Inorganic Chemistry

Alan G Sharpe

University of Cambridge

Copublished in the United States with
John Wiley & Sons, Inc., New York

Longman Scientific & Technical,
Longman Group UK Limited,
Longman House, Burnt Mill, Harlow,
Essex CM20 2JE, England
and Associated Companies throughout the world.

Copublished in the United States with
John Wiley & Sons, Inc., 605 Third Avenue, New York, NY 10158

First published 1981
Reprinted 1982
Second edition 1986
Reprinted 1988, 1989
Third edition 1992

British Library Cataloguing in Publication Data

Sharpe, A.G.
Inorganic chemistry.—3rd ed.
I. Title
546

ISBN 0-582-05913-5 pbk

Library of Congress Cataloging-in-Publication Data

Sharpe, A. G.
Inorganic chemistry / Alan G. Sharpe. — 3rd ed.
p. cm.
Includes bibliographical references and index.
1. Chemistry, Inorganic. I. Title.
QD151.2.S48 1991
546—dc20

Produced by Longman Singapore Publishers Pte Ltd
Printed in Singapore.

Contents

Preface to the third edition

The aim of this book continues to be that stated in the preface to the first edition: to provide, within a single volume of moderate size, a critical introduction to modern inorganic chemistry. It is written primarily to help the student rather than to impress the teacher. Its characteristic features include:

(i) extensive cross-referencing (the fundamental principles developed in the early chapters are for use rather than for ostentation);
(ii) emphasis on thermodynamics (extents and rates of reaction are governed by the energy differences between reactants and products, and between reactants and transition states, respectively);
(iii) a somewhat sceptical attitude towards simple valence theories (they often provide useful frameworks for the memory, but they usually explain rather than predict, and the way inorganic chemists switch between electrostatic, electron-pair repulsion, valence bond, molecular orbital and band theories provides support for the cynical view that the subject is as much a collection of disconnected theories as one of disconnected facts).

The most significant change between this edition and the last one is that outline solutions to all problems are now given. Increased emphasis on the chemistry of the solid state and of the environment will be evident throughout the book, and the treatment of both aqueous and non-aqueous solutions has been extended; space for these changes has been made by judicious pruning in the chapters dealing with the descriptive chemistry of the elements. Many minor additions, corrections and improvements have also been made. The references for further reading (essential for advanced undergraduate and postgraduate students who wish to follow recent research on, for example, cluster compounds, magnetic and electronic properties, bioinorganic chemistry, organometallic chemistry and inorganic spectroscopy) have been brought up to date and provide a full bibliography of inorganic chemistry .

Dr Hazel Rossotti and Professor Peter Edwards have again generously read and criticised parts of the text. To them, and to Dr Michael Rodgers, I offer my heartfelt thanks for their help and encouragement.

A.G.S.

Cambridge, 1991

Acknowledgements

We are grateful to the following for permission to reproduce copyright material:

Addison-Wesley Publishing Company, Menlo Park, CA, USA, for Figs 23.3 from Fig. 11.10 from Porterfield, W.W. (1984), *Inorganic Chemistry*, 3.4 from Fig. 2.17 from Kneen, W.R., Rogers, M.J.W. & Simpson, P. (1972), *Chemistry*, p. 43; Athlone Press, London, for Fig. 11.2 from a Figure from Dewar, M.J.S. (1965) *An Introduction to Modern Chemistry*, p. 16; the author, W.E. Dasent for Fig. 6.16 from Fig. 3 from Dasent, W.E. (1970) *Inorganic Energetics*, p. 47, Penguin Books Ltd, London; Harper and Row, New York, for Fig. 4.2 from Fig. 3.21 from Huheey, J.E. (1972) *Inorganic Chemistry*, p. 99; the author, K.B. Harvey for Figs 6.4, 6.5, 6.6 & 6.7 from Figs 7.11, 7.12, 7.13 & 7.14 from Harvey, K.B. & Porter, G.B. (1963) *Introduction to Physical Inorganic Chemistry*, pp. 242–4, Addison-Wesley Publishing Company, Menlo Park, CA, USA; John Wiley and Sons, Inc., New York, for Figs 6.17 & 25.10 from Figs 2.14 & a Figure, from Cotton, F.A. and Wilkinson, G. (1972) *Advanced Inorganic Chemistry*, 3rd edition, 12.1 from Figs 2.1(a) & 2.3 from Muetterties, E.L. (1967) *The Chemistry of Boron and its Compounds*; Longman Group UK Ltd, Harlow, Essex, for Figs 6.8, 25.3, 25.4, 25.5 & 25.6 from Figs 1.8, 5.2, 5.3, 5.4, 5.8 & 5.9 from Parish, R.V. (1977) *The Metallic Elements*; Marcel Dekker Inc., New York, for Fig. 9.1 from Fig. 9.1 from Lagowski, J.J. (1973) *Modern Inorganic Chemistry*, p. 174; Oxford University Press, Oxford, for Figs 6.9, 6.10, 6.11, 12.7, 13.1, 14.4 & 25.7 from Figs 3.35b, 3.38a, 6.4, 6.5a, 6.13, 24.6, 11.8, 17.1 & 21.3 from Wells, A.F. (1975) *Structural Inorganic Chemistry*, 4th Edition; The Chemical Society, London, for Fig. 13.3 from a Figure, from Ives, D.J.G. (1960) *Principles of the Extraction of Metals*, p. 21; The Open University Press, Milton Keynes, for Fig. 1.3 from Figs 8.14 & 8.15 from The Open University Press (1976) *S 304 Unit 8*; Walter de Gruyter & Co., Berlin, for Fig. 13.4 from Fig. 10.6 from Steudel, R. (1976) *Chemistry of the Non-Metals* p. 354; W.W. Norton & Co. Inc., New York, for Figs 2.5 & 2.6 from Figs 5.8 & 5.9 from Gucker, F.T. & Seitert, R.L. (1967) *Physical Chemistry* p. 126–7.

Whilst every effort has been made to trace the owners of copyright material, in a few cases this has proved impossible and we take this opportunity to offer our apologies to any copyright holders whose rights we may have unwittingly infringed.

1 *Nuclear chemistry*

Introduction • Nuclear binding energy • Radioactivity and nuclear reactions • Nuclear fission and nuclear fusion • Spectroscopic techniques based on nuclear properties • *Ortho-* and *para*-hydrogen • The separation of stable isotopes • The separation of unstable isotopes • The applications of isotopes

1.1 *Introduction*

Most chemistry is concerned with extranuclear atomic structure rather than with the nucleus. Nevertheless, there are two reasons for beginning this book with a brief account of nuclear chemistry. First, there is a logical progression from the study of nuclei to that of atoms and thence to that of assemblies of atoms. Second, there are many aspects of nuclear science that are important to chemistry, and an account of even a restricted selection of these can provide, within a single chapter, a good indication of the scope of modern inorganic chemistry.

The elucidation of atomic structure in the last hundred years is one of the great achievements of science, but it belongs primarily to physics. The following paragraphs summarise only the minimum of information concerning atomic structure that is needed for the understanding of this and the following chapters.

An atom of any element consists of a positively charged *nucleus* surrounded by one or more negatively charged *electrons*, the whole atom being electrically neutral. Nearly all the mass of an atom is concentrated in the nucleus, which, however, has a radius of only about 10^{-15} m, i.e. 10^{-5} times that of the atom; the density of the nucleus is therefore enormous, more than 10^{12} times that of lead. The nucleus can be regarded as consisting of positively charged *protons* and electrically neutral *neutrons*; these particles are known collectively as *nucleons*. The *atomic number*, Z, of an atom of an element is the number of protons in the nucleus, and is equal to the ordinal number of the element in the periodic classification; it defines the chemical identity of an atom. The *mass number*, A, of an atom is the integer nearest to the relative atomic mass

Table 1.1
Properties of the proton, the neutron, and the electron.

Particle	Symbol	m/m_u	m/kg	Charge/C
Proton	^1_1p	1.007 277	$1.672\,65 \times 10^{-27}$	$+1.602 \times 10^{-19}$
Neutron	^1_0n	1.008 665	$1.674\,95 \times 10^{-27}$	0
Electron	e	0.000 548	$9.109\,5 \times 10^{-31}$	-1.602×10^{-19}

(see below) and is equal to the number of nucleons in the nucleus; it follows that the number of neutrons in the same nucleus is $A-Z$.

A particular *nuclide* (i.e. a specific kind of atom) is characterised by its mass number and the atomic number or the name of the element concerned (sometimes by both). Carbon-12, for example, is the name given to atoms of carbon ($Z = 6$) which have $A = 12$; their nuclei contain six protons and six neutrons. Carbon-13 atoms also have $Z = 6$, but their nuclei, for which $A = 13$, contain six protons and seven neutrons. Nuclides of the same element of different mass number are called *isotopes* of that element; their importance is discussed at length later in this chapter. The three isotopes of carbon, for example, can be symbolised as $^{12}_{6}C$, $^{13}_{6}C$, and $^{14}_{6}C$; since all species having $Z = 6$ are carbon atoms, the subscript value of Z is often omitted.

The *isotopic mass*, m, of a nuclide is usually expressed in terms of the *atomic mass unit*, m_u, which is defined as one-twelfth of the mass of an atom of carbon-12 and has the value $1.660\,57 \times 10^{-27}$ kg. The choice of carbon-12 as standard derives from the fact that all accurate values of isotopic masses are now determined by mass spectrometry: the vast number of available volatile carbon compounds of different molecular mass facilitates comparisons of isotopes of unknown mass and carbon-containing species of known mass. The isotopic masses of the proton, the neutron, and the electron are given, together with the charges of these particles, in Table 1.1.

For an element which consists of a mixture of isotopes, the relative abundances of the isotopes (also determined by mass spectrometry) must be known in order to calculate the *relative atomic mass*, A_r (or, as it is often loosely called, the *atomic weight*) of the element. Naturally occurring chlorine, for example, contains 75.53 per cent of ^{35}Cl and 24.47 per cent of ^{37}Cl, the accurate isotopic masses of which are 34.969 and 36.966 m_u respectively; its relative atomic mass is therefore given by

$$A_r = (34.969\, m_u \times 75.53 + 36.966\, m_u \times 24.47)/100\, m_u$$
$$= 35.458$$

1.2 Nuclear binding energy

The mass of an atom of ^1H is equal to the sum of the masses of a proton and an electron. For all other atoms, the atomic mass is less than the sum of the masses of the protons, neutrons and electrons present. This *mass defect* is a measure of the *binding energy* of the protons and neutrons in the nucleus, loss of mass and liberation of energy being related by Einstein's equation

$$\Delta E = \Delta m c_0^2$$

where ΔE is the energy liberated, Δm the loss of mass, and c_0 the velocity of light in a vacuum. (It would be more logical to derive nuclear binding energies from nuclear masses and the sums of the masses of protons and neutrons present, but accurate values of nuclear, as distinct from atomic, masses are known only for elements of low atomic number all of whose electrons can be removed in a mass spectrometer.) For a ^{7}Li atom, for example, which has an isotopic mass of $7.016\,01\ m_u$, the mass defect is

$$[(3 \times 1.007\,277) + (4 \times 1.008\,665) + (3 \times 0.000\,548) - 7.016\,01]m_u$$
$$= 0.042\,13 m_u$$

Since

$$1\ m_u = 1.660\,57 \times 10^{-27}\ \text{kg and } c_0 = 2.998 \times 10^8\ \text{m s}^{-1}$$
$$\Delta E = 0.042\,13 \times 1.660\,57 \times 10^{-27} \times (2.998 \times 10^8)^2\ \text{kg m}^2\ \text{s}^{-2}$$
$$= 6.29 \times 10^{-12}\ \text{kg m}^2\ \text{s}^{-2} \text{ or } 6.29 \times 10^{-12}\ \text{J}$$

This quantity is the binding energy of a single ^{7}Li nucleus; for comparison with the energy changes involved in conventional chemical reactions it must be multiplied by the Avogadro constant, N_A, 6.022×10^{23} mol^{-1}. Thus the molar nuclear binding energy of ^{7}Li is no less than 3.79×10^{12} J mol^{-1}, approximately ten million times the energy liberated by burning one mole (12 g) of carbon to carbon dioxide.

In comparing the binding energies of different nuclei, it is more useful to consider the average binding energy per nucleon; for ^{7}Li this is $(6.29 \times 10^{-12}$ J)/7 nucleons $= 8.98 \times 10^{-13}$ J nucleon^{-1}. The variation in average binding energy per nucleon with mass number is shown in Fig. 1.1.

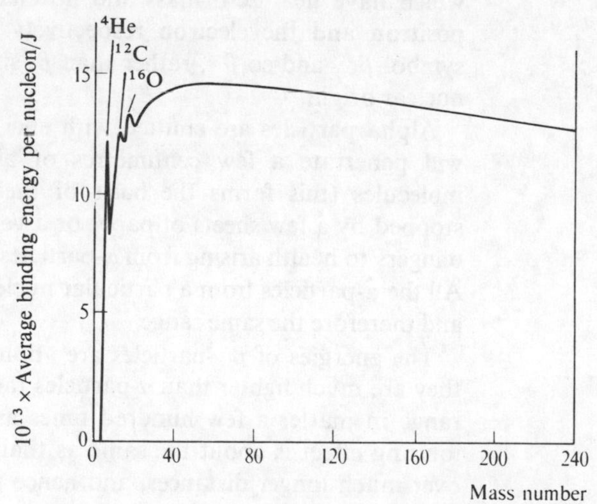

Figure 1.1
Variation in average binding energy per nucleon with mass number.

Three features of interest in Fig. 1.1 may be noted. First, nuclei with mass numbers around 60 have the highest average binding energies per nucleon, and it is elements with mass numbers in this region (e.g. iron, nickel) that are believed to constitute the bulk of the earth's core. Second, species of mass numbers 4, 12 and 16 have high binding energies per nucleon, implying that the nuclei ^{4}He, ^{12}C and ^{16}O are particularly stable; these nuclei are the ones most used as projectiles in the synthesis of the heaviest nuclei. Third, the binding energy per nucleon decreases appreciably above mass number 100.

The form of the relationship between average binding energy per nucleon and mass number indicates that very heavy nuclei would release mass (and therefore energy) on division into two nuclei of medium mass, and also that the lightest nuclei would release mass (and hence energy) on fusion to form heavier nuclei. These processes (*nuclear fission* and *nuclear fusion* respectively) are discussed further after the treatment of radioactivity and nuclear reactions which follows.

1.3 Radioactivity and nuclear reactions

Nuclear transformations are usually very slow (i.e. they have very high activation energies), but spontaneous changes of many heavy nuclides (e.g. $^{238}_{92}U$ and $^{232}_{90}Th$ have been known since the 19th century. In such changes, three types of emission were then recognised: *α-particles* (now known to be helium nuclei, i.e. $^{4}He^{2+}$ ions), *β-particles* (electrons of nuclear origin having high kinetic energies), and *γ-radiation* (high energy X-rays). More recent work has shown that in the decay of some nuclei three other particles may be emitted: the *positron*, of mass equal to, and charge opposite to, that of the electron, and the *neutrino* and *antineutrino*, which have near zero mass and no charge, and which accompany the positron and the electron respectively. Positrons are denoted by the symbol β^{+}, and so β^{-}, rather than β, should be used for an electron of nuclear origin.

Alpha particles are emitted with energy about $(6\text{–}16) \times 10^{-13}$ J; they will penetrate a few centimetres of air, causing ionisation of some molecules (this forms the basis of their determination), but they are stopped by a few sheets of paper or a very thin metal foil; the particular dangers to health arising from α-particles come from the risk of ingestion. All the α-particles from a particular nucleus usually have the same energy and therefore the same range.

The energies of β^{-}-particles are about $(0.03\text{–}5.0) \times 10^{-13}$ J, but since they are much lighter than α-particles they travel much faster and have a range in matter a few hundred times as great as the latter. Their total ionising effect is about the same as that of α-particles, but it is effective over much longer distances, and hence the degree of ionisation per unit length of path is much lower. Particles of different energies can be distinguished by their different penetrating powers. The energies of β^--particles from a particular nuclide show a continuous distribution up to a maximum value; it was this observation, which was surprising because nuclei have discrete energy levels, that led to the postulate that another particle of variable energy (the antineutrino) was emitted simultaneously.

Gamma radiation is very short wavelength (and therefore very high energy) radiation which often accompanies α- or β^--particle emission. After such emission the resulting nucleus (daughter nucleus) is often in an excited (high energy) state; when the nucleons in it rearrange themselves to give the lowest energy level of the nucleus, energy is emitted in the form of a γ-ray. The energies of γ-radiations are in the same range as those of β^--particles, but γ-rays have even greater penetrating power and are stopped only by, for example, several centimetres of lead. Heavy shielding is therefore necessary in experimental work involving γ-radiation.

The chemical consequences of radioactive decay may be summarised as follows. Emission of an α-particle (a helium nucleus) lowers the atomic number by two and the mass number by four; emission of a β^--particle (an electron of nuclear origin) raises the atomic number by one and leaves the mass number unchanged; emission of γ-radiation affects neither atomic number nor mass number. Thus a sequence of an α-emission followed by two β^--emissions produces a nuclide which is isotopic with the original one, e.g.

$$^{238}_{92}\text{U} \xrightarrow{-\alpha} {}^{234}_{90}\text{Th} \xrightarrow{-\beta^-} {}^{234}_{91}\text{Pa} \xrightarrow{-\beta^-} {}^{234}_{92}\text{U}$$

Nuclides having the same mass number but different atomic numbers, such as $^{234}_{90}\text{Th}$, $^{234}_{91}\text{Pa}$ and $^{234}_{92}\text{U}$, are known as *isobars*. In the long natural radioactive decay series from $^{238}_{92}\text{U}$ to $^{206}_{82}\text{Pb}$ (one of four such series), a total of eight α- and six β^--particles is emitted; a slightly simplified version of this series is given in Table 1.2.

For a single radioactive decay process, the number of nuclei disintegrating in a short period of time is proportional to the number present and is independent of temperature:

$$\frac{dn}{dt} = -\lambda n$$

where λ, which has the dimension (time)$^{-1}$, is the *rate* (or *decay*) *constant* of the process. Integration of this equation gives

$$\frac{n}{n_0} = e^{-\lambda t}$$

where n is the number of nuclei remaining out of an original number n_0 after a time t has elapsed. Instead of quoting λ it is common practice to refer to the *half-life*, $t_{\frac{1}{2}}$, of a nuclide. This is the time taken to reduce the number of nuclei to one-half of the original number, and its relationship

Table 1.2
The natural radioactive decay series from $^{238}_{92}U$ to $^{206}_{82}Pb$.

Element	Symbol	Particle emitted	Half-life
Uranium	$^{238}_{92}U$	α	4.5×10^9 y
Thorium	$^{234}_{90}Th$	β^-	24.1 d
Protactinium	$^{234}_{91}Pa$	β^-	1.18 m
Uranium	$^{234}_{92}U$	α	2.48×10^5 y
Thorium	$^{230}_{90}Th$	α	8.0×10^4 y
Radium	$^{226}_{88}Ra$	α	1.62×10^3 y
Radon	$^{222}_{86}Rn$	α	3.82 d
Polonium	$^{218}_{84}Po$	α	3.05 m
Lead	$^{214}_{82}Pb$	β^-	26.8 m
Bismuth	$^{214}_{83}Bi$	β^-	19.7 m
Polonium	$^{214}_{84}Po$	α	1.6×10^{-4} s
Lead	$^{210}_{82}Pb$	β^-	19.4 y
Bismuth	$^{210}_{83}Bi$	β^-	5.0 d
Polonium	$^{210}_{84}Po$	α	138 d
Lead	$^{206}_{82}Pb$		Non-radioactive

to λ is obtained thus:

$$e^{-\lambda t} = \frac{n}{n_0} = \tfrac{1}{2}$$

$$-\lambda t_{\frac{1}{2}} = \ln \tfrac{1}{2}$$

$$\therefore t_{\frac{1}{2}} = \frac{\ln 2}{\lambda} = \frac{0.693}{\lambda}$$

Values of $t_{\frac{1}{2}}$ range from millions of years (e.g. 4.5×10^9 y for $^{238}_{92}U$) to small fractions of seconds (e.g. 10^{-4} s for $^{214}_{84}Po$). Expressions for the variation in radioactivity with time when two or more successive decay stages are under consideration may be derived by standard methods of chemical kinetics.

The SI unit of radioactivity is the becquerel, equal to one nuclear disintegration per second. The radioactivity of a sample of a nuclide undergoing disintegration is also often expressed in curies. One curie is defined as the amount of radioisotope to give 3.7×10^{10} disintegrations per second (the activity associated with 1 g of radium-226, which has a half-life of 1600 years).

There is no essential difference between natural radioactivity as described above and nuclear changes resulting from bombardment of nuclei with positively charged particles or neutrons (the latter are particularly effective, since, being uncharged, they are not subject to electrostatic repulsion by nuclei). Such changes involve conservation of atomic number and mass number, and may be illustrated by the examples:

$$^{27}_{13}Al + ^{4}_{2}He \rightarrow ^{30}_{15}P + ^{1}_{0}n \text{ or } ^{27}_{13}Al(\alpha, n)^{30}_{15}P$$

and

$$^{32}_{16}S + ^{1}_{0}n \rightarrow ^{32}_{15}P + ^{1}_{1}H \text{ or } ^{32}_{16}S(n, p)^{32}_{15}P$$

The first of these reactions is brought about by bombardment of aluminium foil with α-particles which have been given high energies in an accelerating machine such as a cyclotron. The $^{30}_{15}P$ formed is itself radioactive, and decays according to the equation

$$^{30}_{15}P \rightarrow ^{30}_{14}Si + \beta^{+}$$

with $t_{\frac{1}{2}} = 3.2$ min. The second reaction involves the action of 'fast' neutrons produced in the fission of $^{235}_{92}U$ (see the next section); again the product is radioactive, decaying according to the equation

$$^{32}_{15}P \rightarrow ^{32}_{16}S + \beta^{-}$$

with $t_{\frac{1}{2}} = 14.3$ days.

A particularly important process for the production of artificial radioactive isotopes is the (n, γ) reaction brought about by bombardment of many nuclei with 'slow' neutrons, i.e. neutrons formed by fission of ^{235}U whose kinetic energy on formation has been reduced by elastic collisions with low atomic number nuclei (such as $^{12}_{6}C$ or $^{2}_{1}H$) during passage through graphite or deuterium oxide. These reactions are typified by

$$^{31}_{15}P + ^{1}_{0}n \rightarrow ^{32}_{15}P + \gamma \quad \text{or} \quad ^{31}_{15}P(n, \gamma)^{32}_{15}P$$

and

$$^{55}_{25}Mn + ^{1}_{0}n \rightarrow ^{56}_{25}Mn + \gamma \quad \text{or} \quad ^{55}_{25}Mn(n, \gamma)^{56}_{25}Mn$$

In them, it may be noticed, the nuclide produced (which in each instance is a β^{-}-emitter) is isotopic with the target. Reference is made later to the separation of product and target in such cases.

Different nuclei show very wide variation in their susceptibility to neutron absorption when subjected to the action of neutrons of a given kinetic energy, and similar variations are found for the probabilities of other nuclear reactions. Such probabilities are expressed in terms of the *cross-section* of a nucleus for a particular reaction. For thermal neutron capture $^{12}_{6}C$, $^{1}_{1}H$ and $^{2}_{1}H$ have very low cross-sections, $^{10}_{5}B$ and $^{113}_{48}Cd$ very high ones.

The preparation of isotopes of the transuranic (post-uranium) elements neptunium and plutonium by β^{-}-decay of $^{239}_{92}U$, which is made by the action of slow neutrons on uranium-238, is described in the next section; further examples of nuclear reactions are given later in this chapter.

1.4 Nuclear fission and nuclear fusion

Mention was made at the end of Section 1.2 of the possibility of generating large amounts of energy by the fission of very heavy nuclei.

For example, the action of neutrons on $^{235}_{92}U$ results in a variety of reactions, a typical one being

$$^{1}_{0}n + ^{235}_{92}U \rightarrow ^{95}_{39}Y + ^{138}_{53}I + 3^{1}_{0}n$$

followed by β^--decay of the yttrium and iodine nuclei. The yield of different nuclei in the fission of ^{235}U brought about by thermal neutrons is shown in Fig. 1.2; the average number of neutrons released per nucleus undergoing fission is about 2.5, and the energy liberated (2×10^{13} J mol^{-1} of ^{235}U) is about two million times that obtained by burning an equal weight of coal. Since each neutron produced can initiate a further nuclear reaction, a branching chain reaction is possible, and if this takes place in a quantity of ^{235}U larger than a certain critical amount (so that very few neutrons escape), a violent explosion, with enormous liberation of energy, ensues. This is the principle underlying the fission type of nuclear bomb; it has obvious implications for the handling of ^{235}U on an industrial scale.

For the controlled production of energy by nuclear fission the starting material is natural or 235-enriched uranium, usually in the form of its dioxide. In a *thermal reactor* neutrons from ^{235}U fission lose most of their kinetic energy by passage through a moderator (graphite or deuterium oxide) and then undergo one of two nuclear reactions. The first is capture by ^{235}U, leading to further fission; the second is capture by ^{238}U, followed by β^--decay of the product, leading to the production (or *breeding*) of the transuranic elements neptunium and plutonium.

$$^{238}_{92}U(n, \gamma) \longrightarrow ^{239}_{92}U \xrightarrow{-\beta^-} ^{239}_{93}Np \xrightarrow{-\beta^-} ^{239}_{94}Pu$$

The occurrence of a branching chain reaction is prevented by controlling the neutron concentration in the nuclear reactor by the insertion as

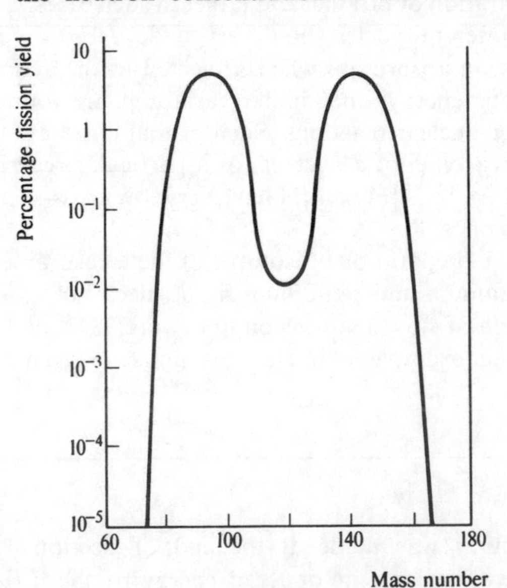

Figure 1.2
Percentage fission yields in slow neutron fission of ^{235}U.

necessary of control rods of boron-containing steel or boron carbide; the isotope ^{10}B has a very large cross-section for neutron capture.

In a *fast reactor*, high energy neutrons are used without a moderator to bring about fission of ^{235}U in a highly enriched uranium fuel; even more effectively, ^{239}Pu can be used as the fissile isotope, and if mixed with ^{238}U the action of some of the neutrons produced on this latter isotope regenerates ^{239}Pu. However, because of the very great toxicity of ^{239}Pu and the technical difficulties associated with making fast reactors as safe as thermal reactors, the overall desirability of a nuclear technology based on the use of the fast breeder reactor is at present open to question.

Fusion reactions leading to the formation of, for example, helium from hydrogen, deuterium (^2_1H), or tritium (^3_1H) are in principle also capable of generating immense amounts of energy, and compared with fission reactions they have the advantage that large amounts of strongly radioactive nuclides (safe storage of which is a serious environmental problem) are not obtained as by-products. However, the activation energies for fusion reactions are very high, and up to the present time it has been possible to produce a fusion reaction, in this case

$$^3_1H + ^2_1H \rightarrow ^4_2He + ^1_0n$$

only if a fission bomb is used to supply the necessary activation energy. This is the principle underlying the hydrogen or thermonuclear bomb, though since tritium is expensive and inconvenient (because of its own short half-life of 12 years) it is prepared *in situ* from lithium deuteride. A fusion explosion generated by compression of a few kilograms of plutonium in a suitable form (e.g. a hollow sphere) brings about reactions such as:

$$^2_1H + ^2_1H \rightarrow ^3_1H + ^1_1H$$

$$^2_1H + ^2_1H \rightarrow ^3_2He + ^1_0n$$

$$^1_0n + ^6_3Li \rightarrow ^4_2He + ^3_1H$$

$$^3_1H + ^2_1H \rightarrow ^4_2He + ^1_0n$$

Fusion reactions are believed to take place in the sun and stars at temperatures above 10^7 K, and the following processes have been suggested as the chief source of the sun's energy:

$$^1_1H + ^1_1H \rightarrow ^2_1H + \beta^+ + \text{neutrino}$$

$$^1_1H + ^2_1H \rightarrow ^3_2He + \gamma$$

$$^3_2He + ^3_2He \rightarrow ^4_2He + 2^1_1H$$

or, in sum,

$$4^1_1H \rightarrow ^4_2He + 2\beta^+ + 2\ \text{neutrinos}$$

Intensive research is now under way to produce controlled fusion by lasers or in a plasma (an ionised gas at a very high temperature). Claims made in 1989 that fusion of deuterium atoms in palladium had been achieved at ordinary temperatures received wide publicity, but are no longer accepted as valid.

1.5 *Spectroscopic techniques based on nuclear properties*

It was believed for a long time that the properties of nuclei were not influenced by their chemical environments, but this is not always true. Certain nuclei have magnetic properties, and when placed in a magnetic field can interact with it so as to assume one of a number of different energy levels, the separation of which varies very slightly according to the chemical environment of the nucleus. Transitions between such energy levels can be induced by irradiation with electromagnetic radiation of the appropriate frequency, and observation of these frequencies is *nuclear magnetic resonance spectroscopy*, now one of the most important techniques in chemistry. One other nuclear spectroscopic technique (*Mössbauer spectroscopy*) will also be mentioned briefly.

Nuclear magnetic resonance spectroscopy

As we shall see in the following chapters, a spinning electron has a magnetic moment which interacts with an applied magnetic field. The same is true of a proton, though the magnetic moment associated with it is only about $\frac{1}{2000}$ of that associated with an electron. The neutron also, however, has spin and gives rise to a magnetic moment, presumably because, although it has no net charge, there is an uneven charge distribution within it. The magnetic moments of nuclei containing both protons and neutrons may be finite or zero according to how the spins of the constituent nucleons are aligned, the resultant amount of spin being expressed by a *nuclear spin quantum number*, *I*. In this book we shall be concerned mainly with nuclei having $I = \frac{1}{2}$ (e.g. ${}^{1}_{1}H$, ${}^{13}_{6}C$, ${}^{19}_{9}F$, and ${}^{31}_{15}P$). If we place such a nucleus in a magnetic field, the nuclear magnetic moment can have only two allowed orientations, parallel and antiparallel to the field. The former is of lower energy, but the difference between them is very small indeed (0.025 J mol^{-1} for a hydrogen nucleus in a field of 1.409 tesla (14 090 gauss)) and so the two orientations are almost equally common. Nevertheless, transitions between them may be observed in the radiofrequency region by the use of a resonance effect. In the case of the proton the resonance frequency at a field of 1.409 tesla is 60×10^{6} s^{-1} (or 60 MHz, the unit of frequency being the hertz, Hz or 1 cycle per second).

The resonance condition specified in the last paragraph holds only for nuclei with no associated electrons. In an atom, the applied field B_0 induces a circulation of electrons round the nucleus which sets up a field proportional to and opposing the field B_0, so that the nucleus feels the effect of a diminished field $B_0(1-\sigma)$, where σ is called the shielding constant. For an actual atom the resonance condition is given by

$$\nu = \frac{\gamma_N B_0(1-\sigma)}{2\pi}$$

where γ_N is a constant called the magnetogyric ratio for the nucleus under discussion; ν is slightly different for atoms of a given element in different chemical environments since these have different values of σ. The range in values for σ increases as the number of electrons in the atom increases, being a minimum for ^1H. Since resonance frequencies observed depend upon field strength, we need to specify both ν and B_0 in order to describe an absorption precisely. But it is very difficult to measure B_0 accurately, and it is therefore universal practice to measure the difference in ν for energy absorption from that for a reference compound; this is tetramethylsilane (TMS), $(CH_3)_4Si$, for both ^1H and ^{13}C nuclear magnetic resonance. The difference in ν is still proportional to B_0, and in order to obtain a parameter which is independent of B_0 the *chemical shift* parameter, δ, is introduced. This is defined by

$$\delta = \frac{\nu(\text{sample}) - \nu(\text{TMS})}{\nu(\text{TMS})} \times 10^6 \text{ ppm}$$

Tetramethylsilane is a convenient standard because it gives a single sharp absorption for both ^1H and ^{13}C, well away from those for these nuclei in other compounds; it is also volatile, soluble in nearly all organic solvents, and very unreactive. In water the salt $(CH_3)_3SiCD_2CD_2CD_2SO_3^- Na^+$ is used.

Under the normal conditions of measurement of ^1H, ^{19}F and ^{31}P magnetic resonance spectra (i.e. spectra of nuclei of high relative abundance), the intensity ratio of the absorptions is proportional to the number of nuclei giving rise to the absorptions. Thus for ethanol, CH_3CH_2OH, the proton magnetic resonance spectrum measured under low resolution contains three peaks at $\delta = 5.2$, 3.7 and 1.2 ppm of relative intensity 1:2:3 assigned to the OH, CH_2 and CH_3 protons respectively. There are well-established rules correlating the values of proton chemical shifts and environments in organic compounds; but such rules are less useful in inorganic chemistry (few inorganic molecules contain protons in several different environments), and we shall not discuss them here beyond remarking that whilst the chemical shifts of the CH_3 and CH_2 protons are constant, that of the OH proton is affected by the extent of hydrogen bonding in which it is involved.

The proton magnetic resonance spectrum of ethanol is a highly instructive example which we shall now consider further. For very carefully purified ethanol under high resolution the spectrum (Fig. 1.3(*a*)) contains CH_3 and OH proton peaks each consisting of a triplet in which the relative intensities of the fine structure peaks are 1:2:1, and a CH_2 proton peak which is an octet whose components have relative intensities 1:1:3:3:3:3:1:1. The origin of these splitting patterns may be explained by considering the methyl group triplet. The protons of the methyl group are under the influence not only of the applied field, but also of the spins of the two protons of the CH_2 group, each of which may have its magnetic moment lined up with or against the applied field. Denoting these possibilities by arrows pointing up and down respectively, there are four possible combinations:

$$\uparrow\uparrow \quad \uparrow\downarrow \quad \downarrow\uparrow \quad \downarrow\downarrow$$
$$\text{(i)} \quad \text{(ii)} \quad \text{(iii)} \quad \text{(iv)}$$

Of these, (ii) and (iii) have the same effect on the CH_3 group protons, so that the combined effects of the CH_2 group protons are in the ratio 1:2:1. Another triplet arises from the effect of the CH_2 group protons on that of the OH group. In general, *spin coupling*, as the phenomenon is called, to one proton leads to a doublet, to two protons to a 1:2:1 triplet, to three protons to a 1:3:3:1 quartet, and so on. The octet associated with the CH_2 protons arises from the splitting of the CH_2 proton resonance into a doublet by the OH proton, followed by the splitting of each peak of the doublet into a quartet by the CH_3 protons. Splitting by nuclei having $I > \frac{1}{2}$ is illustrated for diborane in Section 12.3.

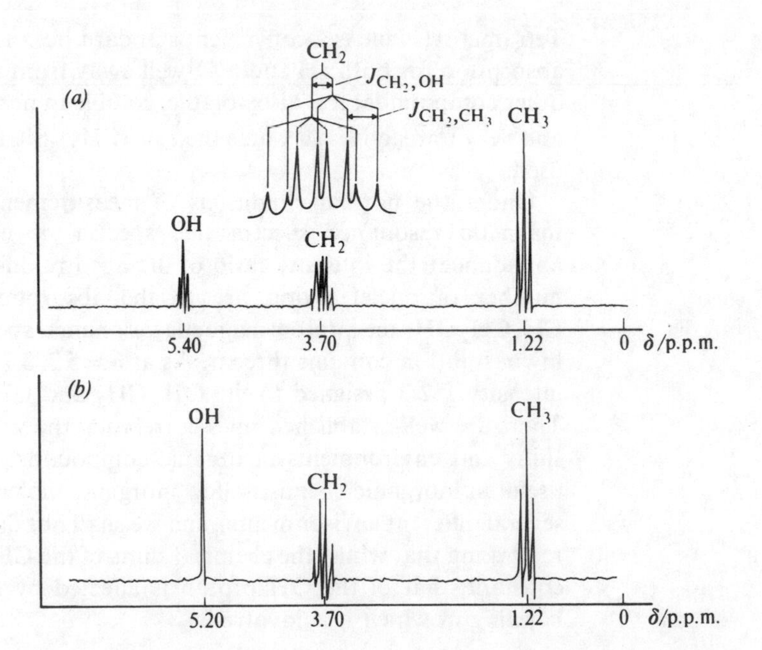

Figure 1.3
High resolution proton magnetic resonance spectrum of: (*a*) pure ethanol; and (*b*) ethanol containing a trace acid

The magnitude of the frequency difference between fine structural peaks in a particular multiplet (the *spin–spin coupling constant*, J) is independent of the strength of the external field or the input frequency used, a fact which makes it possible to distinguish spin–spin splittings from the effects of non-equivalent protons having different chemical shifts. Coupling constants are expressed in Hz. Both of the split signals, when two are split, show the same splitting. Splitting of equivalent proton signals, and splitting by protons other than those on adjacent carbon atoms, are generally not observed.

If a trace of acid is added to pure ethanol the resulting high-resolution spectrum is that normally observed and is shown in Fig. 1.3 (*b*): a singlet (OH), a quartet (CH_2) and a triplet (CH_3). The singlet structure of the OH proton resonance under these conditions can be used to illustrate an aspect of nuclear magnetic resonance spectroscopy that we have not yet mentioned. In order for a proton resonance signal to be observed, the proton must be in a single state for not less than about 10^{-2} s; this is much longer than the time required for proton exchange between different molecules in the presence of a trace of acid, which may be represented by the scheme:

$$C_2H_5OH_a + H_b^+ \rightleftharpoons C_2H_5{-}O^+\begin{matrix} H_a \\ H_b \end{matrix}$$

$$C_2H_5OH_c + C_2H_5{-}O^+\begin{matrix} H_a \\ H_b \end{matrix} \rightleftharpoons C_2H_5{-}O^+\begin{matrix} H_a \\ H_c \end{matrix} + C_2H_5OH_b$$

Because of this proton exchange the average spin state of the OH proton is zero, and it then neither affects, nor is affected by, the CH_2 protons. Such exchange reactions are general for protons bonded to oxygen or nitrogen. Since they are accelerated by a rise in temperature, the observance of spin–spin coupling is sometimes possible only at low temperatures. Conversely, the temperature dependence of a spectrum tells us something about the activation energy and mechanism of the exchange process.

The nucleus ^{12}C has $I = 0$ and hence can give no nuclear magnetic resonance spectrum. The low natural abundance of ^{13}C (1.1 per cent) leads to special difficulties in the observation of ^{13}C spectra, which are overcome by the technique of pulsed Fourier transform (PFT) spectroscopy. In this, all ^{13}C nuclei present are excited by a short pulse, and the frequencies of transitions to lower energy levels are sorted out from the complex spectrum by the mathematical method of Fourier analysis, performed by computer. For spectra measured by this method relative intensities of signals are *not* proportional to numbers of nuclei present in different environments, only qualitative information about the kinds of different ^{13}C atoms being obtainable. Further, because of the low natural abundance of the isotope (so that any molecule is unlikely to contain more than one ^{13}C atom), ^{13}C spin–spin coupling is normally observed only for compounds made from ^{13}C-enriched materials.

The following examples illustrate some of the applications of nuclear resonance spectroscopy in inorganic chemistry.

A number of hydrated cations in aqueous solution undergo exchange with solvent water at rates slow enough for coordinated and solvent water containing the isotope ^{17}O (which has a nuclear magnetic moment) to be distinguished, and from intensity ratios hydration numbers can be obtained: Be^{2+} and Al^{3+}, for example, are found to be present as $[Be(H_2O)_4]^{2+}$ and $[Al(H_2O)_6]^{3+}$ respectively.

A reaction such as

$$PCl_3 + P(OEt)_3 \rightleftharpoons PCl_2OEt + PCl(OEt)_2$$

in which the types and numbers of each type of bond remain constant is known as a *redistribution reaction.* This one can be followed by ^{31}P nuclear magnetic resonance spectroscopy, since the ^{31}P chemical shifts are different in the four compounds. Rate data may be obtained by following the variation in relative intensities with time, equilibrium constants (and hence standard free energy changes) by determining the relative intensities when no further change takes place, and standard enthalpy and entropy changes by determining the equilibrium constants at different temperatures. As would be expected, ΔH° for reactions of this kind is almost zero, the redistribution of the groups attached to the phosphorus atoms being caused by an increase in the entropy of the system.

In investigating inorganic structures by high resolution nuclear magnetic resonance spectroscopy, the relatively long time-scale of the method (10^{-1}–10^{-5} s, depending on the nucleus being studied) must always be borne in mind. Especially at higher temperatures (where the activation energy for intramolecular rearrangement is very much more easily acquired), nuclear magnetic resonance methods often indicate more symmetrical structures than spectroscopic or diffraction methods. In these circumstances, however, the information obtained about the energetics of rearrangement (from which something concerning the mechanism can be inferred) may be of major significance, since few methods for the study of such processes are available. The ^{19}F resonance in bromine pentafluoride at ordinary temperatures, for example, consists of two peaks of relative intensities 1:4, the intense line being a doublet, and the weak line a quintet with the relative intensities 1:4:6:4:1. This indicates that the molecule is a square pyramid. At 180 °C the spectrum collapses to a single line, the fluorine atoms then being, for nuclear magnetic resonance purposes, identical. A similar state of affairs holds in the case of sulphur tetrafluoride. The spectrum at −98 °C consists of two triplets each of relative intensities 1:2:1; this shows that there are two non-equivalent pairs of fluorine atoms, and is in agreement with the structure established

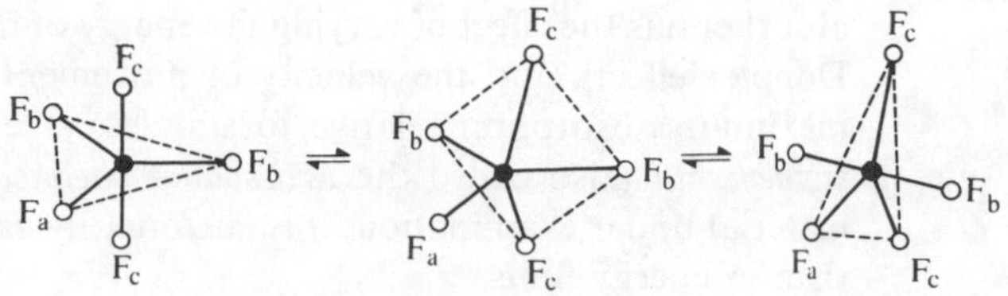

Figure 1.4
Intramolecular exchange of fluorine atoms in PF_5 via a square pyramidal intermediate.

by other methods as a trigonal bipyramid with one of the equatorial positions occupied by an unshared pair of electrons (see Section 5.2 for further details of such structures). At $-58\,°C$ only two broad peaks are seen, and at room temperature there is only a single absorption. For $Fe(CO)_5$ (containing ^{13}C) and PF_5, intramolecular exchange is found to occur even down to $-100\,°C$, showing that the activation energies for exchange must be extremely small. This suggests that exchange involves only a simple molecular vibration of the type shown for PF_5 in Fig. 1.4. In recent years extensive studies by means of ^1H and ^{13}C nuclear magnetic resonance spectroscopy have been made of intramolecular rearrangements of organic ligands coordinated to metal atoms.

Mössbauer spectroscopy

The Mössbauer effect is the emission and resonant absorption of nuclear γ-rays studied under conditions such that the nuclei have negligible recoil velocities when γ-rays are emitted or absorbed; this is achieved by working only with bulky solid samples in which the nuclei are held rigidly in crystal lattices. The energy and hence the frequency of the γ-radiation involved corresponds to the transition between the ground state and the short-lived excited state of the nuclide concerned; the technique is used particularly for ^{57}Fe and ^{119}Sn, but it can be applied to many other nuclides. The magnitude of the excitation energy is influenced very slightly by the electron density round the nucleus; since only *s* electrons have a finite electron density at the nucleus, the Mössbauer effect is, to a first approximation, concerned with the *s*-electron distribution, but occupation of *p*, *d*, or *f* orbitals exerts some screening effect on the *s* electrons and hence has some effect on the spectrum observed.

The method of study of the Mössbauer effect may be illustrated by reference to ^{57}Fe spectroscopy. The basic apparatus includes a radioactive source, a solid absorber containing ^{57}Fe in some form, and a γ-ray detector. The radioactive source is ^{57}Co incorporated in stainless steel; this decays by capture of an extranuclear electron to give an excited state of ^{57}Fe which emits γ-radiation in decaying to the ground state of ^{57}Fe. If the ^{57}Fe is present in the same form in both source and absorber, resonant absorption occurs and no radiation is transmitted. But if the ^{57}Fe is present in two different forms, absorption does not occur, and γ-rays reach the detector. Moving the source at different velocities towards or away from the absorber has the effect of varying the energy of the γ-radiation slightly (the Doppler effect), and the velocity of movement required to bring about maximum absorption relative to stainless steel as arbitrary zero is the *isomer shift* (also called the *Mössbauer chemical shift*), δ, of ^{57}Fe in the material under examination; it is customarily expressed in $mm\,s^{-1}$ rather than in energy units.

A few isomer shifts for ^{57}Fe relative to stainless steel at ordinary temperatures are: $FeCl_2$, +1.3; $FeCl_3$, +0.5; $K_4[Fe(CN)_6]$, −0.05; $K_3[Fe(CN)_6]$, −0.1; K_2FeO_4, −0.9 $mm\,s^{-1}$. Comparison of their isomer shifts with these values has been used to show that the dark blue precipitates formed from Fe^{3+} and $[Fe(CN)_6]^{4-}$ and from Fe^{2+} and $[Fe(CN)_6]^{3-}$ are both iron(III) hexacyanoferrates(II), and oxidation states of iron in naturally occurring materials (e.g. meteorites, moon rock) have also been determined by this method. The closeness of the isomer shifts for $[Fe(CN)_6]^{4-}$ and $[Fe(CN)_6]^{3-}$ suggests that the actual oxidation states of the iron atoms in these ions are less different than their formal oxidation states, the extra electron in the former ion being largely delocalised on to the cyanide ligands. Even for the same oxidation state, different environments lead to different isomer shifts; the existence of two peaks in the Mössbauer spectrum of the carbonyl $Fe_3(CO)_{12}$, for example, provided the first evidence for the presence of two types of iron atom in the structure of this compound.

1.6 Ortho- *and* para-*hydrogen*

The nuclear spins of the two atoms in the molecule of H_2 may be parallel or antiparallel, with a resultant spin of unity or zero. There are thus two *spin isomers* of molecular hydrogen, called (somewhat unsatisfactorily) *ortho-* and *para-*hydrogen respectively. Their interconversion is slow except in the presence of a catalyst on which hydrogen is absorbed in atomic form, or at high temperatures when dissociation into atoms occurs. *Ortho-* and *para-*hydrogen are identical in chemical properties, but they show slight differences in physical properties, e.g. in vapour pressure, thermal conductivity and heat capacity. At 20.4 K, for example, the vapour pressures of *ortho-*H_2, *para-*H_2 and ordinary H_2 are 751, 787 and 760 mm of mercury respectively.

The existence of these two forms of hydrogen was first detected spectroscopically. For reasons which we cannot give here (they are based on the branch of physical chemistry known as statistical mechanics) the *ortho-* and *para-* forms occupy only odd and even (including zero) rotational energy levels (see Section 2.6) respectively. Thus their rotational spectra (observed as fine structure of electronic spectra) are different, and when both forms are present their relative concentrations can be deduced from intensity data. At equilibrium the composition of molecular hydrogen approaches 100 per cent *para*-H_2 (the form of lower internal energy) as the temperature approaches absolute zero; the proportion of *ortho*-H_2 increases with increase in temperature to a limiting value of 75 per cent above 230 K. When ordinary hydrogen (the 3:1 *ortho*-:*para*-mixture) is cooled to liquid air temperature in the presence of charcoal or hydrous iron(III) oxide, it is largely converted into the *para*- form. This may be warmed to the ordinary temperature without change, and it can then be used for the determination of the stationary concentration of hydrogen atoms in a reacting mixture (e.g. H_2+Cl_2), the rate of reconversion to the equilibrium mixture being proportional to the hydrogen atom concentration. Gas chromatographic separation of *ortho*- and *para*-hydrogen, using helium as carrier gas and an aluminium oxide column at 77 K, has been reported.

Catalysis of the *ortho*$\rightarrow$*para* hydrogen conversion is important industrially in connection with the storage of liquid hydrogen. The slow conversion liberates sufficient heat to result in the loss by evaporation of a large fraction of the liquid if a supercooled mixture of the two forms is stored. The presence of hydrous iron(III) oxide when the hydrogen is liquefied, however, greatly accelerates the conversion to almost pure *para*-hydrogen and enables this difficulty to be avoided.

All other diatomic molecules containing atoms having nuclear spin also exist in *ortho*- and *para*- forms, and may be shown spectroscopically to do so; but except in the case of deuterium there are no detectable differences in bulk physical properties.

1.7 The separation of stable isotopes

All methods for the separation of stable isotopes depend upon the difference in isotopic masses, sometimes in a simple way (e.g. the mass spectrographic and diffusion methods), but sometimes only through the influence of the mass difference on the fundamental vibration frequencies of isotopic molecules. Further discussion of this subject is given in Section 1.9, but the following summary should make clear its relevance here.

An extremely important deduction from the quantum mechanical theory of valence is that a molecule has a certain *residual energy* or *zero-point energy*, E_0, even at absolute zero. For a diatomic molecule E_0 per mole is given by

$$E_0 = \tfrac{1}{2}N_A h\nu_0$$

where N_A is Avogadro's constant, h Planck's constant and ν_0 the fundamental vibration frequency. The fundamental vibration frequency is related to the reduced mass μ for the molecule ($\mu = m_a \cdot m_b / m_a + m_b$, where m_a and m_b are the masses of the two atoms in the molecule) and the restoring force constant k for the vibration by the equation

$$\nu_0 = \frac{1}{2\pi}\sqrt{\frac{k}{\mu}}$$

Now the effect of isotopic substitution on the force constant is negligible, so the fundamental vibration frequencies of isotopic species, and hence their zero-point energies, are different. The expression for the reduced mass shows that the effect is greatest for light elements; nevertheless, separation of ^{235}U and ^{238}U by selective excitation and decomposition of $^{235}UF_6$ in a laser beam has been demonstrated.

The mass spectrograph

In this instrument, which is based in principle upon those used by J. J. Thomson and F. W. Aston for the first separations of isotopes, use is made of the fact that particles of different mass/charge ratio move differently when subjected to the influence of electric and magnetic fields. A volatile substance at low pressure is ionised by the action of a beam of electrons and the positive ions of charge ze and mass m so formed are accelerated by a potential difference V to a velocity v. Then

$$\tfrac{1}{2}mv^2 = Vze$$

If a uniform magnetic field of flux density B is applied in a direction perpendicular to that of the ion path, each ion then moves in a circle of radius r such that the centrifugal force balances the force exerted by the field:

$$\frac{mv^2}{r} = Bzev$$

Elimination of v gives

$$\frac{m}{ze} = \frac{B^2r^2}{2V}$$

If V and B are constant, therefore, ions having the same value of ze (e.g. all singly charged ions) can be separated into fractions of different masses. When the method is utilised for the separation of isotopes, these fractions are collected separately. A single operation thus effects complete separation, and has been used, for example, for the separation of ^{235}U and ^{238}U, present as their volatile hexafluorides (fluorine has only one naturally occurring isotope).

The mass spectrograph is not, however, an economically feasible method of isotope separation under normal conditions, and this and related instruments are now used mainly for isotopic analysis and structural work. Amounts of different isotopes can be determined by measurement of the currents carried by them; by adjusting V, ions of the same charge but different mass can be made to follow the same path in succession, and the currents carried at different accelerating potentials can be used to establish isotopic composition. Instruments of this kind are known as *mass spectrometers.* They are important not only in analysis but also in determination of structures of volatile organic and organometallic compounds: there are valuable correlations between the structures of such compounds and the nature of the fragments formed when they are bombarded with electrons of different energies. Mass spectrometry is also used in inorganic chemistry, e.g. in identifying species formed in reactions of the boron hydrides.

Gaseous diffusion

For two isotopic species of molecular mass M_1 and M_2 the ratio of the rates of diffusion through a porous membrane is given by

$$\frac{\text{Rate}_1}{\text{Rate}_2} = \left(\frac{M_2}{M_1}\right)^{\frac{1}{2}}$$

and the separation factor α is defined by the equation

$$\alpha = \frac{n_1'/n_2'}{n_1/n_2}$$

where n_1 and n_2 are the concentrations of the species before, and n_1' and n_2' the concentrations after, diffusion. Unless M_1 and M_2 are very low (e.g. for H_2 and D_2), α is very near unity, and for the method to be useful a large number of units must be used in succession; this has been done industrially for the separation of $^{235}UF_6$ and $^{238}UF_6$, even though the theoretical separation factor is only 1.0043. In practice the degree of separation is always less than the calculated value, since the velocities of individual molecules are spread over a range of values, and bulk flow from higher to lower pressures accompanies molecular diffusion. In the case of the UF_6 separations, several thousand diffusions through a suitable inert porous barrier lead to the separation of 99 per cent pure $^{235}UF_6$. The method has also been used for the separation of deuterium and the enrichment of methane and molecular nitrogen in ^{13}C and ^{15}N respectively.

In a recent modification of the diffusion method, a partial separation is obtained by centrifuging, when the heavier isotope is enriched further away from the axis of the centrifuge.

Thermal diffusion

The theory of this method is very complicated, but the essential principle is that in a vertical tube with an electrically heated wire or thin tube down the middle, heavier molecules tend to diffuse to the cold wall, lighter molecules to the hot surface. This thermal diffusion is reinforced by convection currents, which carry the lighter molecules upward and the heavier molecules downward. If a succession of tubes several metres long and a temperature difference of 500 °C are employed, a good degree of isotopic enrichment is obtained. Although slow, the method requires little attention, and among several applications of it are those to the separation of $H^{37}Cl$, $^{13}CH_4$, $^{15}N_2$ and $^{18}O_2$.

Chemical methods

For the isotopic exchange reaction

$$^{15}NH_3(g) + {}^{14}NH_4^+(aq) \rightleftharpoons {}^{14}NH_3(g) + {}^{15}NH_4^+(aq)$$

the equilibrium constant, K, is 1.031 at 25 °C. The fact that K is not unity arises from a small difference in standard free energy between reactants and products ($\Delta G^\circ = -RT\ln K$); most of this difference is due to differences in zero-point energies. Since $K \neq 1$, the rates of forward and reverse reactions are slightly different, because $K = k_{forward}/k_{backward}$, k being the rate constant; the difference in k values also originates in the difference in zero-point energies (cf. Section 1.9). As the ammonia–ammonium ion equilibrium involves two phases (and, like many other proton-transfer systems, very rapidly reaches equilibrium), it is particularly suitable for isotopic enrichment, and can readily be made the basis of a multistage process.

Other equilibria which have been used as the basis for isotopic enrichments include the following (values of K are at 25 °C):

$$H^{12}CN(g) + {}^{13}CN^-(aq) \rightleftharpoons H^{13}CN(g) + {}^{12}CN^-(aq) \quad K = 1.026$$

$$^{13}CO_2(g) + H^{12}CO_3^-(aq) \rightleftharpoons {}^{12}CO_2(g) + H^{13}CO_3^-(aq) \quad K = 1.012$$

$$HD(g) + H_2O(liq) \rightleftharpoons H_2(g) + HOD(liq) \quad K = 3.70$$

The second and third of these require the use of catalysts to speed up the attainment of equilibrium. The equilibrium

$$^{7}Li(amalgam) + {}^{6}Li^+(methanol) \rightleftharpoons {}^{6}Li(amalgam) + {}^{7}Li^+(methanol)$$

slightly favours the presence of ^6Li in the amalgam and has been used in the separation of ^6Li for the preparation of tritium; ion-exchange has also been used for this purpose.

The electrolytic method

This is important only for deuterium, ^2H or D. It rests upon a complicated theoretical basis, both kinetic and thermodynamic factors being involved since the choice of metal as electrode is important. When an aqueous solution of sodium hydroxide is electrolysed using nickel electrodes, the separation factor (H/D)(gas)/(H/D)(solution) is approximately 6. The electrolysis is continued until about nine-tenths of the liquid has been converted into oxygen and hydrogen; most of the residual liquid is then neutralised with carbon dioxide, and the water is distilled and added to the remainder of the electrolyte. This process can be repeated to give up to 99.9 per cent D_2O. In the later stages of the separation the gas evolved at the cathode is burned to yield partially enriched deuterium oxide that can be electrolysed further. Cheap electric power is, of course, essential for the economic concentration of deuterium oxide by this method.

1.8 *The separation of unstable isotopes*

The production of unstable isotopes by means of radioactive decay or fission of heavy nuclei, and by nuclear reactions involving neutron or charged particle bombardment, has been mentioned in Sections 1.3 and 1.4. Among such reactions (n, γ), (n, p) and (n, d) processes are particularly important because of the effectiveness of the uncharged neutron as a projectile against positively charged nuclei, and they are conveniently carried out in a nuclear reactor, using solids or liquids as the material to be irradiated. For the production of the heaviest elements, however, the use of charged particles such as C^{6+} and O^{8+} nuclei in an accelerating machine is essential, e.g. ${}^{238}_{92}U$ (${}^{16}_{8}O$, 4n) ${}^{250}_{100}Fm$; for such processes solid targets are necessary.

The methods used for separation of the desired isotope depend on whether or not the starting material and the product are the same element. If they are not, as is the case where an isotope has been obtained by fission, α- or β^--decay or a (n, p) or (n, α) reaction, the problem is essentially one of chemical separation of a small amount of one element from much larger amounts of one or more others; volatilisation, electrodeposition, solvent extraction, ion-exchange, or precipitation on a 'carrier' are techniques frequently used.

The radioactive isotope ^{64}Cu, for example, made by a (n, p) reaction using fast neutrons on ^{64}Zn, is easily separated by dissolving the target in dilute nitric acid and depositing the copper electrolytically, there being a difference of more than 1 V between $E^\circ(Cu^{2+}/Cu)$ and $E^\circ(Zn^{2+}/Zn)$. Carbon-14, made by a (n, p) reaction on ^{14}N (present as Be_3N_2) in a nuclear reactor, is liberated as $^{14}CH_4$ by the action of water on the product; this is oxidised to $^{14}CO_2$ and precipitated as $Ba^{14}CO_3$. Astatine (which is the homologue of chlorine, bromine and iodine) is made by the reaction ${}^{209}_{83}Bi(\alpha, 2n){}^{211}_{85}At$ and, being volatile like iodine, may be separated by sublimation. The fissile plutonium isotope ^{239}Pu is obtained, as described in Section 1.4, by the sequence

$$^{238}_{92}U + ^{1}_{0}n \rightarrow ^{239}_{92}U$$

$$^{239}_{92}U \xrightarrow[23\ \text{min}]{-\beta^-} ^{239}_{93}Np \xrightarrow[2.3\ \text{days}]{-\beta^-} ^{239}_{94}Pu$$

After irradiation of the uranium, the targets are set aside for a few weeks, during which the concentrations of ^{239}U and ^{239}Np become negligible. Uranium and plutonium are then dissolved in nitric acid and their nitrates of formula $MO_2(NO_3)_2$ are extracted into tributyl phosphate; plutonium(VI) is much more easily reduced than uranium(VI), and washing the extract with an aqueous solution of a mild reducing agent converts it into plutonium(IV), in which form it is readily transferred back into the aqueous layer, the uranium(VI) remaining in the tributyl phosphate.

If target and product are the same element, as is the case when a (n, γ) reaction is not followed by rapid β^--decay, the concentration of unstable isotope obtained is usually very low. However, the energy of the γ-radiation which is liberated often suffices to break chemical bonds, and then target and product, though isotopic, are present in different chemical forms. This effect (the Szilard–Chalmers effect) may be illustrated for ^{128}I produced by slow neutron bombardment of ^{127}I in the form of ethyl iodide. A significant amount of the ^{128}I formed is liberated as atomic iodine; these atoms may combine to give $^{128}I_2$ or may be allowed to react with added $^{127}I_2$ to form $^{127}I^{128}I$; in either case, molecular iodine containing the ^{128}I is readily separated by shaking with aqueous sodium sulphite solution, which converts iodine into sodium iodide but does not react with ethyl iodide. For such a method to be useful there must, of course, be no rapid exchange reaction between target and product; hence the use of an alkyl halide, rather than an alkali metal halide, in the irradiation.

1.9 The applications of isotopes

Applications of isotopes, both radioactive and stable, are now so numerous that we can do no more here than present a small selection of them, with special reference to inorganic chemistry. Thus we shall mention only briefly such important applications as the detection of flaws or cracks in metallic objects by γ-radiography using the radiation from ^{60}Co, the use of radioactive isotopes to follow movements of sand and of oil, the destruction of malignant growths in the body by irradiation with γ-rays from ^{60}Co or ^{137}Cs, and the elucidation of metabolic and synthetic pathways in organic chemistry and biochemistry. Many of the applications we shall mention here involve only the use of isotopes as 'tracers', in which all isotopes of an element are regarded as chemically equivalent; others, such as those in spectroscopy and kinetics, depend upon the small differences between isotopes of a given element. Uses involving radioactive isotopes often depend upon the readiness with which very small quantities of them can be determined using modern equipment: for an isotope of $t_{\frac{1}{2}} = 14$ days emitting β^--radiation (e.g. ^{32}P), 10^{-16} g is easily detected, making the method about a million times as sensitive as the flame test for sodium. For isotopes with shorter half-lives, even smaller amounts can be detected; but for species of very short half-life it is, of course, difficult to isolate and use the isotope before virtually all of it has changed into its decay product.

Analytical applications

These include determinations of solubilities of sparingly soluble substances, vapour pressures of rather involatile substances, and the investigation of solid solution formation and adsorption by precipitates. The solubility of strontium sulphate, for example, has been determined by preparing and measuring the activity per gram of a uniform mixture of $^{90}SrSO_4$ and the inactive salt, and then measuring the activity of the residue from evaporation of a saturated solution of the mixture; this method is known as *isotope dilution analysis*. The vapour pressure of red phosphorus incorporating ^{32}P has been determined from measurements of its rate of effusion through a small hole into a vacuum (Knudsen's method); by studying the temperature dependence of the vapour pressure, the enthalpy of sublimation has been obtained.

In *activation analysis* an element E is determined by bombardment of the sample under examination (nearly always with neutrons in a reactor) and measurement of the intensity of radioactivity induced; another sample of known E content is treated simultaneously under identical conditions. After irradiation for a period which is usually a few times the half-life of the radioactive isotope being formed, sample and standard are suitably prepared for radioactive assay; the ratio of their activities then gives directly the ratio of the quantities of E present in them. The element E must, of course, have a reasonably high cross-section for neutron capture; the half-life of the product must be long enough for its manipulation and determination, but short enough for its radioactivity to be easily measurable. It is not, for example, feasible to determine carbon by neutron activation analysis owing to the low cross-section of ^{12}C for neutron capture and the long half-life of ^{14}C (5570 years) produced from ^{13}C, but the reaction $^{12}C(d, n)^{13}N$ in a cyclotron can be utilised.

Activation analysis is particularly suitable for the determination of trace impurities or constituents, and it can be used without destroying the sample if the element (or, strictly, the isotope) being determined has a much higher cross-section for neutron capture than anything else present, or if the product from it has a half-life different from those of any products from other constituents. Thus gold in lead is easily determined because of suitable differences in neutron capture cross-sections, and osmium in rhodium can be estimated because the ^{193}Os produced has a half-life (32 hours) much longer than that of the ^{104}Rh (4 min) which is formed simultaneously, so that the activity a few hours after irradiation is all due to the osmium isotope.

Dating applications

Methods for the estimation of the age of objects, and of vegetable and mineral deposits, which involve radioactivity depend on calculation of when the sample under examination was removed from free exchange with its surroundings. This may involve determination of either a species formed during a radioactive decay or of the residual activity of an isotope which is undergoing decay.

The former method may be illustrated by helium dating. Helium present in uranium minerals has almost certainly been formed from α-particles. A gram of uranium in equilibrium with its decay products produces approximately 10^{-7} g of helium per year, so that if the uranium and helium contents of a mineral are known its age can be estimated. Correction must be made if thorium (which also gives rise to helium in a different decay series) is present, and since some helium may have escaped, the age so obtained is likely to be a lower limit.

The latter method is typified by radiocarbon (^{14}C) dating: ^{14}C is formed in the upper atmosphere by the action of cosmic radiation on ^{14}N, and it subsequently undergoes β^--decay to ^{14}N with a half-life of 5730 years. Carbon which has been isolated from the carbon dioxide cycle as bone or wood loses half of its ^{14}C activity every 5730 years, and comparison of its specific β^--activity with that of carbon still in circulation enables the date of isolation of the carbon in the material under investigation to be deduced. It must, of course, be assumed that the rate of production of ^{14}C has been constant over the period of time involved, and the method ceases to be accurate over periods longer than a few times the half-life of ^{14}C. Fluorine dating is mentioned in Section 16.1.

Exchange reactions and tracer studies

Isotopic exchange reactions such as

$$H_2O + D_2O \rightleftharpoons 2HOD$$

are always thermodynamically feasible, since although ΔH° for such processes is very nearly zero, there is always a positive standard entropy of mixing and hence a negative value of ΔG°. The study of whether exchange between isotopes of an element occurs in a particular system therefore really provides information about rates and mechanisms, though correlations with structure may often be made.

Hydrogen attached to oxygen or nitrogen, in organic or in inorganic compounds, always exchanges very rapidly with deuterium in the form of deuterium oxide, whereas hydrogen bonded to carbon exchanges with extreme slowness except in those cases where it is known from independent evidence to be acidic (e.g. in acetylene or nitromethane). The rapid exchange of hydrogen bonded to oxygen is related to the high mobilities of the proton and the hydroxide ion in water as found by conductivity and transport measurements: the low activation energy of both processes arises from the polarity of the O—H bond, which favours the formation of a transition state in which one bond is being broken while another is being made. Hydrogen in the ions $H_2PO_2^-$ and HPO_3^{2-} does not exchange with deuterium when salts containing these ions are crystallised from deuterium oxide; it is therefore deduced that the ions contain no O—H bond and that in them hydrogen is bonded to phosphorus. This conclusion is confirmed by the presence of P—H stretching bands in the infrared spectra of the appropriate sodium salts.

The exchange of ligands in complexes with free labelled ligands in solution has been extensively investigated in recent years. The ions $[Fe(CN)_6]^{3-}$ and $[Fe(CN)_6]^{4-}$ do not exchange with ^{14}C-labelled CN^- in aqueous solution, whereas the corresponding exchange with $[Ni(CN)_4]^{2-}$ is very rapid. This is not because the last species is less stable thermodynamically than the iron-containing complexes (it has in fact a very high formation constant), but because it is in equilibrium with CN^- and the ion $[Ni(CN)_5]^{3-}$:

$$[Ni(CN)_4]^{2-} + CN^- \rightleftharpoons [Ni(CN)_5]^{3-}$$

No comparable species containing an extra cyanide ion is formed by $[Fe(CN)_6]^{3-}$ or $[Fe(CN)_6]^{4-}$, and the absence of exchange shows that dissociation to form $[Fe(CN)_5]^{2-}$ or $[Fe(CN)_5]^{3-}$ does not occur to any significant extent. The rapid 'exchange' of ^{59}Fe between $[Fe(CN)_6]^{3-}$ and labelled $[Fe(CN)_6]^{4-}$ is really a rapid *electron transfer* between the ions.

In net chemical reactions isotopic labelling often provides useful information about the reaction pathway. In the oxidation of aqueous sulphite by chlorate, for example, ^{18}O is shown to be transferred from oxidant to reductant directly, presumably via formation and hydrolysis of an intermediate $O_3SOClO_2^{3-}$, since the reaction is much faster than the exchange of ^{18}O between chlorate ion and water. When hydrogen peroxide labelled with ^{18}O is oxidised by acidified aqueous permanganate, all of the ^{18}O is found in the oxygen evolved and none is present in the solution, showing that the permanganate acts by deprotonating the hydrogen peroxide. In organic chemistry, ^{18}O labelling has been used to show that hydrolysis of carboxylate esters takes place via acyl–oxygen, rather than alkyl–oxygen, fission. As we shall see in Chapter 21, tracer studies have been of immense importance in establishing the nature and mechanism of reactions of transition metal complexes.

Applications in structure determination

The background to this subject is dealt with further in the next chapter, but it is convenient to mention here some of the principal applications of isotopes in structure determination and to make some general points about isotopic syntheses of inorganic compounds.

Interatomic distances and angles in simple molecules are usually determined by electron diffraction (which lies outside the scope of this account) or rotational spectroscopy, either pure rotational (microwave) spectroscopy or from the rotational fine structure of vibrational or electronic spectra. The principle involved is most simply illustrated with reference to pure rotational spectroscopy, which corresponds to absorption in the far infrared and requires that the molecule under investigation possesses a permanent dipole moment. For a linear molecule the difference in frequency between successive lines in the rotational spectrum is given by

$$\Delta\nu = \frac{h}{4\pi^2 I}$$

I here is the moment of inertia about an axis passing through the centre of gravity of the molecule and perpendicular to the linear axis. I is defined by

$$I = \sum_i m_i r_i^2$$

where m_i and r_i are the mass and distance from the centre of gravity of the ith atom in the molecule. For a diatomic molecule $m_1 r_1 = m_2 r_2$, whence it can be shown that

$$I = \frac{m_1 m_2}{m_1 + m_2} r^2$$

Since the masses of the atoms are known, determination of I from $\Delta\nu$ permits evaluation of r, the interatomic distance. For an unsymmetrical linear triatomic molecule (such as carbonyl sulphide, OCS), however, there are two interatomic distances in the expression for the moment of inertia, and clearly it is impossible to obtain values for these two unknowns from a single value of I. If, however, $\Delta\nu'$, and hence I', for an isotopically substituted species is obtained, it is possible to derive values for two interatomic distances if it is assumed that these are not affected by isotopic substitution, i.e. that such substitution affects the moment of inertia only by virtue of a known change in atomic mass. (This assumption can be shown to be true for diatomic molecules.) Thus both bond lengths in carbonyl sulphide may be obtained from the pure rotational spectra of, say, $^{16}O^{12}C^{32}S$ and $^{18}O^{12}C^{32}S$. For more complex molecules many more isotopic substitutions are, of course, necessary.

The effect of isotopic substitution on the fundamental vibration frequency of a diatomic molecule was mentioned in Section 1.7. It is almost universal practice in vibrational spectroscopy to quote not true frequencies in s^{-1} (or Hz) but wave numbers per centimetre (i.e. the number of wavelengths per centimetre), the units of which are cm^{-1}; ν_0 in s^{-1} and $\bar{\nu}_0$ in cm^{-1} are related by the equation

$$\nu_0 = \bar{\nu}_0 c_0$$

where c_0 is the velocity of light (in cm s^{-1}) in a vacuum. The wave number of a vibration is often loosely called its frequency (e.g. 'The fundamental vibration frequencies of $H^{35}Cl$ and $D^{35}Cl$ occur at 2886 and 2091 cm^{-1} respectively') and the symbol ν is often used whether the units are s^{-1} or cm^{-1}; this seldom leads to difficulty in qualitative discussions of spectra, but in this book we shall for consistency distinguish between ν and $\bar{\nu}$.

The effect of isotopic substitution on the positions of vibrational (Raman or infrared) bands is easily calculated from the ratio of the reduced masses of the species concerned, the fundamental relationship being

$$\bar{\nu} = \frac{1}{2\pi c_0}\sqrt{\frac{k}{\mu}}$$

in which k is the restoring force constant and μ the reduced mass, $m_1 \cdot m_2/m_1 + m_2$. If k is taken as the same for both species

$$\frac{\bar{\nu}_1}{\bar{\nu}_2} = \sqrt{\frac{\mu_2}{\mu_1}}$$

For $H^{35}Cl$ and $D^{35}Cl$ the ratio so calculated is 1.395, and the observed ratio 1.380. So long as the other atom, A, in a diatomic hydride is much heavier than hydrogen or deuterium, the ratio of the A–H and A–D frequencies will approach $\sqrt{2}:1$, i.e. 1.41 : 1. In the interpretation of the spectrum of a complex molecule it is therefore sometimes useful to examine the effect of replacing hydrogen by deuterium in order to identify an element – hydrogen vibration; this is particularly true of the study of molecules containing strong hydrogen bonds, for which O—H vibrations often appear at much lower wave numbers than for 'free' OH.

For bonds involving two heavier atoms, e.g. >C=O or —C≡N, the effect of isotopic replacement is, of course, much smaller, but it is still significant, and it is widely employed in assigning spectra of polyatomic species containing two or more functional groups that absorb in the same region.

The methods by which isotopically labelled species are obtained constitute an important area of modern synthetic inorganic chemistry. First, however, it should be said that spectroscopic studies of isotopically substituted molecules do not always involve special preparations: for many elements (e.g. chlorine, silicon, germanium), natural isotopic abundances ensure that ordinary compounds contain several species. In GeH_3Cl, for instance, ^{35}Cl and ^{37}Cl, and ^{70}Ge, ^{72}Ge, ^{74}Ge and ^{76}Ge are all present in proportions sufficient to give rise to observable pure rotational spectra of $^{70}GeH_3{}^{35}Cl$, $^{70}GeH_3{}^{37}Cl$ etc. Where special syntheses are necessary, they should be so designed as to make the best possible use of the isotope to be incorporated. Deuterated ammonia, ND_3, for example, would not be made by exchange between NH_3 and D_2O, as a result of which a large proportion of the deuterium would be wasted by conversion to HOD if replacement of H in NH_3 by D were to be substantially complete: the action of D_2O on magnesium nitride, Mg_3N_2, would be a much better procedure. For the preparation of DCl, the action of D_2O on a readily hydrolysed chloride (e.g. C_6H_5COCl or $AlCl_3$) would be a suitable method. Isotopic carbon is usually supplied as barium carbonate; this may be converted into compounds which are particularly useful in synthesis by the following reactions:

$$Ba{*}CO_3 \xrightarrow{\text{acid}} {*}CO_2 \xrightarrow{\text{C}} {*}CO \xrightarrow{H_2} {*}CH_3OH$$

$$Ba{*}CO_3 \xrightarrow{\text{acid}} {*}CO_2 \xrightarrow{\text{RMgBr}} R{*}CO_2H$$

$$Ba{*}CO_3 \xrightarrow{NH_4Cl,\ K} K{*}CN \quad ({*}C = {}^{13}C \text{ or } {}^{14}C)$$

From $K^{15}NO_3$, $^{15}N_2$ could be obtained by the sequence

$$K^{15}NO_3 \xrightarrow{Hg,\ H_2SO_4} {}^{15}NO \xrightarrow{Cu} {}^{15}N_2$$

and $^{15}N^{14}N$ by heating the $K^{15}NO_3$ with lead, forming $K^{15}NO_2$, and then heating this with $^{14}NH_4Cl$:

$$K^{15}NO_2 + {}^{14}NH_4Cl \rightarrow {}^{15}N^{14}N + KCl + 2H_2O$$

The devising of simple routes to isotopically labelled species often calls for considerable ingenuity; the reader may obtain practice by attempting problem 6 at the end of this chapter.

Further spectroscopic applications of isotopes lie in the field of nuclear magnetic resonance spectroscopy. Syntheses of compounds incorporating a rare isotope of nuclear spin quantum number $I = \frac{1}{2}$ (e.g. ^{13}C) are just the same as those of comparable compounds required for rotational and vibrational spectroscopy; however, syntheses of compounds in which nuclei with $I = \frac{1}{2}$ have been replaced by isotopes which do not give rise to spectra under the same conditions are also important, since by this method spectra may be much simplified. Replacement of H by D is especially important in this connection, as is the use of deuterochloroform, $CDCl_3$, as solvent.

Neutron diffraction should also be mentioned briefly here. Hydrogen atoms do not usually scatter X-rays sufficiently for their positions in crystals to be determined accurately by X-ray diffraction. For the scattering of neutrons, however, deuterium has one of the highest scattering factors of any nuclide. Hence deuterium compounds (e.g. KDF_2, D_2O) are often used in structure determinations by neutron diffraction.

The kinetic isotope effect

Hydrogen and deuterium, or analogous compounds containing them, react at appreciably different rates if the bonds to these atoms are involved in the rate-determining step of the reaction. If, for example, we consider the reactions

$$H_2 + AB \rightarrow HA + HB$$

and

$$D_2 + AB \rightarrow DA + DB$$

it is believed that the absolute energies of the transition states (or activated complexes) H_2AB and D_2AB are very similar; the activation energies for formation of the transition states, however, involve formation from H_2 and D_2 in their lowest energy levels, i.e. when they have only zero-point energies. But, as we have shown earlier, H_2 and D_2 have different zero-point energies because, having different masses, they have different fundamental vibration frequencies; H_2 has the higher zero-point energy, and therefore formation of the transition state H_2AB from H_2 requires less energy than formation of D_2AB from D_2. The relationship between the rate constant, k, of a reaction, the activation energy E and the frequency factor A is given by

$$k = Ae^{-E/RT}$$

A small difference in E therefore makes a significant difference in k. If we compare reactions of hydrogen and tritium, the isotope effect is even more marked; and it is large enough to be detected for reactions involving species containing ^{14}C, ^{15}N and ^{18}O compared with corresponding reactions of species containing ^{12}C, ^{14}N and ^{16}O; for heavier elements the effect is too small to be of any importance.

The rate-determining steps in the photochemical reaction of hydrogen and chlorine and the thermal reaction of hydrogen and bromine both appear to be

$$H_2 + X \rightarrow HX + H$$

where X = Cl or Br. In each case the deuterium reaction is slower by an amount equivalent to a difference in activation energy which is roughly equal to the difference in zero-point energies of H_2 and D_2 (7 kJ mol^{-1}). Conversely, the rates of nitration of benzene and tritiated benzene by nitric acid in sulphuric acid are equal; the C—H or C—T bond is

therefore not involved in the rate-determining step, which is deduced to be addition of the NO_2^+ ion present to benzene, forming $C_6H_6NO_2^+$; this ion subsequently breaks down rapidly to give $C_6H_5NO_2$ and H^+.

PROBLEMS

1 The isotopic masses of ^2_1H and ^4_2He are 2.014 10 and 4.002 60 m_u respectively, and the velocity of light in a vacuum is 2.998×10^8 m s^{-1}. Calculate the quantity of energy (in J) liberated when two moles of ^2_1H undergo fusion to form one mole of ^4_2He.

2 The radioactive isotope $^{60}_{27}Co$, which has now replaced radium in the treatment of cancer, can be made by a (n, p) or (n, γ) reaction. For each reaction indicate the appropriate target nucleus. If the half-life of $^{60}_{27}Co$ is 7 years, evaluate the decay constant in s^{-1}.

3 The high resolution ^{19}F nuclear magnetic resonance spectrum at 60 MHz of chlorine trifluoride at $-50\,°C$ consists of two peaks of intensity ratio 2:1, the more intense peak being a doublet and the less intense peak a triplet. At 60° the spectrum collapses to a single band.
Discuss the interpretation of these observations.

4 A small amount of the radioactive isotope ^{212}Pb was mixed with a quantity of a non-radioactive lead salt containing 0.01 g of lead (atomic weight 207). The whole was brought into aqueous solution and lead chromate was precipitated by addition of a soluble chromate. Evaporation of 10 cm^3 of the supernatant liquid gave a residue having a radioactivity $\frac{1}{24\,000}$ of that of the original quantity of ^{212}Pb. Calculate the solubility of lead chromate in mol dm^{-3}.

5 The β^--activity of 1 g of carbon made from the wood of a recently cut tree is 15.3 counts min^{-1}. If the activity of 1 g of carbon isolated from the wood of an Egyptian mummy case is 9.4 counts min^{-1} under the same conditions, how old is the wood of the mummy case? ($t\frac{1}{2}$ for $^{14}C = 5730$ years.)

6 Suggest preparations for each of the following substances, using deuterium oxide as the only source of deuterium: LiD, D_2O_2, $NaDCO_3$, ND_4Cl, C_6H_5D.

7 If the oxide P_4O_6 is dissolved in an aqueous solution of sodium carbonate a compound **A** of formula Na_2HPO_3 may be crystallised from the solution. The infrared spectrum of **A** contains a band at 2300 cm^{-1}. The corresponding band in the infrared spectrum of **B**, which is obtained by an analogous process from P_4O_6 and a deuterium oxide solution of sodium carbonate, is at 1630 cm^{-1}. On recrystallisation of **A** from deuterium oxide, however, its infrared spectrum is not affected.
Discuss the interpretation of these observations.

REFERENCES FOR FURTHER READING

Banwell, C.N. (1983) *Fundamentals of Molecular Spectroscopy*, 3rd edition, McGraw-Hill, New York. A readable introduction to many branches of spectroscopy.

Butler, I.S. and Harrod, J.F. (1989) *Inorganic Chemistry*, Benjamin/Cummings, Redwood City, California. Chapters 5–8 give a good survey of structural methods with many examples.

Choppin, G.R. and Rydberg, J. (1980) *Nuclear Chemistry*, Pergamon, Oxford. An advanced account of theory and applications.

Cox, P.A. (1989) *The Elements*, Oxford University Press. An account of their origin, abundance and distribution.

Ebsworth, E.A.V., Rankin, D.W.H and Cradock, S. (1987) *Structural Methods in Inorganic Chemistry*, Blackwell, Oxford. The definitive treatment of this subject.

Friedlander, G., Kennedy, J.W., Macias, E.S. and Miller, J.M. (1981) *Nuclear and Radiochemistry*, 3rd edition, Wiley, New York. A general textbook of radiochemistry and its applications.

Greenwood, N.N. and Earnshaw, A. (1984) *Chemistry of the Elements*, Pergamon, Oxford. Chapter 1 gives a brief summary of modern views on the origins of the elements.

Smith, D. (1990) *Inorganic Substances*, Cambridge University Press. Chapter 2 gives an elementary review of structural methods in inorganic chemistry.

The Open University (1976) S304 Units 7 and 8, *Nuclear Magnetic Resonance Spectroscopy*, Open University Press, Milton Keynes. A very clear treatment of nuclear magnetic resonance with special reference to the uses of ^{1}H and ^{13}C nuclear magnetic resonance spectroscopy in organic chemistry.

The Open University (1977) S304 Unit 28, *The Chemical Basis of Nuclear Energy Programmes*, Open University Press, Milton Keynes. An account in simple terms of the nuclear energy industry.

2 *Quantum theory and atomic structure*

2.1 Introduction

In this and the following chapter we shall describe the electronic structures of the elements and some physical properties of isolated atoms that are of special importance in chemistry. No adequate discussion of electronic structures is possible without reference to quantum theory, especially to the more modern branch of it known as wave mechanics. We therefore begin with a brief account of this subject and its applications to chemical problems. The treatment given here is mainly qualitative; for more rigid derivations of mathematical relationships, and for greater detail, the references at the end of the chapter should be consulted.

It is important to understand that the development of quantum theory took place in two stages. In the older quantum theory (1900–25), the electron was treated as a particle, and the achievements of greatest significance to inorganic chemistry were the interpretation of atomic spectra and the assignment of electronic configurations. In the newer quantum theory (1925–) the electron is treated as a wave (hence the name *wave mechanics*), and the chief successes in chemistry are the elucidation of the basis of stereochemistry and a treatment (exact, however, only for species involving light atoms) of the energies of covalent bonds.

Since all the results obtained by the use of the older quantum theory may also be obtained from wave mechanics, it may seem unnecessary to refer to the former, and sophisticated treatments of theoretical chemistry in fact seldom do so. Most chemists, however, often find it easier and more convenient to consider the electron as a particle rather than as a wave, e.g. in discussing atomic spectra, relativistic effects or magnetic properties; and, moreover, in several areas of the subject the nomenclature of the older quantum theory is still employed. On practical as well as on historical grounds, therefore, there is a strong case for introducing the discussion of wave mechanics by a brief account of the older quantum theory, and this approach is adopted here.

Most (though not all) of this chapter is about the hydrogen atom and a few other one-electron species, for which quantum theory leads to exact values for many properties. For systems containing more than one electron, on the other hand, it is rarely satisfactory to use calculated data in place of experimental values, and for atoms of heavy elements only the roughest estimates are possible. In the next few chapters, which are mainly about atoms, molecules, or ions containing several electrons, we shall therefore be using quantum theory mostly to rationalise, rather than to predict, physical data – a distinction of considerable (and not always appreciated) scientific importance. Throughout this book particular attention is paid to making it clear in which way quantum theory is being used.

2.2 The older quantum theory

The energy of the radiation emitted by a hot body is spread over a continuous spectrum which depends on the temperature of that body. At lower temperatures the radiation is mainly of low energy and occurs in the infrared, but as the temperature rises the radiation becomes successively dull red, bright red, and white. All attempts to account for the fact that the proportion of short wavelength (blue) light increases as the temperature is raised failed until Planck in 1901 suggested that energy could be absorbed or emitted only in quanta of magnitude ΔE given by

$$\Delta E = h\nu$$

where ν is the frequency of the radiation and h is Planck's constant, 6.626×10^{-34} J s. Since $\nu = c_0/\lambda$, where c_0 is the velocity of electromagnetic radiation in a vacuum (2.998×10^8 m s^{-1}) and λ is the wavelength of the radiation, this relationship may also be written

$$\Delta E = hc_0/\lambda$$

On the basis of this equation Planck succeeded in deriving a relative intensity/wavelength/temperature relationship which was in very close agreement with the experimental data, though since the derivation is difficult we shall not reproduce it here.

Another early success of the quantum theory was Einstein's treatment of the photoelectric effect in 1905. When light falls on a metal, electrons are ejected if, but only if, the frequency of the light is above a certain minimum value; above this value the number of electrons produced is proportional to the intensity of the light, but their kinetic energy is linearly related only to its frequency. Einstein showed that these results are compatible with the behaviour of the light as a stream of particles (photons) each having energy $h\nu$, the kinetic energy of the ejected electrons being then given by the equation

$$\tfrac{1}{2}mv^2 = h\nu - W$$

where m is the mass and v the velocity of the electrons, and W is the minimum energy (the so-called *work function*) required to remove an electron from the solid metal.

In terms of chemical significance, however, the most important application of the quantum theory was to the interpretation of the atomic spectrum of hydrogen on the basis of Rutherford's model of the atom, in this case a very small nucleus consisting of a positively charged proton, and a negatively charged electron of mass approximately $\frac{1}{1850}$ times that of the proton. According to the laws of classical mechanics such an atom cannot be a static system; if it were, the electron would fall into the nucleus. Nor is a system in which the electron moves round the nucleus satisfactory: the electron would interact with the positive charge and spiral into the nucleus, emitting energy in the form of continuous radiation as it did so. In practice, of course, atoms (other than those of radioactive isotopes) do not emit radiation in the absence of external stimulation, and when they do so they emit only particular frequencies which are characteristic of the atom concerned.

When, for example, an electric discharge is passed through hydrogen, the molecule is dissociated into atoms, in many of which electrons are raised to high energy levels. When such electrons fall back to lower energy levels, a series of spectral lines is emitted. The atomic spectrum of hydrogen contains four lines in the visible region, at 656.28, 486.13, 434.05 and 410.17 nm ($\bar{\nu} = 15\,233$, $20\,565$, $23\,033$ and $24\,373\ \text{cm}^{-1}$ respectively). As long ago as 1885, Balmer pointed out that these wavelengths obey the relationship

$$\bar{\nu} = \frac{1}{\lambda} = R_{\text{H}}\left(\frac{1}{2^2} - \frac{1}{n^2}\right)$$

where $n = 3$, 4, 5 and 6 and R_{H} is the *Rydberg constant* for hydrogen, equal to $109\,678\ \text{cm}^{-1}$. It was later found that other series of lines occur in the ultraviolet (Lyman series) and infrared (Paschen, Brackett and Pfund series). All the lines in all the series fit the general equation

$$\bar{\nu} = \frac{1}{\lambda} = R_{\text{H}}\left(\frac{1}{n_1^2} - \frac{1}{n_2^2}\right)$$

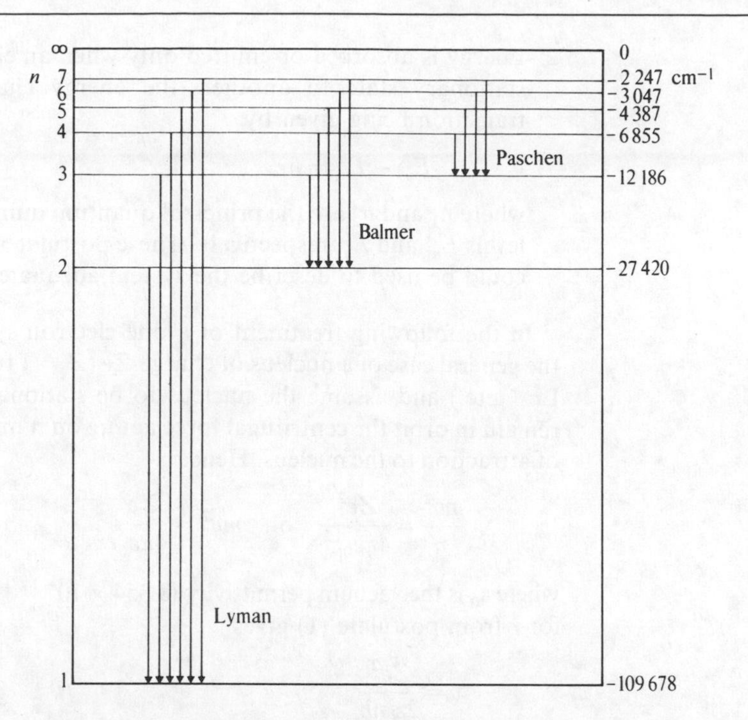

Figure 2.1
Origins of the Lyman, Balmer and Paschen series in the emission spectrum of atomic hydrogen.

where $n_2 > n_1$; for the Lyman series $n_1 = 1$; for the Balmer series, as we have already seen, $n_1 = 2$; and for the Paschen, Brackett and Pfund series $n_1 = 3$, 4 and 5 respectively. The origins of the Lyman, Balmer and Paschen series are shown in Fig. 2.1, and the interpretation of the atomic spectrum of hydrogen is discussed in the following section.

2.3 Bohr's theory of the atomic spectrum of hydrogen

Bohr (1913) combined elements of quantum theory and classical physics in his treatment of the hydrogen atom. He postulated that for an electron in an atom:

1. There exist so-called *stationary states* in which the energy of the electron round the nucleus is constant, such states being characterised by circular orbits in which the electron has angular momentum mvr (where m is the mass and v the velocity of the electron, and r is the radius of the orbit) which is an integral multiple of $h/2\pi$, i.e.

$$mvr = n\left(\frac{h}{2\pi}\right)$$

where n is known as the *principal quantum number*;

2. Energy is absorbed or emitted only when an electron moves from one stationary state to another, the energy change attending such a transition being given by

$$E_{n_2} - E_{n_1} = h\nu$$

where n_1 and n_2 are the principal quantum numbers referring to energy levels E_{n_1} and E_{n_2} respectively. These postulates apart, classical physics could be used to describe the system adequately.

In the following treatment of a one-electron system we shall consider the general case of a nucleus of charge Ze ($Z = 1$ for H^+, 2 for He^{2+}, 3 for Li^{3+} etc.) and assume the nucleus to be stationary. For the electron to remain in orbit the centrifugal force acting on it must be equal to the force of attraction to the nucleus. Hence

$$\frac{mv^2}{r} = \frac{Ze^2}{4\pi\varepsilon_0 r^2} \quad \text{or} \quad mv^2 = \frac{Ze^2}{4\pi\varepsilon_0 r}$$

where ε_0 is the vacuum permittivity ($8.854 \times 10^{-12}\ \mathrm{J^{-1} C^2 m^{-1}}$). Substitution for r from postulate (1) gives

$$v = \frac{Ze^2}{2\varepsilon_0 nh}$$

For an atom in which the electron has angular momentum $n(h/2\pi)$, the total energy of the atom, E_n, is the sum of its kinetic energy ($\frac{1}{2}mv^2$) and its potential energy. The latter is conveniently taken as zero when the electron is removed to infinity, i.e. when the atom is ionised. When the electron is at a distance r from the nucleus the potential energy is then equal and opposite to the work done in removing the electron to infinity against the electrostatic attraction to the nucleus, i.e.

$$-\int_r^\infty \frac{Ze^2}{4\pi\varepsilon_0 r^2}\,\mathrm{d}r = \frac{-Ze^2}{4\pi\varepsilon_0 r}$$

So

$$E_n = \frac{-Ze^2}{4\pi\varepsilon_0 r} + \tfrac{1}{2}mv^2$$

Substitution for mv^2 leads to

$$E_n = \frac{-Ze^2}{8\pi\varepsilon_0 r} = -\tfrac{1}{2}mv^2$$

and further substitution for v to

$$E_n = -\tfrac{1}{2}m\left(\frac{Ze^2}{2\varepsilon_0 nh}\right)^2 = -\frac{me^4Z^2}{8\varepsilon_0^2 h^2}\cdot\frac{1}{n^2}$$

Combination of this formula with postulate (2) gives

$$E_{n_2} - E_{n_1} = h\nu = -\frac{me^4Z^2}{8\varepsilon_0^2h^2}\left(\frac{1}{n_2^2} - \frac{1}{n_1^2}\right)$$

Then $\bar{\nu}$, which is equal to ν/c_0, is given by

$$\bar{\nu} = \frac{me^4Z^2}{8\varepsilon_0^2h^3c_0}\left(\frac{1}{n_1^2} - \frac{1}{n_2^2}\right)$$

In the case of hydrogen $Z = 1$, and comparison with the general equation for the atomic spectrum given at the end of the last section shows that

$$R_H = \frac{me^4}{8\varepsilon_0^2h^3c_0}$$

Substitution of the values for m, e, ε_0, h and c_0 yields $R_H = 109\,738$ cm^{-1}, which, considering the simplicity of the model, is in very good agreement with the experimental value of $109\,678$ cm^{-1}; and the agreement may be made perfect if a correction is made for motion of the nucleus by replacing m, the mass of the electron, by the reduced mass of the atom, $mM/m+M$, where M is the mass of the nucleus.

The Bohr theory may also be used to derive values for the radii of possible orbits of the electron. Rearrangement of

$$\frac{mv^2}{r} = \frac{Ze^2}{4\pi\varepsilon_0r^2}$$

gives

$$r = \frac{Ze^2}{4\pi\varepsilon_0mv^2}$$

and substitution of

$$v = \frac{Ze^2}{2\varepsilon_0nh}$$

then leads to

$$r = \frac{\varepsilon_0h^2}{\pi mZe^2}n^2$$

For the hydrogen atom ($Z = 1$) this gives

$$r = n^2 \times 0.529 \times 10^{-10}\ \text{m} \quad \text{or} \quad n^2 \times 0.529\ \text{Å}$$

Thus for $n = 1$, $r = 0.529$ Å; for $n = 2$, $r = 2.10$ Å; the value for $n = 1$ is often called the *Bohr radius* of the hydrogen atom and given the symbol a_0.

An increase in the principal quantum number from $n = 1$ to $n = \infty$ corresponds to ionisation of the atom, and the *ionisation energy*, *I*, is given by

$$I = -\frac{me^4Z^2}{8\varepsilon_0 h^2}\left(\frac{1}{\infty^2} - \frac{1}{1^2}\right) = 2.180 \times 10^{-18}\ \mathrm{J}$$

Ionisation energies are often expressed in electron volts, eV, and since 1 eV = 1.602×10^{-19} J,

$$I = \frac{2.180 \times 10^{-18}}{1.602 \times 10^{-19}} = 13.61\ \mathrm{eV}$$

The experimental value is 13.58 eV. For a mole of hydrogen atoms

$$I = 2.180 \times 10^{-18} \times 6.022 \times 10^{23} = 1.312 \times 10^6\ \mathrm{J\,mol^{-1}}$$

or $1\,312\ \mathrm{kJ\,mol^{-1}}$

2.4 *The extension of Bohr's theory to systems containing more than one electron*

Impressive as the success of the simple Bohr theory applied to one-electron systems was, extensive modifications were needed to enable it to cope with more complex species. In order to deal with the spectra of such species, which contain many more lines than that of hydrogen, three more quantum numbers had to be introduced. We shall describe these quantum numbers now in order to emphasise that the systematisation of atomic spectra preceded the development of wave mechanics; as we shall see in Section 2.8, however, the modern significance attached to the quantum numbers is different from that given here, though their permitted values remain unchanged.

On the extended Bohr theory, electron orbits may be elliptical or circular; the ratio $(l+1)/n$ gives the ratio of the minor axis of the ellipse to its major axis, and l, the *second (orbital) quantum number*, may have values 0, 1, 2 . . . $n-1$. The splitting of spectral lines in a magnetic field (the Zeeman effect) requires a *third (orbital magnetic) quantum number,* m_l, defining the plane that the electron orbit assumes with respect to the applied field; m_l has the values l, $(l-1)$ . . . 0 . . . $-(l-1)$, $-l$. Finally, a *fourth* (*spin magnetic*) *quantum number*, m_s, is introduced to account for the fine structure of many spectral lines (e.g. the doublets in the spectra of alkali metals). The electron is regarded as spinning whilst it rotates; any electron has *spin quantum number* s equal to $\frac{1}{2}$, but the magnetic field produced by the spinning of the electron can either reinforce or oppose the magnetic field produced by rotation of the electron in its orbit; m_s has the value $+\frac{1}{2}$ or $-\frac{1}{2}$.

For a one-electron system the fact that the spectrum can be accounted for in terms of a single quantum number n means that all orbits associated with a given value of n have the same energy. For atoms containing more than one electron, however, the energies of electrons in orbits of the same n value but different l values are, for reasons we shall give later, substantially different. Electrons in orbits having $l = 0, 1, 2, 3$ are known as *s, p, d, f* electrons respectively, and the same terminology is applied to the orbits. The letters *s, p, d* and *f* stand for sharp, principal, diffuse and fundamental, and refer to the appearance of spectral lines arising from transitions involving electrons in these orbits; an empirical selection rule limits observed transitions to those for which l changes by ± 1. Thus in the atomic spectrum of lithium, transitions from the 3*s*, 4*s* etc. orbits to the 2*p* orbit give rise to the sharp series, transitions from the 2*p*, 3*p*, 4*p* etc. orbits to the 2*s* orbit give rise to the principal series, transitions from the 3*d*, 4*d* etc. orbits to the 2*p* orbit give rise to the diffuse series, and transitions from the 4*f*, 5*f* etc. orbits to the 3*d* orbit give rise to the fundamental series. Part of the emission spectrum of the lithium atom is indicated in Fig. 2.2. At a later stage we shall refer to transitions between spectroscopic states rather than between orbits; the states are denoted by capital letters, and for consistency capital letters are therefore used here.

One other important pre-wave mechanics generalisation remains to be introduced at this stage. This is the empirical *exclusion principle* enunciated by Pauli in 1925, and it states that no two electrons in any

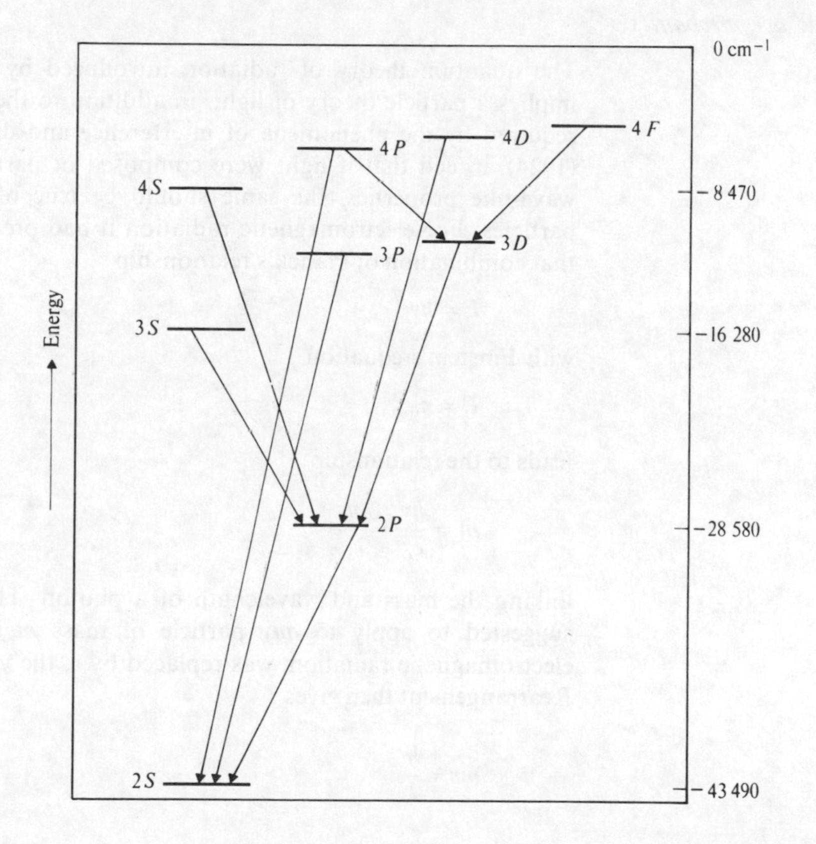

Figure 2.2
Emission of the lithium atom from energy levels with $n = 2$, 3 or 4 (doublet structure is not included).

atom can have the same values for all four of the quantum numbers n, l, m_l, and m_s. Thus for $n = 1$, the only permitted values of l and m_l are both zero, and m_s can be $+\frac{1}{2}$ or $-\frac{1}{2}$; the shell of electrons having $n = 1$ can therefore accommodate only two 1*s* electrons. For $n = 2$, $l = 0$ or 1; for $l = 0$, $m_l = 0$, $m_s = +\frac{1}{2}$ or $-\frac{1}{2}$; for $l = 1$, $m_l = 1, 0,$ or -1, and for each value of m_l, $m_s = +\frac{1}{2}$ or $-\frac{1}{2}$; the shell having $n = 2$ can therefore hold two 2*s* and six 2*p* electrons. For $n = 3$, $l = 0, 1,$ or 2; for $l = 0$ and $l = 1$, m_l and m_s may vary as for $n = 2$; for $l = 2$, $m_l = 2, 1, 0, -1$ or -2, each with $m_s = +\frac{1}{2}$ or $-\frac{1}{2}$; so for $n = 3$ the shell can take two 3*s*, six 3*p* and ten 3*d* electrons, a total of eighteen electrons in all. In a similar way it is easily shown that the shell having $n = 4$ can accommodate two 4*s*, six 4*p*, ten 4*d* and fourteen 4*f* electrons, making a total of thirty-two. The importance of the numbers 2, 6, 10 and 14 in the periodic classification has long been apparent, and although the Pauli principle cannot be verified directly, its value in the systematisation of the chemistry of the elements in terms of electronic configurations provides very powerful evidence for its correctness. We shall return to this topic in the next chapter.

2.5 *Wave mechanics*

The quantum theory of radiation introduced by Planck and Einstein implies a particle theory of light, in addition to the wave theory of light required by the phenomena of interference and diffraction. De Broglie (1924) argued that if light were composed of particles and yet showed wave-like properties, the same should be true of electrons and other particles. For electromagnetic radiation it had previously been accepted that combination of Planck's relationship

$$E = h\nu$$

with Einstein's equation

$$E = mc_0^2$$

leads to the relationship

$$m = \frac{h\nu}{c_0^2} = \frac{h}{c_0\lambda}$$

linking the mass and wavelength of a photon. This equation was now suggested to apply to *any* particle of mass m if c_0, the velocity of electromagnetic radiation, was replaced by v, the velocity of the particle. Rearrangement then gives

$$mv = \frac{h}{\lambda}$$

in which the left- and right-hand sides of the relationship stress particulate and wave behaviour respectively. The greater the mass and velocity of the particle, the shorter its wavelength. Since the wavelength associated with any macroscopic particle is smaller than the dimensions of any physical system, diffraction or other wave phenomena can never be observed with a macroscopic particle; but electrons accelerated to a velocity of 6×10^6 m s^{-1} by a potential of 100 V have a wavelength of approximately 1.2 Å, and the observed diffraction of such electrons by atoms in a crystal provides powerful support for the concept of the electron as a wave.

Let us now consider the Bohr atom again. For the electron to remain in a stationary state, it would have to be a standing wave, i.e. the circumference of the orbit would have to be an integral number of wavelengths:

$$n\lambda = 2\pi r$$

Substitution of $\lambda = h/mv$ gives

$$\frac{nh}{mv} = 2\pi r$$

or

$$mvr = n\left(\frac{h}{2\pi}\right)$$

Thus Bohr's first postulate (the quantisation of angular momentum – see Section 2.3) follows naturally from the wave nature of the electron instead of being introduced arbitrarily.

Another fundamental idea in wave mechanics is expressed by Heisenberg's *uncertainty principle* (1927), which we may illustrate for an electron by considering the problem of determining its position, x, and momentum, mv. The general principles of optics tell us that it is impossible to locate the electron more accurately than to within $\pm\lambda$, the wavelength of the photon used in looking for it. It is therefore essential to use radiation of the shortest wavelength possible; but the action of the photon on the electron results in the transfer of some of the photon's energy to the electron with a resultant increase in its velocity and momentum. Decreasing the wavelength (increasing the energy) of the photon increases the margin of uncertainty in the momentum of the electron, Δ_{mv}; increasing the wavelength of the photon reduces the degree of resolution possible and increases the uncertainty in the position of the electron, Δ_x. The product of these uncertainties is comparable in magnitude to Planck's constant; a rigorous treatment of the problem leads to

$$\Delta_{mv} \cdot \Delta_x \geqslant \frac{h}{4\pi}$$

Thus we can no longer talk of the electron in a hydrogen atom as having definite values for both position and momentum, and as a consequence of specifying precisely the momentum of the electron we must replace the definite orbits of the Bohr theory by more diffuse regions in which the electron may be found and consider only the *probability* of finding the electron there.

It may be noted that the uncertainty principle can also be stated in terms of other variables, for example energy, E, and time, t; the product of the uncertainties in these quantities is similarly given by

$$\Delta_E \cdot \Delta_t \geqslant \frac{h}{4\pi}$$

Thus transitions between excited states of atoms or nuclei of very short half-life give rise to energy changes of considerable indeterminacy and hence diffuse lines in the appropriate spectra. In the remainder of this chapter, however, we shall not be concerned with time as a variable.

2.6 The Schrödinger equation

The general equation for simple harmonic wave motion in one dimension (such as a vibrating string) is

$$\psi = A \sin \frac{2\pi x}{\lambda}$$

where ψ is a quantity whose magnitude is varying in a wave manner with respect to the distance x from the origin (the displacement from the position of rest in the case of a vibrating string), and λ and A are respectively the wavelength and the maximum magnitude (the amplitude) of ψ. Points at which $\psi = 0$ are called *nodes*. Differentiation of ψ with respect to x yields successively

$$\frac{d\psi}{dx} = A \frac{2\pi}{\lambda} \cos \frac{2\pi x}{\lambda}$$

and

$$\frac{d^2\psi}{dx^2} = -\frac{4\pi^2}{\lambda^2} A \sin \frac{2\pi x}{\lambda} = -\frac{4\pi^2}{\lambda^2} \psi$$

For a particle of mass m and velocity v the kinetic energy, T, is $\frac{1}{2}mv^2$; application of the de Broglie relationship $mv = h/\lambda$ gives

$$T = \frac{1}{2m}\left(\frac{h}{\lambda}\right)^2$$

whence

$$\frac{d^2\psi}{dx^2} = -\frac{8\pi^2 m}{h^2} T\psi$$

or

$$T = -\frac{h^2}{8\pi^2 m} \cdot \frac{1}{\psi} \cdot \frac{d^2\psi}{dx^2}$$

This equation applies only to a particle moving in a field-free space, i.e. a space in which V, the potential energy, is constant, and in which the particle can be regarded as having only kinetic energy, T. If the potential energy of the particle does vary, then the total energy, E, is given by $T+V$. Then substitution for T yields

$$E-V = -\frac{h^2}{8\pi^2 m} \cdot \frac{1}{\psi} \cdot \frac{d^2\psi}{dx^2}$$

or

$$\frac{d^2\psi}{dx^2} + \frac{8\pi^2 m}{h^2}(E-V)\psi = 0$$

This is the Schrödinger equation for the motion of a particle in one (the x) dimension. In it, m, the mass of the particle, and V, its potential energy expressed as a function of x, are known; the unknowns to be found by solving the equation are E, the allowed values for the energy of the particle, and ψ, which is called its *wave function*. For electromagnetic radiation ψ^2 measures intensity; for a particle it is a measure of the probability of finding the particle. Thus the probability of finding it between x and $x+dx$ in this instance is $\psi^2 dx$. The wave function ψ itself has no physical meaning, but if ψ^2 is a probability certain properties of ψ follow: (i) ψ must be finite for all values of x; (ii) ψ can have only one value for any value of x; (iii) ψ and $d\psi/dx$ must vary continuously with variation in x. It can be shown that because of these limitations on the values of ψ, there are only a limited number of values of E for which satisfactory equations can be obtained for ψ, and these values of E are related by integers. Thus quantised energy levels and quantum numbers appear naturally from the Schrödinger equation. Satisfactory equations for ψ are called *eigenfunctions*, and the corresponding values of E, *eigenvalues*.

For a particle moving in three-dimensional space, the square of the velocity is equal to the sum of the squares of its component velocities v_x, v_y and v_z in the axial directions x, y and z. Its kinetic energy is given by

$$T = \tfrac{1}{2}mv_x^2 + \tfrac{1}{2}mv_y^2 + \tfrac{1}{2}mv_z^2$$

and the Schrödinger equation for a single particle in three dimensions is

$$\frac{\partial^2\psi}{\partial x^2}+\frac{\partial^2\psi}{\partial y^2}+\frac{\partial^2\psi}{\partial z^2}+\frac{8\pi^2 m}{h^2}(E-V)\psi = 0$$

Then $\psi^2 \mathrm{d}x\mathrm{d}y\mathrm{d}z$ at any point (x, y, z) is a measure of the relative probability of finding the particle within the volume $\mathrm{d}x\mathrm{d}y\mathrm{d}z$ at (x, y, z). Multiplying the equation for ψ by a constant does not change the relative values of ψ at different points, and to maintain the concept of probability the equation obtained for ψ in a given system may have to be *normalised*, i.e. multiplied by a constant factor such that the integral of $\psi^2 \mathrm{d}x\mathrm{d}y\mathrm{d}z$ over all space is unity.

Our ultimate objective in this chapter is to apply the Schrödinger equation to calculate wave functions for the hydrogen atom. Before attempting this rather difficult task, however, we shall first consider some relatively simple problems relevant to the subject matter of the last chapter. (Some of these involve molecules rather than isolated atoms, but it is convenient to deal with them, too, at this stage.) In so doing we shall introduce some more new concepts and derive (or at least show how to derive) some important chemical relationships, and thus give the reader an idea of how fundamental and far-reaching are the applications of wave mechanics to modern chemistry.

2.7 *Applications of wave mechanics to some simple problems*

A particle in a one-dimensional box

The Schrödinger equation for the motion of a particle in one dimension is, as we have already seen,

$$\frac{\mathrm{d}^2\psi}{\mathrm{d}x^2}+\frac{8\pi^2 m}{h^2}(E-V)\psi = 0$$

where E is the total, and V the potential, energy of the particle. The one-dimensional 'box' consists of a line which extends from $x = 0$ to $x = a$, and the particle of mass m can move only on this line. There is no force acting on the particle within its box, and so its potential energy is zero from $x = 0$ to $x = a$; for $x < 0$ and $x > a$, however, the potential energy of the particle is infinite, so it must stay within these limits. The only restriction placed on E is that it must be positive and cannot be infinite. Then outside the box, where $E - V$ is $-\infty$,

$$\frac{\mathrm{d}^2\psi}{\mathrm{d}x^2} = \infty\psi$$

and inside the box

$$\frac{d^2\psi}{dx^2} = -\frac{8\pi^2 mE}{h^2}\psi$$

The limitations on acceptable wave functions given in the last section provide boundary conditions that have to be met by the solutions of this differential equation. If ψ is a continuous function of x (limitation (iii)) and has any value other than zero at $x = 0$ and $x = a$, it will have a value other than zero at points infinitesimally less than zero or greater than a. If ψ has a value other than zero outside the box, then $d^2\psi/dx^2 = \pm\infty$ and $d\psi/dx$ and ψ itself attain infinitely large positive or negative values, a violation of limitation (i). Limitation (i) therefore imposes the boundary condition that $\psi = 0$ at $x \leqslant 0$ and at $x \geqslant a$.

Since E is positive, the coefficient of ψ is a negative constant at any value of E, and the differential equation can be written

$$\frac{d^2\psi}{dx^2} = -k^2\psi \quad \text{where} \quad k^2 = \frac{8\pi^2 mE}{h^2}$$

The solution to this (a well-known general equation) is

$$\psi = A\sin kx + B\cos kx$$

where A and B are integration constants. For $x = 0$, $\sin kx = 0$ and $\cos kx = 1$, so $\psi = B$ at $x = 0$. However, we have already shown that ψ must be zero at $x = 0$, and therefore $B = 0$. Since ψ is also zero at $x = a$, here

$$\psi = A\sin ka = 0$$

Since the probability, ψ^2, that the particle will be at points between $x = 0$ and $x = a$ cannot be zero (the particle must be in the box), A cannot be zero and this equation can be valid only if

$$ka = n\pi$$

where $n = 1, 2, 3 \ldots$ (n cannot be zero, since this would make ψ^2 zero between $x = 0$ and $x = a$, i.e. the particle would not be in the box). Hence

$$\psi = A\sin\left(\frac{n\pi}{a}\right)x$$

and

$$E = \frac{k^2h^2}{8\pi^2 m} = \frac{n^2h^2}{8ma^2}$$

where $n = 1, 2, 3 \ldots$ and is the quantum number determining the energy of a particle of mass m confined within a length a. So we see that the limitations placed on the value of ψ lead to quantised energy levels whose spacing is determined by m and a.

The particle must be somewhere between $x = 0$ and $x = a$, i.e. the probability of finding it there is unity. So

$$\int_0^a \psi^2 \mathrm{d}x = A^2 \int_0^a \sin^2\left(\frac{n\pi}{a}\right) x \mathrm{d}x = 1$$

The value of this integral is $a/2$, so $A = (2/a)^{\frac{1}{2}}$.

The values of E for $n = 1$, 2 or 3, and the corresponding values of ψ and ψ^2 between $x = 0$ and $x = a$, are shown in Fig. 2.3. The different forms of the curves showing the variation of ψ^2 with x for different values of n mean that the probability of finding the particle at a given position depends on n and therefore on the energy of the particle; at $x = a/2$ it is, for example, a maximum for $n = 1$ or $n = 3$ but zero for $n = 2$. The number of nodes in the electron wave is given by $n - 1$.

Let us now consider some of the implications of the expression

$$E = \frac{n^2 h^2}{8ma^2}$$

First, as we have already seen, the fact that the energy of the particle can have only certain values corresponding to $n = 1, 2, 3 \ldots$ introduces the concept of a quantum number. Since E can never be zero, the particle in the box has a minimum energy of $h^2/8ma^2$ even at absolute zero – its *residual energy* or, as it is more commonly called, its *zero-point energy*. The spacing of the energy levels for a box of given length a depends inversely on m, i.e. it is much wider for electrons than for nuclei. For given values of a and n, E differs for different isotopes of an element. For a given value of m, the spacing of the energy levels is inversely proportional to a^2. Thus if we compare an electron in a molecule of length a few angströms with a molecule in a one-dimensional box of length a few centimetres, it is

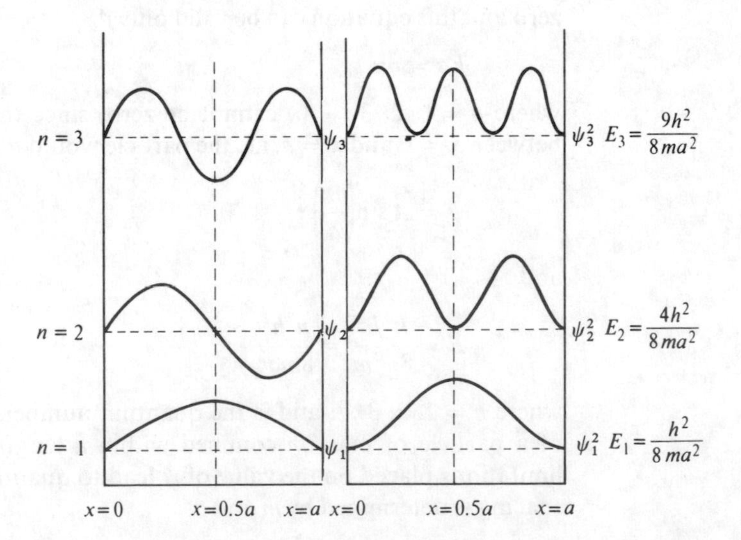

Figure 2.3
Variation of ψ and ψ^2 with x for $n = 1$, 2 and 3 for a particle in a one-dimensional box.

clear that, partly because of the difference in mass but more because of the difference in box length, the energy levels of the latter are very close indeed by comparison with those of the former; thus energy levels of thermal motion of molecules are so close that the kinetic energy of molecules and all larger species appears to vary continuously.

A particle in a cubic box

A similar treatment to the one given above leads to the Schrödinger equation for a constant-energy particle in a field-free cubic box of side a in which V is zero for any value of x, y or z between 0 and a but infinite outside these limits. This is

$$\frac{\partial^2\psi}{\partial x^2}+\frac{\partial^2\psi}{\partial y^2}+\frac{\partial^2\psi}{\partial z^2}+\frac{8\pi^2 m}{h^2}E\psi = 0$$

When ψ is a function of three independent variables x, y and z, it can be shown to be a product of three functions separately dependent on the same variables, i.e.

$$\psi(x, y, z) = \psi_x\psi_y\psi_z$$

where ψ_x, ψ_y and ψ_z are wave functions satisfying

$$\frac{\mathrm{d}^2\psi_x}{\mathrm{d}x^2} = -\frac{8\pi^2 mE_x}{h^2}\psi_x$$

and similar relationships between ψ_y and E_y and between ψ_z and E_z. The total energy E is then given by

$$E = E_x + E_y + E_z$$

Solution of these equations leads to

$$\psi_{n_x,n_y,n_z} = \left(\frac{8}{a^3}\right)^{\frac{1}{2}} \sin\left(\frac{n_x\pi}{a}\right)x \sin\left(\frac{n_y\pi}{a}\right)y \sin\left(\frac{n_z\pi}{a}\right)z$$

and

$$E_{n_x,n_y,n_z} = \frac{h^2}{8ma^2}(n_x^2 + n_y^2 + n_z^2)$$

Three quantum numbers, n_x, n_y and n_z, are now required to specify all permissible solutions of the Schrödinger equation for the system. The zero-point energy is $3h^2/8ma^2$, corresponding to the wave function $\psi_{1,1,1}$. The next energy level, $6h^2/8ma^2$, however, corresponds to any of the wave functions $\psi_{1,1,2}$, $\psi_{1,2,1}$ and $\psi_{2,1,1}$ (which have nodal planes at $z = a/2$, $y = a/2$ and $x = a/2$ respectively); such an energy level is said to be triply *degenerate*. It is easily seen that the next four energy levels, $9h^2/8ma^2$, $11h^2/8ma^2$, $12h^2/8ma^2$ and $14h^2/8ma^2$, have degeneracies of three, three, one, and six respectively.

A particle on a ring

Consider a particle rotating on a circle of radius r, with the potential energy of the system set equal to zero. The x of the one-dimensional box is now replaced by $r\phi$ where ϕ is the angle of rotation, and the Schrödinger equation becomes

$$\frac{d^2\psi}{d\phi^2} = -\frac{8\pi^2 mr^2}{h^2}E\psi$$

or, since I, the amount of inertia, is mr^2

$$\frac{d^2\psi}{d\phi^2} = -\frac{8\pi^2 I}{h^2}E\psi$$

Solutions of this equation for E are given by

$$E = \frac{h^2 M^2}{8\pi^2 I}$$

where $M = 0, 1, 2\ldots$ and is the rotational quantum number for a particle on a ring. This case is of no practical significance, but the treatment of the rotation of a rigid molecule in three dimensions (a rigid rotator), which is analogous to that of a particle in a cubic box, leads to important results. In the particular case of a rigid linear molecule having only one moment of inertia I, the energy of the system is given by

$$E = \frac{h^2}{8\pi^2 I}J(J+1)$$

where J, the rotational quantum number, is 0, 1, 2, 3 ... Further, it may be shown theoretically that for pure rotational spectra to be observed the molecule must have a dipole moment and that the selection rule $\Delta J = \pm 1$ operates; $\Delta J = +1$ corresponds to absorption, $\Delta J = -1$ to emission. Thus the change in energy on going from the Jth rotational level to the $(J+1)$th level is

$$\Delta E = \frac{h^2}{8\pi^2 I}\{(J+1)(J+2) - J(J+1)\}$$

$$= \frac{h^2}{4\pi^2 I}(J+1)$$

Hence the rotational spectrum consists of lines corresponding to $\Delta E = h^2/4\pi^2 I$, $2h^2/4\pi^2 I$, $3h^2/4\pi^2 I\ldots$ and the energies of the rotational levels are 0, $h^2/4\pi^2 I$, $3h^2/4\pi^2 I$, $6h^2/4\pi^2 I$ ... (it should be noticed that there is no zero-point energy of rotation). Thus observation of the rotational spectrum enables I to be determined and hence, as described in Section 1.9, for the one interatomic distance in the case of a diatomic molecule to be found. The quantisation of rotational energy is also important in connection with the heat capacities of gases. In the case of a rigid diatomic gaseous molecule, for example, rotation about its own axis does not occur because, since nearly all the mass of the molecule is in its nuclei, the moment of inertia for such a rotation would be extremely small and the corresponding quantum extremely large; this is why the heat capacity at ordinary temperatures of a rigid diatomic gas at constant volume is $\frac{5}{2}R$, where R is the gas constant, corresponding to three translational and only two rotational degrees of freedom.

The simple harmonic oscillator

For a simple harmonic oscillator the force F tending to restore the particle to its equilibrium position is proportional to its displacement x from that position (Hooke's law):

$$F = -kx$$

where k is the *force constant* (the negative sign arises because the force is in the direction opposite to that of the displacement from the equilibrium position). A force of this type represents the negative gradient of the potential energy–displacement curve. Therefore

$$\frac{dV}{dx} = -F = kx \quad \text{and} \quad V = \tfrac{1}{2}kx^2$$

the integration constant being zero since $V = 0$ at $x = 0$.

The Schrödinger equation now becomes

$$\frac{d^2\psi}{dx^2} + \frac{8\pi^2 m}{h^2}(E - \tfrac{1}{2}kx^2)\psi = 0$$

Solutions for ψ are given by

$$\psi_0 = \left(\frac{2a}{\pi}\right)^{\frac{1}{4}} e^{-ax^2}$$

$$\psi_1 = \left(\frac{2a}{\pi}\right)^{\frac{1}{4}} 2a^{\frac{1}{2}} x e^{-ax^2} \quad \text{etc.}$$

where $a = (\pi/h)(km)^{\frac{1}{2}}$. Hence it may be shown that solutions for E are given by

$$E = (n + \tfrac{1}{2})\frac{h}{2\pi}\sqrt{\frac{k}{m}} \text{ with } n = 0, 1, 2\ldots$$

n in this case being the vibrational quantum number. For a diatomic molecule m is replaced by the reduced mass μ (see Section 1.7) and the energy levels are given by

$$E = (n+\tfrac{1}{2})\frac{h}{2\pi}\sqrt{\frac{k}{\mu}}$$

Thus for a change from the lowest energy level to the next one

$$\Delta E = h\nu_0 = \frac{h}{2\pi}\sqrt{\frac{k}{\mu}}$$

and ν_0, the fundamental vibration frequency, is given by

$$\nu_0 = \frac{1}{2\pi}\sqrt{\frac{k}{\mu}}$$

It is instructive to see how these results compare with those obtained by a treatment of the same problem by classical mechanics. According to the latter the oscillator may have any energy, and it may be at rest and have zero energy. According to the quantum mechanical treatment, the energy of the oscillator is not only quantised but may never be less than $\frac{1}{2}h\nu_0$, the vibrational zero-point energy to which we referred in the last chapter in connection with the separation and uses of isotopes.

The minimum separation of vibrational energy levels is much larger than that of rotational energy levels (vibrational spectra occur in the infrared, pure rotational spectra in the far infrared or in the radio-frequency region), and at ordinary temperatures nearly all molecules of most simple substances are in the lowest vibrational state. However, where μ is large (heavy atoms) or k small (weak bonds) the vibrational quanta are small enough for some molecules to be in higher vibrational levels; thus chlorine, for example, has a molar heat capacity at constant volume which is appreciably greater than that of hydrogen at the same temperature.

In practice, vibrational spectra often show fine structure owing to rotational absorption, and from the same spectrum both the force constant and the interatomic distance may be evaluated in the case of a diatomic species; but for discussion of the details of such a spectrum, a textbook of physical chemistry should be consulted. We should, however, mention here the selection rules for infrared vibrational spectroscopy. These are two in number.

1. $\Delta n = \pm 1$
2. For a vibration to be active in the infrared, it must involve a change in the dipole moment of the molecule.

The first of these is true only for a strictly harmonic oscillator; in practice weak overtone bands corresponding to $\Delta n = \pm 2$, ± 3 etc. are often observed and energy levels become closer together as n increases, the oscillator then being said to be *anharmonic*. The second selection rule, however, is of great importance in the use of infrared spectroscopy and we shall often refer to it.

2.8 The hydrogen atom and other one-electron species

The Schrödinger equation for the motion of a particle in three dimensions is

$$\frac{\partial^2\psi}{\partial x^2}+\frac{\partial^2\psi}{\partial y^2}+\frac{\partial^2\psi}{\partial z^2}+\frac{8\pi^2 m}{h^2}(E-V)\psi = 0$$

In order to simplify the problem, we can to a first approximation consider the nucleus to be the centre of mass and to be at rest. In the general case of a one-electron system the charge on the nucleus is $+Ze$ and the potential energy of an electron at a distance r is given by

$$V = -\frac{Ze^2}{4\pi\varepsilon_0 r}$$

Hence

$$\frac{\partial^2\psi}{\partial x^2}+\frac{\partial^2\psi}{\partial y^2}+\frac{\partial^2\psi}{\partial z^2}+\frac{8\pi^2 m}{h^2}\left(E+\frac{Ze^2}{4\pi\varepsilon_0 r}\right)\psi = 0$$

The spherical symmetry of the system suggests that it would be advantageous to work in the polar coordinates r, θ, and ϕ shown in Fig. 2.4: r is the distance of the electron from the nucleus, θ is the angle which r makes with the z axis, and ϕ is the angle which the projection of r on the xy plane makes with the x axis. The Cartesian and polar coordinates are related by

$$x = r\sin\theta\cos\phi$$

$$y = r\sin\theta\sin\phi$$

$$z = r\cos\theta$$

$$r^2 = x^2+y^2+z^2$$

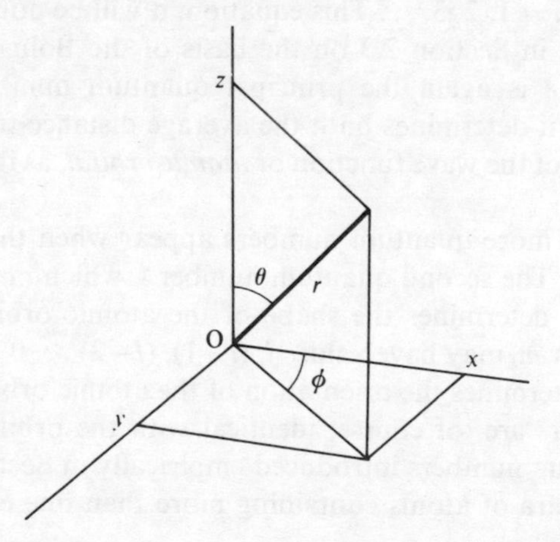

Figure 2.4
Relationship between Cartesian and polar coordinates.

The three-dimensional Schrödinger equation when written in polar coordinates becomes

$$\frac{1}{r^2}\frac{\partial}{\partial r}\left(r^2\frac{\partial\psi}{\partial r}\right)+\frac{1}{r^2\sin\theta}\frac{\partial}{\partial\theta}\left(\sin\theta\frac{\partial\psi}{\partial\theta}\right)+\frac{1}{r^2\sin^2\theta}\frac{\partial^2\psi}{\partial\phi^2}$$
$$+\frac{8\pi^2 m}{h^2}\left(E+\frac{Ze^2}{4\pi\varepsilon_0 r}\right)\psi=0$$

As with the particle in a box problem, limitations are imposed which make it possible to solve the wave equation. The wave function must be single-valued and continuous; it must be zero at infinity (i.e. the species must be finite); and the probability of finding the electron summed over all space must be unity, i.e. the wave function must be normalised.

The difficulties in dealing with this problem appear formidable, and it would seem that in order to represent the dependence of ψ on r, θ and ϕ a four-dimensional graph is necessary. Fortunately, however, it can be shown that the mathematical form of the equation makes it possible to express ψ as a product of three functions, ψ_r, ψ_θ and ψ_ϕ, which depend only on r, θ and ϕ respectively:

$$\psi_{r,\theta,\phi}=\psi_r\psi_\theta\psi_\phi$$

The wave equation can then be broken down into three separate equations expressing the dependence of ψ on r, θ and ϕ respectively; ψ_r is commonly called the *radial part of the wave function* and the product $\psi_\theta\psi_\phi$ the *angular part of the wave function*.

Solutions of the ψ_r part of the wave equation are found only for values of E given by

$$E_n=-\frac{me^4Z^2}{8\varepsilon_0^2h^2}\cdot\frac{1}{n^2}$$

where $n = 1, 2, 3, \ldots$ This equation, it will be noted, is identical with that derived in Section 2.3 on the basis of the Bohr theory of the hydrogen atom; n is again the principal quantum number. For a one-electron system it determines both the average distance from the nucleus and the energy of the wave function or *atomic orbital*, as it will henceforth often be called.

Two more quantum numbers appear when the angular equations are solved. The second quantum number l, which may have values $0, 1, 2 \ldots (n-1)$, determines the shape of the atomic orbital. The third quantum number m_l may have values $l, (l-1), (l-2) \ldots 0 \ldots -(l-2), -(l-1), -l$ and determines the orientation of the atomic orbital. These two quantum numbers are, of course, identical with the orbital and orbital magnetic quantum numbers introduced empirically in Section 2.4 in the discussion of spectra of atoms containing more than one electron. (The need for a spin magnetic quantum number, m_s, with value $\frac{1}{2}$ or $-\frac{1}{2}$, can also be deduced from wave mechanics, but the deduction is beyond the scope of this book. It involves relativity effects and results in wave functions somewhat different from those described here; in particular, they have no nodes.)

Electronic orbitals in atoms, like the electrons in them (cf. Section 2.4) are generally described by giving the principal quantum number and the symbol representing the second quantum number – *s* for $l = 0$, *p* for $l = 1$, *d* for $l = 2$, *f* for $l = 3$; thus we speak of 1*s*, 2*s*, 2*p*, 3*s*, 3*p*, 3*d* etc. orbitals.

The mathematical forms of some of the wave functions that are solutions of the wave equation for hydrogen-like species are given in Table 2.1.

Two features of the radial wave functions are immediately apparent. First, all decay exponentially with increase in *r*, but the decay is slower for $n = 2$ than for $n = 1$; thus the likelihood of the electron being further from the nucleus increases with increase in *n*. Second, there is a node in the 2*s* radial function: at $r = 2a_0/Z$, $\psi_r = 0$ and the value of the radial function changes sign. Inspection of the radial functions for $n = 3$ etc. leads to the generalisation that *ns* orbitals have $n-1$ nodes, *np* orbitals $n-2$ nodes, *nd* orbitals $n-3$ nodes, and so on. Such conclusions are also apparent on examination of Figs. 2.5 and 2.6, which refer to hydrogen ($Z = 1$).

The former is a simple plot of ψ_r against *r*, but the latter calls for additional comment. In discussing the probability of finding the electron at various points in space it is particularly useful to consider the quantity $4\pi r^2\psi_r^2\mathrm{d}r$ ($=P$), which is a measure of the probability of finding the

Table 2.1
The hydrogen-like species wave functions for $n = 1$ and $n = 2$

$a_0 = \dfrac{\varepsilon_0 h^2}{\pi m e^2} = 0.529$ Å

Z is the nuclear charge.

n	l	m_l	ψ_r	$\psi_\theta\psi_\phi$	Symbol
1	0	0	$2\left(\frac{Z}{a_0}\right)^{\frac{1}{2}} \mathrm{e}^{-Zr/a_0}$	$\left(\frac{1}{4\pi}\right)^{\frac{1}{2}}$	1*s*
2	0	0	$\left(\frac{Z}{2a_0}\right)^{\frac{1}{2}}\left(2-\frac{Zr}{a_0}\right)\mathrm{e}^{-Zr/2a_0}$	$\left(\frac{1}{4\pi}\right)^{\frac{1}{2}}$	2*s*
2	1	0	$\frac{1}{\sqrt{3}}\left(\frac{Z}{2a_0}\right)^{\frac{1}{2}}\left(\frac{Zr}{a_0}\right)\mathrm{e}^{-Zr/2a_0}$	$\left(\frac{3}{4\pi}\right)^{\frac{1}{2}} \cos\theta$	$2p_z$
2	1	+1	$\frac{1}{\sqrt{3}}\left(\frac{Z}{2a_0}\right)^{\frac{1}{2}}\left(\frac{Zr}{a_0}\right)\mathrm{e}^{-Zr/2a_0}$	$\left(\frac{3}{4\pi}\right)^{\frac{1}{2}} \sin\theta\cos\phi$	$2p_x$*
2	1	−1	$\frac{1}{\sqrt{3}}\left(\frac{Z}{2a_0}\right)^{\frac{1}{2}}\left(\frac{Zr}{a_0}\right)\mathrm{e}^{-Zr/2a_0}$	$\left(\frac{3}{4\pi}\right)^{\frac{1}{2}} \sin\theta\sin\phi$	$2p_y$*

*The angular parts of these wave functions as quoted are in fact normalised linear combinations of other solutions; this practice is followed because it helps to emphasise that the three 2*p* orbitals differ only in orientation. For a fuller treatment see the references listed at the end of the chapter.

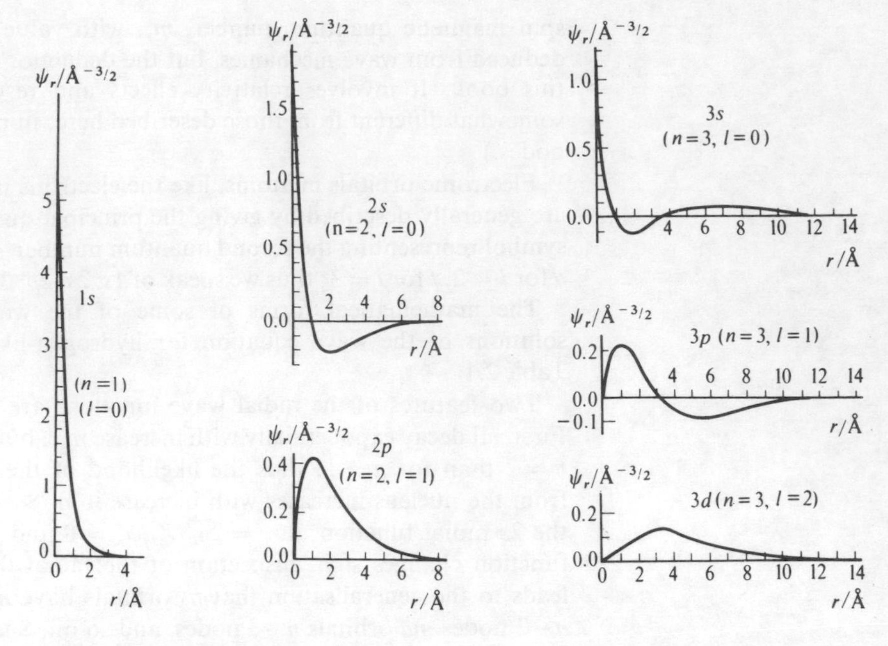

Figure 2.5
Radial parts of wave functions for the hydrogen atom. (After F.T. Gucker and R.L. Seifert (1967) *Physical Chemistry*, English Universities Press, London, p. 126.)

electron within a layer bounded by spheres of radii r and $r+\mathrm{d}r$; $P/\mathrm{d}r$, i.e. $4\pi r^2\psi_r^2$, is known as the *radial distribution function*. The variations of the radial distribution functions with r for several hydrogen orbitals are shown in Fig. 2.6, in which the vertical scale has been normalised to give unit area (probability) under each curve. It is interesting to note that, where the radial distribution function has only one maximum (1s, 2p, 3d etc.) the greatest probability of finding the electron occurs at n^2a_0, i.e. at

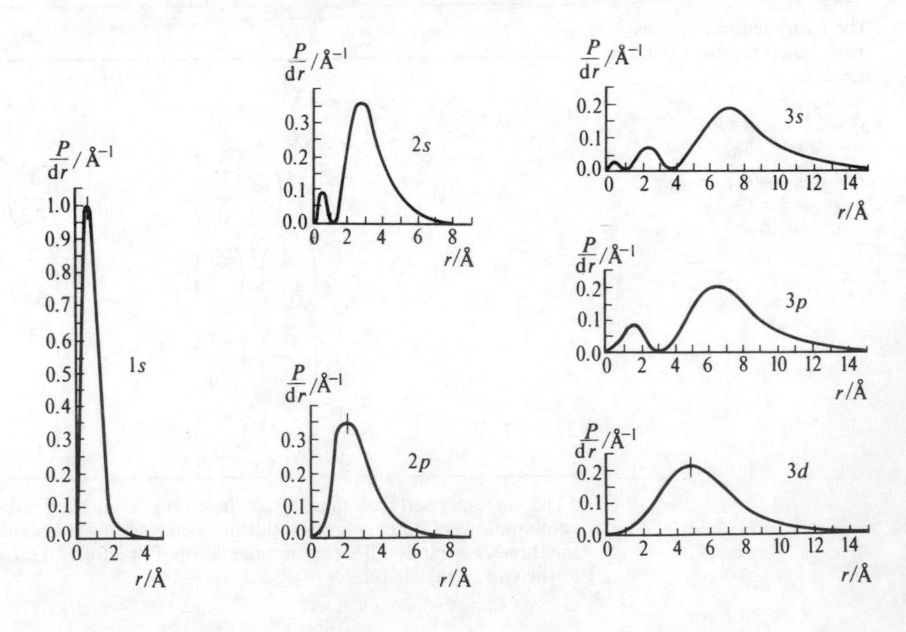

Figure 2.6
Radial distribution functions for the hydrogen atom.

the radii of the orbits of the electron on the simple Bohr theory. Since Z occurs in all expressions for ψ_r, the radial wave functions are, of course, different for H, He^+, Li^{2+} etc.

Now let us look at the angular parts of the wave functions for different types of orbital. These (as is shown in Table 2.1 for s orbitals) are independent of the value of n. Further, for s orbitals, the angular wave function is independent of θ and ϕ and is of constant value. Thus an s orbital is spherically symmetrical. The p_z orbital may be represented as two spheres touching at the origin whose centres lie on the z axis; this is stated in a different manner in Table 2.1, from which it is seen that the angular part of the p_z wave function is independent of ϕ. The p_x and p_y orbitals are similar, but are oriented along the x and y axes respectively.

There is, however, a further important feature of the p orbitals; if we consider the p_z orbital, it will be apparent that $\cos\theta$ is positive for $\theta < 90°$ and $\theta > 270°$, zero for $\theta = 90°$ and $270°$, and negative for $90° < \theta < 270°$. Thus one sphere has a sign different from that of the other. This is true also for the p_x and p_y orbitals.

What we have said so far is, for one-electron systems, mathematically exact, and Table 2.1 contains all there is to be known about the 1s, 2s and 2p orbitals of a hydrogen-like species. Most chemists, however, find mathematical statements of wave functions hard to visualise, and much prefer to think of an orbital as having a single boundary surface, usually chosen so that a certain fraction (commonly 90 per cent) of the electronic charge is contained within it. We have shown some boundary surfaces in Fig. 2.7. In order to emphasise that ψ is a continuous function we have in this book usually extended boundary surfaces up to the nucleus, but for the p orbitals this is not strictly true if we are considering the location of only 90 per cent of the electronic charge; for this reason some authors prefer to represent a p orbital by two spheres which do not touch. Further, the boundary surfaces shown in Fig. 2.7 strictly refer only to 1s and 2p wave functions, whose radial parts contain no nodes; for the 2s and 3p wave functions, there are changes of sign within the envelopes, as is apparent from the form of the ψ_r against r plots in Fig. 2.5.

Just as we have used the radial distribution function to represent the probability of finding an electron at a distance r from the nucleus, we can use the function $\psi_\theta^2\,\psi_\phi^2$ to represent the probability in terms of θ and ϕ. For an s orbital, squaring the angular part of the wave function causes no

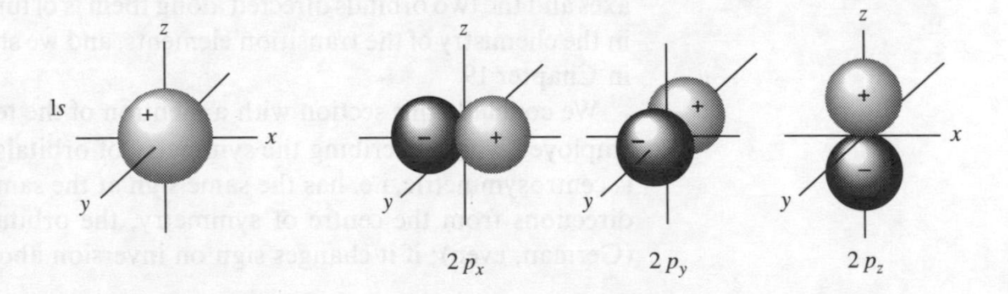

Figure 2.7
Boundary surfaces for angular parts of 1s and 2p wave functions for the hydrogen atom.

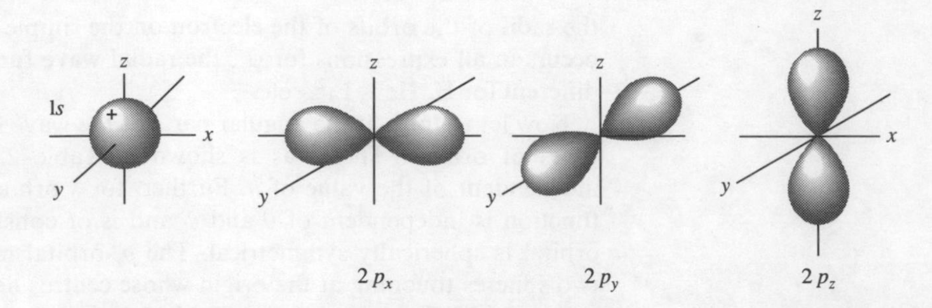

Figure 2.8
Boundary surfaces for angular probability functions of 1*s* and 2*p* orbitals for the hydrogen atom.

change in the spherical symmetry, but for the *p* orbitals the envelope becomes more elongated, as in Fig. 2.8. Note that the positive and negative signs necessarily disappear. (In practice, however, chemists often draw envelopes of shapes more nearly representing probabilities but write on them the signs of the wave functions themselves; these envelopes, as well as the wave functions, are often referred to as orbitals.) The reason for the importance of the signs of the wave functions will become clear when we consider the covalent bond in Chapter 4.

A full representation of electron distribution probability includes both radial and angular probabilities, and requires a three-dimensional model in which variation in ψ^2 is shown by the variation in three dimensions of the density of a fog representing ψ^2. Appreciation of this point may make the reader more tolerant of the admitted shortcomings of the usual two-dimensional representations.

We shall consider the *d* wave functions of hydrogen-like species only briefly, and will not discuss the *f* wave functions here. The *d* radial functions are identical for all five members of a group of *nd* orbitals, but the angular functions vary as shown in Fig. 2.9. The d_{xy}, d_{xz} and d_{yz} orbitals lie in the *xy*, *xz* and *yz* planes respectively, with the lobes midway between the Cartesian axes. The $d_{x^2-y^2}$ orbital lies in the *xy* plane with its lobes along the axes. The d_{z^2} orbital is a normalised linear combination of a $d_{z^2-x^2}$ and a $d_{z^2-y^2}$ orbital (each analogous to the $d_{x^2-y^2}$ orbital). It consists of two main lobes directed along the *z* axis and a cylindrical collar around the *z* axis. The reason for combining the $d_{z^2-x^2}$ and $d_{z^2-y^2}$ orbitals is that although there are six wave functions that can be written for orbitals having the typical four-lobed form, there can be only five *nd* orbitals having any physical reality; it is only by convention that $d_{z^2-x^2}$ and $d_{z^2-y^2}$ are combined. The difference between the three orbitals directed between the Cartesian axes and the two orbitals directed along them is of fundamental importance in the chemistry of the transition elements, and we shall return to this topic in Chapter 19.

We conclude this section with a mention of the terminology commonly employed when describing the symmetry of orbitals. If the wave function is centrosymmetric, i.e. has the same sign at the same distance in opposite directions from the centre of symmetry, the orbital is said to be *gerade* (German, even); if it changes sign on inversion about the centre it is said

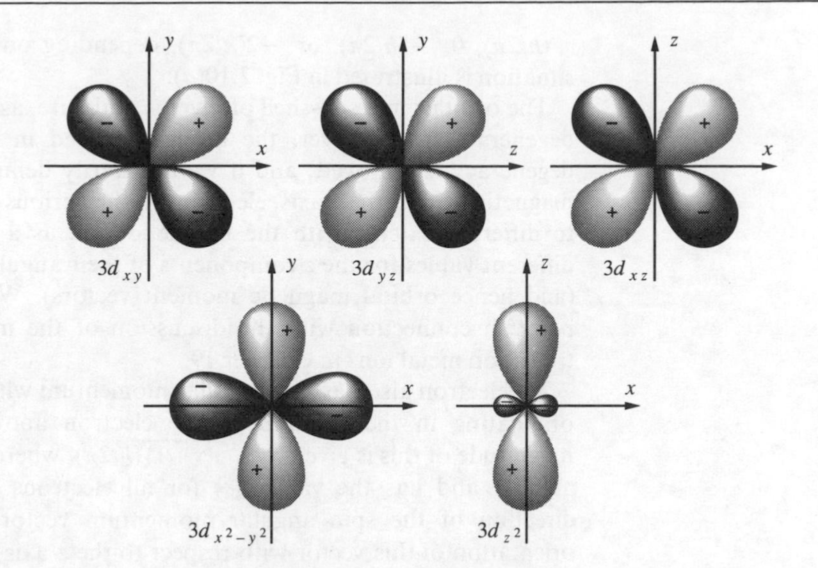

Figure 2.9
Boundary surfaces for angular parts of 3*d* wave functions for the hydrogen atom.

to be *ungerade* (German, uneven). These terms are often abbreviated to *g* and *u*. Thus *s* and *d* orbitals are gerade, *p* orbitals are ungerade.

2.9 *Angular momentum and the inner quantum number* j

The value of the quantum number l determines not only the shape of an electronic orbital, but also the amount of orbital angular momentum associated with an electron in it, this being given by $\sqrt{l(l+1)}(h/2\pi)$. The axis through the nucleus about which the electron (considered in its particulate aspect) can be thought to rotate defines the direction of the orbital angular momentum vector, the magnitude of this vector being that of the orbital angular momentum. This orbital angular momentum gives rise to a magnetic moment whose direction is in the same sense as the angular vector and whose magnitude is proportional to the magnitude of the vector. An electron in an s orbital ($l = 0$) has no orbital angular momentum, an electron in a p orbital ($l = 1$) orbital has angular momentum $\sqrt{2}(h/2\pi)$, and so on. The orbital angular momentum vector has $(2l+1)$ possible directions in space corresponding to the $(2l+1)$ possible values of m_1 for a given value of l.

We are particularly interested in the component of the orbital angular momentum vector along the z axis; clearly this will have a different value for each of the possible different orientations that this vector can take up. The actual magnitude of the z component is given by $m_l(h/2\pi)$. For an electron in a d orbital, for example, $l = 2$, so the orbital angular momentum is $\sqrt{6}(h/2\pi)$ and the z component of this may be $+2(h/2\pi)$,

$+(h/2\pi)$, 0, $-(h/2\pi)$, or $-2(h/2\pi)$, depending on the orientation. This situation is illustrated in Fig. 2.10(*a*).

The orbitals in a sub-shell of given *n* and *l* are, as we have already seen, degenerate. If, however, the atom is placed in a magnetic field this degeneracy is removed; and if we arbitrarily define the direction of the magnetic field as the *z* axis, electrons in the various *d* orbitals will interact to different extents with the magnetic field as a consequence of their different values for the *z* components of their angular momentum vectors (and hence orbital magnetic moment vectors). We shall return to this point in connection with the discussion of the magnetic properties of transition metal ions in Chapter 19.

An electron also has spin angular momentum which can be regarded as originating in the rotation of the electron about its own axis. The magnitude of this is given by $\sqrt{s(s+1)}\,(h/2\pi)$, where *s* is the spin quantum number and has the value $+\frac{1}{2}$ for all electrons. The axis defines the direction of the spin angular momentum vector, but again it is the orientation of this vector with respect to the *z* axis and the magnitude of the component of this vector in the *z* direction that we are particularly interested in. The *z* component is given by $m_s(h/2\pi)$ where m_s is the spin magnetic quantum number that we introduced in Section 2.4; this can be $+s$ or $-s$, i.e. $+\frac{1}{2}$ or $-\frac{1}{2}$. There are thus only two possible orientations of the spin angular momentum vector, and these give rise to *z* components of magnitude $+\frac{1}{2}(h/2\pi)$ and $-\frac{1}{2}(h/2\pi)$, as shown in Fig. 2.10(*b*).

For an electron having both orbital and spin angular momentum, the total angular momentum vector is given by $\sqrt{j(j+1)}\,(h/2\pi)$, where *j* is the

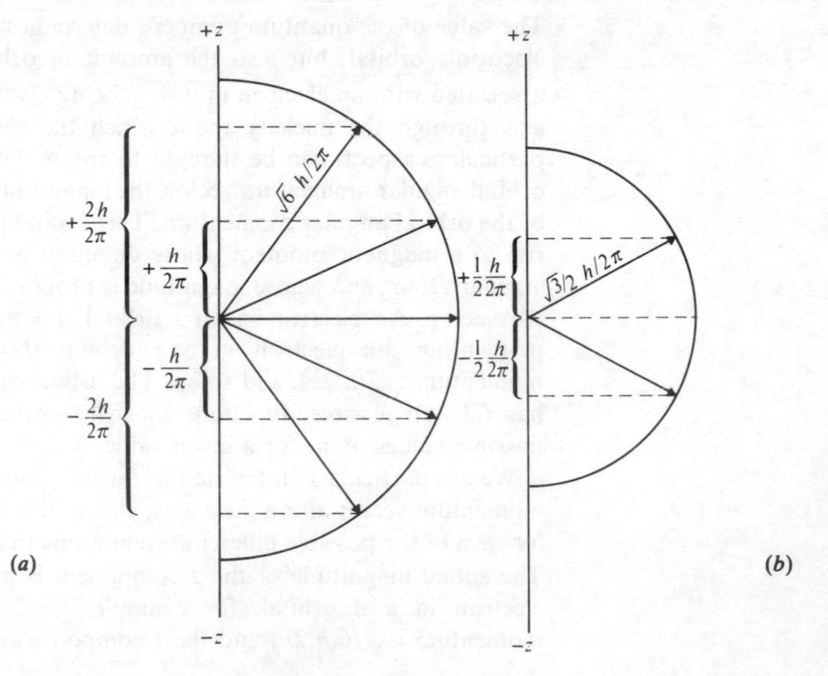

Figure 2.10
(*a*) Possible orientations of the orbital angular momentum vector for a *d* electron ($l=2$). (*b*) Possible orientations of the spin angular momentum vector for an electron ($s=\frac{1}{2}$).

so-called inner quantum number and may be $l+s$ or $l-s$, i.e. $l+\frac{1}{2}$ or $l-\frac{1}{2}$. (When $l = 0$ and the electron has no orbital angular momentum, the total angular momentum is, of course, $\sqrt{s(s+1)}(h/2\pi)$ and $j = s$.) The z component of the total angular momentum vector is now $j(h/2\pi)$ and there are $(2j+1)$ possible orientations in space.

For an electron in a ns orbital ($l = 0$), j can be only $\frac{1}{2}$. When the electron is raised to a np orbital, however, j may be $\frac{3}{2}$ or $\frac{1}{2}$, and the energies corresponding to different j values are not quite equal. In the emission spectrum of sodium, for example, transitions from the $3p_{\frac{3}{2}}$ and $3p_{\frac{1}{2}}$ levels to the $3s_{\frac{1}{2}}$ level therefore correspond to slightly different amounts of energy, and this *spin–orbit coupling* is the origin of the doublet structure of the very strong yellow line in the spectrum. The fine structure of many other spectral lines arises in analogous ways, though the number of possible lines is limited by the selection rule $\Delta j = 0$ or ± 1 and the number actually observed depends, of course, on the difference in energy between states differing only in j value and on the resolving power of the spectrometer. The difference in energy between levels for which $\Delta j = 1$ (the *spin–orbit coupling constant*, λ) increases with the atomic number of the element involved; that between the $np_{\frac{3}{2}}$ and $np_{\frac{1}{2}}$ levels, for example, is 0.23, 11.4 and 370 cm^{-1} for lithium, sodium and caesium respectively.

Throughout this section we have been concerned only with l, s and j for a single electron; we shall go on to consider L, S and J, the corresponding quantum numbers for systems of two or more electrons, in the next chapter, where the importance of Fig. 2.10 will be illustrated further.

2.10 Many-electron atoms

Most of this chapter has been devoted to species such as H, He^+ and Li^{2+} containing a single electron whose energy depends only on n, i.e. is the same in all possible (s, p, d etc.) orbitals having the same value of n; the spectra of such species (as on the Bohr theory) contain only a few lines associated with changes in the value of n. It is only for such species that the Schrödinger equation has been solved exactly.

In the next simplest system, the helium atom, there are three interactions to be considered (the attraction of each electron for the nucleus and inter-electronic repulsion) and exact solution of the wave equation has not yet been possible. When one of the two $1s$ electrons of the helium atom is promoted to an orbital having $n = 2$, the energy of the system depends on whether the electron goes into a $2s$ or a $2p$ orbital; if promotion to a $n = 3$ orbital is involved, different amounts of energy are required in the cases of a $3s$, a $3p$ and a $3d$ orbital (though there is no difference in energy between the three np ($n \nless 2$) orbitals or between the five nd ($n \nless 3$) orbitals). Thus the spectrum of atomic helium contains many more lines than that of atomic hydrogen. For atoms containing more than two electrons the number of interactions to be considered increases rapidly, making an accurate solution of the wave equation even more difficult.

Two methods of tackling this problem have been employed. In the Hartree–Fock or self-consistent field method, a reasonable wave function is assumed for each electron except one, the effect of the field of the nucleus and of the other electrons on the chosen electron is calculated, and then a wave function for the chosen electron, including the effects of the field of the other electrons, is calculated. This procedure is repeated until the wave functions for all the electrons have been improved, and the whole cycle is repeated until no further appreciable improvement occurs. At this point the wave functions are self-consistent and are a fairly accurate description of the atom. With the availability of very powerful computers it has become possible to obtain *ab initio* values (i.e. values involving no assumptions) for wave functions and energies for many-electron atoms (though not yet for atoms of heavy elements). Only for few-electron atoms, however, are values yet reliable enough to be chemically useful.

Results obtained by this method show that the angular parts of the wave functions are about the same as for one-electron species and only the radial parts change appreciably. (It will be remembered that even for one-electron systems the radial parts of the wave functions involve Z.) So *s*, *p* and *d* orbitals have roughly the same shapes and have the same relative orientations throughout chemistry.

PROBLEMS

1 Why do corresponding lines in the atomic spectra of hydrogen and deuterium appear at slightly different wavelengths?

The ionisation energy of hydrogen is 13.6 eV and the first ionisation energy of helium is 24.6 eV. How much energy would be evolved when a helium nucleus combines with two electrons to form a ground-state helium atom?

2 Calculate the frequency in Hz and the wavelength in nm of the first line of the Lyman spectrum of atomic hydrogen. Calculate also the energy of this radiation in kJ per Avogadro's number of quanta.

3 The molecule $^{12}C^{16}O$ has its lowest energy microwave absorption ($J = 0 \rightarrow J = 1$) at 1.5×10^{11} Hz, and its fundamental vibration occurs at 2170 cm^{-1}.

Calculate the bond length and force constant for $^{12}C^{16}O$ in Å and N m^{-1} respectively.

4 How would Fig. 2.7 have to be modified to show boundary surfaces for the 2*s* and 3*p* wave functions of a one-electron species?

The probability of finding the electron of a hydrogen atom in its ground state at a distance r from the proton is a maximum at $r = 0.529$ Å. Explain why this statement is compatible with the occurrence of the maximum in the value at ψ_r at $r = 0$.

REFERENCES FOR FURTHER READING

Atkins, P.W. (1990) *Physical Chemistry*, 4th edition, Oxford University Press. A fuller, more rigid, and more mathematical treatment of the ground covered in this chapter.

Bockhoff, F.J. (1976) *Elements of Quantum Theory*, 2nd edition, Addison-Wesley, Reading, Mass. A very lucid general introduction to quantum theory.

De Kock, R.L. and Gray, H.B. (1980) *Chemical Structure and Bonding*, Benjamin/Cummings, Menlo Park, California. An advanced treatment.

McWeeny, R. (1979) *Coulson's Valence*, 3rd edition, Oxford University Press, Oxford. A very clear account of atomic structure, introductory to molecular structure, at a slightly more advanced level than in this book.

3 Electronic configurations and some physical properties of atoms

Introduction • The periodic table • Hund's rules and state symbols for free atoms and ions • Ionisation energies • Electron affinities • Atomic dimensions • Relativistic effects

3.1 Introduction

In the last chapter we introduced the four quantum numbers n (1, 2, 3, ...), l (0, 1, 2 ... $(n-1)$), m_l (l, $(l-1)$, ... 0 ... $-(l-1)$, $-l$) and m_s ($+\frac{1}{2}$ or $-\frac{1}{2}$) in connection with the interpretation of atomic spectra, and later showed how n, l and m_l also arise from wave mechanics. Taken in conjunction with Pauli's principle (no two electrons in the same atom may have the same values for each of the four quantum numbers), the quantum numbers lead to the conclusion that in a shell of electrons all having the same value for n there may not be more than two s, six p, ten d and fourteen f electrons. Hence for $n = 1$, 2, 3 and 4 the maximum total numbers of electrons are 2, 8, 18 and 32 respectively.

Let us now consider the lowest energy electronic configurations (or *ground states*) of isolated atoms of all the elements. These are given in Table 3.1. It is important to appreciate that they are obtained experimentally, nearly always by the detailed analysis of atomic spectra. Most atomic spectra are unfortunately too complex for discussion here, and we shall take the results of these spectroscopic studies on trust; some idea of how electronic configurations are deduced from them may, however, be obtained from the treatment of symbols for electronic states in Section 3.3.

The electronic configurations of hydrogen and helium are $1s^1$ and $1s^2$ respectively, the $1s$ orbital (and therefore the only orbital corresponding to $n = 1$) then being full. In the next two elements, lithium and beryllium, the filling of the $2s$ orbital takes place, and then from boron to neon the $2p$ orbitals are filled; neon has the electronic configuration $1s^22s^22p^6$ or, as it is

Table 3.1
Ground state electronic configurations of the elements.

Z	Element	Electronic configuration	Z	Element	Electronic configuration
1	H	$1s^1$	53	I	$[Kr]4d^{10}5s^25p^5$
2	He	$1s^2$	54	Xe	$[Kr]4d^{10}5s^25p^6$
3	Li	$[He]2s^1$	55	Cs	$[Xe]6s^1$
4	Be	$[He]2s^2$	56	Ba	$[Xe]6s^2$
5	B	$[He]2s^22p^1$	57	La	$[Xe]5d^16s^2$
6	C	$[He]2s^22p^2$	58	Ce	$[Xe]4f^15d^16s^2$
7	N	$[He]2s^22p^3$	59	Pr	$[Xe]4f^36s^2$
8	O	$[He]2s^22p^4$	60	Nd	$[Xe]4f^46s^2$
9	F	$[He]2s^22p^5$	61	Pm	$[Xe]4f^56s^2$
10	Ne	$[He]2s^22p^6$	62	Sm	$[Xe]4f^66s^2$
11	Na	$[Ne]3s^1$	63	Eu	$[Xe]4f^76s^2$
12	Mg	$[Ne]3s^2$	64	Gd	$[Xe]4f^75d^16s^2$
13	Al	$[Ne]3s^23p^1$	65	Tb	$[Xe]4f^96s^2$
14	Si	$[Ne]3s^23p^2$	66	Dy	$[Xe]4f^{10}6s^2$
15	P	$[Ne]3s^23p^3$	67	Ho	$[Xe]4f^{11}6s^2$
16	S	$[Ne]3s^23p^4$	68	Er	$[Xe]4f^{12}6s^2$
17	Cl	$[Ne]3s^23p^5$	69	Tm	$[Xe]4f^{13}6s^2$
18	Ar	$[Ne]3s^23p^6$	70	Yb	$[Xe]4f^{14}6s^2$
19	K	$[Ar]4s^1$	71	Lu	$[Xe]4f^{14}5d^16s^2$
20	Ca	$[Ar]4s^2$	72	Hf	$[Xe]4f^{14}5d^26s^2$
21	Sc	$[Ar]3d^14s^2$	73	Ta	$[Xe]4f^{14}5d^36s^2$
22	Ti	$[Ar]3d^24s^2$	74	W	$[Xe]4f^{14}5d^46s^2$
23	V	$[Ar]3d^34s^2$	75	Re	$[Xe]4f^{14}5d^56s^2$
24	Cr	$[Ar]3d^54s^1$	76	Os	$[Xe]4f^{14}5d^66s^2$
25	Mn	$[Ar]3d^54s^2$	77	Ir	$[Xe]4f^{14}5d^76s^2$
26	Fe	$[Ar]3d^64s^2$	78	Pt	$[Xe]4f^{14}5d^96s^1$
27	Co	$[Ar]3d^74s^2$	79	Au	$[Xe]4f^{14}5d^{10}6s^1$
28	Ni	$[Ar]3d^84s^2$	80	Hg	$[Xe]4f^{14}5d^{10}6s^2$
29	Cu	$[Ar]3d^{10}4s^1$	81	Tl	$[Xe]4f^{14}5d^{10}6s^26p^1$
30	Zn	$[Ar]3d^{10}4s^2$	82	Pb	$[Xe]4f^{14}5d^{10}6s^26p^2$
31	Ga	$[Ar]3d^{10}4s^24p^1$	83	Bi	$[Xe]4f^{14}5d^{10}6s^26p^3$
32	Ge	$[Ar]3d^{10}4s^24p^2$	84	Po	$[Xe]4f^{14}5d^{10}6s^26p^4$
33	As	$[Ar]3d^{10}4s^24p^3$	85	At	$[Xe]4f^{14}5d^{10}6s^26p^5$
34	Se	$[Ar]3d^{10}4s^24p^4$	86	Rn	$[Xe]4f^{14}5d^{10}6s^26p^6$
35	Br	$[Ar]3d^{10}4s^24p^5$	87	Fr	$[Rn]7s^1$
36	Kr	$[Ar]3d^{10}4s^24p^6$	88	Ra	$[Rn]7s^2$
37	Rb	$[Kr]5s^1$	89	Ac	$[Rn]6d^17s^2$
38	Sr	$[Kr]5s^2$	90	Th	$[Rn]6d^27s^2$
39	Y	$[Kr]4d^15s^2$	91	Pa	$[Rn]5f^26d^17s^2$
40	Zr	$[Kr]4d^25s^2$	92	U	$[Rn]5f^36d^17s^2$
41	Nb	$[Kr]4d^45s^1$	93	Np	$[Rn]5f^46d^17s^2$
42	Mo	$[Kr]4d^55s^1$	94	Pu	$[Rn]5f^67s^2$
43	Tc	$[Kr]4d^55s^2$	95	Am	$[Rn]5f^77s^2$
44	Ru	$[Kr]4d^75s^1$	96	Cm	$[Rn]5f^76d^17s^2$
45	Rh	$[Kr]4d^85s^1$	97	Bk	$[Rn]5f^97s^2$
46	Pd	$[Kr]4d^{10}$	98	Cf	$[Rn]5f^{10}7s^2$
47	Ag	$[Kr]4d^{10}5s^1$	99	Es	$[Rn]5f^{11}7s^2$
48	Cd	$[Kr]4d^{10}5s^2$	100	Fm	$[Rn]5f^{12}7s^2$
49	In	$[Kr]4d^{10}5s^25p^1$	101	Md	$[Rn]5f^{13}7s^2$
50	Sn	$[Kr]4d^{10}5s^25p^2$	102	No	$[Rn]5f^{14}7s^2$
51	Sb	$[Kr]4d^{10}5s^25p^3$	103	Lr	$[Rn]5f^{14}6d^17s^2$
52	Te	$[Kr]4d^{10}5s^25p^4$	104	Rf	$[Rn]5f^{14}6d^27s^2$

often written, $[He]2s^22p^6$. At this stage the shell having $n = 2$ is complete. The filling of the 3*s* and 3*p* orbitals takes place in an analogous sequence from sodium to argon, so that the latter element has the electronic configuration $[Ne]3s^23p^6$.

With potassium and calcium, successive electrons go into the 4*s* orbital, calcium having the electronic configuration $[Ar]4s^2$. Then, however, the pattern changes. To a first approximation we may say that in the following series of ten elements (scandium, titanium, vanadium, chromium, manganese, iron, cobalt, nickel, copper and zinc), the next ten electrons enter the 3*d* orbitals, giving zinc the electronic configuration $[Ar]3d^{10}4s^2$. (The ground states of chromium and copper are actually $[Ar]3d^54s^1$ and $[Ar]3d^{10}4s^1$ rather than $[Ar]3d^44s^2$ and $[Ar]3d^94s^2$ respectively; we return to this minor irregularity later.) Thereafter, from gallium to krypton, the 4*p* orbitals are filled, and krypton has the electronic configuration $[Ar]3d^{10}4s^24p^6$.

From rubidium to xenon the general sequence of filling of orbitals is analogous to that from potassium to krypton; strontium ($[Kr]5s^2$), cadmium ($[Kr]4d^{10}5s^2$) and xenon ($[Kr]4d^{10}5s^25p^6$) are the homologues of calcium, zinc and krypton respectively. Once more there are minor irregularities in the distribution of electrons between *d* and *s* orbitals.

In the following period of thirty-two elements electrons go into *f* orbitals for the first time. The first three elements have electronic configurations analogous to the corresponding members of the previous period: caesium is $[Xe]6s^1$, barium $[Xe]6s^2$ and lanthanum $[Xe]5d^16s^2$. Then the configurations change from $[Xe]4f^15d^16s^2$ for cerium to $[Xe]4f^{14}5d^16s^2$ for lutetium though in these *lanthanide elements* the 5*d* orbital is not usually occupied. After lutetium, successive electrons occupy the remaining 5*d* orbitals as the electronic configuration builds up from $[Xe]4f^{14}5d^26s^2$ for hafnium to $[Xe]4f^{14}5d^{10}6s^2$ for mercury, the homologue of cadmium; yet again, minor departures from a steady increase in the number of *d* electrons occur. Finally, the period is completed with the successive occupation of the 6*p* orbitals from thallium, $[Xe]4f^{14}5d^{10}6s^26p^1$, to radon, $[Xe]4f^{14}5d^{10}6s^26p^6$.

The period which begins after radon is, of course, incomplete, and some of the later elements are too unstable for detailed investigations to be possible. Francium ($[Rn]7s^1$), radium ($[Rn]7s^2$) and actinium ($[Rn]6d^17s^2$) have electronic configurations analogous to those of caesium, barium and lanthanum. Thorium has the configuration $[Rn]6d^27s^2$ rather than $[Rn]5f^16d^17s^2$ or $[Rn]5f^27s^2$ (cf. cerium in the preceding period); thereafter, however, occupation of the 5*f* orbitals begins, and in later elements the electronic configurations are usually of the type $5f^n6s^2$ with no electrons in the 6*d* orbitals until the 5*f* orbitals are all filled at nobelium. Then lawrencium ($[Xe]5f^{14}6d^17s^2$) is the homologue of lutetium, and rutherfordium ($[Xe]5f^{14}6d^27s^2$) is that of hafnium. The metals thorium to lawrencium inclusive are often known as the *actinide elements*.

We should at this point mention the ground states of the ions that are obtained by removal of electrons from the elements. When a gaseous ground state iron atom of electronic configuration $[Ar]3d^6 4s^2$ loses an electron the Fe^+ ion formed has its minimum energy in the configuration $3d^7$, although the ground state manganese atom with which it is isoelectronic has the configuration $3d^5 4s^2$. Similarly, the ground states of the Fe^{2+} and Fe^{3+} ions are $[Ar]3d^6$ and $[Ar]3d^5$ rather than $[Ar]3d^5 4s^1$ and $[Ar]3d^3 4s^2$, which are the ground states of atomic chromium and vanadium respectively. Evidently the differences in nuclear charge between Fe^+ and Mn, Fe^{2+} and Cr, and Fe^{3+} and V are important in determining which orbitals the electrons occupy. Fortunately it turns out that along a series of ions carrying the same charge, the electronic configurations often change much more regularly than the electronic configurations of the atoms from which the ions are derived. Thus for dipositive ions from Sc^{2+} to Zn^{2+} the ground state electronic configuration changes regularly from $[Ar]3d^1$ to $[Ar]3d^{10}$, and for the tripositive ions there is a similar regular change from Sc^{3+}([Ar]) to Zn^{3+}($[Ar]3d^9$). For tripositive ions of the lanthanide elements there is a regular change from Ce^{3+}($[Xe]4f^1$) to Lu^{3+}($[Xe]4f^{14}$). Since the chemistry of all these elements is mostly that of their ions, the regularities in the ground states of the ions are much more important than the irregularities in the ground states of the neutral atoms.

It will be obvious from the foregoing discussion that there is no one sequence which represents accurately the occupation of different sets of orbitals with increase in atomic number. The following scrics is *roughly* true for the relative energies of orbitals in most neutral atoms: $1s < 2s < 2p < 3s < 3p < 4s < 3d < 4p < 5s < 4d < 5p < 6s < 5d \sim 4f < 6p < 7s < 6d \sim 5f$. This is sometimes known as the *aufbau* (building up) order. But it is more important to grasp the fact that the relative energies of orbitals change significantly with increase in atomic number and with change in state of ionisation than to remember the details of the sequence at the high energy end of any one series. The energies of different orbitals are in any case close together for high values of n.

Although it is not possible to calculate the dependence of the energies of orbitals on atomic number with the degree of accuracy necessary to obtain agreement with the data given in Table 3.1, some helpful theoretical comments can be made on the basis of the different *screening effects* of electrons in different orbitals. The radial distribution functions shown in Fig. 2.6 show, for example, that the $2s$ and $2p$ orbitals for the hydrogen atom both have substantial values at distances from the nucleus at which the radial distribution function of the $1s$ orbital is considerable, i.e. the $2s$ and $2p$ orbitals *penetrate* the $1s$ orbital. Accurate calculations show that the penetrating effect of the $2s$ orbital is the greater. Thus if we assume hydrogen-like wave functions, when the $1s$ orbital is filled in an atom of atomic number greater than that of helium, an electron in a $2s$ or a $2p$ orbital experiences the effective charge of a nucleus partly shielded by the 1*s* electrons, and a 2*p* electron is shielded more than a 2*s* electron since it penetrates the 1*s* orbital less. Thus we have the energy sequence $2s < 2p$. Similar considerations lead to the sequences $3s < 3p < 3d$ and $4s < 4p < 4d < 4f$. As we move to atoms of elements of higher atomic number, however, energy differences between orbitals of the same value of *n* become smaller, the validity of assuming hydrogen-like wave functions becomes more doubtful, and predictions of ground states become much less reliable.

There is, nevertheless, a useful set of empirical rules for estimating the effective nuclear charges experienced by electrons in different orbitals. *Slater's rules* are based on experimental data for electron promotion energies and ionisation energies: the effective nuclear charge Z^* acting on a given electron is obtained by subtracting from the atomic number *Z* the screening (or shielding) constant *S* estimated as follows:

1. Write out the electronic configuration of the element in the following order and groupings: (1*s*), (2*s*, 2*p*), (3*s*, 3*p*), (3*d*), (4*s*, 4*p*), (4*d*), (4*f*), (5*s*, 5*p*) etc.
2. Electrons in any group higher in this sequence than the electron under consideration contribute nothing to *S*.
3. Then for an electron in an *ns* or *np* orbital
   (a) All other electrons in the (*ns*, *np*) group contribute $S = 0.35$ each.
   (b) All electrons in the $(n-1)$ shell contribute $S = 0.85$ each.
   (c) All electrons in $(n-2)$ or lower shells contribute $S = 1.00$ each.
4. For an electron in an *nd* or *nf* orbital, all electrons in the same group contribute $S = 0.35$ each; those in groups lying lower in the sequence than the (*nd*) or (*nf*) group contribute $S = 1.00$ each.

As an example of the application of Slater's rules let us consider potassium, the electronic configuration of which is $1s^2 2s^2 2p^6 3s^2 3p^6 4s^1$. The effective nuclear charge experienced by the 4*s* electron of potassium is

$$Z^* = Z - S = 19 - [(0.85 \times 8) + (1.00 \times 10)] = 2.20$$

If the electronic configuration of potassium had been $1s^2 2s^2 2p^6 3s^2 3p^6 3d^1$ the 3*d* electron would have experienced an effective nuclear charge

$$Z^* = Z - S = 19 - (1.00 \times 18) = 1.00$$

Thus the electron in the 4*s* orbital is under the influence of the greater effective nuclear charge and hence in the ground state it is this orbital that is occupied. Slater's rules have been used in the estimation of ionisation energies, ionic radii, and electronegativities. More accurate effective nuclear charges have been calculated by Clementi and Raimondi using self-consistent field methods; they give much higher Z^* values for *d* electrons.

3.2 The periodic table

The great generalisation stated by Mendeléev in 1869 and by Lothar Meyer in the following year, that *the properties of the elements can be represented as periodic functions of their atomic weights*, was shortly afterwards set out by both authors in the form of a *periodic table*. The original form of the periodic table has since been extensively modified as a result of the discovery of the noble gases and the actinides and of the elucidation of the structure of the atom. Furthermore, it is now recognised that the periodic law is essentially the consequence of the periodic variation in electronic configuration summarised in the last section. Modern representations of the periodic table therefore emphasise the blocks of two, six, ten and fourteen elements in which the *s*, *p*, *d* and *f* sub-shells respectively are being filled, though helium is usually placed with neon, argon, krypton, xenon and radon. A typical modern form of the table is shown in Fig. 3.1 (but see below for a recent development).

Many different systems of nomenclature for the various types of elements have been used. In this book we shall refer to hydrogen, helium and the groups of elements headed by lithium, beryllium, boron, carbon, nitrogen, oxygen, fluorine and neon as *typical elements*, *main group elements*, *non-transition elements*, or *s* and *p block elements*; the elements from scandium to zinc inclusive, and their homologues from yttrium to cadmium and from lanthanum to mercury, we shall refer to as *transition elements* or *d block elements*; and the elements from cerium to lutetium inclusive and their homologues from thorium to lawrencium we shall call

Period	Group I	Group II	Transition elements										Group III	Group IV	Group V	Group VI	Group VII	Group VIII or O
1	1 H																	2 He
2	3 Li	4 Be											5 B	6 C	7 N	8 O	9 F	10 Ne
3	11 Na	12 Mg											13 Al	14 Si	15 P	16 S	17 Cl	18 Ar
4	19 K	20 Ca	21 Sc	22 Ti	23 V	24 Cr	25 Mn	26 Fe	27 Co	28 Ni	29 Cu	30 Zn	31 Ga	32 Ge	33 As	34 Se	35 Br	36 Kr
5	37 Rb	38 Sr	39 Y	40 Zr	41 Nb	42 Mo	43 Tc	44 Ru	45 Rh	46 Pd	47 Ag	48 Cd	49 In	50 Sn	51 Sb	52 Te	53 I	54 Xe
6	55 Cs	56 Ba	57 La	72 Hf	73 Ta	74 W	75 Re	76 Os	77 Ir	78 Pt	79 Au	80 Hg	81 Tl	82 Pb	83 Bi	84 Po	85 At	86 Rn
7	87 Fr	88 Ra	89 Ac	104 Rf														

Inner transition elements

58 Ce	59 Pr	60 Nd	61 Pm	62 Sm	63 Eu	64 Gd	65 Tb	66 Dy	67 Ho	68 Er	69 Tm	70 Yb	71 Lu
90 Th	91 Pa	92 U	93 Np	94 Pu	95 Am	96 Cm	97 Bk	98 Cf	99 Es	100 Fm	101 Md	102 No	103 Lr

Figure 3.1
The periodic table.

Table 3.2
Alternative and trivial names for groups of elements.

Name	Elements
Group IA*, alkali metals	Li, Na, K, Rb, Cs, Fr
Group IIA*	Be, Mg, Ca, Sr, Ba, Ra
Alkaline earth metals	Ca, Sr, Ba, Ra
Group IIIA*	Sc, Y, La, Ac
Rare earth elements	La–Lu incl., sometimes also Sc and Y
Transuranic elements	Elements after U
Group IVA*	Ti, Zr, Hf, Rf
Group VA*	V, Nb, Ta
Group VIA*	Cr, Mo, W
Group VIIA*	Mn, Tc, Re
Group VIII*	Fe, Co, Ni; Ru, Rh, Pd,; Os, Ir, Pt
Platinum metals	Ru, Rh, Pd; Os, Ir, Pt
Group IB*, coinage metals	Cu, Ag, Au
Group IIB*	Zn, Cd, Hg
Group IIIB*	B, Al, Ga, In, Tl
Group IVB*	C, Si, Ge, Sn, Pb
Group VB*, pnictides	N, P, As, Sb, Bi
Group VIB*, chalcogens	O, S, Se, Te, Po
Group VIIB*, halogens	F, Cl, Br, I, At
Inert, noble, or rare gases	He, Ne, Ar, Kr, Xe, Rn

*In Mendeléev and similar 'short' forms of periodic table.

inner transition elements or *f block elements*. We should, nevertheless, draw attention to the fact that some authors regard as transition elements only those which, in their ground states as neutral atoms, have partly filled *d* or *f* shells; others also include elements which have partly filled *d* or *f* shells in any common oxidation state. Neither of these definitions, however, fits well into the natural classification in terms of permitted combinations of quantum numbers (2, 6, 10 and 14) and we shall therefore avoid them completely. Nor shall we refer to A and B subgroups,

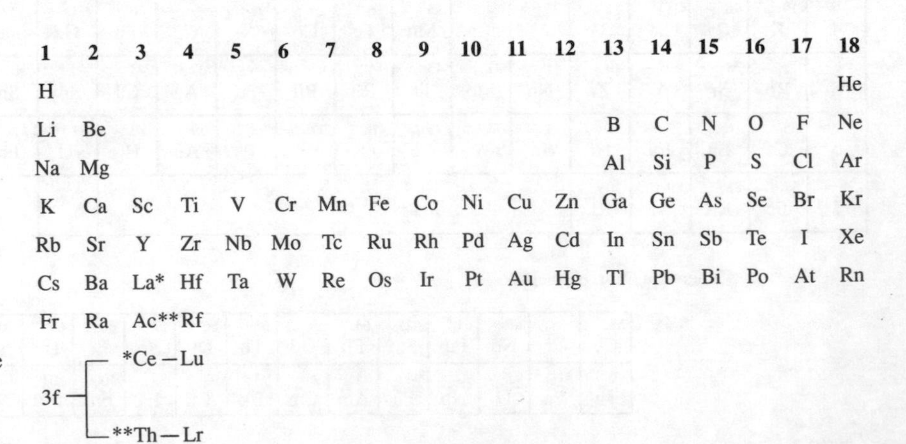

Figure 3.2
International Union of Pure and Applied Chemistry recommended designations of groups of elements.

a terminology based on older forms of the table, though since it is still encountered in the chemical literature we reproduce the essentials of it, along with various trivial names for groups of elements, in Table 3.2.

A new set of designations recommended by the International Union of Pure and Applied Chemistry, and hence likely to be much used in future, is shown in Fig. 3.2.

Among many other forms of periodic table we shall mention only one, a modified form of a version first proposed by Longuet-Higgins to emphasise the order of filling orbitals in relation to their relative energies. This is shown in Fig. 3.3. No simple and easily remembered form of the table can, of course, portray all of the detail given in Section 3.1; and any

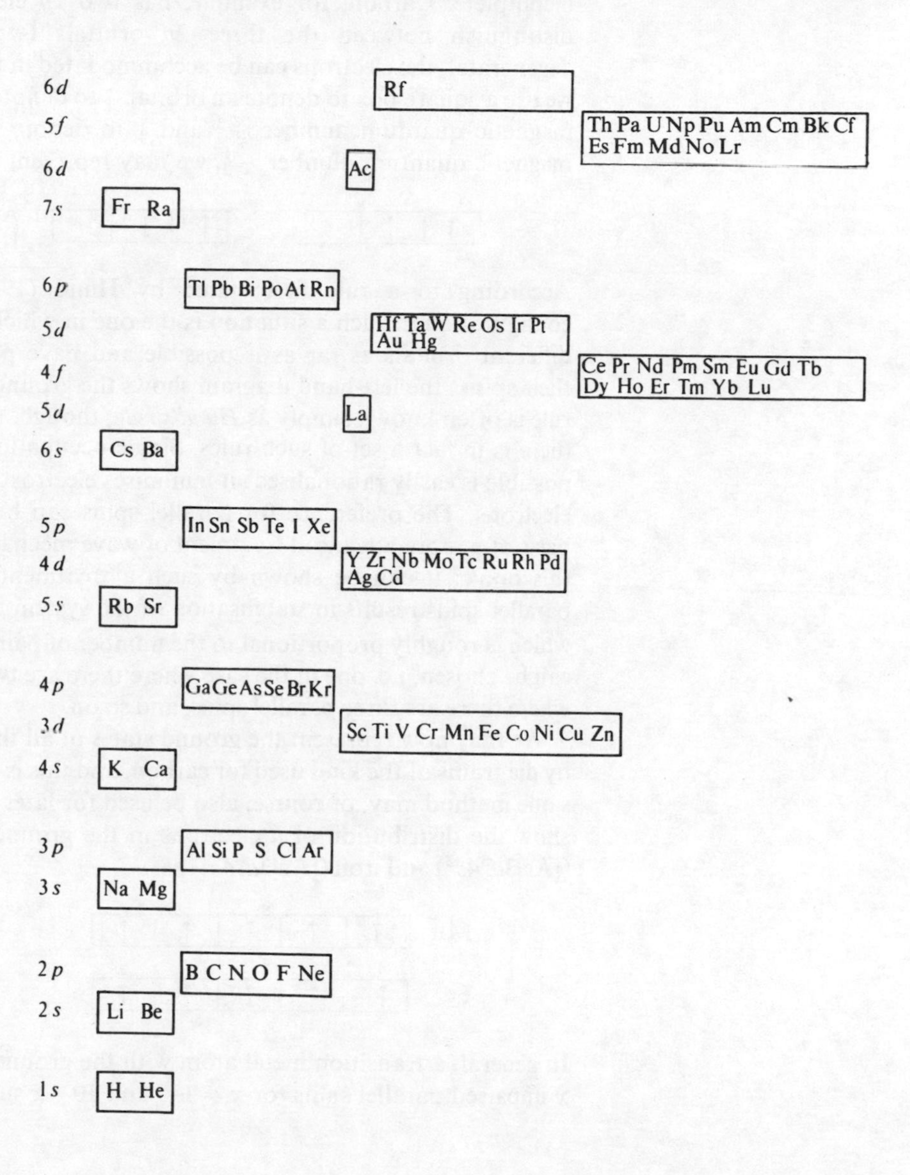

Figure 3.3
Periodic table emphasising relative energy levels.

version of it must be regarded only as a very successful broad generalisation about properties in relation to the overall pattern of electronic configurations.

3.3 *Hund's rules and state symbols for free atoms and ions*

The electronic configurations given in Table 3.1 provide no details of the occupation of orbitals when sub-shells of *p*, *d* and *f* electrons are incomplete. Carbon, for example, has two 2*p* electrons; if we do not distinguish between the three 2*p* orbitals (which are, of course, degenerate), the electrons can be accommodated in three different ways. If we use a square box to denote an orbital, ↑ to denote an electron with spin magnetic quantum number $+\frac{1}{2}$ and ↓ to denote an electron with spin magnetic quantum number $-\frac{1}{2}$, we may represent these as

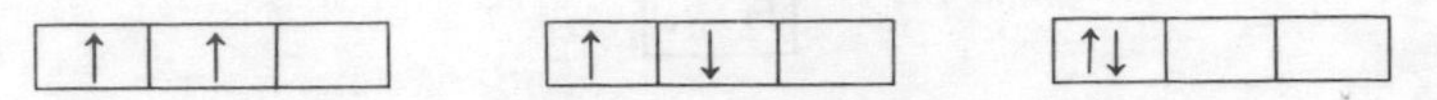

According to a rule first stated by Hund (1925), the most stable configuration in such a situation is the one in which the electrons occupy different orbitals as far as is possible and have parallel orientations of their spins: the left-hand diagram shows the ground state of carbon. This rule is often known simply as *Hund's rule*, though, as we shall see shortly, there is in fact a set of such rules. Single occupation of orbitals as far as possible is easily rationalised: it minimises electrostatic repulsion between electrons. The preference for parallel spins can be justified only on the basis of a more advanced treatment of wave mechanics than that given in this book; it can be shown by such a treatment that the presence of parallel spins results in stabilisation of the system by an *exchange energy* which is roughly proportional to the number of pairs of parallel spins that can be chosen, i.e. one in the case where there are two parallel spins, three where there are three parallel spins, and so on.

We may now represent the ground states of all the elements up to neon by diagrams of the kind used for carbon, and this is done in Table 3.3. The same method may, of course, also be used for later elements; thus we can show the distribution of *d* electrons in the ground states of manganese ($[Ar]3d^5 4s^2$) and iron ($[Ar]3d^6 4s^2$) as:

Mn	↑	↑	↑	↑	↑
Fe	↑↓	↑	↑	↑	↑

In general, a transition metal atom with the ground state $(n-1)d^x ns^2$ has x unpaired parallel spins for $x = 1$–5 and $10-x$ such spins for $x = 6$–10.

Table 3.3
Box diagrams for ground states of elements H–Ne.

Element	Orbitals 1*s*	2*s*	2*p*		
H	↑				
He	↑↓				
Li	↑↓	↑			
Be	↑↓	↑↓			
B	↑↓	↑↓	↑		
C	↑↓	↑↓	↑	↑	
N	↑↓	↑↓	↑	↑	↑
O	↑↓	↑↓	↑↓	↑	↑
F	↑↓	↑↓	↑↓	↑↓	↑
Ne	↑↓	↑↓	↑↓	↑↓	↑↓

The chromium atom in its ground state of $[Ar]3d^5 4s^1$ has six unpaired parallel spins. Similar considerations apply to free ions.

The discussion in the two preceding paragraphs is adequate for many purposes, but when we said that we did not distinguish between the three degenerate $2p$ orbitals we were ignoring the fact that these orbitals *are* distinguishable to some extent: all have $l = 1$, but m_l may be $+1$, 0, or -1. So for a p^2 configuration with two unpaired electrons each having $s = +\frac{1}{2}$, there are three possible pairs of values for m_l for individual electrons: $+1$ and 0, $+1$ and -1, and 0 and -1.

The question of which pair of values of m_l corresponds to the ground state of carbon, and a number of related questions, can be answered only by considering the influence of electrons on each other, primarily by means of the interaction or *coupling* of the magnetic fields generated by their spin and orbital motions: hence the importance of spin and orbital angular momentum, which we discussed for a one-electron system in Section 2.9.

As we have already seen, for any system containing more than one electron, the energy of an electron in a shell of principal quantum number n depends on the value of l for the orbital in which it is present, and the value of l also determines the orbital angular momentum of the electron, the latter quantity being given by $\sqrt{l(l+1)}(h/2\pi)$. By analogy, it is assumed that the energy of a polyelectronic system and its orbital angular momentum is determined by a ***resultant orbital quantum number L*** which may be obtained directly from the values of l for the individual electrons; since the orbital angular momentum not only has magnitude, but also has $(2l+1)$ possible directions in space (the number of possible values of m_l), *vectorial* summation of individual l values is necessary. However, since the value of m_l for any electron denotes the component of its orbital angular momentum along the z axis, $m_l(h/2\pi)$, algebraic summation of m_l values for individual electrons gives the resultant orbital magnetic quantum number M_L and the component of the resultant orbital angular momentum along the z axis, $M_L(h/2\pi)$. Just as m_l may have the $(2l+1)$ values $l, (l-1) \ldots 0 \ldots -(l-1), -l$, so M_L may have the $(2L+1)$ values $L, (L-1) \ldots 0 \ldots -(L-1), -L$; and if, for a polyelectron system, we can find all the possible values of M_L this tells us the value of L. Energy states for which $L = 0, 1, 2, 3, 4 \ldots$ are known as $S, P, D, F, G \ldots$ states respectively, the letters corresponding to $s, p, d, f, g \ldots$ used for orbitals for which $l = 0, 1, 2, 3, 4 \ldots$ respectively in the one-electron case. The resultant orbital angular momentum is $\sqrt{L(L+1)}(h/2\pi)$.

The resultant spin quantum number, S, which denotes the value of the resultant spin angular momentum $\sqrt{S(S+1)}(h/2\pi)$, is analogous to L and is obtained by algebraic summation of the m_s values for individual electrons. One electron with $s = \frac{1}{2}$ obviously also has $S = \frac{1}{2}$ with M_S (the same as m_s) $= +\frac{1}{2}$ or $-\frac{1}{2}$. Two electrons lead to $S = 0$ (with $m_s = +\frac{1}{2}$ and $-\frac{1}{2}$ and therefore $M_S = 0$) or $S = 1$ (with $m_s = +\frac{1}{2}$ and $+\frac{1}{2}$, or $+\frac{1}{2}$ and $-\frac{1}{2}$, or $-\frac{1}{2}$ and $-\frac{1}{2}$, therefore $M_S = 1, 0$, or -1). In general, for any value of S there are $(2S+1)$ values of M_S: $S, (S-1) \ldots 0 \ldots -(S-1), -S$. The quantity $(2S+1)$ is known as the *spin multiplicity* of the state; thus in speaking or writing of states for which $2S+1 = 1, 2, 3, 4$ etc. (corresponding to $S = 0, \frac{1}{2}, 1, \frac{3}{2}$ etc.) we call them singlets, doublets, triplets, quartets etc. The use of S for both the resultant spin quantum number and a state for which $L = 0$ (not to mention screening constants) is regrettable, but it is firmly established and in practice rarely causes confusion.

Finally, we have a resultant inner quantum number, J (also called the total angular momentum quantum number, since the total angular momentum is given by $\sqrt{J(J+1)}(h/2\pi)$), compounded vectorially from L and S, i.e. algebraically from M_L and M_S. This can take the values $L+S$, $L+S-1, \ldots \quad \ldots |L-S|$, the last symbol denoting that only the magnitude, and not the sign, of $L-S$ is involved: like j for a single electron, J must be positive or zero. There are thus $(2S+1)$ possible values of J for $S < L$ and $(2L+1)$ possible values for $L < S$.

This method of obtaining J from L and S is based on LS (or *Russell–Saunders*) *coupling*. Although it is the only form of coupling of orbital and spin angular momentum that we shall be concerned with in this book, it is not valid for all elements (especially for those of high atomic number). In an alternative method of coupling, l and s for all individual electrons are first combined to give j (as in Section 2.9) and the individual j values are combined in *jj coupling*; but we shall not discuss this method further.

If we know $2S+1$, L and J for an energy state we can write the full *state symbol* (or *term symbol* as it is also sometimes called). This is done by writing the symbol for the value of L with the value of $2S+1$ as a left superscript and the value of J as a right subscript. The ground state of carbon, for example, is $^3P_0$, in speech 'triplet-P-nought', denoting $L = 1$, $2S+1 = 3$ (i.e. $S = 1$) and $J = 0$. There are, however, other possible values for $2S+1$, L and J for a carbon atom of configuration $2s^22p^2$, and these we shall soon consider further. Inorganic chemists frequently omit the value of J and refer to ^{2S+1}L as a state and $^{2S+1}L_{J_1}$, $^{2S+1}L_{J_2}$ etc. as *components* of a state; since, for given values of L and S, the value of J often makes very little difference to the energy, we shall usually follow that practice in this book.

Let us now consider systematically the ground states of elements of atomic number lower than that of carbon. For a hydrogen atom with electronic configuration $1s^1$ the only electron has $l = 0$ so L must be 0; $S = \frac{1}{2}$ so $2S+1 = 2$; the only possible value of J is $\frac{1}{2}$, so the state symbol is $^2S_{\frac{1}{2}}$. For helium, $1s^2$, both electrons have $l = 0$, so $L = 0$; two electrons both with $n = 1$ and $l = 0$ must have $m_s = +\frac{1}{2}$ and $-\frac{1}{2}$, so $S = 0$ and $2S+1 = 1$; J must be 0, and the state symbol is $^1S_0$. Thus the ns^2 configuration, having $L = 0$, $S = 0$, and $J = 0$, will contribute nothing to the state symbol in lithium and later atoms. (It is easily seen that for the np^6 configuration ($l = 1$ for each electron; possible values for m_l are 1, 0, and -1, and for each of these $m_s = \pm\frac{1}{2}$) the same is true.) For lithium, $1s^22s^1$, and beryllium, $1s^22s^2$, the state symbols are therefore the same as for hydrogen and helium, $^2S_{\frac{1}{2}}$ and $^1S_0$ respectively.

For boron, $1s^22s^22p^1$, we are concerned only with the p electron. For this $l = 1$, so $L = 1$; $S = \frac{1}{2}$ so $2S+1 = 2$; J may be $L+S$ or $L-S$, i.e. $\frac{3}{2}$ or $\frac{1}{2}$, so the state symbol for boron may be $^2P_{\frac{3}{2}}$ or $^2P_{\frac{1}{2}}$. For carbon, $1s^22s^22p^2$, each of the two electrons $l = 1$ and $m_l = +1$, 0 or -1; L (the algebraic sum of m_l for individual electrons) may therefore be 2, 1 or 0 (D, P or S states respectively); S may be 1 or 0; and it might seem that J could be 3, 2, 1 or 0. This, however, is not so. If, for example, the two electrons each have $n = 2$, $l = 1$ and $m_l = 1$ (giving $L = 2$) they cannot both have $m_s = +\frac{1}{2}$, which would be a breach of Pauli's principle. The only permitted combinations of m_l and m_s, and the corresponding values of M_L and M_S, for two p electrons having the same value of n are shown in Table 3.4. Such combinations are sometimes known as *microstates*. Inspection of Table 3.4 shows that the fifteen entries may be grouped in three sets. First, a set of five with $M_L = 2, 1, 0, -1, -2$, all having $M_S = 0$ and thus having $M_J = 2, 1, 0, -1, -2$, can be assigned to the $^1D_2$ ($L = 2$, $S = 0$, $J = 2$) state. A set of nine with $M_L = 1, 0$ or -1 and $M_S = 1, 0$ or -1 can be assigned to the state ^3P ($L = 1$, $S = 1$), and further examination shows these may be subdivided into a set of five with $J=2$ ($^3P_2$), a set of three with $J = 1$ ($^3P_1$) and a single entry with $J = 0$ ($^3P_0$). The remaining entry, the third one for which $M_L = 0$ and $M_S = 0$, corresponds to the state $^1S_0$. We have, of course, no method of deciding which entry with $M_L = 0$ and $M_S = 0$ should be assigned to which state, and it is, indeed, not meaningful to do so. Thus in the right-hand column of Table 3.4 we have placed more than one state symbol against some entries (in the interests of simplicity J values have been omitted). Of the five possible states for the

Table 3.4
Resultant M_L, M_S and M_J values for various configurations of np^2, together with possible state symbols.

$m_l =$ +1	0	−1	M_L	M_S	M_J	State
↑↓			2	0	2	^1D
	↑↓		0	0	0	$^1D, ^3P, ^1S$
		↑↓	−2	0	−2	^1D
↑	↑		1	1	2	^3P
↑		↑	0	1	1	^3P
	↑	↑	−1	1	0	^3P
↓	↓		1	−1	0	^3P
↓		↓	0	−1	−1	^3P
	↓	↓	−1	−1	−2	^3P
↑	↓		1	0	1	$^1D, ^3P$
↓	↑		1	0	1	$^1D, ^3P$
↑		↓	0	0	0	$^1D, ^3P, ^1S$
↓		↑	0	0	0	$^1D, ^3P, ^1S$
	↑	↓	−1	0	−1	$^1D, ^3P$
	↓	↑	−1	0	−1	$^1D, ^3P$

electronic configuration np^2 ($^1D_2$, $^3P_2$, $^3P_1$, $^3P_0$, and $^1S_0$) it turns out that the one of lowest energy is $^3P_0$ and this is the ground state of carbon. We shall return to this point in a later paragraph.

A similar treatment of the nitrogen atom shows that the $2p^3$ configuration gives rise to ^4S, ^2P and ^2D states. For the $2p^4$ configuration we can introduce a useful simplification by considering it as $2p^6$ plus two positrons to annihilate two of the electrons. Since positrons differ from electrons only in charge, the states arising from the np^4 and np^2 configurations are the same; similarly, np^5 is equivalent to p^1, nd^9 to d^1, nd^8 to d^2, and so on. This positron or *positive hole* concept is therefore a very useful one, and we shall meet it again later in connection with the interpretation of the electronic spectra of complexes of the transition metals.

We conclude this section by returning to the question of the relative energies of the states corresponding to a given electronic configuration, and stating all of Hund's rules in a formal way. It is found from analysis of spectroscopic data that, provided Russell–Saunders coupling holds:

1. The state having the highest multiplicity (and hence the highest value of S) is most stable.
2. If two or more states have the same value of S, the state having the higher value of L is more stable.

3. For all states having the same values of S and L, the component with the lowest value of J is the most stable if the sub-shell is less than half filled, and the component with the highest value of J is most stable if the sub-shell is more than half filled. (If the sub-shell is half filled and S has the highest possible value, L must be zero and $J = S$.)

Thus for the states corresponding to the electronic configuration np^2 that of lowest energy is ^3P and the component of lowest energy is $^3P_0$. For the configuration np^3, the highest possible value of S is $\frac{3}{2}$; L is then zero, so J must be $\frac{3}{2}$ and the full ground state symbol for nitrogen is $^4S_{\frac{3}{2}}$. It should be noted that Hund's rules, though universally valid for ground states, do not always apply to excited states.

We saw in the last chapter that for the sodium atom in what we should now write as the state ^2P there is only a very small difference in energy between the components $^2P_{\frac{1}{2}}$ and $^2P_{\frac{3}{2}}$, though the difference is large enough for spectral lines involving this state to show doublet structure. When transitions involve two states each of which has $J \neq 0$, the multiplicities of spectral lines increase further (though only to an extent limited by the selection rule $\Delta J = 0$ or ± 1); such fine structure is very useful in deducing information about energy states of atoms and ions. For light elements it is generally true that for given values of S and L, variation in the value of J makes only a little difference in energy. For heavier elements, however, this is no longer true, and with increase in atomic number the quantum number J becomes increasingly important.

3.4 Ionisation energies

The ionisation energy of the hydrogen atom and its derivation from the atomic spectrum of hydrogen were mentioned in Section 2.3. We turn now to a general discussion of ionisation energies. The exact definition of the ionisation energy, I, of a species A is the molar internal energy change of the reaction

$$A(g) \rightarrow A^+(g) + e(g)$$

at absolute zero. This is not quite equal to the corresponding enthalpy change at 298 K, the latter quantity being greater by an amount equal to

$$\int_0^{298} [C_p(A^+) + C_p(e) - C_p(A)]dT$$

where C_p denotes a molar heat capacity at constant pressure and T the absolute temperature. If A, A^+, and e are all ideal monatomic gases, their molar heat capacities are $\frac{5}{2}R$ (where R is the ideal gas constant) and for $T = 298$ K the correcting term is thus approximately 6.2 kJ mol^{-1}. This correction is small relative to most ionisation energies, and is very often omitted; in comparing one element with another, of course, its omission introduces no error at all. We should note in passing that, by convention, the enthalpy of the gaseous electron is taken as zero at all temperatures.

Ionisation energies of many elements other than hydrogen have been determined from convergence limits in atomic spectra, but several other methods are available. In the most important group of methods, the principle involved is direct measurement of the minimum potential difference through which it is necessary to accelerate electrons in order for them to ionise atoms with which they collide. This potential difference is often called the *ionisation potential* of the element; the corresponding energy is, of course, this potential times the charge on the electron. This energy refers to a single atom, and for conversion into the usual energy units employed in inorganic chemistry it must first be converted into joules and then multiplied by the Avogadro constant; overall, 1 eV per atom $\equiv$ 96.5 kJ mol^{-1}. Ionisation energies are often tabulated in eV; but in this book we are particularly concerned with their relationship to other thermochemical quantities, and hence will nearly always quote them in kJ mol^{-1}. We should emphasise particularly that all ionisation energies quoted in this book are experimentally determined quantities.

Species having more than one electron can undergo successive ionisations, and the terms *first*, *second*, *third* etc. *ionisation energies* (I_1, I_2, I_3 etc.) refer to the changes

$$A(g) \rightarrow A^{+}(g) + e(g)$$
$$A^{+}(g) \rightarrow A^{2+}(g) + e(g)$$
$$A^{2+}(g) \rightarrow A^{3+}(g) + e(g) \quad \text{etc.}$$

respectively. Ionisation energies always increase as successive electrons are removed, since the same positive charge on the nucleus remains to attract a decreasing number of extranuclear electrons. The relative magnitudes of successive ionisation energies, however, provide interesting evidence for the ideas of electronic structure discussed in the preceding section. The three ionisation energies of lithium, for example, are 520, 7297, and 11 810 kJ mol^{-1}, corresponding to removal of the single 2*s* electron followed by the much more difficult removal of the two 1*s* electrons. The first eleven values for magnesium are 738, 1450, 7731, 10 546, 13 629, 17 997, 21 702, 25 664, 31 643, 35 450 and 169 960 kJ mol^{-1}; the very large proportional increases between the second and third, and between the tenth and eleventh, ionisation energies again find a simple explanation in terms of the electronic configuration of magnesium, $1s^2 2s^2 2p^6 3s^2$; the fact that the next largest increase occurs between the eighth and ninth ionisation energies should also be noted.

Ionisation energies are discussed in connection with the chemistry of groups or periods of elements in several later chapters, but this is the appropriate point to consider the overall pattern of the variation in ionisation energy (taking here the first ionisation energy as an example) with atomic number. This variation is shown in Fig. 3.4.

Figure 3.4
First ionisation energies of the elements.

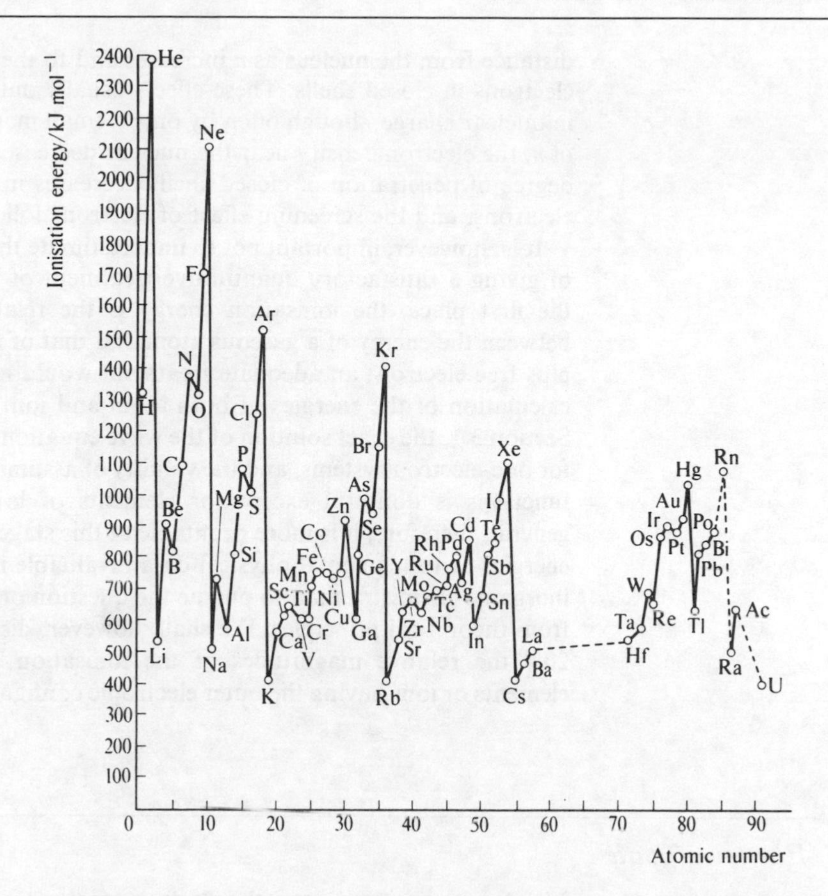

Four general features of Fig. 3.4 call for comment. First, the maxima and minima occur at the noble gases and the alkali metals respectively, indicating that there is a special stability associated with the closed shell electronic configurations of the former elements. Second, there is a general (though somewhat irregular) increase in ionisation energy along each period from the alkali metal to the noble gas. Third, among a group of analogous elements such as the noble gases or the alkali metals, the ionisation energy nearly always decreases with increase in atomic number. Fourth, the variation in ionisation energy along a series of transition or inner transition elements is much less than that along a period of non-transition elements. In addition, we may note the drop in ionisation energy after the elements having the electronic configurations s^2 and s^2p^3, though more detailed examination shows that this is observed only in the earlier periods.

Some helpful comments on these generalisations can be made on the basis of the wave-mechanical model of the atom, just as we discussed the observed ground states of the elements on this basis in Section 3.1. The decrease in ionisation energy down a group may reasonably be attributed to the increasing probability of finding, say, an *ns* electron at a greater distance from the nucleus as n increases and to the screening effect of the electrons in closed shells. These effects usually outweigh that of increase in nuclear charge, though often by only a small margin. For a given value of n, the electron density near the nucleus decreases as l increases, i.e. the degree of penetration of closed shells decreases in the order $s > p > d > f$ electrons, and the screening effect of electrons follows the same order.

It is, however, important not to underestimate the difficulties in the way of giving a satisfactory quantitative treatment of ionisation energies. In the first place, the ionisation energy is the relatively small difference between the energy of a gaseous atom and that of the system gaseous ion plus free electron; an adequate treatment would necessitate the accurate calculation of the energies of both atom and ion. Second, as we said in Section 3.1, the exact solution of the wave equation has been possible only for one-electron systems, and the validity of assuming hydrogen-like wave functions is doubtful except for elements of low atomic number. In general, therefore, it is more profitable at this stage to consider ionisation energies as fundamental physical data invaluable in the interpretation of inorganic chemistry than to pursue the question of how to calculate them from theoretical principles. We shall, however, discuss further in Section 20.4 the relative magnitudes of the ionisation energies of successive elements or ions having the outer electronic configuration p^n or d^n.

3.5 Electron affinities

Like the ionisation energy, the electron affinity E of a species A is defined as the internal energy change of a reaction at absolute zero, in this case

$$A(g) + e(g) \rightarrow A^-(g)$$

As with ionisation energy, little error is introduced by considering the enthalpy change at 298 K instead of the internal energy change at 0 K. For an overall process such as

$$O(g) + 2e \rightarrow O^{2-}(g)$$

the stages

$$O(g) + e \rightarrow O^-(g)$$

$$O^-(g) + e \rightarrow O^{2-}(g)$$

correspond to the first and second electron affinities (E_1, E_2) respectively. It will be noted that for an atom which liberates energy when it combines with an electron, the electron affinity as we have defined it is a negative quantity; this is contrary to older usage, according to which an element that liberates energy on capturing an electron has a positive electron affinity, and for this reason it has been suggested that the term *electron affinity* should be replaced by *enthalpy of electron attachment*. But until such usage becomes general it is desirable when quoting numerical values for electron affinities to include the reminder 'ΔH for $A(g)+e(g) \rightarrow A^-(g)$' in parentheses, i.e. to follow the general thermodynamic practice of accompanying any stated value of ΔH by an equation showing the reaction to which it refers. It may be noted that a similar ambiguity exists for a bond energy if we consider, for example, the combination of two hydrogen atoms rather than the dissociation of a H_2 molecule.

Some electron affinities have been measured directly. Such a measurement is most simply made by a study of the gaseous equilibrium

$$X + e \rightleftharpoons X^-$$

over a range of temperatures. If X is a halogen, the concentration of atoms at a given temperature can be evaluated from a separate study of the equilibrium

$$X_2 \rightleftharpoons 2X$$

and [e], the electron concentration, can also be determined independently if a hot metal wire is used as the source of electrons; a distinction between electrons and halide ions as current carriers can be made on the basis of their different charge/mass ratios, and hence K, given by

$$K = \frac{[X^-]}{[X][e]}$$

can be obtained; a plot of $\ln K$ against $1/T$ gives a straight line of slope $-\Delta H/R$. In this way ΔH for $X + e \rightarrow X^-$ is found to be -338, -355, -331 and -302 kJ mol^{-1} for X = F, Cl, Br and I respectively. Other methods used for the halogens include the application of the Born–Haber cycle (see Section 6.8) and the ultraviolet spectroscopy of halide ions, i.e. measurement of the ionisation energy of X^-, the reverse of electron attachment by the halogen atom.

For oxygen it is found by direct measurement that $E_1 = -142$ kJ mol^{-1}; $E_1 + E_2$, however, estimated from the Born–Haber cycle on the assumption that the alkaline earth metal oxides contain the O^{2-} ion, is $+630$ kJ mol^{-1}: so although O^{2-} has a noble gas electronic configuration, 770 kJ mol^{-1} of energy has to be supplied in order to bring about the gaseous reaction

$$O^- + e \rightarrow O^{2-}$$

This is easily understandable, since to bring up an electron to a negatively charged species must obviously involve doing work. The role of lattice energy in stabilising O^{2-} in crystal structures is discussed in Section 6.8.

Estimates, sometimes approximate, have been made of the electron affinities of several other elements, but the discussion of the procedures involved lies beyond the scope of this book.

The mean of the *numerical* value of the first ionisation energy of an element and its electron affinity (e.g. for Cl, of 1253 and 355 kJ mol^{-1}) was suggested by Mulliken as a quantitative measure of the *electronegativity* of the element, i.e. its tendency to attract electrons when it is in a molecular environment rather than present as isolated atoms in the gas phase. Since several other definitions of electronegativity have been suggested, however, and since some of them are based on observed molecular properties, we shall defer consideration of this property until after the treatment of the covalent bond in Chapter 5.

3.6 Atomic dimensions

The *atomic volume* of an element is the volume occupied by an Avogadro number of atoms in the solid state; the periodic variation in this property with change in atomic mass was recognised soon after the periodic classification was first put forward.

It should be noted, however, that the size of an *isolated* atom is not a meaningful quantity; as we saw in the last chapter, wave functions fall away exponentially with increasing distance from the nucleus, and there is thus no 'atomic radius' at which an atom suddenly ends. Only in the liquid and solid states can the volume occupied by an atom (necessarily in contact with other atoms) be calculated. Even there, comparisons are complicated by the fact that different elements have very different structures. In the solid state, for example, most metals crystallise in infinite symmetrical structures in which each atom has eight or twelve nearest neighbours; non-metals, however, crystallise in a great variety of structures, e.g. the noble gases as single atoms, the halogens as diatomic molecules, sulphur as S_8 rings, and carbon as two infinite macromolecules of totally different geometries.

In practice, therefore, there is little point in discussing atomic volumes in detail, and trends in size among groups or periods of elements will be treated under the structures of the elements concerned. But we should not leave this subject without mentioning the so-called *van der Waals' radii*. In an elementary account of the types of interaction between molecules in condensed phases it is customary to divide the types of bonding into ionic, covalent, metallic and van der Waals' (hydrogen bonding is sometimes added as a special category). The weak residual attractions between molecules which are not covalently bonded, for example, are often grouped together under the heading of van der Waals' forces, though it is now recognised that three types of interaction are involved: dipole–dipole, dipole–induced dipole, and dispersion (or London) forces. The first two of these terms are self-explanatory; dispersion forces arise from the motion of electrons relative to the nucleus, which leads to the presence of a rapidly varying dipole that can in turn induce similar dipoles in neighbouring molecules. All of these forces of attraction tend to offset the repulsive forces that result from the overlapping of the electron clouds around the atoms.

In a crystal of a solidified noble gas, the attraction between molecules arises entirely from dispersion forces; in, say, solid hydrogen chloride, all three types of interaction occur. The closest intermolecular approach of atoms in a solid in which molecules are bonded only by van der Waals' interactions is the sum of the van der Waals' radii of the atoms concerned. In solid argon, for example, each molecule of the monatomic gas is 3.82 Å from twelve other molecules, so the van der Waals' radius of argon is 1.91 Å; the corresponding values for neon, krypton and xenon are 1.60, 1.97 and 2.14 Å. For other atoms the van der Waals' radius naturally varies somewhat according to the chemical environment in which it is measured, as we shall see in Chapter 16 when we discuss the structures of the halogens.

3.7 *Relativistic effects*

The importance of the Schrödinger equation to chemistry will now be obvious. Nevertheless, it is not the only approach to wave mechanics, and it does not provide the basis for understanding all chemical phenomena. It has already been noted in Section 2.8 that it does not predict the existence of the spin quantum number m_s, and it gives no clue as to why, for a state of given values of L and S, the value of J makes little difference to the energy in the case of light atoms but a large difference in the case of heavy atoms (Sections 3.3 and 19.6). Amongst many generalisations about heavy elements that will be encountered in later chapters of this book are two which must clearly depend upon quantum theory for their explanation: the ionisation energies of the 6*s* electrons are anomalously high, leading to the marked stabilisation of Hg(0), Tl(I), Pb(II) and Bi(III) compared with Cd(0), In(I), Sn(II) and Sb(III); and whereas bond energies usually decrease down a group of main group elements, they often increase down a group of transition elements, both in the elements themselves and in their compounds.

All these observations can be accounted for (though often far from simply) if Einstein's theory of relativity is combined with quantum theory, in which case they are attributed to *relativistic effects*. We shall confine attention here to the chemical generalisations. According to the theory of relativity, the mass m of a particle increases from its rest mass m_0 when its velocity v approaches the speed of light, c_0, m being then given by

$$m = \frac{m_0}{\sqrt{1 - \left(\frac{v}{c_0}\right)^2}}$$

Now it was shown in Section 2.3 that for a one-electron system the Bohr model of the atom (which, despite its shortcomings, gives the correct value for the ionisation energy) leads to the expression for the velocity of the electron

$$v = \frac{Ze^2}{2\varepsilon_0 nh}$$

For $n = 1$ and $Z = 1$, v is only about $(1/137)c_0$, but for $Z = 80$, v/c_0 becomes about 0.58, leading to $m = 1.2m_0$ approximately. Since the radius of the Bohr orbit is given by

$$r = \frac{Ze^2}{4\pi\varepsilon_0 mv^2}$$

this increase in m results in a contraction in the radius of the 1s ($n = 1$) orbital of about 20% – the *relativistic contraction*. Other s orbitals are affected similarly and in consequence when Z is high they have diminished overlap with orbitals of other atoms. A detailed treatment shows that p orbitals, having a low electron density near the nucleus, are less affected, and that d orbitals, being more effectively screened from the nuclear charge by the contracted s and p orbitals, undergo a relativistic *expansion* – hence their better overlap with orbitals of other atoms. An analogous argument applies to f orbitals. The relativistic contraction of the s orbitals means that for an atom of high Z there is an extra energy of attraction between s electrons and the nucleus, made manifest in higher ionisation energies for the 6s electrons in such atoms; and this is of particular chemical significance for elements for which the 6s electrons are in the valence shell – Hg, Tl, Pb and Bi, for which $Z = 80$–83. We return to this matter in Section 12.1.

Relativistic effects have been invoked to explain, in whole or in part, many other properties of the heavy elements; more examples are given in some of the references for further reading.

PROBLEMS

1 Use Slater's rules to calculate the effective nuclear charge experienced by (a) one of the 4s electrons (b) one of the 3d electrons in the atom of vanadium, the electronic configuration of which is $1s^2 2s^2 2p^6 3s^2 3p^6 3d^3 4s^2$. What do these calculations suggest about the ground state of the V^+ ion?

2 With the help of Hund's rules work out the state symbols for the ground states of sulphur, chlorine, titanium, chromium and nickel.

Why would an excited carbon atom having the electronic configuration $1s^2 2s^2 2p^1 3p^1$ give rise to a more complex atomic spectrum than a ground state carbon atom?

3 Discuss each of the following observations:
(a) The energy difference between the $1s^2 2s^1\ ^2S_{\frac{1}{2}}$ and $1s^2 2p^1\ ^2P_{\frac{1}{2}}$ states for Li is 14 900 cm^{-1}, whereas for Li^{2+} the energy difference between the $2s^1\ ^2S_{\frac{1}{2}}$ and $2p^1\ ^2P_{\frac{1}{2}}$ states is only 2 cm^{-1}.
(b) The first ionisation energies of boron, carbon, nitrogen, oxygen, fluorine and neon are 800, 1086, 1402, 1313, 1680, and 2080 kJ mol^{-1} respectively.
(c) More energy is liberated when a gaseous chlorine atom combines with an electron than when a gaseous fluorine atom does so.

REFERENCES FOR FURTHER READING

Lagowski, J.J. (1973) *Modern Inorganic Chemistry*, Marcel Dekker, New York. A general account of atomic properties

Phillips, C.S.G. and Williams, R.J.P. (1965) *Inorganic Chemistry*, Oxford University Press. Includes a more advanced discussion of the periodic table and ionisation energies.

Pitzer, K.S. (1979) *Accounts of Chemical Research*, **12**, 271. An introduction to relativistic effects in chemistry.

Porterfield, W.W. (1984) *Inorganic Chemistry*, Addison-Wesley, Reading, Massachusetts. A more advanced and speculative discussion of atomic properties.

Wulfsberg, G. (1987) *Principles of Descriptive Inorganic Chemistry*, Brooks/Cole, Monterey, California. The last chapters of this original treatment give a useful summary of periodic trends and include a discussion of relativistic effects.

4 *Electronic configurations of molecules*

4.1 Introduction

The foundations of modern theories of chemical bonding were laid in 1916–20 by G.N. Lewis and Langmuir, who suggested that polar species such as sodium chloride are formed by electron transfer and non-polar molecules such as carbon tetrachloride by electron sharing. In some cases, e.g. $H_3N—H^+$ and $(CH_3)_3N—BF_3$, both shared electrons are provided by one atom, but once formed the bond, sometimes called a coordinate bond, is indistinguishable from a covalent bond. The treatment of the ionic bond is much simpler than that of the covalent bond, but since the discussion of the energetics of ionic compounds involves a knowledge of the energies of covalent bonds, we shall postpone the account of the former subject until Chapter 6, first considering theories of covalent bonding and then, in the next chapter, some physical properties of molecules.

Modern views of atomic structure are, as we have seen in the last two chapters, based very largely on the applications of wave mechanics to atomic systems. Modern views of molecular structure are, correspondingly, based on the results of applying wave mechanics to molecules, since these studies have provided answers to the questions of how and why atoms combine. The Schrödinger equation can be written down to describe the behaviour of electrons in molecules, but it can be solved only by means of approximation techniques. Two methods have been used to approach this problem: the molecular orbital (MO) theory associated with the names of Hund and Mulliken, and the valence bond (VB) theory developed by Heitler and London, and by Pauling. As its name implies, the molecular orbital theory allocates electrons to molecular orbitals similar in many respects to atomic orbitals, but each containing two or more nuclei. The valence bond theory, on the other hand, treats the formation of a molecule as arising from the bringing together of complete atoms which, although they interact, to a large extent retain their original character. Although the molecular orbital theory is conceptually simpler and is now more popular for the discussion of inorganic compounds, some acquaintance with valence bond theory is also essential, and brief accounts of both treatments are therefore given in the following sections. In later chapters of this book we shall use whichever theory is more appropriate to the topic under consideration. As in Chapter 2, we shall emphasise qualitative aspects of the theories; more mathematical treatments will be found in the references cited at the end of the chapter.

4.2 *Molecular orbital theory: homonuclear diatomic molecules*

In molecular orbital theory we begin by placing the nuclei of a given molecule in their equilibrium positions and calculating the orbitals spread over the whole molecule (i.e. the molecular orbitals) that a single electron might occupy. Each of these corresponds to an atomic orbital in an atom, i.e. $\psi^2 dv$ is proportional to the probability of locating the electron in a given volume dv. Each molecular orbital corresponds to a definite energy and the sum of the individual energies of the electrons in their orbitals, after correction for electron interaction, gives the total energy of the molecule. To derive the ground state of the molecule, the electrons are placed in the available molecular orbitals, beginning with the orbital of lowest energy; each molecular orbital can accommodate a maximum of two electrons, provided their spins are opposed (Pauli's principle extended to molecules); and by filling as few orbitals as possible, and those of the lowest energy possible, we arrive at the electronic ground state.

Let us begin by discussing the case of molecular hydrogen. An approximate description of the molecular orbitals in H_2 can be obtained by considering them as Linear Combinations of Atomic Orbitals (LCAOs). Let us label the nuclei A and B. The lowest energy orbital associated with each nucleus is the 1*s* orbital, and each of these atomic orbitals may be represented by a wave function ψ_A or ψ_B. Now each molecular orbital may also be represented by a wave function which is a suitable linear combination of atomic orbitals; since A and B are identical atoms their atomic orbitals obviously contribute equally to molecular orbitals. In this case we can write down two molecular orbitals derived by combining the 1*s* atomic orbitals, these being represented by:

$$\psi_m = N[\psi_A + \psi_B]$$
$$\psi_m^* = N^*[\psi_A - \psi_B]$$

In these expressions N and N^* normalise the molecular wave functions so that for each orbital the probability of finding an electron somewhere in it is unity; they are given by

$$N = \frac{1}{\sqrt{2(1+S)}} \qquad N^* = \frac{1}{\sqrt{2(1-S)}}$$

where S is the so-called overlap integral;

$$S = \int \psi_A \psi_B \, dv$$

The overlap integral is a measure of the extent to which the two wave functions coincide; it is substantially less than unity and is often neglected in the normalisation procedure, in which case

$$N \approx N^* \approx \frac{1}{\sqrt{2}}$$

The molecular wave functions ψ_m and ψ_m^* are *bonding* and *antibonding* molecular orbitals; orbitals of this type, in which the line joining the two nuclei is a symmetry axis for the electron distribution, are known as σ orbitals if bonding and σ^* orbitals if antibonding, so we may alternatively denote them σ_{1s} and σ_{1s}^* orbitals. When the two 1*s* wave functions are added, they reinforce one another everywhere, and especially in the region between the two nuclei; the build-up of electron density there diminishes the internuclear repulsion and a strong bond results. When one of the two 1*s* wave functions is subtracted from the other, they exactly cancel in a plane midway between the nuclei, and the molecular wave function changes sign at this nodal plane. This lack of electron density raises the internuclear repulsion, the total energy becomes higher, the two nuclei are not bonded together, and the orbital is described as antibonding.

Pictorial representations of this LCAO method are given in Figs. 4.1, 4.2 and 4.3. In the first of these, the relative energies of 1*s* molecular orbitals and their constituent atomic orbitals are shown; note that the antibonding orbital is correctly shown as somewhat more destabilised relative to the atomic orbitals than the bonding orbital is stabilised (N is not strictly equal to N^*, but is rather smaller, as we mentioned earlier).

In the hydrogen molecule in its ground state, both electrons occupy the σ_{1s} orbital; in the hydrogen molecule ion, H_2^+, formed by the action of an electric discharge on hydrogen at low pressures, only a single electron is in this orbital and the total bonding effect is smaller, though the bond is still a strong one, the binding energy being 269 kJ mol^{-1} compared with 458 kJ mol^{-1} for H_2. The bond in H_2^- (which has not yet

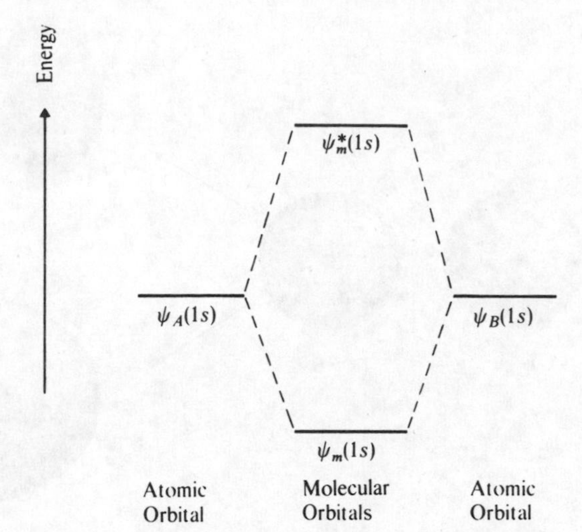

Figure 4.1
The relative energy levels of molecular orbitals and their constituent atomic orbitals for H_2.

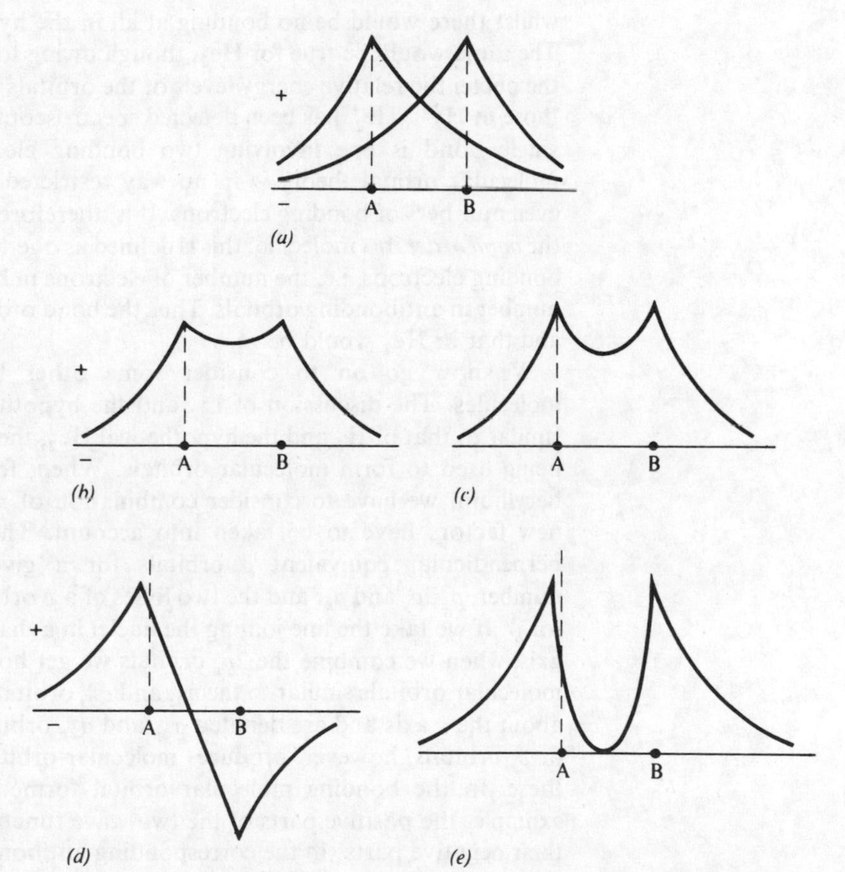

Figure 4.2
The formation of molecular orbitals for H_2.
(*a*) ψ_A and ψ_B for individual atoms. (*b*) ψ_m.
(*c*) Probability function for the bonding orbital, $(\psi_m)^2$.
(*d*) ψ_m^*. (*e*) Probability function for the antibonding orbital, $(\psi_m^*)^2$.

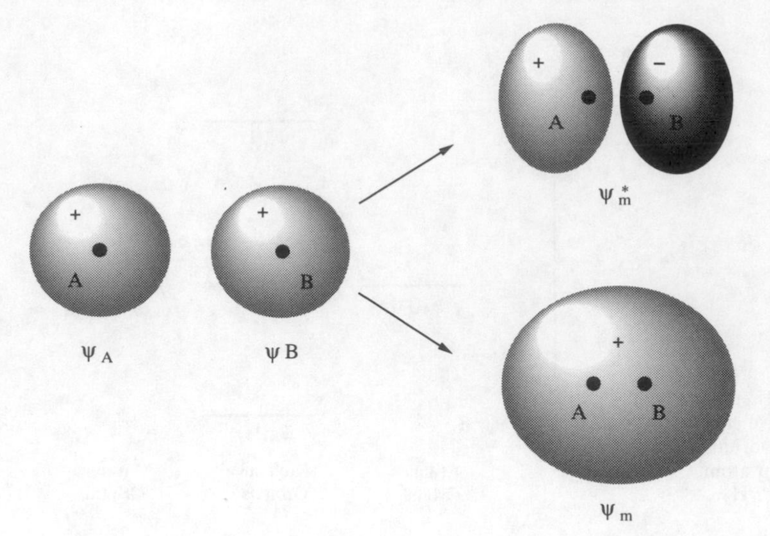

Figure 4.3
Another representation of the formation of molecular orbitals for H_2. Since the σ and σ^* orbitals are respectively centrosymmetric and non-centrosymmetric these orbitals may also be denoted by the symbols σ_g and σ_u^* (cf. Section 2.7).

been characterised) would be expected to be weaker than that in H_2^+, whilst there would be no bonding at all in the hypothetical species H_2^{2-}. The same would be true for He_2, though owing to the different charge on the nuclei the relative energy levels of the orbitals would be different from those in H_2^{2-}; He_2^+ has been detected spectroscopically. The conventional single bond is one involving two bonding electrons (as in H_2), but molecular orbital theory is in no way restricted to situations involving even numbers of bonding electrons. It is, therefore, often useful to refer to the *bond order* in a molecule; this is defined as one-half of the net number of bonding electrons, i.e. the number of electrons in bonding orbitals less the number in antibonding orbitals. Thus the bond order in H_2^+ and He_2^+ is 0.5 and that in He_2 would be zero.

We now go on to consider some other homonuclear diatomic molecules. The discussion of Li_2 and the hypothetical Be_2 is essentially similar to that of H_2 and the hypothetical He_2, the 2*s* atomic orbitals now being used to form molecular orbitals. When, for the elements beyond beryllium, we have to consider combination of *p* atomic orbitals, some new factors have to be taken into account. There are three mutually perpendicular equivalent *p* orbitals for a given principal quantum number, p_x, p_y and p_z; and the two lobes of a *p* orbital have opposite signs for ψ. If we take the line joining the nuclei in a diatomic molecule as the *z* axis, when we combine the $2p_z$ orbitals we get bonding and antibonding molecular orbitals similar to the σ_{2s} and σ_{2s}^* orbitals; they are symmetrical about the *z* axis and are denoted σ_{2p} and σ_{2p}^* orbitals. Combination of p_x or p_y orbitals, however, produces molecular orbitals quite different from these. In the bonding molecular orbital formed from p_x orbitals, for example, the positive parts of the two wave functions overlap, and so do their negative parts; in the corresponding antibonding molecular orbital, parts of the wave functions having opposite signs do not overlap but repel one another. These molecular orbitals are known as π_{2p_x} and $\pi^*_{2p_x}$ orbitals respectively. The corresponding π_{2p_y} and $\pi^*_{2p_y}$ orbitals are perpendicular to them, and are degenerate with the π_{2p_x} and $\pi^*_{2p_x}$ orbitals respectively. These four orbitals collectively are known simply as the π_{2p} and π^*_{2p} orbitals. Thus combination of the $2p$ atomic orbitals leads to a σ_{2s}, a σ^*_{2s}, two π_{2p} and two π^*_{2p} orbitals. All the π orbitals, it may be noted, are ungerade; the π^* orbitals are gerade. A pictorial representation of the combination of the p orbitals is given in Fig. 4.4.

The relative energy levels of the molecular orbitals are determined by two factors: the energies of the atomic orbitals used in forming a molecular orbital and the extent of overlap between these atomic orbitals. The greater the overlap, the more the bonding orbital is lowered and the antibonding orbital is raised in energy relative to the atomic orbitals which are combined. The energies of atomic orbitals, as we have already seen, increase in the sequence $1s < 2s < 2p$, and it might be expected that for molecular orbitals derived from the degenerate $2p$ orbitals the overlap would be greater for the σ than for the π orbitals, thus leading to the overall sequence of increasing energy:

$$\sigma_{1s}, \sigma^*_{1s}, \sigma_{2s}, \sigma^*_{2s}, \sigma_{2p}, \pi_{2p_x} = \pi_{2p_y}, \pi^*_{2p_x} = \pi^*_{2p_y}, \sigma^*_{2p}$$

In filling these orbitals, we must bear in mind two generalisations carried forward from atomic structure, Pauli's principle and Hund's first rule;

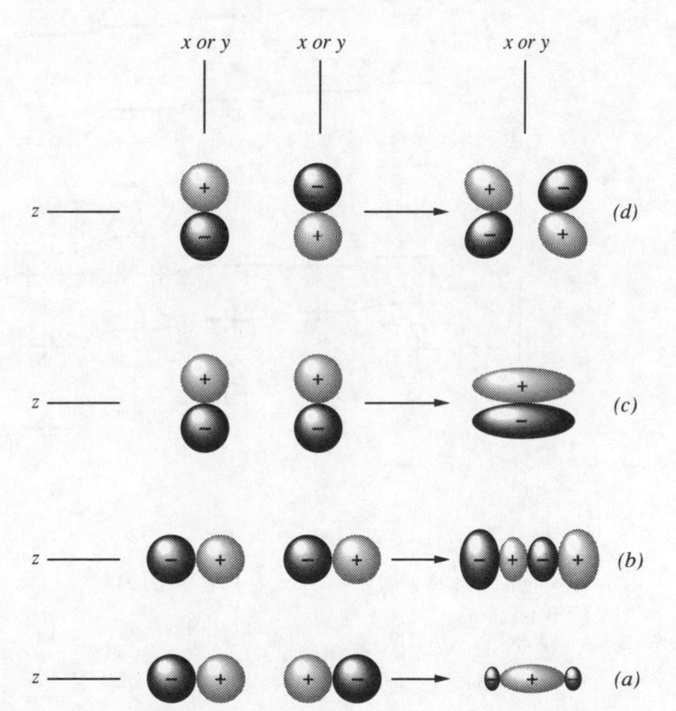

Figure 4.4
The combination of $2p$ orbitals, showing formation of: (*a*) σ_{2p} bonding; (*b*) σ^*_{2p} antibonding; (*c*) π_{2p} bonding; and (*d*) π^*_{2p} antibonding orbitals.

according to the latter two electrons in, say, the degenerate π_{2p} orbitals occupy each orbital singly rather than one doubly.

The actual electronic configurations of molecules are nearly always determined experimentally, particularly by photoelectron spectroscopy, in which electrons in different orbitals are distinguished by their ionisation energies and the effects of their removal on the vibrational spectrum; it is found that whereas the sequence given holds for O_2 and F_2, for B_2, C_2 and N_2 the π_{2p} orbitals are of slightly lower energy than the σ_{2p} orbital; this is also true for Li_2 and Be_2, but in the ground states of these molecules the σ_{2p} and π_{2p} orbitals are, of course, all empty. (This change in energy sequence arises indirectly from the fact that the 2*s* and 2*p* atomic orbitals are much closer in energy for the earlier members of the period lithium–neon than for the later members; in the case of N_2, for example, the molecular orbitals are really only mainly 2*s* or mainly 2*p* in origin.) The ground states of the known molecules Li_2, B_2, C_2, N_2, O_2 and F_2, together with the hypothetical ground states for the non-existent species Be_2 and Ne_2 (in which there would be no net bonding) are represented in Fig. 4.5; Fig. 4.6 is an example of another form of diagram, and shows the molecular orbitals and their occupancy for the oxygen molecule.

Also shown in Fig. 4.5 are the bond lengths, bond energies and bond orders of the diatomic species. Since the nuclear charges are changing along the series, we should not expect all bonds of order unity to have the

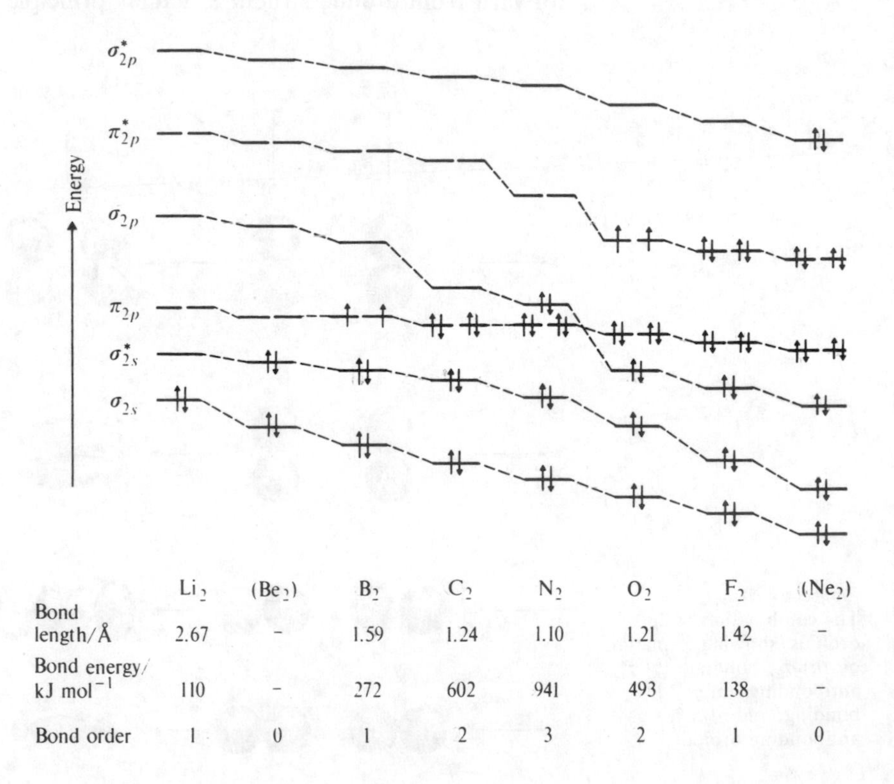

	Li_2	(Be_2)	B_2	C_2	N_2	O_2	F_2	(Ne_2)
Bond length/Å	2.67	–	1.59	1.24	1.10	1.21	1.42	–
Bond energy/ kJ mol^{-1}	110	–	272	602	941	493	138	–
Bond order	1	0	1	2	3	2	1	0

Figure 4.5
Ground states of some homonuclear diatomic molecules.

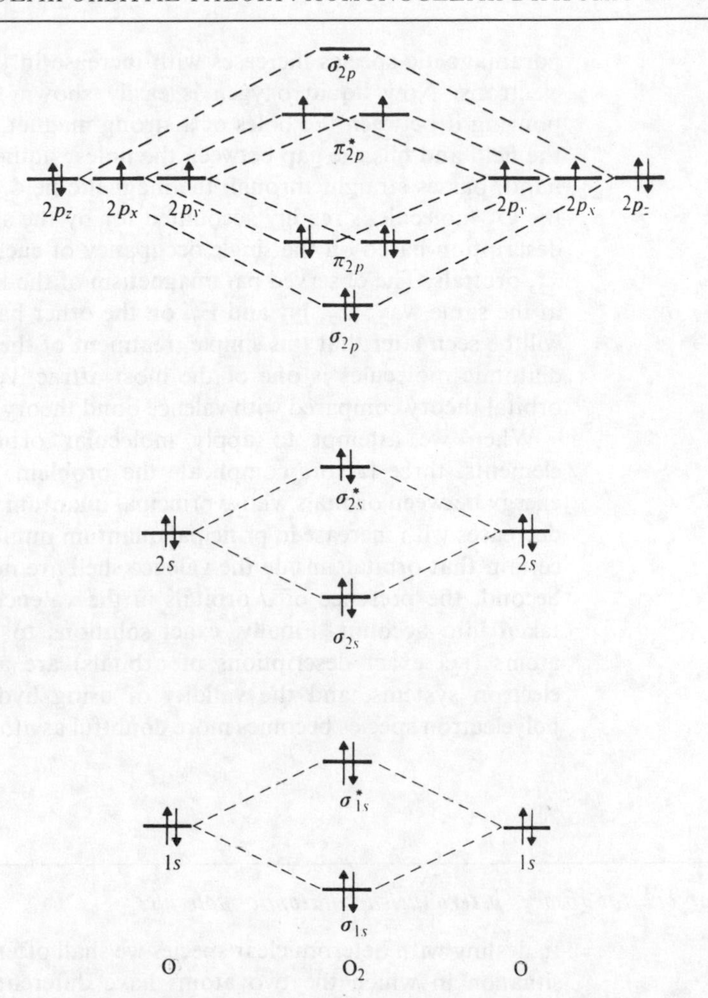

Figure 4.6
The ground state of the O_2 molecule.

same bond energy; but the general relationship between the order, the energy and the shortness of the bond is unmistakable. Further, whilst removal of an antibonding electron from the O_2 molecule (ionisation energy 1170 kJ mol^{-1}) produces an ion O_2^+ in which the bond energy is 625 kJ mol^{-1} compared with 493 kJ mol^{-1} in O_2, removal of a bonding electron from the N_2 molecule (ionisation energy 1510 kJ mol^{-1}) results in a decrease in bond energy from 941 kJ mol^{-1} in N_2 to 613 kJ mol^{-1} in N_2^+. We shall make substantial use of such considerations in later chapters.

Before we leave the subject of homonuclear diatomic molecules we should add a few words about their magnetic properties. Magnetochemistry is discussed in more detail in Chapter 19; for the present it is necessary to know only that the presence of one or more unpaired electrons in an ion or molecule makes it paramagnetic, i.e. attracted by a magnetic field, whereas species which contain no unpaired electrons are diamagnetic, i.e. repelled by a magnetic field; the magnetic moment of a paramagnetic species increases with increase in the number of unpaired electrons. Now liquid oxygen is easily shown to be paramagnetic by pouring it between the poles of a strong magnet, when it is attracted by the field and fills the gap between the poles; liquid nitrogen, on the other hand, passes straight through the magnetic field. The paramagnetism of the O_2 molecule is readily accounted for by the simple molecular orbital description based on the single occupancy of each of the two degenerate π^*_{2p} orbitals. The observed paramagnetism of the B_2 molecule is explained in the same way; C_2, N_2 and F_2, on the other hand, are diamagnetic. It will be seen later that this simple treatment of the magnetic properties of diatomic molecules is one of the most attractive features of molecular orbital theory compared with valence bond theory.

When we attempt to apply molecular orbital theory to heavier elements, three factors complicate the problem. First, the difference in energy between orbitals whose principal quantum number differs by unity decreases with increase in principal quantum number, and it is not always certain that orbitals inside the valence shell are not involved in bonding. Second, the presence of *d* orbitals in the valence shell may have to be taken into account. Finally, exact solutions to the wave equation for atoms (i.e. exact descriptions of orbitals) are available only for one-electron systems, and the validity of using hydrogen-like orbitals for polyelectron species becomes more doubtful as atomic number increases.

4.3 *Molecular orbital theory: heteronuclear diatomic molecules*

In dealing with heteronuclear species we shall often be concerned with the situation in which the two atoms have different orbitals available for possible bond formation. We have seen already that the essential condition for bonding is the overlapping of wave functions of the same sign, and that the overlapping of wave functions of different signs leads to antibonding. A third possibility is that the two wave functions either do not overlap at all (e.g. p_x and p_y) or that some positive overlapping is counteracted by negative overlapping (e.g. a sideways-on or π type of interaction between an *s* orbital and a *p* orbital). These examples are illustrated in Fig. 4.7; the molecular orbitals which result are known as *non-bonding* orbitals. A glance at Fig. 4.7(*b*) makes it immediately clear that in a molecule in which a bonding orbital is to be obtained by combination of the hydrogen *s* orbital and a *p* orbital, the former will most advantageously overlap the p_z orbital.

We can now state the general conditions for the most effective combination of atomic orbitals in a molecule AB. These are that the atomic orbitals involved should: (i) have similar energies; (ii) overlap as much as possible; and (iii) have the same symmetry with respect to the *z*

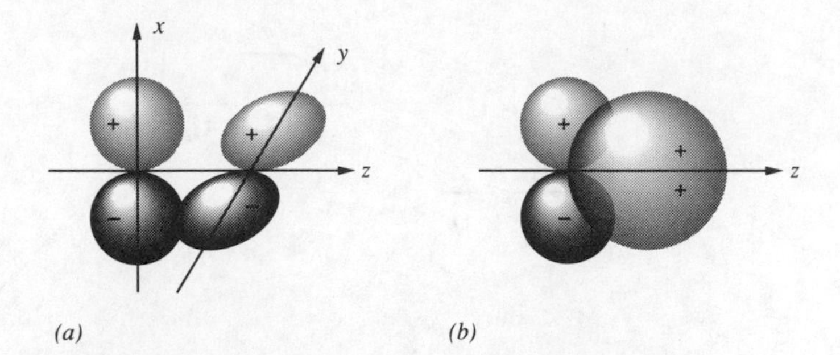

Figure 4.7
Atomic orbital arrangements leading to the formation of non-bonding molecular orbitals: (*a*) p_x–p_y; (*b*) s–p_x.

(i.e. internuclear) axis. With these conditions in mind let us now consider a few simple heteronuclear species.

In a heteronuclear molecule AB the choice of atomic orbitals for combination is guided by information from atomic spectroscopy. Thus in hydrogen fluoride the fluorine atomic orbitals of energy nearest to that of the hydrogen 1*s* orbital are the 2*p* orbitals, though (as is shown by the higher ionisation energy of fluorine) these are considerably lower in energy than the hydrogen 1*s* orbital; as we have just seen, the $2p_z$ orbital is the one to give the best bonding interaction. Where two orbitals of different energy are combined to give molecular orbitals the basic diagram shown in Fig. 4.1 is amended to that given in Fig. 4.8, and the molecular wave functions are now given by

$$\psi_m = N[\psi_A + \lambda\psi_B]$$

and

$$\psi_m^* = N^*[\psi_B - \lambda\psi_A]$$

where B is the element contributing the lower energy orbital (in this case fluorine) and λ is a constant (sometimes called the mixing coefficient) reflecting the extent to which the ψ_B orbital contributes to the molecular wave function. For the case of HF, $\lambda > 1$, i.e. the $2p_z$ wave function for fluorine contributes more than the hydrogen 1*s* wave function to the molecular wave function, and the electron density is greater around the

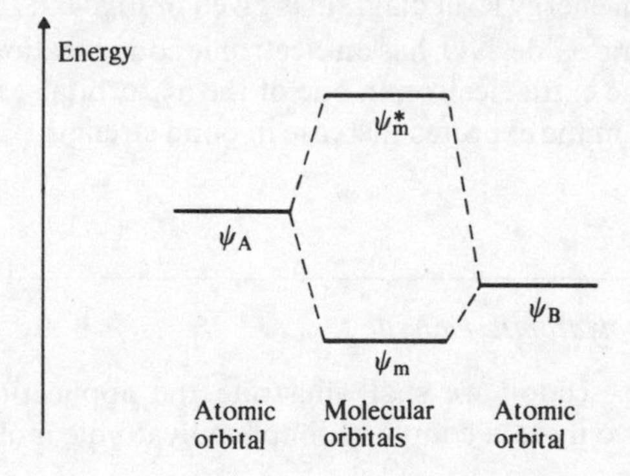

Figure 4.8
The relative energy levels of molecular orbitals and their constituent atomic orbitals for a heteronuclear molecule AB.

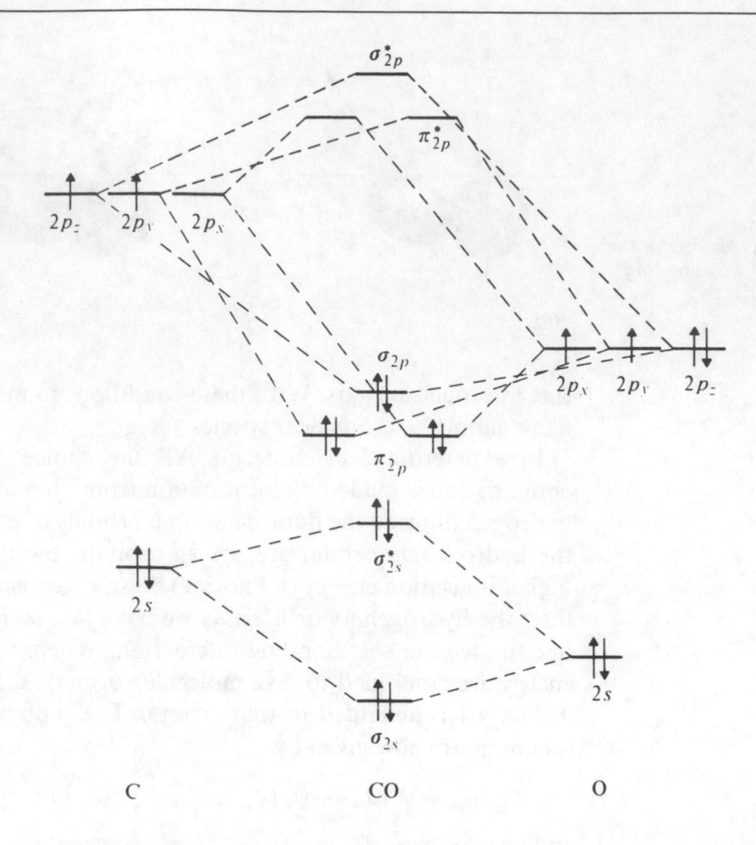

Figure 4.9
Electronic configuration (not to scale) of the CO molecule.

fluorine nucleus than around the proton. The value of λ is thus an indication of the *polarity* of the H—F bond. In the case of a homonuclear diatomic molecule, of course, $\lambda = 1$ and is usually omitted. For the gaseous molecule LiF, on the other hand, λ is much greater than unity, i.e. the bond is a very polar one.

The electronic structure of the carbon monoxide molecule is, to a first approximation, like that of molecular nitrogen (Fig. 4.5), with which it is isoelectronic. A more accurate representation shows all atomic orbitals for oxygen as lower in energy than the corresponding orbitals for carbon; such an energy level diagram is given in Fig. 4.9.

Nitric oxide, NO, has an electronic configuration which is like that of N_2 with the extra electron in one of the π^*_{2p} orbitals; ionisation to yield NO^+ results in the expected increase in bond strength.

4.4 *Molecular orbital theory: polyatomic molecules*

In this section we shall illustrate the application of molecular orbital theory only to a couple of simple polyatomic molecules, $BeCl_2$ and BCl_3, assuming them to contain two and three single bonds respectively (we choose these rather than hydrides because BeH_2 and BH_3 both polymerise, and neither has been observed as a stable species).

In general the combination of *n* (where *n* is an even number) atomic orbitals yields *n*/2 bonding and *n*/2 antibonding molecular orbitals. The molecule $BeCl_2$ is linear, so the beryllium atom has two suitable atomic orbitals (2*s* and $2p_z$), and each chlorine atom has one ($3p_z$). We may represent the bonding and antibonding orbitals formed pictorially by Fig. 4.10, mathematically by the expressions (i)–(iv) below (in which, however, we have omitted to take into account the difference in energy of the atomic orbitals of the two elements), and energetically by Fig. 4.11 (from which we have omitted the $2p_x$ and $2p_y$ orbitals of the beryllium atom and the $3p_x$ and $3p_y$ orbitals of the chlorine atoms). With two bonding molecular orbitals filled and no electrons in antibonding orbitals, the Be—Cl bond order is obviously unity.

$$\text{(i)} \quad \sigma_s = \psi_{\mathrm{Be}(2s)} + (\psi_{\mathrm{Cl_a}} + \psi_{\mathrm{Cl_b}})$$

$$\text{(ii)} \quad \sigma_s^* = \psi_{\mathrm{Be}(2s)} - (\psi_{\mathrm{Cl_a}} + \psi_{\mathrm{Cl_b}})$$

$$\text{(iii)} \quad \sigma_p = \psi_{\mathrm{Be}(2p_z)} + (\psi\psi_{\mathrm{Cl_a}} - \psi_{\mathrm{Cl_b}})$$

$$\text{(iv)} \quad \sigma_p^* = \psi_{\mathrm{Be}(2p_z)} - (\psi_{\mathrm{Cl_a}} - \psi_{\mathrm{Cl_b}})$$

The observant reader will notice that although, as we showed in Section 2.8, the 2*s* orbital contains a node and the outer portion of the wave function is negative, we have marked the *s* orbital of Fig. 4.10 positive. This practice of marking bonding wave functions of the central atom positive is general in the use of such diagrams, which purport to show only the envelope of most of the angular part of the wave function; its justification is that it avoids changing signs on bonding 2*s*, 4*s*, 6*s* (as compared with 1*s*, 3*s*, 5*s*, 7*s*) wave functions and on parts of other wave functions which overlap them.

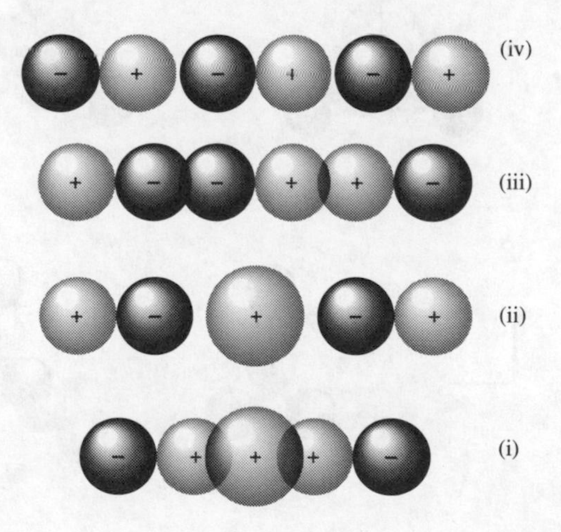

Figure 4.10
Molecular orbitals for $BeCl_2$.

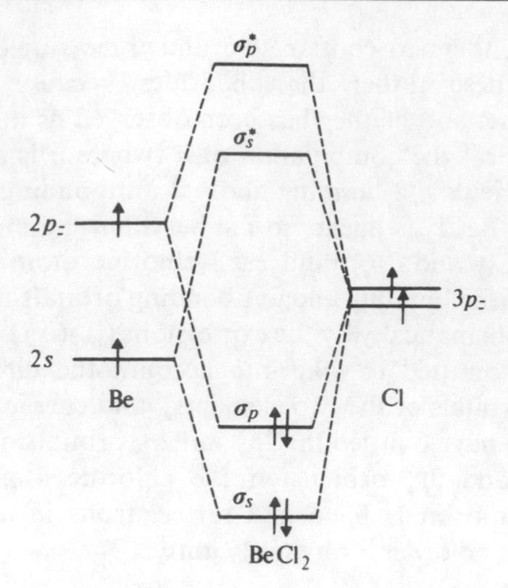

Figure 4.11
Molecular orbital energy levels for $BeCl_2$. For simplifications made see text.

For the planar molecule BCl_3 we choose the z axis as the axis of highest symmetry, i.e. the one through the boron atom and perpendicular to the plane containing the chlorine atoms. We now have four atoms and six suitable atomic orbitals ($2s$, $2p_x$ and $2p_y$ for the boron atom and a $3p$ orbital for each chlorine atom), and there are three bonding molecular orbitals (σ_s, σ_{p_x} and σ_{p_y}) and three antibonding molecular orbitals (σ_s^*, $\sigma_{p_x}^*$ and $\sigma_{p_y}^*$). The formation of these orbitals is represented pictorially by Fig. 4.12 and energetically by Fig. 4.13. A close inspection of Fig. 4.12 will reveal that formation of σ_p bonding orbitals does not result in equal overlapping by all three chlorine orbitals; and accurate expressions for the wave functions would have to take this fact, as well as the inequality in energy levels of all the atomic orbitals, into consideration. We see,

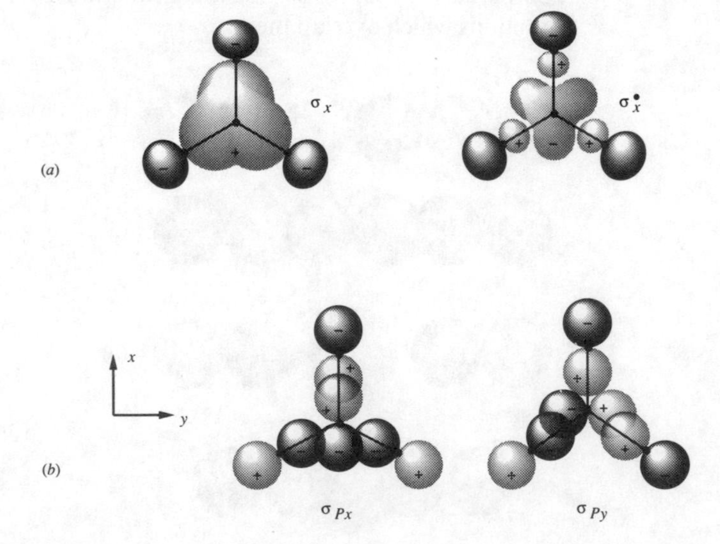

Figure 4.12
(*a*) The σ_s and σ_s^* orbitals.
(*b*) The σ_{p_x} and σ_{p_y} orbitals for BCl_3. The antibonding σ_p orbitals can be obtained by exchanging the signs on the boron p orbital.

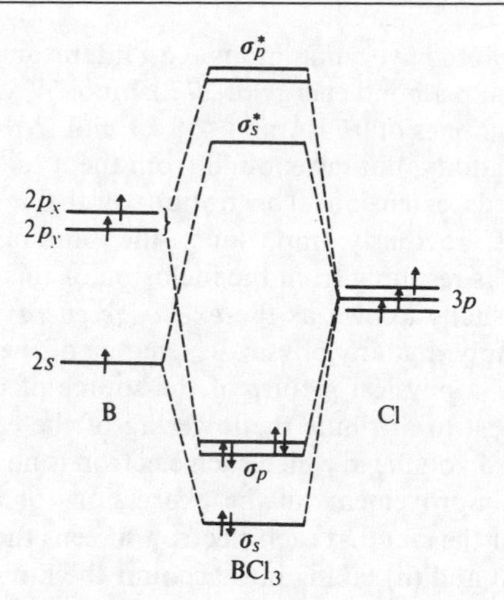

Figure 4.13
Molecular orbital energy levels for BCl_3 (simplified).

therefore, that for a trigonal planar molecule involving two different elements the quantitative aspects of molecular orbital theory have already become much more complicated than for the homonuclear diatomic species with which we began our discussion of this approach to covalent bond formation, and we shall not pursue the subject here.

4.5 *Valence bond theory: diatomic molecules*

In contrast to the molecular orbital theory, valence bond theory begins with the separate atoms and considers the interaction which ensues when they are brought together. Let us begin again by discussing the hydrogen molecule, starting from two hydrogen atoms whose nuclei are labelled A and B and whose electrons are labelled 1 and 2; when the atoms are so far apart that there is no appreciable interaction between them, electron 1 is exclusively associated with nucleus A and electron 2 with nucleus B.

This system can be described by a wave function ψ_{I}. Since, however, in a hydrogen molecule the electrons are indistinguishable, there is an equally good description of the system in which nucleus A is associated with electron 2 and nucleus B with electron 1; this state may be described by a wave function ψ_{II}. As the atoms are brought together, these wave functions interact and two new wave functions, linear combinations of ψ_{I} and ψ_{II} and containing normalising factors N_+ and N_-, result:

$$\psi_+ = N_+(\psi_{\text{I}} + \psi_{\text{II}})$$

$$\psi_- = N_-(\psi_{\text{I}} - \psi_{\text{II}})$$

Calculations of the energies associated with these states show that ψ_- represents a repulsive state; on the other hand the energy curve for the ψ_+

state is found to have a minimum at an interatomic distance of 0.80 Å and corresponds to a bond energy of 277 kJ mol^{-1}, values near enough to the experimental ones of 0.74 Å and 458 kJ mol^{-1} to suggest that the theory has some validity, but far enough from them to show that the expression for ψ_+ needs extension. The notion of the indistinguishability of the electrons is obviously important; the binding energy which can be considered as resulting from the inclusion of this concept in valence bond theory is usually known as the *exchange energy*. This term must not be taken to imply that any physical exchange of the electrons takes place; if we wish for a physical picture of the source of the exchange energy it is probably best to attribute the lowering of the energy of the molecule to the increased volume in which each electron is now free to move.

Further improvements in the expression for ψ_+ can be made by: (i) allowing for the fact that each electron screens the other from the nuclei to some extent and (ii) taking into account the finite (though relatively low) probability that both electrons will be associated with the same nucleus at any one time. We can allow for the latter factor by considering two additional wave functions ψ_{III} and ψ_{IV} corresponding respectively to both electrons being attached to atom A and both electrons being attached to atom B. Then the total wave function becomes

$$\psi_+ = N[\psi_I + \psi_{II} + \lambda(\psi_{III} + \psi_{IV})]$$

or, as it is more conveniently written

$$\psi_+ = \psi_{cov} + \lambda\psi_{ion}$$

since ψ_I and ψ_{II} correspond to covalent structures and ψ_{III} and ψ_{IV} to the ionic structures $H_A^- H_B^+$ and $H_A^+ H_B^-$; λ is again the mixing coefficient, the degree to which the ionic structures contribute to the bonding. Calculations based on this expression for ψ_+ give 0.77 Å and 360 kJ mol^{-1} for the bond length and bond energy; further modifications, which we cannot discuss here, lead to an expression (containing fifty terms, however) from which values in exact agreement with the experimental ones can be obtained.

The wave function of the hydrogen molecule, and thus its structure, can therefore be expressed in terms of the wave functions, and thus the structures, of the configurations

$$H_A.1 \quad H_B.2 \quad (I) \qquad H_A.2 \; H_B.1 \quad (II)$$

$$H_A.1,2 \quad H_B \quad (III) \qquad H_A \quad H_B.1,2 \quad (IV)$$

where, as before, H_A and H_B are nuclei and 1 and 2 are electrons. The actual state of the molecule is often described (in a terminology which is somewhat unfortunate, but which is too firmly established to be eradicated) as a *resonance hybrid* of all four contributing structures, or is said to *resonate* between them; the structures are known as *canonical forms*. Resonance between structures I and II on the one hand, and between III and IV on the other, is known as *ionic–covalent resonance*, and the extra bonding energy resulting from this as *ionic–covalent resonance energy*. We shall not make extensive use of the concept of resonance in this book, but it is desirable to state here the general conditions for the occurrence of resonance or, more accurately, the general conditions in valence bond theory for the making of an appreciable contribution to the total wave function of a molecule by the wave functions associated with simpler structures. These are that the contributing structures should have similar energies, approximately the same relative positions of the nuclei, and the same numbers of unpaired electrons. Further mention of the resonance concept is made later in this chapter, in Section 4.7.

Valence bond theory describes bonds essentially in terms of the pairing of electrons (one from each atom) whose spins are antiparallel (ψ_- corresponds to parallel spins). Thus the number of unpaired electrons in an atom, whether in its ground state or in an accessible excited state, indicates how many bonds it will form. Electrons in the valence shell which are already paired constitute the lone pairs in the electronic configuration of the molecule formed. Thus helium ($1s^2$ in its ground state) does not form He_2, lithium ($2s^1$) forms Li_2, and beryllium ($2s^2$) does not form Be_2. Carbon ($2s^22p^2$), nitrogen ($2s^22p^3$), and fluorine ($2s^22p^5$) form doubly bonded C_2, triply bonded N_2, and singly bonded F_2 respectively; neon ($2s^22p^6$) forms no Ne_2. These descriptions are in essential agreement with those given in Section 4.2 on the basis of molecular orbital theory. Valence bond theory, however, has no satisfactory comment to make on the paramagnetism of B_2 and O_2, for which the unpaired electrons would be expected to pair, and single and double bonded structures respectively to result; attempts to account for the paramagnetism by postulating the presence of special one- and three-electron bonds in these molecules are, to say the least, unsatisfying. Nor has valence bond theory anything to say about the relative strengths of the bonds in, say, O_2^+ and O_2, a subject for which the molecular orbital treatment is simple and convincing.

The polarity of bonds in heteronuclear molecules is dealt with in molecular orbital theory by emphasising the greater contribution from one atomic wave function to the molecular wave function, e.g. in the case of HF, from the $2p_z$ wave function of the halogen, and we then write

$$\psi_{HF} = N[\psi_H + \lambda\psi_F]$$

with $\lambda > 1$. The valence bond description of HF would be an ionic–covalent resonance hybrid of

$$H—F \quad \text{and} \quad H^+F^-$$

(the structure H^-F^+ would be of much higher energy and is therefore disregarded). The increase in bond strength when NO ionises to give NO^+

presents a difficulty similar to the increase in bond energy on the ionisation of O_2; but, broadly speaking, simple qualitative molecular orbital and valence bond descriptions of most diatomic species are, essentially, not very different.

4.6 *Valence bond theory: polyatomic molecules*

We shall begin this section by considering the valence bond theory descriptions of the same two molecules as we choose to illustrate the application of molecular orbital theory to polyatomic species, $BeCl_2$ and BCl_3.

If we consider the formation of two Be—Cl bonds from a beryllium atom in its ground state ($1s^2 2s^2$), we must first promote an electron to raise beryllium to the excited state $1s^2 2s^1 2p^1$. It might then be expected that overlap of the beryllium $2p_z$ orbital with a chlorine $3p_z$ orbital would result in formation of a well-defined σ bond, and overlap of the spherical beryllium $2s$ orbital with the $3p_z$ orbital of another chlorine would lead to a bond of different strength and of uncertain direction.

It is, however, firmly established that in beryllium chloride both bonds are equivalent and that they are colinear; the metal atom can therefore be considered to be using not simple $2s$ and $2p_z$ orbitals but a combination of these. We can solve the wave equation for the beryllium atom and obtain suitable solutions ψ_{2s} and ψ_{2p_z}, each of which describes an orbital capable of holding two electrons with opposed spins. But valid solutions are also given by linear combinations of solutions, so we can describe other possible orbitals by combined or hybridised wave functions. If, for example, the two solutions ψ_{2s} and ψ_{2p_z} contribute equally, we get two new equivalent wave functions describing two linearly directed orbitals which we term hybrid *sp* orbitals:

$$\psi_{sp}\,(\text{i}) = \frac{1}{\sqrt{2}}(\psi_{2s} + \psi_{2p_z})$$

$$\psi_{sp}\,(\text{ii}) = \frac{1}{\sqrt{2}}(\psi_{2s} - \psi_{2p_z})$$

in which $1/\sqrt{2}$ is the normalisation factor. These new orbitals have strong directional characteristics, and each one extends further along the z axis than the original contributing p_z orbital; because of this they provide more effective overlapping and lead to the formation of stronger bonds. Their formation is represented pictorially in Fig. 4.14, and the $BeCl_2$ molecule which results from the overlap of the *sp* hybrid orbitals of the beryllium atom with the $3p_z$ orbitals of two chlorine atoms is represented in Fig. 4.15. The hybridised excited state $1s^2(2s2p)^2$, or, as it is usually

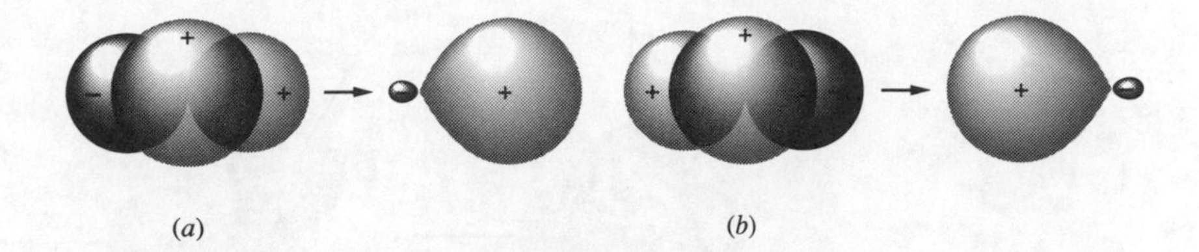

Figure 4.14
The formation of (*a*) ψ_{sp}(i) (*b*) ψ_{sp}(ii) from one *s* and one *p* wave functions.

written, $(sp)^2$, is known as the *valence state* of beryllium; the energy required to raise the beryllium atom from its ground state to its valence state must obviously be more than compensated by that liberated in the formation of two Be—Cl bonds.

We must emphasise that although it is an extremely useful concept, the valence state has no existence outside valence bond theory, and it is not, for example, observable spectroscopically. Its energy can therefore only be estimated theoretically.

For the formation of BCl_3, the boron atom (ground state $(2s^22p^1)$ must first be raised to the hybridised valence state $(2s2p_x2p_y)^3$ corresponding to three coplanar orbitals making an angle of 120° with one another. In this case the wave functions for the hybrid orbitals are given by

$$\psi_{sp^2}\text{(i)} = \frac{1}{\sqrt{3}}\psi_{2s} + \sqrt{\frac{2}{3}}\psi_{2px}$$

$$\psi_{sp^2}\text{(ii)} = \frac{1}{\sqrt{3}}\psi_{2s} - \frac{1}{\sqrt{6}}\psi_{2px} + \frac{1}{\sqrt{2}}\psi_{2py}$$

$$\psi_{sp^2}\text{(iii)} = \frac{1}{\sqrt{3}}\psi_{2s} - \frac{1}{\sqrt{6}}\psi_{2px} - \frac{1}{\sqrt{2}}\psi_{2py}$$

and the formation of these orbitals is represented in Fig. 4.16. The boron–chlorine σ bonds are then formed by overlap of the sp^2 hybrid orbitals of the boron atom and the $3p$ orbitals of the chlorine atoms.

Because of the frequency with which reference is made to it, we shall also mention here the hybridised sp^3 valence state of carbon. In this case, linear combination of the $2s$ and three $2p$ orbitals leads to the formation of four tetrahedrally disposed hybrid orbitals:

$$\psi_{sp^3}\text{(i)} = \tfrac{1}{2}(\psi_{2s} + \psi_{2px} + \psi_{2py} + \psi_{2pz})$$

$$\psi_{sp^3}\text{(ii)} = \tfrac{1}{2}(\psi_{2s} + \psi_{2px} - \psi_{2py} - \psi_{2pz})$$

$$\psi_{sp^3}\text{(iii)} = \tfrac{1}{2}(\psi_{2s} - \psi_{2px} + \psi_{2py} - \psi_{2pz})$$

$$\psi_{sp^3}\text{(iv)} = \tfrac{1}{2}(\psi_{2s} - \psi_{2px} - \psi_{2py} + \psi_{2p_z})$$

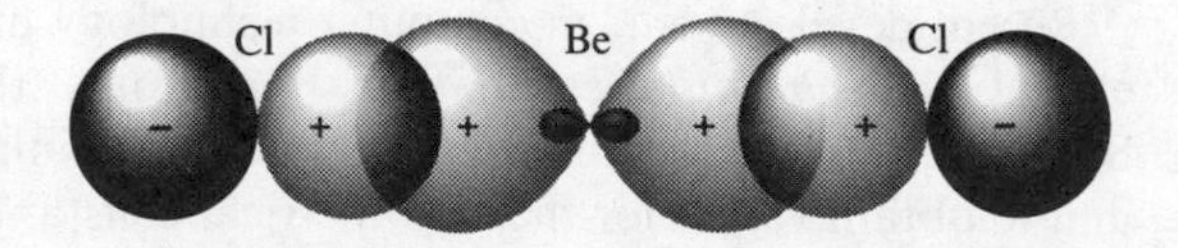

Figure 4.15
The overlap of the sp_z hybrid orbitals of the Be atom with the p_z orbitals of two chlorine atoms.

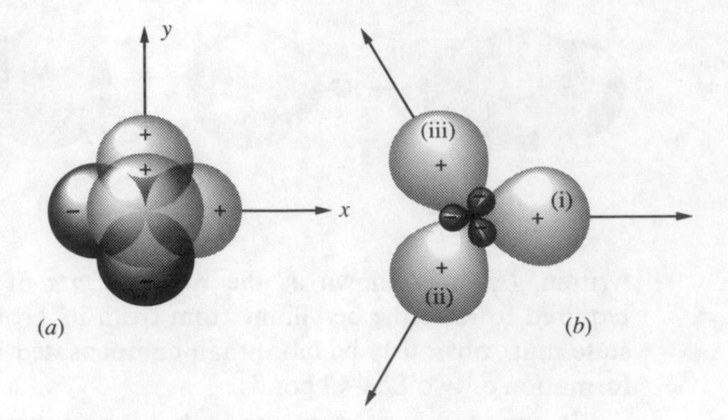

Figure 4.16
The formation of hybridised sp^2 orbitals.
(*a*) Unhybridised $2s$, $2p_x$ and $2p_y$ orbitals. (*b*) The three sp^2 orbitals.

In molecules such as CH_4 and CCl_4 these hybrid orbitals then overlap the $1s$ and $3p$ orbitals of hydrogen and chlorine respectively.

For more complicated geometries it has long been usual to involve *d* wave functions, and we shall not reproduce more mathematical expressions here. The most important cases are as follows:

$s+p_x+p_y+p_z+d_{z^2}$
trigonal bipyramidal hybridisation (e.g. in PF_5).

$s+p_x+p_y+p_z+d_{x^2-y^2}$
square pyramidal hybridisation (e.g. in IF_5).

$s+p_x+p_y+d_{x^2-y^2}$
square planar hybridisation (e.g. in ICl_4^-).

$s+p_x+p_y+p_z+d_{x^2-y^2}+d_{z^2}$
octahedral hybridisation (e.g. in SF_6).

The *extent* of *d* wave function involvement has, however, often been the subject of controversy. As we shall see later, we can often do without *d* orbital participation if we accept that covalent bonds in the species concerned are weak (e.g. in the molecule XeF_4 or the ion I_3^-, where this is certainly true); for an extremely stable molecule such as SF_6, on the other hand, any valence theory treatment which indicates weak bonding is completely at odds with the experimental evidence. Real molecules, of course, do not have to conform to our simple theories of valence and have integral values for the ratios of s:p:d character of their bonds. For a triatomic species AH_2 involving only *s* and *p* orbitals, all bond angles between 90° (pure *p* character) and 180° (*sp* character) are possible; in water, for example, with a bond angle of 104°, we can describe the 0—H bonds as $s^{0.20}p^{0.80}$.

Recent developments in computer technology have made it possible to extend the *ab initio* calculations for atoms that we mentioned in Section 2.10 to simple molecules (i.e. those containing only a few electrons), and to obtain values for their geometrical constants and energies that are in good agreement with the experimental ones. Two factors unfortunately limit the usefulness to a basic treatment of inorganic chemistry of results so far obtained: their restriction to few-electron systems, and the difficulty of expressing them in terms in which the chemist is accustomed to talk and think. At the level of this book and at the present time, therefore, the use of simple molecular orbital and valence bond theories remains essential. We shall, however, occasionally refer to the results of these *ab initio* calculations in the same way as we refer to conventional experimental data.

4.7 Multiple bonding in polyatomic molecules

We shall discuss multiple bonding in polyatomic species mainly by reference to ethylene, acetylene and benzene, partly because these are the most familiar examples, but partly also because some knowledge of the electronic structures of these compounds will be valuable when we describe some organometallic compounds of transition metals in Chapter 23. The discussion will include elements of both valence bond and molecular orbital theories.

Ethylene and acetylene are molecules which contain localised multiple bonds, i.e. it is only between the pair of carbon atoms in each molecule that we need consider the formation of a bond of order higher than unity; this is because the hydrogen atom has no orbital other than the 1*s* orbital that is anywhere near the 2*p* orbitals of carbon in energy. The problem of multiple bonding in these molecules is therefore very like that in molecular nitrogen, and can be discussed entirely in terms of the orbitals of the carbon atoms.

All six atoms in the molecule of ethylene are found experimentally to be coplanar, and the interbond angles are close to 120°. We can therefore consider each carbon atom as making use of its 2*s* orbital and two of its 2*p* orbitals to form three hybrid coplanar sp^2 orbitals; one of these overlaps the corresponding orbital of the other carbon atom, forming a σ bond, and the others form σ bonds by overlapping the 1*s* orbitals of the hydrogen atoms. The remaining *p* orbital on each carbon atom is used to form two π orbitals just as shown in Fig. 4.4 (*c*) and (*d*); since only two electrons have to be accommodated, both go into the bonding π orbital. The lateral overlap is not as great as the end-on type of overlap, and hence a C=C bond, though much stronger than a C—C bond, is not twice as strong; the carbon–carbon bond energy terms in ethylene and ethane are respectively 598 and 346 kJ mol^{-1}. Any attempt to twist the molecule results in diminution of the extent of overlap, and the observed high activation energy for rotation of the CH_2 groups about the C=C bond is thus accounted for. Only the bonding orbitals in ethylene, it should be noted, are shown in Fig. 4.17, which is the customary representation of

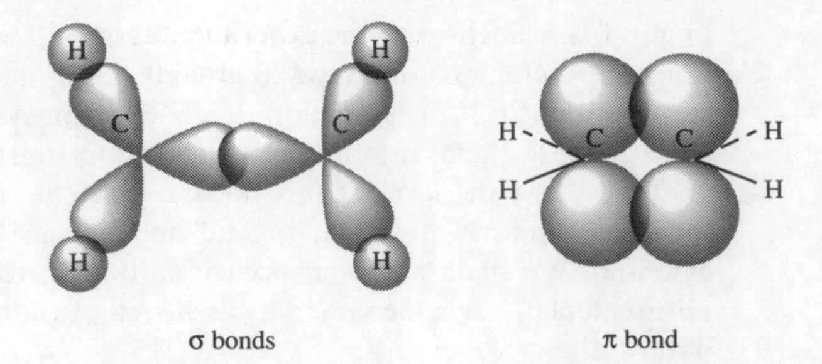

Figure 4.17
The σ and π bonds in ethylene.

the molecule. The compound N_2F_2 (Chapter 14) affords an inorganic example of a molecule with an electronic structure similar to that of ethylene.

In the case of acetylene, in which all the atoms are colinear, each carbon uses hybrid *sp* orbitals for σ bond formation, and two *p* orbitals of each carbon atom are available for π bond formation. There are thus two bonding π orbitals at right angles to one another, each doubly occupied by electrons, and a C≡C bond is formed which is stronger than the C=C bond in ethylene (its energy is 813 kJ mol^{-1}).

It is, of course, possible to form π bonds between two different types of orbital; we shall later encounter several examples in which a *p* orbital overlaps a suitable *d* orbital, e.g. a p_x or p_y orbital a d_{xy} orbital. In general, however, very strong π bonds are formed only between atoms of first period elements, e.g. C≡C, C≡O, C≡N, and N≡N. This is certainly due in part at least to the increasing size of the envelope containing a given fraction of the electron density as the principal quantum number of an orbital increases. Thus one of the main differences between silicon and carbon, the absence of silicon analogues of molecular C_2H_4, C_2H_2, CO, CO_2 and HCN, is simply accounted for. Weaker π bonds involving second period elements, e.g. Si=Si and P=P, do, however, occur in molecules sterically protected against polymerisation (Section 16.1).

Unlike ethylene and acetylene, benzene contains non-localised multiple bonds. All twelve atoms are coplanar, but there is only one carbon–carbon bond length (1.40 Å), which is intermediate between those in ethylene (1.33 Å) and ethane (1.54 Å). The valence bond representation of the benzene molecule is a resonance hybrid with canonical forms

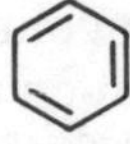   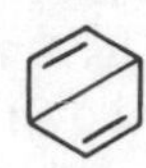 and 

of which the first two are by far the most important; the extent to which benzene is more stable than the hypothetical molecule cyclohexatriene would be expected to be is the resonance energy of benzene. On molecular orbital theory, six *p* orbitals (one to each carbon atom) are available for π orbital formation, and so three bonding and three antibonding π orbitals must be formed. The most strongly bonding π orbital is the one in which

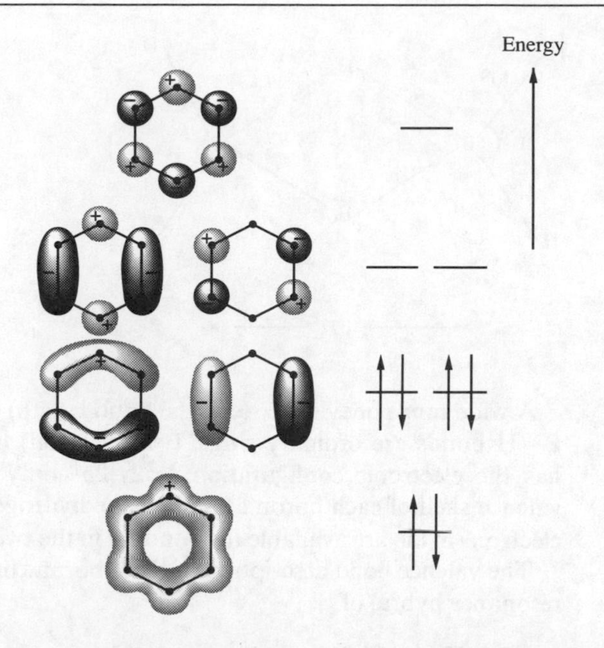

Figure 4.18
The π molecular orbitals in benzene and their relative energy levels. The orbitals are as seen from above the ring of carbon atoms; signs of the lobes below the ring are opposite to those shown.

all six *p* orbitals overlap (i.e. the wave functions on one side of the ring all have the same sign) and this is often shown as *the* π orbital. Like any other molecular orbital, however, it can accommodate only two electrons, and two other (less strongly bonding) π orbitals must therefore also be occupied. A schematic representation of the π molecular orbitals of benzene is given in Fig. 4.18. In molecular orbital terminology the margin by which benzene is more stable than the hypothetical molecule cyclohexatriene would be expected to be is its *delocalisation energy*, which is, of course, equivalent to its resonance energy in valence bond theory terminology.

As we have seen, valence bond theory meets with difficulties in describing the electronic structure of molecular oxygen; it encounters similar difficulties when applied to non-cyclic systems containing non-localised multiple bonds; and in the treatment of these (e.g. buta-1,3-diene, nitrous oxide), molecular orbital theory is generally found to be more satisfactory.

4.8 Multicentre bonding

So far we have considered only molecules in which there are enough electrons for there to be, at least, one electron pair between each two atoms that are bonded to one another. There are, however, a number of species for which this is not so, the best-known being diborane, B_2H_6 which has the structure shown in Fig. 4.19.

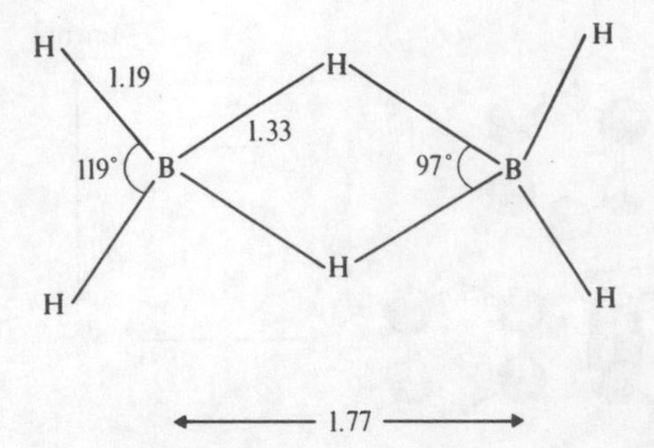

Figure 4.19
The structure of diborane (bond lengths in ångströms).

A wide range of evidence (e.g. the bond length) shows that the terminal B—H bonds are ordinary single (two-electron) bonds, and since boron has the electronic configuration $1s^2 2s^2 2p^1$ only one electron from the valence shell of each boron atom and the hydrogen $1s$ electrons, i.e. four electrons in all, are available for bonding in the two BHB bridges.

The valence bond description of the diborane molecule would then be a resonance hybrid of

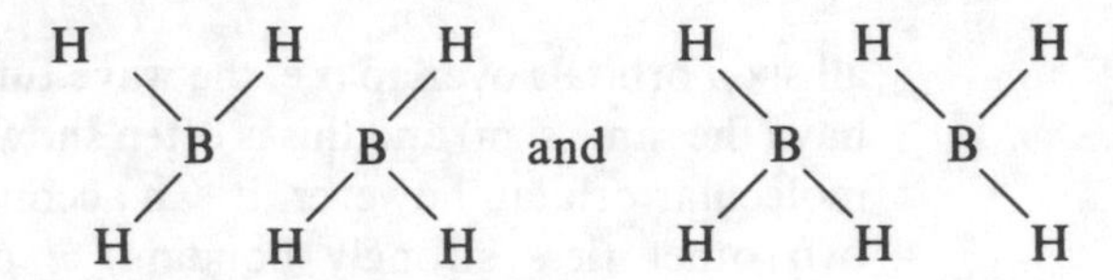

with smaller contributions from structures such as

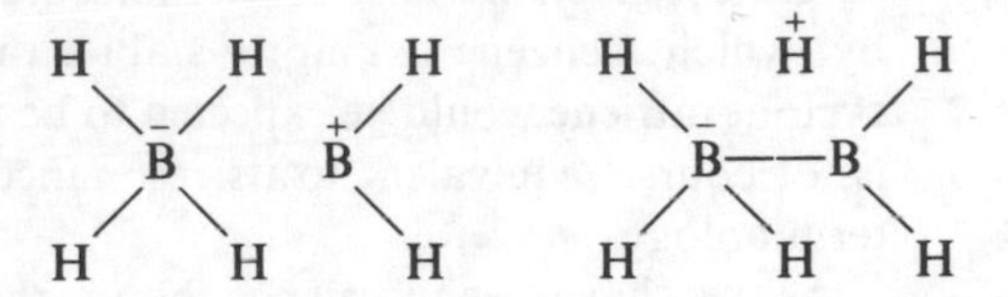

involving the so-called bond–no bond resonance in the bridge.

Alternatively, we can give a simple partial molecular orbital description as follows. We begin with two BH_2 units in which the $2s$ and $2p$ orbitals of a boron atom have been hybridised to form sp^3 orbitals; two of these are used to form B—H terminal bonds, and the BH_2 units are then brought together so that all six atoms are coplanar. If the hydrogen atoms are then placed in the positions they occupy in the actual molecule, each of the $1s$ hydrogen orbitals overlaps two sp^3 boron orbitals, one belonging to each atom, and thereby forms a bonding orbital shaped like a banana and extending over all three atoms. This orbital contains only two electrons, so it may be called a three-centre two-electron orbital; the bonding between the hydrogen atom and a boron atom is thus only about half as strong as in a conventional two-centre two-electron terminal B—H bond.

We may with advantage develop this treatment more formally by writing down the expressions for the three molecular orbitals which result

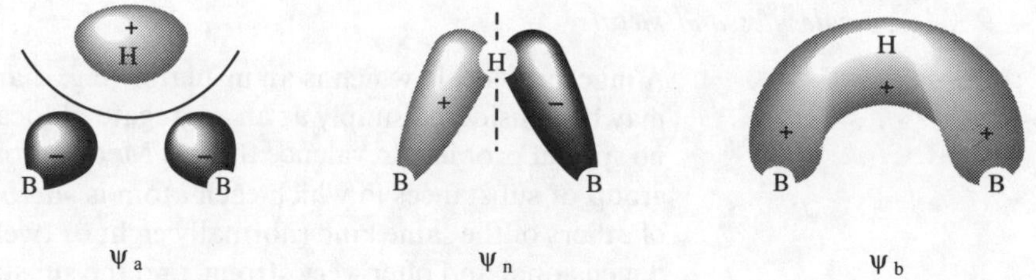

Figure 4.20
Three-centre molecular orbitals in a B—H—B bridge.

from combination of the appropriate wave functions for sp^3 hybrid boron orbitals ψ_{B1} and ψ_{B2} and the 1s hydrogen orbital ψ_H. These are:

$$\psi_b = \tfrac{1}{2}\psi_{B1} + \tfrac{1}{2}\psi_{B2} + \frac{1}{\sqrt{2}}\psi_H$$

$$\psi_n = \frac{1}{\sqrt{2}}\psi_{B1} - \frac{1}{\sqrt{2}}\psi_{B2}$$

$$\psi_a = \tfrac{1}{2}\psi_{B1} + \tfrac{1}{2}\psi_{B2} - \frac{1}{\sqrt{2}}\psi_H$$

Here ψ_b is a bonding orbital (the same as that described in the preceding paragraph), ψ_n a non-bonding orbital, and ψ_a an antibonding orbital. These are represented pictorially in Fig. 4.20, and the corresponding energy level diagram is given in Fig. 4.21.

More sophisticated descriptions can be given in terms of sp^2 hybridised boron atoms and of molecular orbitals extending over the whole molecule, but we shall not make use of them in this book. We shall, however, make considerable use of the concept of a three-centre orbital.

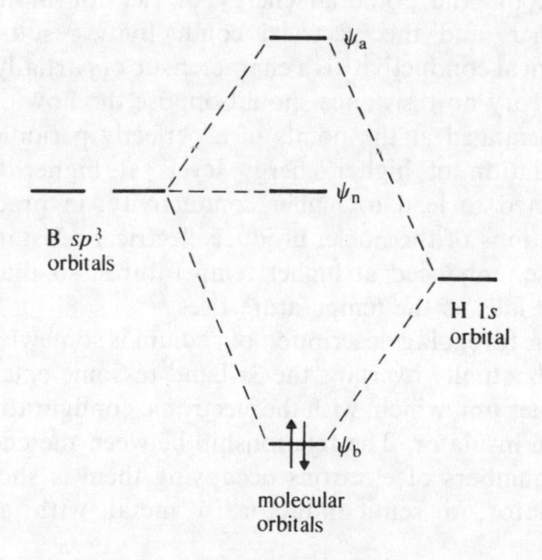

Figure 4.21
Formation of a B—H—B three-centre bond.

4.9 Macromolecules and metals

A macromolecule which is an insulator (e.g. diamond or red phosphorus) may be considered simply as an aggregate of localised bonds, and presents no special problem in valence theory. Metals, however, form an important group of substances in which each atom is surrounded by a large number of others of the same kind (normally eight or twelve); the bonding is three-dimensional and often very strong, and the substance is characterised by a high electrical conductivity which decreases with rise in temperature.

There are not enough valence shell orbitals or electrons for each metal atom to form two-centre two-electron bonds with all its neighbours, so we must consider multicentre orbitals, and the electronic mobility which is the electrical conductivity shows that these multicentre orbitals must extend over the whole crystal. In general, when N atoms combine to form a N-atomic molecule, N molecular orbitals will be obtained, each having an associated energy value; the lowest energy orbital will be completely bonding, the highest energy orbital completely antibonding. For an electron free to move over a whole crystal, as distinct from over only a single small molecule, the separation of the energy levels becomes extremely small (cf. the particle in a box problem), so small indeed that effectively a *band* of finite width of energy values results. In the *band theory of metals* we consider the separation and the extent of filling of these bands.

Each orbital (energy level) in a band can accommodate two electrons. If all the levels in the highest occupied band are filled and the next highest band is at a much higher energy, the substance is an insulator. If, on the other hand, only some of the energy levels in a given band are occupied (e.g. the bottom half of the levels in the 3s band in the case of sodium at very low temperatures), electrons can move from the highest filled levels into higher levels within the same band under the influence of an electric field. In these higher levels the energy is sufficient for the electron to overcome the potential energy barrier for motion from one atom to another, and the electrical conductivity results. On this view, then, electrical conductivity is a characteristic of partially filled bands of orbitals. In theory no resistance should oppose the flow of a current if the nuclei are arranged at the points of a perfectly periodic lattice, and increased population of higher energy levels at higher temperatures might be expected to lead to higher conductivity; in practice, however, thermal vibrations of the nuclei produce electrical resistance and this effect is, of course, enhanced at higher temperatures, so that the conductivity of a metal falls as the temperature rises.

The foregoing description of sodium is somewhat oversimplified; the 3p band actually overlaps the 3s band to some extent. This is true also for magnesium, which, with the electronic configuration $3s^2$, would otherwise be an insulator. The relationship between the energies of the bands and the numbers of electrons occupying them is shown in Fig. 4.22 for an insulator, a semiconductor, a metal with a half-filled band not

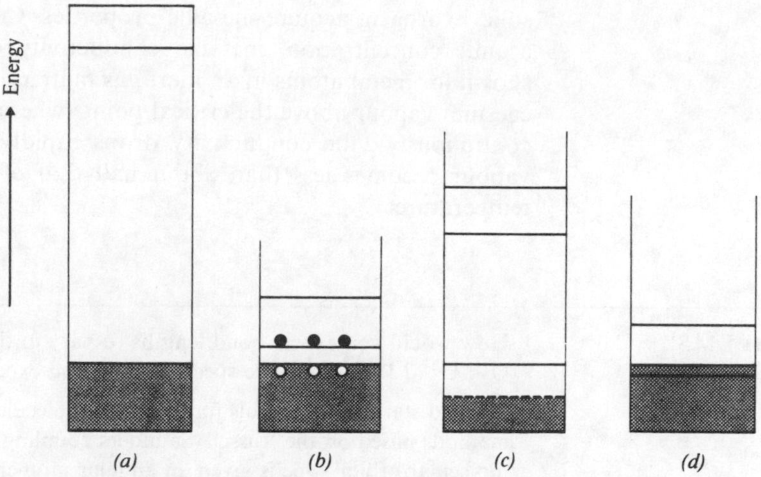

Energy

(a) (b) (c) (d)

Figure 4.22
Energies of filled and empty bands in: (*a*) an insulator; (*b*) a semiconductor in which a few electrons have been promoted to the conduction band; (*c*) a metal with a half-filled band not overlapping a higher energy band; and (*d*) a metal with overlapping bands.

overlapping a higher energy band, and a metal with bands overlapping. For a semiconductor the gap between the bands is very small, so that thermal energies suffice to promote some electrons to the conduction band and conductivity increases as temperature rises.

The property of a solid metal which is of most importance in inorganic chemistry is its energy of atomisation; unfortunately, attempts to use the theory of metals to calculate this quantity have so far not met with much success except in the cases of the lighter alkali metals. Partly for this reason, and partly also because of the difficulty of the subject, we shall not discuss the nature of the metallic bond further, and the physical properties of the metals will be given in later chapters without theoretical comment. We shall, however, return to the subject of band theory when we discuss solid inorganic compounds and semiconductors in Chapter 6.

We should, nevertheless, ask here why some elements are metals and others are not. It is clearly not just a matter of ionisation in the solid state to give a structure in which cations are present in a 'sea' of electrons, since although the alkali metals have relatively low ionisation energies, some transition metals have relatively high ones. Some light may perhaps be thrown on the question by considering the allotropy of the elements: the transition from non-metal to metal (e.g. from grey tin to white tin at 13 °C) is generally accompanied by an increase in density. Attempts have therefore been made to relate the incidence of metallic status in the solid state to elemental density and to some independently measurable or calculable properties of isolated gas-phase atoms, e.g. polarisability (which can be obtained from the refractive index), and it has been shown that any solid should become metallic at a high enough density. Thus it is now clear that the dividing line between metals and non-metals is no longer a very sharp one: by forcing atoms into closer proximity so that their wave functions overlap to a greater extent (i.e. by subjecting elements to very high pressures), many elements, including iodine, silicon and even solid hydrogen, acquire metallic properties. Conversely, below a certain atomic concentration, metallic conductivity disappears. This may be shown for metal atoms in an inert gas matrix at low temperatures, or for caesium vapour above the critical point (where the density can be varied continuously): the conductivity drops rapidly when the density of the vapour becomes less than about half that of the solid metal at room temperature.

PROBLEMS

1 How would you expect bond lengths to vary in the dioxygen species O_2, O_2^+, O_2^-, O_2^{2-}? Which of these species would you expect to be paramagnetic?

2 Ground state term symbols for diatomic molecules can be determined by methods based on the Russell–Saunders coupling scheme for atoms. Each molecular orbital type is given an angular momentum quantum number m_l, zero for σ orbitals, ± 1 for π orbitals. The definition of a term is $^{2S+1}M_L$, where S is the sum of the spin quantum numbers and M_L is the sum of the m_l values. The term symbols representing $M_L = 0$, ± 1 and ± 2 are Σ, Π and Δ respectively.
Derive the ground state terms for H_2^+, H_2, C_2, O_2, and NO.

3 The bond energy of molecular chlorine is 242 kJ mol^{-1} and the first ionisation energies of atomic and molecular chlorine are 1250 and 1085 kJ mol^{-1} respectively. Calculate the enthalpy of formation of the chlorine molecule-ion, Cl_2^+, from a Cl atom and a Cl^+ ion, and comment on its relationship to the bond energy of molecular chlorine.

4 Discuss each of the following observations.
(a) The carbon–carbon distance in cyanogen, C_2N_2, is about 10 per cent shorter than that in ethane.
(b) The NO stretching vibration occurs as 2313 cm^{-1} in $NO[ClO_4]$ compared with 1876 cm^{-1} in nitric oxide.
(c) The energy gaps between the highest filled band and the lowest empty band in diamond, silicon, germanium, grey tin and lead are 530, 110, 70, 10 and 0 kJ mol^{-1} respectively. Diamond is an insulator, lead a metallic conductor; the other three elements are semiconductors and their conductivity increases with increase in temperature.
(d) Beryllium hydride at ordinary temperatures is an infinite chain polymer, $(BeH_2)_n$, in which each beryllium atom is bonded to four hydrogen atoms distributed tetrahedrally around it.
(e) Molecules of the type $XYC = C = CXY$ are optically active.

REFERENCES FOR FURTHER READING

Atkins, P.W. (1990) *Physical Chemistry*, 4th edition, Oxford University Press, Oxford. A good short mathematical introduction to valence theory.

De Kock, R.L. and Gray, H.B. (1980) *Chemical Structure and Bonding*, Benjamin/Cummings, Menlo Park, California. A general treatment of valence theory.

Edwards, P.P. and Sienko, M.J. (1983) *J. Chem. Education*, **60**, 691. On the occurrence of metallic character.

McWeeny, R. (1979) *Coulson's Valence*, 3rd edition, Oxford University Press, Oxford. A general treatment of chemical bonding, decidedly more formal than in this book.

Murrell, J.N., Kettle, S.F.A. and Tedder, J.M. (1985) *The Chemical Bond*, 2nd edition, Wiley, Chichester. A more formal and theoretical treatment.

The Open University (1976) *Photoelectron Spectroscopy*, S 304 unit 20. A critical introduction to this branch of spectroscopy and its relevance to valence theory.

Webster, B (1990) *Chemical Bonding Theory*, Blackwell, Oxford. A very clear introduction to valence theory.

5 Some physical properties of molecules

Introduction • The shapes of molecules and ions of non-transition elements • Symmetry • Bond energies • Force constants • Bond lengths • Bond polarities and electronegativity

5.1 Introduction

In discussing the electronic structures of molecules we have several times referred to the directional properties, energies, lengths and polarities of covalent bonds. We now consider these and some related properties in more detail. In the last chapter we cited only a few examples of molecular structures and bond energies, mainly to illustrate successful applications of theories of valence; in this one, we shall be concerned primarily with the experimental data now available for inorganic species, though we shall also comment critically on some of the theories put forward to account for them.

The first part of this chapter is concerned with structure rather than with energetics, and we shall consider ions as well as molecules, since the geometry of both is determined chiefly by the numbers of electrons they contain. For reasons which will become apparent later, the estimation of bond energies in ions is a difficult task, and so our discussion of the energetics of the covalent bond in the second part of the chapter will be confined to molecules. We shall limit the discussion here to compounds and ions of the non-transition elements, reserving the treatment of the structures of complexes of the transition elements (which are often much more complicated) to Chapters 18 and 19. Much further information on the structures and energetics of molecules and ions of the non-transition elements will be found in Chapters 9–17, which cover the systematic chemistry of these elements.

5.2 The shapes of molecules and ions of non-transition elements

As we have already seen, the molecules $BeCl_2$ and BCl_3 are linear and planar respectively; on the other hand, OCl_2 is bent, and NCl_3 is pyramidal. The central atom in each of these species is a first-row element which cannot contain more than eight electrons in its valence shell. Molecules and ions in which the central atom is not restricted to an octet of electrons exhibit more complex geometries: thus ClF_3 is T-shaped; SeF_4 is irregular tetrahedral; and the $[BrF_4]^-$ ion is planar. Nearly all molecules and ions of formula AB_6 (e.g. SF_6, SiF_6^{2-}) are octahedral, but for AB_5 species there are two common structures: the trigonal bipyramid (e.g. PF_5) and the square pyramid (e.g. BrF_5).

When we considered the bonding in $BeCl_2$, BCl_3, and CH_4 on the basis of either molecular orbital theory or valence bond theory, we began by stating the molecular geometry. In principle it ought to be possible to calculate what structure should be most stable for a particular molecule, but for an interesting species containing several polyelectron atoms (such as ClF_3) this is not yet the case, and so the discussion of molecular structure in terms of hybrid orbitals of the central atom is usually retrospective explanation rather than prediction. A simple theory which can be used predictively is therefore of major interest to inorganic chemists.

As long ago as 1940, Sidgwick and Powell, in a review of the molecular structures then known, suggested that, for a molecule or ion AB_n containing only single bonds, the structure can be derived by assuming that the pairs of electrons in the valence shell of A are as far apart as possible, whether or not they are involved in bonding. Thus for two pairs of electrons in the valence shell of A their distribution is linear; for three

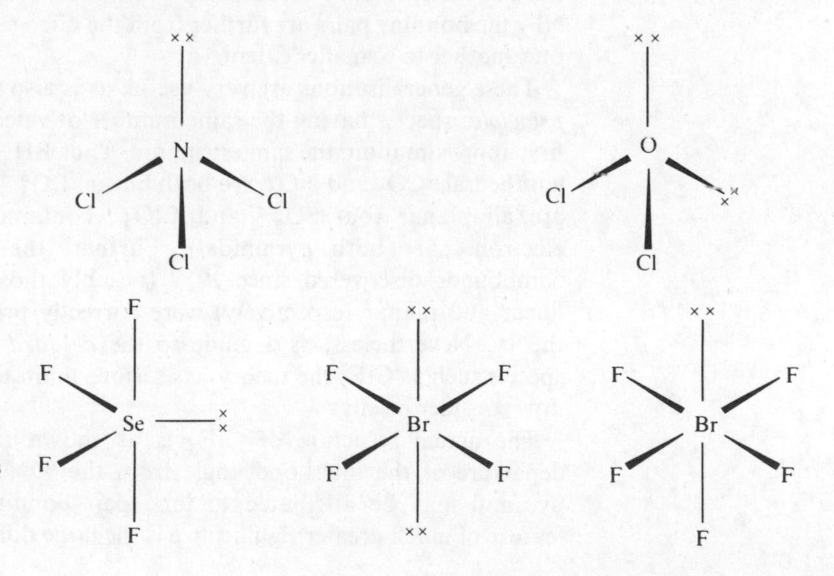

Figure 5.1
Structure of some molecules and ions having unshared pairs of electrons on the central atom.

pairs, triangular; for four pairs, tetrahedral; for five pairs, trigonal-bipyramidal; and for six pairs, octahedral. The structures of species mentioned earlier in this section which have no unshared pairs of electrons are thus simply accounted for; those of NCl_3, OCl_2, SeF_4, $[BrF_4]^-$ and BrF_5 may be represented as shown in Fig. 5.1.

The structure of ClF_3 was not known in 1940; when it was later established as approximately T-shaped, a difficult question arose. Why, since the chlorine atom in this molecule appears to have a valence shell of ten electrons, is it not planar, i.e. a trigonal bipyramid with the two lone pairs at the apices? Further, why does the actual interbond angle decrease steadily along the series CH_4 (109.5°), NH_3 (107.3°) and OH_2 (104.5°), and why is the interbond angle in NF_3 (102°) less than that in NH_3? These and many other experimental observations were rationalised by Nyholm and Gillespie in 1957 in an extension of Sidgwick and Powell's suggestion which has since become known as the *Valence Shell Electron Pair Repulsion* (VSEPR) theory.

This theory may be summarised as follows.

1. Each valence shell electron pair of an atom is stereochemically significant, and repulsions between them determine molecular shape.
2. Repulsions decrease in the sequence lone pair–lone pair > lone pair–bonding pair > bonding pair–bonding pair.
3. Repulsions also decrease in the sequence triple bond–single bond > double bond–single bond > single bond–single bond.
4. Repulsions between the bonding pairs in a species AB_n depend upon the difference in electronegativity between A and B, and decrease as the electronegativity of B increases.

We shall say more about electronegativity later in this chapter; but, however we define it, fluorine is universally accepted as being more electronegative than hydrogen, and so it is reasonable to suppose that in NF_3 the bonding pairs are further from the nitrogen atom and hence repel one another to a smaller extent.

These generalisations are very useful, as is also the related *isoelectronic principle*: species having the same number of valence electrons have, to a first approximation, the same structure. Thus BH_4^-, CH_4 and NH_4^+ are all tetrahedral; CO_2 and NO_2^+ are both linear; BO_3^{3-}, CO_3^{2-}, NO_3^- and SO_3 are all planar (but SO_3^{2-} and ClO_3^-, containing two more valence electrons, are both pyramidal). Further, the structures of several compounds discovered since 1957 (notably those of XeF_2 and XeF_4, linear and planar respectively) were correctly predicted by the VSEPR theory. Nevertheless, as a guide to the *detailed* structures of 'difficult' species such as ClF_3 the theory has serious limitations, and these we shall now consider briefly.

The actual structure of ClF_3 is as shown in Fig. 5.2. The small departure of the interbond angle from the 90° expected for a trigonal pyramid may be attributed to lone pair–bonding pair repulsion; the feature of much greater significance is the large difference in bond lengths,

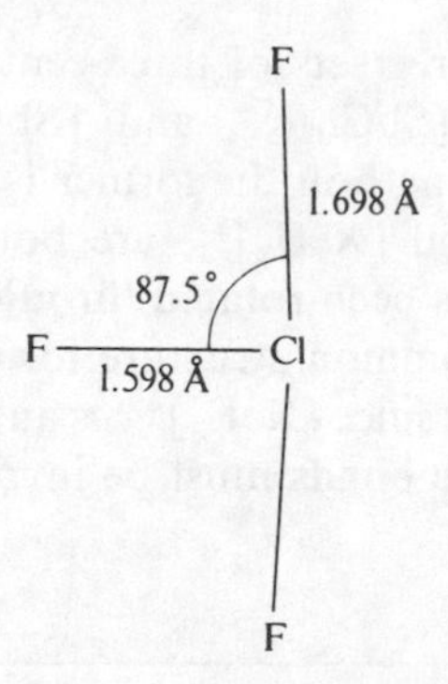

Figure 5.2
The structure of ClF_3.

which implies a substantial difference in bond order between the axial bonds and the equatorial bond in the trigonal bipyramid. Thus in ClF_3 we ought really not to talk about lone pair–bonding pair repulsions but about lone pair–bonding (1.698 Å) pair and lone pair–bonding (1.598 Å) pair repulsions; it is impossible to know accurately the relative values for these different repulsions, so we cannot *predict* the structure of ClF_3. Bromine trifluoride has a structure exactly analogous to that of ClF_3. Similar, though often smaller, differences between axial and equatorial bonds are found for all other molecules whose structures are trigonal bipyramidal or are based on the trigonal bipyramid with one or two positions occupied by electron pairs. In SF_4, for example, the axial and equatorial bond lengths are 1.65 and 1.55 Å; in PF_5 they are 1.58 and 1.53 Å respectively. A similar difference in bond lengths is found in the molecule of BrF_5, a square pyramid with the bromine atom a little below the plane of the four basal fluorine atoms, though here there are three different bond lengths: 1.68 Å (axial) and 1.75 and 1.82 Å (equatorial).

Modern views of the structures of these and a number of related species therefore involve bonds of more than one kind, some of which may be of order other than unity; they are discussed further in Chapters 15, 16 and 17. Incidentally, we may note that the description of a molecule such as PF_5 in terms of hybridised $sp^3d_{z^2}$ orbitals of the phosphorus atom is also somewhat unsatisfactory, since the bonds formed are clearly not all equivalent.

At the present time there are only a few known species which appear to contain more than twelve electrons in the valence shell of a non-transition element. Among the fourteen-electron group, IF_7 has a pentagonal bipyramidal structure; and the $[IF_6]^-$ ion and molecular XeF_6 are certainly not regular octahedra; the ion $[Sb(C_2O_4)_3]^{3-}$ has a pentagonal bipyramidal configuration round the antimony atom, with an axial position occupied by a lone pair. On the other hand the ions $[SeCl_6]^{2-}$, $[SeBr_6]^{2-}$, $[TeCl_6]^{2-}$, $[TeBr_6]^{2-}$ and $[BrF_6]^-$ (in their alkali metal salts) are regular octahedra; this may indicate that in the valence shell of their central atoms two electrons occupying a spherically symmetrical *s* orbital are to be distinguished from the other twelve, but an alternative explanation in terms of three sets of three-centre two-electron bonds is also possible. The ions $[SbCl_6]^{3-}$ and $[SbCl_6]^-$ are *both* regular octahedra, but the bond length in the former is 0.23 Å more than in the latter. The ions $[IF_8]^-$ and $[XeF_8]^{2-}$ are both square antiprisms, i.e. cubes in which one face has been rotated through 45° relative to the face opposite to it. This is a common structure for eight-coordination in an isolated molecule or ion but since $[XeF_8]^{2-}$ is an eighteen-electron system, an inert pair or multicentre bonds must be involved for this ion.

5.3 Symmetry

In describing atomic orbitals we have already mentioned symmetry properties. In this section we shall give a brief introduction to molecular symmetry; although in this book we shall usually describe molecules by adjectives (such as *tetrahedral*, *octahedral*) rather than by their so-called *point groups*, the use of the latter often has the advantage of greater precision. Further, mathematical group theory may be applied to symmetry elements and hence to molecules; this powerful tool is the key to the understanding of the relationship between structure and spectroscopy. It is, for example, the source of our knowledge that a regular tetrahedral molecule or ion of formula AB_4 should have four normal modes of vibration, all Raman-active but only two of them infrared-active; on the other hand a square planar molecule or ion of the same formula type should have seven normal modes of vibration: three Raman-active, a different three infrared-active, and one inactive. Hence we can easily distinguish between tetrahedral and planar structures on the basis of Raman and infrared spectra. The derivation of such results lies outside the scope of this book, but partly because it is helpful in reading scientific literature to understand point group (or *Schonflies*) symbols such as C_{2v} and D_{3h}, and partly because the study of symmetry elements is a powerful encouragement to three-dimensional thinking in chemistry, an elementary account of molecular symmetry is given here. It should be mentioned in passing that solid state chemists, who are interested in crystals in which symmetries must exist in repeating patterns to permit three-dimensional packing, use a different set of symbols, the so-called *space group* (or *Hermann–Mauguin*) symbols. We shall not use the latter in this book.

Symmetry, when considered more rigidly than we have so far considered it, consists of two parts, *symmetry operations* and *symmetry elements*. A symmetry operation is an operation performed on an isolated molecule or ion which leaves the species in a configuration which is indistinguishable from, and superimposable on, the original configuration. If, for example, we consider a water molecule (Fig. 5.3), and rotate it through an angle of 180° (π radians) about an axis bisecting the HOH angle in the plane of the molecule, the resulting configuration is

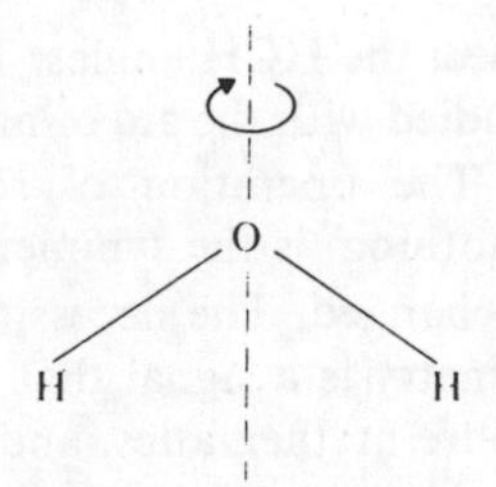

Figure 5.3
The C_2 axis of symmetry of the water molecule.

superimposable on the original one. The rotation of this molecule is a symmetry operation, and the symmetry element is the axis around which the molecule is rotated. This is said to be an *n*-fold axis of symmetry if rotation through $2\pi/n$ brings the molecule into a configuration superimposable on the original one; thus the axis shown, which is the axis of highest symmetry for the water molecule, is a twofold one.

There are five possible types of symmetry operation, and five corresponding basic symmetry elements.

1. *Rotation* (the symmetry operation) *about an n-fold axis of symmetry* C_n (the symmetry element). Thus the molecule of water possesses one twofold axis of symmetry (C_2); the planar BF_3 molecule possesses one threefold axis (C_3) perpendicular to the plane and three twofold axes (C_2) each along a B—F bond.
2. *Reflection through a plane of symmetry* (or *mirror plane*), σ. If reflection of all parts of a molecule through a plane produces an indistinguishable configuration, the plane is a plane of symmetry. The water molecule, for example, possesses two planes of symmetry each containing the C_2 axis (the axis of highest symmetry) and termed *vertical planes* and given the subscript v; the σ_v plane is perpendicular to the plane of the molecule and the σ_v' plane is the plane of the molecule (the individual atoms, of course, are symmetrical about the plane of the molecule). The molecule of BF_3 has three vertical planes of symmetry containing the C_3 axis.
3. *Reflection through a centre of symmetry* (or *centre of inversion*), *i*. If reflection of all parts of a molecule through the centre of the molecule produces an indistinguishable configuration, the centre is a centre of symmetry (or a centre of inversion); this is essentially a three-dimensional analogue of the two-dimensional plane of symmetry. The planar $[ICl_4]^-$ ion and the staggered form of ethane have a centre of symmetry, whilst the tetrahedral CH_4 molecule does not.
4. *Rotation about an axis, followed by reflection through a plane perpendicular to this axis,* S_n. If rotation through $2\pi/n$ about an axis, followed by reflection through a plane perpendicular to the axis, yields an indistinguishable configuration, the axis is an *n*-fold rotation–reflection axis. Such an axis is sometimes known as an *improper rotational axis* to distinguish it from a *proper* or *simple rotational axis.* The CH_4 molecule, for instance, has three S_4 axes coincident with the

three C_2 axes which bisect the HCH angles. These symmetry elements should preferably be studied with the aid of molecular models.

5. *Identity operation*, *E*. The operation of rotation through 2π (or alternatively of doing nothing) is the symmetry element of identity; it leaves the molecule unchanged. The necessity for its inclusion in the rigid treatment of symmetry is a signal that we are now moving into the abstract world of pure mathematics, and we shall not have much occasion to consider the identity operation further in this book.

The number and nature of the symmetry elements in a given molecule are denoted by its *point group* (or *symmetry group*), which has a label such as C_2, C_{2v}, D_{3h}, O_h, T_d etc. These point groups are conveniently divided for further description into *C* groups, *D* groups and special groups.

Point groups given the symbol *C* have a rotational axis of symmetry, though if this is only a onefold axis (i.e. the molecule must be rotated through 360° before it is superimposable on its original configuration) and the point group is C_1, this is tantamount to saying that the only symmetry element is that of identity, *E*; CHFClBr provides an example of a molecule belonging to the C_1 point group. Where, in addition to the identity element *E* (which we shall not mention again, since all molecules possess it) there is only a plane of symmetry, the point group symbol is C_s, e.g. for HOCl or SO_2ClF; where there is only a centre of symmetry, the symbol is C_i, e.g. for staggered CHClF.CHClF.

Point groups denoted by the general symbol C_n all really possess a rotational axis of symmetry. The molecule of hydrogen peroxide has an 'open-book' configuration with the oxygen atoms on the spine and the hydrogen atoms on the pages (see Fig. 5.4).

The O—O bond and the only symmetry element, the C_2 axis, lie in a plane with one hydrogen above and the other below the plane; the point group is C_2. If a molecule, in addition to a rotational axis C_n, also has *n* vertical planes of symmetry (i.e. planes containing the C_n axis), the point group is C_{nv}. As we have already seen, the water molecule belongs to the point group C_{2v}; so does CH_2Cl_2. Ammonia and chloroform belong to the point group C_{3v}, BrF_5 to the point group C_{4v} and the unsymmetrical linear molecules HCl, HCN and OCS to the point group $C_{\infty v}$. If there is,

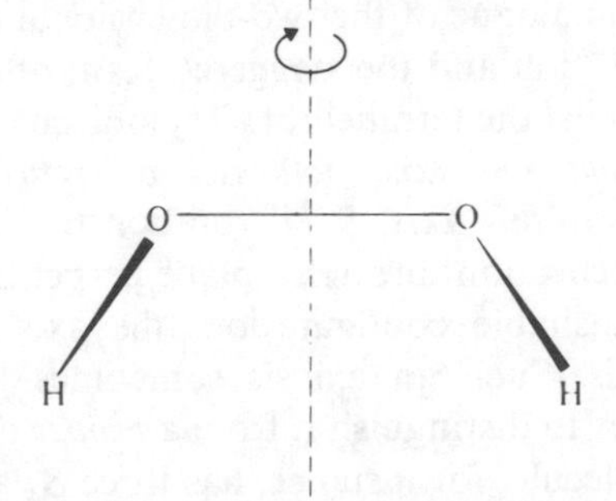

Figure 5.4
The molecular structure of hydrogen peroxide, point group C_2.

in addition to the rotational axis C_n, a plane of symmetry perpendicular to the C_n axis (a *horizontal* plane, symbol σ_h), the point group is C_{nh}; *trans*-CHCl=CHCl belongs to the point group C_{2h}, planar boric acid, $B(OH)_3$ (Chapter 12), to the point group C_{3h}.

A molecule which has a C_n axis of symmetry and nC_2 axes perpendicular to the C_n axis belongs to the point group D_n; these point groups are uncommon, but an example of a species belonging to the D_3 group is the ion $[Co(en)_3]^{3+}$ where en = ethylenediamine (1,2-diaminoethane), the stereochemistry of which is discussed in Chapter 18. A molecule which has, in addition to a C_n axis and nC_2 axes perpendicular to the C_n axis, a horizontal plane of symmetry (i.e. a plane perpendicular to the C_n axis), belongs to the D_{nh} point group. (Such a molecule also has n vertical planes of symmetry and a centre of symmetry, but it is the recognition of the horizontal plane of symmetry that is of overriding importance in allocating it to a D_{nh} point group.) Examples of species belonging to these point groups are B_2H_6 (D_{2h}), BCl_3 (D_{3h}), $[BrF_4]^-$ (D_{4h}) and benzene (D_{6h}), all of whose structures have been described earlier in this book. A symmetrical linear molecule (e.g. CO_2 or Cl_2) belongs to the point group $D_{\infty h}$.

If, in addition to a C_n axis of symmetry and nC_2 axes perpendicular to the C_n axis, there are n planes of symmetry which contain the C_n axis and bisect the angles between the C_2 axes (*dihedral planes*, symbol σ_d), the point group is D_{nd}. Other elements of symmetry are also present. Ethane in its staggered configuration belongs to the point group D_{3d}.

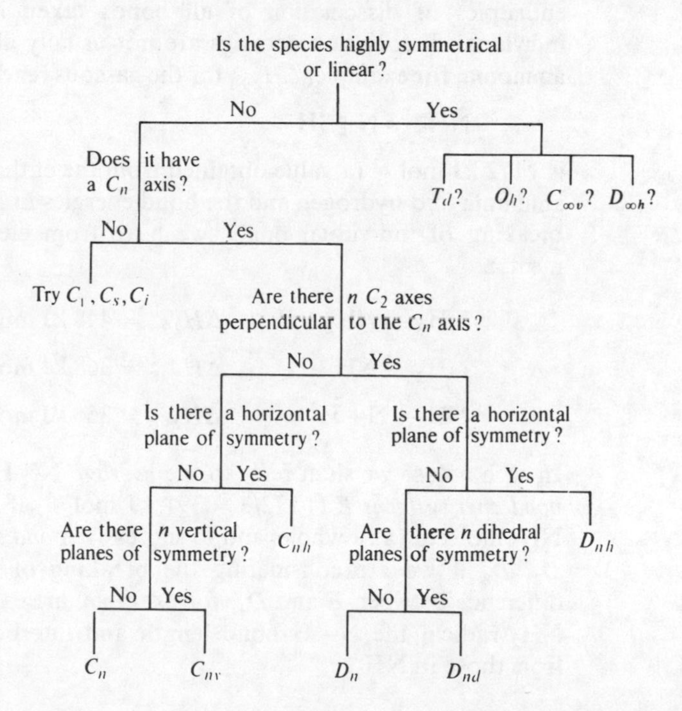

Figure 5.5
Scheme for the derivation of point groups of molecules and ions.

Finally there are a few special point groups that denote highly symmetrical molecules. The regular tetrahedron (e.g. CH_4) belongs to the point group T_d: it has a set of four C_3 axes, a set of three S_4 axes, and a set of six dihedral planes of symmetry (the S_4 axes are coincident with the C_2 axes that bisect the HCH angles). The cube and the regular octahedron (e.g. the $[PaF_8]^{3-}$ ion mentioned in Chapter 18 and SF_6) belong to the point group O_h. This contains a large number of symmetry elements, among which are a set of three C_4 axes (which are also S_4 axes), a set of four C_3 axes, a set of six planes each containing one C_4 axis and bisecting two of the angles made by the other C_4 axes, and a centre of symmetry.

It should be emphasised that many other point groups are known. Schemes for classifying molecules into point groups necessarily vary in complexity according to the number of point groups considered. At the level of this book the scheme given in Fig. 5.5 should prove adequate, though often analogy with species of known point group will suffice.

5.4 Bond energies

In this book we shall define the bond energy in a diatomic molecule as the enthalpy change for dissociation of the gaseous molecule at 298.16 K (25 °C) into ground state gaseous atoms at the same temperature. For a polyatomic gaseous molecule AB_n the enthalpy of dissociation into ground state atoms must obviously be the same as the sum of the enthalpies of dissociation of all bonds taken individually, but these individual dissociation energies are not usually all equal. In the case of ammonia, for example, ΔH_{298} for the gaseous reaction

$$NH_3 \rightarrow N + 3H$$

is 1 172 kJ mol^{-1} (a value obtained from the enthalpies of combustion of ammonia and hydrogen and the bond energies in N_2 and H_2), but for the breaking of individual bonds we have from electron-impact measurements:

$$NH_3 \rightarrow NH_2 + H \qquad \Delta H_{298} = 448 \text{ kJ mol}^{-1}$$

$$NH_2 \rightarrow NH + H \qquad \Delta H_{298} = 368 \text{ kJ mol}^{-1}$$

$$NH \rightarrow N + H \qquad \Delta H_{298} = 356 \text{ kJ mol}^{-1}$$

In such a case we shall refer to the *average* N—H *bond energy* or N—H *bond energy term* E (1 172/3 = 391 kJ mol^{-1}) if we are considering the NH_3 molecule as a whole, and to successive *bond dissociation energies* D_1, D_2, D_3 if we are considering the breaking of individual bonds. The difference between E and D_1, for example, arises from the fact that in a NH_2 radical the N—H bond length and interbond angle are different from those in NH_3.

In defining ionisation energy and electron affinity in Chapter 3, we mentioned that to be exact we should consider internal energy changes at absolute zero, but that the error introduced by the use of enthalpy changes at 298 K is very small and is usually disregarded. The same is true of bond energies, but it is instructive to see just what is being neglected. First, as we showed in Section 2.6, an actual molecule of, say, hydrogen can never have a total energy less than the zero-point energy $\frac{1}{2}h\nu_0$, where ν_0 is the fundamental vibration frequency, and the value of the internal energy change for the reaction

$$H_2(g) \rightarrow 2H(g)$$

at absolute zero could therefore be considered to be less than the total bond energy in H_2 by $\frac{1}{2}N_A h\nu_0 = 25\ \text{kJ mol}^{-1}$. The zero-point energy of heavier molecules is much less than this, since ν_0 decreases with increase in mass; partly for this reason, and partly because for many molecules ν_0 is not known, consideration of zero-point energy is set aside. Then for conversion from an internal energy change at 0 K to an enthalpy change at 298 K we also take into account

$$\int_0^{298} [2C_p(\text{H}) - C_p(\text{H}_2)]\text{d}T$$

(cf. Section 3.4). At 298 K, C_p for H is $5/2R$ and C_p for H_2 is almost exactly $7/2R$ (nearly all molecules of H_2 are in their lowest vibrational level at 298 K); thus the enthalpy change at 298 K for the dissociation of H_2 is $3/2RT = 3.7\ \text{kJ mol}^{-1}$ greater than the internal energy change at 0 K. It would, of course, be impossible to convert enthalpy changes at 298 K (which correspond very much more nearly to what is usually determined experimentally) to internal energy changes at 0 K without a knowledge of the heat capacities of all the species involved; and in general the magnitude of the difference between these quantities is probably little, if any, greater than the uncertainty in the value of a bond energy at 298 K.

For hydrogen and the halogens it is generally accepted that the diatomic molecules contain single bonds (though see Chapter 16 for a comment on the bond energy for F_2); the bond energies in these molecules, obtained from the convergence limit in the electronic spectrum (allowance being made for the fact that one atom is produced in an excited state) or from the temperature-dependence of $\log K$ for the reaction $X_2 \rightarrow 2X$ investigated manometrically, are:

H—H	F—F	Cl—Cl	Br—Br	I—I	
436	159	243	193	151	kJ mol^{-1}

For the hydrogen halides, these values may be combined with standard enthalpies of formation from gaseous halogens to yield:

H—F	H—Cl	H—Br	H—I	
566	431	366	299	kJ mol^{-1}

We shall discuss these values further in Section 5.7 in connection with electronegativities.

As we have seen in Chapter 4, the bonds in molecular oxygen and nitrogen are not single bonds: the triple-bonded formulation N≡N is satisfactory for the latter, but there is no adequate simple bond diagram for oxygen. Although the dissociation energy of nitrogen (946 kJ mol^{-1}) is therefore taken as the N≡N bond energy, we cannot validly interpret the dissociation energy of oxygen (498 kJ mol^{-1}) as the O═O or the O—O bond energy. We cannot give a value for the former, but the latter quantity is obtained from the enthalpies of formation of gaseous hydrogen peroxide and gaseous water and the bond energies in molecular hydrogen and oxygen:

$$H_2O_2 \equiv O{-}O + 2O{-}H$$
$$H_2O \equiv 2O{-}H$$

If we assume that the O—H bond energy is the same in hydrogen peroxide and water, the difference between the enthalpies of dissociation of the two gaseous compounds into ground state atoms gives the O—O bond energy as 146 kJ mol^{-1}. The similarity of the O—H bond lengths in the two molecules, which differ by less than 0.01 Å, indicates that this assumption is a reasonable one; nevertheless, it *is* an assumption, and results in a slight uncertainty in the value of the O—O bond energy. A similar process based on data for N_2H_4 and NH_3 leads to the value of 160 kJ mol^{-1} for the N—N bond energy in N_2H_4.

In order to obtain a value for the average C—H bond energy in methane, the standard enthalpy of formation of methane is determined from the enthalpies of combustion of methane, carbon (graphite) and hydrogen, and this is combined with the enthalpy of atomisation of carbon and the bond energy of hydrogen to yield E(C—H) = 416 kJ mol^{-1}. It would, of course, be of great interest to know accurately the enthalpy of dissociation of methane into sp^3 hybridised carbon atoms and ground state hydrogen atoms, but unfortunately sp^3 hybridised carbon atoms are not observable spectroscopically, and only approximate theoretical estimates for the excitation energy of carbon to this valence state are available. Data for the average C—F and C—Cl bond energies (485 and 327 kJ mol^{-1} in CF_4 and CCl_4 respectively) may be obtained by methods analogous to those used for the C—H bond.

For an element that exists in a range of valence states, the energies of bonds formed with another given element depend on the valence state, partly because of the different energies of excitation to the valence state and partly, no doubt, because of differences in orbital overlap. Thus for BrF, BrF_3 and BrF_5 the average bond energies are 250, 201 and 187 kJ mol^{-1} respectively.

The evaluation of the C—C, C═C and C≡C bond energy terms requires a knowledge of the enthalpies of combustion of ethane, ethylene (ethene) and acetylene (ethyne) in addition to the quantities needed for obtaining the C—H bond energy in methane; further, it has to be assumed that the C—H bond energy is constant. This is most unlikely to be true, since the C—H bonds in ethylene and acetylene are 0.02 and 0.04 Å respectively shorter than that in alkanes. In the context of using bond energy terms to estimate enthalpies of formation, however, this point is of only minor practical importance, since the likely error in *E*(C—H) is largely compensated by one in the value of *E*(C=C) or *E*(C≡C) used. Values compatible with *E*(C—H) = 416 kJ mol^{-1} are *E*(C—C), (C=C) and (C≡C) = 346, 598 and 813 kJ mol^{-1} respectively.

In organic chemistry, thermochemical data are frequently used to provide evidence for electron delocalisation: the enthalpy of atomisation of gaseous benzene, for example, is greater than the sum of 6*E*(C—H) + 3*E*(C=C) + 3*E*(C—C). Their use for this purpose in inorganic chemistry is very limited owing to the relatively small numbers of compounds to be considered, the difficulty of determining accurate enthalpies of formation of many inorganic compounds, and the lack of reference values. Thus although it would be interesting to compare the enthalpy of atomisation of borazine, $B_3N_3H_6$ (which is isostructural with benzene) with that for a hypothetical system containing isolated B=N, B—N, B—H and N—H bonds, it is difficult to estimate the B—H bond energy term, and at the present time no simple compound is known in which a B=N bond can safely be taken to be present. Nevertheless, consideration of bond energy data can often shed light on inorganic problems, and they will frequently be referred to in later chapters.

5.5 *Force constants*

The force constant *k* for the stretching of a diatomic molecule has been discussed in Sections 1.7, 1.9 and 2.6; if the fundamental vibration frequency $\bar{\nu}_0$ is expressed in cm^{-1}, *k* is given by

$$k = 4\pi^2 c_0^2 \bar{\nu}_0^2 \mu$$

where c_0 is the velocity of light in cm s^{-1} and μ the reduced mass of the molecule. Values for the fundamental vibration frequencies of several molecules, corresponding to the excitation of the molecule from vibrational level $v = 0$ to $v = 1$, and of the corresponding force constants, are given in Table 5.1.

In general, it will be seen that as the bond energy increases, the force constant increases also (though the value for F_2 is anomalous in this respect). There is, however, no proportionality between bond energy and force constant, nor should one be expected: the force constant is a measure of resistance to small displacement from equilibrium configuration, whilst the bond energy corresponds to complete separation into ground state atoms.

Table 5.1
Fundamental vibration frequencies $\bar{\nu}_0$ and force constants k of some diatomic molecules*.

Molecule	$\bar{\nu}$/cm^{-1}	k/Newton m^{-1}
H_2	4159	510
D_2	2990	510
F_2	892	450
Cl_2	557	320
Br_2	321	240
I_2	213	170
HF	3958	880
HCl	2886	480
HBr	2559	380
HI	2230	290
O_2	1556	1140
N_2	2331	2260
CO	2143	1870

*The values given in this table are observed vibration frequencies and the force constants derived from them; more accurate values are obtained if correction is made for the fact that the vibrations are not exactly simple harmonic motion, but since such corrections can be made only for very simple molecules we have disregarded them throughout this book.

In polyatomic molecules, there are often strong interactions between the various bonds, and the evaluation of force constants is much more difficult and uncertain. Nevertheless, for a series of molecules in which atomic masses are approximately constant, comparison of values of $\bar{\nu}_0$ can shed some useful light on relative bond strengths in situations where no thermochemical data are available. In the tetrahedral species $[Fe(CO)_4]^{2-}$, $[Co(CO)_4]^{-}$ and $Ni(CO)_4$, for example, the progressive increase in $\bar{\nu}$(CO) (1 788, 1 918 and 2 121 cm^{-1} respectively) is convincing evidence for an increase in the carbon–oxygen bond strength from the $[Fe(CO)_4]^{2-}$ ion to the $Ni(CO)_4$ molecule; the reason for this change in bond strength is discussed in Chapter 21. Measurements of force constants have also provided valuable information concerning the nature of the bonding in polyhalide anions (Chapter 16).

5.6 Bond lengths

Structure determinations by diffraction and spectroscopic methods have resulted in the accumulation of a very large body of data on internuclear distances in inorganic molecules, and some bond lengths have already been quoted, notably in Section 4.5. In many ways the discussion of covalent bond lengths is analogous to that of covalent bond energies. Species which are generally accepted as containing homonuclear single covalent bonds include H_2, F_2, Cl_2, Br_2, I_2, H_2O_2, N_2H_4, P_4, As_4, C_2H_6, Si, Ge, S_8 and Se_8. Halving the internuclear distances in these molecules

Table 5.2
Some single covalent radii of non-metals (Å).

H	C	N	O	F
0.37	0.77	0.74	0.74	0.72
	Si	P	S	Cl
	1.17	1.10	1.04	0.99
		As	Se	Br
		1.22	1.17	1.14
				I
				1.33

or macromolecules gives the value for the single covalent radii of non-metallic elements that appear in Table 5.2.

From values for standard multiple bonds we may quote C═C in monoalkenes (1.33 Å), C≡C in monoalkynes (1.20 Å), and N≡N in N_2 (1.10 Å) as examples. Clearly, a double bond not only has a higher bond energy than a single bond; it is also shorter. In general, for bonds between two given atoms, the higher the bond order, the shorter the bond and the greater the bond energy and the force constant.

For bonds between different elements, it is often found that the internuclear distance corresponds fairly closely to the sum of the covalent radii given in Table 5.2. The C—Cl distances in a wide range of saturated aliphatic compounds (e.g. CCl_4, CBr_2Cl_2, C_2Cl_6, CH_3Cl, CF_3Cl) are all 1.76 ± 0.01 Å, for example. Among the hydrogen halides, however, the bond lengths (HF, 0.92; HCl, 1.27; HBr, 1.41; HI, 1.60 Å) are all substantially less than the sum of the covalent radii, that for HF especially so. It is, indeed, found that bonds involving fluorine (and, to a smaller extent, those involving oxygen or nitrogen) very often show shortenings like this. Sometimes, e.g. for SiF_4 (Si—F = 1.56 Å, sum of covalent radii = 1.89 Å), this may be attributed to a bond order greater than unity, in this case a σ bond being reinforced by π bonds formed by overlap of filled *p* orbitals of the fluorine atoms and the empty *d* orbitals of the silicon atom. A similar effect occurs in BF_3, in which the boron atom formally has an empty *p* orbital available for p_π—p_π bonding. But in the case of CF_4 (C—F = 1.32 Å, sum of covalent radii = 1.49 Å), such an explanation is untenable since there is no empty *p* orbital and there are no *d* orbitals in the valence shell of carbon; the shortening is then attributed to the partial ionic character of the bond and is discussed further in the next section.

Most elements, of course, form no molecules in which two atoms of the element are joined by what can reasonably be taken to be a single bond. For a metal such as titanium the best we can do is to subtract the covalent radius of chlorine from the observed Ti—Cl distance in gaseous $TiCl_4$; for lead, the Pb—C bond length in $Pb(CH_3)_4$ may be used as the source of the covalent radius of lead. Strictly, we thus obtain covalent radii of titanium(IV) and lead(IV); the covalent radius varies slightly according to the valence state, the Pb—Cl distances in gaseous $PbCl_2$ and $PbCl_4$, for example, being 2.46 and 2.43 Å respectively. Differences are also found between the S—F distances in SF_4 and SF_6 and between the P—F distances in PF_3 and PF_5, though in each case the situation is complicated by the occurrence of two different bond lengths in the molecule whose structure is a trigonal bipyramid or is derived from one. In organic chemistry, the variation in the C—H bond length among the molecules C_2H_6, C_2H_4 and C_2H_2 (1.10, 1.08 and 1.06 Å respectively) corresponds to a change in the hybridisation of the carbon atom bonded to the hydrogen (sp^3, sp^2 and *sp* respectively).

It will be apparent from the foregoing discussion that there are no adequate absolute standards for the discussion of most bond lengths, and explanations of properties nearly always involve *comparisons* of interatomic distances. Provided such comparisons are made carefully, however, consideration of bond lengths (which are by far the most readily available properties of covalent bonds) can make a very valuable contribution to the discussion of the chemistry of molecular species.

5.7 *Bond polarities and electronegativity*

The polarities of covalent bonds cannot be measured directly, nor can they often be derived unambiguously from measurable parameters. In this section, therefore, we are mostly concerned with attempts to infer something about electronic distributions in molecules by indirect methods whose reliabilities can be no greater than those of the various assumptions they involve. Before we go on to consider the different ways of attempting to estimate the electronegativity of an element (defined as the power of an atom *in a molecule* to attract electrons to itself), we should, however, mention briefly some techniques that are in principle capable of telling us something about the electron density in a molecule.

X-ray scattering is dependent upon electron density, and it is now possible to obtain accurate electron density maps for simple species (as will be described further in the next chapter); but little information is at present available for molecular species containing many electrons. General correlations sought between nuclear magnetic resonance chemical shifts or ionisation energies of inner electrons (determined by X-ray photoelectron spectroscopy) and electronic distributions in inorganic molecules are not yet firmly established. Dipole moments, which were once thought to provide reliable information about bond polarities, are now recognised as being properties of whole molecules rather than of bonds: for example, the fact that ammonia and nitrogen trifluoride, both of which are pyramidal, have moments of 1.5 and 0.2 debye units (1 debye unit $= 3.3 \times 10^{-30}$ C m) respectively is believed to indicate that the lone pair of electrons on the nitrogen atom contribute substantially to the overall moment, reinforcing the N—H bond moments in the case of ammonia and opposing the N—F bond moments and the fluorine lone pair moments in the case of nitrogen trifluoride. Thus, attempts to calculate the degree of ionic character of the bonds in the hydrogen halides from their observed dipole moments and the moments calculated on the basis of complete charge separation neglect the effects of the lone pairs of electrons on the halogen atom, and so are of doubtful validity. Finally, Mössbauer chemical shifts depend upon electron density (due to electrons in *s* orbitals) at the nucleus, but they are also influenced by other factors such as oxidation state and d_π back-bonding.

There are several definitions of electronegativity in general use at the present time (a fact in itself sufficient to show that the concept of electronegativity is not a very satisfactory one). In the simplest of these, originally due to Mulliken, the electronegativity is taken as the mean value of the first ionisation energy and the first electron affinity, both quantities being given a positive value if loss of an electron involves absorption of energy and gain of an electron involves release of energy. These are, of course, properties of isolated ground state atoms; in subsequent modifications of Mulliken's treatment estimated valence state ionisation energies and electron affinities have been used. Since, however, hybridised valence states are not observed spectroscopically, electronegativities so derived lack a firm experimental basis. We may also note in passing that, as we saw in Section 3.5, not all electron affinities can be determined experimentally; further, the variation in electron affinity among the halogens is so small that their Mulliken electronegativities are determined almost entirely by the ionisation energies.

If we compare the energies of single bonds between different atoms, E(A—B), with those in the reference molecules A_2 and B_2, it is found that E(A—B) is usually greater, and often much greater, than the mean of E(A—A) and E(B—B). This may be seen to be true for the hydrogen halides, for example, from the data given in Section 5.4. This extra bond energy, ΔE, defined by

$$\Delta E = E(\text{A—B}) - \tfrac{1}{2}[E(\text{A—A}) + E(\text{B—B})]$$

has the following values: HF, 268; HCl, 92; HBr, 46; HI, 5 kJ mol^{-1}. Pauling in 1932 suggested that ΔE is a measure of the extra bond energy arising from a contribution A^+B^- or A^-B^+ to the pure covalent structure (i.e. arising from ionic–covalent resonance as defined in Section 4.5) and is a function of the difference in electronegativity, Δx, between A and B. He converted ΔE values into eV mol^{-1} (by dividing by 96.5) in order to make Δx small, and showed empirically that an *approximately* self-consistent set of values for Δx could be obtained by taking

$$\Delta x = (\Delta E)^{\frac{1}{2}}$$

An element of intuition is involved in deciding whether A or B has the higher value of x, and in order to avoid giving any element a negative value of x, x_H was taken as 2.1, leading to the electronegativities of the halogens as: F, 4.0; Cl, 3.0; Br, 2.8; I, 2.4. It is not customary to append units to electronegativity values, but it may be noted that those of x are actually $(\text{eV mol}^{-1})^{\frac{1}{2}}$, whereas on Mulliken's scale values are in eV mol^{-1}.

During the period since Pauling introduced this approach, many new thermochemical data have appeared; the most recent of these have been used to calculate the Pauling electronegativities given in Table 5.3. It should be noted that values of x in this table depend slightly on what compounds were used in the compilation of thermochemical data; some dependence of x on the valence state of the element would be expected, since bond energies depend on valence states; but the concept of electronegativity is not precise enough to justify further consideration of this point now.

For many metals, particularly transition and inner transition metals, neither species known to contain M—M single covalent bonds nor simple volatile molecular compounds exist. For these elements, Pauling electronegativities have been obtained by using metal–metal bond energies in solid metals in place of M—M single bond energies. Values so derived, however, cannot be regarded as entirely consistent with those given in Table 5.3, and we shall not make use of them in this book.

The third definition of electronegativity is that proposed by Allred and Rochow, who chose as a measure of this quantity the electrostatic force exerted by the effective nuclear charge Z^* on the valence electrons, which are assumed to reside at a distance r from the nucleus equal to the covalent radius. Z^* is calculated using Slater's rules (Section 3.1). The attractive force is then proportional to Z^*/r^2. A plot of Pauling electronegativities against Z^*/r^2 gives an approximately straight line, and the slope and intercept indicate that on the Pauling scale Allred–Rochow electronegativities x' (whose units are Å^{-2}) are given by the numerical relationship

$$x' = 0.359\frac{Z^*}{r^2} + 0.744$$

Since, however, Slater's rules are partly empirical and covalent radii are unavailable for some elements, the Allred–Rochow scale is no more rigid or complete than the Pauling one.

Table 5.3
Pauling electronegativities of some elements.

H 2.1	B 2.0	C 2.5	N 3.0	O 3.4	F 4.0
	Al 1.6	Si 1.9	P 2.2	S 2.6	Cl 3.2
	Ga 1.8	Ge 2.0	As 2.2	Se 2.6	Br 3.0
		Sn 2.0	Sb 2.1	Te 2.0	I 2.7
		Pb 1.9	Bi 2.0		

Despite the somewhat dubious scientific basis of all three methods, the relative values of electronegativities obtained by them are roughly in agreement. The most useful of them in inorganic chemistry is probably the Pauling scale, which, being based empirically on thermochemical data, can reasonably be used to predict similar data; if, for example, the electronegativities of two elements X and Y are known from the single covalent bond energies of HX, HY, X_2, Y_2 and H_2, we can estimate the actual bond energy in kJ mol^{-1} in XY by means of the equation

$$E(\text{X—Y}) = \tfrac{1}{2}[E(\text{X—X}) + E(\text{Y—Y})] + 96.5(\Delta x)^2$$

with a fair degree of reliability. In this book we shall avoid the use of the concept of electronegativity as far as possible, and try to base the systematisation of descriptive inorganic chemistry on rigidly defined and independently measured thermochemical quantities such as ionisation energies, electron affinities, bond energies, lattice energies and hydration energies; but some mention of electronegativity is unavoidable, and this is why the foregoing rather full and critical discussion of the subject has been given.

Whatever the precise significance of electronegativity, the fact remains that bonds involving very 'electronegative' elements (particularly oxygen and fluorine) are, as we saw in Sections 5.4 and 5.6, considerably stronger and shorter than would be expected on the basis of additivity relationships; and this generalisation is often useful in discussing bond energies and bond lengths. An empirical correlation between bond energies and bond lengths has been given by Schomaker and Stevenson, who proposed the equation

$$r_{\text{A—B}} = r_{\text{A}} + r_{\text{B}} - 0.09\Delta x$$

in which $r_{\text{A—B}}$ is the actual interatomic distance in a molecule AB_n, r_A and r_B are the single covalent radii of A and B, and Δx is the difference in Pauling electronegativity values between A and B. Considering its simplicity, this equation provides a fairly good representation of bond lengths in many inorganic compounds.

PROBLEMS

1 Predict the structures of the following molecules or ions: BeF_4^{2-}, NF_4^+, $SbCl_5$, H_3O^+, ClO_2, I_3^-, ICl_2^+, BrF_6^+, IF_6^-.

2 Assign point group symbols to each of the following species: ClF_3, C_2H_2, SF_5Cl, C_2H_4, $B_3N_3H_6$ (which has the benzene structure, C atoms being replaced by B and N alternately), $SiCl_2Br_2$, PF_5, BFClBr, $H_2C{=}C{=}CH_2$.

3 The bond energies in N_2 and F_2 are 946 and 159 kJ mol^{-1} respectively, and the standard enthalpy of formation of gaseous NF_3 is -113 kJ mol^{-1}. Calculate the average N—F bond energy in NF_3.

The standard enthalpy of formation of gaseous N_2F_4 is $+8$ kJ mol^{-1}. Assuming that the average N—F bond energy is the same in NF_3 and N_2F_4, estimate the N—N bond energy in N_2F_4.

4 The fundamental vibration frequency of H_2 is 4159 cm^{-1}. Calculate the force constant and molar zero-point energy for this molecule. How would you expect these values to be related to the corresponding values for D_2?

5 The standard enthalpies of formation of gaseous XeF_2, XeF_4 and XeF_6 are -108, -216, and -294 kJ mol^{-1} respectively, and the bond energy in F_2 is 159 kJ mol^{-1}. Calculate the average Xe—F bond energy in each of these compounds, and use the value for XeF_2 to obtain a value for the electronegativity of xenon on the Pauling scale, assuming $x_F = 4.0$

What do the average bond energies in the xenon fluorides imply about the valence states of the xenon in these molecules?

REFERENCES FOR FURTHER READING

Cotton, F.A. (1971) *Chemical Applications of Group Theory*, 2nd edition, Interscience, New York. A more mathematical treatment of symmetry and its importance to chemistry.

Cottrell, T.A. (1958) *The Strengths of Chemical Bonds*, 2nd edition, Butterworths Scientific Publications, London. Though now old, it remains the best critical treatment of bond energies and bond dissociation energies.

Gillespie, R.J. (1970) *J. Chem. Education*, **47**, 18. A fuller account of the Valence Shell Electron Pair Repulsion Theory.

Huheey, J.E. (1983) *Inorganic Chemistry*, 3rd edition, Harper, New York. Contains a fuller treatment of electronegativity and related topics than this book.

Kettle, S.F.A. (1985) *Symmetry and Structure*, Wiley, Chichester. A very clear discussion of symmetry with detailed treatment of a few examples.

Shriver, D.F., Atkins, P.W. and Langford, C.H. (1990) *Inorganic Chemistry*, Oxford University Press. Contains a very clear introduction to symmetry and symmetry-related properties.

Wells, A.F. (1984) *Structural Inorganic Chemistry*, 5th edition, Oxford University Press. The definitive work on the structures of inorganic compounds.

6 The structures and energetics of inorganic solids

6.1 Introduction

In Chapters 4 and 5 we discussed at length the nature and properties of covalent bonds. Most readers of this book will already be familiar also with ionic, van der Waals', and metallic bonding, but for the sake of completeness we begin this chapter with a brief summary of the principal characteristics of each of these.

The *ionic bond* is the strong all-directional electrostatic interaction between charged species such as Na^+ and Cl^- in the solid state; the electron density between the ions at some point approximates to zero, and each ion is surrounded symmetrically by ions of opposite sign; the force between the ions is inversely proportional to the square of their distance apart. Ionic solids are insulators.

Van der Waals' bonding typically involves the weak so-called *dispersion forces* which are responsible for the interaction of non-polar molecules such as the noble gases and the halogens in the liquid and solid states. These arise from momentary asymmetric electron distributions which produce temporary dipole moments that in turn induce dipole moments in neighbouring molecules; the forces are thus often said to result from induced dipole–induced dipole interaction. These dispersion forces are all-directional and give rise to a binding energy proportional to α^2/r^6 where α is the polarisability of the atom or molecule and r the intermolecular separation. The polarisability is the proportionality constant in the equation

$$\mu = 4\pi\varepsilon_0\alpha E$$

where μ is the dipole moment induced by an electric field E and ε_0 is the permittivity of a vacuum; it can be obtained from measurements of the relative permittivity (dielectric constant) or the refractive index of the substance in question. It increases rapidly with increase in atomic or molecular size, i.e. large atoms or molecules give rise to much larger induced dipoles and hence have much larger interatomic or intermolecular forces; this is shown convincingly in the series F_2, Cl_2, Br_2 and I_2, for example. Molecules which have permanent dipole moments attract one another additionally by dipole–dipole interaction, and thus have stronger van der Waals' bonding than non-polar molecules of comparable size and polarisability. Van der Waals' bonding is also present in ionic solids, but there it is much smaller than electrostatic interactions.

Metallic bonding is characterised by strong all-directional cohesive forces and high electrical conductivity. Such conductivity implies the presence of bonding electrons delocalised over the whole crystal. The theory of the metallic bond thus involves an approach, closely related to the molecular orbital method, in which the whole crystal is regarded as a large molecule; a brief outline of the band theory was given in Section 4.9.

Hydrogen bonding is sometimes considered as a fifth fundamental type of chemical interaction. The strengths of hydrogen bonds are intermediate between those of van der Waals' bonds and ordinary covalent bonds. In this book, however, we discuss hydrogen bonding separately, and an account of it will be found in Chapter 9.

Solids in which the strongest interaction is ionic, metallic or van der Waals' bonding typically have structures which exhibit three-dimensional symmetry (e.g. NaCl, Cu and solid Ar respectively). Solids in which covalent bonding is the strongest interaction are conveniently classified as molecular or macromolecular according to whether the covalent bonding is localised within discrete species or extends indefinitely in one, two or three dimensions. In solid chlorine, for example, each chlorine atom has one nearest neighbour at a distance of 1.99 Å, and several others at distances of 3.3–3.8 Å; the enthalpy of sublimation of chlorine is 29 kJ mol^{-1}, but its enthalpy of dissociation is 242 kJ mol^{-1}, the former corresponding to overcoming all the 3.3–3.8 Å interactions and the latter to breaking the bonds in the molecules of Cl_2. Chlorine is thus an example of a molecular solid (so is argon, but in this case the molecule is monatomic). In diamond, on the other hand, each carbon atom is bonded to four other carbon atoms distributed around it tetrahedrally at a distance of 1.54 Å, and no discrete molecule can be discerned in the solid; diamond has a three-dimensional macromolecular structure. In graphite, the covalent bonding is confined within infinite planes, and there are only weak van der Waals' interactions between the planes; graphite has a two-dimensional macromolecular structure.

Fibrous sulphur, silicon disulphide, and α-palladium chloride have infinite chain (or one-dimensional) macromolecular structures.

Although ionic solids have structures which exhibit three-dimensional symmetry, it does not follow that structures exhibiting three-dimensional symmetry are necessarily ionic: diamond, for example, is certainly not. We shall see later in this chapter that in a very large number of compounds which we should normally describe as ionic salts, the cation is surrounded octahedrally by six anions; but, as we saw in Chapter 5, octahedral symmetry is also very common in AB_6 type molecules and ions. There is often, in fact, evidence to show that the bonding in compounds having the sodium chloride structure is not purely ionic in character; for species in which each atom has only four nearest neighbours in an infinite three-dimensional structure (e.g. ZnS), evidence for bonding intermediate between pure ionic and pure covalent is even stronger. We shall return to this subject in Section 6.7.

At the level of this book we shall not, in general, be much concerned with details of the packing of the molecules in a molecular solid whose volatility is convincing evidence for the weakness of intermolecular interaction. The molecules in such a solid usually (though not invariably) have the same structure as in the vapour phase, but the symmetry properties of crystals are more complex than those of individual molecules (as well as being expressed in a different notation), and for descriptions of them the references at the end of the chapter should be consulted. Nor shall we devote much space to the detailed description of the structures of ionic crystals other than those of a few fundamental examples; to the chemist, the main interest in an ionic structure is usually in the *coordination* (or environment) of the individual ions and the interionic distances. Even at an elementary level, however, some feeling for the three-dimensional nature of solid state structures is essential, and the reader is urged to supplement the descriptions given in this chapter by the examination of crystal structure models.

The fundamental concept in solid state structure is that of the *unit cell*, the smallest repeating unit of the structure that, stacked randomly, reproduces the structure of the whole solid. The choice of a unit cell is somewhat arbitrary, but it is customary to choose one with as many atoms as possible at the corners; all the unit cells shown in this chapter are of that form. Thus the unit cell of sodium chloride, illustrated in Fig. 6.1, is a cube of side 5.64 Å with sodium ions at the centre and at the mid-points of the edges, and with chloride ions at the corners and at the mid-points of the faces of the cube (positions of the ions may, of course, be interchanged). Inasmuch as the space in the unit cell is filled by the ions, the representation in Fig. 6.1(*a*) is more realistic, but it hides most of the structure, and the more open representation in Fig. 6.1(*b*) is therefore generally employed. An atom within the unit cell belongs to it alone, one at the mid-point of a face is shared between two unit cells, one at the mid-point of an edge between four unit cells, and one at a corner between eight

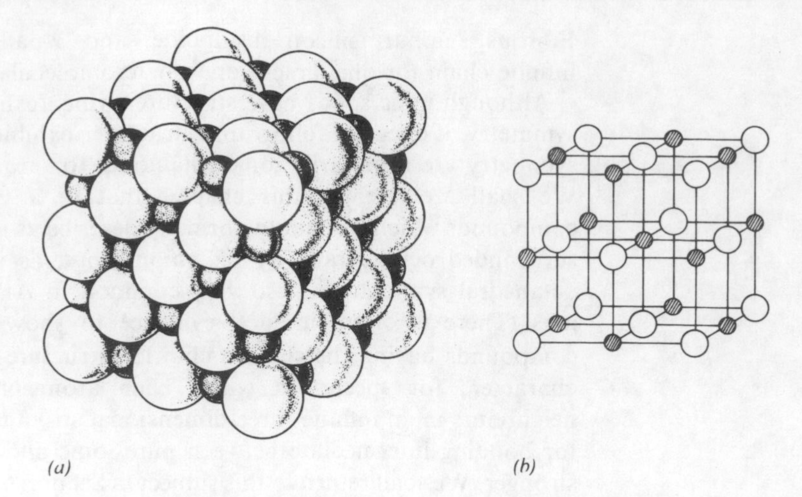

Figure 6.1
Two representations of the structure of sodium chloride: (*a*) shows occupation of space; (*b*) is the conventional representation.

unit cells. It is thus readily shown that for sodium chloride the number of NaCl molecules (in this case, more strictly, of formula units) in the unit cell is four.

The size and shape of a unit cell is denoted by the lengths (a, b, c) of three intersecting edges and the angles (α, β, γ) between pairs of edges (b and c, a and c, a and b respectively) as shown in Fig. 6.2. There are seven basic types of unit cell, and these are described in Table 6.1. Most of these basic types (or crystal systems) may be subdivided, e.g. a primitive (or simple) cubic unit cell contains units (atoms or molecules) only at the corners, a body-centred cubic unit cell contains them at the centre and at the corners, and a face-centred cubic unit cell contains them at the mid-points of the faces and at the corners. These are shown in Fig. 6.3, from which it will be clear that the sodium chloride structure shown in Fig. 6.1 consists of a face-centred cubic lattice of Na^+ ions and an inter-penetrating face-centred cubic lattice of Cl^- ions.

A close examination of the crystal systems shows that the symmetry elements allowed in crystals are different from those allowed in isolated molecules. Thus whereas individual molecules with C_5 and C_7 axes of

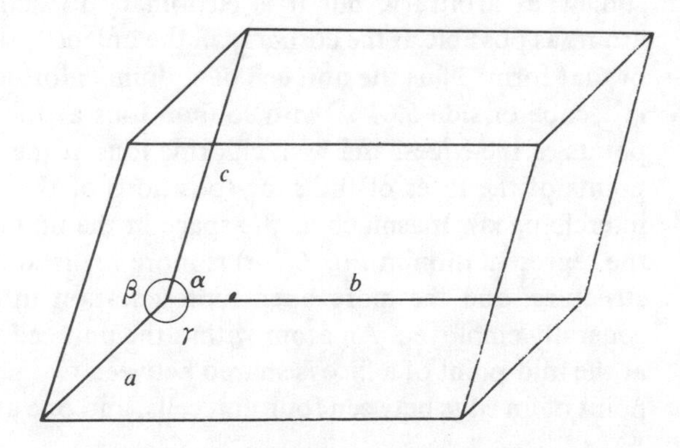

Figure 6.2
The unit cell.

Table 6.1
Types of unit cell.

System	Cell edges	Cell angles
Cubic	$a = b = c$	$\alpha = \beta = \gamma = 90°$
Tetragonal	$a = b \neq c$	$\alpha = \beta = \gamma = 90°$
Orthorhombic	$a \neq b \neq c$	$\alpha = \beta = \gamma = 90°$
Rhombohedral (trigonal)	$a = b = c$	$\alpha = \beta = \gamma \neq 90°$
Hexagonal	$a = b \neq c$	$\alpha = \beta = 90°$ $\gamma = 120°$
Monoclinic	$a \neq b \neq c$	$\alpha = \gamma = 90°$ $\beta \neq 90°$
Triclinic	$a \neq b \neq c$	$\alpha \neq \beta \neq \gamma \neq 90°$

(a)

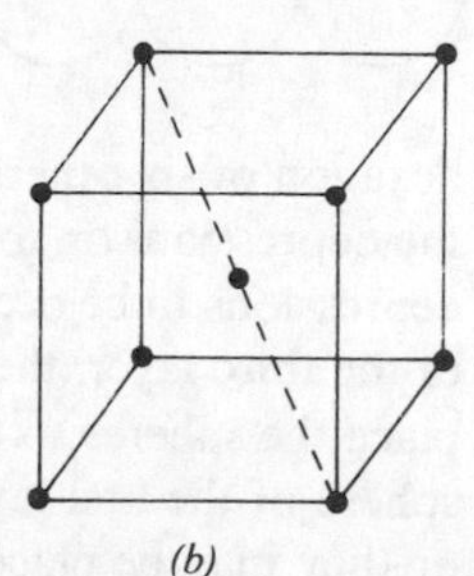
(b)

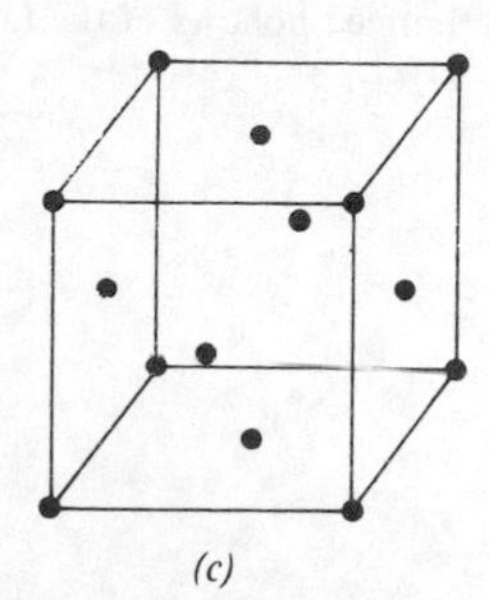
(c)

Figure 6.3
(*a*) Primitive cubic
(*b*) body-centred cubic
(*c*) face-centred cubic, unit cells.

symmetry are well known, these axial symmetries do not occur in crystals because it is impossible to fill all of the space with figures of these symmetries. On the other hand, in a crystal having an infinitely extended pattern in space, new symmetry operations involving translations are possible in addition to most of the operations permitted for individual molecules. This is why the discussion of crystal symmetry is much more difficult than that of molecular species and why the total symmetry of a crystal structure, called its *space group*, is a topic beyond the scope of this book.

6.2 The close packing of spheres

In many crystals in which van der Waals' or metallic bonding plays the dominant role, and also in some ionic solids, the structure adopted can be simply accounted for in terms of the most efficient packing possible. The units involved are often atoms or roughly spherical molecules (e.g. noble gases, metals, SiF_4), and it is therefore interesting to study the geometry of the most effective ways of packing hard spheres of the same size in three-dimensional structures.

As may be seen from Fig. 6.4, spheres forming a single layer and packed as closely as possible arrange themselves so that their centres are at the corners of an equilateral triangle and each sphere is surrounded by a

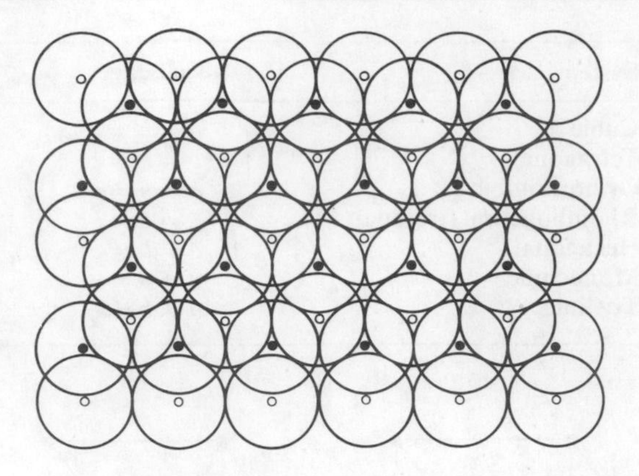

Figure 6.4
Two layers of close-packed spheres. Small open circles denote the positions of the centres of spheres of the lower layer, small closed circles those of the centres of spheres of the upper layer. Tetrahedral holes lie immediately above each sphere of the lower layer and immediately below each sphere of the upper layer. Octahedral holes are the unfilled hollows of the first layer.

hexagon of six others. The second layer is obtained by placing spheres in the depressions on top of the first layer, but there is room for only half the depressions to be occupied in this way. When we come to add the spheres of the third layer, there are thus two ways of placing them. Either we may place the spheres so that each sphere of the third layer lies directly over a sphere of the first layer (thus making the first and third layers identical), or they may be placed over the unoccupied depressions of the first layer (i.e. over neither of the two previous layers of spheres). These arrangements are further illustrated in Figs. 6.5 and 6.6, from which the hexagonal and cubic symmetry of the arrangements is apparent.

We may describe the layer sequences as ababab ... and abcabc ... for *hexagonal* and *cubic close packing* (*hcp* and *ccp*) respectively. In each structure any given sphere is surrounded by twelve other spheres all at the

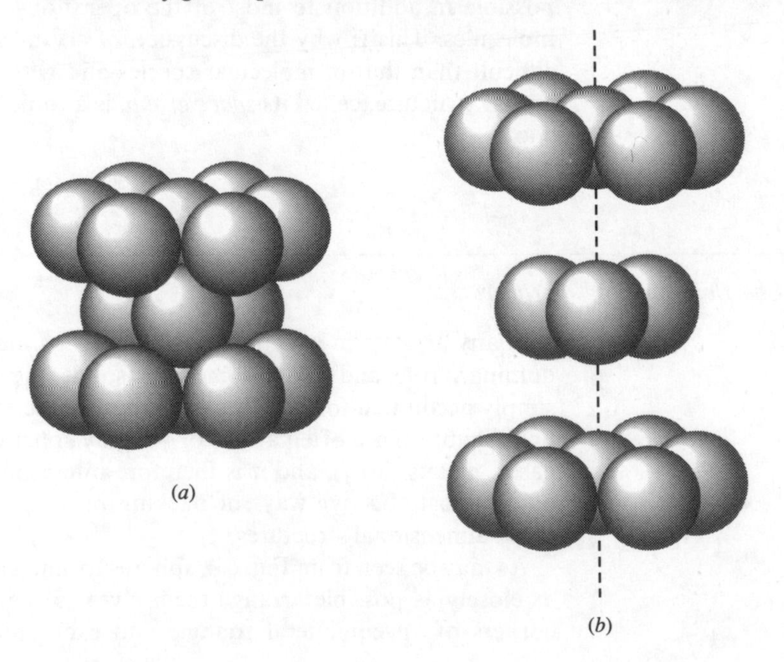

Figure 6.5
Hexagonal close packing of spheres (*a*) normal and (*b*) exploded views.

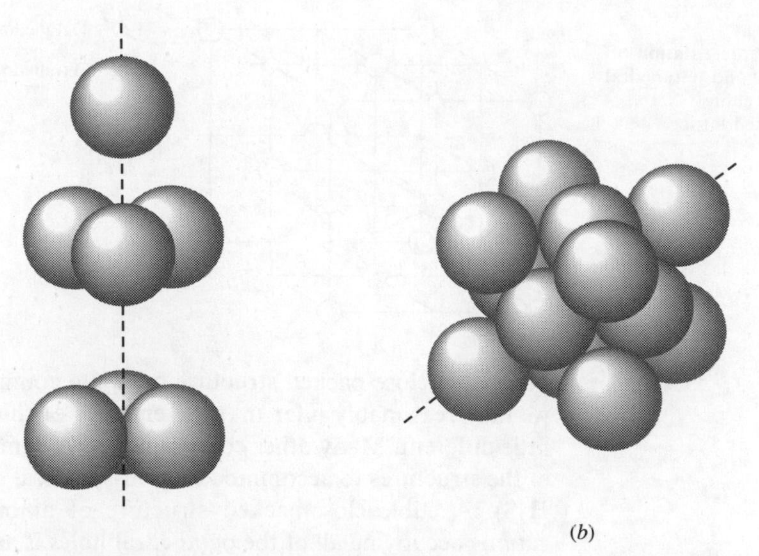

Figure 6.6
Cubic close packing of spheres: (*a*) exploded view to show the relative positions of spheres in the three-layer unit; (*b*) normal view.

same distance, six in the same plane and three in each adjacent layer, and the fraction of the total space occupied by the spheres is 0.74, but the symmetry of the coordination, and hence the overall symmetry, is different. It may be seen that cubic close packing is in fact equivalent to the face-centred cubic structure that was described in the last section.

Another feature of interest in close-packed structures is the occurrence of *interstitial sites*, or *octahedral* and *tetrahedral holes*. In Fig. 6.4 and Fig. 6.7 we can distinguish two types of site or hole, one surrounded octahedrally by six spheres and the other tetrahedrally by four. The octahedral holes, of which there is one to every sphere, are the larger, and can each accommodate a sphere of radius up to 0.41 times that of the larger sphere without causing distortion of the structure. There are two tetrahedral holes per sphere, but these are much smaller, and can accommodate only spheres of radius up to 0.23 times that of the close-packed spheres. Octahedral and tetrahedral sites in a cubic close-packed (face-centred cubic) lattice are also shown in Fig. 6.8.

The noble gases crystallise with the cubic close-packed structure; most metals (as we shall see in Section 6.9) have either the cubic or the

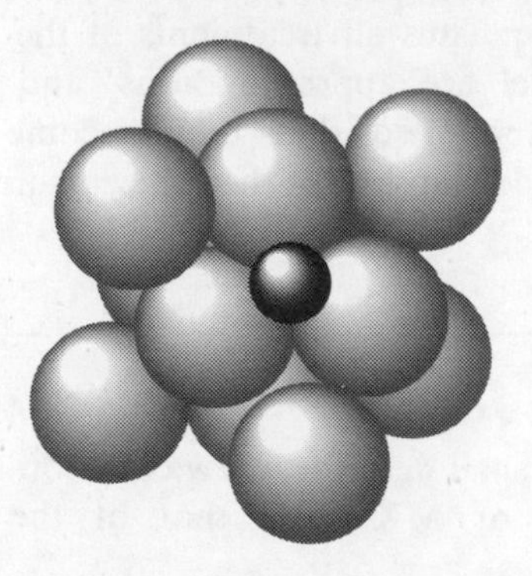

Figure 6.7
Cubic close-packed structure with corner atom removed to show a smaller atom in a tetrahedral hole.

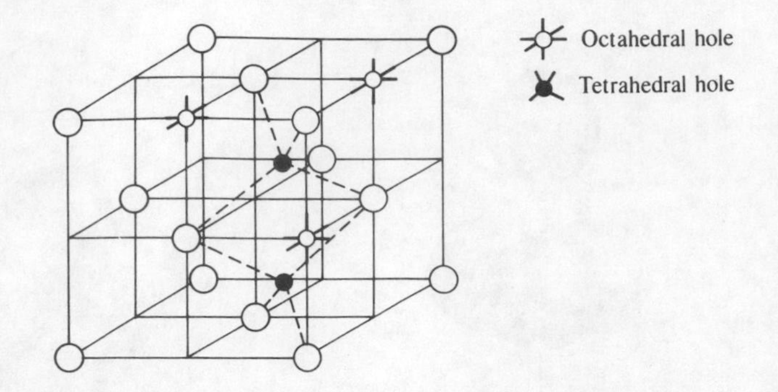

Figure 6.8
Another representation of octahedral and tetrahedral holes in a cubic close-packed lattice.

hexagonal close-packed structure or, quite commonly, both, from which we may reasonably infer that the energies of the two structures are very little different. Many other compounds have somewhat distorted versions of the structures to accommodate non-spherical molecules (e.g. Cl_2, HCl, H_2S). A cubic close-packed structure of anions with relatively small cations occupying all of the octahedral holes is another description of the sodium chloride structure, whilst the same anion structure with one quarter of the tetrahedral holes filled by cations describes the calcium fluoride structure. These geometrical descriptions are often employed by crystallographers, but they are not really satisfactory for salts such as potassium fluoride and barium fluoride (which are isostructural with sodium chloride and calcium fluoride respectively) in which the radii of cations and anions are almost the same; we shall not use them in this book except for species in which one atom or ion is much larger than the other, such as the hydrides, carbides, nitrides, and oxides of certain transition metals. Structures derived by inserting cations into octahedral holes in hexagonal close-packed arrays of anions are best not discussed for the present; that of nickel arsenide, NiAs, is, however, described in Section 14.4.

Useful as the hard sphere model is in acquiring a basic grasp of common crystal structures, it must be clearly understood that it is completely at odds with the modern quantum theory of atoms and molecules. As we saw in Chapter 2, the wave function of an electron does not suddenly drop to zero with increasing distance from the nucleus, and in a close-packed, or indeed any other, crystal there is a finite electron density everywhere. Thus all treatments of the solid state based on the hard sphere model are approximations, and we should not be too surprised or disappointed that they become less satisfying as our knowledge and understanding of the subject increase.

6.3 *The structure of ionic solids*

In this section we shall describe a few common structures of compounds of formula MX or MX_2 and that of the double oxide $CaTiO_3$

(*perovskite*). These structures are nearly always determined by X-ray diffraction; since the scattering powers of different ions depend upon the total number of electrons they contain, however, the use of X-ray methods is subject to two important limitations. First, they are not suitable for the location of light atoms in the presence of much heavier ones: neutron diffraction (Section 1.9) is then necessary as a complementary technique. Second, they are seldom capable of identifying the state of ionisation of the species present: only for a few substances has the electron density distribution been determined with sufficient accuracy for this purpose. One of these is sodium chloride, concerning which it is known that, to within a small margin of uncertainty, ten and eighteen electrons are associated with the sodium and chlorine nuclei respectively, showing that the units present are Na^+ and Cl^-. Much more often the ionic nature of a compound is inferred from the possession of a symmetrical three-dimensional structure and evidence from a variety of other sources (e.g. hardness, conductivity, spectroscopic and thermochemical data). We shall be concerned here mostly with compounds for which the ionic model is generally accepted as valid, but mention will be made of some substances for which it is unsatisfactory.

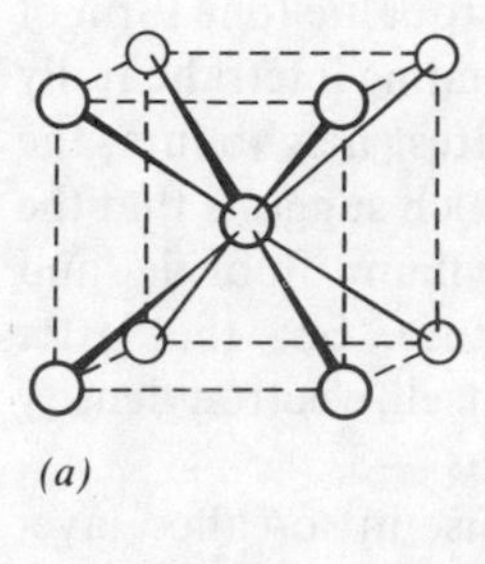
(*a*)

There are four basic structures in which salts of formula type MX commonly crystallise: those of sodium chloride (illustrated already in Section 6.1), caesium chloride, zinc sulphide (*zinc blende*) and zinc sulphide (*wurtzite*); the last three of these are shown in Fig. 6.9. The sodium chloride, caesium chloride and zinc blende structures have cubic symmetry; wurtzite has hexagonal symmetry. For our present purposes, however, what we are particularly interested in is the environments of the individual ions. In the caesium chloride structure, each ion is surrounded by eight nearest neighbours of opposite charge at the corners of a cube; this structure is relatively uncommon, but occurs in caesium chloride, bromide, and iodide, thallium(I) chloride and bromide, and ammonium chloride and bromide at ordinary temperatures. In the sodium chloride structure, each ion has six nearest neighbours at the corners of a regular octahedron; this is the structure of most alkali metal halides, the alkaline earth metal oxides, silver fluoride, chloride and bromide, potassium cyanide (owing to rotation of the anion at the ordinary temperature), and ammonium chloride and bromide above 184 and 138 °C respectively. In the zinc blende and wurtzite structures, each ion is surrounded by four nearest neighbours at the corners of a regular tetrahedron, and the structures differ only in the way in which the tetrahedra are oriented relative to one another: copper(I) chloride, bromide and iodide have the zinc blende structure; silver iodide, zinc oxide, ammonium fluoride and aluminium nitride have the wurtzite structure. The fact that diamond may also be described as having the zinc blende structure (with both zinc and sulphur atoms replaced by carbon) shows that these four-coordination structures are not necessarily purely ionic ones: a regular tetrahedral environment is equally compatible with a purely electrostatic interaction

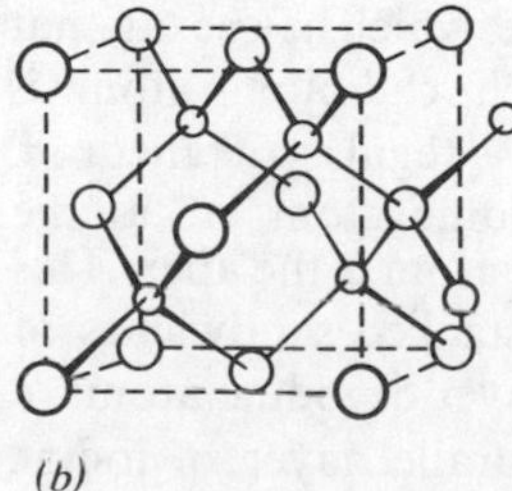
(*b*)

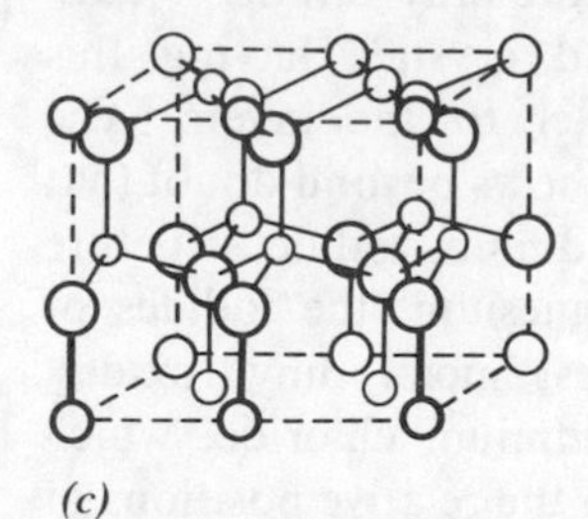
(*c*)

Figure 6.9
(*a*) The caesium chloride structure. (*b*) The zinc blende structure. (*c*) The wurtzite structure.

with four surrounding ions and with covalent bonding involving sp^3 hybrid orbitals. We shall return to this point in Section 6.7.

In salts of formula MX_2, for which the three most important regular three-dimensional structures are illustrated in Fig. 6.10, the coordination number of X must obviously be half that of M. In the cubic calcium fluoride (*fluorite*) structure the cation has eight nearest neighbours at the corners of a cube and the anion four at the corners of a regular tetrahedron; substances which possess this structure include the alkaline earth metal fluorides, barium chloride, and the dioxides of the lanthanide and actinide elements. If the compound has the formula M_2X (e.g. the alkali metal monoxides and sulphides) the positions of cations and anions in the fluorite structure are interchanged and the substance is said to have the *antifluorite* structure. In the tetragonal structure of *rutile* (one form of titanium dioxide), the coordination numbers of the cation and anion are six (octahedral) and three (equilateral triangular) respectively; magnesium fluoride, manganese difluoride, zinc fluoride, and the dioxides of tin, lead, and manganese have this structure. In β-cristobalite (one form of silica) each silicon atom has four oxygen atoms surrounding it tetrahedrally and each oxygen atom has two silicon atoms as nearest neighbours; the angle Si—O—Si is somewhat less than 180°, a fact which suggests that the structure is not a purely electrostatic one; beryllium fluoride and germanium dioxide have the β-cristobalite structure, but the latter compound also exists in the rutile structure. In rutile itself, electron density measurements give the actual charge on the cation as +3.

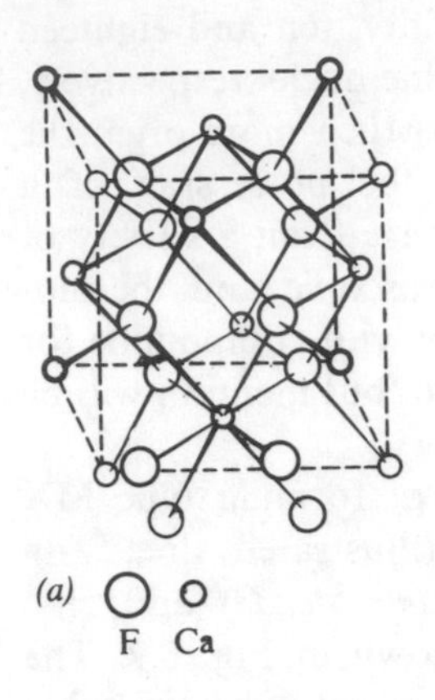

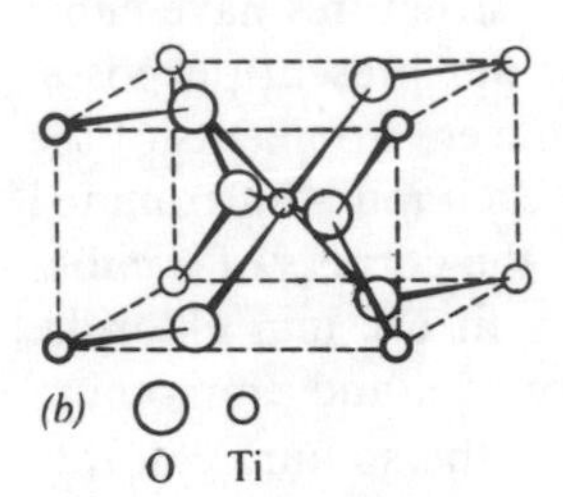

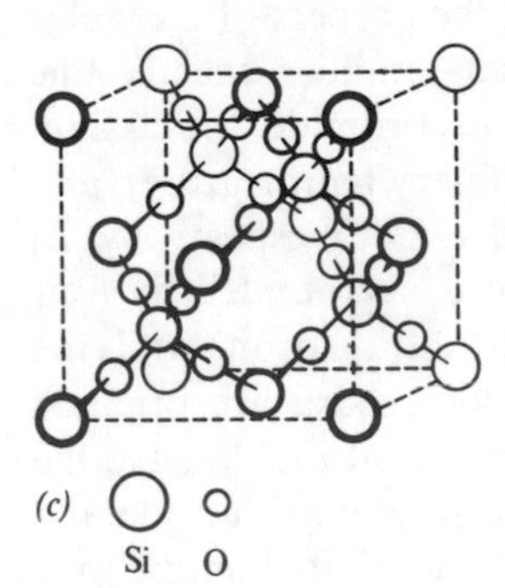

Figure 6.10
(*a*) The fluorite structure.
(*b*) The rutile structure.
(*c*) The β-cristobalite structure.

Many other compounds of formula MX_2 crystallise in so-called layer structures, a typical one being that of cadmium iodide, which has hexagonal symmetry, and is illustrated in Fig. 6.11. Each cadmium atom is surrounded by six iodine atoms at the corners of a regular octahedron, but the three cadmium atoms nearest to each iodine atom are at the corners of the base of a pyramid of which the iodine atom is the apex. This structure may thus be described as consisting of stacked sandwiches of composition CdI_2, each sandwich consisting of a layer of iodine atoms, a parallel layer of cadmium atoms, and another parallel layer of iodine atoms, and thus being electrically neutral. There are only van der Waals' forces operating between the sandwiches, and crystals having this structure show a pronounced cleavage parallel to the layers. The unsymmetrical environment of the iodine atoms shows beyond doubt that the structure is not a purely ionic one. The cadmium iodide structure occurs also in the bromide and iodide of magnesium, the iodides of calcium and many transition metals, and most metal dihydroxides. Another common layer structure is that of cadmium chloride, which differs from the cadmium iodide structure only in the relative positions of the layers. Magnesium chloride and most transition metal dichlorides have the cadmium chloride structure.

The last common structure we have chosen for illustration and discussion here is that of perovskite, $CaTiO_3$. Because of the similarity of

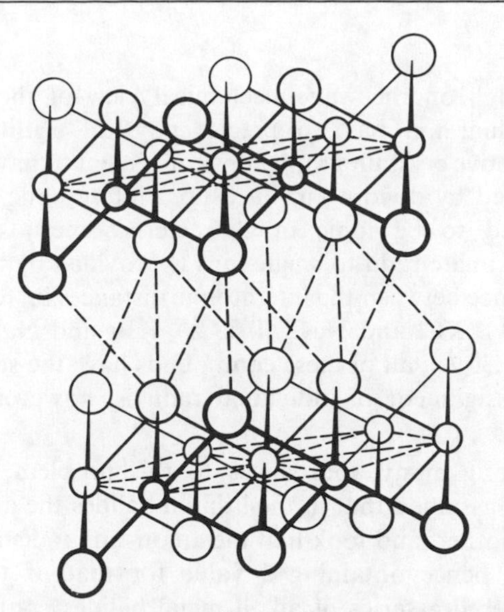

Figure 6.11
The cadmium iodide structure.

its formula to that of calcium carbonate it might be thought that perovskite is calcium titanate(IV), but on examination of the cubic unit cell of this compound, shown in Fig. 6.12, no titanate anion is discernible, and the correct name is calcium titanium(IV) oxide. In Fig. 6.12 (*a*), the titanium atom is at the centre of the cube, the calcium atoms are at the corners and oxygen atoms occupy the mid-points of all the faces. Thus the calcium atom has twelve oxygen atoms as nearest neighbours, their arrangement being that of cubic close packing; the titanium atom has six oxygen atoms, octahedrally distributed around it, as nearest neighbours; and each oxygen atom has two titanium atoms next to it, the coordination being linear. An alternative representation is shown in Fig. 6.12 (*b*) . The reader should check that in each case the content of the unit cell shown is one formula unit of $CaTiO_3$. Among other compounds having this structure (or a slightly deformed version of it) are many double oxides or fluorides such as $BaTiO_3$, $SrFeO_3$, $NaNbO_3$, $KMgF_3$ and $KZnF_3$; the structures of some high-temperature superconductors are also related to that of perovskite. Another mixed oxide structure is that of *spinel*, $MgAl_2O_4$; this structure is of special interest in transition metal chemistry and is discussed further in Section 12.8.

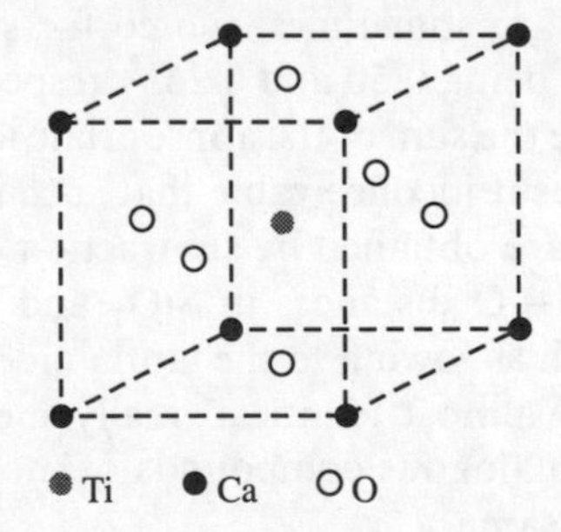

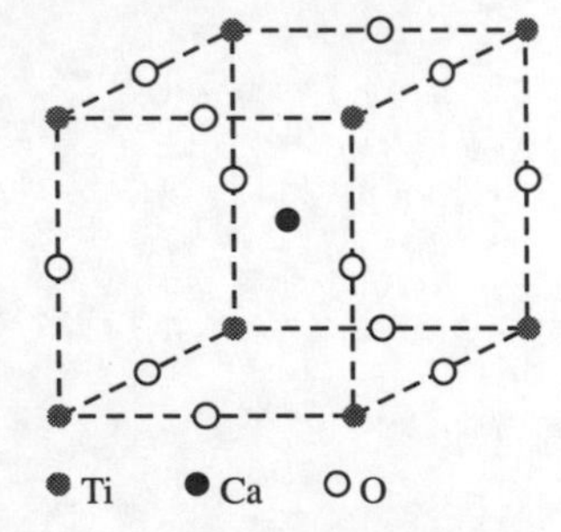

Figure 6.12
The perovskite structure. (*a*) ● Ti ● Ca ○ O (*b*) ● Ti ● Ca ○ O

6.4 Ionic radii

Although on the wave-mechanical view of the atom the radius of an individual ion has no precise physical significance, for purposes of descriptive crystallography it is convenient to have a compilation of values obtained by dividing up measured interatomic distances in compounds believed to be ionic in character. Some justification for allocating approximate radii to single ions is provided by the near-constancy of the difference between the interatomic distances in, for example, KF and NaF (0.35 Å), KCl and NaCl (0.33 Å), KBr and NaBr (0.32 Å), and KI and NaI (0.30 Å); all of these compounds have the sodium chloride structure. The assignment of individual radii is, however, a somewhat arbitrary process.

Among many approaches to this problem we may mention three. Landé assumed that in the lithium halides the anions are in contact with one another, and took half the anion–anion distances as the radius of the anion, hence obtaining a value for that of the lithium ion. Pauling considered a series of alkali metal halides containing isoelectronic ions (NaF, KCl, RbBr, and CsI) and assumed the radius of each ion to be inversely proportional to its actual nuclear charge less an estimated screening effect calculated as described in Section 3.1. Goldschmidt and, more recently, Shannon and Prewitt concentrated on the analysis of experimental data (mostly for fluorides and oxides) with the objective of obtaining a set of values which, when added, reproduced the observed values as closely as possible. In view of the approximate nature of the concept of ionic radius, no great importance should be attached to small differences in quoted values so long as self-consistency is maintained in any one set of values. Further, the fact that there are interionic distances at all implies a balance between attractive and repulsive forces, and some dependence of ionic size on coordination number would be expected unless both forces were affected to the same degree by change in the number of nearest neighbours; evidence is presented in Section 6.6 which suggests that they are not. The representative values for sixfold coordinated ions given in Table 6.2, derived from the compilation by Shannon and Prewitt, must be increased slightly for the same ions in eightfold coordination and reduced slightly for fourfold coordination. Values for some other ions are given in later chapters.

Ionic radii are sometimes quoted for species such as Si^{4+} and Cl^{7+}, typical values being 0.50 and 0.26 Å respectively. These values are highly artificial, since the sum of the appropriate ionisation energies of silicon and chlorine makes it inconceivable that such ions are really present in stable species; they are obtained by subtraction of the O^{2-} ion radius from the Si—O and Cl—O distances in SiO_2 and ClO_4^- respectively. Similarly a statement such as 'owing to the lanthanide contraction the radii of Mo^{6+} and W^{6+} are almost identical' really means only that the interatomic distances in analogous compounds of molybdenum(VI) and tungsten(VI) are nearly the same.

Table 6.2
Some representative ionic radii (Å) for sixfold coordination.

Li^+			O^{2-}	F^-			
0.74			1.40	1.33			
Na^+	Mg^{2+}	Al^{3+}	S^{2-}	Cl^-			
1.02	0.72	0.53	1.85	1.81			
K^+	Ca^{2+}	Ga^{3+}		Br^-			
1.38	1.00	0.62		1.96			
Rb^+	Sr^{2+}	In^{3+}		I^-			
1.49	1.13	0.80		2.20			
Cs^+	Ba^{2+}	Tl^{3+}					
1.70	1.36	0.88					
V^{2+}	Cr^{2+}	Mn^{2+}	Fe^{2+}	Co^{2+}	Ni^{2+}	Cu^{2+}	Zn^{2+}
0.79	0.82	0.82	0.77	0.74	0.70	0.73	0.75
Sc^{3+}	Ti^{3+}	V^{3+}	Cr^{3+}	Mn^{3+}	Fe^{3+}	Co^{3+}	Ni^{3+}
0.73	0.67	0.64	0.62	0.64	0.64	0.61	0.60

We should mention here that in the few instances in which the variation in electron density in a crystal has been accurately determined, the minimum electron density does not usually occur at distances from nuclei indicated by the ionic radii in general use; in LiF, NaCl and KBr, for example, the minima are found at 0.92, 1.18 and 1.57 Å from the nucleus of the cation, whereas the appropriate cationic radii given in Table 6.2 are 0.74, 1.02 and 1.38 Å.

These considerations make it clear that the discussion of the occurrence of crystal structures in terms of radius ratios is at best only a rough guide, and for this reason only a brief account of this subject is given in the next section. In the discussion of lattice energetics, however, only the total interionic distance is involved, and hence lattice energetics is a more satisfactory subject and is treated much more fully in the later sections of this chapter.

6.5 Radius ratio rules

If, despite the uncertainties about ionic radii discussed in the last section, we continue to consider ions as rigid spheres, the structures of many ionic crystals can be accounted for *to a first approximation* by postulating that they are determined by the relative numbers and relative sizes of the ions present. It is easy to calculate from geometrical considerations how many ions of a given size can be in contact with a single ion of smaller radius. For monatomic ions, cations are usually smaller than anions (loss of one or more electrons strengthens the hold of the nucleus on the remaining electrons; gain of one or more electrons results in an increase in size arising from interelectronic repulsion). It is customary, therefore, to

Table 6.3
The influence of radius ratios on the coordination of ions.

r_+/r_-	Permitted coordination of cation by anions	
	Number	Geometry
<0.15	2	linear
0.15–0.22	3	triangular
0.22–0.41	4	tetrahedral
0.41–0.73	6	octahedral
>0.73	8	cubic

consider the radius ratio r_+/r_- and the permitted coordination numbers and the arrangements of anions around a cation for different values of r_+/r_- are shown in Table 6.3.

In the discussion of the distribution of the structures of ionic salts in terms of radius ratios, it is assumed that maximum stability is attained when every ion is surrounded by as many as possible ions of opposite sign. This assumption is somewhat less simple to apply than it at first appears to be. In the structure of a salt of formula MX, for example, the coordination numbers of the two ions must be the same; increase in the coordination number of M^+, giving it more X^- ions as nearest neighbours, therefore inevitably leads to its having more M^+ ions as next nearest neighbours, and an increase in electrostatic attraction is thus largely cancelled by a subsequent increase in electrostatic repulsion. Further, the very high coordination numbers predicted on the basis of radius ratio considerations do not occur in practice. If M^+ and X^- are the same size it is possible to pack twelve X^- ions round a M^+ ion, but it is impossible to do this so that each X^- ion is then surrounded by twelve M^+ ions. Both of these arguments show that for a satisfactory understanding of ionic structures it is essential to consider not only the immediate environment of one ion, but also the infinite three-dimensional character of the structure; we return to this point in the following section.

Radius ratio considerations thus provide a rough guide to what is possible rather than to what actually occurs. Among the alkali metal halides, for example, only caesium chloride, bromide and iodide have the caesium chloride structure under ordinary conditions; caesium fluoride and all the other halides (including those of lithium, some of which would certainly be expected to contain only tetrahedrally coordinated cations) have the sodium chloride structure. However, when caesium chloride is sublimed on to an amorphous surface, it is obtained in the sodium chloride structure, and the interionic distance decreases from 3.57 to 3.47 Å; on the other hand rubidium chloride, bromide and iodide adopt the caesium chloride form when subjected to high pressures. Zinc oxide (wurtzite structure) adopts the sodium chloride structure under high pressures. When, on the rigid sphere model, r_+/r_- is not too different from 0.73 or 0.41, polymorphism is thus found to occur.

A good illustration of the dependence of structure on the relative sizes of the ions is provided by the fluorides BeF_2 (β-cristobalite), MgF_2 (rutile), CaF_2, SrF_2 and BaF_2 (all fluorite). In the series of oxides CO_2 (linear molecule), SiO_2 (β-cristobalite, quartz, and several other structures in all except one of which the silicon atom is fourfold coordinated), GeO_2 (dimorphic, β-cristobalite or rutile), and SnO_2 and PbO_2 (rutile), a similar increase in the coordination number of the Group IV element occurs (though solid carbon dioxide is certainly not an ionic compound). For molecular species like CO_2, and for halides having the cadmium chloride and cadmium iodide layer structures, it is often said that the ionic structure has been modified as a consequence of polarisation (the distortion of the electronic charge density round the larger anion by the smaller, more highly charged, cation). This description suffers from two limitations. First, although polarisability can be defined and measured (as we mentioned in Section 6.1), polarising power is defined in different ways by different authors and cannot be measured directly at all; the discussion of polarisation in relation to structures of ionic crystals therefore lacks a quantitative basis. Second, although considerations of polarisation may help in understanding why an ionic structure is not adopted by a particular compound, they do not help at all in trying to predict just what structure will be assumed. In this book we shall therefore make very little use of the concept of polarisation in the discussion of the structures of inorganic solids.

6.6 *Lattice energy*

Let us begin by considering a salt MX which has the sodium chloride structure with ions carrying charges z_+e and z_-e where e is the electronic charge and z_+ and z_- are integers (necessarily the same in the case of a salt of this formula, but it is desirable to develop the argument in a general way). A study of the geometry of the structure (Fig. 6.1) shows that each M^{z+} ion is surrounded by six X^{z-} ions at a distance r, twelve M^{z+} ions at $\sqrt{2}r$, eight X^{z-} ions at $\sqrt{3}r$, six M^{z+} ions at $\sqrt{4}r$, twenty-four X^{z-} ions at $\sqrt{5}r$, and so on. The coulombic energy change when a M^{z+} ion is brought from infinity to its position in the lattice is therefore given by

$$-\frac{e^2}{4\pi\varepsilon_0}\left(\frac{6}{r}z_+z_- - \frac{12}{\sqrt{2}r}(z_+)^2 + \frac{8}{\sqrt{3}r}z_+z_- - \frac{6}{\sqrt{4}r}(z_+)^2 + \frac{24}{\sqrt{5}r}z_+z_- - \cdots\right)$$

$$= -\frac{z_+z_-e^2}{4\pi\varepsilon_0 r}\left(6 - \frac{12z_+}{\sqrt{2}z_-} + \frac{8}{\sqrt{3}} - \frac{6z_+}{\sqrt{4}z_-} + \frac{24}{\sqrt{5}} - \cdots\right)$$

where ε_0 is the permittivity of a vacuum. The ratio of the charges on the ions (z_+/z_-) is constant for a given type of structure (unity for NaCl, two for CaF_2 etc.) so the series in parentheses (which slowly converges and may be summed algebraically) is a function only of the crystal geometry and, for a particular structure, is independent of z_+, z_-, and r; it is called the Madelung constant, A (after its first evaluator). Values of Madelung constants are discussed later. The coulombic energy of interaction of a mole of cations with all other ions in the crystal is thus

$$-\frac{N_A A z_+ z_- e^2}{4\pi\varepsilon_0 r}$$

The same expression gives the energy of interaction of a mole of anions with all other ions in the crystal, but we must not take it into account again, since to do so would be to count each interaction twice.

Since r is a finite quantity, the coulombic interaction must be opposed by a repulsive force which, once the ions come into contact, increases very rapidly as the internuclear distance is decreased. The simplest expression for the repulsive energy is $+N_A B/r^n$, where B is the repulsion coefficient, which is proportional to the number of nearest neighbours, and n is the so-called Born exponent, which can be evaluated from compressibility data, and generally lies between 7 and 10 but is lower if both ions have the helium electronic configuration and higher if both ions have the xenon electronic configuration (this is tantamount to saying that n shows some degree of dependence on the sizes of the ions). Thus the total interaction energy is given by

$$-\frac{N_A A z_+ z_- e^2}{4\pi\varepsilon_0 r} + \frac{N_A B}{r^n}$$

and the lattice energy (defined as the internal energy required at 0 K to convert one mole of the salt into infinitely separated ions) is

$$U_0 = \frac{N_A A z_+ z_- e^2}{4\pi\varepsilon_0 r} - \frac{N_A B}{r^n}$$

Since $\mathrm{d}U_0/\mathrm{d}r = 0$ at the equilibrium distance $r = r_0$,

$$0 = -\frac{N_A A z_+ z_- e^2}{4\pi\varepsilon_0 r_0^2} + \frac{n N_A B}{r_0^{n+1}}$$

whence

$$B = \frac{A z_+ z_- e^2 r_0^{n-1}}{4\pi\varepsilon_0 n}$$

and

$$U_0 = \frac{N_A A z_+ z_- e^2}{4\pi\varepsilon_0 r_0}\left(1 - \frac{1}{n}\right)$$

This is the Born–Landé expression for lattice energy. As for ionisation energies and electron affinities (Sections 3.4 and 3.5), if we are to be accurate corrections should be made when we consider enthalpy changes at standard temperature in place of internal energy changes at absolute zero. In practice, however, relatively little error is involved by taking U_0 as equal to U_{298} and neglecting the term $2RT$ (for a salt containing two ions) that should be added to U_{298} if we require ΔH° at 298 K for the process

$$MX(s) \rightarrow M^+(g) + X^-(g)$$

This is particularly so when, as is frequently the case, we are comparing one lattice energy with another. In the remainder of this chapter, therefore, we shall use U as the symbol for both lattice internal energy and lattice enthalpy and refer simply to lattice energy.

Because of its simplicity, the Born–Landé expression for lattice energy is the chemist's usual one; many chemical problems involve estimated lattice energies of hypothetical compounds or salts of unknown structures (or of structures for which Madelung constants are not yet available). In these circumstances it is common practice to assume $n = 9$; an error of one or two in n makes only a small difference in $(1 - 1/n)$, and this may disappear altogether if differences in lattice energies are being considered. Nevertheless, for the accurate evaluation of lattice energies several improvements may be made to the Born–Landé expression. The most important of these arises by replacing B/r^n by $Be^{-r/\rho}$, a change which reflects the fact that wave functions show exponential dependence on r. This leads to the Born–Mayer expression

$$U_0 = \frac{N_A A z_+ z_- e^2}{4\pi\varepsilon_0 r_0}\left(1 - \frac{\rho}{r_0}\right)$$

where ρ is a constant that can be expressed in terms of the compressibility. It has a value of 0.35 Å for all the alkali metal halides, but it should be noted that r_0 now appears in the repulsion energy term. Further refinements in lattice energy calculations include the incorporation of terms for the dispersion energy and the zero-point energy. In the case of sodium chloride, for which $A = 1.748$ and $r_0 = 2.82$ Å, the contributions to the total lattice energy (766 kJ mol^{-1}) made by electrostatic attraction, electrostatic repulsion, London dispersion energy and zero-point energy are +860, −99, +12 and −7 kJ mol^{-1} respectively; clearly, the error introduced by neglecting the last two terms is very small.

Lattice energies derived from Madelung constants, formal ionic charges, and interionic distances are often referred to as 'calculated' values to distinguish them from lattice energies obtained by means of thermochemical cycles. It should, however, be appreciated that values of r_0 obtained by X-ray diffraction are themselves experimental quantities, and may conceal departures from ideal ionic behaviour. In addition, the actual charges on ions may well be less than their formal charges, and the calculation of the non-electrostatic part of the lattice energy may be open to question. Nevertheless, the concept of lattice energy is of immense importance in inorganic chemistry, and it will very often be used in the later chapters of this book.

We must now return briefly to the Madelung constant. As we have stated, for the sodium chloride structure $A = 1.748$. For the caesium chloride structure $A = 1.763$. When we remember the difference in coordination number for the two structures, the similarity in their Madelung constants may seem surprising, but it is simply the consequence of the infinite nature of the structures: although the first (attraction) term in the expression for A is greater by a factor of $\frac{8}{6}$ for the caesium chloride structure, the second (repulsion) term is also greater, and so on. Values for a number of Madelung constants are given in Table 6.4. We should note that values of A for MX_2 structures are about 50 per cent higher than those for MX structures; we shall comment further on this point in Section 6.8.

To conclude this section let us look again at the polymorphism of certain alkali metal halides, taking as an example the case of caesium chloride. The question of what structure represents the most stable ionic arrangement for caesium chloride could be answered completely if we could calculate the standard free energies for different possible structures. Since the standard entropies of the different possible structures would be almost identical, it would suffice to calculate the appropriate lattice energies. The higher Madelung constant for the caesium chloride structure would favour the adoption of this rather than the sodium chloride structure, and since the dispersion energy would be larger for a structure involving more nearest neighbours this factor would also favour the eight-coordination structure. Experiment shows, however, that caesium chloride has a smaller interionic distance in the sodium chloride structure (3.474 Å) than in its normal form (3.566 Å) and this factor almost cancels the effect of the difference in Madelung constant. It thus appears that the caesium chloride structure is preferred by a small margin because of the greater dispersion energy for this structure.

The significance of the foregoing discussion is that it suggests that two forms of a salt may well have very nearly equal thermodynamic stability;

Table 6.4
Some Madelung constants.

Structure	A
CsCl	1.763
NaCl	1.748
ZnS (zinc blende)	1.638
ZnS (wurtzite)	1.641
CaF_2	2.519
TiO_2	2.408*
CdI_2	2.191*

*For these structures values vary very slightly according to the ratio of the lattice constants for the unit cell.

this has been confirmed experimentally in a few instances. If, therefore, we use lattice energy considerations to estimate the standard enthalpy or free energy of a compound, we need not worry too much about whether we have postulated the correct structure for it. This conclusion is a great encouragement to the use of lattice energy considerations in the interpretation or prediction of inorganic chemistry.

6.7 The Born–Haber cycle

The lattice energy of a salt may be related to several other quantities by means of the Born–Haber thermochemical cycle. If the anion present is a halide, all of the other quantities in the cycle have been determined independently, and we shall therefore first discuss the cycle for a metal halide of formula MX_n.

In this cycle, shown in Fig. 6.13, L is the enthalpy of sublimation of the metal, $\sum I$ the sum of its first n ionisation energies, E the electron affinity of X (or, as we defined it in Section 3.5, the enthalpy of electron capture by X), D the dissociation energy of the gaseous halogen molecule, $\Delta H_f^\circ(MX_n)$ the standard enthalpy of formation of the metal halide, and U the lattice energy of MX_n; it is assumed that the halogen is gaseous, but if it is bromine or iodine the necessary enthalpy of conversion into the gaseous state can easily be incorporated. Then

$$U + \Delta H_f^\circ(MX_n) = L + \frac{n}{2}D + \sum I + nE$$

If we take sodium chloride as a typical example, $L = 109$, $I = 496$, $D = 243$, $E = -356$ and $\Delta H_f^\circ(\text{NaCl})$, determined calorimetrically, is -411 kJ mol^{-1}, whence $U = 786$ kJ mol^{-1}. The value calculated, using the experimental value of r_0, from the Born–Mayer formula is 761 kJ mol^{-1}; a more refined calculation, the basis of which was indicated in the last section, gives 768 kJ mol^{-1}.

The agreement between the 'calculated' and 'cycle' lattice energies is as close as this for all the alkali metal halides (including those of lithium)

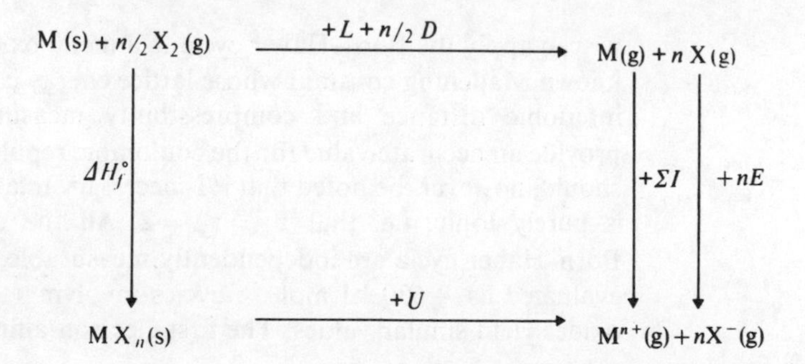

Figure 6.13
The Born–Haber cycle for an ionic halide MX_n.

and for the alkaline earth metal fluorides. Such agreement, whilst not a rigid proof that all these compounds are completely ionic in nature, at least establishes that the ionic model provides a very satisfactory basis for discussing their thermochemistry. For compounds having layer lattices the position is very different; for cadmium iodide, for example, the calculated and cycle values for the lattice energy are 1986 and 2435 kJ mol^{-1}, showing that the ionic model for this substance is unsatisfactory (a conclusion we reached earlier on the grounds that the coordination of the iodine atoms in the solid is unsymmetrical). The ionic model is similarly found to be unsatisfactory for the copper(I) halides having the zinc blende structure and for silver iodide, which has the wurtzite structure; the cycle values for the lattice energies of the silver halides become increasingly larger than the calculated values as we go from AgF to AgI, for which the extra (non-coulombic) contributions to the total lattice energy are approximately AgF, 27; AgCl, 70; AgBr, 80; AgI, 110 kJ mol^{-1}. This extra energy stabilising the solid, which increases from AgF to AgI, is the origin of the diminution in solubility of the silver halides in water as the atomic number of the halogen increases.

6.8 *Applications of lattice energetics*

In this section we shall consider a few typical applications of values of lattice energies, which may be either calculated from known structures or estimated on the basis of postulated structures and interionic distances. Many other examples will be found in later chapters.

Estimation of electron affinities and related quantities

Electron affinities can be measured only for the conversion of atoms into singly charged ions. For the estimation of the sum of the first two electron affinities of the oxygen atom, i.e. ΔH° for the gas-phase reaction

$$O + 2e \rightarrow O^{2-}$$

we can apply the Born–Haber cycle to a metal oxide having a structure of known Madelung constant whose lattice energy can be obtained from the interionic distance and compressibility measurements (the latter to provide an accurate value for the coulombic repulsion term), e.g. MgO. It should, however, be noted that it is necessary to assume that the structure is purely ionic, i.e. that $z_+ = z_- = 2$. All the other quantities in the Born–Haber cycle are independently measurable, and hence ΔH° can be evaluated as $+630$ kJ mol^{-1}; cycles involving the alkaline earth metal oxides yield similar values. The first electron affinity of oxygen (ΔH° for the reaction $O+e \rightarrow O^-$ in the gas phase) is -142 kJ mol^{-1}, so it is evident that addition of the second electron *absorbs* about 770 kJ mol^{-1}. It seems, in fact, that the only reason the O^{2-} ion exists at all is the high lattice energy which is liberated when it is incorporated into a solid ionic oxide; the lattice energy of MgO, for example, is calculated to be approximately 3900 kJ mol^{-1}.

A somewhat similar method may be used to estimate the fluoride ion affinity of boron trifluoride, i.e. ΔH° for the gas-phase reaction

$$BF_3 + F^- \rightarrow BF_4^-$$

from the known lattice energy of KF, the calculated lattice energy of KBF_4 (based on the assumption that the anion can be treated as a hollow sphere whose charge resides at its centre, an assumption supported by the fact that the high-temperature form of KBF_4 has the CsCl structure) and ΔH° for the reaction

$$KBF_4(s) \rightarrow KF(s) + BF_3(g)$$

determined from the temperature variation of the dissociation pressure of solid KBF_4. In this way the fluoride ion affinity of boron trifluoride is found to be about -360 kJ mol^{-1}.

Estimation of proton affinities

As an example let us consider the estimation of the proton affinity, P, of ammonia, i.e. ΔH° for the gas-phase reaction

$$NH_3 + H^+ \rightarrow NH_4^+$$

This may be done from the cycle shown in Fig. 6.14, for which we need the lattice energy of ammonium chloride, its enthalpy of formation from ammonia and hydrogen chloride, the dissociation energy of hydrogen chloride, the ionisation energy of hydrogen and the electron affinity of chlorine. If we calculate the lattice energy of ammonium chloride from structural data to be 653 kJ mol^{-1}, the proton affinity of ammonia is

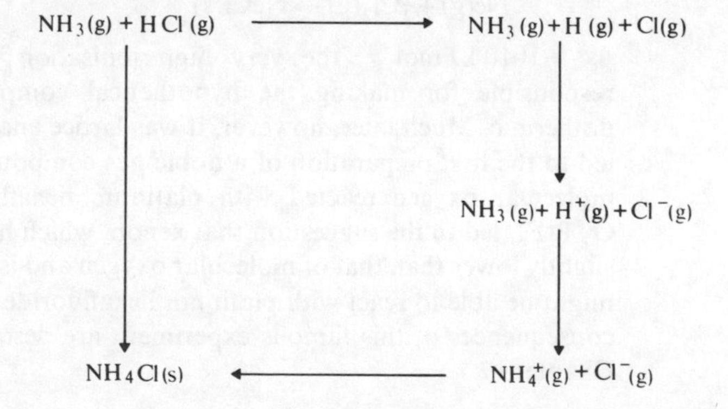

Figure 6.14
Cycle for the estimation of the proton affinity of ammonia.

found to be -895 kJ mol^{-1}, similar values being obtained from analogous cycles involving ammonium bromide or iodide. A cycle based on ammonium fluoride, however, yields a different value; as we mentioned in Section 6.3, ammonium fluoride has the wurtzite structure, presumably because of strong N—H—F hydrogen bonding, which makes a substantial extra contribution to the lattice energy. The proton affinity of phosphine (-816 kJ mol^{-1}) may be obtained similarly from the lattice energy of phosphonium iodide.

The proton affinity of water is obtained from an analogous cycle based on $H_3O^+ClO_4^-$, the corresponding cycle for $NH_4^+ClO_4^-$ (which is isomorphous and almost isodimensional with the former compound), and the proton affinity of ammonia determined as above. This method leads to a value of -762 kJ mol^{-1}. It is interesting to note that this value is lower than that for phosphine, which suggests that the reason for the decomposition of phosphonium salts by water lies in the more negative enthalpy of further hydration of the H_3O^+ ion in aqueous solution.

Estimation of enthalpies of formation and disproportionation

It is seldom the case that for a well-known ionic compound the lattice energy is known and the standard enthalpy of formation is not, so the main application of the Born–Haber cycle in this context is to the estimation of the enthalpies of formation of unknown compounds on the basis of reasonable estimates of what their lattice energies would be. The earliest example of this method was to the question of whether it was conceivable that neon would form a saline chloride Ne^+Cl^-; on the assumption that Ne^+ would be a little smaller than Na^+ and that the chloride, if formed, would have the sodium chloride structure, its lattice energy was estimated as about 840 kJ mol^{-1}. Combination of this value with the ionisation energy of neon (2080 kJ mol^{-1}), half the dissociation energy of chlorine, and the electron affinity of chlorine, gives ΔH° for the reaction

$$Ne(g) + \tfrac{1}{2}Cl_2(g) \rightarrow NeCl(s)$$

as $+1010$ kJ mol^{-1}, the very high ionisation energy of neon being responsible for making the hypothetical compound so strongly endothermic. Much later, however, it was lattice energy considerations that led to the first preparation of a noble gas compound. The discovery that molecular oxygen reacted with platinum hexafluoride to form a salt $O_2^+PtF_6^-$ led to the suggestion that xenon, which has an ionisation energy slightly lower than that of molecular oxygen and is of about the same size, might be able to react with platinum hexafluoride to form $Xe^+PtF_6^-$; the consequences of this famous experiment are described in some detail in Chapter 17.

As a final example we may consider the possibility of forming a saline monofluoride of calcium. In this case the simple Born–Haber cycle is not helpful, since the compound is not unstable with respect to decomposition into calcium and fluorine, but with respect to *disproportionation* into calcium and calcium difluoride:

$$2CaF(s) \rightarrow Ca(s) + CaF_2(s)$$

The cycle to be considered is shown in Fig. 6.15, from which it can be seen that the quantities involved are the enthalpy of sublimation of calcium (which is relatively small), both ionisation energies, the known lattice energy of CaF_2, and the estimated lattice energy of CaF. The first and second ionisation energies of calcium are 590 and 1146 kJ mol^{-1}, so ΔH° for the process

$$2Ca^+(g) \rightarrow Ca(g) + Ca^{2+}(g)$$

will be +556 kJ mol^{-1}; gaseous Ca^+, therefore, will not disproportionate, and the possibility of obtaining Ca^+ in a solid fluoride depends mainly upon the relative magnitudes of the lattice energies of CaF and CaF_2. That of the latter will be the greater for three reasons: z_+ is twice as large as in the former compound; Ca^{2+} is smaller than Ca^+ would be; and the Madelung constants for MX_2 structures are about one and a half times those for MX structures (cf. Table 6.4). Thus the lattice energy of CaF would be rather less than one-third of that of CaF_2, which is 2 610 kJ mol^{-1}; the extra lattice energy of CaF_2 over that of 2CaF therefore outweighs the effects of the values of the ionisation energies. It is not difficult to see that CaI would be less unstable with respect to disproportionation than CaF.

Attempts have been made to utilise the fact that the Madelung constants for MX and MX_2 structures are roughly in the ratio of 2:3 (i.e. are proportional to the number of ions in the formula) to produce a general equation for lattice energies which could be used in estimating unknown values. The best known of these is Kapustinskii's equation

$$U_0 = \frac{0.874 N_A \nu z_+ z_- e^2}{4\pi\varepsilon_0(r_+ + r_-)}\left(1 - \frac{1}{n}\right)$$

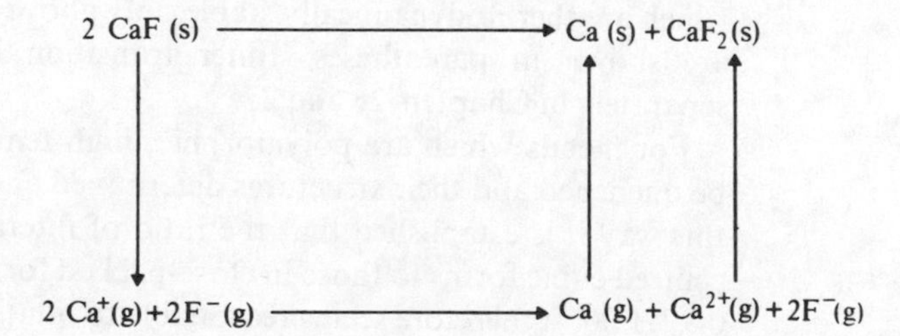

Figure 6.15
Cycle for the disproportionation of calcium monofluoride.

or, taking $n = 9$ (the value for sodium chloride), and expressing U_0 in kJ mol^{-1},

$$U_0 = \frac{1071 v z_+ z_-}{r_+ + r_-}$$

In the former equation, 0.874 is half the Madelung constant for the sodium chloride structure, and in both equations v is the number of ions in the formula of the salt; r_+ and r_- are empirical radii for sixfold coordination – not always the radii derived from X-ray diffraction data, since their purpose is to enable U to be calculated for unknown structures; they are known as 'thermochemical radii'. An equation of this kind is often useful in estimating unknown lattice energies, but in using it the assumptions underlying its formulation are apt to be forgotten, and the results to be accepted as more reliable than they really are. In general, therefore, the use of such an equation at the level of this book is to be avoided as far as possible.

6.9 Metals

Most metals crystallise in the cubic close-packed, the hexagonal close-packed, or the body-centred cubic structure; many are polymorphic and are known in more than one structure. As we have stated earlier, on the basis of the hard sphere model the close-packed structures represent the most efficient utilisation of space, with a common packing efficiency of 74 per cent. The body-centred cubic structure is not much less efficient in packing terms, for although the coordination number of each atom is only eight, there are six more neighbours only 15 per cent further away, and the packing efficiency is 68 per cent. Among the few metals having other structures are mercury (the only liquid metal at normal temperature), zinc and cadmium at ordinary temperatures, gallium and indium, tin, and antimony and bismuth. In all of these the coordination number is below eight, and it may be noted that they are some of the most volatile metals. Manganese at ordinary temperatures also has an atypical structure, but in this case the complex lattice contains atoms in three different environments all having coordination numbers of twelve or more. The structures of most of the metallic elements are summarised in Table 6.5, where forms which are thermodynamically stable only above the ordinary temperature are shown in parentheses. Inner transition elements are dealt with separately in Chapters 26 and 27.

For metals which are polymorphic, high-temperature forms can often be quenched and their structures determined at ordinary temperatures. In this way it is established that the ratio of interatomic distances in body-centred cubic forms to those in close-packed forms of the same elements is 0.97:1.00. It therefore seems reasonable to multiply the metallic radii (half

Table 6.5
Structures of some metallic elements*.

Typical elements

Li	Be		
bcc	hcp		
(ccp, hcp)			
Na	Mg	Al	
bcc	hcp	ccp	
(ccp, hcp)			
K	Ca	Ga	
bcc	ccp	X	
	(hcp)		
Rb	Sr	In	Sn
bcc	ccp	X	X
	(hcp)		
Cs	Ba	Tl	Pb
bcc	bcc	hcp	ccp

Transition elements

Sc	Ti	V	Cr	Mn	Fe	Co	Ni	Cu	Zn
hcp	hcp	bcc	bcc	X	bcc	ccp	ccp	ccp	X
(bcc)	(bcc)			(bcc, ccp)	(hcp)	(hcp)			(hcp)
Y	Zr	Nb	Mo	Tc	Ru	Rh	Pd	Ag	Cd
hcp	hcp	bcc	bcc	hcp	hcp	ccp	ccp	ccp	X
(bcc)	(bcc)								(hcp)
La	Hf	Ta	W	Re	Os	Ir	Pt	Au	Hg
hcp	hcp	bcc	bcc	hcp	hcp	ccp	ccp	ccp	X
(ccp, bcc)	(bcc)								

*(bcc = body-centred cubic; hcp = hexagonal close-packed; ccp = cubic close-packed; X = atypical metal structure).

the interatomic distances) of metals known only in the body-centred cubic structure by 1.00/0.97 to arrive at the set of metallic radii for twelve-coordination shown in Table 6.6; in this table, of course, no value can be given for mercury. The very close similarity in values for metals of the second transition series and their analogues in the third transition series should be noted; the reasons for, and consequences of, this close similarity are discussed in Chapter 25.

Table 6.6
Some metallic radii for twelve-coordination (Å).

Li	Be										
1.57	1.12										
Na	Mg										
1.91	1.60										
K	Ca	Sc	Ti	V	Cr	Mn	Fe	Co	Ni	Cu	Zn
2.35	1.97	1.64	1.47	1.35	1.29	1.37	1.26	1.25	1.25	1.28	1.37
Rb	Sr	Y	Zr	Nb	Mo	Tc	Ru	Rh	Pd	Ag	Cd
2.50	2.15	1.82	1.60	1.47	1.40	1.35	1.34	1.34	1.37	1.44	1.52
Cs	Ba	La	Hf	Ta	W	Re	Os	Ir	Pt	Au	Hg
2.72	2.24	1.87	1.59	1.47	1.41	1.37	1.35	1.36	1.39	1.44	—

The metals in Table 6.6 which have the largest radii (implying the weakest bonding) are the alkali and alkaline earth metals, which have much lower melting- and boiling-points than most transition metals. Further, the values for the enthalpies of sublimation of the metals at 25 °C (often referred to as enthalpies of atomisation, since the vapours are monatomic) shown in Fig. 6.16 show that these metals are the ones which are vaporised with the minimum energy absorption; since the sublimation enthalpy appears not only in the Born–Haber cycle but also in cycles for the formation of cations in solution (Section 7.11), it is clear that this fact is extremely important in accounting for the reactivity of the metals. In general, there appears to be a rough correlation between the enthalpy of sublimation and the number of unpaired electrons (or of electrons that can become unpaired if one is promoted to produce a relatively low-lying excited state of the atom). Thus in any long period of elements (if we exclude manganese, which has an atypical structure, from consideration) the maximum values are reached at around the middle of the transition series: before this there are not enough valence electrons, after this too many occupy antibonding orbitals and weaken the bonding. As we explained in Section 4.9, however, the theory of the metallic bond has not

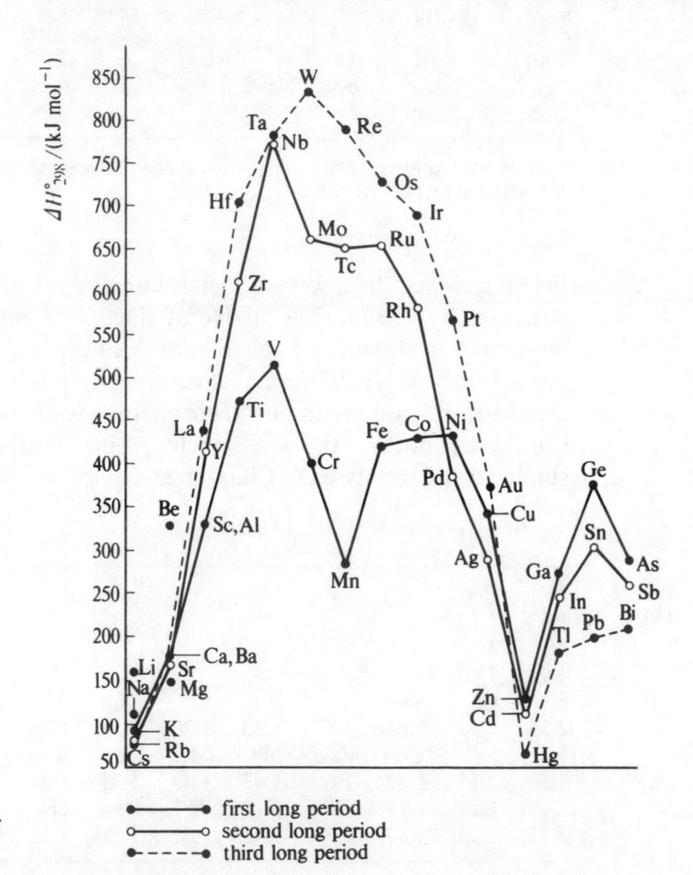

Figure 6.16
Enthalpies of sublimation of metals.

yet been developed to a state at which it is possible to predict in detail the observed pattern of behaviour.

6.10 The defect solid state

So far in this chapter we have assumed implicitly that all the solids considered have ideal lattices in which every site is occupied by the right kind of atom or ion. Such a state of affairs, however, represents equilibrium only at absolute zero. Above this temperature, defects are always present: although energy has to be supplied to create them, this is more than compensated by the resulting increase in the entropy of the structure. Defect formation frequently leads to the development of semiconducting properties; since the concentration of defects increases rapidly with increase in temperature, so does semiconductivity. In this way it is distinguished from metallic conductivity; according to the modern theory of metals, a perfect metallic conductor at absolute zero should have zero resistance, since resistance arises from the scattering of electrons by lattice vibrations and imperfections; these increase in concentration, owing to the thermal motions of the atoms, as the temperature is raised, and increase in temperature is accompanied by a decrease in metallic conduction. In the case of a semiconductor, these effects of increase in temperature are usually far outweighed by the increase in the carrier concentration.

There are two types of defect which are inherent in all infinite lattices. A *Schottky defect* consists of a vacant atom site (for a metal), a vacant cation site and a vacant anion site (for a salt of formula type MX), or a vacant cation site and two vacant anion sites (for a salt of formula MX_2); in each case the atom or ion is removed to the surface of the crystal. In a *Frenkel defect* an atom or ion occupies a normally vacant interstitial site, leaving its proper lattice site vacant. Frenkel defects are commonest in ionic crystals in which the anion is much larger than the cation. The formation of both types of defect is illustrated in Fig. 6.17.

Among several methods that may be used to study the occurrence of Schottky and Frenkel defects in stoichiometric crystals (i.e. crystals of ideal composition) the simplest in principle is extremely accurate density measurements. Low concentrations of Schottky defects lead to a decrease in pycnometrically determined density relative to that calculated from X-ray data on the size and content of the unit cell; low concentrations of Frenkel defects, on the other hand, leave the density unchanged. In this way it is found that, for example, metallic copper, silver and gold have Schottky defects and silver chloride and bromide have Frenkel defects.

There are many inorganic compounds that have variable composition and exhibit maxima or minima in physical properties (e.g. melting-point,

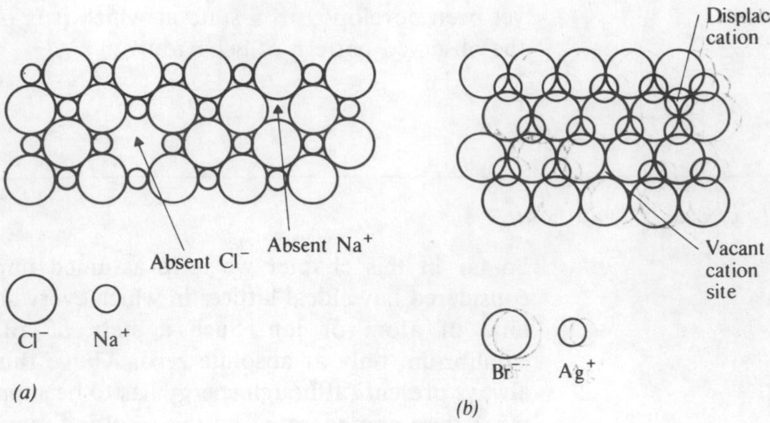

Figure 6.17
(*a*) A Schottky defect in a layer of sodium chloride.
(*b*) A Frenkel defect in a layer of silver bromide.
Such defects need not be restricted to one layer as shown here.

conductivity, or lattice order) at atomic proportions which are not simple ratios; such species are known as *non-stoichiometric compounds*. The incorporation of an excess of an element B in a phase of ideal composition AB_n can be accomplished in three ways: substitution of B atoms for A atoms, found only when both A and B are metals or metalloids, so that ionic repulsions are not involved; interstitial incorporation of extra B atoms, which is possible only when B is considerably smaller than A; and subtractive incorporation, the leaving vacant of some A lattice sites whilst all B lattice sites are occupied. Interstitial incorporation of an ion results in a pycnometric density which is higher than that calculated on the basis of X-ray data: when zinc oxide is heated, for example, it turns yellow owing to the loss of a little oxygen and the interstitial incorporation of the zinc atoms that are produced, the resulting composition at 800 °C being $Zn_{1.00007}O$. Iron(II) oxide and iron(II) sulphide provide examples of subtractive incorporation: both have variable composition, but density measurements show them to be deficient in iron rather than rich in sulphur, and their composition ranges are therefore written $Fe_{0.84-0.94}O$ and $Fe_{0.88-1.00}S$. Cadmium oxide resembles zinc oxide by losing oxygen when heated, but in this case the product, $Cd_{0.9995}O$ at 650 °C, is deficient in metal; this may well be due to the fact that cadmium oxide has the sodium chloride structure, in which there is less room for interstitial cations than in the zinc blende structure; further, the Cd^{2+} ion is appreciably larger than the Zn^{2+} ion.

Such deviation from the stoichiometry of an ideal crystal must necessarily involve the presence of the metal in two different oxidation states in order to keep the whole crystal electrically neutral. A sample of iron(II) oxide of composition $Fe_{0.90}O$, for example, is really $Fe^{2+}_{0.70}Fe^{3+}_{0.20}O$. The presence of cations of the same element in different oxidation states and crystallographically equivalent positions leads to electronic semiconduction. Where conductivity arises from excess of negative charge, the substance is called a n-type semiconductor; where there is a deficit of negative charge or an excess of positive charge, it is a p-type or positive hole semiconductor. In each case the conductivity, like that of other semiconductors, increases exponentially with rise in temperature. Investigation of the variation in conductivity with pressure of oxygen over the solid at constant temperature provides an alternative to density measurements for determining the nature of non-stoichiometry. For zinc oxide, increase in oxygen pressure lowers the conductivity, which means that it is due to the presence of species with more electrons than the Zn^{2+} ion; for copper(I) oxide, increase in oxygen pressure increases the conductivity. Thus zinc oxide and copper(I) oxide are n-type and p-type semiconductors respectively. For the p-type semiconductor $Li_xNi_{1-x}O$, made by heating a mixture of Li_2O and NiO in air at 1 200 °C, Li^+, Ni^{2+} and Ni^{3+} are all of about the same size, and a range of compositions, giving rise to a range of specific conductivity from 1 to 10^{-10} ohm^{-1} cm^{-1}, can be obtained. The relatively high temperature superconductors containing Cu^{2+} and Cu^{3+} (e.g. $La_{2-x}Ba_xCuO_{4-x}$ and $YBa_2Cu_3O_7$) are much more complex in structure; they are described in Section 24.10.

Defect structures which result from the planned presence of impurities are of outstanding importance in solid-state electronic technology. A crystal of very pure germanium or silicon, for example, can be 'doped' with a trace of Group III element, giving a p-type extrinsic semiconductor, or with a trace of a Group V element, giving a n-type extrinsic semiconductor. Similar semiconductors may be made from 'doped' gallium arsenide, GaAs, a Group III–Group V (or, more succinctly, III–V) compound. The whole of transitor technology rests on combinations of n- and p-type semiconductors.

6.11 Band theory of inorganic compounds

When we introduced band theory in Section 4.9 we did so for metals, for which neither an ionic nor a covalent model is satisfactory. The band model is of little use in the treatment of thermodynamic (and therefore of chemical) properties, but it provides the best basis for the discussion of the electronic properties of many (though not all) inorganic solids, and it is mainly for this reason that we return to it now.

Metals are characterised by a band structure in which the valence band (the highest energy occupied band) is only partly full. The energy level of the highest occupied orbital at absolute zero is called the *Fermi level*; it lies at the centre of the band. In the metal at temperatures above absolute zero some levels just below this are vacant and some just above it are occupied. Electrons in singly occupied levels close to the Fermi level are the cause of metallic conductivity. In lithium, the 2*s* band is half full and is the origin of the conductivity. In beryllium, a filled 2*s* band would make the metal an insulator if the 2*p* band were of much higher energy or a semiconductor if the energy gap between the bands were small; in reality, however, the 2*s* and 2*p* bands (both broad) overlap to some extent, and beryllium is metallic.

In the macromolecular structures of diamond, silicon, germanium and α-tin each atom is in the tetrahedral sp^3 hybridised state, and hybrid orbitals on two atoms can interact to form either a bonding or an antibonding orbital. Overlapping of the bonding orbitals throughout the structure leads to formation of the (full) valence band; overlapping of the antibonding orbitals gives the (empty) conduction band. The energy required to promote an electron across the band gap may be determined spectroscopically; it is approximately 530, 110, 70 and 10 kJ mol^{-1} for C, Si, Ge and Sn respectively. The very large band gap for diamond makes it an insulator; for the other three elements, some electrons have sufficient thermal energy to cross the gap and the elements are *intrinsic semiconductors*, the extent of occupation of the conduction bands increasing with increase in temperature. Electrons in the conduction band are the origin of n-type semiconductivity; the positive holes left in the valence band are the origin of p-type semiconductivity. The 1:1 compounds formed by aluminium, gallium or indium with phosphorus, arsenic or antimony have the zinc blende structure and thus are similar to diamond both structurally and in the average number of valence electrons per atom. They are intrinsic semiconductors with band gaps which decrease with increase in atomic number, e.g. AlP, 240; GaAs, 140; InSb, 20 kJ m^{-1}. The uses of these compounds in light-emitting or -absorbing devices involving the conversion of electrical energy into light, or vice versa, are mentioned in Section 12.9.

In sodium chloride (for which, as we have seen earlier, the ionic model holds) both ions have noble gas configurations. The valence band, formed from the 3*s* and 3*p* orbitals of the Cl^- ions, is full; the conduction band, formed from the 3*s* orbitals of the Na^+ ion, is empty. The band gap is very large (about 800 kJ mol^{-1}) and the solid is an insulator; it is also white, because the amount of energy needed to promote an electron from the valence band to the conduction band is available only in the ultraviolet. The other alkali metal halides and alkaline earth halides are similar, though the band gap shows a general decrease with increase in interionic distance: caesium iodide is relatively easily made metallic at high pressure.

Compounds of zinc, cadmium or mercury with sulphur, selenium or tellurium have the zinc blende structure and are intermediate between alkali or alkaline earth metal salts and aluminium phosphide and its homologues; e.g. for CdS, CdSe and CdTe the band gap is 240, 180 and 140 kJ mol^{-1} respectively. All are coloured and are intrinsic semiconductors.

The factors which influence the size of the band gap appear to be several in number: the homopolar bond energy in the series C, Si, Ge, Sn; the lattice energy in the alkali metal halides; the difference in electronegativity in other series. No prediction of the magnitude of the band gap is yet possible.

Most transition metal cations have partly filled *d* shells, and the interaction of *d* orbitals can then give rise to the presence of electons in a partly filled *d* band and metallic conductivity: this occurs for the high-temperature form of VO_2 and ReO_3, both containing d^1 cations; the isostructural d^0 compounds TiO_2 and WO_3 are insulators. It is, however, also necessary that the transition metal cations should be near enough together for overlapping of orbitals to occur: salts containing aquo or complex ions, for example, are insulators irrespective of the presence of partly filled *d* orbitals. Further, electrons in different *d* orbitals are not necessarily equivalent in giving rise to metallic conductivity.Observations in the series of transition metal monoxides (all having the NaCl structure) are complicated by the occurrence of non-stoichiometry; nevertheless, the fact that NiO (containing a d^8 cation) is an insulator whilst TiO and VO (d^2 and d^3) are metallic conductors points strongly towards the importance of crystal field effects and orbital ovcrlap in this matter. Further comment on this subject will be made in Sections 19.3 and 24.9.

The properties of doped extrinsic semiconductors can also be interpreted on band theory. In gallium-doped silicon, for example, one in every four Ga—Si bonds can involve only a single electron. The energy level associated with such a bond does not form part of the valence band of silicon, but instead forms a discrete level just above the top of the valence band—an *acceptor level,* since it can accept another electron. The gap between these levels and the top of the valence band is very small (about 10 kJ mol^{-1}) and electrons can be promoted by thermal energies. The acceptor levels remain discrete if the concentration of gallium atoms is low, and in these circumstances the electrons in them do not contribute directly to conductance; however, the positive holes left behind can move and provide the conductivity, which is very much greater than that obtained by promoting electrons right up to the conduction band. In arsenic-doped

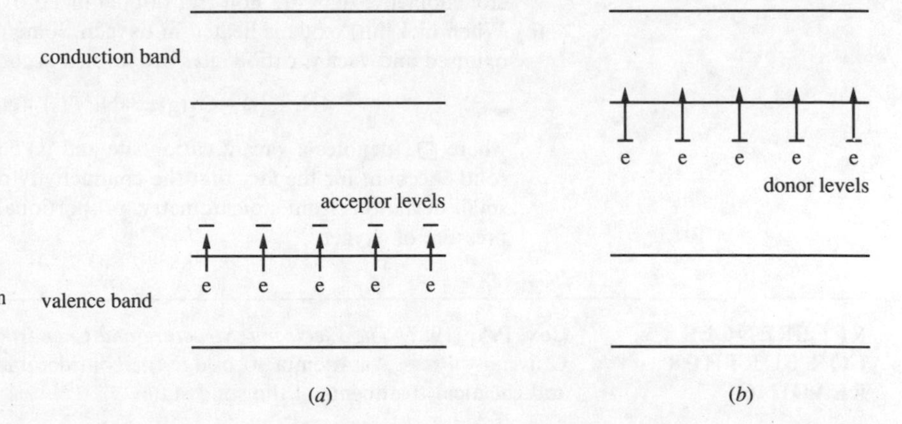

Figure 6.18
Origin of (*a*) p-type extrinsic semiconductivity in gallium-doped silicon, (*b*) n-type extrinsic semiconductivity in arsenic-doped silicon.

silicon, the extra electron occupies a discrete level about 10 kJ mol^{-1} below the bottom of the conduction band – a *donor level*, because thermal energy is now sufficient to move the electron up into the conduction band, where it is free to move. The origin of *extrinsic conductivity* in these examples may be represented by the energy level diagrams shown in Fig. 6.18. Since the numbers of acceptor or donor levels introduced are proportional to the concentration of gallium or arsenic in the silicon, the conductivity of extrinsic semiconductors can be controlled by controlling the dopant concentration.

PROBLEMS

1 How many molecules of CaF_2 are there in the unit cell of fluorite? If the density of fluorite is 3.18×10^3 kg m^{-3} and the interionic distance is 2.37 Å, calculate Avogadro's constant ($CaF_2 = 78.1$).

2 Use the Born–Landé equation to calculate the lattice energy of KBr, for which $r_0 = 3.28$ Å and $n = 9.5$.

3 How would you attempt to estimate:
(i) the electron affinity of hydrogen;
(ii) the standard enthalpy of formation of $Na^{2+}(F^-)_2(s)$;
(iii) the standard enthalpy change of the process CsCl(ordinary form) → CsCl(NaCl form)?

4 Discuss the interpretation of each of the following observations:
(a) ΔH_f° becomes less negative along the series LiF, NaF, KF, RbF, CsF but more negative along the series LiI, NaI, KI, RbI, CsI.
(b) LiF and MgO are isostructural and almost isodimensional, but a crystal of MgO is much harder than one of LiF.
(c) $LiFeO_2$ and Li_2TiO_3 form a continuous range of solid solutions with MgO but not with ZnO.
(d) The thermal stability of the isomorphous sulphates of calcium, strontium and barium with respect to decomposition into metal oxide and sulphur trioxide increases in the sequence $CaSO_4$, $SrSO_4$, $BaSO_4$.

5 (a) Why are transition metal oxides much more frequently non-stoichiometric than are non-transitional metal oxides?
(b) When nickel(II) oxide is heated in oxygen, some of the cations are oxidised and vacant cation sites are formed according to the equation

$$4Ni^{2+}(s) + O_2(g) \rightleftharpoons 4Ni^{3+}(s) + 2\square_+ + 2O^{2-}(s)$$

where $\square_+$ denotes a vacant cation site and (s) denotes an ion in the solid. Account for the fact that the conductivity of the product is, for small deviations from stoichiometry, proportional to the sixth root of the pressure of oxygen.

REFERENCES FOR FURTHER READING

Cox, P.A. (1987) *The Electronic Structure and Chemistry of Solids*, Oxford University Press. An attempt to bridge the considerable gap between physical and chemical treatments of the solid state.

Duffy, J.A. (1990) *Bonding, Energy Levels and Bands in Inorganic Solids*, Longmans. A similar attempt.

Johnson, D.A. (1982) *Some Thermodynamic Aspects of Inorganic Chemistry*, 2nd edition, Cambridge University Press. A rather fuller account of lattice energetics than the one in this book.

Ladd, M.F.C. (1979) *Structure and Bonding in Solid State Chemistry*, Ellis Horwood, Chichester. A more mathematical and physical approach.

Shriver, D.F., Atkins, P.W. and Langford, C.H. (1990) *Inorganic Chemistry*, Oxford University Press. Chapter 18 gives a very good account of the defect solid state.

Wells, A.F. (1984) *Structural Inorganic Chemistry*, 5th edition, Oxford University Press. The definitive account of the geometry of inorganic structures, with a wealth of beautiful illustrations.

West, A.R. (1988) *Basic Solid State Chemistry*, Wiley, Chichester. The best general introductory account of the solid state.

7 Inorganic chemistry in aqueous media

7.1 Introduction

Although it is now recognised that many of the properties of water as a solvent and as a medium for inorganic reactions are to some extent shared by other liquids, the study of aqueous solutions still commands a special place in inorganic chemistry. This is, of course, partly because the availability of water is enormously greater than that of any other solvent; but it is also partly because of the abundance of accurate physicochemical data for aqueous solutions and their relative scarcity for solutions in non-aqueous media. It is therefore logical to discuss inorganic chemistry in aqueous media (a relatively precise branch of science) before turning to the often fascinating, but less exact, subject of non-aqueous solutions of inorganic substances. It should be noted that this chapter is concerned almost entirely with equilibria; the kinetics and mechanisms of reactions in aqueous media are dealt with under individual elements or groups of elements, especially in Chapter 21.

Liquid water is approximately 55 molar H_2O, a fact commonly overlooked in the study of classical physical chemistry, in which the concentration (or activity) of liquid water is by convention taken as unity. Except for those of a few compounds that undergo extensive hydrogen bonding to water, solubilities in this solvent are usually, in mole fraction units, relatively low; even in concentrated solutions most of the molecules present are water molecules. We therefore begin this chapter with a survey of the physical properties of water; some of these, together with corresponding properties of methanol and dimethyl ether, the mono- and di-methyl derivatives of water, are presented in Table 7.1.

The structure of liquid water is still uncertain, but we know that of ice; X-ray diffraction studies do not locate the hydrogen atoms, but this can be done by a neutron diffraction investigation of solid deuterium oxide. In the common form of ice, the oxygen atoms occupy the same positions as the zinc and sulphur atoms in the wurtzite structure (Section 6.3). Each oxygen atom is involved, through its two lone pairs of electrons and its two bonded hydrogen atoms, in four hydrogen bonds. The hydrogen atoms are situated slightly off the lines joining the oxygen atoms, the H—O—H angle being 105°, almost the same as in the gaseous molecule of water. The hydrogen bonding is not symmetrical, the O—H distance being 1.01 Å, a little greater than that in water vapour (0.96 Å); the overall O—H···O distance is 2.76 Å. The wurtzite structure is a very open one (hence the low density of ice); on melting, a partial breakdown of the hydrogen-bonded network occurs, and some individual water molecules are accommodated in the cavities in the structure, with a consequent increase in density. Up to 4 °C (the temperature of maximum density) this effect outweighs that of thermal expansion, but above 4 °C the latter effect prevails and the density falls. Even at boiling point, however, much of the hydrogen bonding remains, as is shown by the high enthalpy and entropy of vaporisation. The strength of the hydrogen bond in ice or water has been estimated as about 25 kJ mol^{-1}. Hydrogen bonds are continually being made and broken; the lifetime of a water molecule in a particular environment is estimated to be only about 10^{-11} s.

It is the hydrogen bonding that is mainly responsible for the very low solubility of non-polar molecules in water. The mixing of two substances is always accompanied by an increase in entropy, and on these grounds is always favoured. In the case of two liquids forming an ideal solution (i.e. one which obeys Raoult's law) there is no enthalpy of mixing; for gases or solids forming ideal solutions in liquids (e.g. nitrogen or naphthalene in hydrocarbons), the only enthalpy change is that associated with conversion of the solute into the liquid phase, and the solubility

Table 7.1
Some physical properties of water, methanol and dimethyl ether.

	H_2O	CH_3OH	$(CH_3)_2O$
Boiling-point/°C	100	64.7	−23.7
Melting-point/°C	0	−97.8	−138.5
Enthalpy of vaporisation/kJ mol^{-1}	40.67	35.23	21.50
Enthalpy of fusion/kJ mol^{-1}	6.02	2.17	5.05
Entropy of vaporisation/J K^{-1} mol^{-1}	109	102	86
Relative permittivity at 25 °C	78.5	32.6	5.0
Dipole moment/debyes	1.84	1.68	1.30

(expressed as a mole fraction) is independent of the solvent chosen and may be calculated fairly accurately by a straightforward application of the second law of thermodynamics. When water acts as a solvent, however, hydrogen bonds have to be destroyed, and unless the cross-interaction between the structural units (molecules or ions) in a substance and water is stronger than both the interaction between the structural units in the substance and the interaction between the molecules in liquid water, dissolution of the substance will be associated with an increase in the enthalpy of the system. Thus most molecular species that are readily soluble in water contain a high proportion of polar bonds (e.g. H—F, O—H, N—H) and take part in hydrogen bonding to solvent molecules. In the case of ionic species, solubility is determined by a delicate balance between lattice energy and hydration energies (resulting from ion–dipole interactions), and is discussed in the following section. The high relative permittivity (dielectric constant) of water plays an important part in making hydration energies of ions large and interionic forces in solution relatively small.

Water is ionised to a small extent according to the equation

$$2H_2O \rightleftharpoons H_3O^+ + OH^-$$

On the classical (Arrhenius) theory, an acid is a compound that produces protons in aqueous solution, a base a compound that produces hydroxide ions. The role of the solvent is more clearly shown in Brønsted's theory of acids and bases, according to which an acid is a species that can act as a proton donor and a base a species that can act as a proton acceptor. For example, when hydrogen chloride (a non-electrolyte) is dissolved in water the following equilibrium is established:

$$HCl + H_2O \rightleftharpoons H_3O^+ + Cl^-$$

In this reaction water is acting as a base; hydrogen chloride is a much stronger acid than water, and the equilibrium position is far to the right. Alternatively, we may say that since water competes successfully with chloride ion for most of the protons in the system it is a much stronger base than chloride ion. In an aqueous solution of ammonia, on the other hand, water acts as an acid in the equilibrium

$$NH_3 + H_2O \rightleftharpoons NH_4^+ + OH^-$$

The difference in acid strengths between H_2O and NH_4^+, or in base strengths between NH_3 and OH^- is, however, relatively small, so that in aqueous solution NH_3 may be described as a weak base or NH_4^+ as a weak acid. Every acid is related to a conjugate base, and every base to a conjugate acid, by the relationship

$$\text{Acid} \rightleftharpoons H^+ + \text{Base}$$

Thus in the ionisation of hydrogen chloride, Cl^- is the conjugate base of HCl and HCl the conjugate acid of Cl^-. The Brønsted theory is applicable to all proton-containing systems, and we shall meet it again (together with still wider definitions of acids and bases) in the following chapter.

In aqueous systems the H_3O^+ ion is further hydrated, and it is then often represented simply as $H^+(aq)$, a procedure we shall follow in this book; this symbol stands for H_3O^+ and all the other hydrated protonic species that may be present in a particular solution, e.g. $H_5O_2^+$, $H_7O_3^+$ and $H_9O_4^+$. The $H_9O_4^+$ ion, a H_3O^+ ion hydrogen-bonded to three water molecules, is a well-established entity of widespread occurrence.

7.2 *Conventions and units in aqueous solution chemistry*

Before we discuss equilibria in aqueous media it is desirable to remind the reader of the physicochemical conventions and units generally employed in the study of aqueous solutions. Failure to appreciate that in some respects these are not the same as those used in most other branches of chemistry is the cause of much misunderstanding.

(i) In most of physical chemistry the standard state of a solid, liquid or ideal gas is taken as the pure substance at 25 °C under 1 bar (approximately atmospheric) pressure, and the activity (or, approximately, the concentration) of any pure substance in its standard state is defined as unity. For species in aqueous solution, however, the standard state is arbitrarily chosen as 1 mol kg^{-1} of water (unit molality); for dilute solutions at ordinary temperature this is not very different from 1 mol kg^{-1} of solution or 1 mol dm^{-1} of solution (unit molarity), though for the sake of consistency in the choice of units the use of 1 mol kg^{-1} of water is much to be preferred. Let us consider a dilute solution of sodium chloride containing 0.585 g NaCl kg^{-1} H_2O. The relative activity of each ion a_i is a dimensionless quantity defined by

$$a_i = m_i \gamma_i / m^\circ$$

where m_i is the molality of the ion in solution, m° the standard molality of 1 mol kg^{-1}, and γ_i the activity coefficient of the ion (in this case very nearly unity). By convention $a_{H_2O} = 1$; so for Na^+ and Cl^-, $a = 0.01$. These values, however, tend to obscure the fact that the solution actually contains not 100 but about 5500 molecules of water to every sodium ion or chloride ion.

As a further example we may discuss the self-ionisation of water. The equilibrium constant K for the reaction

$$2H_2O \rightleftharpoons H_3O^+ + OH^-$$

would normally be written as

$$K = a_{H_3O^+} \cdot a_{OH^-} / a_{H_2O}^2$$

If we write $a_{H_2O} = 1$ we must then use a different symbol for the equilibrium constant; this is K_w, given by

$$K_w = a_{H_3O^+} \cdot a_{OH}$$

and having the value 10^{-14} at 25 °C. We then say $a_{H_3O^+} = a_{OH^-} = 10^{-7}$. If, however, we are considering a reaction in aqueous solution in which OH^- and H_2O are both potential reactants, we must remember that the relative numbers of these species present are not $1:10^7$ but $1:55 \times 10^7$.

(ii) It is impossible to measure the potential, E, of a single electrode, and the universal practice is to express all such potentials relative to that of a standard hydrogen electrode (hydrogen ions at unit activity in equilibrium with hydrogen at unit activity, i.e. atmospheric pressure). This electrode is taken to have potential $E^\circ = 0$ at all temperatures. The half-reaction corresponding to the standard hydrogen electrode is

$$H^+ \ (a = 1, aq) + e \rightleftharpoons \tfrac{1}{2}H_2 \ (1 \ atm)$$

Application of the basic thermodynamic and electrochemical relationships

$$\Delta G = \Delta H - T\Delta S$$

$$\Delta G - \Delta H = T\frac{d(\Delta G)}{dT}$$

and

$$-\Delta G = nFE$$

(in which n is the number of electrons involved in the half-reaction, F is Faraday's constant and the other symbols have their usual meanings) shows that the convention $E^\circ = 0$ and $dE^\circ/dT = 0$ for the standard hydrogen electrode is equivalent to the convention that ΔG°, ΔH° and ΔS° for the corresponding half-reaction are all zero. This is not true, but we cannot yet assign accurate absolute values to ΔG°, ΔH° and ΔS° for the half-reaction; further difficult problems arise from the inclusion of the electron in the system. As was stated in Section 3.4, the enthalpy of the gaseous electron is taken as zero; here, however, the electron is not gaseous but is in a condensed state. It is therefore preferable to eliminate the electron from our discussion, and this can be done in the following way.

Suppose we measure the e.m.f. of the cell:

$$Zn/Zn^{2+} \ (a=1, aq) \| H^+ \ (a=1, aq), \ H_2 \ (1 \ atm)/Pt$$

This gives us ΔG° for the cell reaction:

$$Zn + 2H^+ \ (a = 1, aq) \rightarrow Zn^{2+} \ (a = 1, aq) + H_2 \ (1 \ atm)$$

We can determine ΔH° for this reaction by dissolving zinc in dilute acid (or by measuring dE°/dT in addition to E°), and hence we can obtain ΔS°.

Since the standard enthalpies and free energies of zinc and hydrogen are both zero and the standard entropies of both are known from heat capacity measurements (the so-called third law of thermodynamics method), we thus obtain ΔG°, ΔH° and ΔS° for Zn^{2+} ($a = 1$, aq) relative to the corresponding values for H^+ ($a = 1$, aq), which are all taken as zero. Great care must obviously be exercised in combining thermochemical data obtained in this way with those obtained solely by calorimetric methods (and hence based on the convention that only enthalpies and free energies of elements in their standard states are zero). An example of the need for such care is given later in this chapter in the discussion of factors determining the magnitude of standard electrode potentials. Thermodynamic data for ions in solution are often given relative to the values for the hydrogen ion; where this is so the inclusion of the values for the latter species (even though they are zero) is highly desirable. For many purposes, however, we are interested only in differences between electrode potentials or free energies of half-reactions, and the difficulties associated with the conventions that E° is zero for the standard hydrogen electrode and ΔG°, ΔH° and ΔS° are zero for the hydrogen ion in its standard state in aqueous solution then disappear.

(iii) The equilibrium between a solid salt MX which yields singly-charged M^+ and X^- ions in aqueous solution and the saturated solution may be represented as:

$$MX_{(s)} + aq \rightleftharpoons M^+_{(aq)} + X^-_{(aq)}$$

The *solubility* of the solid at the specified temperature is the mass of solid which dissolves when equilibrium is reached in the presence of an excess of the solid divided by the mass of solvent: it is thus a dimensionless quantity; when divided by the molar mass of the solid it gives the molality of the saturated solution, m_{sat}. If the saturated solution is dilute, little error is generally introduced by taking activity coefficients as unity, and then the activities of the ions are given by

$$a_{sat} = m_{sat}/m^\circ$$

where m° is the standard molality 1 mol kg^{-1}. Thus for copper(I) iodide, which has a molar mass of 0.191 kg mol^{-1} and 3.08×10^{-5} g of which dissolve in 100 g of water at 25 °C, the solubility is 3.08×10^{-7}, the molality of the saturated solution is 1.61×10^{-6} mol kg^{-1} and the relative activity of each of the ions in solution is 1.61×10^{-6}.

The solubility product of MX, K_S, is the equilibrium constant of the reaction given above for the dissolution of MX in water; since the activities of solid MX and of water are taken as unity, it is given by

$$K_S = a^2_{sat}$$

In general, for a substance MX_n which yields M^{n+} and nX^- ions in solution

$$K_S = a_{M^{n+}} \cdot a^n_{X^-}$$

Thus for a saturated solution of calcium fluoride

$$K_S = a_{Ca^{2+}} a_{F^-}^2$$

and since $a_{F^-} = 2a_{Ca^{2+}}$

$$K_S = 4a_{Ca^{2+}}^3$$

From values for K_S, values of ΔG° for dissolution reactions may be obtained, and combination with calorimetrically determined values of ΔH° for dissolution (or precipitation) leads to values of ΔS° for dissolution. When other than very dilute solutions are involved, activity coefficients must be taken into consideration. This is often quite a difficult matter, and in one group of solution equilibria, complex formation by metal ions, it is common practice to make measurements in a relatively concentrated solution of a supporting electrolyte (commonly 3 molar sodium perchlorate) and to assume all reacting species behave as if they are at infinite dilution in that medium. Thermodynamic data obtained under these conditions are, of course, not strictly comparable with those obtained by making measurements in the absence of added electrolyte and then incorporating any necessary corrections for activity coefficients which are not unity.

7.3 *The hydration of ions and the solubilities of salts*

We may consider the equilibrium between a solid salt MX and its ions in saturated aqueous solution in terms of the following thermodynamic cycle:

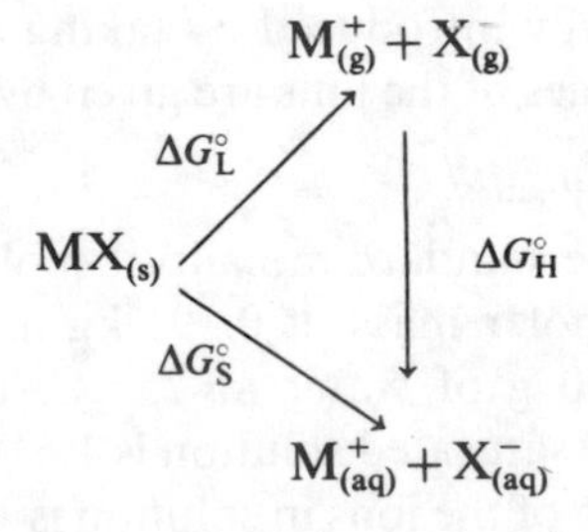

In this cycle ΔG°_L is the standard free energy change for conversion of the solid salt into infinitely separated gaseous ions (which differs from the lattice energy only by the inclusion of a small term to take into account the entropy of vaporisation of the ions); ΔG°_H is the sum of the standard free energies of hydration of the ions; and ΔG°_S is the standard free energy change for the dissolution of the salt in water, which is related to the solubility product by the equation

$$\Delta G^\circ_S = -RT \ln K_S$$

It should in principle be possible to predict solubility products from values of ΔG°_L and ΔG°_H, but there are two difficulties in the way of doing so. First, ΔG°_S is the small difference between two much larger quantities, neither of which is usually accurately known, and the situation is made worse by the exponential relationship between ΔG°_S and K_S. Second, hydration energies are not very accessible quantities. We must therefore precede the discussion of solubility with a brief account of the thermodynamics of the hydration of ions.

We showed in the last section how it is possible to determine ΔG°, ΔH° and ΔS° for the conversion of a solid metal into its ion at unit activity in aqueous solution. Similar methods may be used to obtain data for the formation of aqueous anions from non-metallic elements. In each case, of course, the electrode involved must be reversible; and, as we have emphasised earlier, the values obtained are all relative to those for the conversion of hydrogen into hydrogen ion under standard conditions.

In an alternative approach, we can consider the dissolution of a salt MX of known solubility product K_S and known enthalpy of solution ΔH°_S; since K_S is related to ΔG°_S, we can then obtain ΔS°_S. The value of ΔH°_L corresponds to the lattice energy of the salt, and ΔS°_L may be obtained from the entropies of the gaseous ions (which, provided the ions are monatomic, may be calculated from statistical mechanics) and the third-law entropy of MX: hence ΔG°_L may be derived, and so the values of ΔH°_H, ΔG°_H and ΔS°_H for the pair of ions $M^+_{(g)}$ and $X^-_{(g)}$ are known. Further, consideration of all the data for, say, NaCl and NaBr enables us to obtain accurate values for the differences in ΔH°_H, ΔG°_H and ΔS°_H between $Cl^-_{(g)}$ and $Br^-_{(g)}$, and these differences agree very closely with those obtained from data for the $\frac{1}{2}Cl_2/Cl^-$ and $\frac{1}{2}Br_2/Br^-$ standard electrodes. The problem of assigning an absolute value to one ion, however, remains.

We should mention here that in principle the value of ΔG°_H for an ion of charge ze and radius r can be calculated on the basis of electrostatics by means of the equation

$$\Delta G^\circ_H = -\frac{Nz^2e^2}{8\pi\varepsilon_0 r}\left(1 - \frac{1}{\varepsilon_r}\right)$$

where ε_0 is the permittivity of free space, and ε_r the relative permittivity of the solvent; but in practice the use of this equation gives unsatisfactory results since it is certain that the bulk relative permittivity of the solvent is changed in the vicinity of an ion, and the only available values for r are those of ions in crystals rather than in solution.

The simplest way of obtaining thermodynamic functions of hydration of individual ions rests on the assumption that the very large ions Ph_4As^+ (tetraphenylarsonium) and BPh_4^- (tetraphenylborate) have the same values. This is the so-called TATB assumption. From the appropriate data for the salts $Ph_4As^+BPh_4^-$, $Ph_4As^+Cl^-$, and $K^+BPh_4^-$ (lattice energies being estimated in each case) we can then, for example, obtain absolute values for K^+ and Cl^-. More sophisticated approaches based on the estimation of values for the gaseous proton lead to results in fair agreement (± 20 kJ mol^{-1} or better) with those based on the TATB assumption. Some representative data are assembled in Table 7.2; some values for transition metal ions are given in Chapter 20. As we have mentioned earlier, the relative values are certainly more reliable than the absolute ones.

A survey of the values in Table 7.2 reveals several features of interest. Highly charged ions have much more negative enthalpies and entropies of hydration than ions which carry a single charge. The more negative enthalpy term is readily understood in terms of simple electrostatic attraction; the more negative entropy of hydration may be thought of as arising from the introduction of more order into the system when ions of high charge orient more water molecules in fixed configurations around them. For ions of a given charge, some dependence of enthalpy and entropy of hydration on size (in the crystallographic sense) may be noted: smaller ions have more negative values of both ΔH°_{H} and ΔS°_{H}; the variation in ΔH°_{H}, however, outweighs that in $T\Delta S^{\circ}_{H}$, and ΔG°_{H} is much more negative for smaller ions of the same charge or more highly charged ions of the same size. It is noteworthy that for monatomic ions of the same crystal radius (e.g. K^{+} and F^{-}), anions are more strongly hydrated than cations.

The entropies of monatomic ions in solids or in the gaseous state are found to vary only slightly from one species to another, and it is therefore not far from exact to discuss the solubilities of ionic salts in terms of differences between lattice energies and free energies of hydration of gaseous ions. These differences are, as we have already pointed out, small differences between large quantities; furthermore, change in ionic size along a series of related compounds affects both quantities, so that changes in them tend to cancel. Nevertheless, some helpful conclusions

Table 7.2
Absolute values of ΔH°_{H}, ΔS°_{H} and ΔG°_{H} for some ions at 25 °C.

Ion	ΔH°_{H}/kJ mol^{-1}	ΔS°_{H}/J K^{-1} mol^{-1}	ΔG°_{H}/kJ mol^{-1}
H^{+}	−1100	−130	−1060
Li^{+}	−528	−140	−485
Na^{+}	−413	−110	−379
K^{+}	−330	−70	−308
Rb^{+}	−305	−70	−285
Cs^{+}	−280	−60	−262
Mg^{2+}	−1940	−320	−1845
Ca^{2+}	−1595	−230	−1525
Sr^{2+}	−1465	−220	−1400
Ba^{2+}	−1325	−200	−1265
Al^{3+}	−4700	−530	−4540
La^{3+}	−3300	−430	−3170
F^{-}	−513	−150	−468
Cl^{-}	−37	−90	−344
Br^{-}	−339	−70	−317
I^{-}	−394	−50	−280

can be drawn. The lattice energy of a series of salts of similar structure is proportional to $1/(r_+ + r_-)$, and hence there will be only a small variation in this quantity if $r_- \gg r_+$; if this condition holds, however, the free energy of hydration of the cation will be very much more negative than that of the anion for all values of r_+, and hence the sum of the free energies of hydration of the ions will be very nearly the same as the free energy of hydration of the cation alone and will be roughly proportional to $1/r_+$. Now if, along a series of salts, the lattice energy remains nearly constant whilst the total free energy of hydration of the ions becomes less negative, solubility will decrease. This is the situation which prevails among the alkali metal perchlorates and hexafluorophosphates, and similar considerations also apply to the chloroplatinates(IV) and hexanitrocobaltates(III). It may be noted that for a very small anion, such as fluoride, the variation in solubilities is reversed, with lithium fluoride being the least soluble; the minimum solubility among the chlorides occurs at potassium chloride. A more detailed mathematical treatment along these lines of the solubility of salts supplies a theoretical justification of an old experimental generalisation, that the best precipitant for a large complex ion is an ion of equal but opposite charge and of the same size.

All of the foregoing discussion relates to compounds for which the ionic model is valid. Many marked variations in solubilities have, of course, quite different origins. Thus among the silver halides there is a substantial increase in the extra non-electrostatic contribution to the lattice energy as we go along the series AgF, AgCl, AgBr, AgI; this progressively stabilises the solid more with respect to the ions in solution. Hence although Ag^+ is of about the same crystallographic size as Na^+ in the fluorides, the overall patterns of solubilities among the halides of the two elements are quite different.

We have so far said very little about the interpretation of the hydration of ions on a molecular basis. It is generally accepted that the interaction of a simple cation such as Na^+ and the solvent is mainly ion–dipole in nature, the water molecules nearest to the ion having their oxygen atoms adjacent to it; for an anion the orientation of the water molecules corresponds to ion–dipole interaction augmented by hydrogen bonding, the structure being approximately planar:

```
           H                          
          /                           
M+       O        and     X-   H—O    
          \                        \  
           H                        H 
```

These configurations have been established for dilute solutions of lithium and sodium chlorides by detailed neutron diffraction studies. In concentrated solutions, however, the plane of the water molecule makes an angle of up to 50° with the M^+—O axis, implying interaction of the cation with a lone pair of electrons rather than with the dipole moment as a whole. For both cations and anions, there are six water molecules in the primary solvation layer. Spectroscopic studies suggest that the hydration of other halide ions is similar to that of chloride, but for more complex anions very little is known. For a limited number of hydrated cations, however, tracer methods and electronic and nuclear magnetic resonance spectroscopy now provide reliable information about coordination number and stereochemistry.

Tracer studies show, for example, that the $[Cr(H_2O)_6]^{3+}$ ion, whose presence in solid chromium(III) salts is shown by X-ray diffraction, does not undergo exchange with solvent water labelled with ^{18}O. In this case, and also in a number of other cases in which the hydrated ion is too labile for the tracer method to be applied (e.g. $[Co(H_2O)_6]^{2+}$), comparison of solid-state and solution spectra shows the same species to be present in both phases. For cations with a relatively low rate of exchange of water molecules between the immediate coordination sphere and the bulk solvent, separate ^{17}O nuclear magnetic resonance signals can be observed for the two types of water molecule; hydration numbers calculated from relative intensities have proved the existence of $[Be(H_2O)_4]^{2+}$, $[Mg(H_2O)_6]^{2+}$ and $[Al(H_2O)_6]^{3+}$ in solution. None of these methods, however, can be applied successfully to the study of the hydration of anions.

7.4 *The ionisation of acids in aqueous solution*

The strengths of different acids in aqueous solution are extensively discussed in elementary textbooks of inorganic and organic chemistry, though often only on a qualitative basis. In the case of the hydrogen halides, however, an exact treatment in terms of independently measurable thermodynamic quantities is *almost* possible, and we therefore begin by considering the aqueous equilibrium

$$HX \rightleftharpoons H^+ + X^-$$

The factors which influence the degree of ionisation are apparent from examination of the following cycle:

$$\begin{array}{ccccc} HX_{(g)} & \xrightarrow{(2)} & H_{(g)} & + & X_{(g)} \\ \uparrow & & (3)\downarrow & & (4)\downarrow \\ (1) & & H^+_{(g)} & & X^-_{(g)} \\ & & (5)\downarrow & & (6)\downarrow \\ HX_{(aq)} & \rightarrow & H^+_{(aq)} & + & X^-_{(aq)} \end{array}$$

Stages (3) and (5) are common to all the halides. Stage (2) is the dissociation of the hydrogen halide into atoms, stage (4) the capture of an electron by the gaseous halogen atom, and stage (6) the hydration of the gaseous halide ion; for each of these stages ΔH° and ΔS° are known.

Stage (1), however, which is the reverse of the dissolution of the gaseous hydrogen halide in water to form an unionised solution, presents more difficulty. Since hydrogen fluoride is a weak acid in dilute aqueous solution, it might seem that ΔH° and ΔS° can be obtained directly for this compound, but infrared spectroscopy shows that the species present in solution is the strongly-bonded ion pair F^-—HOH_2^+; for the other hydrogen halides values for stage (1) have to be estimated from somewhat unsatisfactory comparisons with the noble gases and the methyl halides. There is clearly a strong resemblance between this cycle and that given in Section 6.8 for the protonation of a gaseous base by a hydrogen halide, though the latter cycle has the advantage of not including anything analogous to stage (1), and variations in ΔS° in it may be disregarded – a simplification which is never permissible when we are considering processes that involve the formation of ions in aqueous solution.

Since some of the data needed for the cycle under discussion here are estimates, we shall not tabulate them all, but merely summarise the conclusions regarding the strengths of the halides other than hydrogen fluoride ($pK_a = -7$, -9 and -10 for HCl, HBr and HI respectively). Hydrogen chloride, bromide and iodide are estimated to have about the same value for stage (1), and the chloride is marginally the weakest acid because the much greater strength of the H—Cl bond outweighs the effects of electron affinity and free energy of hydration. It will be noticed that the discussion of this thermodynamic cycle does not involve electronegativity or the inductive effect, less precise concepts than the variables we have taken into consideration; but it will also be noticed that, since four different factors enter into the relative ionisation constants, this superficially simple problem is really rather a difficult one. Similar cycles may be devised for the two series H_2S, H_2Se and H_2Te and NH_3, PH_3, AsH_3 and SbH_3, in both of which ionisation constants increase with increase in atomic number; for these, however, more data have to be estimated. It is, nevertheless, clear that the decrease in bond strength with increasing atomic number plays an important part in accounting for what is often thought to be the puzzling observation that as, down a periodic table group of elements, the element becomes more metallic in character, its hydride becomes more acidic. If we consider successive ionisations of an acid like H_2S, in which both protons are attached to the same atom, it is always found that $K_2 \ll K_1$; in this instance $pK_1 = 7$, $pK_2 = 19$.

The largest group of oxo-acids are the organic carboxylic acids, many of which have pK_a values in the range 0–5 (e.g. HCOOH, 3.8; CH_3COOH, 4.8; $ClCH_2COOH$, 2.9; $Cl_2CHCOOH$, 1.3; CCl_3COOH, 0.7). It is common practice in organic chemistry to attribute these variations to the inductive effects of CH_3 and Cl, electron-repelling and electron-attracting substituents respectively; but inductive effects are not independently measurable, and since we know very little about the electron affinities of RCOO radicals or the hydration energies of $RCOO^-$ ions, a quantitative treatment of the ionisation of carboxylic acids is impossible. It is, however, interesting to note that studies of the temperature dependence of pK_a values show that for the ionisation of HCOOH, CH_3COOH and CCl_3COOH, ΔH° is in each case very close to zero, and that the variation in pK_a therefore arises from a variation in the entropy of ionisation, which becomes less negative along the series CH_3COOH, HCOOH, CCl_3COOH; apparently withdrawal of electrons from the carboxylate end of the anion results in less orientation of surrounding water molecules.

For inorganic oxo-acids, which represent a variety of elements in different oxidation states, there is again no adequate thermodynamic treatment, though there are certain useful empirical generalisations about pK_a values. The best known of these is Bell's rule that for an acid of formula $EO_n(OH)_m$ (where E is any element) the first ionisation constant is given approximately by

$$pK_a = 8 - 5n$$

Typical values are Cl(OH), 7.2 and $B(OH)_3$, 9.2; ClO(OH), 2.0 and NO(OH), 3.3; $ClO_2(OH)$, -1 and $NO_2(OH)$, -1.4; and $ClO_3(OH)$, -10. For polybasic acids it is often found that successive pK_a values differ by about 4 or 5. This increase in acid strength with increase in the number of oxygen atoms attached to the central atom is generally attributed to the greater possibility of delocalisation of negative charge on to the oxygen atoms, and this seems to be the best comment on the variation in pK_a values that can be made at the present time. It is worth noting that three acids which appear to be out of line (telluric(VI) acid, periodic acid and carbonic acid) are not really exceptions to Bell's rule; the first two are not H_2TeO_4 and HIO_4 but $Te(OH)_6$ and $IO(OH)_5$, and the last has a true pK_a value of 3.9. In an aqueous solution of carbon dioxide most of the solute is present as CO_2 rather than as H_2CO_3, but the pK_a value is usually evaluated as 6.0 on the assumption that it is all present as H_2CO_3 or the HCO_3^- ion.

Aquated cations may also function as acids, e.g.

$$[M(H_2O)_6]^{n+} \rightleftharpoons [M(H_2O)_5OH]^{(n-1)+} + H^+$$

This is the reason for the acidity of solutions of salts of many cations, especially those of high *n*. The pK_a values of $Li^+_{(aq)}$, $Mg^{2+}_{(aq)}$ and $Al^{3+}_{(aq)}$, for example, are 14, 11.4 and 5.0 respectively; $Al^{3+}_{(aq)}$ is as strong an acid as CH_3COOH.

7.5 Complex formation

The formation of complexes in aqueous solution may be recognised by a number of different methods, of which the classical test of modification of chemical properties is only one, and a somewhat unreliable one at that: all reactions have equilibrium constants, and chemical tests are often only investigations of relative values of equilibrium constants. In a solution of a silver salt saturated with ammonia, for example, nearly all the silver ion is present as the complex $[Ag(NH_3)_2]^+$, and addition of a chloride-containing solution produces no precipitate; addition of an iodide-containing solution, however, results in the precipitation of silver iodide. Silver iodide is much less soluble than silver chloride, the values for their solubility products being 10^{-16} and 10^{-10} respectively, so what these experiments indicate is that the overall equilibrium constant for the reaction

$$Ag^+ + 2NH_3 \rightleftharpoons [Ag(NH_3)_2]^+$$

is large enough for silver chloride to be soluble in a saturated solution of ammonia whilst silver iodide is unaffected.

Physical methods (such as investigations of colligative properties, electronic or vibrational spectra, solubility, conductivity, or, for preference, electrode potentials) provide more reliable evidence as well as, in favourable circumstances, leading to values of equilibrium constants for complex formation.

The examples of complexes that are encountered early in the study of chemistry usually contain only one metal ion, which is combined with one or more anionic or neutral *ligands*. Complexes containing anionic ligands include the hexacyanoferrate(II) and hexacyanoferrate(III) ions, $[Fe(CN)_6]^{4-}$ and $[Fe(CN)_6]^{3-}$ respectively, and the neutral bright red nickel(II) dimethylglyoximate, $[Ni(CH_3C({=}NO)C({=}NOH)CH_3)_2]$, the formation of which in weakly alkaline solution is a sensitive test for nickel. The complexes $[Ag(NH_3)_2]^+$ and $[Co(H_2NCH_2CH_2NH_2)_3]^{3+}$ contain uncharged ligands. Complexes containing more than one cation are by no means uncommon, however: partially hydrolysed magnesium salts, for example, contain ions such as $[Mg_2(OH)_3]^+$, and when silver iodide is dissolved in a saturated solution of silver nitrate the ions Ag_2I^+ and Ag_3I^{2+} are produced.

Neutral complexes are usually only sparingly soluble in water but are often readily soluble in organic solvents, e.g. iron(III) acetylacetonate, $[Fe(CH_3COCHCOCH_3)_3]$ (or, as it is often written, $[Fe(acac)_3]$), is a red solid (m.p. 179 °C) which is readily extracted from aqueous media by benzene or chloroform. The anion in this substance is formed by the deprotonation of the weak acid acetylacetone, and the formation of the compound in aqueous solution therefore involves both of the equilibria

$$CH_3COCH_2COCH_3 \rightleftharpoons CH_3COCHCOCH_3^- + H^+ \qquad K = 10^{-10}$$

and

$$Fe^{3+} + 3CH_3COCHCOCH_3^- \rightleftharpoons [Fe(CH_3COCHCOCH_3)_3] \qquad K = 10^{26}$$

The quantity of complex formed therefore depends upon the pH of the solution: if it is too low, H^+ ions compete with Fe^{3+} ions for the anion; if it is too high, the iron is precipitated as hydroxide, for which K_s is about 10^{-37}. There is thus an optimum pH for the extraction of iron(III) from aqueous media by acetylacetone and chloroform. Since most ligands are bases in the Brønsted sense, accurate pH control is of great importance in studies of complex formation. Solvent extraction is now important in the separation, both analytical and industrial, of many metals.

7.6 *Formation constants of complexes*

Let us consider a metal ion M and a ligand L (charges being omitted for the sake of simplicity) forming a series of complexes ML, ML_2,... ML_n in aqueous solution. Although each stage in complex formation really represents successive replacement of water from the hydration shell of the metal ion, on the convention that the activity of water in such a solution is unity we can disregard it as part of a reactant or as a product. For the formation of successive complexes we then have, provided equilibrium is attained in each case,

$$M + L \rightleftharpoons ML \qquad K_1 = \frac{a_{ML}}{a_M a_L}$$

$$ML + L \rightleftharpoons ML_2 \qquad K_2 = \frac{a_{ML_2}}{a_{ML} a_L}$$

$$ML_{n-1} + L \rightleftharpoons ML_n \qquad K_n = \frac{a_{ML_n}}{a_{ML_{n-1}} a_L}$$

The equilibrium constants K_1, K_2 ... K_n are called stepwise formation (or stability) constants. Alternatively we may consider a series of overall equilibrium constants, for which the general symbol β is used. Thus

$$M + L \rightleftharpoons ML \qquad \beta_1 = \frac{a_{ML}}{a_M a_L}$$

$$M + 2L \rightleftharpoons ML_2 \qquad \beta_2 = \frac{a_{ML_2}}{a_M a_L^2}$$

$$M + nL \rightleftharpoons ML_n \qquad \beta_n = \frac{a_{ML_n}}{a_M a_L^n}$$

Obviously

$$\beta_n = K_1 K_2 \ldots K_n$$

For the sake of consistency, it may be noted, we should regard the equilibrium between an acid and its ionisation products as the complexing of the proton by the conjugate base, e.g.

$$H^+ + L^- \rightleftharpoons HL$$

but this representation is seldom employed, most chemists preferring to start from the acid and consider its dissociation.

If it is shown that only one complex, of known formula, is present in a solution containing known total activities (or, approximately, concentrations) of M and L, the formation constant of this complex may be obtained directly from a determination of the concentration of uncomplexed M or L in that solution. The measurement may be made by polarographic or e.m.f. measurements if a suitable reversible electrode exists, by pH measurements if the ligand is the anion of a weak acid, or by ion-exchange, spectrophotometric, and distribution methods. In general, however, more than one complex is present in a given solution, and then measurements over a wide range of relative concentrations, followed by computational treatment of the experimental data, are necessary for the evaluation of all the consecutive formation constants; details of these operations are given in some of the references listed at the end of the chapter. From measured equilibrium constants, standard free energies of complex formation may be calculated. By making measurements at different temperatures, enthalpies and entropies of complex formation may also be obtained, though since enthalpies of complexing in aqueous solution are often small and are not quite independent of temperature, it is preferable to obtain them by direct calorimetry.

It is generally found that stepwise formation constants show a progressive decrease as n increases; for the Ni^{2+}–NH_3 system in 2 molar NH_4NO_3 solution at 30 °C, for example, a pH study using a glass electrode gave $K_1 = 10^{2.79}$, $K_2 = 10^{2.26}$, $K_3 = 10^{1.69}$, $K_4 = 10^{1.25}$, $K_5 = 10^{0.74}$, $K_6 = 10^{0.03}$, leading to an overall formation constant of the $[Ni(NH_3)_6]^{2+}$ ion $\beta_6 = 10^{8.74}$. Thus the principal complex present in an ammoniacal solution of a nickel(II) salt depends upon the $[NH_3]:[Ni^{2+}]$ ratio; only at high concentrations of ammonia is the pentammine the main species present, and even in the presence of very high concentrations of the ligand, formation of the hexammine complex is incomplete. For the Ni^{2+}–NH_3 system, the enthalpies of successive stages in complex formation are all about -17 kJ mol^{-1} and the relative values of successive formation constants agree very roughly with what might be expected on the basis of a statistical replacement of water by ammonia with the coordination number of the metal ion constant. In other cases, however, an abrupt variation or even reversal in value of K occurs. Thus in the Ag^+–NH_3 system $K_2 > K_1$, suggesting that coordination of one molecule of ammonia converts the hydration shell round the cation into a monosubstituted tetrahedron or octahedron, but that formation of the second complex produces a total change of structure with formation of the linear $[Ag(NH_3)_2]^+_{(aq)}$ ion.

The detailed discussion of the thermodynamics of complex formation in aqueous solution lies beyond the scope of this book, but we should touch briefly on entropies of complex formation and the so-called *chelate effect*. As we saw in Section 7.3, ions carrying high charges have very negative entropies of hydration because of their effect in orienting water molecules around them. When complex formation occurs between highly charged cations and anions, with a resulting partial or total cancellation of charges, ΔH° is substantially negative and ΔS° substantially positive, and very stable complexes are formed; instances of such behaviour are provided by the interaction of many di- and tri-positively charged metal ions and sulphate, phosphate or the ethylenediaminetetraacetate anion $[(OOCCH_2)_2NCH_2CH_2N(CH_2COO)_2]^{4-}$ ($EDTA^{4-}$).

There is another source of increase in entropy that is important in this connection. Even when we are dealing with comparable uncharged ligands, it is generally found that ligands which can coordinate at two or more positions (*polydentate*, i.e. many-toothed, ligands) form more stable complexes with a given ion than ligands which can coordinate at only a single position. Coordination by polydentate ligands leads to the formation of ring structures in which the metal is incorporated, and the process is known as *chelation* (a term derived from the Greek for a crab's claw). Typical chelated species are the oxalate and ethylenediamine (en) complexes of many metals. If we compare complexing of the Cd^{2+} ion by methylamine and ethylenediamine it is found although ΔH° for the reactions

$$Cd^{2+} + 4CH_3NH_2 \rightleftharpoons [Cd(CH_3NH_2)_4]^{2+}$$

and

$$Cd^{2+} + 2en \rightleftharpoons [Cd(en)_2]^{2+}$$

is almost identical (-57.3 and -56.5 kJ mol^{-1} respectively), β_4 for the first reaction is $10^{6.5}$ whilst β_2 for the second reaction is $10^{10.6}$, corresponding to values of ΔG° of -37.2 and -60.7 kJ mol^{-1} respectively. Thus the second reaction has a much more positive standard entropy change, $+14.1$ J K^{-1} mol^{-1} compared with -67.3 J K^{-1} mol^{-1}. Most of this can be accounted for if we remember that complexing is really displacement of water; when the chelating ligand acts as the complexing agent, there is an increase in the number of unbound molecules (water or ligand) and hence in the translational entropy of the system. The greater stability of the chelated complex compared with the most closely analogous non-chelated complex is termed the *chelate effect*. In the example given it appears to be purely an entropy effect (though this may be the chance result of the cancellation of any difference in metal–ligand bond energies by differences in hydration energies of the ligands or the complexes – we know only that the overall values of ΔH° are almost identical). When we are comparing less closely similar ligands (e.g. ammonia and ethylenediamine) ΔH° as well as $T\Delta S^\circ$ certainly contributes to the greater stability of the chelate complex. Macrocyclic and bicyclic (cryptand) polyethers form still more stable complexes. The enhancement of complex stability when a macrocyclic ligand replaces a comparable acyclic ligand, e.g. when cyclo- $(CH_2CH_2O)_6$ replaces $CH_3O(CH_2CH_2O)_5CH_3$ in its potassium complex, with an increase in the formation constant of 10^4, is called the *macrocyclic effect*. Like the chelate effect, this may be due mainly to entropy or partly also to enthalpy, but detailed discussion of the macrocyclic effect is complicated by factors such as different possible conformations of the ligand, steric strain, and the compatibility or otherwise of the size of the gap in the middle of the molecule and that of the ion to be accommodated. Further discussion of macrocylic and polymacrocyclic ligands is therefore deferred until we consider complexes of the alkali metals in Section 10.7.

All the systems discussed so far in this section involve only equilibria which are rapidly attained and hence are amenable to study by the standard physicochemical methods. There are, however, some systems for which this is not so, e.g. the formation of hexacyanoferrate(III):

$$Fe^{3+} + 6CN^- \rightleftharpoons [Fe(CN)_6]^{3-}$$

Both formation and dissociation of the $[Fe(CN)_6]^{3-}$ ion are extremely slow. For such a system, direct measurement of the overall formation constant is impossible, and it has to be obtained from ΔG°, in turn obtained from a lengthy series of measurements leading to values for ΔH° and ΔS°. The study of the thermodynamics of complexing then presents great difficulty, and it is not surprising that few accurate data are available for such non-labile systems.

7.7 *Factors affecting the stabilities of complexes containing only monodentate ligands*

Although there is no single generalisation relating values of formation constants of complexes of different cations with the same ligand, or of the same cation with different ligands, a number of useful correlations exist, and we shall now present the most important of them.

The stabilities of complexes of non-transition metal ions of a given charge normally decrease with increasing size (in the crystallographic sense) of the cation (e.g. among the alkaline earth metal ions the order of stability is $Ca^{2+} > Sr^{2+} > Ba^{2+}$). Analogous behaviour is found for the series of lanthanide tripositive ions; for the transition metal ions, however, an additional effect (crystal or ligand field stabilisation) is usually involved, and data for these ions are therefore discussed later in Chapter 20. For a given ionic size, increase in charge almost invariably results in a substantial increase in complex stability (e.g. along the series $Li^+ < Mg^{2+} < Al^{3+}$). If a metal exists in two different oxidation states, the more highly charged ion is the smaller, and the two effects reinforce one another; metals in their higher oxidation states nearly always form more stable complexes than in the lower oxidation states. Occasional exceptions to this rule do occur, for example in the 1,10-phenanthroline and 2,2′-bipyridyl complexes of Fe(II) and Fe(III); but where this is so there are strong reasons for believing that the nature of the ligand (in particular the electron density on it) is not quite the same in complexes of the metal in different oxidation states. It is noteworthy that all ligands that are prone to form stable complexes with metals in low oxidation states (carbon monoxide and cyanide ion, for example, in addition to the two heterocyclic species mentioned already) have relatively low-energy antibonding orbitals available to accept electrons back from the metal ion; they thus differ essentially from the simple inorganic ligands like H_2O, NH_3 and the halide ions, with which we have so far been mainly concerned.

Where their acceptor behaviour towards different ligands is concerned, cations fall into two classes, though the distinction between them is not always clear-cut. Most quantitative data on complex formation refer to aqueous or aqueous organic systems with halide ions or species containing oxygen or nitrogen as the donor atom as the ligands; quantitative data for complexes containing phosphorus, sulphur or other atoms as the donor atom in the ligand are extremely scarce. For halide ions as ligands, it is found that for the lighter *s* and *p* block cations, metals at the beginnings of the transition series, lanthanides and actinides the usual order of complex stability is fluoride > chloride > bromide > iodide. For tellurium, polonium, the platinum metals, copper, silver, gold, cadmium, mercury and thallium this order is reversed. On the basis of mainly qualitative evidence, cations of the first group also appear to form more stable complexes with amines than with phosphines, and with ethers than with sulphides; these orders are reversed for cations of the second group.

Cations of the first and second groups were called class (a) and class (b) acceptors by Ahrland, Chatt and Davies, and these non-committal terms are widely used. In the wider context of general electron donor (*Lewis base*)–electron acceptor (*Lewis acid*) interactions, class (a) cations have been termed 'hard' acids, and class (b) cations 'soft' acids by Pearson, who has also classified ligands as 'hard' or 'soft' bases according to whether they are small and their valence electrons are tightly held (e.g. F^-, H_2O, NH_3) or are large and easily polarised (e.g. I^-, CN^-, $(CH_3)_2S$, $(CH_3)_3P$). 'Hard' acids bind more strongly to 'hard' bases, 'soft' acids more strongly to 'soft' bases – the so-called principle of hard and soft acids and bases.

This principle certainly has some usefulness in chemistry, but it lacks a satisfactory quantitative basis and its value in the study of equilibria in aqueous solution, in particular, is doubtful. If, for example, we consider the equilibrium

$$MF_6^{2-} + 6I^- \rightleftharpoons MI_6^{2-} + 6F^-$$

in aqueous solution by means of a thermochemical cycle, we see that the position of equilibrium depends not only on the relative values of the M^{4+}–F^- and M^{4+}–I^- interaction energies in the gas phase complex ions, but also on the hydration energies of all the species concerned. Since fluoride ion has a substantially more negative free energy of hydration than iodide ion, it is quite feasible for M^{4+} to form a more stable complex with iodide ion in aqueous media even though the gas phase M^{4+}–F^- interaction is stronger. The position of the equilibrium

$$HgCl_2 + 2Br^- \rightleftharpoons HgBr_2 + 2Cl^-$$

under different conditions can be evaluated from thermochemical data for the vaporisation and dissolution of the $HgCl_2$ and $HgBr_2$ (both stable compounds) and data for their formation from Hg^{2+} and Cl^- and Br^- ions. It lies to the left-hand side in the gas phase, to the right-hand side in aqueous solution, and is intermediate in acetonitrile: this is clear proof of the role of the solvent in such systems.

It has recently been suggested that for the gas-phase equilibrium

$$RX_n + nY^- \rightleftharpoons RY_n + nX^-$$

where Y is a halogen of higher atomic number than X, the position of equilibrium lies to the left whatever the nature of R, e.g. whether it is a metal ion or an alkyl group. The possibility of displacing the position of equilibrium to the right then depends on whether solvation effects are great enough to overcome the differences in R—X and R—Y bond energy terms. For water as solvent, hydration effects are sufficient if R is a class (b) cation or 'soft' base, but not if it is a class (a) cation or 'hard' base.

Comparable effects to those on equilibria have been observed in kinetics. For a bimolecular reaction such as

$$CH_3Br + F^- \rightarrow CH_3F + Br^-$$

the transition state $[BrCH_3F]^-$, being much larger than F^-, is relatively weakly solvated, both alkyl halides are very weakly solvated, and the activation energy in solution incorporates desolvation of the fluoride ion. No such factor enters into the activation energy for the gas-phase reaction, which has a rate constant at 25° greater than that in aqueous solution by a factor of 10^{18}, the activation energies being 1 and 105 kJ mol^{-1} respectively.

7.8 Redox processes

In this and the following section we are concerned with equilibria involving a change in oxidation state* in aqueous media, i.e. with transfer of electrons. An example of such a process is the reduction of zinc ions at unit activity in aqueous solution to metallic zinc:

$$Zn^{2+}\ (a = 1) + 2e \rightleftharpoons Zn$$

The standard free energy change, ΔG°, associated with this half-reaction is related to the standard redox potential, E°, of the Zn^{2+} $(a = 1)/Zn$ electrode by the equation

$$\Delta G^\circ = -nE^\circ F$$

where n is the number of electrons transferred and F is Faraday's constant. As we explained in Section 7.2, ΔG° and E° are relative to the corresponding quantities for the half-reaction

$$H^+\ (a = 1) + e \rightleftharpoons \tfrac{1}{2}H_2\ (1\ \text{atm})$$

and the standard hydrogen electrode respectively; E° is found to be -0.76 V. For the cell conventionally written

$$Zn/Zn^{2+}\ (a=1) \,\vdots\vdots\, H^+\ (a=1)/H_2\ (1\ \text{atm}),\ Pt$$

the chemical reaction is

$$Zn + 2H^+ \rightarrow Zn^{2+} + H_2$$

The e.m.f. of this cell is $+0.76$ V, the standard free energy change is

$$\Delta G^\circ = -2E^\circ F$$

and the equilibrium constant is given by

$$\log K = \frac{2E^\circ F}{2.303\ RT} = 26$$

A number of standard redox potentials are given in Table 7.3. Most of these have been obtained directly from e.m.f. measurements, but a few values have been calculated from data obtained by calorimetric methods; these are for systems which cannot be investigated in aqueous media because of decomposition of the solvent (e.g. $\tfrac{1}{2}F_2/F^-$) or for those in which equilibrium is established only very slowly, so that the electrode is not reversible (e.g. $\tfrac{1}{2}O_2$, $2H^+/H_2O$). The more positive the value of E°, the better is the system (with the species on both sides of the half-reaction at unit activity) as an oxidising agent relative to the $H^+/\tfrac{1}{2}H_2$ system.

* Readers not familiar with the oxidation state concept will find a summary of it in Problem 7 at the end of this chapter.

Table 7.3
Some standard redox potentials in aqueous solution at 25 °C.

Electrode reaction	$E°$/V
$\frac{1}{2}F_2 + e \rightleftharpoons F^-$	+2.9
$\frac{1}{2}S_2O_8^{2-} + e \rightleftharpoons SO_4^{2-}$	+2.01
$Ag^{2+} + e \rightleftharpoons Ag^+$	+1.98
$Co^{3+} + e \rightleftharpoons Co^{2+}$	+1.93
$MnO_4^- + 8H^+ + 5e \rightleftharpoons Mn^{2+} + 4H_2O$	+1.51
$\frac{1}{2}Cl_2 + e \rightleftharpoons Cl^-$	+1.36
$\frac{1}{2}Cr_2O_7^{2-} + 7H^+ + 3e \rightleftharpoons Cr^{3+} + \frac{7}{2}H_2O$	+1.33
$\frac{1}{2}O_2 + 2H^+ + 2e \rightleftharpoons H_2O$	+1.23
$[Fe(phen)_3]^{3+} + e \rightleftharpoons [Fe(phen)_3]^{2+}$	+1.12
$\frac{1}{2}Br_2 + e \rightleftharpoons Br^-$	+1.07
$Ag^+ + e \rightleftharpoons Ag$	+0.80
$Fe^{3+} + e \rightleftharpoons Fe^{2+}$	+0.77
$\frac{1}{2}I_2 + e \rightleftharpoons I^-$	+0.54
$[Fe(CN)_6]^{3-} + e \rightleftharpoons [Fe(CN)_6]^{4-}$	+0.36
$Cu^{2+} + 2e \rightleftharpoons Cu$	+0.34
$AgCl + e \rightleftharpoons Ag + Cl^-$	+0.22
$Cu^{2+} + e \rightleftharpoons Cu^+$	+0.15
$H^+ + e \rightleftharpoons \frac{1}{2}H_2$	0
$Cr^{3+} + e \rightleftharpoons Cr^{2+}$	−0.41
$Fe^{2+} + 2e \rightleftharpoons Fe$	−0.44
$Zn^{2+} + 2e \rightleftharpoons Zn$	−0.76
$Mn^{2+} + 2e \rightleftharpoons Mn$	−1.18
$Al^{3+} + 3e \rightleftharpoons Al$	−1.66
$Mg^{2+} + 2e \rightleftharpoons Mg$	−2.37
$Na^+ + e \rightleftharpoons Na$	−2.71
$Ca^{2+} + 2e \rightleftharpoons Ca$	−2.87
$Li^+ + e \rightleftharpoons Li$	−3.04

The way of writing half-reactions used here is now generally accepted; it has the advantage that it gives the correct sign of the potential of the actual electrode relative to that of the standard hydrogen electrode. If the half-reaction is written as an oxidation reaction, e.g.

$$Zn \rightleftharpoons Zn^{2+}(a = 1) + 2e$$

the sign of the standard free energy change is then *necessarily* the opposite of that of the half-reaction

$$Zn^{2+}\ (a = 1) + 2e \rightleftharpoons Zn$$

and hence the sign of $E°$ must change too. Many chemists find that if they are in any doubt as to whether the sign of a $E°$ value is correct, it is best to consider $\Delta G°$, a negative value for which means that the half-reaction under consideration occurs spontaneously with respect to the half-reaction

$$H^+\ (a = 1) + e \rightleftharpoons \tfrac{1}{2}H_2\ (1\ \text{atm})$$

Any half-reaction may be represented in the form

$$aA + bB + \ldots + ne \rightleftharpoons pP + qQ + \ldots$$

The value of the standard redox potention E° then refers to the condition that *all* substances present are at unit activity; under non-standard conditions E relative to the standard hydrogen electrode is given by

$$E = E^\circ - \frac{RT}{nF} \ln \frac{a_P^p a_Q^q}{a_A^a a_B^b} \cdots$$

$$= E^\circ - \frac{0.059}{n} \log \frac{a_P^p a_Q^q}{a_A^a a_B^b} \cdots \text{ at } 25\,^\circ\text{C}$$

(E will be less positive, i.e. ΔG will be less negative and the reduction will be less likely to go spontaneously, in the presence of high concentrations of products.) Thus for the reduction of Zn^{2+}, since the activity of solid Zn is by convention taken as unity,

$$E = E^\circ - \frac{0.059}{2} \log \frac{1}{a_{Zn^{2+}}}$$

For the reduction of MnO_4^- to Mn^{2+} in acidic solution, remembering that the activity of water is taken as unity,

$$E = E^\circ - \frac{0.059}{5} \log \frac{a_{Mn^{2+}}}{a_{MnO_4^-} a_{H^+}^8}$$

Thus MnO_4^- being reduced to Mn^{2+} becomes a more powerful oxidant as the hydrogen ion activity increases: MnO_4^- will not oxidise chloride in neutral solution, for example, but it liberates chlorine from concentrated hydrochloric acid.

The potentials for the reduction of water ($[H^+] = 10^{-7}$) to hydrogen and for the reduction of oxygen to water (the reverse of the oxidation of water to oxygen) are of particular importance in aqueous solution chemistry: they provide general guidance (subject, of course, to the usual limitation of thermodynamics that kinetic factors lie outside its scope) concerning the nature of chemical species that can exist under aqueous conditions. For the reduction

$$H^+ + e \rightleftharpoons \tfrac{1}{2}H_2$$

$$E = E^\circ - 0.059 \log \frac{p_{H_2}^{\frac{1}{2}}}{a_{H^+}}$$

since hydrogen is almost an ideal gas. For neutral water in equilibrium with hydrogen gas at atmospheric pressure, $E = -0.41$ V; at $[H^+] = 10^{-14}$, $E = -0.83$ V. Thus any couple with E° more negative than -0.41 V should reduce water under a pressure of one atmosphere of hydrogen, and any couple with E° more negative than -0.83 V should reduce molar alkali under the same conditions. The potential of -0.83 V for the $H^+/\frac{1}{2}H_2$ electrode in molar alkali is of only limited importance in isolation; many metal ion/metal systems that should reduce water under these conditions are prevented from doing so by formation of a coating of hydroxide or hydrated oxide; others which are less powerfully reducing (e.g. Zn^{2+}/Zn) bring about reduction because they are modified by complexing (e.g. conversion of the Zn^{2+} ions into a stable hydroxo-complex $[Zn(OH)_4]^{2-}$). Some authors make a practice of quoting E values in alkaline solution at $[OH^-] = 1$ relative to $E = -0.83$ V for the hydrogen electrode and denoting them by the symbol E°_b, but we shall not adopt that usage in this book.

For the reduction of oxygen the relevant reaction and standard potential at $[H^+] = 1$ are

$$\tfrac{1}{2}O_2 + 2H^+ + 2e \rightleftharpoons H_2O \qquad E^\circ = +1.23 \text{ V}$$

Then under non-standard conditions

$$E = E^\circ - \frac{0.059}{2} \log \frac{1}{p_{O_2}^{\frac{1}{2}} a_{H^+}^2}$$

For $p_{O_2} = 1$ atm, $E = 1.23 - 0.059$ pH; for neutral water under atmospheric pressure of oxygen E is thus $+0.83$ V, and for molar alkali under the same conditions, $+0.41$ V. So from the standpoint of thermodynamics, oxygen in the presence of water should oxidise any system with E° less positive than $+1.23$ V at pH 0, $+0.83$ V at pH 7, and $+0.41$ V at pH 14. Conversely, any system with E° more positive than 1.23 V should oxidise water at pH 0, and so on. For both the reduction and the oxidation of water, the other system refers, of course, to one in which all species involved in the appropriate half-reaction are present at unit activity (otherwise we would not be considering E° values). In reality, of course, most of experimental chemistry involving changes in oxidation states is carried out by taking an almost pure reductant and allowing it to react with an almost pure oxidant. Thus although a solution combining Cr^{3+} and Cr^{2+} both at unit activity (E° $Cr^{3+}/Cr^{2+} = -0.41$ V) in water under one atmosphere pressure of hydrogen is in equilibrium, addition of a chromium(II) salt to water should lead to liberation of hydrogen. Bench chemistry is seldom carried out under standard conditions, and it is essential to bear this in mind when using only a table of standard potentials in the discussion of real experiments.

7.9 *The stabilisation of oxidation states by complex formation or precipitation*

For the half-reaction

$$Ag^+ (a = 1) + e \rightleftharpoons Ag$$

E° is $+0.80$ V. If the concentration of Ag^+ ions is lowered, E° will become less positive and reduction to the metal will be less easy. We may then say that the oxidation state Ag(I) has been stabilised. This may be brought about by removal of the ion either by formation of a stable complex or by precipitation as a sparingly soluble salt. If K is the overall equilibrium constant for the process which lowers the silver ion concentration, we can calculate the E° value for the new half-reaction from the relationships

$$-\Delta G^\circ = RT \ln K = n(\Delta E^\circ)F$$

Silver chloride, for example, has a solubility product of 10^{-10}, from which we can easily show that for the half-reaction

$$AgCl_{(s)} + e \rightleftharpoons Ag + Cl^- \ (a = 1)$$

E° is +0.22 V: it is harder to reduce silver chloride than the hydrated silver ion. Silver iodide is less soluble than silver chloride, but it is much more soluble in aqueous potassium iodide than silver chloride is in aqueous potassium chloride; thus the iodo complexes of silver are much more stable than the chloro complexes. For the complex AgI_3^{2-} the overall formation constant is found to be 10^{14}. Combination of this value with E° for the Ag^+/Ag electrode gives

$$AgI_3^{2-} + e \rightleftharpoons Ag + 3I^- \qquad E^\circ = -0.02 \text{ V}$$

Again silver(I) has been stabilised against reduction, but this time to a greater degree: silver in the presence of AgI_3^{2-} and I^- both at unit activity is as powerful a reductant as hydrogen in the presence of hydrogen ion at unit activity. If we heat powdered silver with a concentrated solution of hydrogen iodide, evolution of hydrogen occurs.

The modification of the relative stabilities of two oxidation states of a metal, both subject to removal by precipitation or complexing, may be treated in a similar way. For example, manganese(III) is a powerful oxidant:

$$Mn^{3+} + e \rightleftharpoons Mn^{2+} \qquad E^\circ = +1.51 \text{ V}$$

In alkaline medium both cations are precipitated, but Mn^{3+} much more completely than Mn^{2+}, the solubility products of $Mn(OH)_3$ and $Mn(OH)_2$ being 10^{-36} and 10^{-13} respectively, whence at $[OH^-] = 1$

$$Mn(OH)_3 + e \rightleftharpoons Mn(OH)_2 + OH^- \qquad E = +0.15 \text{ V}$$

Thus although oxygen at $[H^+] = 1$ cannot oxidise Mn^{2+}, oxygen at $[OH^-] = 1$, for which $E = +0.41$ V (as we have already seen) can and does oxidise $Mn(OH)_2$. Most transition metals resemble manganese in being more stable in higher oxidation states in alkali than in acid because the hydroxide of the metal in its higher oxidation state is much less soluble than the hydroxide of the metal in its lower oxidation state.

Analogous principles apply when two oxidation states of a metal are both stabilised by complexing: it is usually found that the higher oxidation state is stabilised to a greater degree. Thus comparison of

$$Co^{3+} + e \rightleftharpoons Co^{2+} \qquad E^\circ = +1.93\ V$$

and

$$[Co(NH_3)_6]^{3+} + e \rightleftharpoons [Co(NH_3)_6]^{2+} \qquad E^\circ = +0.10\ V$$

shows that the overall formation constant of the $[Co(NH_3)_6]^{3+}$ ion must be about 10^{31} times as great as that of the $[Co(NH_3)_6]^{2+}$ ion. A similar comparison of

$$Fe^{3+} + e \rightleftharpoons Fe^{2+} \qquad E^\circ = +0.77\ V$$

and

$$[Fe(CN)_6]^{3-} + e \rightleftharpoons [Fe(CN)_6]^{4-} \qquad E^\circ = +0.36\ V$$

leads to the conclusion that the overall formation constant of the $[Fe(CN)_6]^{3-}$ ion is 10^7 times that of the $[Fe(CN)_6]^{4-}$ ion. The action of a few organic ligands, notably 1,10-phenanthroline and 2,2′-bipyridyl, in preferentially stabilising lower oxidation states of transition metals was mentioned in Section 7.7.

There are some reactions in which an element in one oxidation state is converted into two other oxidation states, one higher than that in the starting material and the other lower; examples of such *disproportionations* (as they are usually called) are

$$2Cu^+ \rightleftharpoons Cu + Cu^{2+}$$

and

$$3MnO_4^{2-} + 4H^+ \rightleftharpoons 2MnO_4^- + MnO_2 + 2H_2O$$

The former of these may be seen to take place when copper(I) sulphate (prepared by interaction of copper(I) oxide and dimethyl sulphate) is added to water; the latter occurs when acid is added to a solution of potassium manganate(VI). Equilibrium constants for such disproportionations may be calculated from potential data (most simply via ΔG° values); the reaction

$$2Cu^+ \rightleftharpoons Cu + Cu^{2+}$$

for example, for which $K = 10^6$, is the sum of

$$Cu^+ \rightleftharpoons Cu^{2+} + e$$

and

$$Cu^+ + e \rightleftharpoons Cu$$

Species which are unstable with respect to disproportionation, such as Cu^+ in aqueous solution, may, like other oxidation states, be stabilised under suitable conditions – in this case by precipitation as a sparingly soluble species (e.g. as CuCl) or by complexing (e.g. as $[Cu(CN)_4]^{3-}$). In the case of MnO_4^{2-} all that is necessary is to remove the H^+ ions involved in bringing about disproportionation by keeping the solution alkaline. But even this may be considered as yet another example of complexing: neutralisation of an acid in aqueous solution is only hydroxide ion complexing of the proton, for which, on the convention that the activity of the water is unity, $K = 10^{14}$.

Many further examples of aqueous equilibria involving acid–base behaviour, redox potentials, complexing and precipitation will be given later in this book, and a firm grasp of the relationships between them is invaluable in developing an understanding of an important part of modern inorganic chemistry.

7.10 Potential diagrams and oxidation state diagrams

For an element which exists in several different oxidation states in aqueous solution, many half-reactions need to be tabulated to give a clear picture of its solution chemistry. Manganese, for example, exhibits well-defined oxidation states of II, III, IV, VI and VII in aqueous media. The following standard potentials may be determined experimentally:

$$Mn^{2+} + 2e \rightleftharpoons Mn \qquad E^\circ = -1.18\ V$$

$$Mn^{3+} + e \rightleftharpoons Mn^{2+} \qquad E^\circ = +1.51\ V$$

$$MnO_2 + 4H^+ + 2e \rightleftharpoons Mn^{2+} + 2H_2O \qquad E^\circ = +1.23\ V$$

$$MnO_4^- + 8H^+ + 5e \rightleftharpoons Mn^{2+} + 4H_2O \qquad E^\circ = +1.51\ V$$

$$MnO_4^- + e \rightleftharpoons MnO_4^{2-} \qquad E^\circ = +0.56\ V$$

These potentials may be used (care being taken to remember that different numbers of electrons are involved in many of the half-reactions) to derive E° for other half-reactions such as

$$MnO_4^- + 4H^+ + 3e \rightleftharpoons MnO_2 + 2H_2O \qquad E^\circ = +1.69\ V$$

In seeking to obtain an overall view of the relative stabilities of the different oxidation states, it is useful to present all these data on a diagram whose main features are apparent at a glance, and to this end the so-called potential diagrams and oxidation state diagrams have been devised.

The potential diagram for manganese at $[H^+] = 1$ is shown in Fig. 7.1. It shows immediately that MnO_4^{2-} being reduced to MnO_2 is a more powerful oxidant (more negative value of ΔG°) than MnO_4^- being reduced

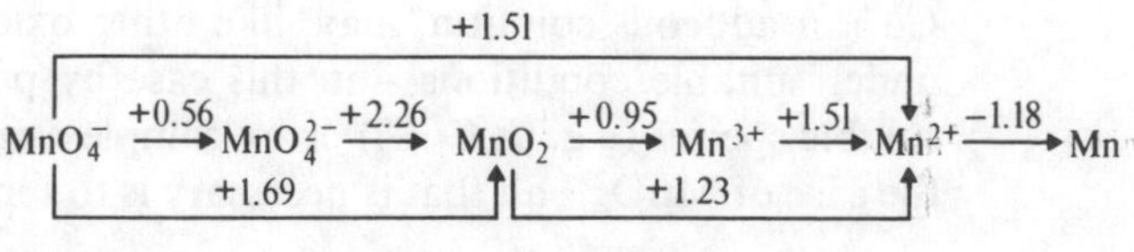

$$MnO_4^- \xrightarrow{+0.56} MnO_4^{2-} \xrightarrow{+2.26} MnO_2 \xrightarrow{+0.95} Mn^{3+} \xrightarrow{+1.51} Mn^{2+} \xrightarrow{-1.18} Mn$$

($MnO_4^- \rightarrow Mn^{2+}$: +1.51; $MnO_4^- \rightarrow MnO_2$: +1.69; $MnO_2 \rightarrow Mn^{2+}$: +1.23)

Figure 7.1
Potential diagram for manganese at $[H^+] = 1$.

to MnO_4^{2-} and hence MnO_4^{2-} will not accumulate during the reduction of MnO_4^- to MnO_2; alternatively, it shows that MnO_4^- being reduced to MnO_2 is a better oxidant than MnO_4^- being reduced to MnO_4^{2-}; each of these statements is tantamount to saying that MnO_4^{2-} at pH 0 is unstable with respect to disproportionation into MnO_4^- and MnO_2. Similarly Mn^{3+} is unstable with respect to disproportionation into MnO_2 and Mn^{2+} at pH 0. But because the protons that are involved in some stages are not explicitly shown, there is a dangerous probability that the reader will forget that they enter into many redox equilibria. At pH values other than zero, those potentials which relate to half-reactions involving protons will change to extents depending on the number of protons involved, whilst the others will remain constant. Thus for an element forming insoluble oxides and oxo-anions, a new potential diagram is needed for every change in hydrogen ion concentration. The incorporation of data for systems involving complexes or insoluble salts is usually not attempted.

In the commonest graphical method of summarising redox relationships the standard free energy change ΔG° for the formation of M(Z) from M(0), where Z is the oxidation state, is plotted against increasing Z. This form of plot, it should be noted, is often used even where half-reactions are written, as in this book,

$$\text{oxidised form} + ne \rightleftharpoons \text{reduced form}$$

To conform to general chemical practice ΔG° should be in the customary units of kJ mol^{-1}; in order to be able to work with small numbers, however, it is usual to replace ΔG° by the quantity $\Delta G^\circ/F$, sometimes called the volt-equivalent, which is equal to $ZE^\circ(\mathrm{M}(Z)/\mathrm{M})$. Since E° is relative to the standard hydrogen electrode, ΔG° is relative to that for the half-reaction corresponding to the standard hydrogen electrode under the same conditions.

The oxidation state diagram for manganese at $[H^+] = 1$ is given in Fig. 7.2. The lowest point on it (Mn(II)) represents the most stable oxidation state. Those halt-reactions which have negative ΔG° are represented by downward slopes, those with positive ΔG° by upward slopes; if $\Delta G^\circ/F$ has been plotted against Z, the slope gives the value of E°. Any state represented by a 'convex' point is thermodynamically unstable with respect to the states on either side of it; the reverse is true of a state represented by a 'concave' point. For example Mn^{3+} has a higher standard free energy than $\frac{1}{2}(Mn^{2+} + MnO_2)$ whilst Mn^{2+} has a lower standard free energy than $(\frac{1}{3}Mn + \frac{2}{3}Mn^{3+})$. The reader should verify that the information contained in Fig. 7.2 is the same as that contained in Fig. 7.1 or in the potential data set out at the beginning of this section. It must again be emphasised that, since protons are involved in many of the changes represented, the diagram would look quite different at other hydrogen ion concentrations, and hence the relative ease with which it may be committed to memory constitutes a potential danger to the unwary. Partly for this

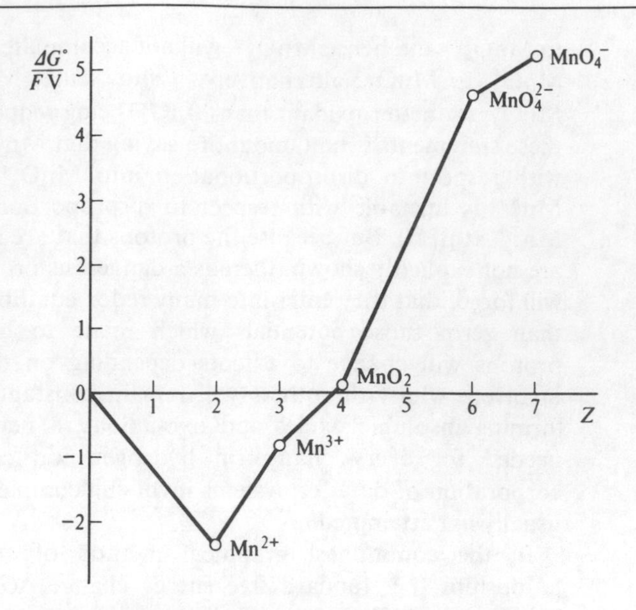

Figure 7.2
Oxidation state diagram for manganese at $[H^+] = 1$.

reason, but mainly because of the desirability of dealing with complex formation and precipitation (which would also necessitate modification of the diagram) at the same time as oxidation states, we shall not make much use of oxidation state diagrams in the discussion of solution chemistry in this book.

7.11 Factors influencing the magnitudes of standard redox potentials

It is interesting to correlate values for standard redox potentials with those for other independently determined thermodynamic quantities and thus to see why, for example, sodium is so much more reactive than silver in aqueous media. The standard potential of the Na^+/Na electrode ($E^\circ = -2.7$ V) can be determined experimentally by a rather elaborate set of experiments involving sodium amalgam electrodes; that of the Ag^+/Ag electrode ($E^\circ = +0.8$ V) is, of course, easily measurable.

We can represent the general half-reaction

$$M^+_{(aq)} + e \rightleftharpoons M_{(s)}$$

as taking place in stages:

$$M^+_{(aq)} \xrightarrow{(1)} M^+_{(g)} \xrightarrow{(2)} M_{(g)} \xrightarrow{(3)} M_{(s)}$$

Of these, (1) is the dehydration of the metal ion, (2) the reverse of the ionisation of the metal atom, and (3) the liberation of energy equivalent to

Table 7.4
Factors determining the magnitude of the Na^+/Na and Ag^+/Ag potentials in aqueous solution: values of ΔH°/kJ mol^{-1}.

Stage	(1)	(2)	(3)	Sum
$Na^+ \rightarrow Na$	413	−494	−109	−190
$H^+ \rightarrow \frac{1}{2}H_2$	1100	−1310	−218	−428
$Ag^+ \rightarrow Ag$	480	−728	−280	−528

the energy of atomisation of the solid metal. Since all standard potentials are relative to that for the half-reaction

$$H^+_{(aq)} + e \rightleftharpoons \tfrac{1}{2}H_{2(g)}$$

we must also consider the stages

$$H^+_{(aq)} \xrightarrow{(1)} H^+_{(g)} \xrightarrow{(2)} H_{(g)} \xrightarrow{(3)} \tfrac{1}{2}H_{2(g)}$$

in which (1) and (2) are again the reversal of hydration and ionisation respectively, and (3) is the liberation of half the bond energy in molecular hydrogen. In exact treatment we should consider ΔG° values for all these stages, but to a first approximation we may disregard entropy changes (which largely cancel one another here) and consider only enthalpy changes. These are tabulated in Table 7.4; conversion into units of potential yields

$$Na^+ + e \rightleftharpoons Na \qquad E^\circ = -2.5\ \text{V}$$

$$Ag^+ + e \rightleftharpoons Ag \qquad E^\circ = +1.0\ \text{V}$$

in good agreement with the experimental values. It should be noted that the difference in enthalpies of atomisation of sodium and silver is almost as large as the difference in their ionisation energies. A similar analysis of the factors determining the difference in the values of E° for the Cu^{2+}/Cu and Zn^{2+}/Zn electrodes shows that it can be attributed mainly to the difference in enthalpies of atomisation of the metals. Thus what is often regarded as a purely 'physical' property may be very important in influencing chemical behaviour.

A similar analysis of the factors influencing the values of the standard potentials for the $\frac{1}{2}X_2/X^-$ electrodes (where X is a halogen) shows that in this case the variation in hydration energy plays the major part. This subject is considered further in Section 16.8.

PROBLEMS

1 What are: (a) the conjugate bases of the acids HF, HSO_4^-, $[Fe(H_2O)_6]^{3+}$ and NH_4^+; (b) the conjugate acids of the bases HSO_4^-, PH_3, NH_2^- and $VO(OH)^+$?

2 Discuss the interpretation of each of the following observations.

(a) Copper(II) oxide dissolves in concentrated aqueous ammonia; addition of sodium sulphide to the solution results in precipitation of copper(II) sulphide.

(b) Magnesium oxide is more soluble in aqueous magnesium chloride solution than in pure water.
(c) A solution of titanium(IV) chloride in aqueous hydrochloric acid forms an orange peroxo compound on treatment with hydrogen peroxide; the colour is destroyed by addition of ammonium fluoride.
(d) The freezing-point of a solution of potassium iodide rises when mercury(II) iodide (which is practically insoluble in pure water) is dissolved in it.

3 Given the overall formation constant of the $[Fe(CN)_6]^{4-}$ ion as 10^{35} and the standard potentials for the half-reactions

$$Fe^{3+} + e \rightleftharpoons Fe^{2+} \qquad E^\circ = +0.77\ V$$

$$[Fe(CN)_6]^{3-} + e \rightleftharpoons [Fe(CN)_6]^{4-} \qquad E^\circ = +0.36\ V$$

calculate the overall formation constant of the $[Fe(CN)_6]^{3-}$ ion.

4 Although the Cu^{2+}/Cu^{+} and $\frac{1}{2}I_2/I^-$ standard potentials are +0.15 and +0.54 V respectively, copper(II) salts liberate iodine from potassium iodide solution. In the presence of sodium tartrate, iodine oxidises copper(I) iodide to a copper(II)-containing solution, but if excess of dilute sulphuric acid is then added, the reaction is reversed.
Suggest an explanation for all these observations.

5 The following potential diagram at pH 0 summarises the results of electrochemical studies of the aqueous solution chemistry of a metal M.

$$MO_2^{2+} \xrightarrow{+0.06} MO_2^{+} \xrightarrow{+0.58} M^{4+} \xrightarrow{-0.63} M^{3+} \xrightarrow{-1.80} M$$

($MO_2^{2+} \rightarrow M^{4+}$: +0.32)

Use the information given in the diagram to deduce as much as possible about the chemistry of M.

6 In the presence of a little hydrochloric acid, sulphur dioxide reduces iron(III) to iron(II), but in concentrated hydrochloric acid solution this reduction takes place to only a slight extent.
Suggest an explanation of these observations, and describe how you would attempt to show that your explanation is correct.

7 Oxidation states (for oxidation numbers) are the charges (often fictitious) assigned to atoms in molecules or ions according to the following abbreviated set of rules. Atoms in elements are assigned the oxidation state zero. Combined fluorine is always assigned the oxidation state −1 (i.e. it is assumed for these purposes always to be present as F^-). Combined hydrogen is assigned the oxidation state +1 in compounds with non-metals and −1 in compounds with metals. Combined oxygen, except in compounds with fluorine and in hydrogen peroxide and its derivatives, is assigned the oxidation state −2. The sum of the oxidation numbers of the atoms in a molecule or ion must be equal to the electrical charge on that species. Oxidation states are commonly shown by Roman numerals to distinguish them from ionic charges.

(a) Apply these rules to work out the oxidation states of
 (i) oxygen in H_2O_2
 (ii) chromium in $K_2Cr_2O_7$
 (iii) phosphorus in H_3PO_2
 (iv) nitrogen in HN_3
 (v) carbon in CH_2F_2.

(b) When HN_3 and H_3PO_2 react with excess of iodine (which is converted into iodide) the molar ratios of reactants are $2HN_3{:}I_2$ and $H_3PO_2{:}2I_2$ respectively. What are the oxidation states of nitrogen and phosphorus in the nitrogen-containing and phosphorus-containing products?

REFERENCES FOR FURTHER READING

Bard, A.J., Parsons, R. and Jordan, J. (1985) *Standard Potentials in Aqueous Solution*, Marcel Dekker, New York. A critical compilation of values, the successor to Latimer's famous treatment of this subject.

Burgess, J. (1978) *Metal Ions in Solution*, Ellis Horwood, Chichester, and Halstead Press, New York. A thorough treatment of most aspects of metal ions in both aqueous and non-aqueous solutions.

Gillespie, R.J. (1975) 'Proton acids, Lewis acids, hard acids, soft acids and superacids', **Chap. 1** pp. 1–30 in Caldin, E. and Gold, V. (eds.), *Proton-Transfer Reactions*, Chapman and Hall, London, and Halstead Press, New York. A comparative account of different definitions of acid and base and assessments of acidity.

Hartley, F.R., Burgess, C. and Alcock, R. (1980) *Solution Equilibria*, Ellis Horwood, Chichester and Halstead Press, New York. The fullest recent account of the study of complexes in solution with reference to analytical chemistry and biochemistry as well as inorganic chemistry.

Johnson, D.A. (1982) *Some Thermodynamic Aspects of Inorganic Chemistry*, 2nd edition, Cambridge University Press, Cambridge. Contains a very useful discussion of solubility and redox potentials.

Jolly, W.L. (1984) *Modern Inorganic Chemistry*, McGraw-Hill, New York. Contains an account of the chemistry of the electron in aqueous solutions, and complements the treatment of redox potentials given in this chapter by discussing mainly systems involving non-metallic elements.

Marcus, Y. (1985) *Ion Solvation*, Wiley, New York. The definitive work on this subject.

Parish, R.V. (1977) *The Metallic Elements*, Longman, London and New York. Makes extensive use of oxidation state diagrams in the descriptive chemistry of the metals.

Rossotti, H. (1978) *The Study of Ionic Equilibria in Aqueous Solution*, Longman, London and New York. An introduction to quantitative aspects of complex formation, solubility and redox processes with special reference to experimental methods.

Sharpe, A.G. (1990) *J. Chem. Education*, **67**, 309. A short review of the solvation of halide ions and its chemical significance.

8 Inorganic chemistry in non-aqueous media

Introduction • Ammonia • Hydrogen fluoride • Sulphuric acid and fluorosulphuric acid • Bromine trifluoride • Dinitrogen tetroxide

8.1 Introduction

In this chapter we aim to extend the discussion of solubility, complex formation (including acid–base relationships), and redox processes to non-aqueous solutions. The extent to which this is possible is severely limited by the fact that no other solvent is nearly as good as water for dissolving inorganic substances; further, many non-aqueous solvents are highly reactive, thus restricting the range of chemical studies possible in them. Quantitative data are scarce and, in solvents of relative permittivity (dielectric constant) lower than that of water, difficult to interpret because of ion-association. Although, therefore, we shall make some general observations in this section, no integrated treatment of inorganic chemistry in non-aqueous media is yet possible, and much of our discussion will appear under headings of specific solvents.

A few molecular inorganic solids (e.g. iodine) which are insoluble in water dissolve in a wide range of non-aqueous solvents, often forming ideal non-conducting solutions. Most inorganic compounds, however, are either insoluble in all non-aqueous solvents or dissolve only in those of high relative permittivity. As we mentioned in the last chapter, the standard free energy of solvation of an ion should, on a purely electrostatic theory of solvation, be proportional to $(1-1/\varepsilon_r)$ where ε_r is the relative permittivity of the solvent in the neighbourhood of the ion. This is not the same as the bulk value, but may be supposed to be related to it, and hence it is not surprising that the organic liquids which are the best solvents for salts are those which have high relative permittivities, e.g. dimethyl sulphoxide (47) and nitromethane (36). In accounting for solubilities in general, however, certain molecular properties of solvents must also be taken into consideration. Ion–solvent interaction is favoured by the use of a solvent with a large dipole moment, but for maximum effect the solvent molecule should also be small, and both ends of it should be able to interact with ions in the same way as water interacts with cations at the oxygen end of the molecules and with anions at the hydrogen ends. Thus ammonia (relative permittivity 23 and dipole moment 1.5 debyes) is a better solvent for salts than dimethyl sulphoxide and nitromethane, although both of these have dipole moments in the range 3–4 debyes. Attempts have been made to place the donor properties of solvents in a general scale, but with only limited success: donor properties depend markedly on what acceptor is involved. (We shall see that this is true also of gas-phase complex formation when we discuss donor–acceptor complexes of the boron halides, borane and boron trimethyl in Chapter 12.) In practice, of course, ion–dipole interaction in solution is often difficult to distinguish from specific complex formation resulting from the use of a lone pair of electrons on one atom of the solvent for covalent bond formation, such as certainly occurs in ammine complexes of transition metal ions. We must therefore be prepared to consider many different factors in accounting for the solvent powers of different liquids, and no comprehensive quantitative treatment of solubility in non-aqueous media is yet possible.

There are, nevertheless, sufficient accurate data for the transfer of simple ions from water to a few organic solvents of high specific permittivity to enable some interesting conclusions to be drawn. Since most organic liquids are soluble in water to some extent, and a few are completely miscible with it, enthalpies and free energies of transfer are usually obtained from separate measurements on different solvents as for water, and the results are combined: the difference between two enthalpies of solution, for example, gives the enthalpy of transfer. Enthalpies and free energies of transfer of tetraphenylarsonium and tetraphenylborate ions from water to all organic solvents are by convention taken as being equal – the TATB assumption we met in the last chapter. Results are usually left as enthalpies, etc., of transfer because of the uncertainties in the absolute values of enthalpies, etc., of solvation in water.

We shall confine attention here to enthalpies and free energies of transfer from water to methanol, formamide, dimethylformamide and acetonitrile, for which the specific permittivities and dipole moments (the latter, in debyes, in parentheses) are 33 (2.9), 111 (3.3), 37 (3.9) and 37 (3.5) respectively. The data are presented in Tables 8.1 and 8.2.

The large non-polar ions Ph_4As^+ and BPh_4^- are more solvated in the organic solvents than in water, enthalpy and entropy effects both contributing in the same direction. Alkali metal ions exhibit no simple pattern of behaviour except insofar as ΔH°_{trans} and ΔG°_{trans} are nearly all less positive than those for halide ions. Amongst the latter, however, there is a marked difference between transfer to a solvent capable of strong hydrogen bonding (methanol or formamide) and one which is not

Table 8.1
$\Delta H^\circ_{transfer}$/kJ mol^{-1} from water to other solvents.

Ion	Methanol	Formamide	Dimethylformamide	Acetonitrile
F^-	12	20	—	—
Cl^-	8	4	18	19
Br^-	4	−1	1	8
I^-	−2	−7	−15	−8
Li^+	−22	−6	−25	—
Na^+	−20	−16	−32	−13
K^+	−19	−18	−36	−23
$Ph_4As^+ = BPh_4^-$	−2	−1	−17	−10

(dimethyl or acetonitrile): not only are free energies of transfer more positive for aprotic solvents, but the variation in values amongst the halide ions is much greater. This leads us to conclude that because hydrogen bonding to the solvent is impossible, fluoride and chloride ions, in particular, are much less strongly solvated in aprotic solvents than in water or protic solvents. This is the origin of their greater reactivity in aprotic solvents, a well-known generalisation in organic chemistry; e.g. in the bimolecular reaction

$$CH_3Br + X^- \rightarrow CH_3X + Br^-$$

the rate increases from X = F to X = I in aqueous solution, but decreases in dimethylformamide. Fluoride ion in aprotic solvents is sometimes described as 'naked', but this term is misleading; in dimethylformamide, for example, it still has a solvation free energy of about −400 kJ mol^{-1} (60 kJ mol^{-1} less negative than in water), and it is still very much less reactive than it is in the gas phase.

The thermodynamics of metal complex formation is also affected by change of solvent, though this subject has been much less investigated

Table 8.2
$\Delta G^\circ_{transfer}$/kJ mol^{-1} from water to other solvents*.

Ion	Methanol	Formamide	Dimethylformamide	Acetonitrile
F^-	20	25	~60	71
Cl^-	13	14	48	42
Br^-	11	11	36	31
I^-	7	7	20	17
Li^+	4	−10	−10	25
Na^+	8	−8	−10	15
K^+	10	−4	−10	8
$Ph_4As^+ = BPh_4^-$	−23	−24	−38	−33

*Concentrations from which $\Delta G^\circ_{transfer}$ is evaluated are customarily on the molar (mol dm^{-3} of solution) scale and are therefore not strictly comparable with those (mol kg^{-1} of water) used in Chapter 7. The difference does not affect the qualitative discussion in this section, however.

than reaction kinetics. We have, however, mentioned the $HgCl_2$–Br^- reaction in Section 7.7, and as a further example we may cite halide complexes of the Cd^{2+} ion, the stability sequence of which is chloride < bromide < iodide in aqueous solution but iodide < bromide < chloride in dimethyl sulphoxide (which in solvating properties behaves very like dimethylformamide).

Acid–base behaviour in protonic solvents is a relatively favourable subject for treatment, since we are always dealing with the same acceptor, the proton. Non-aqueous solvents which are good proton acceptors (e.g. ammonia) naturally encourage acids to ionise in them; thus in a basic solvent all acids become strong ones, and the solvent is said to exert a *levelling effect*. The reverse is true for acidic solvents (e.g. acetic acid, sulphuric acid) in which all bases become strong ones, and in which most acids are relatively weak, and some even ionise as bases. Acidic solvents exert a *differentiating effect* on the ionisation of acids. In acetic acid, for example, hydrogen chloride, bromide and iodide (all very strong acids in aqueous solution) are weak electrolytes, though the extent of ionisation increases with increase in the atomic number of the halogen. In sulphuric acid, even perchloric acid is practically a non-electrolyte, and nitric acid ionises according to the equation:

$$HNO_3 + 2H_2SO_4 \rightleftharpoons NO_2^+ + H_3O^+ + 2HSO_4^-$$

This may be regarded as the summation of the processes:

$$HNO_3 + H_2SO_4 \rightleftharpoons H_2NO_3^+ + HSO_4^-$$

$$H_2NO_3^+ \rightleftharpoons NO_2^+ + H_2O$$

$$H_2O + H_2SO_4 \rightleftharpoons H_3O^+ + HSO_4^-$$

It is, of course, the presence of the nitronium ion, NO_2^+, in the solution that is responsible for use of nitric acid–sulphuric acid in the nitration of aromatic compounds. In *super-acidic* media such as mixtures of antimony pentafluoride and hydrogen fluoride or fluorosulphuric acid ($HOSO_2F$), even hydrocarbons may be protonated, i.e. act as bases.

The ionisation of water according to the equation

$$2H_2O \rightleftharpoons H_3O^+ + OH^-$$

is the transfer of a proton from one solvent molecule to another. Liquid ammonia ionises to a small extent by an analogous process:

$$2NH_3 \rightleftharpoons NH_4^+ + NH_2^-$$

Thus by analogy with aqueous media, an acid in liquid ammonia may be described as a substance which produces NH_4^+ ions and a base as one which produces NH_2^- ions; NH_4Cl and KNH_2, for example, react to form KCl and NH_3. Such a solvent-oriented definition may be widened to include behaviour in any solvent which undergoes self-ionisation; an acid is a substance which produces the cation, and a base a substance which produces the anion, characteristic of the solvent. Liquid dinitrogen tetroxide, for example, ionises slightly according to the equation:

$$N_2O_4 \rightleftharpoons NO^+ + NO_3^-$$

In it, nitrosonium salts (e.g. $NO^+ClO_4^-$) are acids, and metal nitrates are bases. In some ways the use of this acid–base terminology for systems not involving protons at all results in an unfortunate addition to the meanings attached to these overworked words (we have already met Arrhenius, Brønsted, Lewis, and hard and soft acids and bases); but it has certainly been helpful in suggesting lines of investigation in the study of non-aqueous media, and its use will probably continue in the future.

As we saw in Section 7.8, the range of oxidising and reducing agents that can persist in aqueous solution is limited by oxidation or reduction of the solvent. Other solvents have quite different susceptibilities to oxidation and reduction. Liquid ammonia, for example, is ionised to a very much smaller extent than water, i.e. the concentration of solvated protons in it is much lower. It is therefore more difficult to reduce than water and, as we shall see, partly for this reason and partly because of the lower boiling-point of ammonia, electrons can survive in ammonia solution for a considerable time; hence arises the use of sodium in liquid ammonia as an excellent reducing agent for organic compounds – sodium in water would be useless for these purposes since the water would be reduced preferentially. At the other end of the scale, hydrogen fluoride is extremely resistant to oxidation, and so in this solvent many oxidations are possible which could not be carried out in water. Non-aqueous solvents sometimes stabilise oxidation states that are unstable in water: copper(I), for example is stable with respect to disproportionation in acetonitrile solution. The acceptor properties of a metal ion may also be affected by a change of solvent: copper(I), a class (b) acceptor in aqueous solution, is a class (a) acceptor in acetonitrile or pyridine. Such observations add further support to the view taken in the last chapter, that complex formation in solution is a competitive process in which the relative strengths of the interactions of the solvent with all the species present are involved.

We now turn to some individual inorganic solvents, beginning with proton-containing species (ammonia, hydrogen fluoride, sulphuric acid and fluorosulphuric acid) and then going on to aprotic systems (bromine trifluoride and dinitrogen tetroxide). The solvents chosen here represent only a few of those which have been investigated. Any liquid which has even a small electrical conductivity must contain cations and anions, and species which produce these ions in solution and which can interact to form a salt and the solvent can often be found. Thus in nitrosyl chloride (which we shall not discuss in detail) the ionisation equilibrium is

$$NOCl \rightleftharpoons NO^+ + Cl^-$$

and the interaction of $NO^+[AlCl_4]^-$ and $Me_4N^+Cl^-$, leading to the formation of $Me_4N^+[AlCl_4]^-$ and the solvent, is a neutralisation process that can be followed by conductivity measurements. The solvents we have chosen for more detailed study all present some special features of interest in addition to the demonstration of the

$$\text{acid} + \text{base} \rightarrow \text{salt} + \text{solvent}$$

reaction in them.

One omission from the list of solvents discussed in some detail here calls for comment. The solvent-based definition of acids and bases was first put forward for sulphur dioxide, for which the self-ionisation equilibrium

$$2SO_2 \rightleftharpoons SO^{2+} + SO_3^{2-}$$

was suggested. Unlike suggested methods of ionisation for other solvents, this one requires the separation of doubly charged species, and on these grounds alone must be considered improbable; its correctness is further called into question by the fact that thionyl chloride, $SOCl_2$, the only reported acid in the solvent, does not exchange ^{35}S or ^{18}O with the solvent. So although sulphur dioxide is a useful inert solvent in many reactions and, considering its low relative permittivity (15), quite a good ionising medium (e.g. for triphenylchloromethane), we have not judged it to be one of the solvents most suitable for detailed discussion here.

Much of our knowledge of inorganic chemistry in non-aqueous media has been obtained by classic physicochemical methods such as the study of colligative properties and conductance, potentiometric, and spectroscopic measurements. In addition to the examples given in this chapter, important applications of non-aqueous solvents include the separation of uranium and plutonium in nuclear technology (Sections 1.8 and 27.6) and the analytical separation of many metals (e.g. iron; cf. Sections 7.5 and 24.7). Mixed aqueous-organic media are frequently used for the determination of formation constants of metal complexes of organic ligands which are sparingly soluble in water, though their use introduces considerable complications into the interpretation of experimental data. The use of fused salts as reaction media is a recent development of potentially far-reaching significance, since it greatly extends the temperature range over which solution measurements can be made; for an account of this subject, however, the references at the end of the chapter should be consulted.

8.2 Ammonia

Ammonia is a liquid in the temperature range −78 to −33 °C under atmospheric pressure; it has a relative permittivity of 23 at its boiling-point and a specific conductivity of approximately 10^{-11} ohm^{-1} cm^{-1};

from the latter value and the molar conductivities of the NH_4^+ and NH_2^- ions (142 and 166 ohm^{-1} cm^2 respectively), the ionisation constant is 5×10^{-27}. Its lower boiling-point suggests that hydrogen bonding in liquid ammonia is much weaker than that in water, a conclusion supported by the fact that values for the enthalpy and entropy of vaporisation are also lower. Another consequence of the less polar nature of the N—H bond is that, unlike H_3O^+ and OH^- in water, which conduct by a proton-switch mechanism and are much more mobile than other singly-charged ions, NH_4^+ and NH_2^- in ammonia have ionic mobilities approximately equal to those of alkali metal and halide ions.

Salts are generally less soluble in liquid ammonia than in water, partly as a consequence of the lower relative permittivity. There are, however, some noteworthy exceptions to this generalisation; thus ammonium salts, iodides and nitrates are usually readily soluble. Silver iodide, for example, which is very sparingly soluble in water, dissolves easily in liquid ammonia, a fact which indicates that both the silver ion (which forms an ammine complex) and the iodide ion interact strongly with the solvent. Changing solubility patterns lead to the possibility of carrying out interesting precipitations in ammonia; thus whereas in aqueous solution barium chloride and silver nitrate react to precipitate silver chloride, in ammonia solution silver chloride and barium nitrate react to precipitate barium chloride. Most chlorides (and nearly all fluorides) are practically insoluble in ammonia. Molecular organic compounds are generally more soluble than in water.

As we have previously mentioned, ammonium salts and metal amides react in ammonia, e.g.

$$NH_4^+Br^- + K^+NH_2^- \rightarrow K^+Br^- + 2NH_3$$

This reaction may be followed by conductivity or potentiometry, or by the use of an indicator such as phenolphthalein (which is colourless but is deprotonated to a red anion by a strong base like NH_2^-, as it is by OH^- in water). Solutions of ammonium halides in ammonia may be used as acids, e.g. in the preparation of silane and arsine:

$$Na_3As + 3NH_4Br \rightarrow AsH_3 + 3NaBr + 3NH_3$$

$$Mg_2Si + 4NH_4Br \rightarrow SiH_4 + 2MgBr_2 + 4NH_3$$

A saturated solution of ammonium nitrate in ammonia (which has a vapour pressure of less than one atmosphere even at the ordinary temperature) dissolves many metal oxides and even some metals; when the latter dissolve some of the nitrate is reduced to nitrite. Metals which form insoluble hydroxides form insoluble amides; some of these (e.g. $Zn(NH_2)_2$) are soluble in metal amide solutions as a result of a reaction analogous to that which occurs when zinc hydroxide dissolves in aqueous alkali:

$$Zn(NH_2)_2 + 2NH_2^- \rightarrow [Zn(NH_2)_4]^{2-}$$

$$Zn(OH)_2 + 2OH^- \rightarrow [Zn(OH)_4]^{2-}$$

Metal nitrides on the ammonia system are analogous to metal oxides; and many other analogies may be drawn.

All of the alkali and alkaline earth metals except beryllium dissolve in liquid ammonia to form metastable solutions from which the alkali metals may be recovered unchanged and the alkaline earth metals as hexammoniates. Dilute solutions of the metals are bright blue, the colour being due to the short wavelength tail of a broad and intense absorption band in the infrared. The visible-region spectra of solutions of all the metals are the same; the colour must therefore be due to the presence of a common species, which is the solvated electron. The solutions occupy a much greater volume than the sum of the volumes of metal and solvent, so it is suggested that the electrons occupy cavities of radius 3–4 Å. Very dilute solutions of the metals are paramagnetic, the susceptibility corresponding with that calculated for the presence of one free electron per metal atom.

As the solutions are made more concentrated, the molar conductivity at first decreases, reaching a minimum at about 0.05 molar; thereafter, it increases again until in saturated solutions it is comparable to that of a metal. Such saturated solutions are no longer blue but are diamagnetic and bronze-coloured, like a molten metal. They have been described as expanded metals.

The blue solutions of alkali metals in ammonia decompose very slowly with liberation of hydrogen (i.e. reduction of the solvent):

$$NH_3 + e \rightarrow NH_2^- + \tfrac{1}{2}H_2$$

Decomposition is accelerated by the presence of many transition metal compounds, e.g. by stirring the solutions with a rusty iron wire. Ammonium salts, strong acids in this medium, suffer immediate decomposition:

$$NH_4^+ + e \rightarrow NH_3 + \tfrac{1}{2}H_2$$

Among other reducing actions we may note the following:

$$GeH_4 + e \rightarrow GeH_3^- + \tfrac{1}{2}H_2$$

$$O_2 + e \rightarrow O_2^-$$

$$O_2 + 2e \rightarrow O_2^{2-}$$

$$[Ni(CN)_4]^{2-} + 2e \rightarrow [Ni(CN)_4]^{4-}$$

$$Fe(CO)_5 + 2e \rightarrow [Fe(CO)_4]^{2-} + CO$$

Further reference is made to these reactions under the element concerned.

Reduction potentials for the reversible conversion of metal ions in ammonia solution into metals in general follow the same sequence as for the corresponding ions in aqueous solution, as may be seen from the values Li^+/Li, −2.24; K^+/K, −1.98; Na^+/Na, −1.85; Zn^{2+}/Zn, −0.53; $H^+/\frac{1}{2}H_2$, 0; Cu^{2+}/Cu, +0.43; Ag^+/Ag, +0.83 V. Information deduced from these values and from lattice energies and solubilities indicates that the proton and transition metal ions have more negative absolute standard free energies of solvation in ammonia than in water, whereas those for alkali metal ions are about the same in the two solvents. This is, of course, in agreement with the experimental observation that hydrogen and transition metal ions are complexed by ammonia in aqueous solution whilst alkali metal ions are not. Potential data for oxidising systems cannot be obtained in liquid ammonia owing to the ease with which the solvent is oxidised. Lattice energy and solubility data, however, reveal the interesting fact that whilst ΔG_S° is 30 kJ mol^{-1} less negative for chloride ion in ammonia than in water, the values for iodide ion are the same in the two solvents.

8.3 Hydrogen fluoride

Hydrogen fluoride attacks glass and has found applications as a non-aqueous solvent only comparatively recently; it can be handled without excessive inconvenience in fluorine-containing plastics (such as polytetrafluoroethylene) or, if absolutely dry, in copper or stainless steel vacuum lines. It is liquid from −89 to 19.5 °C; it has a relative permittivity of 84 at 0 °C, and its specific conductivity at this temperature is approximately 10^{-6} ohm^{-1} cm^{-1}. The equilibrium constant for its ionisation according to the equation

$$3HF \rightleftharpoons H_2F^+ + HF_2^-$$

is about 10^{-12} at 0 °C. Hydrogen fluoride is strongly hydrogen-bonded, but, having only one H per molecule, it forms only chains and rings of various sizes; some of these, in particular cyclic $(HF)_6$, persist in the vapour, thus accounting for the relatively low boiling-point (though it should be noted that the other hydrogen halides, which are not hydrogen-bonded, are much more volatile).

Most salts are converted into fluorides by the action of liquid hydrogen fluoride, and only a few of these are soluble. Alkali, alkaline earth, silver, and thallium(I) fluorides dissolve to form acid fluorides, e.g. $K[HF_2]$, $K[H_2F_3]$; fluorine was first isolated by the electrolysis of fused $K[HF_2]$. Inorganic and organic acids (together with many other organic compounds) are usually protonated, e.g. acetic acid forms $CH_3C(OH)_2^+HF_2^-$; a few molecular fluorides, however, act as fluoride ion acceptors and lead to formation of the H_2F^+ cation and very strongly acidic solutions, e.g.

$$2HF + SbF_5 \rightarrow H_2F^+[SbF_6]^-$$

$$2HF + AsF_5 \rightarrow H_2F^+[AsF_6]^-$$

Phosphorus pentafluoride and boron trifluoride form $H_2F^+[PF_6]^-$ and $H_2F^+[BF_4]^-$ only to slight extents, i.e. they are weak acids in this medium. Carbon and silicon fluorides are insoluble. Proteins react immediately with liquid hydrogen fluoride, and it produces very serious burns if it is allowed to come into contact with skin.

Electrolysis in liquid hydrogen fluoride is an important route to the preparation of both inorganic and organic fluorine compounds. Thus anodic oxidation of ammonium fluoride yields NFH_2, NF_2H and NF_3; of H_2O yields OF_2; and of CH_3COOH, $(C_2H_5)_2O$ and $(CH_3)_3N$ yields CF_3COOH, $(C_2F_5)_2O$ and $(CF_3)_3N$ respectively.

8.4 *Sulphuric acid and fluorosulphuric acid*

Sulphuric acid, which melts at 10 °C and boils at 320 °C, is a widely used acidic non-aqueous solvent, though its viscosity and high enthalpy of vaporisation (both the consequence of extensive hydrogen bonding) sometimes make the isolation and purification of products troublesome. It has a relative permittivity of 100 and a specific conductivity of 10^{-2} ohm^{-1} cm^{-1} at 25 °C. For the primary self-ionisation

$$2H_2SO_4 \rightleftharpoons H_3SO_4^+ + HSO_4^-$$

the equilibrium constant is 2.7×10^{-4} at this temperature; other equilibria, such as

$$2H_2SO_4 \rightleftharpoons H_3O + HS_2O_7^-$$

are also involved to a small extent, but we shall disregard them here. Sulphuric acid is, of course, a highly acidic solvent, and most other acids are neutral or basic in it; as we have seen for nitric acid, however, when another acid is protonated the resulting species often loses water and a H_3O^+ ion is also formed:

$$HNO_3 + 2H_2SO_4 \rightarrow NO_2^+ + H_3O^+ + 2HSO_4^-$$

The nature of such reactions may be investigated by an ingenious combination of conductivity and cryoscopic measurements. Cryoscopy tells us ν, the total number of particles present per molecule of solute; conductivity in sulphuric acid is almost entirely due to the presence of $H_3SO_4^+$ and/or HSO_4^- ions (which carry current by proton-switching mechanisms, thus avoiding the need for migration through the viscous solvent) and tells us γ, the number of these ions produced per molecule of solute. For a solution of acetic acid in sulphuric acid, $\nu = 2$ and $\gamma = 1$, corresponding to the reaction

$$CH_3COOH + H_2SO_4 \rightarrow CH_3C(OH)_2^+ + HSO_4^-$$

For nitric acid, on the other hand, $\nu = 4$ and $\gamma = 2$. For a solution of boric acid, it is found that $\nu = 6$ and $\gamma = 2$:

$$H_3BO_3 + 6H_2SO_4 \rightarrow [B(HSO_4)_4]^- + 3H_3O^+ + 2HSO_4^-$$

For the anion $[B(HSO_4)_4]^-$ to be formed, the acid $H[B(HSO_4)_4]$ must be a strong one even in sulphuric acid solution. A solution of it can be made by dissolving boric acid in oleum:

$$H_3BO_3 + 2H_2SO_4 + 3SO_3 \rightarrow H_3SO_4^+ + [B(HSO_4)_4]^-$$

and can be titrated conductimetrically against a solution of a strong base such as $KHSO_4$ to produce a solution containing $K[B(HSO_4)_4]$ and the solvent.

Fluorosulphuric acid, FSO_2OH, is a liquid between -89 and 163 °C; it is much less viscous than sulphuric acid. It does not attack glass appreciably. Its self-ionisation

$$2HSO_3F \rightleftharpoons H_2SO_3F^+ + SO_3F^-$$

has an equilibrium constant of about 10^{-8}. The particular importance of fluorosulphuric acid lies in the fact that in combination with antimony pentafluoride it forms the strongest acids known. A typical equilibrium (one of several, for the system is a complex one) is:

$$2HSO_3F + SbF_5 \rightleftharpoons H_2SO_3F^+ + [SbF_5(OSO_2F)]^-$$

Such a mixture, which is often diluted with sulphur dioxide as an inert solvent, is an even more powerful protonating agent than hydrogen fluoride and antimony pentafluoride; it will protonate not only a very wide range of organic compounds but also such exceedingly weak bases as the phosphorus trihalides, which are converted into substituted phosphonium cations of general formula PHX_3^+, and carbonic acid, which forms the unstable cation $C(OH)_3^+$.

8.5 Bromine trifluoride

Bromine trifluoride, a pale yellow liquid freezing at 9 °C and boiling at 126 °C, is the only aprotic solvent for which the postulated self-ionisation has been substantiated by the isolation and characterisation by X-ray diffraction of acids and bases, and by conductimetric titration of them. Bromine trifluoride fluorinates everything which dissolves in it, and a great many other substances too; but massive quartz is kinetically stable towards the reagent and vessels of quartz, or of metals (e.g. nickel) that become protected by a thin layer of fluoride, may be used for handling the compound.

The specific conductivity of bromine trifluoride is 8×10^{-3} ohm^{-1} cm^{-1} at 25 °C; its relative permittivity is 107. Self-ionisation is believed to occur according to the equation

$$2BrF_3 \rightleftharpoons BrF_2^+ + BrF_4^-$$

Alkali metal, barium, and silver(I) fluorides combine with the solvent to form polyhalides containing the planar BrF_4^- ion, e.g. $KBrF_4$, $Ba[BrF_4]_2$, $AgBrF_4$. Antimony(V), tin(IV) and gold(III) fluorides also combine with it; the product from antimony pentafluoride has been shown to be $(BrF_2)^+[SbF_6]^-$, with a bent cation and a regular octahedral anion, and by analogy the compounds formed from the other halides are formulated $(BrF_2^+)_2[SnF_6]^{2-}$ and $(BrF_2^+)[AuF_4]^-$. Conductivity measurements on solutions containing $(BrF_2)[SbF_6]$ and $AgBrF_4$ or $(BrF_2)_2[SnF_6]$ and $KBrF_4$ show minima at 1:1 and 1:2 molar ratios of reactants, thus supporting the formulation of the neutralisation reactions

$$(BrF_2)^+[SbF_6]^- + Ag^+BrF_4^- \rightarrow Ag^+[SbF_6]^- + 2BrF_3$$

$$(BrF_2^+)_2[SnF_6]^{2-} + 2K^+BrF_4^- \rightarrow K_2^+[SnF_6]^{2-} + 4BrF_3$$

Since bromine trifluoride fluorinates antimony, silver, tin and potassium chloride, the complex fluorides produced in the above reactions may conveniently be prepared by treating a 1:1 mixture of antimony and silver, or a 1:2 mixture of powdered tin and potassium chloride, with the reagent and removing the products and excess of bromine trifluoride by evaporation *in vacuo*.

Many other fluoro complexes may be obtained by analogous methods, e.g.

$$Ag + Au \rightarrow Ag[AuF_4]$$

$$SnCl_4 + 2NOCl \rightarrow (NO)_2[SnF_6]$$

$$SbCl_5 + N_2O_4 \rightarrow (NO_2)[SbF_6]$$

$$VCl_3 + KCl \rightarrow K[VF_6]$$

$$AgBO_2 \rightarrow Ag[BF_4]$$

These reactions may also involve ionic intermediates, but it is not necessarily true that all do so. Some of the complexes thus prepared are also obtainable by fluorinations using elemental fluorine, but this often requires the use of a higher temperature; thus silver(I) salts of complex fluoro acids, which often show remarkable solubility in organic solvents and hence are important reagents in organic and organometallic chemistry, are not obtained by the latter method, silver(II) fluoride being formed.

8.6 Dinitrogen tetroxide

Although this compound has a liquid range only from −12 to 21 °C and a relative permittivity of only 2.4 (so that it is a poor solvent for most inorganic compounds), its preparative uses justify the allocation of a separate section in this chapter.

The proposed self-ionisation of dinitrogen tetroxide according to the equation

$$N_2O_4 \rightleftharpoons NO^+ + NO_3^-$$

must occur to an extremely small extent since the specific conductivity is only 2×10^{-13} ohm^{-1} cm^{-1} at 17 °C. The presence of nitrate ions in the liquid solvent is indicated by a rapid exchange of nitrate between liquid dinitrogen tetroxide and tetraethylammonium nitrate (which is soluble owing to its very low lattice energy). Metals such as lithium and sodium react with the liquid to liberate nitric oxide, e.g.

$$Li + N_2O_4 \rightarrow LiNO_3 + NO$$

Less reactive metals may react rapidly if nitrosyl chloride, tetraethylammonium nitrate, or an organic donor molecule such as acetonitrile or ethyl acetate is present. Nitrosyl chloride can be considered a very weak acid in liquid N_2O_4 and its action is explained on this basis. Tetraethylammonium nitrate is, of course, a base, and its action on metals such as zinc and aluminium arises from the formation of nitrato complexes (analogous to hydroxo complexes in aqueous solution) by reactions such as

$$Zn + 2Et_4NNO_3 + 2N_2O_4 \rightarrow (Et_4N)_2[Zn(NO_3)_4] + 2NO$$

Organic donor molecules appear to act by increasing the degree of self-ionisation of the solvent by coordination with the NO^+ cation. Thus acetonitrile or ethyl acetate–dinitrogen tetroxide mixtures readily dissolve copper, iron and zinc with formation of the acids $NO[Cu(NO_3)_3]$, $NO[Fe(NO_3)_4]$ and $(NO)_2[Zn(NO_3)_4]$, e.g.

$$Cu + 3N_2O_4 \rightarrow NO[Cu(NO_3)_3] + 2NO$$

The presence of the NO^+ cation in these substances is shown by a characteristic infrared absorption at about 2 300 cm^{-1}. Analogous derivatives of other metals are obtained by the action of dinitrogen tetroxide on metal carbonyls, e.g.

$$Mn_2(CO)_{10} + 8N_2O_4 \rightarrow 2(NO)_2[Mn(NO_3)_4] + 4NO + 10CO$$

The structures of the $[M(NO_3)_4]^{2-}$ anions are of considerable interest: in them the nitrate ion functions as a bidentate ligand and the metal atom is eight-coordinated. Decomposition of $NO[Cu(NO_3)_3]$ etc. yields the otherwise inaccessible anhydrous transition metal nitrates, volatile solids that are readily soluble in organic solvents.

PROBLEMS

1 Discuss the following observations:
(a) Zinc dissolves in a solution of sodium amide in liquid ammonia with liberation of hydrogen; careful addition of ammonium iodide to the resulting solution produces a white precipitate which dissolves if excess of ammonium iodide is added.
(b) For a solution of N_2O_4 in sulphuric acid it is found that $\nu = 6$ and $\gamma = 3$.
(c) The alkene $Ph_2C{=}CH_2$ forms a conducting solution in liquid hydrogen chloride; when such a solution is titrated conductimetrically with one of boron trichloride in liquid hydrogen chloride, a sharp end-point is obtained at the molar ratio $Ph_2C{=}CH_2{:}BCl_3 = 1{:}1$.

2 How would you attempt to show that arsenic trichloride ionises slightly according to the equation

$$2AsCl_3 \rightleftharpoons AsCl_2^+ + AsCl_4^-$$

and that there exist acids and bases on the aresenic trichloride solvent system?

3 Early in the study of chemistry in liquid ammonia it was noticed that nitrogen compounds behave in ammonia as the solvent in a manner similar to that of analogous oxygen compounds in aqueous media: thus in the so-called nitrogen system of compounds $NH_4^+Cl^-$ is analogous to $H_3O^+Cl^-$ and $K^+NH_2^-$ to K^+OH^-. What would be corresponding compounds in the nitrogen system of compounds to H_2O_2, HgO, HNO_3, CH_3OH, H_2CO_3 and $[Cr(H_2O)_6]Cl_3$?

REFERENCES FOR FURTHER READING

Lagowski, J.J. (1973) *Modern Inorganic Chemistry*, Marcel Dekker, New York. Gives a good comparative account of water, ammonia and hydrogen fluoride as solvents.

Marcus, Y. (1985) *Ion Solvation*, Wiley, New York. The definitive work on solvation by organic liquids.

Nicholls, D. (1979) *Inorganic Chemistry in Liquid Ammonia*, Elsevier, Amsterdam. A general account of chemistry in the most important non-aqueous solvent.

Purcell, K.F. and Kotz, J.C. (1977) *Inorganic Chemistry*, W.B. Saunders, Philadelphia. Contains a good account of protonic solvents, including some quantitative data on liquid ammonia solutions.

Sharpe, A.G. (1990) *J. Chem. Education*, **67**, 309. A review of the solvation of halide ions and its chemical significance.

Waddington, T.C. (1969) *Non-aqueous Solvents*, Nelson, London. A slightly fuller and more advanced treatment than that given in this book, and includes a chapter on reactions in molten salts.

9 Hydrogen

Introduction • Isotopes of hydrogen • The physical properties of hydrogen • The preparation and chemical properties of hydrogen • Hydrogen bonding • Classification and general properties of hydrides

9.1 Introduction

The hydrogen atom and molecule are so important in theoretical chemistry, and hydrogen compounds are so numerous and so significant in the systematisation of inorganic chemistry, that many aspects of the chemistry of hydrogen have already been discussed in this book. These include:

(i) the nuclear spin quantum numbers of hydrogen and deuterium (Section 1.5), and the existence of *ortho*- and *para*-hydrogen (Section 1.6);
(ii) chemical applications of deuterium and tritium (Section 1.9);
(iii) the atomic spectrum and ionisation energy of hydrogen (Section 2.3);
(iv) the solution of the wave equation for the hydrogen atom (Section 2.7);
(v) the molecular orbital (Section 4.2) and valence bond (Section 4.5) theories of the bonding in the hydrogen molecule;
(vi) the electronegativity of hydrogen (Section 5.7);
(vii) proton affinities (Section 6.8);
(viii) the properties of water (Section 7.1);
(ix) the hydrogen ion in aqueous solution (Sections 7.2, 7.4, 7.8 and 7.11);
(x) the hydrogen ion in non-aqueous media (Sections 8.1, 8.2 and 8.3).

Most of these topics will not be considered again in this chapter, but it is now desirable to extend somewhat the treatment of the isotopes of hydrogen, and to bring together the thermochemical and related data for atomic and molecular hydrogen in order to provide a basis for the discussion of the chemistry of the element. Other subjects covered later include hydrogen bonding and the general properties of the different types of hydrides.

9.2 Isotopes of hydrogen

The three known isotopes of hydrogen, ^1H, ^2H and ^3H (isotopic masses 1.0078, 2.0141 and 3.0161 m_u respectively) exhibit greater differences in physical and chemical properties than isotopes of any other element, and the heavier isotopes are commonly known as deuterium and tritium and given the symbols D and T; H, D and T have nuclear spin quantum numbers of $\frac{1}{2}$, 1, and $\frac{1}{2}$ respectively.

Deuterium is present to the extent of about 0.015 per cent in ordinary hydrogen. Where cheap electrical power is available, the isotope is obtained by fractional electrolysis (Section 1.7). A dilute solution of sodium hydroxide and nickel electrodes are used, the residues being distilled from time to time to free them from the accumulated alkali. Partial enrichment may be obtained by fractional distillation (HOD and D_2O are slightly less volatile than H_2O) or by means of a chemical exchange process (Section 1.7) based on the equilibrium

$$HSD(g) + H_2O(l) \rightleftharpoons H_2S(g) + HOD(l)$$

For the final concentration to 99.8 per cent D_2O, however, the electrolytic process is required.

The origin of all the differences between hydrogen and deuterium, or between their compounds, lies in the difference in mass, which in turn affects their fundamental vibration frequencies and hence their zero-point energies. As we showed in Section 1.9, for two isotopically different molecules of reduced masses μ_1 and μ_2, the ratio of the fundamental vibrations is given by

$$\frac{\bar{\nu}_1}{\bar{\nu}_2} = \sqrt{\frac{\mu_2}{\mu_1}}$$

Thus for H_2, HD and D_2 the fundamental vibrations occur at 4159, 3630, and 2990 cm^{-1} respectively. When a much heavier atom is bonded to hydrogen or deuterium the ratio $\bar{\nu}_1/\bar{\nu}_2$ approaches $\sqrt{2}$:1, i.e. 1.41, a relationship which is often used in identifying O—H etc. vibrations in infrared spectra. From their fundamental vibration frequencies the zero-point energies of H_2 and D_2 may be obtained as 26.0 and 18.4 kJ mol^{-1} respectively. The total electronic binding energies for these molecules (represented by the overlap of their atomic wave functions) are the same, so their dissociation energies (the differences between the molecules in their lowest energy states and the isolated atoms) differ by 7.6 kJ mol^{-1}. Differences in rates of reaction arise because although the absolute energies of analogous transition states involving hydrogen or deuterium are nearly the same, they have to be produced from H_2 and D_2 in their actual lowest energy states; thus reactions of D_2 are slower.

The variations in properties between hydrogen and deuterium compounds are illustrated for H_2O and D_2O (heavy water) in Table 9.1; the higher boiling-point of D_2O implies stronger hydrogen bonding. The determination of the deuterium content of a partially deuterated substance can be made by mass spectrometry, density measurement (after conversion into water) or infrared spectroscopy. Examples of the preparation and uses of deuterium compounds in vibrational spectroscopy, nuclear magnetic resonance spectroscopy, the location of hydrogen atoms in crystals by means of neutron diffraction, and as a tracer have all been given in Section 1.9. The major industrial use of deuterium oxide is as a moderator in nuclear reactors; deuterium has a much lower cross-section for neutron capture than hydrogen, and hence can be used for reducing the energies of fast neutrons produced in fission without appreciably diminishing the neutron flux.

Tritium occurs in very small amounts in natural hydrogen, being formed by the reaction ${}^{14}_{7}N + {}^{1}_{0}n \rightarrow {}^{12}_{6}C + {}^{3}_{1}T$ brought about by neutrons arriving in the upper atmosphere from outer space. It was first obtained synthetically by the bombardment of deuterium compounds such as $(ND_4)_2SO_4$ with fast deuterons (${}^{2}D + {}^{2}D \rightarrow {}^{3}T + {}^{1}H$); it is now prepared by irradiation of lithium enriched in ${}^{6}Li$ in the form of Li/Mg alloy or LiF with slow neutrons in a reactor:

$$ {}^{6}Li + {}^{1}n \rightarrow {}^{3}T + {}^{4}He $$

The tritium formed is separated from helium by conversion into UT_3 (which is decomposed at a higher temperature) or by burning to T_2O. Tritium is a weak β^--emitter with a half-life 12.4 years; it is extensively used as a tracer, not only in chemical but also in biochemical studies, in which its weak radioactivity, rapid excretion, and failure to concentrate in

Table 9.1
Some properties of H_2O and D_2O.

	H_2O	D_2O
Melting-point/°C	0.00	3.82
Boiling-point/°C	100.00	101.42
Temperature of maximum density/°C	4	11.6
Relative permittivity at 20 °C	82	80.5
Density at 20 °C/g cm^{-3}	0.998 2	1.105 9
Ionic product K_w at 25 °C	1×10^{-14}	2×10^{-15}
Ionic mobilities at 18 °C: K^+	64.2	54.5
/ohm^{-1} cm^2 mol^{-1} Cl^-	65.2	55.3
H^+	315.2	213.7
Symmetric O—H stretching $\bar{\nu}_1$ (gas)/cm^{-1}	3 657	2 671

vulnerable organs make it one of the least toxic radioactive isotopes. Its roles in the generation of nuclear energy by a fusion reaction ($^3T + ^2D \rightarrow ^4He + ^1n$) and in the elucidation of reaction mechanisms have been mentioned in Sections 1.4 and 1.9 respectively.

9.3 *The physical properties of hydrogen*

Hydrogen is a colourless, odourless gas, sparingly soluble in all solvents, which at ordinary temperature and pressure conforms very closely to the ideal gas laws. Solid hydrogen, which has the hexagonal close-packed structure, melts at 14 K, and the liquid boils at 20.4 K. The dissociation energy (i.e. ΔH° for $H_2 \rightarrow 2H$) at 25 °C is 436 kJ mol^{-1}, and the standard entropy is 131 J K^{-1} mol^{-1} at the same temperature. The bonding in H_2 is very strong for a molecule containing only a single bond, and the observed heat capacity ($C_p = 29$ J K^{-1} mol^{-1} at 25 °C) shows that, as is then to be expected for a molecule of very small reduced mass, only the lowest vibrational level is occupied at the ordinary temperature. The covalent radius of hydrogen (i.e. half the H—H distance in H_2) is 0.37 Å.

The ionisation energy of atomic hydrogen is 1310 kJ mol^{-1}, a value high enough to preclude the existence of free H^+ under ordinary conditions. The ion H_3O^+, however, is a well-defined species which occurs in several crystals; ΔH° for the gas-phase reaction

$$H^+ + H_2O \rightarrow H_3O^+$$

may be obtained from the estimated lattice energy of $H_3O^+ClO_4^-$ and other data, by the method described in Section 6.8, as -762 kJ mol^{-1}; mass spectrometric studies of the competition for protons by different bases lead to the somewhat different value $\Delta H^\circ = -690$ kJ mol^{-1}; the source of the discrepancy is uncertain. In addition to H_3O^+, the ions $H_5O_2^+$ and $H_9O_4^+$ (H_3O^+ to which one and three water molecules respectively are hydrogen-bonded through the protons of the cation) have also been identified in crystalline acid hydrates.

For the solvation of the gaseous proton by water, ΔH° for the reaction

$$H^+(g) + aq \rightarrow H^+(aq)$$

is -1100 kJ mol^{-1} (Section 7.3), whence ΔH° for the reaction

$$H^+(aq) + e \rightarrow \tfrac{1}{2}H_2(g)$$

is -428 kJ mol^{-1}. This last reaction is, of course, that involved in the standard hydrogen electrode; the difference between the absolute thermochemical value and the electrochemical convention $\Delta H^\circ = \Delta G^\circ = 0$ for this reaction, a source of much confusion, was emphasised throughout Chapter 7. If we take the proton affinity of water as -762 kJ mol^{-1}, the hydration enthalpy of the H_3O^+ ion is $-(1100-762)$, i.e. -338 kJ mol^{-1}, about the same as that for the K^+ ion.

The electron affinity of the hydrogen atom (ΔH° for the gas phase reaction $H + e \rightarrow H^-$) has been determined by the method used for the halogens, and described in Section 3.5, as -77 kJ mol^{-1}. All the alkali metal hydrides crystallise in the sodium chloride structure; if we subtract from the metal–hydrogen distance the ionic radii for the alkali metals given in Table 6.2, the apparent radius of the H^- ion is found to vary from 1.30 Å in LiH to 1.54 Å in CsH, i.e. to be about the same as that of the F^- ion; evidently interelectronic repulsion when an electron is added to a hydrogen atom causes a large increase in size and is responsible for the electron affinity being so near to zero. The lower value derived from the structure of LiH may signify some degree of covalent bonding in this compound, but there is no independent evidence to support this view; calculated and cycle values for the lattice energies of all the alkali metal hydrides are in good agreement. Combination of the value for the electron affinity with half that for the dissociation energy of molecular hydrogen leads to $\Delta H^\circ = +148$ kJ mol^{-1} for the gas-phase reaction

$$\tfrac{1}{2}H_2 + e \rightarrow H^-$$

When we compare this value with those for the analogous reactions of fluorine and chlorine (-259 and -234 kJ mol^{-1} respectively), it is at once apparent why, since H^- is about the same size as F^-, ionic hydrides are relatively unstable species with respect to dissociation into the elements, and salt-like hydrides of metals in high oxidation states are most unlikely to exist.

Electronegativities on the Pauling scale (Section 5.7) are assessed relative to an arbitrarily chosen value of 2.1 for hydrogen. Thus bonds between hydrogen and elements of electronegativity about 2.1 are believed to be non-polar or nearly so, whilst those between hydrogen and elements of electronegativity substantially greater or less than 2.1 are believed to be appreciably polar; bonds in the former category include H—B, H—Si and H—P; those in the latter category include H—F, H—O and H—N, and it is these bonds which are particularly involved in hydrogen bonding, which is discussed further in Section 9.5.

9.4 *The preparation and chemical properties of hydrogen*

In the laboratory, small quantities of hydrogen are conveniently made by the action of dilute acid on metals, e.g. zinc and iron, or of alkali on metals that form amphoteric hydroxides, e.g. zinc and aluminium. The alkali and alkaline earth metals liberate hydrogen from water, but many other metals that, on thermodynamic grounds, would be expected to do so are made kinetically inert by the presence of a thin film of an insoluble oxide.

On an industrial scale, hydrogen for use *in situ* (because the low density and very low boiling-point make transport uneconomic) in the Haber process or for the hydrogenation of fats is now made mainly from methane or other low molecular weight alkanes present in natural gas or produced in the 'cracking' of crude petroleum. The hydrocarbons and steam are passed over a nickel catalyst at about 800 °C, when reactions typified by the following occur:

$$CH_4 + H_2O \rightarrow CO + 3H_2$$

$$C_3H_6 + 3H_2O \rightarrow 3CO + 6H_2$$

The hot gases are cooled to 400 °C by addition of steam, and then passage over an iron–copper catalyst results in the conversion of carbon monoxide into the dioxide:

$$CO + H_2O \rightarrow CO_2 + H_2$$

The carbon dioxide is absorbed in potassium carbonate or ethanolamine ($HOCH_2CH_2NH_2$) solution, from which it may be regenerated by the action of heat. Large quantities of hydrogen are also produced in the petroleum industry (in the dehydrogenation of alkanes) and as a by-product in the manufacture of chlorine. Electrolysis of dilute alkali is used in places remote from the large production centres in order to avoid transportation costs.

Depletion of resources of fossil fuels will at some future time make hydrogen, either for use directly by combustion and electrochemically in fuel cells (as now used in space craft, where the water formed is drunk) or indirectly via hydrogenation of coal, the major alternative to nuclear energy; hence arises the current interest in the so-called *hydrogen economy*. Production of hydrogen from water inevitably requires a net input of energy; this would come from nuclear or solar sources. Energy collected with photovoltaic cells, for example, could be used to electrolyse water. A *thermochemical* cycle for hydrogen production involves at least one element that can exist in two different oxidation states; a typical cycle is

$$6FeCl_2 + 8H_2O \xrightarrow{600\,°C} 2Fe_3O_4 + 12HCl + 2H_2$$

$$2Fe_3O_4 + 3Cl_2 + 12HCl \xrightarrow{150\,°C} 6FeCl_3 + 6H_2O + O_2$$

$$6FeCl_3 \xrightarrow{400\,°C} 6FeCl_2 + 3Cl_2$$

Since water is transparent to light the *photolytic* production of hydrogen from it requires a catalyst; if the oxidised form of this is represented by A(ox) and the reduced form by A(red), the steps involved are

$$H_2O + h\nu + 2A(ox) \rightarrow 2A(red) + \tfrac{1}{2}O_2 + 2H^+$$

$$2A(red) + 2H^+ \rightarrow 2A(ox) + H_2$$

Various devices may be employed to get round the difficulty created by the simultaneous production of the two gases; photocatalysts may be semi-conductors, photosensitive dyes, or complexes such as $[Ru(bipyridyl)_3]^{3+}$ (Section 25.8). In biochemical (*biomass*) methods, the standard green-plant conversion of carbon dioxide and water into carbohydrate and oxygen (which is tantamount to a photolysis of water followed by reduction of carbon dioxide by hydrogen) is modified so that some of the hydrogen is liberated; certain blue-green algae are effective for this purpose. All of these new methods for hydrogen production are still at the experimental stage, but obviously they have great potential importance and will be the subject of a vast amount of research in years to come.

Hydrogen is not very reactive under ordinary conditions, though the lack of reactivity is kinetic rather than thermodynamic in origin, and originates in the strength of the H—H bond. Fluorine reacts explosively even at low temperatures. The complicated photochemical reaction with chlorine involves dissociation of the latter (which contains a weaker bond) as the initial step:

$$Cl_2 \xrightarrow{h\nu} 2Cl$$

$$Cl + H_2 \rightarrow HCl + H$$

$$H + Cl_2 \rightarrow HCl + Cl \text{ etc.}$$

Reactions with bromine and iodine, which occur only at higher temperatures, also involve initial dissociation of the halogen molecules, though the reaction with iodine is not a chain reaction. The explosive branching-chain reaction with oxygen (the basis of the use of hydrogen as a rocket fuel), initiated by sparking, involves the steps:

$$H_2 + O_2 \rightarrow HO_2 + H$$

$$H_2 + HO_2 \rightarrow OH + H_2O$$

$$H_2 + OH \rightarrow H + H_2O$$

$$H + O_2 \rightarrow OH + O$$

$$O + H_2 \rightarrow OH + O$$

Many metals form hydrides when heated in hydrogen; metal oxides and halides are often reduced to the metal or to compounds of the metal in a low oxidation state.

By the action of an electric discharge, hydrogen is partly dissociated into atoms, particularly at low pressures. Atomic hydrogen so obtained is much more reactive than ordinary hydrogen, and combines with many elements (e.g. Sn, As) that are unaffected by the latter. The bond energy liberated when the atoms recombine (in the presence of a third body to absorb the heat) is made use of in the hydrogen torch for the welding of metals in a strongly reducing atmosphere.

Most of the important reactions of hydrogen involve heterogeneous catalysis, which lowers activation energies for reactions by first weakening or breaking the H—H bond at a metal surface. Among them are the Haber process for the formation of ammonia from nitrogen and hydrogen over an iron catalyst at 400–540 °C, the reduction of carbon monoxide to methanol over a copper–zinc catalyst at 300 °C, and the hydrogenation of enormous numbers of unsaturated organic compounds on various nickel, palladium, or platinum catalysts under a wide variety of conditions. Within recent years, however, homogeneous catalysis has become increasingly important. The hydroformylation reaction

$$RCH{=}CH_2 + H_2 + CO \rightarrow RCH_2CH_2CHO$$

is catalysed by cobalt carbonyl, $Co_2(CO)_8$, via the hydride $HCo(CO)_4$; and many specific reductions of organic compounds can be effected by the use of hydrogen and suitable transition metal complexes such as $[Co(CN)_5]^{3-}$ and $Rh(PPh_3)_3Cl$ (Section 23.7).

9.5 Hydrogen bonding

There are many compounds in which evidence of various kinds indicates that a hydrogen atom interacts with two other atoms (usually, though not invariably, of nitrogen, oxygen or fluorine) to an extent greater than corresponds to normal van der Waals' bonding. Such interaction is called ***hydrogen bonding*** and, as we stated in Section 6.1, is sometimes regarded as a fifth basic type of chemical bonding additional to ionic, covalent, metallic and van der Waals' bonding. It is now recognised, however, that the term hydrogen bonding covers more than one kind of interaction. Where, as is most commonly the case, the hydrogen atom is asymmetrically placed with respect to the two atoms to which it is bonded (e.g. O—H ... O in the dimer of acetic acid), the stronger bond is only slightly longer than the comparable bond in the absence of hydrogen bonding; ΔH for hydrogen bond formation is then only slightly negative (approximately -30 kJ mol^{-1}) and the interaction is mostly simply looked upon as a weak electrostatic attraction between a covalently bonded hydrogen atom carrying a fractional positive charge and a lone pair of electrons on a neighbouring atom. Where the hydrogen bond is symmetrical (e.g. in the $[F{-}H{-}F]^-$ ion), appreciable stretching of the original single covalent bond occurs, and ΔH for hydrogen bond formation may be as great as about -200 kJ mol^{-1}, i.e. greater than ΔH for the formation of the covalent bond in molecular iodine; in these circumstances it is more appropriate to treat the hydrogen bond as a case of multicentre covalent bonding (cf. Section 4.8).

The simplest evidence for the existence of hydrogen bonding comes from comparisons of melting- and boiling-points in series of hydrides;

data for the hydrides of carbon, nitrogen, oxygen and fluorine, together with data for the noble gases, are presented in Fig. 9.1. In general, melting- and boiling-points of analogous molecular species increase with increase in molecular size owing to the increase in the dispersion forces; thus the high values for ammonia, water and hydrogen fluoride show that some extra intermolecular attraction must be involved in these substances. A similar pattern can be seen in the enthalpies of vaporisation of the liquid hydrides, shown in Fig. 9.2. Some caution must, however, be exercised in reaching quantitative conclusions about the relative strengths of hydrogen bonds from the data in Figs. 9.1 and 9.2; it would be unsound to deduce that the hydrogen bonding in water is stronger than

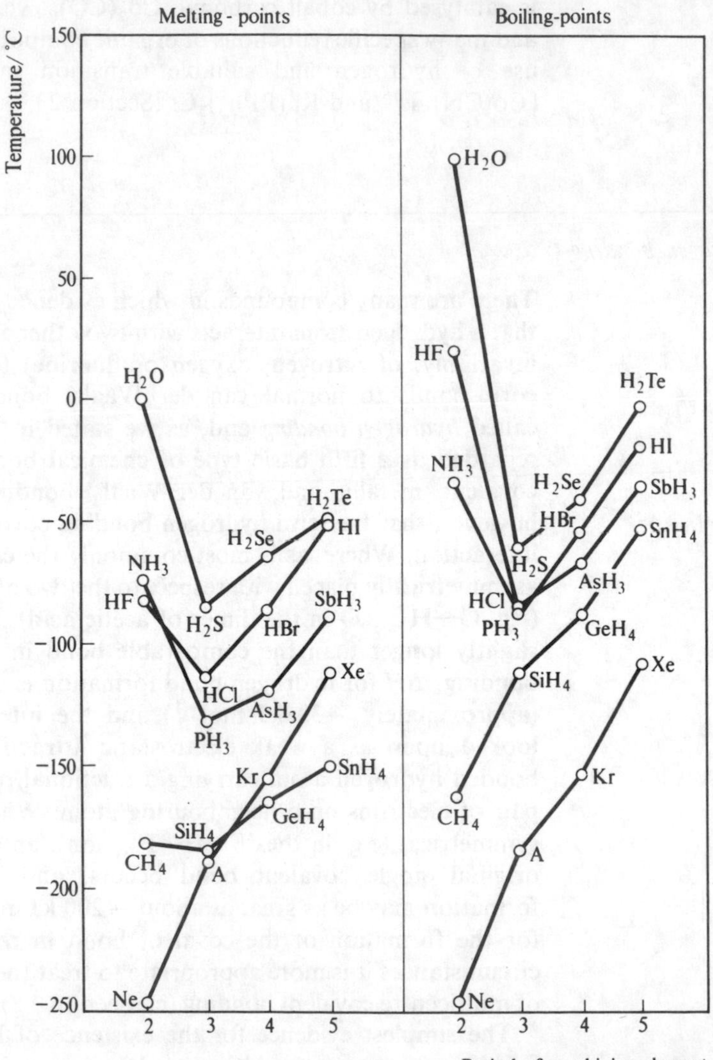

Figure 9.1
Melting- and boiling-points of the molecular hydrides and the noble gases.

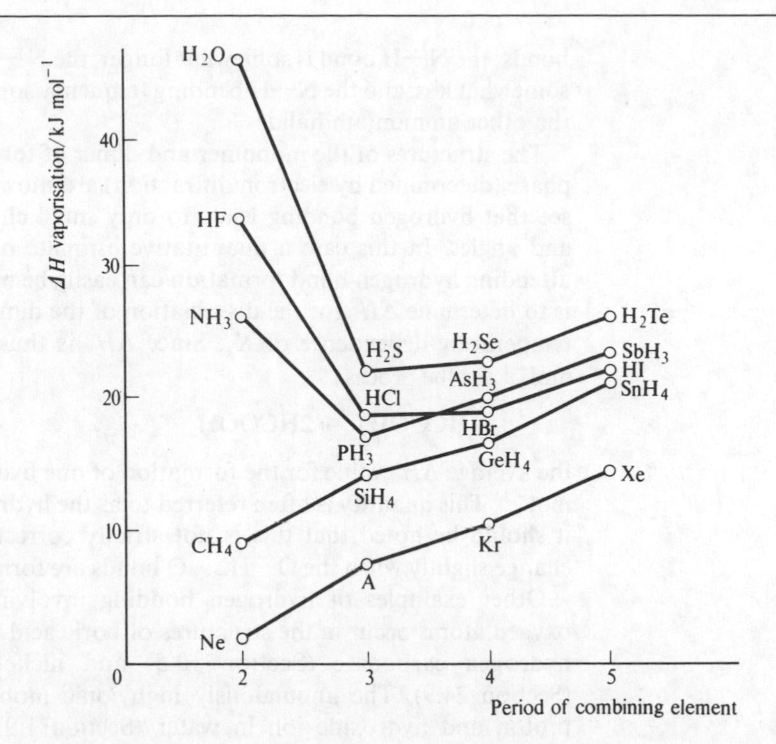

Figure 9.2
Enthalpies of vaporisation of molecular hydrides and noble gases at their boiling-points.

that in hydrogen fluoride, since boiling-points and enthalpies of vaporisation are concerned with differences between the liquid and gaseous states, and there is independent evidence from vapour pressure measurements to show that whilst water is hydrogen-bonded in the liquid but not in the vapour, hydrogen fluoride is strongly hydrogen-bonded in both.

The structure of ice, determined by X-ray and neutron diffraction, provides a classic example of hydrogen bonding: the common form of ice has a structure closely related to that of wurtzite (Section 6.3); the oxygen atoms occupy the sites of both the zinc and the sulphur atoms in wurtzite, and the hydrogen atoms are situated just off lines joining the oxygen atoms. The O—O distance is 2.74 Å at 0 °C; the O—H bond length, 1.01 Å, is slightly greater than that in water vapour (0.96 Å); correspondingly, the symmetric O—H stretching frequency in ice (3 400 cm^{-1}) is lower than that in water vapour (3 657 cm^{-1}). Each oxygen atom thus has two near and two distant hydrogen atoms as neighbours. Consideration of the possible ways of building up a whole crystal on the basis of such units shows that ice should have a disordered structure even at absolute zero, i.e. should have a non-zero entropy at absolute zero; this prediction is confirmed by a discrepancy between the values for the standard entropy of water vapour calculated from the third law of thermodynamics (i.e. on the assumption that $S^\circ = 0$ at 0 K) and from statistical mechanics. Ammonium fluoride also has the wurtzite structure with, in this case, four N—H···F hydrogen bonds; the N—H bond is somewhat longer, the N—H stretching frequency somewhat less, and the N—H bending frequency somewhat greater, than in the other ammonium halides.

The structures of the monomer and dimer of formic acid in the vapour phase (determined by electron diffraction) are shown in Fig. 9.3; again we see that hydrogen bonding leads to only small changes in bond lengths and angles. In this case a quantitative estimate of the enthalpy change attending hydrogen bond formation can easily be made: all we have to do is to determine ΔH° for the dissociation of the dimer by investigating the temperature-dependence of K_p. Since ΔH° is thus found to be $+60$ kJ mol^{-1} for the process

$$(HCOOH)_2 \rightarrow 2HCOOH$$

the average ΔH° value for the formation of one hydrogen bond is -30 kJ mol^{-1}. This quantity is often referred to as the hydrogen bond energy, but it should be noted that this is not strictly correct: all the other bonds change slightly when the O—H . . . O bonds are formed.

Other examples of hydrogen bonding involving a proton and two oxygen atoms occur in the structures of boric acid (Section 12.8), sodium hydrogen carbonate (Section 10.5) and nickel dimethylglyoximate (Section 24.9). The anomalously high ionic mobilities of the solvated proton and hydroxide ion in water (Section 1.9) and of the solvated proton in some non-aqueous solvents (Section 8.4) also arise from O—H . . . O hydrogen bonding. Such bonding and N—H . . . O hydrogen bonding (the source of the high solubility of ammonia in water) are also of immense importance in biological systems.

In addition to the relatively low volatility of hydrogen fluoride, the structure of the solid, which consists of planar zig-zag chains as shown in

O
1.217
1.106
H — C
123°
1.361
O — H
0.984

(a)

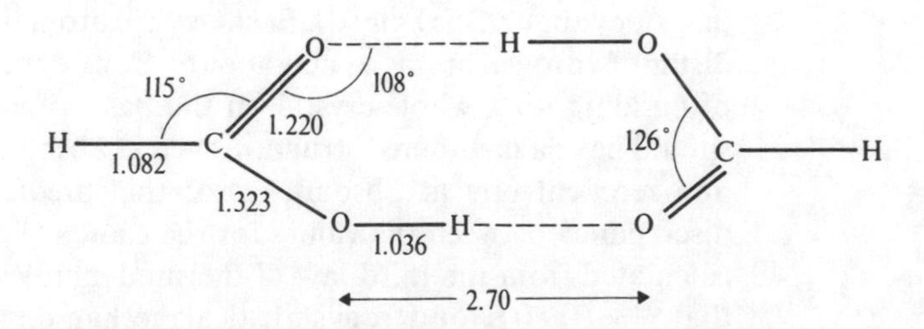

(b)

Figure 9.3
The structures of (*a*) the monomer, (*b*) the dimer, of formic acid (interatomic distances in ångströms).

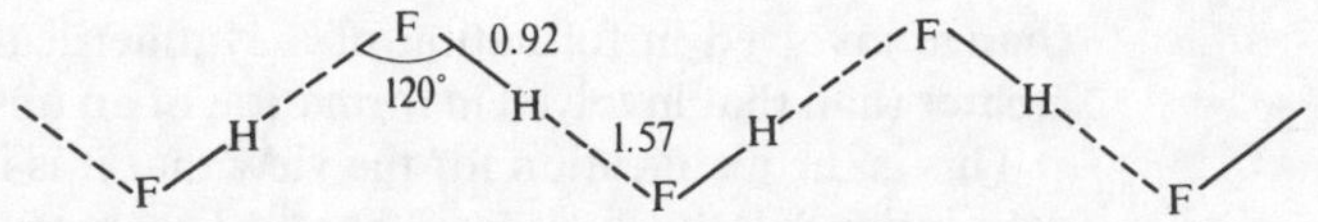

Figure 9.4 The structure of solid hydrogen fluoride (interatomic distances in ångströms).

Fig. 9.4, also establishes the presence of hydrogen bonds in this compound, though the positions of the hydrogen atoms are not known with certainty. Very strong hydrogen bonding occurs in the alkali metal hydrogen fluorides of formula $M[HF_2]$; in KHF_2, for example, an X-ray diffraction study, together with a neutron diffraction study of the deuterium-containing analogue, shows that there is a linear symmetrical anion having an overall F—H—F distance of 2.26 Å, which may be compared with the H—F bond length of 0.92 Å in hydrogen fluoride monomer. Thus the stretching of the H—F bond in the anion is 0.21 Å, compared with a stretching of 0.05 Å of the O—H bond in ice or formic acid dimer. For the gas-phase reaction

$$HF + F^- \rightarrow HF_2^-$$

an estimate of ΔH° may be made by considering the thermochemical cycle shown in Fig. 9.5, in which the other quantities involved are the lattice energies of $M[HF_2]$ and MF and the enthalpy change for the combination of MF(s) and HF(g), most readily obtained from the temperature-dependence of K_p for the reverse reaction. The difficulty in this operation lies in the estimation of the lattice energy of $M[HF_2]$, since the anion is linear (so that salts of this formula do not have simple structures) and the charge distribution in it is uncertain; the best estimate of ΔH° yet given is about -240 kJ mol^{-1}. An ingenious attempt to avoid this difficulty has been made by determining ΔH°_{diss} for the thermal dissociation of a series of tetraalkylammonium hydrogen fluorides; as the size of the cation increases, the difference between the lattice energies of $R_4N[HF_2]$ and R_4NF may be supposed to approach zero and hence the value of ΔH°_{diss} approaches that for dissociation of HF_2^-(g) to gaseous F^- and HF. The value of ΔH°_{diss} for $(CH_3)_4N[HF_2]$ indicates ΔH° is about -160 kJ mol^{-1} for the combination of gaseous F^- and HF. Although this result is not in very good agreement with that obtained via the estimated lattice energy of $K[HF_2]$, it does serve to establish beyond doubt that the energy

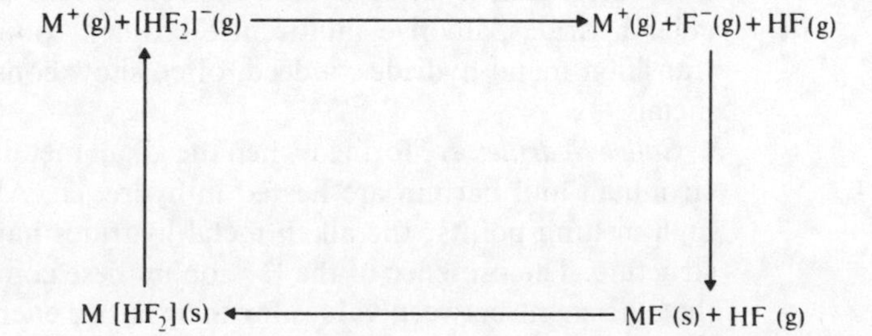

Figure 9.5 Thermochemical cycle for the estimation of ΔH° for the gas-phase reaction $HF + F^- \rightarrow [HF_2]^-$.

change involved in formation of a symmetrical hydrogen bond is much greater than that involved in formation of an unsymmetrical one.

This is the justification for the view that it is impossible to give a single description of the hydrogen bond. The most electronegative elements form bonds with hydrogen which are significantly polar, e.g. $—O^{\delta-}—H^{\delta+}$. Covalently bonded hydrogen differs from other covalently bonded elements in the fact that there are no non-bonded electrons to shield the nucleus from interaction with other species. Electrostatic attraction to another oxygen atom, for example, then suffices to account for formation of a weak O—H . . . O bond. For the symmetrical hydrogen bonds in $[HF_2]^-$ ions, however, a molecular orbital description is more satisfactory. The energy level diagram is basically like that shown in Fig. 4.21, but in this case the orbitals involved are the two p_z orbitals of the fluorine atoms and the *s* orbital of the hydrogen atom. These give rise to a bonding orbital, a non-bonding orbital and an antibonding orbital, and the four electrons involved (the number is more clearly seen if we think of the $[HF_2]^-$ ion as constructed from a proton and two fluoride ions) occupy the first two of these, giving a bond order of one-half for each bond in the anion. As we shall see in Chapter 16, this description closely resembles that for polyhalide ions such as I_3^-.

It should not be thought that hydrogen bond formation is absolutely restricted to the elements nitrogen, oxygen and fluorine. The linear chain structure of solid HCN, the formation of a 1:1 complex between acetone and chloroform and the existence of unstable salts of large cations with the ion $[HCl_2]^-$, for example, show that carbon and chlorine can also form hydrogen bonds, though only weak ones; but the importance of hydrogen bonding in the chemistry of other elements is small compared with that in the chemistry of the three elements with which most of this section has been concerned.

9.6 *Classification and general properties of hydrides*

A convenient first classification of the hydrides of the elements is into saline hydrides (believed to contain the H^- ion), molecular hydrides and macromolecular hydrides. A few molecular hydrides, and all macromolecular hydrides, involve multicentre covalent bonding; macromolecular transition metal hydrides, indeed, often show considerable similarities to metals.

Saline hydrides are formed when the alkali metals, magnesium, calcium, strontium and barium are heated in hydrogen. All are white solids with high melting-points; the alkali metal hydrides have the sodium chloride structure. The presence of the H^- ion in these compounds is indicated by the agreement between values for their lattice energies obtained from the

Born–Haber cycle and from X-ray and compressibility data; further, when molten lithium hydride is electrolysed, hydrogen is liberated at the anode. Among the alkali metal halides ΔH_f° becomes less negative, and reactivity increases, with increase in atomic number (and ionic size) of the alkali metal; hydride ion is about the same size as fluoride and, as with the alkali metal fluorides, the lattice energy of the salt decreases more rapidly than the ionisation energy of the metal as we go down the group. Thus ΔH_f° is -91 kJ mol^{-1} for LiH and -50 kJ mol^{-1} for CsH. Saline hydrides are decomposed immediately by protonic solvents such as water, ammonia and ethanol with formation of hydrogen and the hydroxide, amide and ethoxide respectively, showing that the hydride ion is an extremely strong base. Important reactions of these hydrides include those of lithium hydride with aluminium chloride in ether to form lithium tetrahydridoaluminate, $LiAlH_4$, and of sodium hydride with trimethyl borate at 250 °C to form the analogous boron compound $NaBH_4$ (Section 12.3); both of these substances, and sodium hydride itself, are widely used as reducing agents.

Molecular hydrides are formed by the non-metals and by a few other non-transition elements such as tin, arsenic and antimony. Those of the halogens, sulphur and nitrogen are obtained by the action of hydrogen on the element under appropriate conditions; others are made by the hydrolysis of metal salts with water, aqueous acid, or ammonium bromide in liquid ammonia, e.g.

$$Ca_3P_2 \xrightarrow{H_2O} Ca(OH)_2 + PH_3$$

$$FeS \xrightarrow{\text{dil HCl}} FeCl_2 + H_2S$$

$$Mg_2Si \xrightarrow[\text{in } NH_3]{NH_4Br} MgBr_2 + SiH_4$$

A third method of widespread applicability is the action of lithium tetrahydridoaluminate on a halide in ethereal solution:

$$4BCl_3 + 3LiAlH_4 \rightarrow 2B_2H_6 + 3LiCl + 3AlCl_3$$

$$SnCl_4 + LiAlH_4 \rightarrow SnH_4 + LiCl + AlCl_3$$

Most molecular hydrides are very volatile and have simple structures which are easily accounted for on the basis of the Valence Shell Electron Pair Repulsion Theory (Section 5.1); the simplest hydride of boron, however, is not BH_3 but B_2H_6, and is an electron-deficient compound whose structure involves multicentre covalent bonding (Section 4.8). There is very wide variation in properties among the molecular hydrides, and it is impossible to summarise them satisfactorily; considerable detail is therefore given under individual elements.

Macromolecular hydrides are formed by beryllium, aluminium and many transition metals. Those of beryllium and aluminium, obtained by thermal decomposition of beryllium alkyls and by the action of lithium tetrahydridoaluminate on aluminium chloride in ether, are white polymeric solids which probably contain four-coordinated beryllium and six-coordinated aluminium respectively. If, for example, BeH_2 has the structure

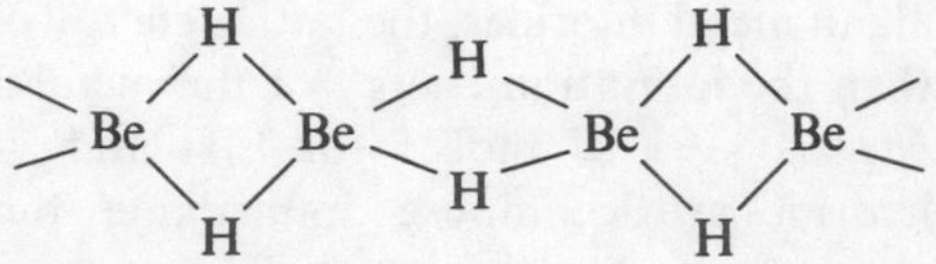

```
       H             H
      / \     H     / \
  \ Be   Be /   \ Be   Be /
  /   \ / \  H  /   \ / \
       H             H
```

it must involve multicentre bonding. Transition metal hydrides are usually made by direct combination or electrolytically; typical limiting compositions are TiH_2, ZrH_2, NbH_2. None of these formulae, it should be noted, corresponds to a stable oxidation state of the metal concerned. Transition metal hydrides are often dark grey solids similar in appearance and reactivity to the parent metals; many of them are metallic conductors. The dihydrides have a structure which, on the basis of atomic positions, can be described either as a cubic close-packed metal with hydrogen atoms in the tetrahedral holes or as a fluorite lattice containing metal cations and hydride anions (cf. Section 6.2), but because of the physical properties of the compounds the latter description must be rejected. On the basis of the former description these hydrides have often been called *interstitial hydrides*, and since they have only relatively small enthalpies of formation (e.g. -170 kJ mol^{-1} for ZrH_2) it has sometimes been suggested that only weak bonding is involved. This cannot be so, however; for a hydride with zero enthalpy of formation, the bonding must compensate the dissociation energy of hydrogen and some expansion of the metal lattice. It seems, therefore, likely that some of the electrons of the metal and those of the hydrogen atoms are involved in multicentre covalent bonding, the other metal electrons remaining in a conduction band extending throughout the structure.

Not all hydrides, it should be said, fit neatly into this classification; some which do not are those of palladium, copper and the lanthanides and actinides. Palladium reversibly absorbs large amounts of hydrogen or deuterium (but no other gas – a fact of great importance in the separation of hydrogen from mixtures of gases). The hydrogen has a high mobility within the metal. It is not known how the hydrogen is present in the solid, the limiting composition of which is about $PdH_{0.7}$.

PROBLEMS

1 How would you attempt to prepare a sample of pure HD and to establish the purity of the product?

2 Discuss the interpretation of each of the following observations:

(a) The rate of conversion of *para*-H_2 into *ortho*-H_2 at high temperature is proportional to [*para*-H_2][total H_2]$^{\frac{1}{2}}$.

(b) The infrared spectrum of a 0.01 molar solution of tertiary butanol in carbon tetrachloride shows a sharp peak at 3610 cm^{-1}; the spectrum of a similar 1.0 molal solution shows this peak much diminished in intensity, but a very strong broad peak centred at 3330 cm^{-1} is now observed.

(c) Caesium chloride, but not lithium chloride, absorbs hydrogen chloride at low temperatures.

(d) Ammonia is more soluble in aqueous ammonium chloride than in pure water.

(e) Solid ammonium fluoride and ice are miscible in all proportions.

3 How would you attempt to determine the standard enthalpies of formation of LiH and TiH_2? What use would the value for TiH_2 be in the discussion of the nature of the bonding in TiH_2?

REFERENCES FOR FURTHER READING

Emsley, J. (1980) *Chem. Soc. Rev.*, **9**, 91. A comprehensive review of very strong hydrogen bonds.

Jolly, W.L. (1984) *Modern Inorganic Chemistry*, McGraw-Hill, New York. Contains chapters on reactions of molecular hydrogen and hydrogen compounds.

McAuliffe, C.A. (1980) *Hydrogen and Energy*, Macmillan, London. A short and readable account of present and future energy sources.

Wells, A.F. (1984) *Structural Inorganic Chemistry*, 5th edition, Oxford University Press, Oxford. Full treatments of the structures of ice, water, other hydrogen-bonded compounds and hydrides.

10 The alkali metals

Introduction • The metals • Halides • Oxides and hydroxides • Salts of oxo-acids • Aqueous solution chemistry • Complexes • Organometallic compounds

10.1 Introduction

The alkali metals illustrate, more clearly than any other group of elements, the influence of increase in atomic and ionic size on physical and chemical properties. Thus they appear prominent in any discussion of general principles, and we have already discussed several aspects of alkali metal chemistry, e.g. atomic spectra (Section 2.4), ionisation energies (Section 3.4), structures and lattice energies of halides (Sections 6.3 and 6.6), solvation energies of ions and solubilities of salts (Sections 7.3 and 8.1), standard electrode potentials (Section 7.8), solutions in liquid ammonia (Section 8.2) and hydrides (Section 9.6).

Hitherto, however, we have mentioned only lithium, sodium, potassium, rubidium and caesium. The heaviest alkali metal, francium (atomic number 87, symbol Fr), has been little investigated because its longest-lived isotope, ^{223}Fr, has a half-life of only 21 minutes, and is only a minor product in the radioactive decay of a rather inaccessible isotope of actinium, ^{227}Ac. Precipitation and ion-exchange studies show that francium behaves as the homologue of rubidium and caesium, but since no quantitative data are available for the element and its compounds we shall not consider it further.

Some important physical properties of alkali metal atoms and ions are assembled in Table 10.1. With increase in atomic number, the atoms become larger and the strength of the metallic bonding decreases. The effect of increasing size evidently outweighs that of increasing nuclear charge, since the ionisation energies decrease from lithium to caesium. As we have seen earlier, enthalpies and entropies of hydration both become

Table 10.1
Some properties of the alkali metals and their ions.

Element	Li	Na	K	Rb	Cs
Atomic number	3	11	19	37	55
Electronic configuration	$[He]2s^1$	$[Ne]3s^1$	$[Ar]4s^1$	$[Kr]5s^1$	$[Xe]6s^1$
Atomic radius[a]/Å	1.52	1.86	2.27	2.48	2.65
ΔH° atomisation/kJ mol^{-1}	161	108	90	82	78
M.p./°C	180	98	64	39	29
B.p./°C	1326	883	756	688	690
First ionisation energy/ kJ mol^{-1}	520	496	419	403	375
Second ionisation energy/ kJ mol^{-1}	7297	4561	3069	2650	2420
Ionic radius[b]/Å	0.74	1.02	1.38	1.49	1.70
ΔH° hydration of M^+/ kJ mol^{-1}	−528	−413	−330	−305	−280
ΔS° hydration of M^+/ J K^{-1} mol^{-1}	−140	−110	−70	−70	−60
ΔG° hydration of M^+/ kJ mol^{-1}	−485	−379	−308	−285	−262
E° $M^+(aq) + e \rightleftharpoons M$/V	−3.04	−2.71	−2.92	−2.99	−3.02

[a]In body-centred cubic metal.
[b]For six-coordination.

less negative as ionic size increases, but the variation in free energy of hydration follows fairly closely that in enthalpy of hydration. Standard electrode potentials are related to energy changes attending atomisation, ionisation and hydration; among the alkali metals these energy changes almost cancel one another, and E° for the different M^+/M systems is nearly constant. The lower reactivity of lithium towards water is therefore kinetic rather than thermodynamic in origin: as a harder and less easily melted metal, lithium is less rapidly dispersed and hence reacts more slowly than the other metals.

The second ionisation energies of all the alkali metals are so high that the formation of M^{2+} ions under chemically realisable conditions is not possible. A few covalently bonded organic compounds are formed, particularly by lithium, and evidence for a compound of Na^- has recently been obtained (see Section 10.7). In general, however, the chemistry of the alkali metals is dominated by their low first ionisation energies and is overwhelmingly that of the M^+ ions, and lattice energies calculated on the electrostatic model provide us with a very satisfactory understanding of most of it. This understanding extends to many of the features of the chemistry of lithium in which the element shows similarities to magnesium (e.g. the instability of peroxides and relative instabilities of many salts of oxo-acids, in each case with respect to the metal oxide and other products); these, and certain similarities in solubilities of salts (e.g. LiF and MgF_2, unlike the other halides of both elements, are sparingly soluble in water) may all be attributed to the fact that Li^+ and Mg^{2+} have similar ionic radii, 0.74 and 0.72 Å respectively.

Sodium and potassium are both abundant elements (sodium especially so), their principal sources being *rock salt* (NaCl) and natural brines and sea water, *sylvite* (KCl and NaCl) and *carnallite* ($KMgCl_3.6H_2O$). Rock salt is almost pure sodium chloride, and sodium chloride is the major component of dissolved solids in the sea to a degree sufficient for it to be readily obtainable in warm climates by evaporation of sea water. Potassium chloride is considerably less soluble than sodium and magnesium chlorides at ordinary temperatures, and is separated by fractional crystallisation.

Lithium, rubidium and caesium occur chiefly in relatively small amounts in various silicate minerals. For the extraction of lithium, *spodumene* ($LiAlSi_2O_6$) is heated with the calculated quantity of calcium oxide, and the lithium hydroxide formed is extracted with water. The other two metals are obtained in small amounts as by-products of this process.

Sodium chloride is extensively used as a preservative and as the source of other sodium salts. Potassium is one of the major nutrients needed by plants, and potassium compounds are therefore important in fertilisers. Both sodium and potassium ions are involved in various electrophysiological functions in higher animals; the $[Na^+]:[K^+]$ ratio is different in intra- and extra-cellular fluids, and the concentration gradients of these ions across cell membranes are the origin of the trans-membrane potential difference that, in nerve and muscle cells, is responsible for the transmission of nerve impulses. A balanced diet therefore includes both sodium and potassium salts. Lithium carbonate has recently found application in the treatment of manic-depressive disorders, but large amounts of lithium salts damage the central nervous system.

10.2 The metals

Sodium, economically much the most important of the alkali metals, is always manufactured by electrolysis of the fused chloride, normally at about 600 °C in a $NaCl–CaCl_2$ eutectic which melts at 505 °C, some 300 °C below the melting-point of pure sodium chloride; the anode is carbon and the cathode steel. Under these conditions the potential for the discharge of sodium is lower than that for the discharge of calcium; further, any calcium liberated is insoluble in molten sodium, which, being less dense than the molten chloride, rises to the surface and is run off.

The electrolytic preparation of potassium is difficult owing to the volatility of the metal and the fact that on electrolysis of mixtures of molten salts potassium is discharged last. The modern method is by the action of sodium vapour on molten potassium chloride in a countercurrent fractionating tower; this yields a sodium–potassium alloy which can be separated into its components by distillation. Lithium is made by electrolysis of a fused mixture of lithium and potassium chlorides. Small quantities of all the metals except lithium may be obtained by thermal decomposition of their azides *in vacuo*; lithium combines with nitrogen to form the red-brown nitride Li_3N and hence is not obtained under these conditions. Rubidium and caesium are also made by reduction of their chlorides with calcium at 800°C.

Lithium, sodium, potassium and rubidium are silvery-white, but caesium has a golden-yellow cast. All are soft metals, lithium the least so, in line with its higher melting-point. The first three metals, though stored under a hydrocarbon solvent, can be handled in air, provided undue exposure is avoided, but for rubidium and caesium handling in an inert atmosphere is necessary. Lithium reacts quickly with water; sodium reacts vigorously; and the other three metals react violently with ignition of the hydrogen which is liberated. All the metals react with the halogens and form hydrides when heated in hydrogen, but only lithium combines with nitrogen. With oxygen, lithium forms only Li_2O, sodium forms Na_2O_2 together with a little NaO_2, and the heavier metals form mainly KO_2, RbO_2, and CsO_2; the temperature necessary for reaction decreases down the group. The thermodynamics of metal hydride formation are essentially like the thermodynamics of metal fluoride formation (i.e. are expressed in the Born–Haber cycle); the rather different thermodynamics of the various oxide systems are discussed in Section 10.4.Heated with carbon, lithium and sodium yield acetylides; the others form intercalation compounds (see Section 13.3).

All of the alkali metals dissolve in liquid ammonia, forming solutions which contain solvated cations and electrons; their properties have been described in some detail in Section 8.2. Similar solutions may be prepared in some amines and ethers. The metals also dissolve in mercury; sodium amalgam, which is a liquid only when poor in sodium, is a useful reducing agent in both organic and inorganic chemistry; it can be used in aqueous media because there is a large overpotential for the discharge of hydrogen on mercury. Sodium and lithium are also used for many purposes in synthetic organic chemistry, and liquid sodium finds employment as a heat exchanger in certain nuclear reactors. A major but decreasing demand for sodium is in the preparation of alkyl derivatives of lead (from lead–sodium alloy) for use as antiknock additives to petrol. The sodium–sulphur battery consists of a molten sodium anode and a liquid sulphur cathode separated by a solid β-alumina electrolyte at 300–350 °C; the cell reaction is

$$2Na(l) + 5S(l) \rightarrow Na_2S_5(l)$$

and is reversed on recharging. The e.m.f. is 2.0 V.

10.3 Halides

The distribution of the sodium chloride and caesium chloride structures among the alkali metal halides (all of which are solids of high melting-point) has been discussed at length in Chapter 6, together with the evidence from electron density and lattice energy data that suggests they are all well represented by the simple electrostatic model. All of the halides have strongly negative standard enthalpies of formation, but the actual values (shown in Table 10.2) show two different trends; ΔH_f° for fluorides becomes less negative as we go down the group, whilst the reverse is true, and to an increasing extent, for ΔH_f° for chlorides, bromides and iodides. For a given metal, ΔH_f° always becomes less negative as we go from the fluoride to the iodide.

These generalisations may easily be explained in terms of the Born–Haber cycle. Suppose we consider the formation of MF and MI from the elements in their standard states. For MF the variable quantities are the enthalpies of atomisation and ionisation of M, and the lattice energy of MF. Combination of the values for the first two of these quantities gives ΔH_f° for the Li^+, Na^+, K^+, Rb^+ and Cs^+ gaseous ions as 687, 610, 514, 490 and 459 kJ mol^{-1} respectively. The lattice energy values given in Table 10.3 show that the variation in ΔH_f° for formation of the gaseous cation is less than the variation in lattice energy for the fluorides but greater than the variation in lattice energy for the chlorides, bromides and iodides. This is, of course, because lattice energy is proportional to $1/r_0$ or $1/(r_+ + r_-)$ and so the variation in lattice energy is greatest when r_- is smallest and least when r_- is largest. For a given metal, the small change in $\Delta H_f^\circ(X^-)$ (-259, -234, -217, -197 kJ mol^{-1} for X = F, Cl, Br, and I respectively) is outweighed by the decrease in the lattice energy of MX, though it may be noted that the difference between the enthalpies of formation of the fluoride and the iodide decreases substantially as the size of the alkali metal cation increases. As we mentioned in Section 6.8, complex halide formation such as that represented by the equations

$$MF + BF_3 \rightarrow M[BF_4]$$

and

$$MI + ICl \rightarrow M[ICl_2]$$

Table 10.2
Standard enthalpies of formation of alkali metal halides.

M	$-\Delta H_f^\circ(MX)$/kJ mol^{-1}			
	MF	MCl	MBr	MI
Li	617	409	351	270
Na	575	411	361	288
K	568	436	394	328
Rb	557	443	393	333
Cs	555	435	405	348

Table 10.3
Lattice energies of alkali metal halides.

M	U(MX)/kJ mol^{-1}			
	MF	MCl	MBr	MI
Li	1035	845	800	740
Na	908	770	736	690
K	803	703	674	636
Rb	770	674	653	615
Cs	720	644	623	590

is favoured by increase in the size of the cation, the decrease in lattice energy on going from the simple halide to the complex halide diminishing as r_+ increases. Complex halide formation has been further discussed in Section 9.5 in connection with the enthalpy of formation of the $[HF_2]^-$ ion, and it will be mentioned again in the treatment of polyhalide anions in Chapter 16.

The solubilities of the alkali metal halides in water are determined by a delicate balance between lattice energies and hydration free energies (Section 7.3); lithium fluoride is only sparingly soluble, but solubility relationships among the other halides call for too detailed a treatment for a discussion to be given in this book.

Lithium chloride, bromide and iodide, and also sodium iodide, are readily soluble in some oxygen-containing organic solvents; both iodides are very soluble in, and form stable complexes with, ammonia. It is likely that in all these cases there is strong complexing of the alkali metal cation by the organic solvent or ammonia; the unstable complex $[Na(NH_3)_4]I$ has been isolated and shown to contain a tetrahedrally coordinated sodium ion. Solubility in oxygen-containing organic solvents should therefore not be taken as a proof of covalent character of the bonding in the solid halide.

In the vapour phase, alkali metal halides are present mainly as ion-pairs, but measurements of bond lengths and dipole moments suggest that under these conditions electron-sharing is also involved to a considerable extent, especially for the lithium halides.

10.4 Oxides and hydroxides

When the alkali metals are heated in excess of air the principal products are the oxide Li_2O, the peroxide Na_2O_2 and the superoxides KO_2, RbO_2 and CsO_2 respectively. The monoxides Na_2O, K_2O, Rb_2O and Cs_2O can be obtained impure by the use of a limited amount of air, but they are better prepared by thermal decomposition of the peroxides and superoxides. Partial oxidation of rubidium and caesium at low temperatures yields suboxides such as Rb_9O_2 and Cs_3O which involve metal–metal bonding.

A peroxide of lithium may be obtained by the action of hydrogen peroxide on an ethanolic solution of lithium hydroxide, but it decomposes when gently heated. Sodium peroxide is manufactured by heating sodium metal on aluminium trays in air; when pure, it is colourless, and the faint yellow colour of the usual product arises from the presence of a small amount of the superoxide NaO_2. All the superoxides are yellow or orange in colour and are paramagnetic (with moments close to 1.73 μ_B) because of the presence of an unpaired electron; the peroxides, on the other hand, are diamagnetic. These properties, and the increase in bond lengths along the series O_2, O_2^-, O_2^{2-} (1.21, 1.26 and 1.49 Å respectively), are readily interpreted on the basis of the simple molecular orbital theory outlined in Section 4.2. Oxygen has an unpaired electron in each of two degenerate π^* orbitals, and the extra electrons in O_2^- and O_2^{2-} are paired with these, leading to progressive loss of paramagnetism and lower bond order.

All three types of oxide react with water:

$$M_2O + H_2O \rightarrow 2MOH$$

$$M_2O_2 + 2H_2O \rightarrow 2MOH + H_2O_2$$

$$2MO_2 + 2H_2O \rightarrow 2MOH + H_2O_2 + O_2$$

Sodium peroxide is widely used as an oxidising agent in inorganic chemistry (e.g. for $Cr^{3+} \rightarrow CrO_4^{2-}$); its reactions with organic compounds are dangerously violent. With carbon dioxide it yields sodium carbonate and oxygen, and hence may be used for the purification of air in confined spaces such as submarines; potassium superoxide is even better for these purposes.

The dependence of the thermal stabilities of the alkali metal peroxides and superoxides on cation size may be discussed in terms of thermochemical cycles such as that shown in Fig. 10.1, which relates to the general reaction

$$M_2O_2(s) \rightarrow M_2O(s) + \tfrac{1}{2}O_2(g)$$

for which the entropy change is very nearly independent of the identity of M. From Fig. 10.1 it is readily seen that the enthalpy change of this reaction is determined by the difference between the lattice energies of

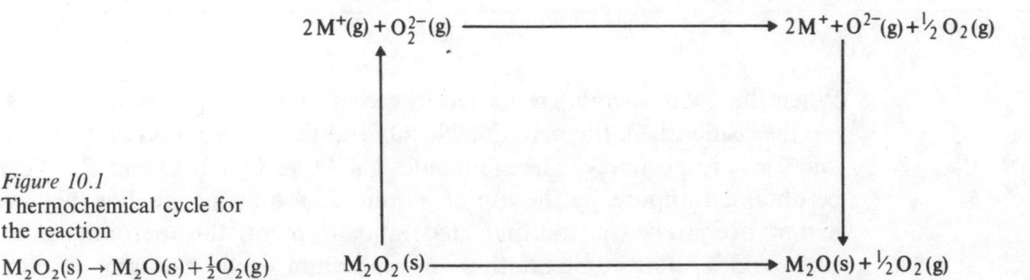

Figure 10.1
Thermochemical cycle for the reaction
$M_2O_2(s) \rightarrow M_2O(s) + \tfrac{1}{2}O_2(g)$

M_2O_2 and M_2O and ΔH° for the process

$$O_2^{2-}(g) \rightarrow O^{2-}(g) + \tfrac{1}{2}O_2$$

If we assume that the Madelung constant A_1 is the same for the structures of all the peroxides, their lattice energies will be proportional to

$$\frac{A_1}{r(M^+) + r(O_2^{2-})}$$

Similarly, we may take the lattice energies of the monoxides as proportional to

$$\frac{A_2}{r(M^+) + r(O^{2-})}$$

where A_2 is the Madelung constant for the structures of the monoxides. Since O^{2-} is obviously smaller than O_2^{2-}, the lattice energy of M_2O is always greater than that of M_2O_2, but the difference between them decreases as the radius of the cation increases. With respect to decomposition into monoxide and oxygen, therefore, Li_2O_2 is the least, and Cs_2O_2 the most, stable peroxide. The process

$$2MO_2 \rightarrow M_2O_2 + O_2$$

can be treated in an analogous way.

Sodium hydroxide is now made almost entirely as a by-product in the manufacture of chlorine by electrolysis of aqueous sodium chloride. In one type of cell a steel cathode and a graphite or platinum-coated titanium anode, separated by an asbestos diaphragm, are employed; sodium hydroxide accumulates in the cathode compartment and the solution is concentrated by evaporation. In the other common type of cell, a mercury cathode is used and is pumped away before the sodium which dissolves in it can react with the aqueous electrolyte; the sodium amalgam is hydrolysed separately, giving a clean, concentrated solution, and the mercury is returned to the electrolytic cell. Owing to the danger of pollution by mercury lost from the cell, however, the use of such cells is likely to diminish in the future. Solid sodium hydroxide is obtained as a white translucent solid of m.p. 318 °C by evaporation of the solution; it is fused and cast into sticks or made into flakes or pellets. It forms several hydrates and is very soluble in water, in which it dissolves with considerable evolution of heat. Sodium hydroxide is used throughout organic and inorganic chemistry wherever a cheap alkali is needed. Potassium hydroxide closely resembles the sodium compound in preparation and properties, but is more soluble in ethanol, in which it produces a low concentration of $OC_2H_5^-$ ions by the equilibrium

$$OH^- + C_2H_5OH \rightleftharpoons OC_2H_5^- + H_2O$$

Hence arises the special use of ethanolic potassium hydroxide in organic chemistry.

The crystal structures of the alkali metal hydroxides are usually complicated, but a high-temperature form of KOH has the sodium chloride structure with the hydroxide ion undergoing rotation in the lattice so that it is pseudo-spherical; potassium cyanide, it may be noted, has a similar structure.

The reactions of the hydroxides with acids and acidic oxides call for no special mention here; it is, however, worth noting that the anhydrous compounds react with carbon monoxide to produce formates, e.g.

$$NaOH + CO \xrightarrow{180\,°C} HCO_2Na$$

Many non-metals disproportionate on treatment with aqueous alkali, e.g. chlorine gives chlorate or hypochlorite and chloride, sulphur gives sulphide and a mixture of oxo-anions, and white phosphorus gives hypophosphite ($H_3PO_2^-$) and phosphine. Non-metals which do not form stable hydrides, together with amphoteric metals, give only hydrogen and oxo-anions, e.g. aluminium and silicon give hydrogen together with sodium aluminate and sodium silicate respectively.

10.5 *Salts of oxo-acids*

Alkali metal salts of most oxo-acids are mentioned under the acid concerned, since their properties are almost entirely those of the anion present. There are, however, a few aspects of the chemistry of these salts that call for mention here.

Sodium carbonate is manufactured by the Solvay (ammonia–soda) process, in which sodium hydrogencarbonate (or bicarbonate as it is still commonly called) is precipitated from a solution of sodium chloride by the successive action of ammonia and carbon dioxide, and is then heated to produce the carbonate and regenerate some of the carbon dioxide. The remainder of the carbon dioxide needed is produced by heating limestone, the other product of the reaction (calcium oxide) being used, after conversion into the hydroxide, to regenerate the ammonia from the ammonium chloride that is the other product of the precipitation reaction. The essential reactions are:

$$2NaCl + 2CO_2 + 2NH_3 + 2H_2O \rightarrow 2NaHCO_3 + 2NH_4Cl$$

$$2NaHCO_3 \rightarrow Na_2CO_3 + CO_2 + H_2O$$

$$CaCO_3 \rightarrow CaO + CO_2$$

$$CaO + H_2O \rightarrow Ca(OH)_2$$

$$Ca(OH)_2 + 2NH_4Cl \rightarrow 2NH_3 + CaCl_2 + 2H_2O$$

The whole process therefore amounts to the change

$$2NaCl + CaCO_3 \rightarrow Na_2CO_3 + CaCl_2$$

and calcium chloride is the only by-product; unfortunately it is a by-product for which no major use exists. In certain parts of the world (notably in the USA), the compound $Na_2CO_3.NaHCO_3.2H_2O$ (sesqui-carbonate) occurs as the mineral *trona*; thermal decomposition of this mineral there replaces the ammonia–soda process for the manufacture of sodium carbonate. Elsewhere the sesquicarbonate, which is a milder alkali than the carbonate and is much used in washing powders, is made by cooling a warm solution containing equimolar concentrations of the carbonate and the bicarbonate to 35 °C. Pure sodium bicarbonate for use in baking powders is made by passing carbon dioxide into a solution of the carbonate.

Lithium carbonate is only sparingly soluble in water and no bicarbonate of lithium has been isolated. Potassium carbonate cannot be made by a method analogous to the ammonia–soda process, since potassium bicarbonate is too soluble to be precipitated; the carbonate is therefore prepared from the hydroxide (produced electrolytically) and carbon dioxide.

The thermal stabilities of the alkali metal carbonates with respect to the reaction

$$M_2CO_3 \rightarrow M_2O + CO_2$$

increase with increasing size of the cation. As in the decomposition of peroxides to oxides and oxygen, lattice energy considerations play a dominant role. Such a gradation in stability is, indeed, common to all series of oxo-salts of the alkali metals.

The structures of sodium and potassium bicarbonates provide interesting examples of hydrogen bonding. The potassium salt contains a dimeric anion of the structure shown in Fig. 10.2(*a*). In the sodium salt, however, the HCO_3^- anions form an infinite chain as shown in Fig. 10.2(*b*). The hydrogen bonds are in each case unsymmetrical.

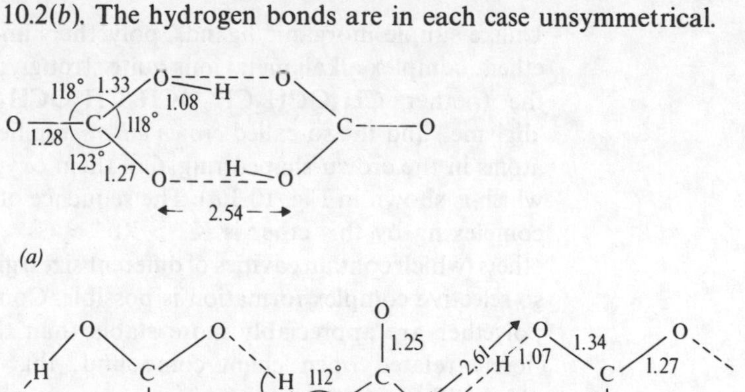

Figure 10.2
Structures of the anions in (*a*) $KHCO_3$ (*b*) $NaHCO_3$ (interatomic distances in ångströms).

10.6 *Aqueous solution chemistry*

All the alkali metal cations are hydrated in aqueous solution. Neutron diffraction studies of concentrated aqueous solutions of sodium and lithium chlorides show that the cations there are surrounded by six water molecules.

The thermodynamics of hydration of the ions and the standard M^+/M electrode potentials have been discussed already (Sections 7.8 and 10.1); since the metals all exhibit only the unipositive oxidation state, there is little more to be said about their standard potentials. Many lithium salts of acids which have small anions are sparingly soluble in water, e.g. LiF, Li_2CO_3; for large anions, on the other hand, it is the potassium, rubidium and caesium salts that are insoluble, e.g. $KClO_4$, $K_2[PtCl_6]$, $K_3[Co(NO_2)_6]$ and the analogous rubidium and caesium salts. A discussion of this behaviour in terms of lattice and hydration energies has been given in Section 7.3.

Although, as we shall see in the next section, the alkali metal cations are now known to form stable complexes, they very rarely do so in dilute aqueous solution; in so far as complexes with the usual anions are concerned, the normal order of formation constants (e.g. with $P_2O_7^{4-}$ or $EDTA^{4-}$) is $Li^+ > Na^+ > K^+ > Rb^+, Cs^+$. For adsorption on an ion-exchange resin, however, the order of the strength of adsorption is usually $Li^+ < Na^+ < K^+ < Rb^+ < Cs^+$, suggesting that the hydrated ions are adsorbed, since hydration energies decrease along that series and the total interaction (i.e. primary hydration plus secondary interaction with more water molecules) is greatest for lithium and least for caesium.

10.7 *Complexes*

Unlike simple inorganic ligands, polyethers and, even more, cyclic polyethers, complex alkali metal ions quite strongly; examples are provided by the triether $CH_3OCH_2CH_2OCH_2CH_2OCH_3$ (commonly known as 'diglyme') and the so-called *crown ethers*, typified by 18-crown-6 ether (18 atoms in the crown-shaped ring, 6 of them oxygen atoms), the formula of which is shown in Fig. 10.3(*a*). The sequence of equilibrium constants for complexing by this ether is $K^+ > Rb^+ > Cs^+, Na^+ > Li^+$; other crown ethers (which contain cavities of different sizes) give different sequences, and so selective complex formation is possible. Complexes formed by a cyclic polyether are appreciably more stable than those formed by the most closely related open chain compound; this is the *macrocyclic effect* (Section 7.6).

Because of the very large size and hydrophobic character of the complexed cation, crown ether complexing can lead to salts becoming soluble even in organic solvents which contain no donor atoms, e.g.

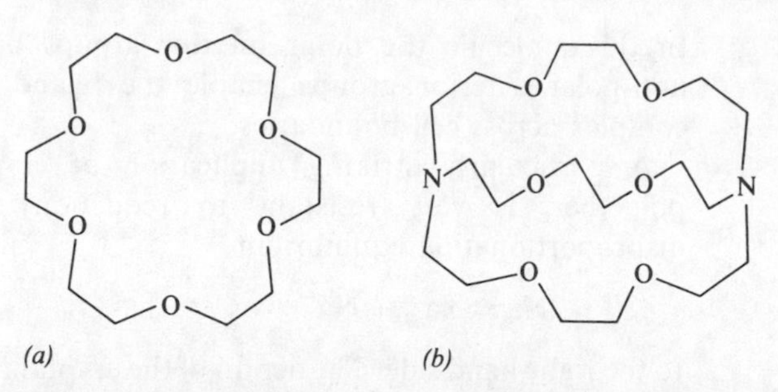

Figure 10.3
Macrocyclic and cryptand ligands.

[K(crown-6)]OH and [K(crown-6)]MnO_4 in toluene and benzene respectively. This phenomenon is very useful in preparative organic chemistry.

Three-dimensional equivalents of crown ethers contain nitrogen atoms to obtain branching and to provide extra donor sites; they are called *cryptands* (because they hide the cation). A typical one, shown in Fig. 10.3(*b*), is cryptand-222, so-called because each of the chains joining the nitrogen atoms contains two $-CH_2OCH_2-$ groups. All the nitrogen and oxygen atoms act as donors, the conformations at the nitrogen atoms being *endo*, i.e. the lone pairs inwards. Cryptands shield the cations that they complex even more effectively than crown ethers, and they show similar selectivities, cryptands-211, -221, and -222 forming their most stable alkali metal complexes with Li^+, Na^+, and K^+ respectively. These stabilities are essentially enthalpy-based; entropies of complexing, even in aqueous solution, are very small. Sodium and potassium cryptates are interesting models for the biologically occurring materials involved in the transfer of those ions across cell membranes. One of these is the polypeptide valinomycin, commonly formulated as

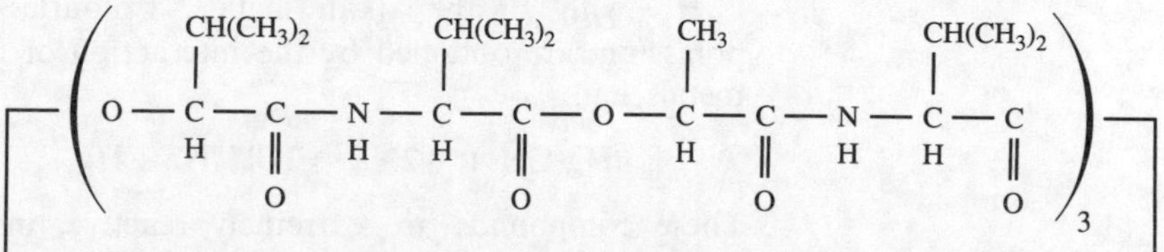

$$\left(-O-\underset{H}{\overset{CH(CH_3)_2}{\overset{|}{C}}}-\underset{\underset{O}{\|}}{C}-\underset{H}{N}-\underset{H}{\overset{CH(CH_3)_2}{\overset{|}{C}}}-\underset{\underset{O}{\|}}{C}-O-\underset{H}{\overset{CH_3}{\overset{|}{C}}}-\underset{\underset{O}{\|}}{C}-\underset{H}{N}-\underset{H}{\overset{CH(CH_3)_2}{\overset{|}{C}}}-\underset{\underset{O}{\|}}{C} \right)_3$$

but more faithfully represented by the symmetrical cyclic structure

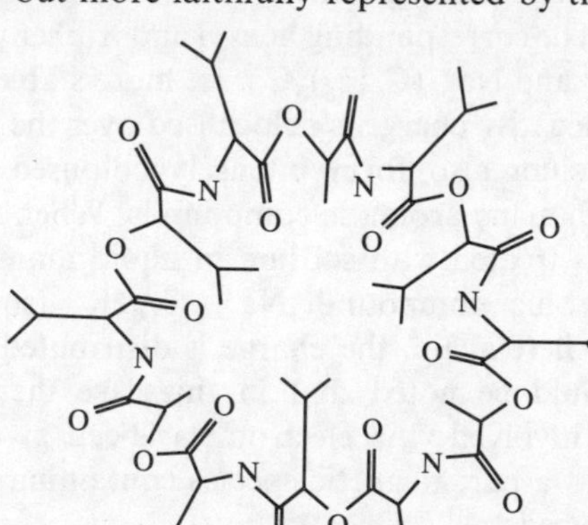

In this molecule the polar interior groups bind the cations and the non-polar exterior groups enable the ligand to transport the whole complex across cell boundaries.

A particularly striking application of cryptand-222 (which, as a polyether, is very resistant to reduction) is the shifting of the disproportionation equilibrium

$$2Na \rightleftharpoons Na^+ + Na^-$$

to the right-hand side. The action of the cryptand on sodium in ethylamine solution is to produce a diamagnetic golden-yellow solid of structure $[Na(\text{cryptand-222})]^+Na^-$. Crystallographic studies show that the effective radius of the Na^- ion in this compound is about 2.3 Å, i.e. about the same as that of I^-. Similar *alkalides* containing the ions K^-, Rb^-, and Cs^- have also been prepared. In these reactions, 1 mol of cryptand reacts with 2 mol of alkali metal. If the reaction is carried out using more cryptand, paramagnetic black *electrides*, e.g. $[Cs(\text{cryptand-222})_2]^+e^-$, may be isolated. In these the electron is trapped in a cavity of radius about 2.4 Å.

10.8 Organometallic compounds

Organic compounds which contain relatively acidic hydrogen atoms (e.g. alkynes, cyclopentadiene) form salts such as Na^+ $RC{\equiv}C^-$ and Na^+ $C_5H_5^-$ with all the alkali metals. Colourless alkyls of sodium and potassium are obtained by the interaction of mercury dialkyls and the metals, e.g.

$$Hg(CH_3)_2 + 2Na \rightarrow 2CH_3Na + Hg$$

These compounds are extremely reactive, insoluble in most organic solvents, and, when stable enough with respect to thermal decomposition, have fairly high melting-points; potassium methyl has the nickel arsenide structure. The corresponding benzyl and triphenylmethyl compounds Na^+ $C_6H_5CH_2^-$ and Na^+ $(C_6H_5)_3C^-$ are more stable and are red in colour; in these the negative charge is delocalised over the aromatic system. Sodium and potassium also form intensely coloured salts containing radical anions with many aromatic compounds. When naphthalene ($C_{10}H_8$), for example, is treated with sodium in liquid ammonia or tetrahydrofuran, the deep blue compound Na^+ $C_{10}H_8^-$ (sodium naphthalenide) is produced; here again, the charge is distributed over the aromatic rings, but it should be noted that in this case there is no displacement of hydrogen involved; the electron has been given to the neutral hydrocarbon and a paramagnetic species containing an unpaired electron has been produced.

Lithium alkyls and aryls are more stable thermally than the corresponding compounds of the other alkali metals (though they ignite spontaneously in air), and mostly differ from them in being soluble in hydrocarbons and other non-polar organic solvents and in being liquids or solids of low melting-point. The alkyls are polymeric both in solution and in the solid state; lithium methyl, for example, is tetrameric, and has a structure which may be described as a tetrahedron of lithium atoms with a methyl group just outside the middle of each face of the tetrahedron; each lithium atom has three carbon atoms, and each carbon atom three lithium and three hydrogen atoms, as nearest neighbours. Short intermolecular distances imply some bonding between the tetrameric units. There is disagreement on whether Li_4Me_4 is an ionic aggregate or whether multicentre covalent bonding is involved.

Organolithium compounds are now of great importance as synthetic reagents; in their chemistry they show a general similarity to the Grignard reagents, but they are more reactive. Butyl lithium, which is commercially available, is best made by the action of the metal on n-butyl chloride or bromide in a hydrocarbon solvent (it is decomposed by ether, forming butane, ethylene and lithium ethoxide). It is a liquid which reacts with many molecular chlorides and bromides with transfer of the alkyl group, e.g.

$$3LiBu + BCl_3 \rightarrow BBu_3 + 3LiCl$$

and with organic halides or hydrocarbons to give other lithium derivatives, e.g.

$$LiBu + CH_2{=}CHBr \rightarrow BuBr + CH_2{=}CHLi$$

$$LiBu + C_6H_6 \rightarrow BuH + C_6H_5Li$$

Lithium alkyls are important catalysts in the synthetic rubber industry for the stereospecific polymerisation of alkenes.

PROBLEMS

1 Comment on each of the following observations.

(a) The mobilities of the alkali metal ions in aqueous solution are $Li^+ < Na^+ < K^+ < Rb^+, Cs^+$.

(b) Lithium is the only alkali metal to form a stable nitride.

(c) The solubility of sodium sulphate in water increases up to 32 °C, and thereafter decreases.

(d) E° for $M^+_{(aq)} + e \rightleftharpoons M$ is nearly constant for the alkali metals.

2 (a) Predict the outcome of heating a mixture of lithium iodide and sodium fluoride.

(b) For the replacement of chlorine in organic compounds by fluorine by the autoclave reaction

$$\gt C{-}Cl + MF \rightarrow \gt C{-}F + MCl$$

potassium fluoride is a better reagent than sodium fluoride. Why?

REFERENCES FOR FURTHER READING

Burgess, J. (1988) *Ions in Solution*, Ellis Horwood, Chichester. Contains detailed accounts of macrocyclic and cryptand complexes.

Dye, J.L. (1977) *J. Chem. Education*, **54**, 333. The formation of Na^-.

Johnson, D.A. (1982) *Some Thermodynamic Aspects of Inorganic Chemistry*, 2nd edition, Cambridge University Press. Illuminating account of many aspects of the chemistry of the alkali metals.

Parker, D. (1983) *Adv. Inorg. Chem. Radiochem.*, **28**, 1. Metal cryptates.

Powell, P. (1988) *Principles of Organometallic Chemistry*, 2nd edition, Chapman and Hall, London. Contains a full account of alkali metal organometallic compounds.

Thompson, R. (editor) (1977) *The Modern Inorganic Chemicals Industry*, The Chemical Society, London. Contains informative accounts of the modern alkali and other chemical industries.

Wells, A.F. (1984) *Structural Inorganic Chemistry*, 5th edition, Oxford University Press. A much fuller account than that given here of the structures of alkali metal compounds, illustrated by a wealth of beautiful diagrams.

11 *Beryllium, magnesium and the alkaline earth metals*

11.1 Introduction

The relationship between the elements of this group is very like that between the alkali metals, though beryllium stands apart from the other metals to a greater extent than lithium from its homologues. This is mainly because the relative sizes of the beryllium and magnesium atoms (in the metals), or of the Be^{2+} and Mg^{2+} ions, differ by factors greater than those for the relative sizes of the lithium and sodium atoms, or of the Li^{+} and Na^{+} ions. Thus whereas analogous lithium and sodium salts usually crystallise in the same structure, this is not true for beryllium and magnesium salts; the typical coordination number for beryllium is four (e.g. in BeF_2 and BeO) whilst for magnesium it is six (e.g. in MgF_2 and MgO); for the alkaline earth metals it is six (e.g. in the oxides) or eight (e.g. in the fluorides).

Among aspects of the chemistry of beryllium, magnesium, calcium, strontium, barium and radium that we have discussed earlier in this book are successive ionisation energies of magnesium and their correlation with the electronic structure of the magnesium atom (Section 3.4), the non-existence of the molecule Be_2 (Section 4.2), the application of the molecular orbital and valence bond theories to account for the linear structure of the isolated $BeCl_2$ molecule (Sections 4.4 and 4.5), the structures of many halides (Sections 6.3 and 6.5), the lattice energy treatment of the solid-state disproportionation of CaF into Ca and CaF_2 (Section 6.8), the thermodynamics of the hydration of the Mg^{2+}, Ca^{2+}, Sr^{2+} and Ba^{2+} ions (Section 7.3), solutions of calcium, strontium and barium in liquid ammonia (Section 8.2), and the hydrides of all these metals (Section 9.6). These topics will be mentioned only briefly, or will not be mentioned at all, in the present chapter.

Table 11.1 contains values for some physical properties of beryllium, magnesium and the alkaline earth metals and their ions. The intense radioactivity of radium makes it impossible to obtain all the data for this element, but we should note that its first two ionisation energies are greater than those of barium – the first manifestation of the 6*s inert pair effect*. The quoting of a value for the radius of Be^{2+} depends upon the assumption, which is open to doubt, that this ion is present in the oxide and fluoride. It will be noted that the variation in some atomic properties (e.g. melting-points, boiling-points and enthalpies of atomisation) is somewhat irregular, a fact for which no simple explanation can be given. For the ionisation energies and for properties of the ions themselves, however, much more regular variations are seen, and the discussion of these aspects of the chemistry of the elements follows closely that of the alkali metals. Thus the near-constancy of E° for the M^{2+}/M systems where M = Ca, Sr, Ba or Ra, for example, is exactly similar to that found for the M^+/M systems where M = Li, Na, K, Rb or Cs, and arises from a similar balance between atomisation, ionisation and hydration energies.

The very high third ionisation energies preclude the formation of M^{3+} ions under chemical conditions, but the question of why the elements form M^{2+} rather than M^+ ions is so important that, although it has been mentioned in Section 6.8, it merits further discussion here. For a solid-state disproportionation such as

$$2CaF \rightarrow Ca + CaF_2$$

Table 11.1
Some properties of beryllium, magnesium and the alkaline earth metals and their ions.

Element	Be	Mg	Ca	Sr	Ba	Ra
Atomic number	4	12	20	38	56	88
Electronic configuration	[He]$2s^2$	[Ne]$3s^2$	[Ar]$4s^2$	[Kr]$5s^2$	[Xe]$6s^2$	[Rn]$7s^2$
Atomic radius[a]/Å	1.12	1.60	1.97	2.15	2.24	—
ΔH° atomisation/kJ mol^{-1}	326	149	177	164	178	130
M.p./°C	1278	651	851	767	707	700
B.p./°C	2770	1107	1437	1366	1637	1140
First ionisation energy/kJ mol^{-1}	898	736	589	548	503	508
Second ionisation energy/kJ mol^{-1}	1762	1449	1144	1060	960	975
Third ionisation energy/kJ mol^{-1}	14850	7730	4940	4150	3440	—
Ionic radius[b]/Å	0.31	0.72	1.00	1.13	1.36	1.48
ΔH° hydration of M^{2+}/kJ mol^{-1}	−2500	−1940	−1595	−1465	−1325	—
ΔS° hydration of M^{2+}/J K^{-1} mol^{-1}	−300	−320	−230	−220	−200	—
ΔG° hydration of M^{2+}/kJ mol^{-1}	−2410	−1845	−1525	−1400	−1265	—
E° $M^{2+}(aq) + 2e \rightleftharpoons M$/V	−1.85	−2.37	−2.87	−2.89	−2.90	−2.92

[a] For twelve-coordination.
[b] For four-coordination in the case of Be^{2+}, six-coordination for the others.

we assume that the Ca^+ ion would be somewhat larger than the Ca^{2+} ion and that CaF would have the NaCl structure. Three factors then lead to a lattice energy for CaF_2 rather more than three times that estimated for the salt CaF: the double charge on the cation, the smaller size of the cation and the higher Madelung constant for the CaF_2 structure (2.52 compared with 1.75). Thus although the second ionisation energy of calcium is about twice the first one, the lattice energy factor (reinforced by the liberation of the atomisation energy of solid calcium) brings about the disproportionation; however, the margin of instability of CaF is not great and a very unstable species of this formula (of unknown structure) can be obtained at high temperatures; it disproportionates on cooling. In solution, other factors must obviously be involved. In this case the critical quantity is the free energy of solvation which, as we saw in Section 7.3, is roughly proportional to z^2/r_+ where z is the ionic charge and r_+ is the crystallographic radius of the ion. The solvation energy of Ca^{2+} should on this basis be rather more than four times that of Ca^+; again, therefore, the effect of the higher second ionisation energy is overcome.

We discussed the similarity between lithium and magnesium in Section 10.1; beryllium shows a similar 'diagonal relationship' to aluminium. The two metals have similar standard electrode potentials; both are rendered passive by nitric acid; both form electron-deficient hydrides, volatile chlorides and amphoteric hydroxides; and both form carbides that give methane on hydrolysis. As in the case of lithium and magnesium, the fact that, in atomic and ionic size, the first member of the group is nearer to the second member of the following group than to its own homologues is doubtless the main factor underlying the relationship.

Beryllium occurs principally as the mineral *beryl*, $Be_3Al_2(Si_6O_{18})$. Magnesium, calcium, strontium and barium are all widely distributed in minerals and in the sea, particularly important minerals being *dolomite* ($CaMg(CO_3)_2$), *carnallite* ($KMgCl_3.6H_2O$), chalk, limestone and marble ($CaCO_3$), *strontianite* ($SrSO_4$) and *barytes* ($BaSO_4$). Radium is formed as ^{226}Ra, an α-emitter of half-life 1 600 years, in the ^{238}U decay series; it may be separated by co-precipitation with barium sulphate. Its use in the treatment of cancer has been superseded by that of other isotopes, and since it is now of only minor chemical importance we shall not specifically mention radium compounds again in this chapter; their general properties may be inferred by extrapolation from those of the corresponding calcium, strontium and barium compounds. The isotope ^{90}Sr, a β^--emitter, is a relatively long-lived ($t_{\frac{1}{2}} = 28$ years) fission product of uranium; the danger that, via contamination of grass and milk, it may be incorporated with calcium phosphate in bone is the cause of great anxiety when it is distributed as a result of a mishap in a nuclear energy plant.

Both calcium and magnesium are catalysts for the diphosphate–triphosphate transformations (Section 14.9) by means of which energy obtained from the oxidation of organic compounds is stored and utilised. Magnesium is also an essential constituent of chlorophylls (Section 11.6).

Beryllium compounds, on the other hand, are extremely toxic, probably because beryllium displaces magnesium in enzymes and prevents their functioning.

11.2 The metals

The only one of the metals to be manufactured on a very large scale is magnesium: dolomite, $CaMg(CO_3)_2$, is decomposed by heat to a mixture of the oxides, and this is reduced by ferrosilicon in nickel vessels, the effective reaction being represented by the equation

$$2MgO + 2CaO + Si \xrightarrow{1200\ ^\circ C} 2Mg + Ca_2SiO_4$$

The magnesium is distilled out *in vacuo*. The metal can also be obtained by reduction of the oxide with carbon at 2000 °C or by electrolysis of a fused mixture of magnesium, calcium and sodium chlorides; it is used in light alloys (particularly with aluminium), as a reducing agent for the production of other metals, and for the preparation of the Grignard reagents. The extraction of magnesium from sea water by the action of calcined dolomite depends upon the fact that magnesium hydroxide is much less soluble than calcium hydroxide and may be filtered off after completion of the reaction

$$Ca(OH)_2 + Mg^{2+} \rightarrow Mg(OH)_2 + Ca^{2+}$$

Beryllium is obtained from beryl by heating it with sodium fluorosilicate, extracting the beryllium fluoride formed, and precipitating the hydroxide. The element is made by reduction of the fluoride with magnesium. The metal is a poor absorber of electromagnetic radiation, hence its use in the windows of X-ray tubes. Its high melting-point and low cross-section for neutron capture also make it a useful material in the nuclear energy industry. The alkaline earth metals are obtained by electrolysis of their fused halides or by reduction of their oxides with aluminium.

Beryllium and magnesium are greyish metals which are kinetically inert to oxygen and water because of the formation of a surface film of oxide. Magnesium amalgam liberates hydrogen from water, since no coherent oxide film forms on its surface, and magnesium metal decomposes steam. Both metals dissolve readily in non-oxidising acids, but only magnesium is attacked by nitric acid. Only beryllium, which forms an amphoteric hydroxide, liberates hydrogen from aqueous alkali. The alkaline earth metals are soft and silvery; they are somewhat less reactive than sodium, but in general they closely resemble that metal. This similarity extends to the formation of blue solutions in liquid ammonia, from which, in the case of the alkaline earth metals, ammines such as $Ca(NH_3)_6$ can be isolated.

All the Group II metals combine with oxygen, nitrogen, sulphur and halogens when heated. Calcium, strontium and barium also combine when heated with hydrogen, but magnesium does so only under pressure. These metals form saline hydrides; beryllium hydride is a polymeric species formed by decomposition of beryllium alkyls. Beryllium combines with carbon at high temperatures to form Be_2C, which has the antifluorite structure and reacts with water to liberate methane; the other metals form carbides of formula MC_2 which yield acetylene on hydrolysis and contain the $[C{\equiv}C]^{2-}$ ion.

11.3 Halides

Beryllium fluoride is obtained as a very soluble glass (m.p. 803 °C) by the thermal decomposition of $(NH_4)_2[BeF_4]$, which may be made by addition of ammonia to a solution of beryllium oxide in excess of aqueous hydrofluoric acid. In the molten state it is almost a non-conductor of electricity, and the fact that the solid has the cristobalite structure (Section 6.3) indicates that it is a polymeric solid, not a salt.

The fluorides of the other metals are all solids of high melting-points which are sparingly soluble in water, the solubility increasing slightly with increase in cation size. Magnesium fluoride has the 6:3-coordinated rutile structure; the others all have the 8:4-coordinated fluorite structure. Calcium fluoride occurs naturally as *fluorspar*. Neither magnesium fluoride nor alkaline earth metal fluorides form complexes with alkali metal fluorides. The bent (100°) structures of SrF_2 and BaF_2 in the vapour phase are surprising; they imply *sd* rather than *sp* hybrid bonding.

Anhydrous beryllium chloride (m.p. 405 °C) is made by the action of carbon tetrachloride on the oxide at 800 °C, a standard method for metal chlorides which cannot be obtained by thermal decomposition of hydrates isolated from aqueous solution. Its vapour at 750 °C consists of linear $BeCl_2$ molecules with Be—Cl = 1.77 Å, but at lower temperatures a dimer is present, and the solid has an infinite chain structure in which each metal atom is surrounded tetrahedrally by four chlorine atoms at 2.02 Å:

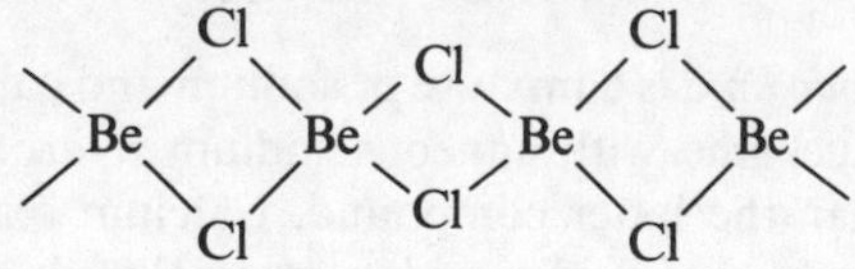

```
          Cl              Cl
   \     /    \   Cl  /     \    /
    Be          Be     Be      Be
   /     \    /   Cl  \     /    \
          Cl              Cl
```

The bond length data show the polymer has three-centre two-electron bonds. Beryllium chloride is a Friedel–Crafts catalyst (like aluminium chloride), and forms complexes with many oxygen- or nitrogen-containing organic compounds. It also forms complexes of formula $M_2[BeCl_4]$ with alkali metal chlorides, but these are decomposed by water; in aqueous media beryllium forms a stronger complex with fluoride than with chloride.

Magnesium halides (other than the fluoride) crystallise from aqueous solution as hydrates which undergo partial hydrolysis when heated, so the anhydrous halides are made from the elements. Chlorides, bromides and iodides of the alkaline earth metals can be obtained by dehydration of hydrates; anhydrous calcium chloride, made from the by-product in the ammonia–soda process for the manufacture of sodium carbonate, is widely used as a cheap drying agent. Most of these halides are somewhat soluble in polar oxygen-containing organic solvents, complexing of the cation doubtless being involved. Many of them have layer lattices, but the structures are complicated and will not be described here. Mixed chloride hydrides of formula MClH which have layer lattices (the different anions occupying different layers) are formed when the hydrides and chlorides of the alkaline earth metals are heated together, e.g.

$$CaH_2 + CaCl_2 \rightarrow 2CaHCl$$

11.4 Oxides and hydroxides

Beryllium oxide, made by ignition of the metal or its compounds in oxygen, is an insoluble white solid (m.p. 2570 °C) which has the wurtzite structure. The hydroxide, which may be precipitated from aqueous solutions of beryllium salts, is amphoteric.

The oxides of the other metals are usually made by thermal decomposition of the carbonates. All have the sodium chloride structure, and the regular fall in melting-point along the series (MgO, 2800 °C; CaO, 1730 °C; SrO, 1635 °C; BaO, 1475 °C) reflects the decrease in lattice energy as the radius of the cation increases. Magnesium oxide (a useful refractory) is converted into the sparingly soluble hydroxide (the antacid *milk of magnesia*) only very slowly, but the other oxides react readily with water and absorb carbon dioxide from the atmosphere. The conversion of calcium oxide (*quicklime*) to the hydroxide (*slaked lime*) is strongly exothermic:

$$CaO + H_2O \rightarrow Ca(OH)_2 \qquad \Delta H^\circ = -65 \text{ kJ mol}^{-1}$$

Soda lime is a mixture of sodium and calcium hydroxides made by slaking quicklime with aqueous sodium hydroxide; it is much easier to handle than the latter compound. Calcium oxide is used in metallurgy for the removal of acidic oxides (especially silica) in the extraction of metals, and in water-softening. The hydroxide finds numerous applications: in the manufacture of sodium carbonate, in the treatment of 'acid' soils, in the preparation of mortar (a mixture of the hydroxide and sand, which is slowly converted into a mixture of the carbonate and sand), and in the production of calcium hydrogensulphite (used in paper manufacture) and of bleaching powder (a mixture of the hydroxide, chloride and hypochlorite). The solubilities of the hydroxides in water increase from magnesium to barium, and their thermal stabilities with respect to decomposition into oxide and water increase similarly.

Barium oxide reacts with oxygen at 600 °C to form the peroxide; strontium peroxide may be obtained from the oxide at 350 °C under 200 atm pressure of oxygen. Calcium peroxide may be obtained by cautious dehydration of the hydrate obtained from the hydroxide and hydrogen peroxide. Hydrated magnesium peroxide is prepared by an analogous reaction for use in toothpastes; it cannot be dehydrated without decomposition. Thus for the process

$$MO_2 \rightarrow MO + \tfrac{1}{2}O_2$$

the standard free energy change becomes less negative as the size of M^{2+} increases; as for the decomposition of the alkali metal compounds, this can be traced to a diminishing increase in lattice energy on going from $M^{2+}O_2^{2-}$ to $M^{2+}O^{2-}$ as r_+ increases.

11.5 Salts of oxo-acids

As in the section dealing with alkali metal salts of oxo-acids, we shall mention here only a relatively small number of compounds which are of some special interest or importance.

Most beryllium salts of strong oxo-acids crystallise as soluble hydrates. Beryllium carbonate is prone to hydrolysis and can be isolated only by precipitation under an atmosphere of carbon dioxide. Another manifestation of the weakness of beryllium hydroxide as a base is the formation of a basic acetate $Be_4O(CH_3COO)_6$ by the action of acetic acid on the hydroxide. This compound is insoluble in water, but is readily soluble in organic solvents, in which it is unionised. X-ray diffraction shows it to contain a central oxygen atom surrounded by four beryllium atoms, which are linked by chelating acetate groups (Fig. 11.1). Beryllium nitrate,

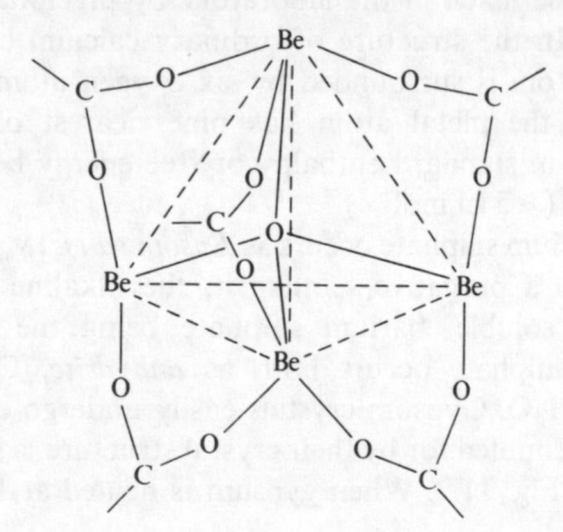

Figure 11.1
The structure of basic beryllium acetate. Methyl groups and one acetate ion have been omitted.

which is obtained by the reaction sequence

$$BeCl_2 \xrightarrow{N_2O_4} (NO)_2[Be(NO_3)_4] \xrightarrow{50\,°C} Be(NO_3)_2$$

decomposes at 125 °C with formation of a basic nitrate $Be_4O(NO_3)_6$ in which the nitrate group acts as a bidentate ligand.

The carbonates of magnesium and the alkaline earth metals are all sparingly soluble in water; their thermal stabilities increase with increasing cationic size; once again lattice energy considerations provide a simple explanation of the pattern of behaviour. All of these carbonates are much more soluble in a solution of carbon dioxide than in water owing to formation of the hydrogencarbonate ion, but no solid metal hydrogencarbonates of elements of this group have been isolated. The 'hardness' of water is due to the presence of magnesium and calcium salts: on addition of soap, these ions have to be precipitated as stearates before a lather forms. 'Temporary' hardness is caused by the presence of the hydrogencarbonates of magnesium and calcium, and may be removed by boiling, which upsets the equilibrium

$$2HCO_3^- \rightleftharpoons CO_3^{2-} + CO_2 + H_2O$$

and results in precipitation of the normal carbonates. It may also be removed by addition of the calculated quantity of calcium hydroxide. 'Permanent' hardness, which is not removed by boiling, is caused by other calcium and magnesium salts, usually the sulphates. Both types of hardness may be removed by exchanging calcium and magnesium for sodium by means of an ion-exchange resin.

Calcium carbonate is used in the ammonia–soda process, in the glass industry, and in making cement, which is a complex mixture of calcium silicates and aluminates obtained by heating a mixture of limestone and clay; mixed with water into a paste, either alone or with sand, it slowly sets to a hard mass. Concrete is a mixture of cement and sand which is used in much the same way as cement.

A metastable form of calcium carbonate (*aragonite*) occurs naturally and may be made in the laboratory by precipitation from hot aqueous solution. In the structure of ordinary calcium carbonate (*calcite*), each calcium atom is surrounded by six oxygen atoms of carbonate ions; in aragonite the metal atom has nine nearest oxygen neighbours. The difference in standard enthalpy or free energy between the two forms is very small (<5 kJ mol^{-1}).

Magnesium sulphate occurs as *Epsom salts*, $[Mg(H_2O)_6]SO_4.H_2O$, and is used as a purgative. Unlike it, the alkaline earth sulphates are all sparingly soluble, barium sulphate being the least soluble of them. Calcium sulphate occurs both as *anhydrite*, $CaSO_4$, and as *gypsum*, $CaSO_4.2H_2O$. Gypsum crystals easily undergo cleavage, a fact which is readily accounted for by their crystal structure, a section through which is shown in Fig. 11.2. When gypsum is heated at 120–130 °C it forms the

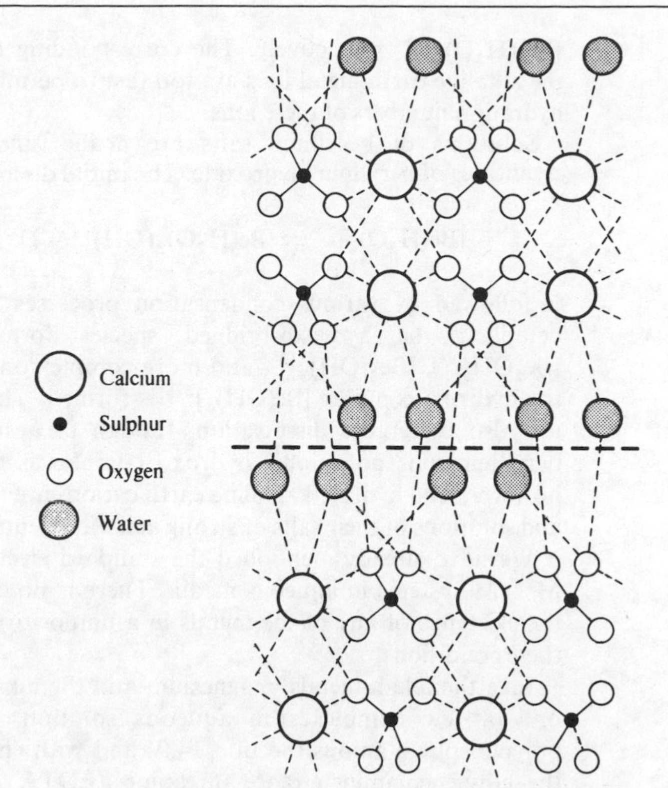

Figure 11.2
The structure of gypsum. The heavy broken line indicates the cleavage, which breaks only hydrogen bonds.

hemihydrate $CaSO_4 \cdot \frac{1}{2}H_2O$ (*plaster of Paris*); if this is mixed with water it evolves heat and solidifies to gypsum, expanding slightly in the process – hence its use in making plaster casts. Anhydrite is used in the manufacture of ammonium sulphate (Section 14.3) and sulphuric acid (Section 15.5). Another calcium salt of great industrial importance is the phosphate: its use in the manufacture of phosphate fertilisers is referred to in Section 14.9. Strontium and barium sulphates are the principal sources of these elements; they are reduced to sulphides by heating them with carbon; treatment with hydrochloric acid then produces the chlorides, from which other salts are readily obtained.

11.6 Aqueous solution chemistry and complexes

For solutions of beryllium and magnesium salts in water containing ^{17}O, exchange between solvent coordinated to the metal and the bulk solvent is sufficiently slow for the two types of water to be distinguished by nuclear magnetic resonance spectroscopy, and in this way the nature of the hydrated cations present is established as $[Be(H_2O)_4]^{2+}$ and

$[Mg(H_2O)_6]^{2+}$ respectively. The corresponding exchange processes for the alkaline earth metal ions are too fast to permit the elucidation of the hydration numbers of these ions.

Solutions of beryllium salts are acidic, and dissolve appreciable quantities of beryllium hydroxide. The initial dissociation of the aquo ion

$$[Be(H_2O)_4]^{2+} \rightleftharpoons [Be(H_2O)_3(OH)]^+ + H^+$$

is followed by various condensation processes involving the hydroxo complex; the hydroxo-bridged species formed include hydrated $[Be_2OH]^{3+}$, $[Be_3(OH)_3]^{3+}$ and more complex ones. In alkaline solution the hydroxo complex $[Be(OH)_4]^{2-}$ is formed. The hydrated magnesium ion also undergoes dissociation, though to a smaller extent than the beryllium ion; no anionic hydroxo complex is formed by magnesium, however. The hydrated alkaline earth cations are not appreciably ionised, and solutions of their salts of strong acids are neutral.

We have already mentioned the standard electrode potentials for the M^{2+}/M systems in aqueous media. There is no convincing evidence for the existence of any of the metals in a unipositive oxidation state under these conditions.

Like the alkali metals, magnesium and the alkaline earth metals form only a few complexes in aqueous solution, notably with various polyphosphate anions (Section 14.9) and with chelating ligands such as the ethylenediaminetetraacetate anion ($EDTA^{4-}$). Such complex formation may be utilised in water-softening. In complexing by $EDTA^{4-}$, the reaction goes mainly because of a large increase in entropy; it should, however, be remembered that most of this arises from the reduction in total ionic charge by reactions such as

$$Ca^{2+} + EDTA^{4-} \rightarrow [Ca(EDTA)]^{2-}$$

for which $\Delta H^\circ = -27\,\text{kJ mol}^{-1}$, $\Delta S^\circ = 113\,\text{J K}^{-1}\,\text{mol}^{-1}$ and $\Delta G = -60\,\text{kJ mol}^{-1}$. The alkaline earth metals also form complexes with macrocyclic ligands of the types described in Section 10.7.

Chlorophylls, the pigments that are involved in the conversion of carbon dioxide and water into organic compounds by green plants, are planar coordination complexes of magnesium. They are derivatives of various substitution products of the dianion derived from the parent molecule porphin (Fig. 11.3(*a*)), the skeleton of which also occurs in the blood pigment haem (Section 24.7). The formula of chlorophyll a is shown in Fig. 11.3(*b*). Because of the extensive conjugation in the ring system (cf. Section 2.7), the molecule can absorb light in the visible region and initiate a complicated series of reactions involving other systems containing manganese or iron; it is the macrocycle, not the magnesium, that is involved in these redox reactions.

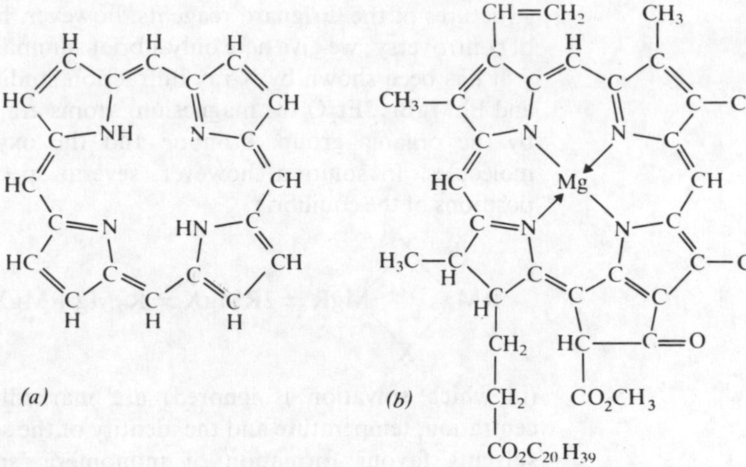

(a) (b)

Figure 11.3
The structures of
(*a*) porphin (*b*) chlorophyll a.

11.7 *Organometallic compounds*

Organometallic compounds of the alkaline earth metals are ionic, highly reactive species like the sodium and potassium derivatives, but they are of little importance. Beryllium and magnesium, on the other hand, both form organic compounds of great interest.

Beryllium alkyls and aryls are best made by reactions such as

$$HgMe_2 + Be \xrightarrow{110\,^\circ C} BeMe_2 + Hg$$

and

$$2LiPh + BeCl_2 \xrightarrow{Et_2O} 2LiCl + BePh_2$$

respectively. They are hydrolysed by water and inflame in air. Beryllium methyl is monomeric and linear in the vapour phase, but in the solid state it has the same structure as the chloride (Section 11.3). Since the methyl group has no unshared electrons available for electron donation, solid beryllium methyl must necessarily be an electron-deficient species, and it is believed that the beryllium and carbon atoms form three-centre two-electron molecular orbitals analogous to those in the bridges in the structure of diborane (Section 4.8). Higher alkyls are progressively less polymerised, and the t-butyl compound is monomeric under all conditions.

Alkyl and aryl magnesium halides (the Grignard reagents) are so well known on account of their uses in synthetic organic chemistry that it is unnecessary to discuss their preparation and reactions here. The structures of the Grignard reagents, however, have long been the subject of controversy; we give here only a brief summary of modern views.

It has been shown by X-ray diffraction studies that in $EtMgBr.2Et_2O$ and $PhMgBr.2Et_2O$ the magnesium atoms are tetrahedrally coordinated by the organic group, bromine and the oxygen atoms of the ether molecules. In solution, however, several species may be present; the positions of the equilibria

$$RMg(\mu\text{-}X)_2MgR \rightleftharpoons 2RMgX \rightleftharpoons R_2Mg + MgX_2 \rightleftharpoons R_2Mg(\mu\text{-}X)_2Mg$$

(in which solvation is ignored) are markedly dependent upon concentration, temperature and the identity of the solvent; strongly donating solvents favour formation of monomeric species, with which they coordinate. Solvent-free species RMgX have not been isolated; dialkyl derivatives of magnesium (e.g. $MgEt_2$) are known and closely resemble the corresponding beryllium compounds in both structure and reactivity. The magnesium derivative obtained by the action of the metal on cyclopentadiene, $Mg(C_5H_5)_2$, is decomposed by water, and hence is often inferred (not necessarily correctly) to be an ionic compound; it has the same sandwich structure as ferrocene (Section 23.5).

PROBLEMS

1 Discuss the variations in ΔH_f° among the following compounds.

	$-\Delta H_f^\circ/\text{kJ mol}^{-1}$			
M	MF_2	MCl_2	MBr_2	MI_2
Mg	1113	642	517	360
Ca	1214	795	674	535
Sr	1213	828	715	567
Ba	1200	860	754	602

2 How would you attempt to estimate

(a) ΔH° for the solid-state reaction

$$MgCl_2 + Mg \rightarrow 2MgCl$$

(b) ΔH° for the reaction

$$CaCO_3\,(\textit{calcite}) \rightarrow CaCO_3\,(\textit{aragonite})?$$

REFERENCES FOR FURTHER READING

Cotton, F.A. and Wilkinson, G. (1988) *Advanced Inorganic Chemistry*, 5th edition, Interscience Publishers, New York. A fuller treatment than that given in this book, especially of organometallic compounds and bioinorganic chemistry.

Hughes, M.N. (1981) *The Inorganic Chemistry of Biological Processes*, 2nd edition, John Wiley, Chichester and New York. The biochemistry of the alkaline earth metals.

Huheey, J.E. (1983) *Inorganic Chemistry: Principles of Structure and Reactivity*, 3rd edition, Harper and Row, New York, Chapter 18 gives an introduction to bioinorganic chemistry.

Wells, A.F. (1984) *Structural Inorganic Chemistry*, 5th edition, Oxford University Press. Full account of the structural chemistry of beryllium, magnesium and the alkaline earth metals.

12 *Boron, aluminium, gallium, indium and thallium*

12.1 Introduction

The elements in this group show a very wide variation in properties: boron is a typical non-metal, aluminium is a metal but shows many chemical similarities to boron, and the remaining elements are almost exclusively metallic in character. Although the tripositive oxidation state is the characteristic one for all members of the group, the unipositive state occurs in compounds of all the elements except boron, and for thallium it is the stable oxidation state. Thallium in fact shows similarities to so many other elements (the alkali metals, silver, mercury and lead) that Dumas called it the duckbill platypus among the elements. A feature of the chemistry of boron is the existence of large numbers of electron-deficient species which pose formidable problems in valence theory; these include not only the hydrides, but also organic and metallic derivatives of the hydrides, the metal borides, boron subhalides and the element itself.

Some physical properties of the elements are presented in Table 12.1. The first point to be noticed in this table is that whilst all the elements have the outer electronic configuration ns^2np^1, with the expected larger differences between the first and second than between the second and

Table 12.1
Some properties of boron, aluminium, gallium, indium and thallium.

Element	B	Al	Ga	In	Tl
Atomic number	5	13	31	49	81
Electronic configuration	$[He]2s^2 2p^1$	$[Ne]3s^2 3p^1$	$[Ar]3d^{10}4s^2 4p^1$	$[Kr]4d^{10}5s^2 5p^1$	$[Xe]4f^{14}5d^{10}6s^2 6o^1$
Atomic radius[a]/Å	—	1.43	1.22	1.62	1.70
ΔH° atomisation/kJ mol^{-1}	565	324	272	244	180
M.p./°C	2250	660	30	157	303
B.p./°C	2550	2500	2400	2100	1475
First ionisation energy/kJ mol^{-1}	800	578	579	558	589
Second ionisation energy/kJ mol^{-1}	2428	1817	1979	1820	1970
Third ionisation energy/kJ mol^{-1}	3650	2745	2962	2705	2880
Fourth ionisation energy/kJ mol^{-}1	25000	11600	6190	5250	4890
Ionic radius of M^{3+}/Å	—[b]	0.53	0.62	0.80	0.90[c]
E° $M^{3+}(aq) + e \rightleftharpoons M$/V	—[b]	−1.66	−0.53	−0.34	+0.72[d]

[a]Only the values for Al, In and Tl are strictly comparable; these metals crystallise in close-packed structures. Boron and gallium have unique structures which are described in Section 11.2, and the value given for Ga is half the distance of the one nearest neighbour; no meaningful value can be given for boron.
[b]There is no evidence for the existence of simple cationic boron under chemical conditions.
[c]Ionic radius of Tl^+ 1.45 Å.
[d]E° $Tl^+(aq) + e \rightleftharpoons Tl$ is −0.34 V.

third ionisation energies, the relationship between the electronic structures of the elements and those of the preceding noble gases are more complex than for the two groups of elements whose chemistry we discussed in Chapters 10 and 11. For gallium and indium the electronic structures of the species left after removal of three electrons are $[Ar]3d^{10}$ and $[Kr]4d^{10}$ respectively, whilst for thallium the species so formed has the configuration $[Xe]4f^{14}5d^{10}$. Thus the fourth ionisations of these three elements do not involve removal of an electron from a noble gas configuration, and the difference between the fourth and the third ionisation energies is not nearly so large as for boron and aluminium. The observed discontinuities in the values of ionisation energies between aluminium and gallium and between indium and thallium originate in the failure of *d* and *f* electrons, which have low screening power, to compensate the increases in atomic number; with thallium, relativistic effects are also involved. The increases in the second and third ionisation energies at gallium and at thallium lead to the marked increases in stability of the unipositive oxidation state at these elements. In the case of thallium (the only saline trihalide of which is the fluoride) this is often described as the 6*s* *inert pair effect*; it is also a feature of the chemistry of lead and bismuth in the following groups. As usual, the discussion of the variation in the stability of covalently bonded species is more difficult, but we can at least say that an increase in the ground state (ns^2np^1) to valence state (ns^1np^2) promotion energy must lead to a decrease in bond energy terms (which, of course, refer to bond formation from ground state atoms) unless there is a compensating increasing in the overlapping of valence state orbitals; with increasing atomic size this is most unlikely.

Most of the elements discussed in this chapter form compounds in which they have a formal oxidation state of two, e.g. B_2Cl_4, $GaCl_2$, GaS. All of these are diamagnetic, however, and structural studies show them to be $Cl_2B.BCl_2$, $Ga^+[GaCl_4]^-$ and $Ga_2^{4+}(S^{2-})_2$ respectively. There is in fact no evidence for a dipositive oxidation state of any of these elements. Nor are compounds containing B=B or analogous bonds yet known.

When boron, aluminium, gallium, indium or thallium forms a molecular trihalide or similar species it is still capable of accepting a pair of electrons, and very large numbers of complexes such as $[BF_4]^-$, $[AlCl_4]^-$, $[GaCl_4]^-$, $[InCl_4]^-$ and $[TlI_4]^-$ are known. Aluminium and the heavier elements are not restricted to an octet of electrons in their valence shells, and for these elements coordination numbers higher than four may be found, e.g. in $[AlF_6]^{3-}$ and $[TlCl_6]^{3-}$. The ability of three-coordinated boron to accept electrons is also indicated by bond lengths in boron trifluoride, various borates and boron nitride, which are much less than in the corresponding four-coordinated species, and (together with other properties) establish that π bonds are formed by donation of lone pairs of the fluorine, oxygen or nitrogen atoms into the empty *p* orbital of the boron atom.

The boron isotopes ^{10}B and ^{11}B have nuclear spin quantum numbers of 3 and $\frac{3}{2}$ respectively; ^{11}B nuclear magnetic resonance spectroscopy is important in the study of many aspects of boron chemistry, as we shall see later in this chapter.

Borax, $Na_2B_4O_7.10H_2O$ or $Na_2[B_4O_5(OH)_4].8H_2O$, is the principal source of boron; it occurs as deposits from hot springs and lakes in volcanic regions. Aluminium is the most abundant metal in the earth's crust, but its isolation from the aluminosilicate minerals in which it most commonly occurs is prohibitively difficult; it is always extracted from *bauxite* (a generic name for various hydrated oxide minerals) and *cryolite*, $Na_3[AlF_6]$. Naturally occurring bauxite is dissolved in hot aqueous sodium hydroxide under pressure to separate it from iron(III) oxide and the solution is seeded with a little $Al_2O_3.3H_2O$ and cooled; crystalline $Al_2O_3.3H_2O$ separates and is converted into the anhydrous oxide by the action of heat. Many applications of boron and aluminium and their compounds are mentioned later. Gallium, indium and thallium occur in traces in sulphide minerals; gallium is also found in traces in bauxite. All three metals may be obtained electrolytically from enriched aqueous solutions of their salts obtained as by-products in the extraction of other elements. Gallium and indium phosphides, arsenides and antimonides (often termed III–V semi-conductors) are used as transitor materials and in the light-emitting devices familiar in pocket calculators, watches, etc.; the colour of the light emitted depends on the band gap (Section 6.11). Thallium compounds have no major industrial uses.

12.2 The elements

Boron of a low degree of purity is obtained by reduction of the oxide with magnesium, followed by washing of the resulting material successively with alkali, hydrochloric acid and hydrofluoric acid. The product is a very hard black solid of low electrical conductivity which is inert towards most acids, but is slowly attacked by concentrated nitric acid or fused alkali; it does, however, combine with many metals. It is used in the production of impact-resistant steels and (because of the high cross-section for neutron capture of ^{10}B) in boron steel control rods for nuclear reactors (Section 1.4).

Pure boron is made by reduction of boron tribromide with hydrogen or by pyrolysis of diborane or boron triiodide. At least four allotropes may be obtained under different conditions, but transitions between different forms are extremely slow.

The simplest structure is that of α-rhombohedral boron, the unit cell of which contains twelve equivalent boron atoms in the form of an icosahedron (a highly symmetrical figure that may be described as two separate pentagonal pyramids with their base planes turned through 36° from one another, i.e. a bicapped pentagonal antiprism) (Fig. 12.1(*a*)). Each boron atom has five neighbours at 1.77 Å located at the corners of the base of the pentagonal pyramid of which it is the apex, so that a C_5 axis of symmetry passes through each atom. Six of the twelve atoms (marked 'r' in Fig. 12.1(*a*)) have also a boron atom of an adjacent B_{12} icosahedron 1.71 Å away on the C_5 axis passing through the atom; the other six atoms of the icosahedron (marked 'e' in Fig. 12.1(*a*)) are also 2.03 Å distant from each of two atoms belonging to different icosahedra. Of the atoms in any one icosahedron, therefore, half have one neighbour at 1.71 Å and five at 1.77 Å, and the other half have five neighbours at 1.77 Å and two at 2.03 Å; the linking of the 'e' atoms only is shown in Fig. 12.1(*b*).

The detailed discussion of such structures in terms of molecular orbital theory is difficult, but the essential features of the description of the structure of α-rhombohedral boron can be indicated briefly. The shortest bonds (intericosahedral bonds formed by 'r' atoms) are about the same length as the B—B bond in B_2Cl_4 and are taken as ordinary single (i.e.

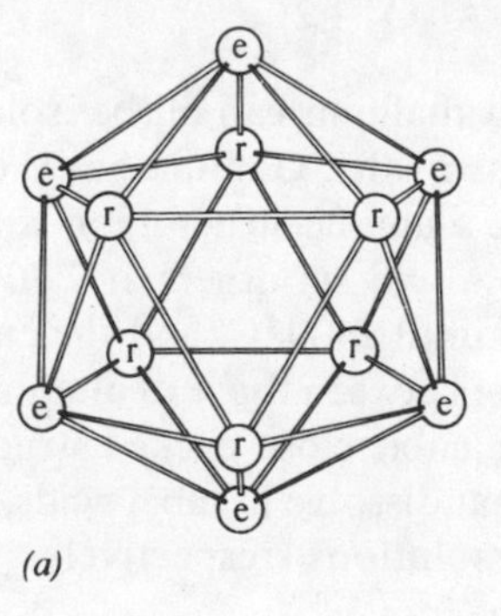

(*a*)

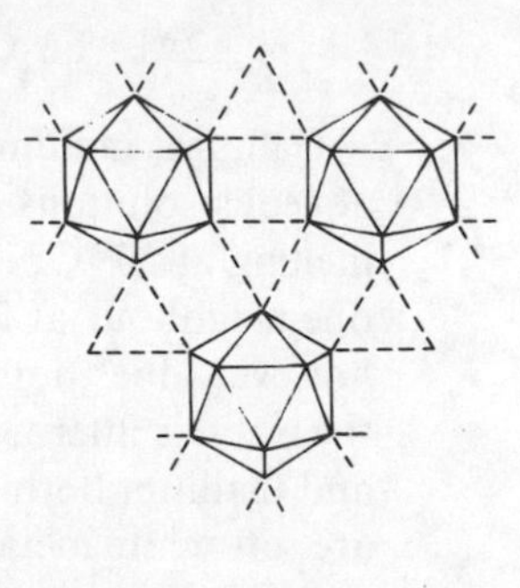

(*b*)

Figure 12.1
(*a*) B_{12} icosahedron.
(*b*) Linking of 'e' atoms in α-rhombohedral boron.

two-centre two-electron) bonds; six such bonds account for six electrons from each B_{12} unit. The longest bonds (those of the B_3 triangles formed between 'e' atoms) are 'closed' three-centre two-electron bonds of the type found in some of the boron hydrides; each B_{12} unit participates in six such bonds, contributing four electrons in all. Since a B_{12} unit contains thirty-six valence electrons, twenty-six are left for intraicosahedral bonding and occupy one multicentre orbital directed from each atom towards the centre of the icosahedron and twelve multicentre orbitals directed between the boron atoms, i.e. thirteen bonding orbitals in all.

Aluminium is manufactured by electrolysis at 950 °C of the oxide in solution in cryolite (to which calcium fluoride has been added to lower the melting-point); graphite-lined steel tanks are used as the cathodes, graphite rods as the anodes. Molten aluminium sinks to the bottom of the cell and is drawn off. The process is expensive in terms of electric power used, and so aluminium production is usually associated with hydro-electric schemes. Aluminium is a hard white metal which has the cubic close-packed structure; it is used in building construction, domestic utensils, drinks containers, electric cables (it has a weight for weight conductance twice that of copper), and, usually in alloys, in ship and aircraft construction. Thermodynamically it should react with both air and water, but in fact it is resistant to corrosion because of coating by a layer (10^{-6}–10^{-4} mm thick) of the oxide. A thicker layer of oxide (about 0.01 mm) can be obtained by making aluminium the anode in the electrolysis of sulphuric acid; the product (anodised aluminium) will take up dyes and pigments to give a strong and decorative finish. If the ordinary surface film of oxide is broken by scratching or amalgamation, the metal is rapidly corroded. Aluminium dissolves in dilute mineral acids, but is made passive (i.e. coated with oxide) by concentrated nitric acid; aluminium hydroxide is soluble in aqueous alkali, and the metal is rapidly attacked by the same reagent, hydrogen being liberated. The metal combines with halogens in the cold and with nitrogen on heating. Aluminium is often used as a reducing agent for the liberation of other metals from their oxides: the *thermit* process, for example, is based upon the reaction

$$2Al + Fe_2O_3 \rightarrow Al_2O_3 + 2Fe$$

Gallium, indium and thallium can all be isolated by electrolysis from aqueous solutions of their salts. Gallium has the remarkable property of melting at 30 °C; it has a unique structure in which each metal atom has one neighbour at 2.43 Å and six others at 2.70–2.79 Å. Liquid gallium, however, does not boil until 2400 °C, and the enthalpy of atomisation of the metal is intermediate between those of aluminium and indium. Indium and thallium both have cubic close-packed structures. All three elements are soft white metals that dissolve in most acids, forming Ga(III)-, In(III)-, and Tl(I)-containing solutions respectively. Only gallium liberates hydrogen from aqueous alkali: indium hydroxide is insoluble and not amphoteric, and although thallium(I) hydroxide is soluble in water, $E°$ for the Tl^+/Tl system is only −0.34 V, i.e. not negative enough for the metal to liberate hydrogen from, say, molar alkali (cf. Section 7.8). All three metals react with halogens at or a little above the ordinary temperature, forming trihalides except in the reactions of thallium with bromine and iodine, in which the products are $TlBr_2$ and Tl_3I_4 respectively.

12.3 Diborane and hydrogen compounds of aluminium, gallium, indium and thallium

The chemistry of the hydrides of boron and their derivatives is now so extensive that we shall devote two sections to hydrides of the Group III elements. For the present, we shall confine attention to the simplest hydride of boron, B_2H_6, and some of its derivatives, and the analogous compounds (in so far as they exist) of the later elements in the group. A few of the higher boron hydrides will be described in the next section.

We have already discussed the structure of diborane and the nature of the bonding in it (Section 4.8). Among several convenient laboratory preparations of diborane, which is now an important reagent in synthetic organic chemistry, are the following in diglyme solution

$$3NaBH_4 + 4BF_3.OEt_2 \rightarrow 2B_2H_6 + 3NaBF_4 + 4Et_2O$$

$$2NaBH_4 + I_2 \rightarrow B_2H_6 + 2NaI + H_2$$

$$(\text{diglyme} = (MeOCH_2CH_2)_2O)$$

The preparation of $NaBH_4$ is discussed later in this section. More direct syntheses include the industrial process based on

$$2BF_3 + 6NaH \xrightarrow{180\,°C} B_2H_6 + 6NaF$$

Diborane is a colourless gas (b.p. −90 °C) which is rapidly decomposed by water with formation of boric acid and hydrogen. Like all other boron hydrides, diborane has a small positive enthalpy of formation (36 kJ mol^{-1}). Mixtures with air or oxygen are liable to inflame or explode. Because the boron hydrides react with stopcock grease, they have to be manipulated in all glass apparatus using special valves and many other ingenious devices; for a brief account of some of the experimental techniques used, the references given at the end of this chapter should be consulted.

The bridged structure of diborane was determined by electron diffraction; vibrational and rotational spectroscopy, X-ray diffraction, and nuclear magnetic resonance spectroscopy now provide confirmatory evidence for it. As an indication of the uses of the last technique in elucidating the structures of the higher boron hydrides we may consider a few features of the ^{1}H and ^{11}B spectra of diborane (signals from the less abundant ^{10}B isotope are usually obscured by those arising from ^{11}B). Hydrogen atoms attached to one ^{11}B ($I = \frac{3}{2}$) atom give rise to four equally spaced lines of equal intensity corresponding to the four possible orientations of the ^{11}B nuclear spin ($\frac{3}{2}, \frac{1}{2}, -\frac{1}{2}, -\frac{3}{2}$). Hydrogen atoms attached to two ^{11}B atoms give rise to a septet of lines with components of relative intensities 1:2:3:4:3:2:1 since the combined nuclear spins of the two ^{11}B atoms can adopt seven orientations (3, 2, 1, 0, −1, −2, −3) in 1, 2, 3, 4, 3, 2, 1 ways respectively. Since there are twice as many terminal protons as bridge protons the ^{1}H spectrum (if we neglect ^{1}H—^{1}H coupling) consists of four lines each of relative intensity 8 (total intensity 32) and seven lines of relative intensity 1:2:3:4:3:2:1 (total intensity 16); this is the spectrum observed at a frequency of 100 MHz. If, whilst the proton resonance spectrum is being recorded, a strong radio frequency field at the ^{11}B resonance frequency (9.63 MHz) is applied (a procedure known as spin–spin decoupling or double resonance), the protons experience only the average of the ^{11}B orientations, and the ^{1}H spectrum then consists only of two peaks of relative intensities 2:1.

The interpretation of ^{11}B nuclear magnetic resonance spectra is simplified by the fact that ^{11}B—^{11}B coupling is not observed in practice. Each ^{11}B in diborane is attached to two ^{1}H terminal atoms of combined nuclear spins 1, 0 or −1, the relative probabilities of these values being 1, 2 and 1. The ^{11}B resonance is thus split into a triplet by the terminal protons; and each peak of the triplet is then split into a further triplet by coupling with the two bridge protons. The overall ^{11}B spectrum thus consists of a triplet of triplets, the relative intensities of the peaks being 1:2:1:2:4:2:1:2:1. In practice, of course, the ease with which the spectra are interpreted depends upon the values of the different spin–spin coupling constants involved and the degree of separation (i.e. the difference in chemical shifts) of identical nuclei in different chemical environments. This point was touched on briefly in Section 1.5, however, and we shall not consider it further here.

A few reactions of diborane which have not been mentioned earlier are shown below:

$$B_2H_6 + HCl \rightarrow B_2H_5Cl + H_2$$

$$B_2H_6 + 6MeOH \rightarrow 2B(OMe)_3 + 6H_2$$

$$B_2H_6 + 2LiH \xrightarrow{Et_2O} 2LiBH_4$$

$$B_2H_6 + 2Me_3N \rightarrow 2Me_3NBH_3$$

$$B_2H_6 + 2Me_3P \rightarrow 2Me_3PBH_3$$

$$B_2H_6 + 2CO \xrightarrow[20\ \text{atm}]{200\ ^\circ C} 2H_3BCO \text{ (borane carbonyl)}$$

$$B_2H_6 + 2Et_2S \rightarrow 2Et_2SBH_3$$

$$B_2H_6 + 2NH_3 \rightarrow [H_2B(NH_3)_2][BH_4] \xrightarrow{\text{heat}} B_3N_3H_6 \text{ (borazine – see Section 12.9)}$$

$$B_2H_6 + 6C_2H_4 \rightarrow 2(CH_3CH_2)_3B$$

$$2(CH_3CH_2)_3B \xrightarrow{H_3O^+_{aq}} 3C_2H_6 + B(OH)_3$$

$$2(CH_3CH_2)_3B \xrightarrow{H_2O_2} 3C_2H_5OH + B(OH)_3$$

Many of these reactions involve the species BH_3, the existence of which has been established by mass spectroscopic and kinetic studies; monoborane, however, cannot be isolated from among the products of the thermal decomposition of diborane, since further reactions to yield hydrogen and higher boranes take place. The best estimate of the enthalpy of dissociation of B_2H_6 into $2BH_3$ is about 150 kJ mol^{-1}. With the aid of this value, we can compare the strength of BH_3 as a Lewis acid with the strengths of the boron halides and the boron alkyls. Towards simple bases like trimethylamine, monoborane is generally, as might be expected from electronegativity considerations, intermediate between boron trihalides and boron trimethyl in acceptor properties. It is, however, alone amongst these Lewis acids in forming stable complexes with carbon monoxide and also with phosphorus trifluoride. For reasons which we shall describe more fully in Chapter 22, it is now generally believed that in their complexes CO and PF_3 are simultaneously electron donors (using the lone pairs of electrons on the carbon and phosphorus atoms respectively) and electron acceptors (using empty π^* and d orbitals respectively). Thus BH_3 must be taken to act in both capacities; its electron donation is ascribed to *hyperconjugation* analogous to that often postulated for a methyl group in organic chemistry. In borane carbonyl, for example, the structure is envisaged on valence bond theory as a resonance hybrid to which the following canonical forms contribute:

$$H_3B{-}C^-{\equiv}O^+ \leftrightarrow H{-}B^-H_2{-}C^+{=}O \leftrightarrow H^+\,H{-}B^-H{=}C{=}O$$

Sodium tetrahydridoborate, $NaBH_4$, the most important salt containing the BH_4^- anion, is conveniently prepared by heating sodium hydride with trimethyl borate in tetrahydrofuran under pressure:

$$4NaH + 4B(OMe)_3 \xrightarrow{250\,°C} NaBH_4 + 3Na[B(OCH_3)_4]$$

It is a white non-volatile crystalline substance, stable in dry air and soluble in water (towards which it is kinetically, rather than thermodynamically, stable). Although it is insoluble in diethyl ether, it dissolves in tetrahydrofuran and polyethers. As we have mentioned already, it is an important reagent for the laboratory preparation of diborane.

Sodium tetrahydridoborate has the sodium chloride structure, and is a typical ionic salt, but derivatives of some other metals have quite different properties. Thus $Be(BH_4)_2$ and $Al(BH_4)_3$, made from the appropriate chlorides and sodium tetrahydridoborate, sublime at 91 °C and boil at 45 °C respectively; both are violently decomposed by water. They have the electron-deficient structures shown in Fig. 12.2.

Aluminium hydride monomer and dimer may be detected when aluminium is slowly vaporised in hydrogen; the solid compound, however, is polymeric and is believed to contain Al—H—Al three-centre two-electron bonds analogous to those in diborane. It is made by the action of aluminium chloride on lithium tetrahydridoaluminate ('lithium aluminium hydride') in an ethereal solvent:

$$3LiAlH_4 + AlCl_3 \rightarrow 3LiCl + 4/n(AlH_3)_n$$

The tetrahydridoaluminate is a white or grey solid obtained by the action of lithium chloride on aluminium chloride (using stoichiometric quantities) in diethyl ether:

$$4LiH + AlCl_3 \rightarrow 3LiCl + LiAlH_4$$

It is readily soluble in many ethers, and the solution is a very important reducing agent in both organic and inorganic chemistry: it is, for example, capable of reducing carboxylic acids to alcohols, and it is the standard reagent for the conversion of molecular halides into hydrides (cf. Section 9.6). Lithium tetrahydridoaluminate decomposes into its elements above 120 °C, and is violently decomposed by water:

$$LiAlH_4 + 4H_2O \rightarrow LiOH + Al(OH)_3 + 4H_2$$

Sodium tetrahydridoaluminate can be made by the direct reaction between sodium, aluminium and hydrogen at 150 °C under pressure in tetrahydrofuran; it may be converted into the lithium salt by interaction with lithium chloride in ether, in which sodium chloride is insoluble:

$$NaAlH_4 + LiCl \rightarrow NaCl + LiAlH_4$$

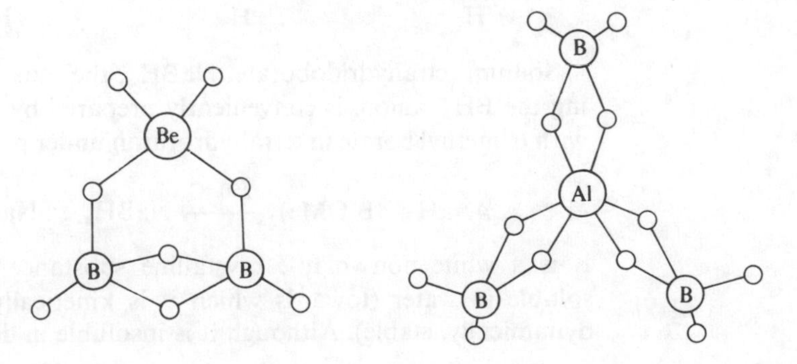

Figure 12.2
The structures of $Be(BH_4)_2$ and $Al(BH_4)_3$

Complexes of aluminium hydride (or alane) and donor molecules may be obtained by reactions such as

$$3LiAlH_4 + AlCl_3 + 4Me_3N \rightarrow 4Me_3NAlH_3 + 3LiCl$$

Some of these compounds are important reducing agents and polymerisation catalysts in organic chemistry.

Unstable complexes $LiMH_4$, where M = Ga, In or Tl, have been made at low temperatures by reactions analogous to that used for the preparation of the tetrahydridoaluminate. Gallium hydride, though very unstable, is now known, but simple hydrides of indium and thallium have not yet been prepared.

12.4 *The higher boranes and the carboranes*

There is now evidence for the existence of more than twenty hydrides containing four or more atoms of boron. Several of these, including B_4H_{10} (b.p. 16 °C), B_5H_9 (b.p. 60 °C), B_5H_{11} (b.p. 63 °C) and $B_{10}H_{14}$ (m.p. 99 °C) were isolated by Stock in his classic researches in the period 1912–36. In this work, a mixture of boron hydrides was made by the action of aqueous hydrochloric acid on magnesium boride; B_2H_6 and B_5H_{11}, which are rapidly hydrolysed, were not obtained in this way, but were prepared by the action of heat on other members of the series. All the higher hydrides are now made by controlled pyrolysis of diborane under rigidly controlled conditions, e.g. B_4H_{10} from B_2H_6 after five hours at 200 atm pressure at 80–90 °C, B_5H_9 from $B_2H_6 + H_2$ by rapid passage through a reactor at 200–240 °C, and $B_{10}H_{14}$ by pyrolysis of B_2H_6 at 150 °C in the presence of dimethyl ether; the mechanisms of these interconversions are obviously very complicated. Pentaborane(11) is best made from B_5H_9, which yields the $B_5H_9^{2-}$ ion on treatment with potassium naphthalenide; protonation then gives B_5H_{11}. There was at one time considerable interest in the possibility of using boron hydrides as high-energy fuels, and $B_{10}H_{14}$ has been made on a ton scale for this purpose. In principle, the combustion of boranes is capable of providing about 30 J per gram more energy than the combustion of hydrocarbon fuels. In practice, however, it is difficult to ensure complete combustion to B_2O_3, and an involatile polymer of empirical formula BO tends to block exhausts; interest in the boron hydrides from this point of view has therefore now faded.

In general, the reactivity of the boron hydrides decreases with increase in the number of boron atoms present; $B_{10}H_{14}$, or decaborane(14), is (kinetically) inert to air and water. Nucleophilic substitution (e.g. with lithium alkyls or sodium cyanide) takes place at boron atoms 6 and 9 (see Fig. 12.3), electrophilic substitution (e.g. iodination or Friedel–Crafts methylation) at boron atoms 1, 2, 3 and 4.

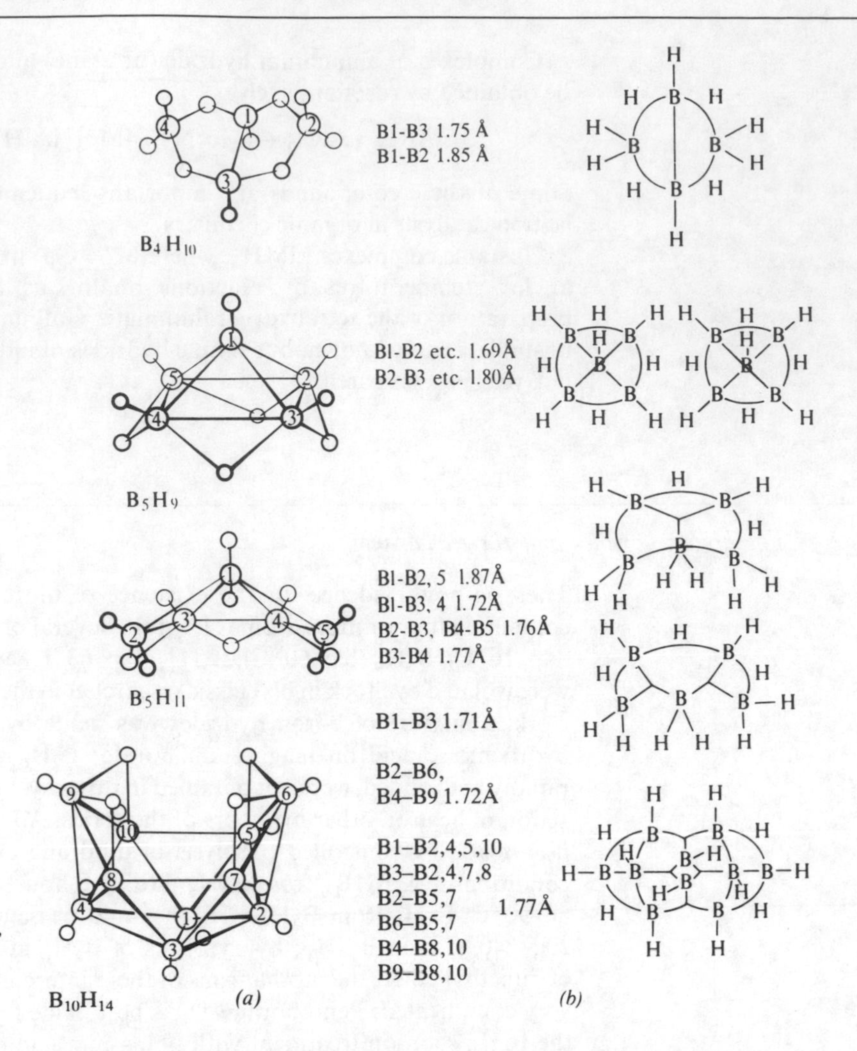

Figure 12.3
(*a*) The structures of B_4H_{10}, B_5H_9, B_5H_{11} and $B_{10}H_{14}$.
(*b*) Representation of the bonding in the hydrides.

The structures of B_4H_{10}, B_5H_9, B_5H_{11} and $B_{10}H_{14}$ (and several others) have been determined by X-ray and electron diffraction and are shown in Fig. 12.3 in the way in which they are usually drawn, which emphasises the spatial arrangements of the boron atoms – *approximately* an octahedron with two adjacent positions empty, a square pyramid (or an octahedron with one position empty), a pentagonal pyramid with one equatorial position empty (or a pentagonal bipyramid with one equatorial and one axial position empty), and an icosahedron with two adjacent positions empty (or, less obviously, an octadecahedron, a polyhedron with eleven vertices and C_{2v} symmetry, with the unique apical position empty) respectively. Figure 12.3(*a*), however, emphasises the geometry of the groupings of boron atoms at the expense of showing which atoms are linked by the strongest (i.e. shortest) bonds; in the structure of B_5H_{11}, for example, the B1–B2 or B5 distance (1.87 Å) is so much greater than the B1–B3 or B4, B2–B3, B3–B4 or B4–B5 distances (all in the range 1.72–1.77 Å) that it may be thought that no bond should be shown between B1 and B2 or B5. On the other hand, the ^{11}B nuclear magnetic resonance signal originating from the B1 atom is a doublet, suggesting that this boron atom has only a single proton attached to it, the other proton shown attached only to B1 being close enough to B2 and B5 (capping the B1 B2 B5 triangular face) to be intermediate in character between bridging and terminal and for some reason not influencing the B1 signal. Since some other fine structure is absent from the ^{11}B spectrum of B_5H_{11}, this reason is evidently extensive exchange of bridging hydrogen atoms. Thus it is obviously very difficult to give a simple bonding diagram for B_5H_{11}.

This difficulty is a fundamental one: structure determinations reveal only atomic positions (strictly, in the case of X-ray diffraction the positions of centres of electron density), from which we infer the presence of bonds. In organic chemistry, where we can safely assume all valencies in the vast majority of compounds, this presents few problems, but in the case of electron-deficient compounds there are often several possible reasonable ways of joining up the atoms. It is then found that no one representation is an adequate description, i.e. in valence bond language the molecule has to be regarded as a resonance hybrid of various canonical forms. Even the writing of the canonical forms, however, requires the use of multicentre orbitals; in addition to ordinary single (two-centre two-electron) B—H and B—B bonds and three-centre two-electron B—H—B bonds as in B_2H_6, two types of three-centre two-electron bonds involving only boron atoms are required. In an *open* three-centre two-electron bond the bridge hydrogen of the B—H—B bond is replaced by a boron atom contributing a *p* orbital; in a *closed* three-centre two-electron bond, in which the boron atoms are at the corners of an equilateral or nearly equilateral triangle, each boron atom is envisaged as contributing a sp^3 hybrid orbital pointing towards the centre of the triangle; a closed three-centre two-electron bond is represented in formulae by three lines from three boron atoms meeting in a point, an open three-centre two-electron bond by a line joining three boron atoms passing through the symbol of the middle one.

In Fig. 12.3(*b*) we give representations of the structures of B_4H_{10}, B_5H_9, B_5H_{11} and $B_{10}H_{14}$ in terms of these five types of bond. Each hydrogen is assumed to contribute a 1*s* orbital and each boron a 2*s* and three 2*p* orbitals appropriately hybridised. One electron from each hydrogen and three from each boron are available; these are used first to form terminal B—H and bridging B—H—B bonds, and then any remaining electrons are allocated to framework orbitals. Tetraborane (10) is easily dealt with: of the 22 electrons available, six pairs are used for terminal B—H bonds, four pairs for B—H—B bridge bonds, and the remaining pair bond B1 and B3, which are 1.72 Å apart. The representation of the bonding in Fig. 12.3(*b*) is completely satisfactory for this compound.

For pentaborane (9), B1 forms four short (1.69 Å) identical bonds; in view of the greater interatomic distances between basal boron atoms (1.80 Å) we assume no B—B bonding there. Then out of 24 electrons, five pairs being used for terminal B—H bonds and four pairs for B—H—B bridge bonds, six electrons remain to bond B1 to B2, B3, B4 and B5 by means of two two-electron B—B bonds and one open three-centre B—B—B bond; since all bonds from B1 to other boron atoms have to be equivalent, resonance between the structure shown in Fig. 12.3(*b*) and an equivalent one in which the three-centre bond involves the other basal boron atoms has to be invoked. Even so, there remains the implication that the bonds between B1 and the other boron atoms will be of order less than unity, although the actual bond length is less than that of the ordinary single B—B bond in B_4H_{10}. In these circumstances, therefore, a molecular orbital treatment involving multicentre orbitals (as in B_{12}) is inherently more satisfactory; unfortunately it is also much more difficult.

In the case of pentaborane (11), out of 26 electrons eight pairs are needed for B—H terminal bonds and three pairs for B—H—B bridges, leaving four electrons to bond the boron atoms, either by means of two closed (B—B—B, with the third B below the centre) bonds or by means of one closed and one open three-centre bond; again resonance has to be invoked, both structures being shown in Fig. 12.3(*b*). For $B_{10}H_{14}$ there are several possibilities and we have simply shown one.

The structures of the boron atom clusters (all of which have triangular faces) in boron hydrides and related species are usefully summarised by *Wade's rules*, which have a foundation in molecular orbital theory but are presented here as an empirical basis for remembering and predicting structures. Any borane or borane anion can be represented by the formula $B_nH_{n+m}^{p-}$; e.g. for B_5H_{11}, $n=5$, $m=6$, $p=0$; for $B_{10}H_{14}^{2-}$, $n=10$, $m=4$, $p=2$. If we count the number of valence electrons in the species $[3n+(n+m)+p]$ and assign $2n$ to n ordinary two-electron B—H bonds, the remaining $2n+m+p$ electrons (or half that number of electron pairs) are used for bonding the cluster. This is: a *closo* (closed) cluster, a complete polyhedron, if the n boron atoms are held together by $n+1$ electron pairs; a *nido* (nest) cluster, a polyhedron with one vertex missing, if there are $n+2$ electron pairs; and an *arachno* (web) cluster, a polyhedron with two vertices missing, if there are $n+3$ electron pairs. An electron count shows, for example, that there are $n+3$, $n+2$, and $n+3$ electron pairs available for boron cluster bondings in B_4H_{10}, B_5H_9, and B_5H_{11} respectively; hence the description of their boron clusters given earlier as an octahedron with two positions vacant, an octahedron with one position vacant (a square pyramid), and a pentagonal bipyramid with two positions vacant, respectively. It should be noted that in cases where there is choice we have not predicted which positions are vacant, nor what the polyhedron will be (though symmetric structures would be expected to be favoured). In practice the polyhedron is found to be a trigonal bipyramid, an octahedron, a pentagonal bipyramid, a dodecahedron, an octadecahedron and an icosahedron for 5, 6, 7, 8, 11 and 12 vertices respectively.

In addition to $[BH_4]^-$, many anions containing more than one boron atom are known. A few preparations are indicated by the equations

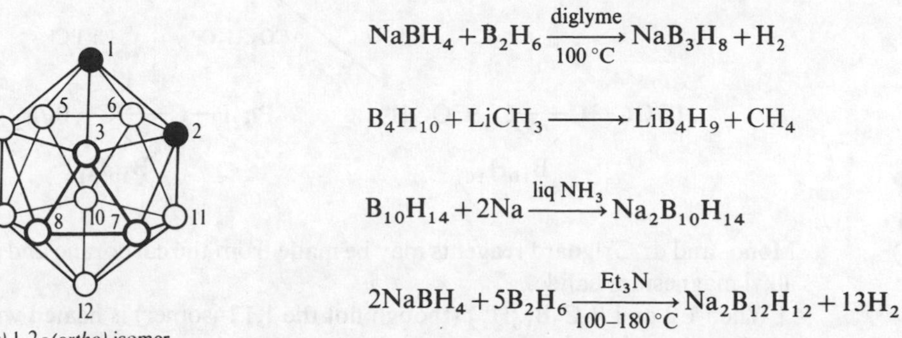

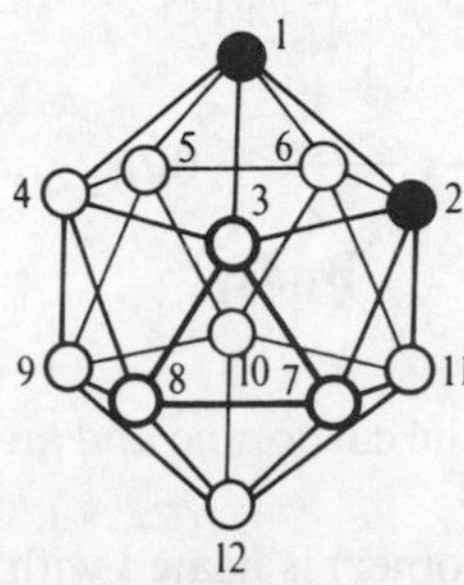

(a) 1,2-(*ortho*) isomer

$$NaBH_4 + B_2H_6 \xrightarrow[100\,°C]{\text{diglyme}} NaB_3H_8 + H_2$$

$$B_4H_{10} + LiCH_3 \longrightarrow LiB_4H_9 + CH_4$$

$$B_{10}H_{14} + 2Na \xrightarrow{\text{liq } NH_3} Na_2B_{10}H_{14}$$

$$2NaBH_4 + 5B_2H_6 \xrightarrow[100\text{–}180\,°C]{Et_3N} Na_2B_{12}H_{12} + 13H_2$$

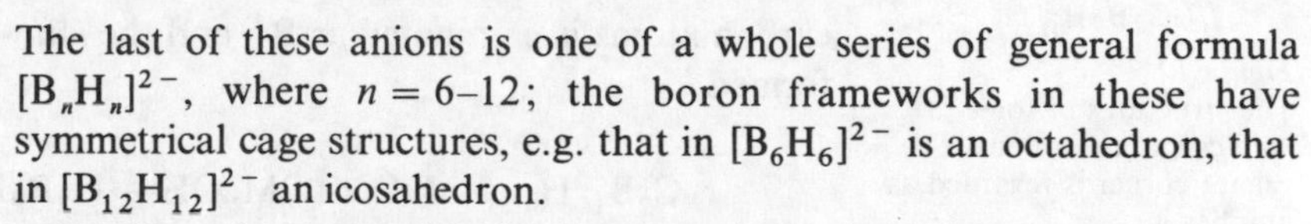

The last of these anions is one of a whole series of general formula $[B_nH_n]^{2-}$, where $n = 6$–12; the boron frameworks in these have symmetrical cage structures, e.g. that in $[B_6H_6]^{2-}$ is an octahedron, that in $[B_{12}H_{12}]^{2-}$ an icosahedron.

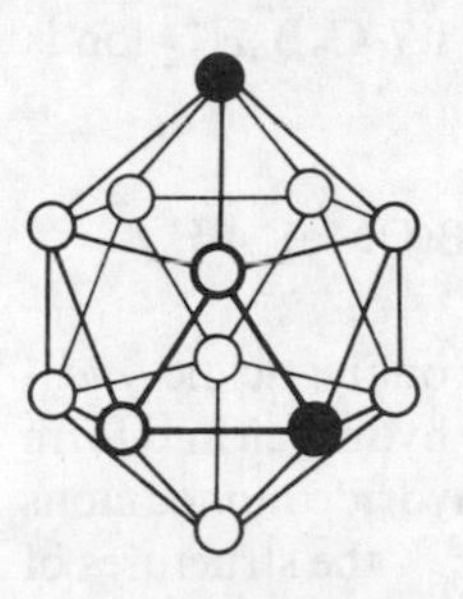

(b) 1,7-(*meta*) isomer

Studies of substitution in $B_{10}H_{14}$ show that the electron density distribution in the boron cage is uneven: Friedel–Crafts alkylation, for example, occurs at B1, B2, B3, and B4; lithium methyl, on the other hand, attacks mainly at B6 and B9. These are standard electrophilic and nucleophilic substitutions respectively.

Carboranes are neutral molecules in which a BH^- unit has been formally replaced by an isoelectronic CH unit. When, for example, decaborane(14) is heated with alkynes in the presence of a dialkyl sulphide as catalyst, the following reactions occur:

$$B_{10}H_{14} + 2R_2S \rightarrow B_{10}H_{12}(R_2S)_2 + H_2$$

$$B_{10}H_{12}(R_2S)_2 + RC{\equiv}CR' \rightarrow B_{10}H_{10}C_2RR' + 2R_2S + H_2$$

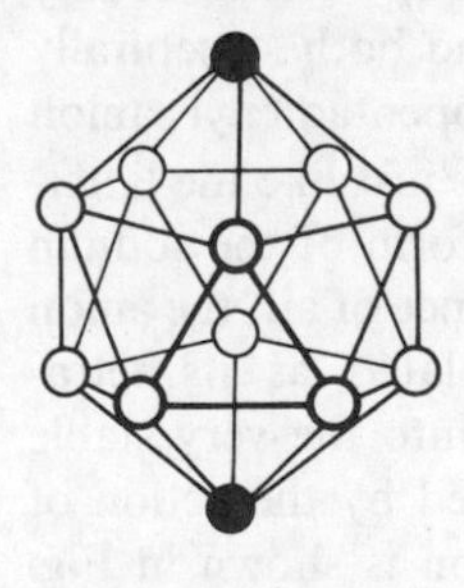

(c) 1,12-(*para*) isomer

Figure 12.4
The three dicarba-*closo*-dodecaboranes $C_2B_{10}H_{12}$.

In the case where R = R′ = H the product has the structure shown in Fig. 12.4(*a*); at 470 °C this rearranges to compound (*b*), and this in turn gives compound (*c*) at 600 °C. These three compounds are known respectively as 1, 2 (or *ortho*), 1, 7 (or *meta*) and 1, 12 (or *para*) dicarba-*closo*-dodecaboranes. The $B_{10}C_2$ icosahedron is electron-deficient, with electron delocalisation extending over the whole framework. It is thus effectively a three-dimensional aromatic unit, and the electron-withdrawing effect of the $B_{10}C_2$ skeleton is to make the hydrogen atoms attached to carbon distinctly acidic. Thus a dilithium derivative may be made by the action of lithium butyl on the 1,2 compound, and this may in turn be converted into a range of other compounds according to the following scheme, in which a self-explanatory abbreviation is used for the carborane:

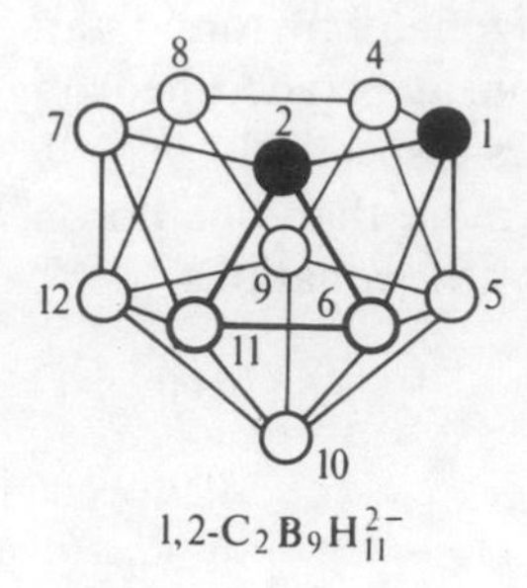

$1,2\text{-}C_2B_9H_{11}^{2-}$

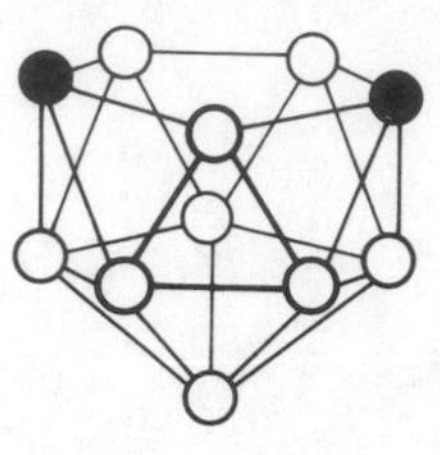

$1,7\text{-}C_2B_9H_{11}^{2-}$

Figure 12.5
The structures of some *nido*-carborane anions. The vacant corner is regarded as number 3.

$$\underset{B_{10}H_{10}}{H\text{—}C\overset{O}{\text{——}}C\text{—}H} \xrightarrow{C_4H_9Li} \underset{B_{10}H_{10}}{Li\text{—}C\overset{O}{\text{——}}C\text{—}Li}$$

$$\xrightarrow{CO_2,H_3O^+} \underset{B_{10}H_{10}}{HOOC\text{—}C\overset{O}{\text{——}}C\text{—}COOH} \qquad \xrightarrow{Ph_2PCl} \underset{B_{10}H_{10}}{Ph_2P\text{—}C\overset{O}{\text{——}}C\text{—}PPh_2}$$

Mono- and di-Grignard reagents may be made from the carborane and an alkyl magnesium halide.

When 1,2 or $1,7\text{-}C_2B_{10}H_{12}$ (though not the 1,12-isomer) is heated with potassium methoxide and methanol, one of the borons linked to two carbon atoms is eliminated as B^+ and the *nido*-1,2 or $1,7\text{-}C_2B_9H_{12}^-$ ion is formed:

$$C_2B_{10}H_{12} + MeO^- + 2MeOH \rightarrow C_2B_9H_{12}^- + B(OMe)_3 + H_2$$

These anions may be protonated to form the strongly acidic *nido*-carboranes 1,2 and $1,7\text{-}C_2B_9H_{13}$, which at 100 °C lose hydrogen and form a *closo*-carborane $C_2B_9H_{11}$. By thc action of sodium hydride monoanions $C_2B_9H_{12}^-$ are deprotonated to give 1,2 and $2,7\text{-}C_2B_9H_{11}^{2-}$, the structures of which are shown in Fig. 12.5 (in which the vacant corner is the original number 3 of the parent $C_2B_{10}H_{12}$ *closo*-carborane). These doubly charged *nido*-anions have an open pentagonal face and both structurally and electronically they closely resemble the cyclopentadienyl anion $C_5H_5^-$, and form complexes with transition metal ions just like the latter. Thus by the action of the $1,2\text{-}C_2B_9H_{11}^{2-}$ ion (in the form of the sodium salt) on iron(II) chloride in tetrahydrofuran in the absence of air, the anion $[(1,2\text{-}C_2B_9H_{11})_2Fe]^{2-}$ is formed, and may be isolated as its tetramethylammonium salt. Aerial oxidation converts it into the very stable $[(1,2\text{-}C_2B_9H_{11})_2Fe]^-$ ion; the change may be reversed by the action of sodium amalgam. The structure of the complex anion is shown in Fig. 12.6. The $C_2B_9H_{11}^{2-}$ anions are known by the trivial name of *dicarbollide* ion (from the Spanish *olla*, a pot, reflecting the potlike shape of the C_2B_9 cage). A large number of other dicarbollide complexes are now known, including some in which one dicarbollide ligand is replaced by $C_5H_5^-$,

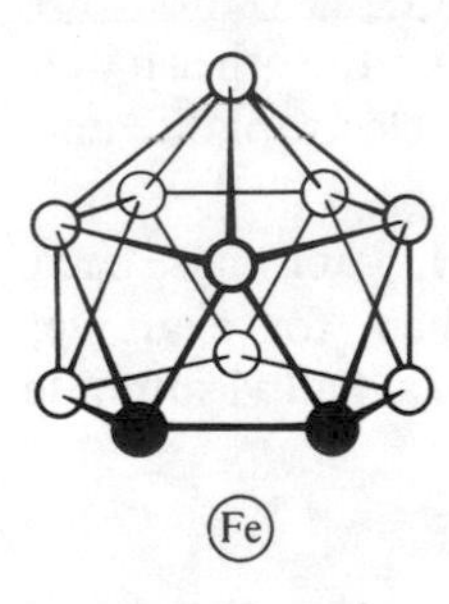

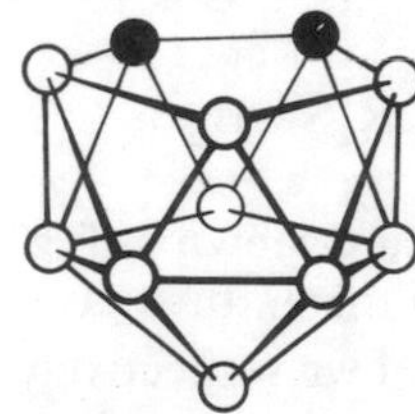

Figure 12.6
The structure of the $[Fe(C_2B_9H_{11})_2]^-$ ion.

C_4Ph_4 or three molecules of carbon monoxide. Complexes derived from other *nido*-carborane anions are also known.

12.5 Metal borides

Most metal borides are very hard substances of high melting-point that contain groups, chains, layers or three-dimensional networks of boron atoms. Their formulae bear no relation to what would be expected on the basis of the normal oxidation states of the metals and a B^{3-} anion. Most of them are made by reduction of a metal oxide by a mixture of boron carbide and carbon, but direct combination of the elements and the action of sodium on a mixture of metal oxide and boric oxide can also be used. Beryllium and magnesium borides are decomposed by dilute acids, but those of other metals are chemically inert.

The hexaborides, typified by CaB_6 and LaB_6, may be described to a first approximation as having metal and B_6^{2-} ions in a caesium chloride type structure (Fig. 12.7): the distances between the boron atoms of adjacent B_6 octahedra are, however, substantially the same as bond lengths within the octahedra (cf. the structure of boron itself), so a description in terms of a polymeric anion is more accurate. In the borides ZrB_{12} and UB_{12} there is a somewhat similar structure, this time approximating to that of sodium chloride, based on metal and B_{12} units.

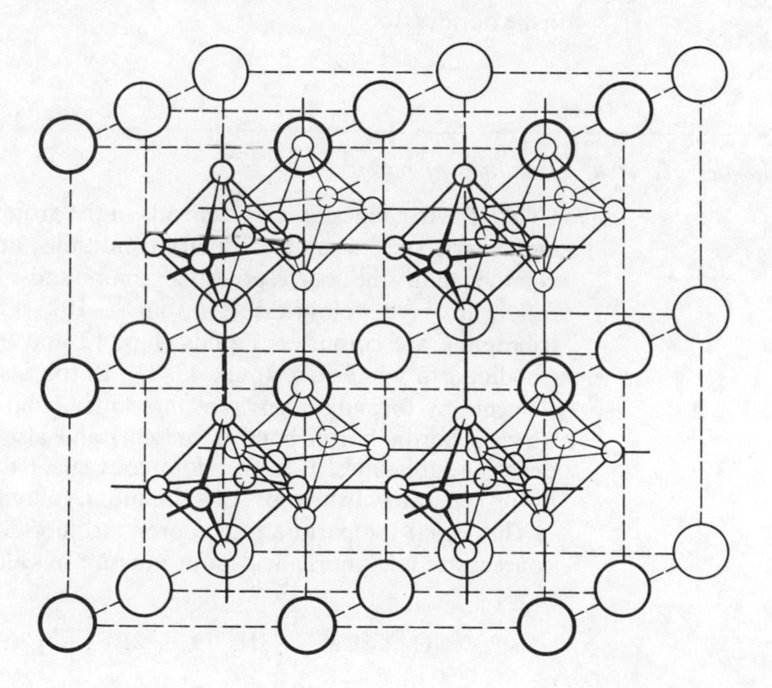

Figure 12.7
The structure of CaB_6.

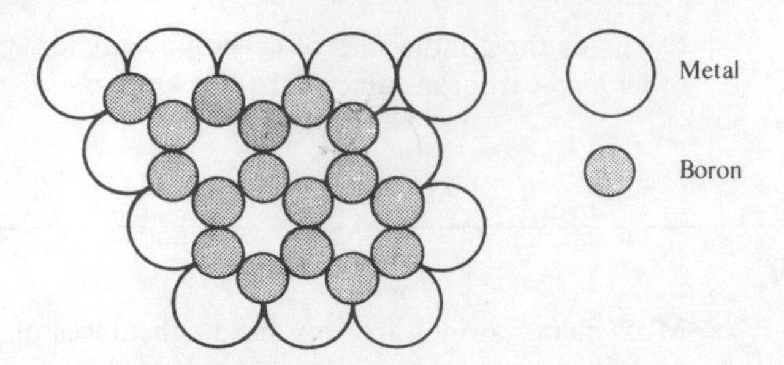

Figure 12.8
Parallel layers of metal atoms and boron atoms in compounds of formula MB_2.

The diborides (e.g. AlB_2, MgB_2, TiB_2 and CrB_2) have a structure in which hexagonal layers of boron atoms alternate with triangular layers of the much larger metal atoms. As may be seen from Fig. 12.8, in which for clarity only two layers are shown, each boron is in contact with six metal atoms (three in each of two layers), and each metal atom is in contact with twelve boron atoms (six in each of two layers); hence the composition MB_2. This is not a layer structure analogous to that of cadmium iodide, since successive planes always consist of different atoms; it is a true macromolecular structure, and thus satisfactorily accounts for the hardness and very high melting-point of borides having it; TiB_2, for instance, has a melting-point more than 1000 °C higher than that of titanium metal. Chains of boron atoms occur in FeB, single boron atoms in Fe_2B, and B_2 units in V_3B_2.

Greatly simplified though this brief account of metal boride structures is, it will be plain that the structural complexity which is a feature of the chemistry of boron and the boron hydrides remains very much in evidence in the borides, too.

12.6 *Halides and complex halides of boron*

Unlike the trihalides of later elements in the group, those of boron are all monomeric species under ordinary conditions, and they are much more volatile than the corresponding compounds of aluminium. Boron trifluoride is a colourless gas (b.p. −101 °C); the trichloride and tribromide are colourless liquids (b.p. 12 and 90 °C respectively); the triiodide is a white solid (m.p. 43 °C). Boron also differs from the later elements by forming a series of molecular halides of formula B_2X_4 (in which a boron–boron bond is present) and also a number of electron-deficient subhalides. It is therefore convenient to discuss the halides of boron separately from those of aluminium, gallium, indium and thallium.

The usual preparation of boron trifluoride is by the action of concentrated sulphuric acid on a mixture of calcium fluoride and boric oxide:

$$B_2O_3 + 3CaF_2 + 3H_2SO_4 \rightarrow 2BF_3 + 3CaSO_4 + 3H_2O$$

Excess of sulphuric acid removes the water formed. Boron trifluoride fumes strongly in moist air, and is partially hydrolysed by excess of water:

$$4BF_3 + 6H_2O \rightarrow 3H_3O^+ + 3BF_4^- + B(OH)_3$$

With small amounts of water at low temperatures, however, the adducts $BF_3.H_2O$ and $BF_3.2H_2O$ are obtained. The tetrahedral $[BF_4]^-$ ion may be made in aqueous solution by the interaction of boric acid and aqueous hydrofluoric acid

$$B(OH)_3 + 4HF \rightarrow H_3O^+ + [BF_4]^- + 2H_2O$$

Unlike boric acid, HBF_4 is a very strong acid; mixtures of hydrogen fluoride and boron trifluoride are extremely strong proton donors, though not quite so strong as those of hydrogen fluoride and antimony pentafluoride. Potassium tetrafluoroborate, KBF_4, is sparingly soluble in water, and so are the corresponding salts of many large cations; silver fluoroborate, made by the action of bromine trifluoride on silver borate (Section 8.5) is soluble in many organic solvents, and is a valuable reagent in organic and organometallic chemistry. The increase in stability of alkali metal fluoroborates with respect to decomposition into fluoride and boron trifluoride as the size of the cation increases has been discussed in terms of lattice energies in Section 10.3.

Boron trifluoride forms a very wide range of complexes with ethers, nitriles and amines; the compound is, in fact, commonly encountered in the form of its 1:1 adduct with dimethyl ether, a liquid boiling at 126 °C which is used as a source of boron trifluoride as a catalyst in many organic reactions among them ester formation, alkene polymerisations and Friedel–Crafts type alkylations and acylations.

The trichloride and tribromide (made by heating together the elements) and the triiodide (made by the action of hydrogen iodide on the trichloride at high temperatures) are all completely and vigorously decomposed by water, and react (with elimination of hydrogen halide) with inorganic and organic compounds containing labile protons. Thus whereas boron trifluoride forms adducts with ammonia, alcohols and amines, boron trichloride reacts with liquid ammonia to give the amide $B(NH_2)_3$, with ethanol to give $B(OEt)_3$ and with aniline to give $B(NHPh)_3$ (the reaction with ammonium chloride to give trichloroborazine is mentioned in Section 12.9). Tetrahalogenoborates containing the $[BCl_4]^-$, $[BBr_4]^-$ and $[BI_4]^-$ ions can be made only in non-aqueous media and isolated only as salts of very large cations. Mixed trihalides can be shown by ^{11}B or ^{19}F nuclear magnetic resonance spectroscopy to be formed from mixtures of BF_3, BCl_3 and BBr_3 or any two of these; BF_3 and BI_3, however, do not react, suggesting that BF_2I and BFI_2 are unstable with respect to disproportionation into the binary halides.

The thermodynamics of complex formation by BF_3, BCl_3 and BBr_3 has been much discussed. In the most direct experimental work on this subject, displacement reactions in the gas phase have shown that, for trimethylamine as Lewis base, the order of adduct stabilities is $BF_3 < BCl_3 < BBr_3$. The same sequence is found for ΔH° of adduct formation by pyridine in nitrobenzene solution, the values being -143, -189 and -217 kJ mol^{-1} respectively for the reaction

$$\text{py(soln)} + BX_3(\text{g}) \rightarrow \text{py}BX_3(\text{soln})$$

This sequence is, of course, opposite to what might have been expected on the basis of the electron-attracting effects of the halogens (Section 5.7), but the paradox is easily resolved by considering the reaction in detail. When a boron trihalide combines with an electron donor the stereochemistry at the boron atom changes from planar to tetrahedral. We may, therefore, consider adduct formation as taking place in two stages, the reorganisation of planar BX_3 into pyramidal BX_3, followed by combination with the donor species. The observed behaviour therefore indicates that the energy needed for the reorganisation of the BX_3 molecule is greatest for X = F and least for X = Br, and that this outweighs an increase in energy liberated from X = F to X = Br in the second stage. It should be noted that since it is impossible to prepare pyramidal BX_3 and examine its properties, we cannot prove that this explanation is correct. Indirect evidence, however, can be cited. The B—X bond energy decreases from X = F to X = Br, and so do both stretching and bending force constants. Further, the increase in the B—X bond length in the trihalides (1.30, 1.76 and 1.87 Å in BF_3, BCl_3 and BBr_3 respectively) is greater than the increase in the conventional single covalent radii of the halogens (0.72, 0.99 and 1.14 Å respectively – see Section 5.6). The bond length data indicate that in BF_3 there is strong p_π–p_π bonding between an empty p orbital of the sp^2 hybridised boron atom and filled p orbitals of the fluorine atoms, and the extent of π interaction decreases from X = F to X = Br. Such π bonding must be destroyed when the configuration changes from planar to pyramidal (it may be noted that the B—F bond length in Me_3NBF_3 is 1.39 Å) and this factor would account for an increase in reorganisation energy from BF_3 to BBr_3. It has, indeed, been suggested further that π bonding of this kind provides a basis for explaining why the boron trihalides are planar monomers whilst those of the later elements are polymers in which the metal is tetrahedrally or octahedrally coordinated: π bonding is always much weaker when other than first-period elements are involved. In contrast to BH_3, boron trihalides form much more stable complexes with trimethylamine than trimethylphosphine, and they do not combine with carbon monoxide.

Diboron tetrachloride (b.p. 65°C) is a colourless unstable liquid obtained by co-condensing boron trichloride and copper vapours on a surface cooled with liquid nitrogen. It may be converted into the corresponding fluoride (b.p. −34 °C) by the action of antimony trifluoride. In the solid state the molecules of B_2F_4 and B_2Cl_4 are planar, whereas in the vapour phase the BCl_2 groups of B_2Cl_4 are mutually perpendicular; the planar configuration in the solid state presumably arises from better packing and more van der Waals' bonding, but it is not clear why this structure is retained in gaseous B_2F_4.

The thermal disproportionation of B_2Cl_4 gives BCl_3 and a number of solid subchlorides, among them pale yellow B_4Cl_4 (the structure of which consists of a tetrahedron of boron atoms with one chlorine atom attached to each boron) and purple B_8Cl_8. The latter contains an irregular B_8 dodecahedron with two different sets of boron–boron bond lengths and one chlorine atom attached to each boron. Electron counts show that if the B–Cl bonds in these substances are ordinary two-electron bonds (as from their lengths they appear to be) the B_4 and B_8 cages must be electron-deficient.

12.7 Halides and complex halides of aluminium, gallium, indium and thallium

All of the trifluorides of these elements are non-volatile solids, best prepared by fluorination of the metal or one of its simple compounds with elemental fluorine; their structures, though geometrically complicated, are those of three-dimensional ionic compounds. *Cryolite*, $Na_3[AlF_6]$, contains an octahedrally coordinated aluminium atom; in the complexes Tl_2AlF_5 and $TlAlF_4$ there are polymeric chains and layers respectively. In the former, AlF_6 octahedra have opposite corners in common; in the latter opposite edges of such octahedra are in common, leading to the observed atomic ratios. Mixed aquo–fluoro complexes occur in aqueous solution.

Thallium trichloride and tribromide are very unstable with respect to the monohalide and free halogen; thallium triiodide is isomorphous with the alkali metal triiodides and hence is really thallium(I) triiodide, $Tl^+I_3^-$. This decrease in stability of the higher oxidation state as we go from the fluoride to the iodide is, of course, a general feature of the chemistry of all metals that show more than one oxidation state; for ionic compounds, it is easily explained by lattice energy considerations, the increase in lattice energy accompanying an increase in oxidation state being greatest for the smallest anion.

The trichlorides, tribromides and triiodides of aluminium, gallium and indium are all obtained by combination of the elements. They are relatively volatile and have layer lattices or lattices containing dimeric molecules; the vapours consist of dimeric molecules, which are also present in solutions of the compounds in organic solvents; only at high temperatures does dissociation of the dimers, with formation of planar monomers, take place. When water is dropped on to solid aluminium chloride, vigorous hydrolysis ensues, but in dilute aqueous solution $[Al(H_2O)_6]^{3+}$ and halide ions are present.

The structures of the halide dimers are almost exactly like that of diborane. In Al_2Cl_6, for example, the Al—Cl distance is 2.06 Å (terminal) and 2.21 Å (bridge). The electronic structure of this dimer (whose structure is nearly tetrahedral at each metal atom) was formerly written

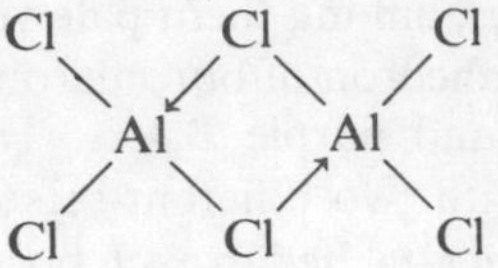

```
Cl        Cl        Cl
   \    ↙    \    /
     Al        Al
   /    \    ↗    \
Cl        Cl        Cl
```

with pairs of unshared electrons from each of two chlorine atoms being used to bring the valence shells of the metal atoms up to that of argon; by analogy with diborane, however, it would now appear more reasonable to write the Al—Cl—Al bridge bonds as three-centre two-electron bonds of lower bond order than the terminal two-centre two-electron ordinary single covalent bonds.

In coordinating solvents such as diethyl ether, aluminium chloride forms 1:1 complexes such as Et_2OAlCl_3. The ion $[AlCl_4]^-$ is important in Friedel–Crafts acylation; the intermediate formed from an acyl chloride and aluminium chloride has the structure $RCO^+[AlCl_4]^-$ at ordinary temperatures. The anion is decomposed by water, but the corresponding species obtained from gallium, indium and thallium trichlorides are only partially hydrolysed under comparable conditions. Indium and thallium exhibit coordination numbers higher than four in complex chlorides: the anion in $(Et_4N)_2[InCl_5]$ is a square pyramid with the indium atom 0.6 Å above the base of the pyramid, in contrast to the usual trigonal bipyramid for five-coordination when the central atom has no unshared electrons (cf. Section 5.2); thallium is six-coordinated in the anions of $K_3[TlCl_6]$ and $Cs_3[Tl_2Cl_9]$, the $[Tl_2Cl_9]^{3-}$ ion consisting of two $TlCl_6$ octahedra sharing a face, i.e. having three chlorine atoms in common.

Aluminium monohalides are formed in the reactions of the trihalides with the metal at 1000 °C followed by rapid cooling; the red monochloride is also obtained from the metal and hydrogen chloride at 900 °C. They all disproportionate at room temperature. Gallium(I) chloride is produced by thermal decomposition of gallium trichloride at 1100 °C, but it has not been isolated pure: gallium dichloride is readily obtained by heating the trichloride with the metal at 180 °C; its diamagnetism and crystal structure show it to be $Ga^+[GaCl_4]^-$. From the system $InCl_3$–In both $InCl_2$ (analogous to the gallium compound) and InCl can be isolated, the latter has a deformed sodium chloride structure.

Thallium(I) halides are stable compounds which in some ways resemble the silver(I) halides. Thus thallium(I) fluoride is very soluble in water whilst the other halides are only sparingly soluble. The fluoride has a deformed sodium chloride structure with pairs of fluorine atoms at 2.59,

2.75 and 3.04 Å; the chloride and bromide have the caesium chloride structure; the iodide is dimorphic, like mercury(II) iodide, existing in a red caesium chloride form above 170 °C and a yellow layer lattice form below that temperature. Owing to the unsymmetrical structure, the lattice energy of TlF has not been calculated, but cycle values for TlCl and TlBr are (like those of the silver salts, though to a less extent) appreciably greater than values calculated from the observed interatomic distances on the basis of an electrostatic model; this extra lattice energy may be the source of the abrupt change in solubility between TlF and the other halides. Since the structure of Tl_2O is unknown it will be apparent that the assignment of the radius of 1.45 Å to the Tl^+ ion is even more of an approximation than the assignment of radii to most other cations. Intermediate halides, e.g. yellow Tl_2Cl_3, red Tl_2Br_3 and black Tl_3I_4, are diamagnetic and contain Tl(I) and Tl(III). Thallium(III) is stabilised by complex halide formation, and when thallium(I) triiodide combines with alkali metal iodides an interesting redox reaction occurs with formation of the tetrahedral $[TlI_4]^-$ ion; this process is discussed further in Section 12.10.

12.8 Oxides, oxo-acids, oxo-anions and hydroxides

As in later groups of elements, basic character increases in Group III with increase in atomic number. Thus the boron oxides are exclusively acidic, those of aluminium and gallium amphoteric, and those of indium and thallium exclusively basic. Thallium(I) oxide is soluble in water and the resulting hydroxide is, in fact, as strong a base as potassium hydroxide.

The principal oxide of boron, B_2O_3, is obtained as a glass by dehydration of boric acid at red heat, but a crystalline form can be made by very slow dehydration. This has a complicated three-dimensional macromolecular network structure built up by sharing of oxygen atoms between planar BO_3 groups with B—O = 1.38 Å. Under pressure at 530 °C a more dense form is produced; this contains BO_4 tetrahedra, three vertices of which are each shared with two other BO_4 units, the fourth vertex being shared with only one other BO_4 unit. Thus each boron atom has a one-third share of three oxygen atoms (at 1.51 Å) and a half share of a fourth (at 1.37 Å), corresponding to the formula $BO_{1.5}$. Trigonal and tetrahedral coordination of boron by oxygen both occur frequently in borate anions.

A polymeric oxide of empirical formula BO is obtained in the incomplete combustion of boranes or by heating B_2O_3 with boron at 1000 °C; since on treatment with water it forms an acid $(HO)_2BB(OH)_2$, together with boric acid and hydrogen, it must contain boron–boron bonds, but its structure is not known.

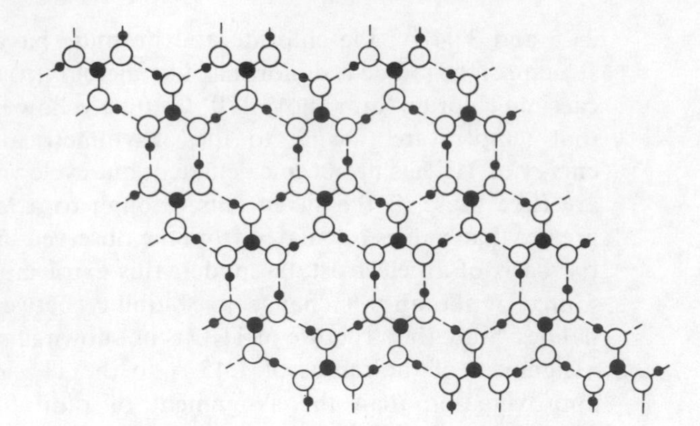

Figure 12.9
The layer structure of H_3BO_3. Broken lines indicate hydrogen bonds.

Fused boron trioxide dissolves many metal oxides with the formation of glasses (metal borates); it reacts with phosphorus pentoxide to give a compound BPO_4 (boron phosphate) which is insoluble in water and dilute acids and can be sublimed unchanged at 1500 °C. This has a silica-like structure in which alternate silicon atoms have been replaced by boron and by phosphorus, and is thus better considered a mixed oxide than a salt of cationic boron; nevertheless, it should be noted that aluminium phosphate has a similar structure. Boron phosphate is a valuable catalyst for the hydration of alkenes and the dehydration of amides to nitriles.

Orthoboric acid, $B(OH)_3$, commonly referred to simply as boric acid, is obtained by addition of mineral acid to a hot solution of borax; it crystallises on cooling. When heated it forms metaboric acid, $B_3O_3(OH)_3$, and then boric oxide. Both boric acids have layer structures in which molecules are linked by hydrogen bonding. The layer structure of $B(OH)_3$, shown in Fig. 12.9, is the cause of the slippery feel of boric acid, which is quite a good lubricant.

Boric acid is a very weak acid (pK_a 9.2); it acts not as a proton donor but as an electron acceptor, the ionisation equilibrium being

$$B(OH)_3 + H_2O \rightleftharpoons [B(OH)_4]^- + H^+$$

Complex formation with polyhydric alcohols containing *cis*-diol groups, however, leads to a substantial increase in acidic strength; the equilibria involved are shown in Fig. 12.10. Many borates occur naturally; others may be made from the metal oxide and boron trioxide. A very wide range of borate anions are found in such salts. Species containing only planar BO_3 groups include BO_3^{3-} (in $Ca_3(BO_3)_2$), $B_2O_5^{4-}$ or $O_2BOBO_2^{4-}$ (in $Co_2B_2O_5$), $B_3O_6^{3-}$ (in $Na_3B_3O_6$), and $(BO_2)_n^{n-}$ (in CaB_2O_4); the

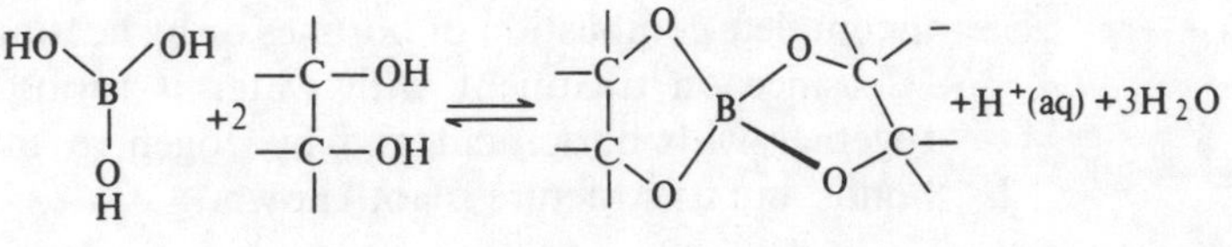

$$\mathrm{B(OH)_3} + 2\,\begin{matrix}-\mathrm{C}-\mathrm{OH}\\ -\mathrm{C}-\mathrm{OH}\end{matrix} \rightleftharpoons \left[\mathrm{B}\left(\mathrm{O_2C_2}\right)_2\right]^- + \mathrm{H^+(aq)} + 3\mathrm{H_2O}$$

Figure 12.10
The equilibrium between boric acid and a 1,2-diol.

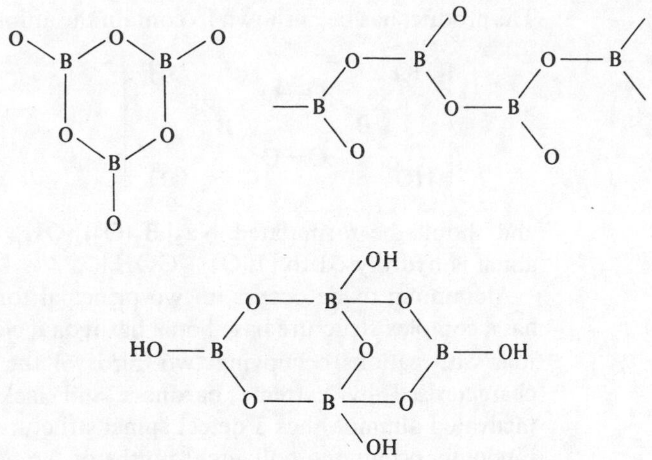

Figure 12.11
The structures of the $B_3O_6^{3-}$, $(BO_2)_n^-$, and $[B_4O_5(OH)_4]^{2-}$ anions.

structures of the last two of these anions are shown in Fig. 12.11. The $[B(OH)_4]^-$ ion is found in $LiB(OH)_4$, and polymeric species containing only four-coordinated boron atoms are also known (e.g. the high pressure form of CaB_2O_4 contains an anion having the silica structure). Among many anions containing both three- and four-coordinated boron is $[B_4O_5(OH)_4]^{2-}$ (in borax, which is commonly written $Na_2B_4O_7.10H_2O$ but should really be formulated $Na_2[B_4O_5(OH)_4].8H_2O$). The structure of this anion is also shown in Fig. 12.11.

For planar BO_3 groups the B—O bond length is usually close to 1.36 Å, but for tetrahedral BO_4 groups the length increases to about 1.48 Å. This increase is very similar to that on going from BF_3 to $[BF_4]^-$, and suggests that in the planar grouping π bonding involving lone pairs of electrons from the oxygen atoms occurs; this π bonding is necessarily lost in the tetrahedral group, in which a lone pair from the extra oxygen atom occupies the previously empty orbital on the boron atom.

All of the foregoing information about borate anions is derived from crystallographic studies. Much less is known about the nature of the anions in solution, though evidence from ^{11}B nuclear magnetic resonance spectroscopy, which serves to distinguish between planar and tetrahedrally coordinated boron, suggests that in aqueous media species containing only three-coordinated boron are unstable and are rapidly converted into species containing some of the boron tetrahedrally coordinated.

Boric acid is a mild antiseptic and is used as a food preservative. Borax and other borates are used in water treatment, timber preservation, glass manufacture, and in the preparation of sodium peroxoborate, an important constituent of washing powders. Sodium peroxoborate is formed when boric acid is treated with sodium peroxide, but on the industrial scale it is made by electrolysis of a solution of sodium borate containing sodium carbonate (the function of which is not understood).

The product has been shown to contain the anion

```
┌HO                OH ┐2−
    \     O—O     /
      B          B
    /     O—O     \
└HO                OH ┘
```

and should be formulated $Na_2[B_2(O_2)_2(OH)_4].6H_2O$. In solution the anion is hydrolysed to $[B(OH)_3OOH]^-$.

Aluminium oxide occurs in two principal forms: α-Al_2O_3 (*corundum*) has a complex structure based on a hexagonal close-packed array of oxide ions with cations occupying two-thirds of the octahedral holes, and is characterised by extreme hardness and lack of reactivity; γ-Al_2O_3 (activated alumina) has a defect spinel structure (see below) with cations randomly occupying only eight-ninths of the octahedral and tetrahedral holes that are occupied in spinel itself. The α- and γ- forms are made by dehydration of hydrous oxides at 1000 °C and 450 °C respectively; the γ-form is widely used as an absorbent and as a catalyst. When ammonia is added to a hot solution of an aluminium salt, AlO(OH) is precipitated; $Al(OH)_3$ is obtained by passing carbon dioxide into a solution of sodium aluminate. Not much is known about solid aluminates: $NaAlO_2$, made from aluminium oxide and sodium oxalate at 1000 °C, is a mixed oxide, and so is spinel, $MgAl_2O_4$, but there is a calcium salt $Ca_9[Al_6O_{18}]$ containing a cyclic anion which is isostructural with the $Si_6O_{18}^{12-}$ ion. This calcium salt (generally called tricalcium aluminate, $Ca_3Al_2O_6$, until the structure was determined) is an important component of Portland cement; the very open structure facilitates the formation of hydrates that constitutes the setting of cement.

What was referred to as β-alumina until the presence of sodium in it was established is obtained by heating sodium carbonate or sodium hydroxide in a closed vessel with either form of alumina or its hydrate at about 1500 °C. Its composition approximates to $NaAl_{11}O_{17}$ or $Na_2O.11Al_2O_3$. The structure is very complicated, but its essential character may be described as chunks of close-packed structure separated by relatively empty planes containing only Na^+ and an equal number of O^{2-} ions; within these planes the Na^+ ions can move easily and the substance is a two-dimensional ionic conductor (used, for example, in the sodium–sulphur battery – see Section 10.2).

We shall often have occasion later in this book to refer to the structure of spinel. This consists of a cubic close-packed array of oxide ions with one-eighth of the tetrahedral holes occupied by magnesium ions and one-half of the octahedral holes by aluminium ions. Many other mixed oxides, e.g. MAl_2O_4 (M = Mn, Fe, Co or Zn), have the spinel structure; others, e.g. $MgFe_2O_4$, have the so-called inverse spinel structure in which the dipositive and half the tripositive cations have exchanged positions; the significance of this structure change is discussed in Section 20.2.

Traces of transition metal ions incorporated into α-Al_2O_3 give substances classified as gem stones: ruby, for example, contains Cr^{3+}; blue sapphire contains Fe^{2+}, Fe^{3+} and Ti^{4+}. Synthetic gems are now produced in large quantities.

Gallium shows its customary resemblance to aluminium by forming two forms of Ga_2O_3, GaO(OH) and $Ga(OH)_3$; indium forms only an oxide and hydroxide, and thallium(III) forms only an oxide. None of these compounds calls for further comment here.

12.9 *Nitrogen derivatives*

The group BN is isoelectronic with C_2, and many boron–nitrogen compounds possess interesting similarities to carbon compounds; no species containing a B≡N group has yet been discovered, however.

Boron nitride, $(BN)_n$, can be made by the action of ammonia on boron at 1000 °C, by passing nitrogen over a mixture of boron trioxide and carbon at a slightly higher temperature, or by heating borax with ammonium chloride. It is chemically rather inert, but is hydrolysed to boric acid and ammonia by the action of hot acid.

The common form of boron nitride, obtained by heating the products of any of the above reactions at 1800 °C, has an ordered layer structure closely resembling that of graphite, but with layers superimposed so that boron and nitrogen atoms lie vertically above one another (Fig. 12.12). The B—N distance within layers is 1.45 Å, appreciably shorter than that

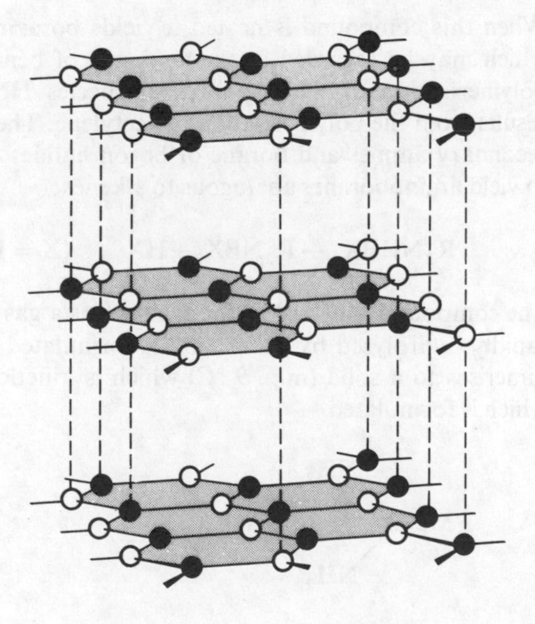

Figure 12.12
The structure of boron nitride, BN.

in the tetrahedrally bonded compound H_3NBH_3 (1.56 Å) and indicative of π bonding involving the lone pair of electrons on the nitrogen and the empty boron *p* orbital. Like graphite, the compound is a good lubricant, there being only van der Waals' interactions between the layers, which are 3.30 Å apart. Unlike graphite, however, it is white and is an insulator; this last fact can be interpreted in terms of the band theory of solids on the basis of an energy gap between filled bonding molecular orbitals and empty antibonding molecular orbitals that is considerably greater than that in graphite; it can be shown to arise from the polarity of the B—N bond.

When boron nitride is heated at 1500–2000 °C and 50 000 atm pressure in the presence of traces of lithium or magnesium nitride as catalyst, it is transformed into a more dense form having the cubic zinc blende structure (i.e. that of diamond with alternate atoms replaced by boron and by nitrogen). In this form of boron nitride, which is as hard as diamond, the B—N bond length is 1.57 Å – further proof of the existence of π bonding in the hexagonal form. Another form of boron nitride, having the wurtzite structure, has been obtained under similar conditions.

Of the other Group III elements, only aluminium reacts directly with nitrogen (at 750 °C); AlN has the wurtzite structure and is hydrolysed to ammonia by hot dilute alkali. The homologous compounds BP, AlP, GaN, GaP and GaAs all have zinc blende or wurtzite structures.

Mention has already been made of the ammonia derivative H_3NBH_3, which may be obtained by the action of ammonium chloride on sodium tetrahydridoborate:

$$NH_4Cl + NaBH_4 \rightarrow H_3NBH_3 + H_2 + NaCl$$

When this compound is heated it yields borazine, $B_3N_3H_6$ (see below), which may be regarded as the analogue of benzene; it results from the polymerisation of the hypothetical species HN$\rightleftarrows$BH just as benzene results from the polymerisation of acetylene. The addition compounds of secondary amines and borane or boron halides decompose when heated to yield aminoboranes analogous to alkenes:

$$R_2NHBX_3 \rightarrow R_2NBX_2 + HX \qquad (X = H, Cl \text{ or } Br)$$

The compound Me_2NBH_2, for example, is a gas (b.p. 1 °C) which is very rapidly hydrolysed by water, and is formulated $Me_2N\rightleftarrows BH_2$; it readily dimerises to a solid (m.p. 9 °C) which is kinetically stable to water and which is formulated

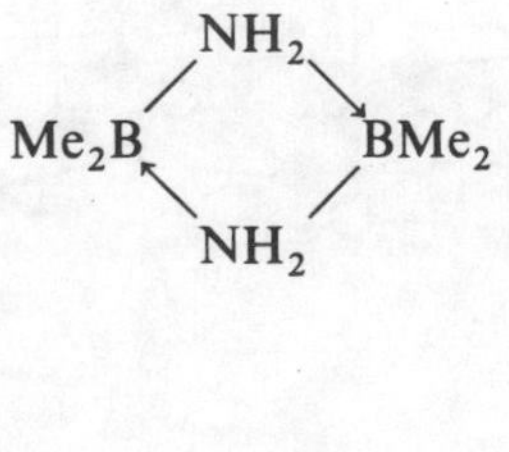

$$\begin{array}{ccc} & NH_2 & \\ \nearrow & & \searrow \\ Me_2B & & BMe_2 \\ \nwarrow & & \swarrow \\ & NH_2 & \end{array}$$

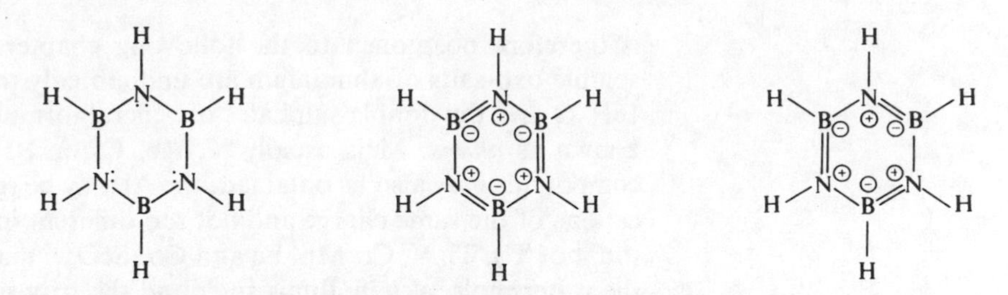

Figure 12.13
Valence bond structures for borazine.

Many other polymers of this and analogous compounds are also known. Phosphinoboranes, it may be noted, are known only as polymers, e.g. $(Me_2PBH_2)_3$, which has a structure like that of cyclohexane.

Borazine, though obtained as described above, is now made by reduction of B-trichloroborazine with sodium tetrahydridoborate, the trichloro derivative being prepared by the reaction

$$3BCl_3 + 3NH_4Cl \xrightarrow[C_6H_5Cl]{150\ °C} B_3N_3Cl_3H_3$$

Use of an alkylammonium halide in this reaction yields a B-trichloro-N-trialkyl borazine, which may be similarly converted into a N-trialkyl borazine. Borazine is a colourless liquid (b.p. 55 °C) with an aromatic odour; in its physical properties it closely resembles benzene, and the molecular structure is also that of a planar hexagon. The B—N distance is 1.44 Å throughout the ring, only 0.01 Å less than that in boron nitride, whereas the C—C distance in benzene is 0.03 Å less than that in graphite; this suggests substantial, but not complete, delocalisation of the lone pairs of electrons on the nitrogen atoms. In valence bond terminology the structure may be written as a resonance hybrid of the structures shown in Fig. 12.13. Hexamethylborazine (though not borazine itself) forms complexes with transition metals analogous to those formed by benzene and alkylbenzenes.

Borazine is much more reactive than benzene (though it should be remembered that benzene is only kinetically inert towards addition of, e.g., H_2O or HCl). With hydrogen chloride the compound $B_3N_3Cl_3H_9$ is formed, chlorine becoming attached to boron as expected; this product can be reduced by $NaBH_4$ to the cyclohexane analogue $B_3N_3H_{12}$, which has the chair conformation. Water and methanol similarly add to borazine, though the products are themselves unstable and disruption of the ring takes place readily.

12.10 *Aluminium, gallium, indium and thallium salts of oxo-acids and aqueous solution chemistry*

Aluminium-containing silicates are of great importance in mineralogy, but they are best discussed along with other silicates, and mention of them is therefore postponed to the following chapter. The most important soluble oxo-salts of aluminium are undoubtedly the sulphate $Al_2(SO_4)_3$.16H_2O and the double sulphates of general formula $MAl(SO)_4)_2$.12H_2O known as *alums*: M is usually K, Rb, Cs or NH_4, but Li, Na and Tl compounds may also be obtained; the Al may be replaced by a number of cations of the same charge and not too different in size, including Ga, In (but not Tl), Ti, V, Cr, Mn, Fe and Co; SeO_4^{2-} may replace SO_4^{2-}. Six of the water molecules in alums surround the tripositive cation; the others (lattice water) form hydrogen bonds between this hydrated cation and the anions. Aluminium sulphate is used in water purification for removal of phosphate (Section 14.9) and of colloidal matter, for the coagulation of which the high charge on the cation is very effective. Aluminium in food, however, is suspected of causing Alzheimer's disease.

All the tripositive aquo ions of these elements are acidic, that of aluminium the least and that of thallium the most so. Thus aqueous solutions of their salts are appreciably hydrolysed, and salts of weak acids (e.g. carbonates and cyanides) cannot exist in contact with water. Nuclear magnetic resonance studies have shown that in acidic solution aluminium is present as the $[Al(H_2O)_6]^{3+}$ ion; as the acidity is decreased, polymeric hydrolysed species such as hydrated $[Al_2(OH)_2]^{4+}$ and $[Al_7(OH)_{16}]^{5+}$ appear, then $Al(OH)_3$ is precipitated, and finally, in alkaline solutions, aluminate anions such as $[Al(OH)_6]^{3-}$, $[Al(OH)_4]^-$ and polymeric species like $[(HO)_3AlOAl(OH)_3]^{2-}$ are formed. The chemistry of gallium is broadly similar to that of aluminium in this respect; indium and thallium(III) hydroxides, however, are not amphoteric.

Redox potential data (Section 12.1) show that Al^{3+}(aq) is much less readily reduced than the other tripositive cations in aqueous solution. This doubtless arises partly from a more negative hydration free energy of the smaller Al^{3+} ion, but another important contributory factor is the increases in ionisation energies between aluminium and gallium, and between indium and thallium; there is relatively little variation in atomisation enthalpies, and the overall variation in E° is therefore quite different from that in the two preceding groups of elements.

In Section 7.5 we discussed the influence of hydrogen ion concentration on the formation of iron(III) acetylacetone in aqueous solution and on the extraction of iron into organic solvents by acetylacetone. Since formation of the aluminium compound, $[Al(acac)_3]$, is an exactly analogous process, we shall not repeat the treatment here. We may, however, note that 8-hydroxyquinoline (oxine) forms similar compounds, and aluminium trisoxinate (a weighing form for the metal in gravimetric analysis) is similarly soluble in organic solvents.

The aqueous solution chemistry of thallium calls for special comment, since this is the first element we have encountered in our systematic survey that has two relatively stable oxidation states in solution and its chemistry provides a good opportunity to illustrate further the principles discussed in Section 7.8. (Indium(I) can be obtained in low concentration by oxidation of an indium anode in dilute perchloric acid, and E° for the system In^{3+}/In^{+} has been measured as -0.42 V, but the solution rapidly evolves hydrogen with formation of indium(III).)

For the half reaction

$$Tl^{3+} + 2e \rightleftharpoons Tl^{+}$$

E° has been found to be $+1.26$ V in molar $HClO_4$; under these conditions Tl(III) is therefore a powerful oxidant. The value of E° is, however, very dependent upon the anion present, because Tl(I) resembles an alkali metal ion by forming few stable complexes in aqueous solution (TlCl, unlike AgCl, is not soluble in aqueous ammonia or potassium cyanide, for example), whereas Tl(III) is very strongly complexed by a variety of anions. Thus at unit chloride ion concentration, although TlCl is fairly insoluble, E° for the system $[TlCl_4]^-/TlCl$ is only $+0.9$ V. Iodide forms a more stable complex than chloride (class (*b*) behaviour – Section 7.7) and at high iodide ion concentrations $[TlI_4]^-$ is a stable species even though E° Tl^{3+}/Tl^{+} is much higher than E° $\frac{1}{2}I_2/I^-$ ($+0.54$ V) and TlI is sparingly soluble. Thus I_3^- $(I^- + I_2)$ in solid TlI_3 can under these conditions oxidise Tl(I) and bring about the reaction

$$Tl^{I}I_3 + I^- \rightarrow [Tl^{III}I_4]^-$$

In alkaline media Tl(I) is also easily oxidised, since Tl(OH) is soluble in water and hydrated Tl_2O_3 (which is in equilibrium with Tl^{3+} and OH^- ions in solution, of course) is very sparingly soluble, with K_{sp} about 10^{-45}.

Soluble thallium(I) compounds are extremely poisonous; they include the fluoride, hydroxide, carbonate and many other salts of oxo-acids. On the other hand the sulphide, chloride, bromide, iodide and chloroplatinate are very sparingly soluble in water. These facts illustrate not only the similarities of thallium to various other metals, but also the difficulties in the way of treating solubilities in water on a quantitative basis (cf. Section 7.3).

12.11 *Organometallic compounds*

The preparations of boron alkyls by the action of diborane, generated *in situ*, on alkenes, and of carboranes from boron hydrides and alkynes, have already been mentioned. Both boron alkyls and boron aryls may be made by the action of the boron trifluoride–dimethyl ether complex on active organometallic compounds (e.g. Grignard reagents or lithium alkyls) in ethereal solution, e.g.

$$2BF_3 + 6RMgX \rightarrow 2BR_3 + 3MgF_2 + 3MgX_2$$

Trialkylboranes are all monomeric and are inert towards water, but are spontaneously inflammable in air; the triaryl compounds are less reactive.

Both sets of compounds contain planar three-coordinated boron and act as strong Lewis bases towards amines or carbanions. Boron triphenyl, for example, reacts with sodium phenyl to form sodium tetraphenylborate, $Na[BPh_4]$; this is soluble in water, but the salts of the larger alkali metal ions are not, and may be used for the gravimetric determination of these elements.

As we mentioned in Section 12.3, boron trimethyl is a weaker Lewis acid than the boron trihalides or monoborane. Although there is no π-bonding to be overcome when the planar molecule becomes tetrahedral in a donor–acceptor complex, the electron-repelling effect of the methyl groups hinders complex formation. No compound is formed with carbon monoxide and compounds with tertiary phosphines are much less stable than those with tertiary amines. As we might expect, steric factors may influence either donor or acceptor properties: thus 2,6-dimethylpyridine does not react with boron trimethyl (though pyridine does so), and boron trimesityl (tris(1,3,5-trimethylphenyl)boron) shows no acceptor properties at all. We may summarise some of the information given in this chapter on donor–acceptor complexes of boron compounds by the following stability sequences:

Me_3N as donor: $BBr_3 > BCl_3 > BF_3 \sim BH_3 > BMe_3$

Me_3P as donor: $BBr_3 > BCl_3 \sim BH_3 > BF_3 \sim BMe_3$

CO as donor: $BH_3 \gg BF_3, BMe_3$

In addition to steric and inductive effects, we have had occasion to invoke halogen to boron π-bonding and hyperconjugation in our explanations of the observed data. Donor–acceptor interactions, even when free from complications arising from the presence of a solvent (cf. Sections 7.7 and 8.1), are therefore far from being determined simply by the ionisation energy of the donor and the electron affinity (measured or presumed) of the acceptor. Attempts have been made to derive general relationships for enthalpies of formation of such complexes, but the equations produced inevitably contain several parameters, and even so are of only limited reliability. Their discussion would accordingly not be appropriate here, and for an account of them recourse should be had to the references at the end of the chapter.

Aluminium alkyls can be prepared from metallic aluminium and mercury alkyls or from the chloride and Grignard reagents; on an industrial scale they are now made by direct synthesis from alkenes, aluminium and hydrogen, e.g.

$$Al + \tfrac{3}{2}H_2 + 3C_2H_4 \rightarrow Al(C_2H_5)_3$$

Interaction of aluminium and alkyl halides gives a mixture of alkyl aluminium halides, e.g.

$$2Al + 3RX \rightarrow R_3Al_2X_3$$

Alkyl aluminium hydrides are obtained from the metal, aluminium alkyls and hydrogen:

$$Al + \tfrac{3}{2}H_2 + 2AlR_3 \rightarrow 3AlR_2H$$

These substances, which are unstable to both air and water, are (together with transition metal compounds) important catalysts for the polymerisation of alkenes and other unsaturated organic compounds.

Aluminium trimethyl (m.p. 15 °C) is a dimer having a structure like that of aluminium chloride with methyl groups replacing the chlorine atoms; Al—C bond lengths are 1.95 Å (terminal) and 2.12 Å (bridge), so that three-centre two-electron bonding must be involved in the Al—C—Al bridges. Trialkyls containing large alkyl groups (e.g. the isopropyl compound) do not dimerise. All are strong Lewis bases and form adducts with amines and alkali metal halides, e.g. R_3NAlR_3 and $K[AlR_3F]$.

Gallium, indium and thallium trialkyls may be made by the action of various metal alkyls on the chlorides; many are polymeric in the solid state or in solution. Thallium trichloride and alkyl or aryl Grignard reagents yield only dialkyl or diaryl thallium halides, however; these substances are saline, and give conducting solutions in water; the $[(CH_3)_2Tl]^+$ cation has been shown to have a linear C—Tl—C skeleton.

PROBLEMS

1 Use Wade's rules to derive the structures of the boron atom clusters in B_4H_{10}, B_6H_{10}, $B_{10}H_{14}$, $B_6H_6^{2-}$ and $B_{10}H_{14}^{2-}$.

2 Suggest syntheses (starting from deuterium oxide as the only source of deuterium) for (a) B_2D_6, (b) $B(OD)_3$, (c) $B_3N_3D_6$, (d) D_3BCO, (e) $B(CH_2CH_2D)_3$.

3 Suggest explanations for the following facts.
 (a) $NaBH_4$ is very much less rapidly hydrolysed by water than is $NaAlH_4$.
 (b) The rate of hydrolysis of diborane by water vapour is proportional to $p^{\frac{1}{2}}_{B_2H_6} \cdot p_{H_2O}$.
 (c) 2,6-Dimethylpyridine, although a stronger base towards protons than pyridine, differs from the latter by not forming an adduct with boron trimethyl; it does, however, combine with aluminium triethyl.
 (d) A saturated aqueous solution of boric acid is neutral to the indicator bromocresol green (pH range 3.8–5.4), and a solution of potassium hydrogen difluoride is acidic to this incidator; when, however, excess of boric acid is added to a solution of potassium hydrogen difluoride, the solution becomes alkaline to bromocresol green.

4 Comment on each of the following observations:
 (a) The enthalpies of dimerisation of the aluminium alkyls AlR_3 are -41, -35 and -16 kJ mol^{-1} for R = Me, Et and iso-Bu respectively.
 (b) Aluminium trifluoride is almost insoluble in anhydrous hydrogen fluoride, but dissolves if potassium fluoride is also present. Passage of boron trifluoride through the solution results in the reprecipitation of aluminium trifluoride.

(c) The Raman spectra of germanium tetrachloride, a solution of gallium trichloride in concentrated hydrochloric acid and fused gallium dichloride contain the following lines:

$GeCl_4$	134 cm^{-1}	172 cm^{-1}	396 cm^{-1}	453 cm^{-1}
$GaCl_3$/HCl	114	149	346	386
$GaCl_2$	115	153	346	380

(d) When TlI_3, which is isomorphous with the alkali metal triiodides, is treated with aqueous sodium hydroxide, hydrated Tl_2O_3 is quantitatively precipitated.

(e) At 25 °C the ^1H nmr spectrum of $Al(BH_4)_3$ exhibits only a single resonance.

5 Pyrolysis of diborane under controlled conditions gives two new boranes **A** and **B**. The ^{11}B nmr spectrum of **A** exhibits a triplet of triplets and a poorly resolved doublet of relative intensities 1:1. Treatment of **A** with hot water produces eleven moles of hydrogen *per* mole of **A** and the mass spectrum exhibits a parent ion at $m/e = 54$. Borane **B**, which contains 85.7 per cent boron, reacts with alcohols ROH to produce twelve moles of hydrogen *per* mole. Friedel–Crafts alkylation of **B** yields a derivative **C** which liberates eleven moles of hydrogen *per* mole with base. On heating **C** at *ca* 200° for several hours, a new product **D** which has the same molecular formula as **C** is produced. The ^{11}B nmr spectrum of **C** may be taken to indicate only two boron environments whereas that of **D** indicates that there are four different environments. Identify compounds **A**–**D**.

REFERENCES FOR FURTHER READING

Cotton, F.A. and Wilkinson, G. (1988) *Advanced Inorganic Chemistry*, 5th edition, Interscience, New York. A full account of the chemistry of all the Group III elements.

Greenwood, N.N. and Earnshaw, A. (1984) *The Chemical Elements*, Pergamon Press, Oxford. Chapters 6 and 7 give the best available account of the chemistry of B, Al, Ga, In and Tl and technical applications of their compounds.

Housecroft, C.E. (1990) *Boranes and Metalloboranes*, Ellis Horwood, Chichester. A very clear and well-illustrated introduction to the structures, bonding and reactions of boranes and related compounds.

Powell, P. and Timms, P.L. (1974) *The Chemistry of the Non-Metals*, Chapman and Hall, London. A treatment of boron chemistry in which more emphasis is placed on comparison with other non-metallic elements than in this book.

Purcell, K.F. and Kotz, J.C. (1977) *Inorganic Chemistry*, W.B. Saunders, Philadelphia. Contains a full and critical account of attempts to rationalise donor–acceptor interactions.

Wade, K. (1971) *Electron Deficient Compounds*, Nelson, London. A classic account of the boron hydrides and related electron-deficient compounds.

Wells, A.F. (1984) *Structural Inorganic Chemistry*, 5th edition, Oxford University Press. Full account of the structural chemistry of all the Group III elements.

13 Carbon, silicon, germanium, tin and lead

Introduction • The elements • Intercalation compounds of graphite • Hydrides • Carbides and silicides • Halides and complex halides • Oxides and oxo-acids of carbon • Oxides, oxo-acids and hydroxides of silicon, germanium, tin and lead • Silicates • Silicones • Sulphides • Cyanogen, its derivatives and silicon nitride • Aqueous solution chemistry and oxo-acid salts of tin and lead • Organometallic compounds

13.1 Introduction

Many of the general trends in the comparative chemistry of the elements of this group resemble those seen in the chemistry of boron and its homologues: there is a gradation from the non-metals carbon and silicon through germanium and tin to lead, which, though its oxides are amphoteric, is mainly metallic in behaviour. All members of the group exhibit an oxidation state of four, but they also form species containing the elements in the dipositive state. These increase in stability with increase in atomic number: dipositive carbon-containing species (carbenes) exist only as reaction intermediates and silicon dihalides are stable only at high temperatures, but dipositive germanium and tin are well established, and for lead the lower oxidation state is the stable one. In this respect lead resembles thallium in the preceding group, and bismuth in the following one; the inertness of the 6*s* electron pair is a general feature of the chemistry of all three elements.

The variation in ionisation energies (Table 13.1) as we go from carbon to lead is very like that from boron to thallium: note in particular the relatively large increases between the second and third ionisation energies, and the increases in the third and fourth ionisation energies at germanium and lead. Whether cations carrying an actual charge of +4 occur in any compounds of these elements is, owing to the magnitudes of the sums of

Table 13.1
Some properties of carbon, silicon, germanium, tin and lead.

Element	C	Si	Ge	Sn	Pb
Atomic number	6	14	32	50	82
Electronic configuration	$[He]2s^22p^2$	$[Ne]3s^23p^2$	$[Ar]3d^{10}4s^24p^2$	$[Kr]4d^{10}5s^25p^2$	$[Xe]4f^{14}5d^{10}6s^26p^2$
Atomic radius[a]/Å	0.77	1.17	1.22	1.40	1.75
ΔH° atomisation/kJ mol^{-1}	715	452	372	301	197
M.p./°C	>3550[b]	1410	937	232	327
B.p./°C	4827	2355	2830	2260	1744
First ionisation energy/kJ mol^{-1}	1086	786	760	707	715
Second ionisation energy/kJ mol^{-1}	2360	1575	1540	1415	1450
Third ionisation energy/kJ mol^{-1}	4620	3220	3310	2950	3090
Fourth ionisation energy/kJ mol^{-1}	6220	4350	4420	3930	4080
Ionic radius M^{2+}/Å	—	—	—	0·93	1.18
E° $M^{2+}(aq) + 2e \rightleftharpoons M$/V	—	—	—	−0.14	−0.13
E° $M^{4+}(aq) + 2e \rightleftharpoons M^{2+}$/V	—	—	—	+0.15	+1.7
Pauling electronegativity	2.5	1.9	2.0	2.0	1.9

[a] The data are for C, Si, Ge and Sn in diamond-type structures and are therefore also covalent radii for four-coordination for the elements; the value for Pb relates to the cubic close-packed metal; the covalent radius for four-coordinated Pb is 1.45 Å. There are also metallic forms of Ge and Sn in which the atomic radii are 1.24 and 1.51 Å respectively.
[b] Diamond.

Table 13.2
Some covalent bond energy terms for compounds of carbon, silicon, germanium and tin in which these elements are four-coordinated ($kJ\ mol^{-1}$).

C—C	C—H	C—F	C—Cl	C—O
346	416	485	327	359
Si—Si	Si—H	Si—F	Si—Cl	Si—O
226	326	582	391	466
Ge—Ge	Ge—H	Ge—F	Ge—Cl	Ge—O
186	289	465	342	350
Sn—Sn	Sn—H		Sn—Cl	
151	251		320	
			Pb—Cl	
			244	

the first four ionisation energies, open to considerable doubt: neither SnF_4 nor PbF_4, though both are non-volatile solids, has a symmetrical three-dimensional crystal structure, and whilst SnO_2 and PbO_2 have the rutile structure the fact that the latter is brown is a point against the simple formulation $Pb^{4+}(O^{2-})_2$; the agreement between Born cycle and calculated lattice energies (cf. Section 6.7) is good for SnO_2 but poor for PbO_2. Cations containing singly-charged carbon (e.g. Ph_3C^+) are stable only when delocalisation of the positive charge over aromatic rings occurs, but the salts [Si(acac)$_3$]Cl and [Ge(acac)$_3$]Cl (acac = $CH_3COCHCOCH_3^-$) are known. Aqueous solution chemistry of cations of elements in this group is restricted to those of tin and lead.

Some important experimental values for covalent bond energy terms are given in Table 13.2; these are, of course, also the source of the Pauling electronegativity values given in Table 13.1. In addition to the values in Table 13.2 we may note the enthalpies of atomisation of the elements in Table 13.1, the bond energy terms for C=O and Si=O in gaseous CO_2 and SiO_2 (806 and 642 $kJ\ mol^{-1}$ respectively), and the bond energy terms for C=C and C≡C in hydrocarbons (598 and 813 $kJ\ mol^{-1}$ respectively). When we try to interpret the chemistry of the elements on the basis of these bond energy terms, caution is necessary for two reasons. First, many reactions which are thermodynamically possible (i.e. have negative values for ΔG° or, if we neglect the effect of $T\Delta S^\circ$, of ΔH°) are subject to kinetic control. Hydrocarbons, for example, are much more stable than silicon hydrides, both with respect to decomposition into elements in their standard states (ΔG_f° is -75 and $+62$ $kJ\ mol^{-1}$ for CH_4 and SiH_4 respectively) and with respect to combustion to the dioxide and water; nevertheless, the striking difference between silane and methane, that the former is spontaneously inflammable in air, is kinetic in origin, as a moment's reflection will show – methane and air explode when the necessary activation energy is supplied by a spark. Second, in order to use bond energy terms successfully it is necessary to consider complete reactions. Thus although the C—H bond is stronger than the C—Cl bond, chlorine does react with methane; in the process

$$\gt C\text{—H} + \text{Cl—Cl} \rightarrow \gt C\text{—Cl} + \text{H—Cl}$$

the difference in bond energy between Cl—Cl (243 kJ mol^{-1}) and H—Cl (431 kJ mol^{-1}) is also involved and overrides the unfavourable difference between C—H and C—Cl.

Carbon atoms show a much greater tendency to form compounds containing chains or rings than those of any other element, and in this the strength of the C—C bond certainly plays an important part. Again, however, we must stress that kinetic as well as thermodynamic considerations may be involved. Unfortunately any detailed discussion of kinetic factors is subject to two complications: even when carbon–carbon bond breaking is the rate-determining step in a reaction, it is the bond dissociation energy (Section 5.4) rather than the bond energy term that is involved; and reactions are often bimolecular processes in which bond making takes place simultaneously with bond breaking. Where this is the case the rate of the reaction may bear no relationship to the difference between the bond energy terms for reactants and products.

Carbon, unlike later elements in the group, is limited to eight electrons in its valence shell. Accordingly carbon tetrafluoride forms no complexes analogous to the fluorosilicates (salts containing $[SiF_6]^{2-}$ ions) and, though thermodynamically unstable with respect to hydrolysis by water, is inert towards it, presumably because no five-coordinated intermediate can be formed. Silicon and later elements have *d* orbitals available in the valence shell, and it is noteworthy that not only do silicon halides readily form complexes and react with water, but also that whereas the Si—Si and Si—H bonds are weaker than the corresponding bonds involving carbon, the Si—F, Si—Cl, and Si—O bonds are stronger. This last observation calls to mind the 'extra' strengths of B—F, B—Cl, and B—O bonds involving three-coordinated boron, and suggests that donation of electrons from fluorine, chlorine or oxygen into *d* orbitals of silicon occurs, giving rise to p_π–d_π π bonding as distinct from the p_π–p_π π bonding that is an important feature of boron chemistry. Bond length data, and the planar Si_3N skeleton and non-basic character of trisilylamine, $(SiH_3)_3N$ (Section 13.4), provide further evidence for the occurrence of p_π–d_π bonding in silicon compounds.

On the other hand p_π–p_π double or triple bonding, which is so important in carbon chemistry, is relatively unimportant for silicon and later elements of the group. An analogous state of affairs is found in the nitrogen and oxygen groups; on the views of the nature of π-bonding outlined in Chapter 4, relatively poor overlap would result from interaction of a 3*p* orbital of an atom of a second period element and a 2*p* orbital of an atom of a first period element or of two 3*p* orbitals of two atoms of second period elements. Thus there is no silicon analogue of graphite, benzene, ethylene, acetylene, carbon monoxide or hydrogen cyanide; and silicon dioxide, silicates and silicones are structurally very different from carbon dioxide, carbonates and ketones. Nevertheless, it should be noted that molecular O=Si=O has recently been characterised in an inert matrix at 20 K, and $R_2Si = SiR_2$, where R = 1,3,5 trimethylphenyl, is stable at the ordinary temperature in the absence of oxygen, the bulky aryl groups preventing polymerisation and perhaps also stabilising the molecule by conjugation with the Si=Si bond. It should also be noted that although ethylene, for example, is apparently a stable species, it is thermodynamically unstable with respect to polymerisation to polyethylene, which involves conversion of each double bond into two single ones. Because of the marked differences between the chemistry of carbon and that of later elements in the group, it will often be convenient to discuss carbon compounds separately, though our coverage will be limited to the element and its simpler derivatives. In this book we shall not deal with organometallic compounds of the main group elements except under individual groups of elements, but carbonyls and other organometallic derivatives of transition metals will be considered in general terms in Chapters 22 and 23.

The isotope ^{13}C (which is present to the extent of 1.1 per cent in natural carbon) has nuclear spin quantum number $I = \frac{1}{2}$, and nuclear magnetic resonance studies involving this isotope are very important in organic and organometallic chemistry; ^{13}C is also used as a non-radioactive tracer. The β^--active isotope ^{14}C ($t_{\frac{1}{2}} = 5730$ years), made by a (n,p) reaction on $Li_3{}^{14}N$ in a reactor, is also an important tracer; its use in radiocarbon dating has been mentioned in Section 1.9. The tin isotope ^{119}Sn is suitable for Mössbauer spectroscopy.

Carbon occurs naturally as diamond and graphite. Carbon compounds are ubiquitous in living matter, and occur also as coal, natural gas, petroleum, atmospheric carbon dioxide and many minerals, e.g. *chalk*, *limestone* and *marble* ($CaCO_3$), *magnesite* ($MgCO_3$) and *dolomite* ($CaMg(CO_3)_2$). Silicon is, after oxygen, the most abundant element in the earth's crust, occurring mainly as silicon dioxide (*quartz*, *flint* and *sand*) and many silicates. By comparison, germanium, tin and lead are relatively rare elements, though the last two have been known since antiquity because of the ease with which the metals are isolated. Germanium occurs in traces in coal, and accumulates in flue dust as GeO_2, tin as *tinstone* or *cassiterite* (SnO_2), and lead mostly as *galena* (PbS). The isolation of the elements and the uses of them and their compounds are mentioned in later sections.

13.2 The elements

Diamond and graphite, the two well-defined crystalline forms of carbon, occur naturally; graphite is also manufactured by heating powdered coke (high-temperature carbonised coal) with silica at 2500 °C. Carbon black (amorphous carbon) for use in the manufacture of synthetic rubbers is made by burning oil in a limited supply of air. Many other amorphous and crystalline forms are also known and will be mentioned later. Impure silicon and germanium are made by reduction of silica with carbon or calcium carbide in an electric furnace, and by reduction of germanium dioxide with hydrogen or carbon respectively. For electronic work, very pure silicon and germanium are obtained by conversion into the tetrachlorides or tetraiodides, fractional distillation of these to remove boron- and phosphorus-containing impurities, and decomposition by pyrolysis or reduction with hydrogen; alternatively, the pure hydrides may be prepared and pyrolysed. The solid elements are then further purified by zone-refining: a bar of silicon or germanium is traversed with a molten zone by means of a moving radio-frequency heater; impurities, which are more soluble in the melt than in the solid, are thus concentrated at one end of the bar, which is removed. In this way silicon and germanium containing less than 10^{-9} per cent of impurity atoms may be prepared. Tin is obtained by reduction of the dioxide with carbon. Lead occurs mainly as sulphide ores, which cannot be reduced by carbon (carbon disulphide has a positive standard free energy of formation): the sulphide can, however, be converted into lead monoxide by heating it in air, and this can be reduced by carbon. Alternatively, the controlled action of atmospheric oxygen on the sulphide may be used to effect the successive reactions:

$$2PbS + 3O_2 \rightarrow 2PbO + 2SO_2$$

$$PbS + 2PbO \rightarrow 3Pb + SO_2$$

Both metals are refined electrolytically.

Silicon, germanium and α-tin (or grey tin – the stable form of tin below 13 °C) all crystallise with the same structure as diamond, which may be described as that of zinc blende (Section 6.3) with the zinc and sulphur atoms both replaced by carbon. In diamond each carbon atom has four nearest neighbours tetrahedrally disposed around it at 1.54 Å, and the carbon–carbon bond energy term is 346 kJ mol^{-1}; both of these values are the same as those for carbon–carbon bonds in saturated hydrocarbons. It is this strong three-dimensional infinite structure that is responsible for the hardness of diamond; the bond energies in silicon, germanium and tin (Table 13.2) are much lower, and this fact is reflected in their physical properties.

The variation in electrical conductivity from diamond to grey tin may be accounted for on the basis of a band theory analogous to that already outlined in connection with the electronic structures of metals (Section 4.9). If we consider the solid as a giant molecule containing n atoms each with four valence electrons, $2n$ bonding orbitals (all filled) form one band of energy levels and $2n$ antibonding orbitals (all empty) another. Energy is needed to promote electrons from one band to the other and hence give rise to semiconductivity. For diamond the energy gap is very large (580 kJ mol^{-1}), and pure diamond is colourless and an insulator, but for silicon, germanium and grey tin the energy gap is only 115, 70 and 10 kJ mol^{-1} respectively and these elements are metallic in appearance and are semiconductors. The conductivity of all the elements is increased by the presence of Group III or Group V elements, which give rise to p-type or n-type extrinsic semiconductivity (Sections 6.10 and 6.11) respectively. Starting with very pure silicon or germanium, suitable materials for use in semiconductor devices may be made by addition of the exact amounts of impurities necessary to produce the required properties.

Graphite is the most stable allotrope of carbon by a margin of 3 kJ mol^{-1} under standard conditions. Its structure normally consists of layers of carbon atoms stacked so that atoms of alternate layers lie vertically beneath one another, as shown in Fig. 13.1; in another form of graphite, every third layer is superposed. Each carbon atom in graphite forms bonds to three other atoms at 1.42 Å; the interlayer distance of 3.35 Å indicates that there is only van der Waals' bonding between the layers and accounts for the fact that graphite is a good lubricator. The carbon–carbon distance is a little greater than that in benzene (1.39 Å) and corresponds to a bond order of $1\frac{1}{3}$. Molecular π orbitals extend over each whole layer; all the bonding orbitals are filled, but the energy gap between them and the antibonding orbitals is very small, and the electrical conductivity along the layers approaches that of a metal. The conductivity perpendicular to the layers is smaller by a factor of about 1:5000.

Since graphite is more stable than diamond and has a more open structure (their densities are 2.22 and 3.51 g cm^{-3} respectively), the conversion of graphite into diamond is favoured by increase in temperature and pressure. Further, since the change involves the initial

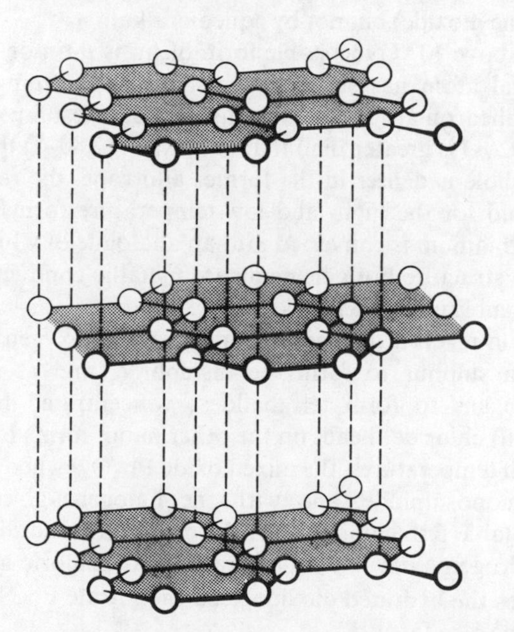

Figure 13.1
The structure of graphite. Atoms in alternate layers are directly under one another.

breaking of a large number of bonds, it has a very large activation energy. Small diamonds for industrial purposes are made by heating graphite at 2000 °C and 10^5 atm pressure in the presence of a transition metal catalyst (e.g. chromium, iron or platinum); the catalyst is believed to act by forming a film of molten metal on the surface of the graphite, dissolving some of it, and reprecipitating it in the form of the less soluble diamond.

Charcoal (made by heating wood or sugar) and animal charcoal (made by charring treated bones) both consist of microcrystalline forms of graphite, in the latter case supported on calcium phosphate. These finely divided forms, possessing extensive internal surfaces and unsatisfied valencies, readily absorb gases and solutes and are very good catalysts, particularly when impregnated with palladium, platinum or other transition metal compounds. Carbon fibres of great tensile strength obtained by pyrolysis of oriented organic polymer fibres at 1500 °C or above contain graphite crystallites oriented with the fibre axis; they are very important in the manufacture of reinforced plastics.

Graphite possesses the remarkable property of forming many compounds by taking up atoms or ions between the layers with an increase in the interlayer distance; these *intercalation compounds* are described in the following section. Other reactions of graphite include oxidation in air above 700 °C and combination with sulphur at 900 °C. Diamond is less reactive, but burns in air at 900 °C.

Silicon is much more reactive than carbon; at elevated temperatures it combines with oxygen, all the halogens, sulphur, nitrogen, carbon and boron. It liberates hydrogen from aqueous alkali, but it is insoluble in acids other than a mixture of HNO_3 and HF. Germanium generally resembles silicon except in that it is attacked by nitric acid (with formation of the dioxide) but not by aqueous alkali.

Above 13 °C the stable form of tin is β-tin or white tin, in which each metal atom has six nearest neighbours forming a somewhat distorted octahedron around it. Although the shortest Sn—Sn distance in white tin (3.02 Å) is greater than that in grey tin (2.80 Å) the packing considered as a whole is denser in the former allotrope, the reverse of what is usually found for the high- and low-temperature forms of the same substance. Germanium is converted into an analogue of white tin under pressure; in this structure both elements are metallic conductors. Lead exists only in the cubic close-packed structure.

Tin reacts at higher temperatures with oxygen to form the dioxide and with sulphur to form the disulphide, and at room temperature with halogens to form tetrahalides; concentrated hydrochloric acid yields tin(II) chloride. Lead, on the other hand, forms only the monoxide (or, at high temperatures, the mixed oxide Pb_3O_4) when heated in air, forms only a monosulphide, and with the halogens gives the tetrafluoride, the unstable tetrachloride, and the dibromide and diiodide; it slowly evolves hydrogen from hot concentrated hydrochloric acid. With nitric acid tin gives the hydrated dioxide, lead the nitrate $Pb(NO_3)_2$, oxides of nitrogen being also formed.

Important alloys of tin and lead include various solders, pewter (mainly tin), bronze (tin and copper), and type metal (lead, tin and antimony). Tin is also used for containers for food and drink, for plating on to other metals to provide a corrosion-free surface and in the float glass process for the manufacture of sheet glass, for which its low melting point (232 °C) and lack of reaction with the glass make it eminently suitable. Metallic lead now finds its chief use in storage batteries; lead compounds are dangerous cumulative poisons, and the use of lead for water pipes has been abandoned because in the presence of air and the absence of carbonate the metal is slowly attacked to form the hydroxide $Pb(OH)_2$, which is slightly soluble in water.

13.3 Intercalation compounds of graphite

Graphite undergoes several reactions in which the carbon layers move further apart and atoms or ions are accumulated between them; the products of such reactions are known as *intercalation* or *lamellar compounds* or, sometimes, merely as *graphitic compounds*. It is now recognised that there are, broadly, two different types of product: (i) those in which the layers become buckled owing to saturation of the carbon atoms and the loss of the π electron system; (ii) those in which the planarity of the layers is retained, but they move further apart. Compounds of class (i) are colourless and non-conductors of electricity, those of class (ii) coloured (though different from graphite in appearance) and conductors. Only formation of class (ii) compounds is reversible.

The most extensively investigated compound of class (i) is the polymeric carbon monofluoride. Although graphite reacts with fluorine at 700 °C to form carbon tetrafluoride, at 450 °C (or at lower temperatures in the presence of hydrogen fluoride) an unreactive hydrophobic grey (or, when pure, white) solid of composition $(CF)_n$ is formed. In this, the carbon–carbon distance within layers is 1.54 Å, and between layers is 8.2 Å; the structure is shown in Fig. 13.2. Like graphite, carbon monofluoride is a lubricating agent because of its layer structure, but it has the advantage of being more resistant to atmospheric oxidation. At high temperatures, however, it disproportionates into carbon and carbon tetrafluoride.

Treatment of graphite with various oxidising agents (e.g. a mixture of nitric and sulphuric acids or potassium permanganate) causes swelling and formation of a brown or yellow product (commonly known as graphite oxide) containing carbon, hydrogen and oxygen. The constitution of this material has not been established, but since it is a non-conductor and the layers of carbon atoms are 6–11 Å apart (according to how dry the product is), a similarity to carbon monofluoride is apparent; chemical evidence, however, suggests that ether, hydroxyl and carbonyl

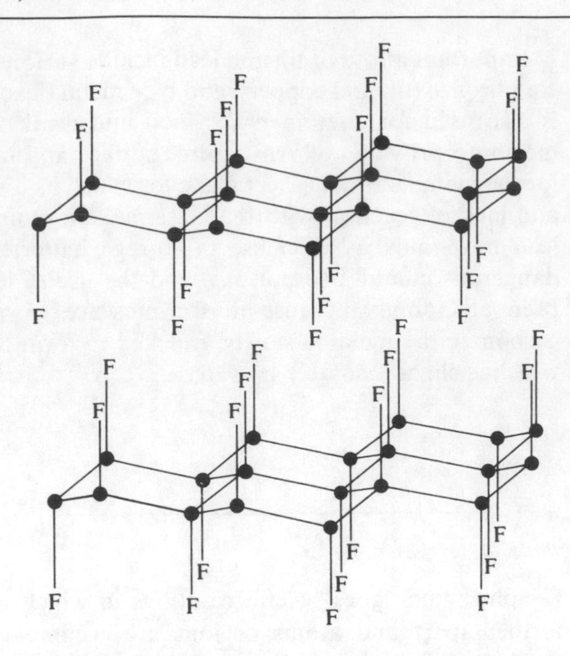

Figure 13.2
The structure of carbon monofluoride.

groups are all present, and it is unlikely that graphite oxide is a single compound.

Class (ii) intercalation compounds include the metallic-looking red or blue compounds formed with potassium and heavier alkali metals (in which the graphite layer acquires a negative charge), and the blue graphite salts formed with strong acids in the presence of small quantities of oxidising agents (in which the layer acquires a positive charge). When graphite is treated with excess of potassium and the excess is washed out with mercury, a paramagnetic copper-coloured residue of composition C_8K remains. In this, which is formulated $C_8^-K^+$, there are alternate layers of carbon and metal atoms, the latter being arranged so that they lie above and below the centres of C_6 rings in two layers; the carbon layers are now vertically above one another, unlike their relative orientation in graphite itself, and 5.40 Å apart. Thermal decomposition of C_8K yields successively blue $C_{24}K$, $C_{36}K$, $C_{48}K$ and $C_{60}K$, in which there are two, three, four and five layers of carbon atoms respectively between layers containing potassium atoms. The extra electrons contributed by the potassium atoms to the conduction band of the graphite give potassium–graphite compounds conductivities higher than that of graphite itself. The potassium–graphite compound is highly reactive, igniting in air and exploding on treatment with water.

Graphite hydrogensulphate has the composition $C_{24}^+HSO_4^-.2H_2SO_4$, and is obtained by the action of concentrated sulphuric acid and a little nitric acid or chromium trioxide on graphite. Analogous products are formed with perchloric and trifluoroacetic acids. In the perchlorate, the separation between the layers of carbon atoms has been shown to be 7.94

Å, with anions and acid molecules between each pair of layers; cathodic reduction or treatment with graphite results in successive formation of species in which there are two, three, four or five layers of carbon atoms between layers of inorganic material. Such behaviour is exactly analogous to that found in the potassium–graphite system. Graphite salts also conduct better than graphite, in this case by a positive hole mechanism.

Other intercalation compounds are formed by graphite and chlorine, bromine, iodine monochloride and many halides, among them KrF_2, UF_6 and $FeCl_3$. With dioxygenyl hexafluoroarsenate, $O_2^+AsF_6^-$, graphite yields the complex fluoride $C_8^+AsF_6^-$.

Graphite intercalation compounds are of practical interest because of their catalytic properties (C_8K, for example, is a good hydrogenation catalyst) or because of catalysis by graphite of relevant reactions (e.g. the combination of hydrogen and bromine or carbon monoxide and chlorine).

13.4 Hydrides

The extensive chemistry of the saturated and unsaturated hydrocarbons lies outside the scope of this book, though their formation by the hydrolysis of metal carbides is considered in the following section, and a few points about them must be made here to illustrate the comparative chemistry of the Group IV elements. We have already drawn attention to the fact that the Si—H bond is weaker than Si—Cl or Si—O, whereas the C—H bond is stronger than C—Cl or C—O. Thus methane is chlorinated with some difficulty and is thermodynamically stable to hydrolysis, whilst monosilane, SiH_4, reacts violently with chlorine and is hydrolysed to silicate and hydrogen by alkali; and although methane is only kinetically stable to atmospheric oxidation at ordinary temperatures, its combustion is much less exothermic than that of monosilane, which is, furthermore, spontaneously inflammable in air.

A mixture of SiH_4, Si_2H_6, Si_3H_8 and Si_4H_{10} (b.p. −112, −14, 53 and 108 °C respectively), plus traces of hydrides containing more silicon atoms, is obtained by decomposing magnesium silicide with aqueous acid in an inert atmosphere or *in vacuo*; the germanium hydrides may be obtained similarly. The hydrides of silicon and germanium, together with monostannane, may also be made by the action of lithium tetrahydridoaluminate on the appropriate chloride in ethereal solution. Monosilane, which has some importance in the preparation of pure silicon, can also be prepared directly from silica by hydrogenation at 175 °C and 400 atm pressure in an aluminium chloride and sodium chloride melt. No hydride of lead has yet been satisfactorily characterised.

Mixed hydrides of silicon and germanium, e.g. H_3SiGeH_3, may be obtained by the action of acid on an intimate mixture of magnesium silicide and magnesium germanide. Pyrolysis under controlled conditions of silane and germane yields hydrides containing up to ten atoms of the element; solid polymeric species of composition $(SiH)_n$ and $(GeH)_n$ may also be obtained.

Stannane decomposes into its elements even at room temperature, and its chemistry has been little examined, but silane and germane give rise to many compounds containing the silyl (SiH_3) and germyl (GeH_3) groups. Aluminium chloride catalyses the reaction

$$SiH_4 + HCl \xrightarrow{100\,°C} SiH_3Cl + H_2$$

Alternatively, $PhSiH_3$ (made by reduction of $PhSiCl_3$ with $LiAlH_4$) may be treated with liquid hydrogen chloride to give silyl chloride:

$$PhSiH_3 + HCl \rightarrow SiH_3Cl + PhH$$

(Neither of these reactions, it may be noted, has an organic counterpart – this is easily interpreted in terms of the points made earlier in this section about bond energies.) Some reactions of SiH_3Cl (b.p. −30 °C) are shown below; the preparation and properties of GeH_3Cl closely resemble those of the silicon compound.

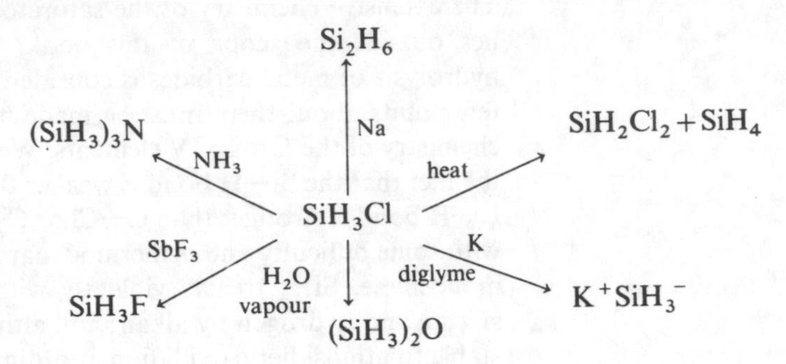

As we mentioned in Section 13.1, trisilylamine has a planar Si_3N skeleton and is not basic, indicating that the unshared pair of electrons formally present on the nitrogen atom is delocalised into 3*d* orbitals of the silicon atoms. A similar delocalisation also occurs, though to a smaller extent, in disilyl ether, in which the SiOSi bond angle is 144° compared with 111° for the COC bond angle in dimethyl ether. Trigermylamine is isostructural with trisilylamine; but trisilylphosphine is pyramidal and basic, and the SiSSi bond angle in disilyl sulphide, $(SiH_3)_2S$, is only 97°, indicating that phosphorus and sulphur, in contrast to nitrogen and oxygen, do not take part in p_π–d_π bonding to silicon.

Trichlorosilane, or silicochloroform, results from the action of hydrogen chloride on silicon at 400 °C:

$$Si + 3HCl \rightarrow SiHCl_3 + H_2$$

It is a volatile liquid which forms SiH_3 and $SiCl_4$ when heated and which adds to alkenes (the *hydrosilation* reaction):

$$RCH{=}CH_2 + SiHCl_3 \rightarrow RCH_2CH_2SiCl_3$$

13.5 Carbides and silicides

Binary compounds containing carbon and any element other than the non-metals of Groups V to VII are generally described as carbides; a consistent basis for their classification is not easily established, but the limiting cases of saline and macromolecular carbides are easily recognised.

The commonest macromolecular carbides are those of boron and silicon, both made by heating the appropriate oxide with excess of carbon. They are extremely hard black substances widely used as abrasives and are very unreactive, though the lack of reactivity is generally kinetic, rather than thermodynamic, in origin. Silicon carbide (*carborundum*) is polymorphic and exists in both the zinc blende and the wurtzite structures. Boron carbide, $B_{13}C_2$, has an immensely complicated structure made up of B_{12} icosahedra and linear CBC groups. Each boron atom is bonded to five other boron atoms in the icosahedron and either to a terminal carbon atom of a CBC group or to a boron atom of a neighbouring icosahedron; three-dimensional networks of B—B and BCBCB bonds extend throughout the crystal.

Saline carbides which give methane on hydrolysis include those of beryllium and aluminium, Be_2C and Al_4C_3 respectively, both made from the elements; it is exceedingly unlikely that anions carrying four negative charges on a single atom are present in these compounds (the interelectronic repulsion energy would be enormous), but it is certain that in both compounds the carbon atoms occur singly, so they may justifiably be considered derivatives of methane. Calcium carbide or acetylide, CaC_2, is manufactured as a grey solid by heating lime with coke at 2000 °C, but when pure it is colourless; it yields acetylene on treatment with water:

$$CaC_2 + 2H_2O \rightarrow Ca(OH)_2 + C_2H_2$$

It has a tetragonally distorted sodium chloride structure, the axis along which the C_2^{2-} ions lie being lengthened relative to the other two axes of the unit cell; the carbon–carbon distance in the anion is 1.19 Å, compared with 1.20 Å in acetylene. Sodium acetylide, Na_2C_2, is made from sodamide and acetylene in liquid ammonia; it also is decomposed by water. Other acetylides, such as those of silver and copper, may be precipitated from aqueous solutions of complex ammines, e.g.

$$2[Ag(NH_3)_2]^+ + C_2H_2 \rightarrow Ag_2C_2 + 2NH_4^+ + 2NH_3$$

Like acetylene itself, silver and copper acetylides are sensitive to heat and mechanical shock.

Carbides of formula MC_2 are formed by several normally tri- and tetrapositive elements, e.g. lanthanum and thorium; the carbon–carbon distances in the anions of these substances are 1.29 and 1.33 Å respectively. Unlike calcium carbide, which is an insulator, these carbides are conductors. The bond length data indicate structures approximating to $La^{3+}\ C_2^{3-}$ and $Th^{4+}\ C_2^{4-}$, with the extra electrons given from the metal to the C_2^{2-} anion being accommodated in antibonding orbitals and thus weakening the bond (cf. Section 4.2); but the conducting properties and diamagnetism show that this description must be an oversimplification and that electron delocalisation into a conduction band occurs. The hydrolytic behaviour of these carbides is complicated, different mixtures of hydrocarbons and hydrogen being formed at different temperatures.

The structures of the so-called interstitial carbides, formed by heating transition metals of atomic radius greater than about 1.3 Å (e.g. Ti, Zr, V, Mo, W) with carbon, may be described as based on a cubic close-packed metal lattice with carbon atoms occupying octahedral holes (cf. Section 6.2). A formula such as V_2C corresponds to occupation of half the octahedral holes, one such as TiC to occupation of all of them. As we explained in discussing interstitial hydrides (Section 9.6), the non-metal has to be atomised in the formation of such a compound, and hence the metal–carbon bonding, which has to compensate this process, must be very strong; interstitial carbides are characterised by extreme hardness and infusibility, tungsten carbide (WC) being one of the hardest substances known. Like the interstitial hydrides, these carbides are now recognised as involving multicentre bonding.

Transition metals of smaller atomic radius than about 1.3 Å (e.g. Cr, Fe, Co, Ni) form carbides with a wide variety of atomic ratios and complicated structures involving carbon–carbon bonding. In Fe_3C, for example, there is a very distorted metal lattice with carbon atoms close enough together (at 1.65 Å) to indicate some interaction between them. Carbides of this group, such as Cr_3C_2 and M_3C (where M = Fe, Co or Ni), are hydrolysed by water or dilute acid to give mixtures of hydrocarbons and hydrogen.

The structures of the metal silicides, all of which are made from the elements, are even more diverse than those of the carbides. Isolated silicon atoms occur in Mg_2Si and Ca_2Si, but Si_2 groups (in U_3Si_2), chains (in CaSi), planar and puckered hexagonal networks (in USi_2 and $CaSi_2$ respectively), and three-dimensional frameworks (in $SrSi_2$) of silicon atoms all occur; these structures, however, are too complicated for discussion here.

Reduction of germanium, tin or lead by sodium in liquid ammonia gives polymetallic anions, e.g. M_5^{2-} and M_9^{4-}, trigonal bipyramidal and capped antiprismatic respectively. Many related anions have recently been characterised.

13.6 Halides and complex halides

The tetrahalides of carbon (CF_4, b.p. -128 °C; CCl_4, b.p. 76 °C; CBr_4, m.p. 93 °C; CI_4, m.p. 171 °C) differ from those of silicon and tetrapositive germanium, tin and lead in their inertness towards water or dilute alkali and in their failure to form complexes with metal halides. Both of these characteristics may be attributed to the absence of *d* orbitals in the valence shell of the carbon atom, though in the case of the inertness towards hydrolysis this comment is not entirely adequate, since it presupposes that the reaction would have to proceed by way of a five-coordinated intermediate (as the hydrolysis of silicon–halogen compounds is believed to do), and it is impossible to establish the mechanism of a reaction that does not go. Certainly, however, carbon tetrafluoride and tetrachloride are thermodynamically unstable towards hydrolysis; it is easily shown from the standard enthalpies of formation and entropies of the reactants and products that, for example, $\Delta G^\circ = -380\,\text{kJ mol}^{-1}$ for the reaction.

$$CCl_4(l) + 2H_2O \rightarrow CO_2 + 4HCl(aq)$$

This value is much more negative than that ($-290\,\text{kJ mol}^{-1}$) for the hydrolysis of silicon tetrachloride, which is very rapidly hydrolysed by water.

Carbon tetrafluoride, the end-product of the fluorination of any organic compound, is an extremely inert substance, conveniently made by fluorination of silicon carbide, the silicon tetrafluoride which is also produced being removed by passage through aqueous sodium hydroxide. A very large number of other carbon fluorides, or *fluorocarbons*, are known. Uncontrolled fluorination of hydrocarbons usually leads to extensive decomposition because of the heat evolved in the reaction

$$\geq C{-}H + F_2 \rightarrow \geq C{-}F + HF; \quad \Delta H^\circ = -480\,\text{kJ mol}^{-1}$$

The preparation of fully fluorinated organic compounds is therefore carried out in an inert solvent (the evaporation of which absorbs the heat liberated), in a reactor packed with gold- or silver-plated copper turnings (which serve the same purpose, though catalysis by formation of a surface layer of AgF_2 or AuF_3 may also be involved), or by the use of cobalt(III) fluoride as a milder fluorinating agent; electrolysis in liquid hydrogen fluoride (Section 8.3) may also be used, though more commonly for the fluorination of oxygen- or nitrogen-containing compounds than for hydrocarbons. Fluorocarbons have boiling-points very close to those of the corresponding hydrocarbons; they are inert towards concentrated alkali or acids, and dissolve only in non-polar organic solvents. Their chief applications are as high-temperature lubricants and (together with the chlorofluoro compounds mentioned below) as refrigerants.

Freons are chlorofluoro-carbons or -hydrocarbons, made by partial replacement of chlorine in the corresponding chlorinated compounds by means of antimony trifluoride in the presence of antimony pentachloride or of hydrogen fluoride in the presence of both these compounds. They are volatile inert compounds which are non-toxic, non-inflammable and non-corrosive, and find extensive applications in refrigerants and aerosols. However, chlorine atoms produced by their solar decomposition in the upper atmosphere catalyse the decomposition of ozone (Section 15.2) and their use is now being discouraged.

Mention may be made here of two important plastics which are manufactured from chlorofluoro compounds. Polytetrafluoroethylene (*Teflon*), the non-stick plastic, is made by polymerisation of tetrafluoroethylene in the presence of water with an organic peroxide catalyst; the fluoroalkene is obtained by the reactions

$$CHCl_3 \xrightarrow[SbCl_5, SbF_3]{HF} CHF_2Cl \xrightarrow{700\ °C} C_2F_4 + HCl$$

Polytetrafluoroethylene is a white solid which is stable to 300 °C, is insoluble in most solvents, and is inert towards both acids and alkalis. Polymerised perfluorovinyl chloride, $(CF_2{=}CFCl)_n$, or *Kel-F*, is made by polymerising the monomer obtained by dechlorination of the compound $CF_2ClCFCl_2$ with zinc and ethanol; it is almost transparent and is relatively easy to fabricate.

Carbon tetrachloride is made by the carefully controlled chlorination of methane at 250–400 °C, or by the chlorination of carbon disulphide:

$$CS_2 + 3Cl_2 \xrightarrow{\text{Fe catalyst}} CCl_4 + S_2Cl_2$$

$$CS_2 + 2S_2Cl_2 \rightarrow CCl_4 + 6S$$

$$6S + 3C \rightarrow 3CS_2$$

It is widely used as a solvent and for the chlorination of inorganic compounds; it is decomposed photochemically or thermally with the production of CCl_3 radicals and chlorine atoms. Carbon tetrabromide and tetraiodide, which are easily decomposed into their elements, are made by the reactions

$$3CCl_4 + 4AlBr_3 \rightarrow 3CBr_4 + 4AlCl_3$$

$$CCl_4 + 4C_2H_5I \xrightarrow{AlCl_3} CI_4 + 4C_2H_5Cl$$

Carbonyl chloride (*phosgene*) is a highly toxic colourless gas (b.p. 8 °C) made by addition of chlorine to carbon monoxide in sunlight or in the presence of active charcoal. It is also produced by the action of water on carbon tetrachloride at high temperatures. The action of antimony trifluoride transforms it into COClF and COF_2. All these halides are unstable to water, and react with ammonia and alcohols to form urea and esters.

Many fluorides and chlorides of silicon are known, but we shall confine the discussion here to one of each. Silicon tetrafluoride, a gas b.p. −86 °C, is conveniently made by the reaction

$$2CaF_2 + 2H_2SO_4 + SiO_2 \rightarrow 2CaSO_4 + SiF_4 + 2H_2O$$

or by fluorination of the tetrachloride, which is obtained by the action of chlorine on the element. The tetrafluoride is only partially hydrolysed by water, the overall reaction being

$$2SiF_4 + 4H_2O \rightarrow SiO_2 + 2H_3O^+ + [SiF_6]^{2-} + 2HF$$

The tetrachloride, however, is completely hydrolysed. Fluorosilicates are best made from the tetrafluoride and metal fluorides in aqueous hydrofluoric acid; the potassium and barium salts are very sparingly soluble. Fluorosilicic acid is a strong acid in aqueous solution, but pure H_2SiF_6 has not been isolated. The other silicon halides do not form complex anions with alkali metal halides, but lattice energy considerations suggest that it might be possible to obtain a hexachlorosilicate of a very large quaternary ammonium cation.

The halides of tetrapositive germanium closely resemble those of silicon except in that salts containing the $[GeCl_6]^{2-}$ ion are known and the ion persists in concentrated hydrochloric acid solution. However, whereas SiF_2 and $SiCl_2$ can be obtained only as unstable species by the action of the tetrahalides on silicon at 1200 °C and polymerise to cyclic products on cooling, the germanium dihalides, made from GeS (Section 13.11) and the appropriate silver halide, are stable solids. The yellow iodide, like many metal iodides, has the cadmium iodide structure.

Tin tetrafluoride, made from the chloride and hydrogen fluoride, sublimes at 700 °C; it has a structure based on octahedral SnF_6 groups which share four fluorine atoms to give an infinite layer, the Sn—F bridging and (*trans*)terminal distances being 2.02 and 1.88 Å respectively; the presence of longer bonds to bridging atoms is general among structures of this kind. Tin tetrafluoride is thermally stable, but lead tetrafluoride, which is isostructural with it, decomposes into the difluoride and fluorine when heated, and has to be made by the action of fluorine or halogen fluorides on lead compounds. Tin difluoride is readily soluble in water and may be made in aqueous media; lead difluoride is only sparingly soluble. One form of PbF_2 has the fluorite structure; SnF_2 contains puckered Sn_4F_8 rings, each Sn atom being bonded to two shared and one unshared F atoms, with pyramidal coordination of the metal atom, as expected from the presence of a lone pair of electrons.

Tin tetrachloride (b.p. 114 °C) and the dichloride (m.p. 246 °C) are obtained from the metal and chlorine and hydrogen chloride respectively; both are partially hydrolysed in water, but react with alkali metal chlorides in the presence of hydrochloric acid to give salts containing the $[SnCl_6]^{2-}$ and $[SnCl_3]^-$ ions. The structures of $SnCl_2$ (in the gas phase) and the $[SnCl_3]^-$ ion are bent and pyramidal respectively; that of the $[SnCl_6]^{2-}$ ion is, of course, octahedral. Lead tetrachloride is obtained as an oily liquid by the action of cold concentrated sulphuric acid on the yellow salt $(NH_4)_2[PbCl_6]$, which is precipitated when chlorine is passed into a saturated solution of lead chloride in aqueous ammonium chloride; it freezes at −19 °C, is hydrolysed by water, and decomposes into the dichloride and chlorine when gently heated. Lead tetrabromide and tetraiodide do not exist. The ease with which the complex chloride of lead(IV) is obtained is a striking instance of the stabilisation of oxidation states by complexing. Lead dichloride is a white solid which is sparingly soluble in water but much more soluble in hydrochloric acid owing to formation of the complex ion $[PbCl_4]^{2-}$. The dichloride has a complicated three-dimensional structure containing a nine-coordinated cation, but the yellow iodide has the cadmium iodide layer structure.

13.7 *Oxides and oxo-acids of carbon*

In this section it is again necessary to consider carbon separately from the other elements, since the latter form no stable volatile monomers analogous to CO and CO_2. An interesting comment on the difference between carbon dioxide and silicon dioxide can be made in the light of the thermochemical data given in Section 13.1, according to which the bond energy term for C=O (806 kJ mol^{-1}) is more than twice that for C—O (359 kJ mol^{-1}) whilst the bond energy term for Si=O (642 kJ mol^{-1}) is less than twice that for Si—O (466 kJ mol^{-1}). In discussing these values we may say that the C=O bond is strengthened relative to Si=O by p_π–p_π bonding or that the Si—O bond is strengthened relative to C—O by d_π–p_π bonding; there is probably considerable justification for both comments. Irrespective of the interpretation of the bond energy terms, however, they indicate that (neglecting enthalpy and entropy changes associated with vaporisation) silica is stable with respect to conversion into O=Si=O molecules, whilst carbon dioxide is stable with respect to conversion into a macromolecule in which each carbon forms four single C—O bonds.

Carbon monoxide, a colourless gas (b.p. −190 °C), is formed when carbon burns in a deficiency of oxygen or, on a laboratory scale, by dehydration of formic acid with concentrated sulphuric acid. On an industrial scale it is prepared by the reduction of carbon dioxide with coke at above 800 °C, or, together with hydrogen, by the action of steam on coke at about 1000 °C. It may be separated by low-temperature fractionation or by reversible absorption in ammoniacal copper(I) chloride solution.

Carbon monoxide is almost insoluble in water under ordinary conditions and does not react with aqueous sodium hydroxide, but at high pressures and elevated temperatures formic acid and sodium formate are obtained. It combines with the halogens (except iodine), sulphur and selenium, and forms metal carbonyls (Chapter 22) when heated with many transition metals or their compounds. The highly poisonous nature of the gas arises from its formation of a stable compound with haemoglobin (Section 24.7), which is thereby inhibited from acting as an oxygen carrier in respiration. Carbon monoxide may be oxidised to the dioxide by a mixture of manganese dioxide, copper(II) oxide and silver(I) oxide at ordinary temperatures (a process made use of in respirators), or by iodine pentoxide at 90 °C:

$$I_2O_5 + 5CO \rightarrow I_2 + 5CO_2$$

The latter reaction may be employed for the determination of carbon monoxide.

The thermodynamics of the oxidation of carbon is of immense importance in metallurgy: most metal oxides may be reduced by carbon if a sufficiently high temperature is employed. We shall not discuss this subject in detail here, but the underlying principles are readily grasped. The reaction

$$C(s) + O_2 \rightarrow CO_2$$

involves no change in the number of gaseous molecules, so ΔS° is very small (actually $+3$ J K^{-1} mol^{-1}) and ΔG° is nearly independent of temperature; the effectiveness of carbon being converted into the dioxide as a reducing agent is therefore almost temperature-independent. For the reaction

$$2CO + O_2 \rightarrow 2CO_2$$

there is a decrease in the number of gaseous molecules, ΔS° is negative (actually -172 J K^{-1} mol^{-1}) and ΔG° increases with increasing temperature, corresponding to a decrease in the power of carbon monoxide as a reducing agent. On the other hand, for the reaction

$$2C(s) + O_2 \rightarrow 2CO$$

the increase in the number of gaseous molecules leads to an increase in entropy ($\Delta S^\circ = 178$ J K^{-1} mol^{-1}), ΔG° becomes more negative as the temperature increases, and carbon being converted into carbon monoxide becomes an increasingly powerful reducing agent. A solid metal combining with gaseous oxygen to give a solid oxide will always result in a decrease in entropy, and the effect will be even greater if, as is often the case at high temperatures, the metal is a liquid or a gas and the oxide is a solid. Thus ΔG° for the reaction

$$\text{metal} + O_2 \rightarrow \text{oxide (s)}$$

always increases with increase in temperature, and there must come a point at which it is more positive than that for the reaction

$$2C(s)+O_2 \rightarrow 2CO$$

This point is attainable in practice for nearly all metal oxides other than those of the alkaline earth metals, aluminium and a few transition elements, though sometimes very rapid cooling is necessary to prevent reversal of the reaction at lower temperatures. The fact that carbon monoxide is a gas is also important for kinetic reasons, gas–solid reactions being much faster than reactions between solids. Thus carbon monoxide can act as a reducing agent by virtue of the reaction

$$2CO+\text{oxide} \rightarrow 2CO_2+\text{metal}$$

and the resulting dioxide can be reconverted into the monoxide by the reaction

$$C(s)+CO_2 \rightarrow 2CO$$

which, because of the increase in entropy, is also favoured at high temperatures. Data for reduction of a wide range of metal oxides are summarised on the so-called *Ellingham diagram* shown in Fig. 13.3.

The physical properties of carbon monoxide are very similar to those of nitrogen, and the electronic structures of the two molecules both involve a σ bond and two π bonds, the only difference being that in carbon monoxide the electrons for one of these bonds must be provided by the oxygen atom. This might suggest a polar molecule $\overset{\times}{_\times}\bar{C}\equiv\overset{+}{O}\overset{\times}{_\times}$, but in fact the dipole moment of carbon monoxide is only 0.1 debye; however, with the effects of the lone pairs and the greater electronegativity of oxygen as factors that may also be involved, it is clear that no simple satisfactory comment on the very low polarity of the molecule can be made. The bond energy in CO is 1075 kJ mol^{-1}, making it the strongest known bond in a stable molecule; the bond length is 1.13 Å.

Carbon dioxide is normally present to the extent of about 0.03 per cent by volume in the atmosphere, from which it is removed by photosynthesis. It is obtained in the laboratory from a carbonate and acid or, more commonly, in the form of *dry ice*, solid carbon dioxide (which sublimes at -78 °C under atmospheric pressure) made by allowing the liquid to expand rapidly. On an industrial scale the dioxide is obtained by heating limestone (cf. Section 10.5) or as a by-product in the manufacture of hydrogen from natural gas (Section 9.4). The C=O bond energy term in carbon dioxide is 806 kJ mol^{-1} and the C=O bond length in the linear molecule is 1.16 Å. A simple molecular orbital description of the molecule is as follows. If the molecular axis is, as usual, taken as the z axis, the s and p_z orbitals of the three atoms are used to form two σ bonding orbitals (each containing two electrons), two non-bonding orbitals (containing the two lone pairs of the oxygen atoms) and two antibonding orbitals (which are empty). Out of the sixteen valence electrons, eight are therefore

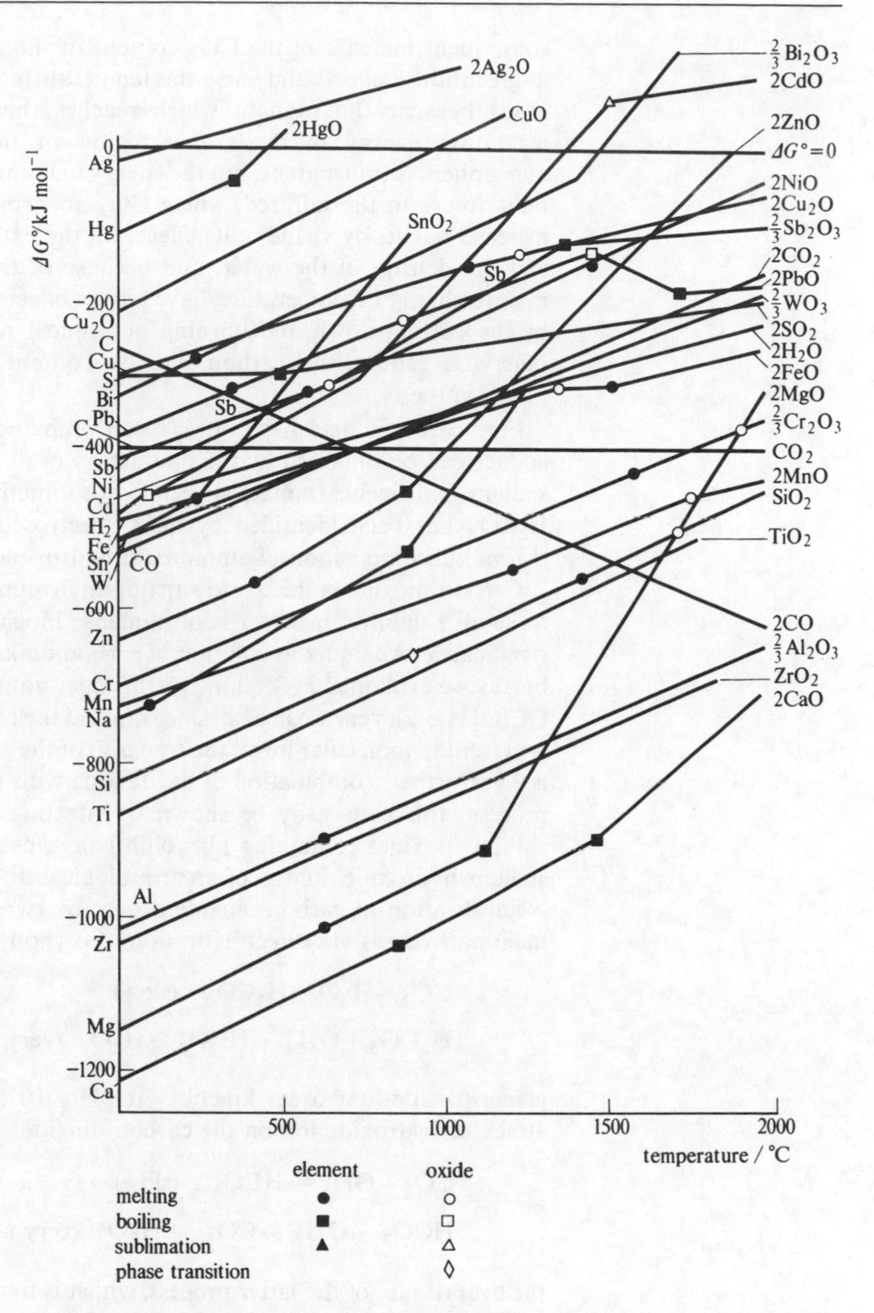

Figure 13.3
Variation of ΔG° with temperature for formation of some oxides per mol of O_2 consumed (Ellingham diagram).

available for π bonding. Each atom contributes its p_x and p_y orbitals to form π orbitals, so we have two filled bonding, two filled non-bonding, and two empty antibodning π orbitals, corresponding to a net bond order of two (one σ and one π) between each pair of atoms.

The increase in combustion of fossil fuels and decomposition of limestone for cement manufacture in recent years have given rise to fears that a consequent increase in the CO_2 content of the atmosphere may lead to a 'greenhouse effect' and raise the temperature of the atmosphere. This arises because the sunlight which reaches the earth's surface has its maximum energy in the visible region of the spectrum where the atmosphere is transparent; but the energy maximum of the earth's thermal radiation is in the infrared, where CO_2 absorbs strongly. Even a small increase might, by virtue of its effects on the extent of the polar ice caps and the density of the water, and because of the sensitivity of reaction rates to change of temperature, have serious effects. The danger is enhanced by the cutting down and burning of tropical rain forests which would otherwise reduce the carbon dioxide content of the atmosphere by photosynthesis.

Free carbonic acid has not yet been prepared, but an unstable ether adduct can be obtained by the interaction of dry hydrogen chloride and sodium hydrogencarbonate suspended in dimethyl ether at $-30\,°C$ and H_2CO_3 has been identified by mass spectroscopy as a product of the thermal decomposition of ammonium hydrogencarbonate at $120\,°C$.

Carbon dioxide is the world's major environmental source of acid and its small solubility in water is of immense biochemical and geochemical significance. An aqueous solution of carbon dioxide is only weakly acidic, but, as we explained in Section 7.4, this does not mean that carbonic acid, $OC(OH)_2$, is a very weak acid, since most of the carbon dioxide in solution is present in molecular form; the true pK_a of the acid for its first ionisation is 3.9. Further, combination of the dioxide with water is a relatively slow process; this can easily be shown by titrating a saturated solution of carbon dioxide, containing phenolphthalein as indicator, with aqueous sodium hydroxide, and is of great biological and industrial importance. Neutralisation of carbon dioxide occurs by two routes. For $pH < 8$ the main pathway is via direct hydration of carbon dioxide

$$CO_2 + H_2O \rightarrow H_2CO_3 \quad \text{(slow)}$$

$$H_2CO_3 + OH^- \rightarrow HCO_3^- + H_2O \quad \text{(very fast)}$$

giving pseudo-first order kinetics. At $pH > 10$ the main pathway is via attack of hydroxide ion on the carbon dioxide.

$$CO_2 + OH^- \rightarrow HCO_3^- \quad \text{(slow)}$$

$$HCO_3^- + OH^- \rightarrow CO_3^{2-} + H_2O \quad \text{(very fast)}$$

the overall rate of the latter process, which is first order in both CO_2 and OH^-, being the greater.

The carbonate ion is planar, with all carbon–oxygen bonds of length 1.29 Å; since the electronic structure on molecular orbital theory indicates net π-bonding corresponding to one filled bonding π orbital embracing all four atoms, and on valence bond theory is a hybrid of three structures each containing one C═O and two C—O^- bonds, this bond length (which is much greater than that in CO_2) is satisfactorily accounted for on the basis of a carbon–oxygen bond order of $1\frac{1}{3}$.

Most metal carbonates other than those of the alkali metals are sparingly soluble in water. The normal and acid carbonates of sodium and potassium have been discussed in some detail in Section 10.5, and the carbonates of the alkaline earth metals in Section 11.5. Potassium peroxodicarbonate, $K_2C_2O_6$, is obtained by electrolysis of a solution of potassium carbonate at -10 °C, using a high current density (a general method for the preparation of peroxo salts). The pale blue salt is believed to contain the anion $[^{-}O(O)C{-}O{-}O{-}C(O)O^{-}]^{2-}$, analogous to the peroxodisulphate ion. Peroxocarbonates are also believed to be intermediates in the reaction of carbon dioxide with superoxides, e.g.

$$4NaO_2 + 2CO_2 \rightarrow 2Na_2CO_3 + 3O_2$$

This reaction is used for the regeneration of oxygen in closed systems such as submarines and rocket capsules.

A third oxide of carbon, the suboxide C_3O_2, may be mentioned briefly. It is a gas (b.p. 6 °C) which is obtained by dehydration of malonic acid with phosphorus pentoxide at 150 °C, and which at room temperature rapidly polymerises. The molecule is linear and its structure is approximately represented by the formulation O=C=C=C=O.

There are, of course, numerous other oxo-acids of carbon, some of which have been mentioned briefly in Section 7.4 in connection with acidity, but they are generally regarded as belonging to the realm of organic chemistry and will not be discussed in this book.

13.8 Oxides, oxo-acids and hydroxides of silicon, germanium, tin and lead

The dioxides of all these elements are involatile solids. Silica, SiO_2, occurs in many different forms, nearly all of which have macromolecular structures built up of SiO_4 tetrahedra sharing all oxygen atoms. At atmospheric pressure there are three polymorphic forms, each stable within a characteristic temperature range, but each also having a low-temperature (α) and a high-temperature (β) modification, the transition temperatures being as shown below:

$$\begin{array}{ccccccc}
\beta\text{-quartz} & \underset{\text{slow}}{\overset{870\,^\circ\text{C}}{\rightleftharpoons}} & \beta\text{-tridymite} & \underset{\text{slow}}{\overset{1470\,^\circ\text{C}}{\rightleftharpoons}} & \beta\text{-cristobalite} & \underset{\text{slow}}{\overset{1710\,^\circ\text{C}}{\rightleftharpoons}} & \text{liquid} \\
573\,^\circ\text{C} \Big\| \text{fast} & & 120\text{–}160\,^\circ\text{C} \Big\| \text{fast} & & 200\text{–}275\,^\circ\text{C} \Big\| \text{fast} & & \\
\alpha\text{-quartz} & & \alpha\text{-tridymite} & & \alpha\text{-cristobalite} & &
\end{array}$$

The cubic β-cristobalite structure was described in Section 6.3; the silicon atoms are arranged as in diamond, and the oxygen atoms are somewhat off the lines joining the silicon atoms. In all of the other species shown above the environment of any one silicon atom is the same as in β-cristobalite, but the ways in which the tetrahedra are joined up are different in the different polymorphs, and the relative orientations of joined tetrahedra are different in the different modifications. Changes between polymorphs require the initial breaking of Si—O bonds and therefore require higher temperatures than changes between modifications of any one polymorph. When liquid silica is cooled slowly it does not crystallise but forms a glass; this is again a giant molecule built up from SiO_4 tetrahedra, but it consists of a random array of chains, sheets and three-dimensional units without the long-range order that characterises a crystal. Only a few oxides form glasses (e.g. B_2O_3, GeO_2, P_2O_5 and As_2O_5 in addition to SiO_2), since for such a random arrangement to occur (i) the coordination number of the element to oxygen must not be two (which would lead to a chain) or more than four (which would lead to too rigid a structure), (ii) only one oxygen atom must be shared by any two atoms of the element (sharing of more than one would also lead to rigidity). When the silica glass has cooled to about 1500 °C, it is plastic and may be worked in an oxy-hydrogen flame to give apparatus that, owing to the low coefficient of thermal expansion of silica, is highly insensitive to thermal shock. Pyrex glass, which has a somewhat lower melting-point, contains 10–15 per cent of boron oxide; in soda glass added alkali has converted some of the Si—O—Si links in the silica network into thermal Si—O groups and the melting-point is lower still.

Several forms of silica are of economic importance. α-Quartz has an interlinked helical chain structure; it shows optical activity arising from the fact that in any one crystal the helices can be either right-handed or left-handed; it is also piezoelectric and hence is very widely used in crystal oscillators and filters for frequency control, and in electromechanical devices such as pickups, microphones and loudspeakers. *Kieselguhr* is an earthy form of silica much used in filtration plants.

Silica gel is obtained from the gel formed when acid is added to an aqueous solution of sodium silicate by drying it until it contains only a few per cent by weight of water. Because of its large surface area : weight ratio it is a good drying agent and heterogeneous catalyst.

No definite silicic acid has yet been characterised, though esters of formula $Si(OR)_4$, obtained from ROH and $SiCl_4$, are well known. Although silica in its normal forms is only very slowly attacked by alkali, silicates are readily formed by fusion with alkali; some of them are discussed in the next section. Silica is not attacked by acids other than hydrofluoric acid, which dissolves it with formation of fluorosilicate. Silica occurs in small quantities in plants; it is not poisonous to animals, but when finely divided it damages the lungs and causes silicosis; finely divided silicate minerals (e.g. asbestos dust) have a similar effect.

In all the forms of silica mentioned so far, the Si—O bond length is approximately 1.61 Å and the Si—O—Si bond angle approximately 144°, nearly the same as in $(SiH_3)_2O$. By heating silica under very high pressures, however, a rutile form, in which the interatomic distance is 1.79 Å, may be obtained; this is much more dense and much less reactive than ordinary forms of silica.

Germanium dioxide closely resembles silica, and exists in both quartz and rutile structures; it dissolves in concentrated hydrochloric acid to form the $[GeCl_6]^{2-}$ ion, and in alkalis to form germanates. Tin dioxide and lead dioxide crystallise only with the rutile structure. Tin dioxide is readily made by oxidation of the metal, but lead dioxide is formed only by the use of powerful oxidising agents (e.g. alkaline hypochlorite) on lead(II) compounds. It is a powerful oxidant whose reactions are discussed further in Section 13.13. Tin dioxide, when freshly precipitated, is soluble in many acids, but lead dioxide shows no basic properties. Stannates and plumbates, e.g. $K_2[Sn(OH)_6]$ and $K_2[Pb(OH)_6]$, may be crystallised from solutions of the dioxides in aqueous potassium hydroxide. Tin dioxide is a widely used opacifier for enamels and glazes.

Germanium monoxide is prepared by dehydration of the yellow hydrate obtained by the action of aqueous ammonia on the dichloride. Tin monoxide, which is very sensitive to oxidation, is best made by thermal decomposition of the oxalate SnC_2O_4; lead monoxide (which exists in two forms, red and yellow) results from heating the metal in air above 550 °C or from dehydration of the white hydroxide $Pb(OH)_2$ precipitated from solutions of lead(II) salts. All three monoxides are amphoteric, but the oxo-anions formed from them are not well characterised.

The structure of GeO is not known, but SnO and red PbO have layer lattices in which each metal atom is bonded to four oxygen atoms at the corners of the base of a square pyramid of which the metal forms the apex; each oxygen atom is bonded to four metal atoms. It is generally believed that the inert electron pair of the metal ion occupies an orbital pointing away from the oxygen atoms and is responsible for this asymmetric structure.

A third oxide of lead, Pb_3O_4 (*red lead*), is obtained by heating the monoxide in air at 450 °C. This is diamagnetic, and hence can contain only Pb(II) and Pb(IV); in its crystal structure two types of lead atoms can be discerned, and it is really a mixed oxide. Nitric acid decomposes it to lead(II) nitrate and lead dioxide; glacial acetic acid reacts to form lead(II) acetate and the tetraacetate $Pb(CH_3COO)_4$, and the latter, an important reagent in organic chemistry, may be separated by crystallisation.

13.9 Silicates

Fusion of silica with metal oxides, hydroxides or carbonates leads to the formation of a wide range of silicates. Sodium silicates, for example, made by heating sand with sodium carbonate at 1300 °C, are soluble in water if the sodium content is high, and the alkaline solution (*water glass*) contains ions such as $[SiO(OH)_3]^-$ and $[SiO_2(OH)_2]^{2-}$; if the sodium content is low, large polymeric ions are present and the products are insoluble, like the silicates of most other metals and the enormous number that occur naturally. Equilibrium between the different anions is attained rapidly at pH > 10, more slowly in less alkaline solutions.

The earth's crust is very largely composed of silica and silicates, which form the principal constituents of all rocks and of the sands, clays and soils that are the breakdown products of rocks. Most inorganic building materials are based on silicate minerals. These include natural silicates such as sandstone, granite and slate, and manufactured materials such as cement, concrete and ordinary glass (which is made by fusing together limestone, sand and sodium carbonate). Clays are used in the ceramics industry, asbestos in fireproofing, zeolites as ion-exchangers and molecular sieves, and mica as an electrical insulator. Some of these materials are too complex to be discussed at the level of this book, but we shall give some indication of the enormous range of silicate structures known and of the relatively simple principles which underlie them. Some of the structures described here will turn up again when we discuss polyphosphates in the next chapter.

First, however, we should mention that it is the universal practice in describing silicate structures to adopt a purely ionic model; silicon is always written as Si^{4+}, even though so high a charge is most unlikely on ionisation energy grounds and is incompatible with the commonly observed Si—O—Si bond angle of about 140°. On the scale of the ionic radii given in Table 6.2, which include those for O^{2-} (1.40 Å), Na^+ (1.02 Å), Ca^{2+} (1.00 Å), Mg^{2+} (0.72 Å) and Al^{3+} (0.53 Å), the radius of Si^{4+} is 0.40 Å, i.e. it is about the same size as Al^{3+}. If, however, Al^{3+} replaces Si^{4+} in a crystal structure (a common occurrence), an extra singly charged cation must be present to maintain electrical neutrality. Thus in the felspar *orthoclase*, $KAlSi_3O_8$, the infinite anion of composition $AlSi_3O_8^-$ has the structure of quartz, with one-quarter of the silicon replaced by aluminium; the K^+ ions are accommodated in the relatively open $AlSi_3O_8^-$ structure. Double replacements such as ($Na^+ + Si^{4+}$) by ($Ca^{2+} + Al^{3+}$) are also common.

A very few silicates are now known to contain six-coordinated silicon (cf. the recent isolation of a rutile form of silica itself), but the overwhelming majority have structures based on SiO_4 tetrahedra; these may occur singly or, by sharing oxygen atoms, in small groups, cyclic groups, infinite chains, infinite layers, or infinite three-dimensional frameworks. Sharing always involves only corners: in silicate structures two SiO_4 units never have two O^{2-} ions, i.e. an edge of the tetrahedra, in common; such a structure would be unstable owing to the close approach of the O^{2-} ions that it would entail. Of the metal ions most commonly occurring in silicates, Be^{2+} is always surrounded by four oxide ions, Al^{3+}

by four or six, Mg^{2+}, Fe^{3+}, Ti^{4+} by six, Na^{+} by six or eight, and Ca^{2+} by eight; radius ratio considerations (Section 6.5), even though they rest on the questionable assumption that all species present are simple ions, therefore provide a very useful guide to silicate structures.

The simplest silicates contain the SiO_4^{4-} ion and are represented by Mg_2SiO_4 (*olivine*) and the synthetic β-Ca_2SiO_4, an important constituent of cement, which sets to a hard mass when finely ground and mixed with water; one of the products of this reaction has the composition $Ca_2[SiO_3(OH)](OH)$. When two SiO_4 units share an oxygen the resulting ion is $Si_2O_7^{6-}$; this occurs in $Sc_2Si_2O_7$ (*thortveitite*). The cyclic anions $Si_3O_9^{6-}$ and $Si_6O_{18}^{12-}$, in which each SiO_4 unit shares two corners, occur in $Ca_3(Si_3O_9)$ (α-*wollastonite*) and $Be_3Al_2(Si_6O_{18})$ (*beryl*) respectively; the cyclic anion $Si_4O_{12}^{8-}$ occurs in a synthetic potassium salt. The structures of most of these ions, together with some more complicated ones, are shown in Fig. 13.4. Conformational details (e.g. the fact that $Si_4O_{12}^{8-}$ has the chair form, are (as is usual in diagrams of silicate structures) omitted.

Short chain polysilicate anions are uncommon, though $Si_3O_{10}^{8-}$ occurs in a few rare minerals. If the SiO_4 units sharing two corners form an infinite chain, the Si:O ratio is again 1:3; this is found in $CaSiO_3$ (β-*wollastonite*) and $CaMg(SiO_3)_2$ (*diopside*, a member of the *pyroxene* group of minerals). Although their structures both contain infinite chain anions, the relative orientations of adjacent SiO_4 units of the chain are different in these two substances. Asbestos consists of a group of fibrous minerals some of which, e.g. $Ca_2Mg_5(Si_4O_{11})_2(OH)_2$ (*tremolite*), contain the double chain anion of formula $Si_4O_{11}^{6-}$. More extended cross-linking of the chains leads to layer anions of composition $Si_2O_5^{2-}$; the rings within the layers may contain four, six, eight or twelve silicon atoms or there may be a mixture of rings of different sizes. Such sheets occur in micas, though OH^- anions are very often also present; *talc*, for example, which is noted for its softness, has the composition $Mg_3(Si_2O_5)_2(OH)_2$, in which the Mg^{2+} ions are sandwiched between composite layers each containing $Si_2O_5^{2-}$ sheets and OH^- ions. The whole sandwich may thus be represented roughly as $[(Si_2O_5)^{2-},(OH^-),(Mg^{2+})_3(OH^-)(Si_2O_5)^{2-}]$; this is electrically neutral, so talc cleaves very readily parallel to the sandwich and is used as a dry lubricant, e.g. in toilet preparations.

Infinite sharing of all four oxygen atoms leads to the composition SiO_2, but partial replacement of silicon by aluminium leads to three-dimensional anions of composition $[AlSi_nO_{2n+2}]^{-1}$ and the felspar *orthoclase*, as we mentioned earlier, contains K^+ ions in an anion framework of composition $AlSi_3O_8^-$; *celsian* similarly contains Ba^{2+} ions in a framework of composition $Al_2Si_2O_8^{2-}$ (the doubled formula is used to emphasise the relationship to orthoclase). In felspars the holes in the structure which accommodate the cations are quite small; in zeolites they are much larger, and can accommodate not only monatomic cations but also larger species such as water, carbon dioxide, methanol and

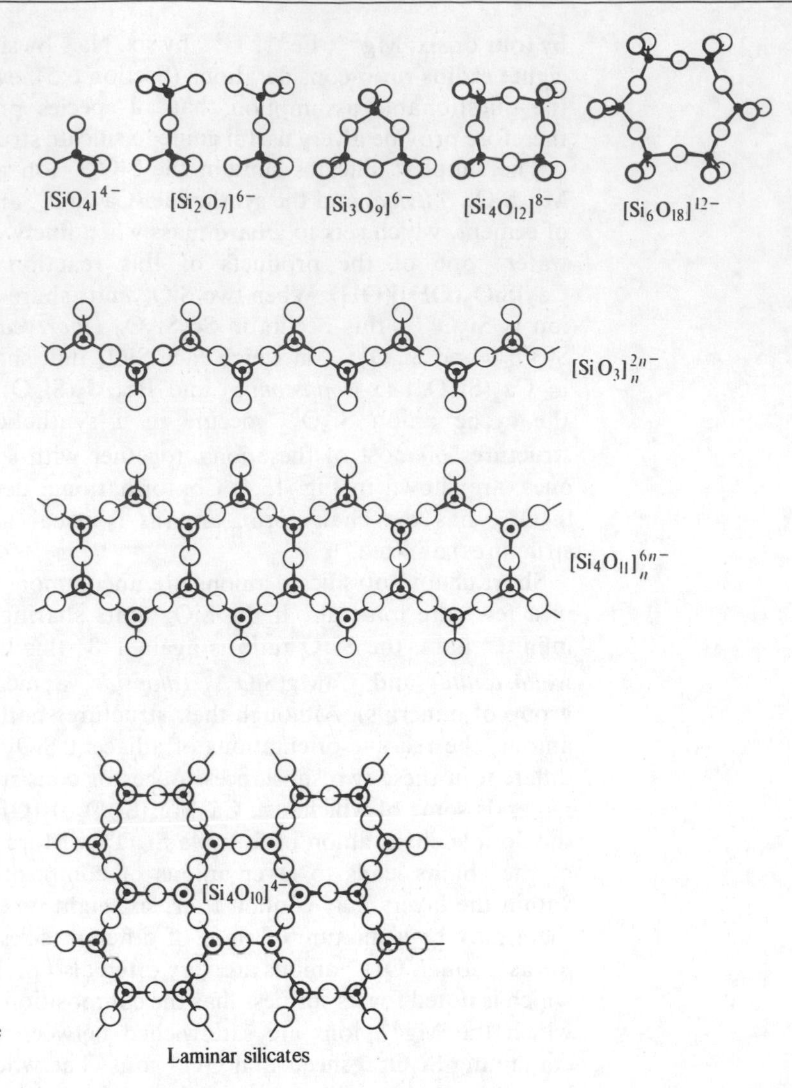

Figure 13.4
Structures of some silicate anions (negative charges are omitted).

hydrocarbons. The positive ions may be replaced by others, e.g. two sodium ions in *thomsonite*, $Na_5[Al_5Si_5O_{20}].6H_2O$ by one calcium ion; the original material may be regenerated by washing the zeolite with a concentrated solution of sodium chloride, and hence is useful in water-softening, though synthetic materials have now largely replaced natural zeolites for this purpose. Molecular sieves are usually synthetic zeolites in which the holes in the structures are capable of accommodating some molecules but not others and which therefore function as selective absorbents; in this way, for example, n-butane may easily be separated from iso-butane. Molecular sieves are also widely used as drying agents and as catalysts.

13.10 *Silicones*

Although the silicones are organometallic compounds they are conveniently mentioned at this point because of their structural similarities to the silicates. Hydrolysis of Me_3SiCl, Me_2SiCl_2 and $MeSiCl_3$ (all of which are obtained by passing methyl chloride over a copper–silicon alloy at 300 °C) would be expected to lead to Me_3SiOH, $Me_2Si(OH)_2$ and $MeSi(OH)_3$ respectively. On the basis of analogies with carbon compounds the first of these would be expected to be stable (except with respect to dehydration to an unsaturated compound at higher temperatures) and the other two to undergo dehydration with formation of $Me_2Si{=}O$ and $MeSiOOH$; as we have seen, however, the Si=O bond is unstable with respect to transformation into singly bonded structures, and in fact the methylsilicon chlorides on hydrolysis yield products containing the tetrahedral groups

$$\mathrm{Me{-}\underset{Me}{\overset{Me}{\overset{|}{\underset{|}{Si}}}}{-}O{-}} \qquad \mathrm{{-}O{-}\underset{Me}{\overset{Me}{\overset{|}{\underset{|}{Si}}}}{-}O{-}} \quad \text{and} \quad \mathrm{{-}O{-}\underset{\underset{|}{O}}{\overset{Me}{\overset{|}{\underset{|}{Si}}}}{-}O{-}}$$

(in which O really represents half an atom of oxygen, since one O is shared by two silicon atoms) respectively. These arise from intermolecular condensation, with elimination of water, from molecules containing Si—OH bonds, e.g.

$$2Me_3SiCl \rightarrow 2Me_3SiOH \rightarrow Me_3SiOSiMe_3 + H_2O$$

Diols formed by hydrolysis of Me_2SiCl_2 can condense to give chains or rings, e.g.

$$\cdots O{-}\underset{Me\ Me}{Si}{-}O{-}\underset{Me\ Me}{Si}{-}O{-}\underset{Me\ Me}{Si}{-}O\cdots \quad \text{and} \quad \text{cyclo-}(Me_2SiO)_3$$

These are the analogues of the chain and ring silicates, an $—O^-$ of the latter being replaced by a —Me group. Hydrolysis of $MeSiCl_3$ leads to a cross-linked polymer.

In practice, mixtures of methylsilicon chlorides are co-hydrolysed, or the initial products of hydrolysis are 'equilibrated' by heating with sulphuric acid, which catalyses the conversion of ring polymers into chain polymers and brings about redistribution of $—OSiMe_3$ end-groups. Thus when linear oligomers of formula

$$HOSiMe_2(OSiMe_2)_nOSiMe_2OH$$

are equilibrated with $Me_3SiOSiMe_3$ (from hydrolysis of Me_3SiCl) methyl silicone polymers of formula

$$Me_3Si(OSiMe_2)_nOSiMe_3$$

are formed. These are useful for waterproofing: the polar Si—O bonds orient themselves near the surface and the organic groups present a water-repelling exterior. They are more thermostable and less viscous than hydrocarbons of comparable molecular weight, and are valuable lubricants and hydraulic fluids. Cross-linking, achieved by co-hydrolysis of Me_2SiCl_2 and $MeSiCl_3$, leads, after heating at 250 °C, to the formation of silicone resins which are hard and inert; a smaller degree of cross-linking gives silicone rubbers.

13.11 *Sulphides*

The disulphides of carbon, silicon, germanium and tin show the gradation in properties that might be expected to accompany the increasingly metallic character of the elements. Lead(IV) is too powerful an oxidising agent to coexist with sulphide ion, and PbS_2 is not known.

Carbon disulphide is a volatile liquid (b.p. 46 °C) with a very disagreeable odour, and is almost insoluble in water; it is made by heating charcoal in sulphur vapour at 900 °C or by passing methane and sulphur vapour over alumina at 675 °C. It is an excellent solvent which is used in the production of rayon and cellophane, but because of its volatility, poisonous character and low ignition temperature it is dangerous to work with. Carbon disulphide is, by a narrow margin, thermodynamically unstable with respect to hydrolysis to carbon dioxide and hydrogen sulphide, but this reaction (like the hydrolysis of carbon tetrahalides) is very slow. When shaken with alkali metal sulphide solutions, carbon disulphide dissolves readily, and thiocarbonates such as Na_2CS_3, analogous to Na_2CO_3, may be isolated. The free acid H_2CS_3 separates as an oil when concentrated hydrochloric acid is added to a solid thiocarbonate. Unlike carbon dioxide, carbon disulphide polymerises under high pressure, forming a black solid of structure

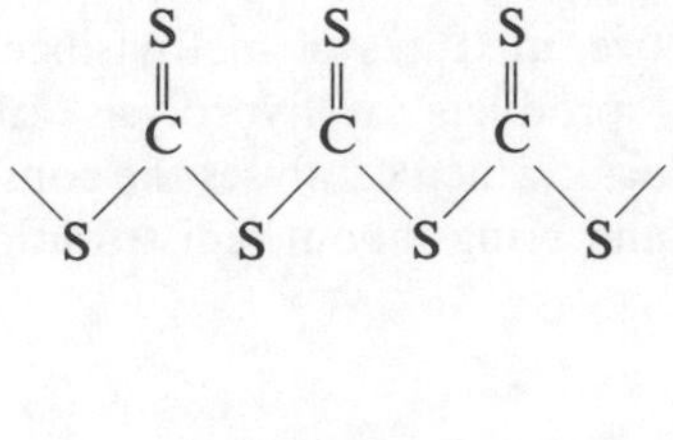

```
        S       S       S
        ‖       ‖       ‖
        C       C       C
  \   /   \   /   \   /   \   /
    S       S       S       S
```

Silicon disulphide, prepared by heating silicon in sulphur vapour, is a colourless fibrous solid in which chains of structure

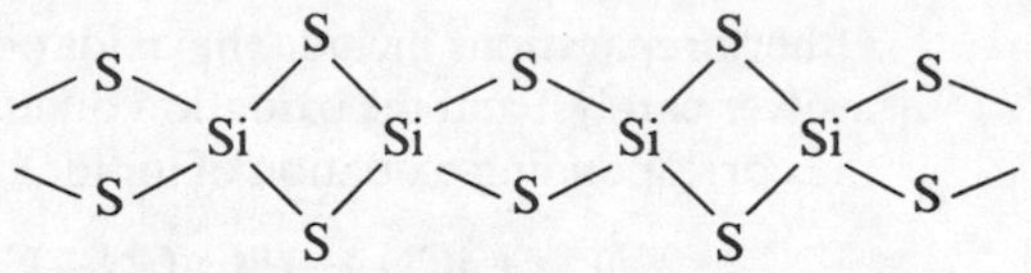

```
         S               S
   /S\  / \  /S\  / \  /S\
     Si    Si     Si    Si
   \S/  \ /  \S/  \ /  \S/
         S               S
```

are present. This sharing of edges of SiS_4 tetrahedra, it may be noted, has no parallel in the oxygen chemistry of silicon. The compound is instantly hydrolysed by water to silica and hydrogen sulphide.

The disulphides of germanium and tin are precipitated when hydrogen sulphide is passed into acidic aqueous solutions of germanium(IV) and tin(IV) compounds respectively. Germanium disulphide at high temperatures has a silica-like structure, but tin disulphide has a cadmium iodide type lattice. Both form thio-salts with alkali metal sulphides.

Germanium, tin and lead monosulphides may all be obtained by precipitation from aqueous media. The first two compounds have layer structures similar to that of black phosphorus, each element being pyramidally coordinated (cf. the effect of the inert pair of electrons on the structure of SnO). Lead sulphide, however, has the sodium chloride structure, though its black colour and extremely low solubility in water (K_{sp} is about 10^{-30}) suggest that it is certainly not a purely ionic compound.

13.12 *Cyanogen, its derivatives and silicon nitride*

Cyanogen, C_2N_2, and related compounds and ions are of great interest in inorganic chemistry. The CN radical is a *pseudo-halogen* (i.e. its chemistry is in many respects like that of a halogen); and although C_2N_2 and HCN are thermodynamically unstable with respect to decomposition into their elements, hydrolysis by water and oxidation by atmospheric oxygen, they and the cyanide anion are kinetically stable enough for them to have been studied extensively. The cyanide ion is an extremely important ligand in transition metal chemistry, and cyano complexes have figured prominently in the development of theories of bonding in metal complexes; moreover, CN^- is isoelectronic with two other very important ligands, CO and NO^+, and comparisons between complexes of these ligands have suggested several fruitful lines of research.

Cyanogen itself is a very poisonous colourless gas (b.p. −21 °C) which when pure is unchanged over long periods at the ordinary temperature, though with $\Delta H_f^\circ = +297$ kJ mol^{-1} it is one of the most endothermic compounds known. It can be made by heating mercury(II) cyanide with the chloride at 300 °C:

$$Hg(CN)_2 + HgCl_2 \rightarrow C_2N_2 + Hg_2Cl_2$$

Other preparations include the oxidation of hydrogen cyanide by air over a silver catalyst and the oxidation of aqueous cyanide by the Cu^{2+} ion (cf. the corresponding oxidation of iodide):

$$2Cu^{2+} + 4CN^- \rightarrow 2CuCN + C_2N_2$$

Cyanogen has a linear structure commonly written N≡C—C≡N, but the shortness of the C—C bond (1.38 Å) indicates considerable electron delocalisation. It burns in air with a very hot flame. Resemblances to the halogens include thermal dissociation into CN radicals at high temperatures and hydrolysis by alkali to form cyanide and cyanate:

$$C_2N_2 + 2OH^- \rightarrow OCN^- + CN^- + H_2O$$

Hydrogen cyanide, HCN, is also extremely poisonous; it is a colourless liquid (b.p. 26 °C) with a very high relative permittivity (107 at 25 °C) which arises from strong hydrogen bonding. It is usually prepared on a small scale by the action of dilute acid on sodium cyanide; industrially, it is obtained from methane and ammonia by the processes

$$2CH_4 + 2NH_3 + 3O_2 \xrightarrow[1000\text{–}1200\,^\circ C]{Pt/Rh} 2HCN + 6H_2O$$

$$CH_4 + NH_3 \xrightarrow[1200\text{–}1300\,^\circ C]{Pt} HCN + 3H_2$$

The pure liquid polymerises readily to the compounds $HC(NH_2)(CN)_2$ and $(H_2N)(NC)C{=}C(CN)(NH_2)$ and polymers of high molecular weight; in the presence of traces of water and ammonia it forms adenine (6-aminopurine), and on reduction it is converted into methylamine. Hydrogen cyanide is thought to have been one of the small molecules in the early atmosphere of the earth, and to have played an important role in the formation of many biologically important compounds. Among its many synthetic reactions may be mentioned the catalytic addition to butadiene to give adiponitrile for Nylon manufacture. In aqueous solution hydrogen cyanide, in contrast to the hydrogen halides, is a very weak acid (pK_a 9.5), and it is slowly hydrolysed to ammonium formate.

Sodium cyanide, much the most important salt of the acid, is now made by neutralisation of the acid. It is a very important reagent in organic chemistry, and is used industrially in the extraction of silver and gold; many cyano complexes are readily made from aqueous cyanide and other transition metal compounds. Mercury(II) cyanide, like mercury(II) chloride, is soluble in water to form a solution of the unionised compound; silver cyanide is sparingly soluble.

Sodium and potassium cyanides crystallise at the ordinary temperature in the sodium chloride structure, the CN^- ion behaving as a freely rotating ion of effective radius about 1.9 Å, i.e. about the same as that of Br^-; at low temperatures, however, transitions to structures of lower symmetry occur.

As we have stated earlier, cyanide is oxidised to cyanogen by mild oxidising agents such as Cu(II); more powerful oxidants (e.g. neutral permanganate) convert it into cyanate, OCN^-. Potassium cyanate may be obtained by heating the cyanide with lead monoxide. Fulminates contain the isomeric anion CNO^-; they can be reduced to cyanides but not made by oxidation of them. Mercuric fulminate, a dangerous initiatory explosive, is made by the action of alcohol and nitric acid on mercury or by the interaction of mercuric chloride and the sodium salt of nitromethane:

$$2CH_2NO_2^- + HgCl_2 \rightarrow Hg(CNO)_2 + 2H_2O + 2Cl^-$$

Among other cyanogen derivatives we shall mention only cyanogen chloride, ClCN (b.p. 13 °C), made by the action of chlorine on sodium cyanide, and the thiocyanates such as KSCN, obtained by fusing KCN with sulphur. Thiocyanate, like cyanide, forms many complexes with transition metal ions, some bonded to the metal through sulphur and others through nitrogen (see Section 18.6).

Silicon tetrachloride reacts with ammonia to form the amide $Si(NH_2)_4$, which when heated gives first the imide $Si(NH)_2$ and then silicon nitride, Si_3N_4. Silicon nitride is usually made, however, by the action of coke and nitrogen on silica at 1500 °C. It is a very hard and chemically inert substance, a very good insulator, which has recently found considerable application as a ceramic. It has an infinite three-dimensional structure in which silicon and nitrogen form four tetrahedral and three approximately planar bonds respectively; the structure is also that of the mineral *phenacite*, Be_2SiO_4, in which two-thirds of the silicon atoms of Si_3N_4 are replaced by beryllium. Another new refractory is Si_2N_2O, obtained by heating silica, silicon and nitrogen at 1450°C. This consists of puckered hexagonal nets of alternating silicon and nitrogen atoms, the nets being linked by Si—O—Si bonds; again the silicon and nitrogen atoms have coordination numbers of four and three respectively.

13.13 *Aqueous solution chemistry and oxo-acid salts of tin and lead*

As implied by the value $E^\circ = +0.15$ V for the half-reaction

$$Sn^{4+}(aq) + 2e \rightleftharpoons Sn^{2+}(aq)$$

tin(II) salts (or *stannous* salts as they are still sometimes called) in aqueous solution are very readily oxidised by oxygen; further, hydrolysis to $[Sn_3(OH)_4]^{2+}$ and other species is extensive. Solutions of tin(II) salts in water are therefore usually acidified, in which case complex anions are present, e.g. $[SnCl_3]^-$ in a solution of tin(II) chloride in dilute hydrochloric acid. In alkaline media an oxo- or hydroxo-anion, probably $[Sn(OH)_3]^-$, is formed.

Tin(IV) compounds (*stannic* compounds) are also extensively hydrolysed in aqueous solution unless sufficient acid is present to complex the tin, e.g. as $[SnCl_6]^{2-}$. From alkaline solutions stannates containing the $[Sn(OH)_6]^{2-}$ ion may be isolated; potassium stannate, for example, has the analytical composition, and is often formulated, $K_2SnO_3.3H_2O$, but is actually $K_2[Sn(OH)_6]$.

Little is known about tin salts of oxo-acids, which can be obtained pure only from non-aqueous media. Tin(IV) nitrate, $Sn(NO_3)_4$, for example, made by the action of dinitrogen pentoxide on the tetrachloride, is a volatile solid in which the metal atom is eight-coordinated, the NO_3^- ions acting as bidentate ligands.

Lead(II) salts are much more stable with respect to both hydrolysis and oxidation. The most important soluble oxo-salts are the nitrate and the acetate; the fact that many insoluble lead(II) compounds are dissolved by a mixture of ammonium acetate and acetic acid shows that lead(II) is very strongly complexed by acetate. Most lead(II) oxo-salts are, like the halides, sparingly soluble in water; lead sulphate, like barium sulphate (with which it is isomorphous) dissolves in concentrated sulphuric acid.

The ion Pb^{4+} does not exist as such in aqueous solution, but the aqueous chemistry of lead(IV) enters into the operation of the familiar lead–acid battery, a group of six cells each capable of producing about 2 V. The electrodes are lead–antimony plates holding spongy lead and lead(IV) oxide alternately, the electrolyte sulphuric acid. The cell reaction during discharge is

$$Pb(s)+PbO_2(s)+2H_2SO_4(aq) \rightarrow 2PbSO_4(s)+2H_2O(l)$$

and is accompanied by a decrease in the specific gravity of the electrolyte; it is, of course, reversed during recharging. The relevant half-reactions are:

$$PbO_2+SO_4^{2-}+4H^++2e \rightleftharpoons PbSO_4+2H_2O \qquad E^\circ = +1.68 \text{ V}$$

$$PbSO_4+2e \rightleftharpoons Pb+SO_4^{2-} \qquad E^\circ = -0.35 \text{ V}$$

For the half-reaction

$$PbO_2+4H^++2e \rightleftharpoons Pb^{2+}+2H_2O \qquad E^\circ = +1.45 \text{ V}$$

the fourth-power dependence of the potential on the hydrogen-ion concentration accounts at once for the way in which the relative stabilities of lead(IV) and lead(II) depend upon the acidity of the solution, e.g. lead dioxide oxidises concentrated hydrochloric acid to chlorine, but chlorine oxidises lead(II) in alkaline media to lead dioxide. It may be noted that thermodynamically PbO_2 should oxidise water at $[H^+]=1$, and the usefulness of the lead–acid battery depends upon a high overpotential for oxygen evolution.

Lead(IV) sulphate may be obtained as yellow crystals by electrolysis of fairly concentrated sulphuric acid using a lead anode; it is (like the tetraacetate and the complex chloride $(NH_4)_2[PbCl_6]$ mentioned in Section 13.6) hydrolysed to lead dioxide by cold water.

Lead dioxide dissolves in a concentrated solution of potassium hydroxide with formation of the hydroxo complex $K_2[Pb(OH)_6]$; on dilution PbO_2 is reprecipitated.

13.14 *Organometallic compounds*

The stability of organometallic compounds of Group IV elements towards all reagents decreases from silicon to lead. The great majority of these compounds are those of the tetravalent element, and these alone will be considered here.

We have already mentioned the direct synthesis of methylsilicon chlorides from copper–silicon alloy and methyl chloride in connection with the preparation of silicones (Section 13.10). Silicon tetraalkyl and tetraaryl derivatives, as well as alkyl- or aryl-silicon halides, may also be obtained by the action of lithium alkyls or aryls or Grignard reagents on silicon halides, the identity of the main products being dependent upon the molar ratio of reactants used. The stability of the alkyl–silicon bond is illustrated by the fact that chlorination of tetraethyl silicon gives $Si(CH_2CH_2Cl)_4$; chlorination of the tetraalkyl derivatives of germanium and tin gives alkyl germanium and tin chlorides. Except in this respect, the tetraalkyl-germanium and -tin compounds closely resemble those of silicon. All three elements now have an extensive organic chemistry.

Trimethyl tin compounds have structures quite different from those of the silicon and germanium compounds and contain planar Me_3Sn^+ ions, though these are usually so placed with respect to the anions (e.g. F^- or ClO_4^-) as to suggest anion bridging to give a trigonally bipyramidally coordinated tin atom. In aqueous solution the compounds are ionised to give the $[Me_3Sn(H_2O)_2]^+$ cation. Organo-tin compounds are employed as polyvinyl chloride stabilisers and agricultural pesticides.

Lead tetraethyl, which is widely used as an anti-knock agent in petrol, is made by treating lead–sodium alloy with ethyl chloride in an autoclave at 80–100 °C:

$$4NaPb + 4EtCl \rightarrow PbEt_4 + 3Pb + 4NaCl$$

Alternatively, electrolysis of solutions of $NaAlEt_4$ or EtMgCl using a lead anode may be employed. Lead tetraethyl begins to decompose at its boiling-point, 110 °C; its vapour, and the vapours of other lead compounds produced from it by the action of ethylene dichloride or dibromide added to prevent formation of a deposit of oxide, are dangerously toxic; the use of lead tetraethyl in petrol, although leading to more efficient use of the available energy of combustion, may give rise to dangerous atmospheric pollution, and is now being phased out.

PROBLEMS

1 Predict the shapes of the following molecules or ions: HCN, OCS, SiH_3^-, $SnCl_5^-$, Si_2OCl_6, $[Ge(C_2O_4)_3]^{2-}$.

2 Comment on each of the following observations:
(a) When an aqueous solution of potassium cyanide is added to one of aluminium sulphate, $Al(OH)_3$ is precipitated.
(b) The carbide Mg_2C_3 liberates the alkyne $CH_3C{\equiv}CH$ on treatment with water.
(c) Magnesium silicide reacts with ammonium bromide in liquid ammonia to form silane.
(d) The mineral *spodumene*, $LiAlSi_2O_6$, is isostructural with *diopside*, $CaMgSi_2O_6$, but when it is heated it is transformed into a polymorph having the quartz structure with the lithium ions in the interstices.
(e)

```
        SiMe2   Me
       /     \ /
  Me2Si       Si
       \     / \
        SiMe2   H
```

is hydrolysed by aqueous alkali at the same rate as the corresponding Si—D compound.

3 What would you expect to happen when:
(a) tin is heated with a concentrated aqueous solution of sodium hydroxide;
(b) sulphur dioxide is passed over lead dioxide;
(c) carbon disulphide is shaken with aqueous sodium hydroxide;
(d) dichlorosilane is hydrolysed by water;
(e) four moles of $ClCH_2SiCl_3$ react with three moles of $LiAlH_4$ in diethyl ether solution?

4 Suggest one method for the estimation of each of the following quantities:
(a) ΔH° for the conversion

$$GeO_2\,(quartz) \rightarrow GeO_2\,(rutile);$$

(b) the Pauling electronegativity value for silicon;
(c) the purity of a preparation of lead tetracetate.

REFERENCES FOR FURTHER READING

Ashcroft, S.J. and Beech, G. (1973) *Inorganic Thermodynamics*, van Nostrand, London. Contains good accounts of the thermochemistry of many of the compounds discussed in this chapter.

Ebert, L.B. and Selig, H. (1980) *Adv. Inorg. Chem. Radiochem.*, **23**, 281. A review of graphite intercalation compounds and their uses.

Greenwood, N.N. and Earnshaw, A. (1984) *Chemistry of the Elements*, Pergamon Press, Oxford. Chapters 8, 9, and 10 give a full account of Group IV elements, including technical applications of their compounds.

Powell, P. (1988) *Principles of Organometallic Chemistry*, Chapman and Hall, London. Chapter 4 gives a very good account of the organometallic chemistry of silicon, germanium, tin and lead.

Powell, P. and Timms, P.L. (1974) *The Chemistry of the Non-Metals*, Chapman and Hall, London. Contains good discussions of many aspects of carbon and silicon chemistry presented so as to emphasise comparisons with other elements.

Shriver, D.F., Atkins, P.W. and Langford, C.H. (1990) *Inorganic Chemistry*, Oxford University Press. Chapter 11 gives good accounts of many technical aspects of the chemistry of Groups III and IV elements.

Wells, A.F. (1984) *Structural Inorganic Chemistry*, 5th edition, Oxford University Press. Chapters 21–23 and 26 describe the structural chemistry of Group IV elements in detail.

14 Nitrogen, phosphorus, arsenic, antimony and bismuth

14.1 Introduction

The rationalisation of the properties of the Group V elements and their compounds is one of the most difficult problems of inorganic chemistry, despite some general similarities between these elements and those of Groups III and IV.

As we go down the group, for example, there are the overall increases in metallic characteristics and stabilities of lower oxidation states that we saw earlier for the boron and carbon groups. Ionisation energies (Table 14.1) increase rather sharply after removal of the *p* electrons; they decrease only slightly between phosphorus and arsenic (behaviour not very different from that found between aluminium and gallium, and between silicon and germanium) and, for removal of the *s* electrons at least, they increase between antimony and bismuth, just as they increased between indium and thallium and between tin and lead – bismuth, like thallium and lead, shows the 6*s* inert pair effect. Further, the enthalpy of atomisation decreases steadily from nitrogen to bismuth, just as it did from boron to thallium and from carbon to lead. Data for most single covalent bond energies (Table 14.2) also follow the Group IV pattern, e.g. nitrogen forms stronger bonds than phosphorus with hydrogen, but weaker bonds with halogens or oxygen. This fact, together with the absence of stable phosphorus-containing analogues of N_2, NO, HCN, N_3^- and NO_2^+, indicates that strong p_π–p_π bonding is more important for

Table 14.1
Some properties of nitrogen, phosphorus, arsenic, antimony and bismuth.

Element	N	P	As	Sb	Bi
Atomic number	7	15	33	51	83
Electronic configuration	$[He]2s^22p^3$	$[Ne]3s^23p^3$	$[Ar]3d^{10}4s^24p^3$	$[Kr]4d^{10}5s^25p^3$	$[Xe]4f^{14}5d^{10}6s^26p^3$
ΔH° atomisation/kJ mol^{-1}	473[b]	315	287	259	207
M.p./°C	−210	590 (red)	817 (grey)	630	271
B.p./°C	−196	280 (white)	615	1380	1560
First ionisation energy/kJ mol^{-1}	1403	1012	947	834	703
Second ionisation energy/kJ mol^{-1}	2855	1903	1800	1590	1610
Third ionisation energy/kJ mol^{-1}	4577	2910	2735	2440	2465
Fourth ionisation energy/kJ mol^{-1}	7473	4955	4830	4250	4370
Fifth ionisation energy/kJ mol^{-1}	9400	6270	6030	5400	5400
Covalent radius[a]/Å	0.74	1.10	1.22	1.41	1.52
Ionic radius M^{3+}/Å	—	—	—	—	1.03
Pauling electronegativity	3.0	2.2	2.2	2.1	2.0

[a] For three-coordinated element.
[b] Note this term refers to production of one mole of gaseous atoms from the element in its standard state; thus for nitrogen it is half the dissociation energy of N_2.

Table 14.2
Some covalent bond energy terms for compounds of nitrogen, phosphorus and arsenic in which these elements are three-coordinated ($kJ\ mol^{-1}$).

N—N	N—H	N—F	N—Cl	N—O
160	391	272	193	201
P—P	P—H	P—F	P—Cl	P—O
209	322	490	319	340
As—As	As—H	As—F	As—Cl	As—O
180	247	464	317	330
			Sb—Cl	
			312	
			Bi—Cl	
			280	

nitrogen, and that p_π–d_π bonding plays an important part in the chemistry of phosphorus. Finally, the non-existence of nitrogen pentahalides, and the slowness of the reaction between nitrogen trifluoride and water, are interpreted in terms of the lack of *d* orbitals in the valence shell of nitrogen; conversely, the existence of phosphorus pentahalides and the reactivity of the phosphorus trihalides arise from the availability of valence shell *d* orbitals to permit the presence of more than eight electrons in the valence shell of phosphorus. In all these ways the comparison between nitrogen and phosphorus follows that between carbon and silicon.

The biggest difference between Groups IV and V lies in the relative strengths of the N≡N (in N_2) and N—N (in N_2H_4) bonds compared with the relative strengths of the C≡C and C—C bonds: *E*(N≡N), 946 kJ mol^{-1}, is nearly six times *E*(N—N), 160 kJ mol^{-1}, whilst *E*(C≡C), 813 kJ mol^{-1}, is only slightly more than twice *E*(C—C), 346 kJ mol^{-1}. There is some uncertainty about *E*(N=N) arising from the difficulty of choosing a reference compound, but it is generally accepted as being about 400 kJ mol^{-1}, i.e. more than twice *E*(N—N); *E*(C=C), on the other hand (598 kJ mol^{-1}) is substantially less than twice *E*(C—C). Thus of the isoelectronic species N_2 and C_2H_2, the former is thermodynamically stable, the latter thermodynamically unstable, with respect to polymerisation to singly-bonded species. Indeed, the very high N≡N bond energy makes many nitrogen compounds endothermic, and nearly all the others only very slightly exothermic. The strength of the bond in P_2, an unstable species observed only at high temperatures, is 490 kJ mol^{-1}; P_2 is P≡P, so this bond is only rather more than twice as strong as P—P, and P_2, like C_2H_2, is unstable with respect to polymerisation. The very high strength of the N≡N bond, one σ plus two π bonds, has already been discussed in Chapter 4, but the weakness of the N—N single bond (compared with the C—C and P—P bonds as well as with the N≡N bond) calls for comment. It is found that O—O and F—F (146 and 159 kJ mol^{-1} in H_2O_2 and F_2) are also very weak bonds, much weaker than S—S and Cl—Cl respectively. Now the N—N, O—O and F—F bonds in N_2H_4, H_2O_2 and F_2 differ from the C—C bond in C_2H_6 by having non-bonding lone pairs on the nitrogen, oxygen and fluorine atoms, and it is believed that these bonds are weakened by repulsion between the lone pairs (in N_2 the lone pairs are, of course, at opposite ends of the molecule and do not interact). When, however, nitrogen is singly bonded to an atom with no lone pairs (e.g. hydrogen) the bond is a strong one (though, as we pointed out in Section 5.7, the extra energies of single covalent bonds above the mean of the single bond energies in the parent elements are often attributed to the partial ionic character of such bonds). Lone pairs on larger atoms (e.g. P, S, Cl) are further apart and lead to less repulsion.

Another difference between nitrogen and later elements that has no counterpart in Group IV is the ability of nitrogen, arising from its high electronegativity, to take part in strong hydrogen bonding. Ammonia, for example, is much less volatile than phosphine or arsine and much more soluble in water; methane, silane and germane, on the other hand, show very much less variation in such properties.

Very little of the chemistry of nitrogen and its homologues is that of simple ions. Although nitrides and phosphides that react with water are usually formulated as containing N^{3-} and P^{3-} ions, electrostatic considerations make it very doubtful whether these formulations are correct. There is no instance of the existence of a monatomic cation of any of the Group V elements in a +v oxidation state, and indeed the only definite case of a tripositive monatomic cation of one of these elements in a chemical environment appears to be that of Bi^{3+}. So nearly all the chemistry of these elements is that of covalently bonded compounds. Not only is the thermochemical basis of the chemistry of such species much harder to establish than that of ionic compounds, but they are also much more likely to be kinetically inert, not only to substitution (e.g. NF_3 to hydrolysis, $H_2PO_2^-$ to deuteration), but also to oxidation or reduction when these processes involve the making or breaking of covalent bonds as well as transfer of electrons. Thus nitrogen, for example, forms a whole range of oxo-acids and oxo-anions, and in aqueous media can exist in all oxidation states from +v to −III (NO_3^-, N_2O_4, NO_2^-, NO, N_2O, N_2, NH_2OH, N_2H_4, NH_3); tables of redox potentials (calculated from thermodynamic data) or oxidation state diagrams are of limited use in summarising the relationships between these species, for although they tell us about the thermodynamics of the possible reactions, they tell us nothing about their rates. Much the same is true of the chemistry of phosphorus. In this book we shall mention only a small fraction of the inorganic compounds of nitrogen and phosphorus that are now known, but even so we shall be more concerned with kinetic factors than we have been in any chapter so far.

Naturally occurring nitrogen, which forms 78 per cent by volume of the atmosphere of the earth, contains 0.36 per cent of ^{15}N, which is useful for isotopic labelling and may be obtained in concentrated form by chemical exchange processes (Section 1.7). The only naturally occurring isotope of phosphorus is ^{31}P, but ^{32}P, made by a (n,p) reaction on ^{32}S, is a suitable radioactive tracer of half-life 14 days; ^{31}P has a nuclear spin quantum number $I = \frac{1}{2}$ and a large nuclear magnetic moment, and has been used extensively in nuclear magnetic resonance spectroscopy. Tracer work involving arsenic, antimony and bismuth is of minor importance.

Because of the availability of nitrogen in the atmosphere and the need for it in living organisms (in which it is present as proteins), the fixing of nitrogen in forms in which it can be assimilated by plants is of great importance. It has not yet proved possible to devise a synthetic process to emulate the action of various bacteria (containing iron and iron–molybdenum proteins) that live on root nodules of leguminous plants, though a number of interesting reactions of molecular nitrogen (described in the next section) have been discovered. The only major natural source of suitable combined nitrogen is crude sodium nitrate (Chile saltpetre) that occurs in rainless areas of South America.

Phosphorus is also an essential constituent of plant and animal tissue; calcium phosphate occurs in bones and teeth, and phosphate esters of nucleotides are of immense biological significance. The most important sources of phosphorus are calcium phosphate in the form of *apatite*, $Ca_5(OH)(PO_4)_3$, and *fluorapatite*, $Ca_5F(PO_4)_3$, and secondary deposits of $Ca_3(PO_4)_2$ itself, formed from apatite by weathering; most of this is used for the manufacture of the more soluble acid phosphate fertilisers, the rest for the production of the element and its compounds. Arsenic, antimony and bismuth occur in small quantities as the sulphides As_4S_4 (*realgar*) and As_2S_3 (*orpiment*), Sb_2S_3 (*stibnite*), and Bi_2S_3 (*bismuthinite*); arsenic is commonly obtained (as arsenic(III) oxide) as a by-product in the extraction of other metals from sulphide ores. All three metals are used in alloys; arsenic compounds, which are highly toxic, are used in weedkillers, sheep-dips, sprays and as poisons for vermin.

14.2 The elements

Nitrogen is prepared industrially by the fractional distillation of liquid air; so obtained, it contains a little argon and traces of oxygen. The latter may be removed by addition of a small amount of hydrogen and passage over a platinum catalyst, or by bubbling through an aqueous solution of chromium(II) chloride. Small quantities of the pure gas may be obtained by thermal decomposition of sodium azide.

Nitrogen is generally unreactive, and its chief use is for the provision of an inert atmosphere. It combines slowly with lithium at ordinary temperatures, and with magnesium, the alkaline earth metals, aluminium and many transition metals when heated. Calcium carbide reacts at 800 °C to form the calcium salt of cyanamide, which, since it is hydrolysed to ammonia by moisture, may be used as a fertiliser:

$$CaC_2 + N_2 \rightarrow CaNCN + C$$
$$CaNCN + 3H_2O \rightarrow CaCO_3 + 2NH_3$$

Small amounts of nitric oxide are formed when nitrogen and oxygen are passed through an electric discharge at about 3000 °C, but as an economic process for the utilisation of atmospheric nitrogen this reaction is much less suitable than the reversible combination with hydrogen under less extreme conditions (the Haber process, discussed in the following section). Among recently discovered reactions of nitrogen which take place at ordinary temperatures are (i) the reduction to hydrazine by vanadium(II) and magnesium hydroxides and (ii) the formation of complexes analogous to the metal carbonyls (N_2 and CO are isoelectronic), e.g. by the reaction

$$CoH_3(PPh_3)_3 + N_2 \rightarrow CoH(PPh_3)_3N_2 + H_2$$

Atomic nitrogen, obtained by passing nitrogen at low pressures through an electric discharge, is very reactive, as would be expected, and combines with many elements (including mercury, sodium and sulphur) that are not affected by molecular nitrogen.

Phosphorus exhibits complicated allotropy; no fewer than eleven forms have been reported, but we shall confine attention here to the three best-characterised ones – white, black and red phosphorus. When calcium phosphate is heated with sand and coke in an electric furnace at 1400 °C, phosphorus vapour distils out and is condensed under water to yield white phosphorus.

$$2Ca_3(PO_4)_2 + 6SiO_2 + 10C \rightarrow P_4 + 6CaSiO_3 + 10CO$$

The lattice of white phosphorus consists of P_4 tetrahedra with an interatomic distance of 2.21 Å and an interbond angle of 60°. This allotrope is a soft waxy solid which becomes yellow on exposure to light; it melts at 44 °C and boils at 287 °C, giving a vapour also consisting of P_4 molecules; this is dissociated to P_2 to an appreciable extent only above 800 °C. White phosphorus is very poisonous; it is practically insoluble in water, but it dissolves in benzene, phosphorus trichloride and, especially, carbon disulphide. When exposed to air it slowly oxidises with a greenish glow and formation of the oxide PO_2 (Section 14.8) and some ozone; the chain reaction involved is an extremely complicated one. White phosphorus inflames in air above 50 °C, producing phosphorus pentoxide, and it combines violently with all the halogens. Concentrated nitric acid slowly oxidises it to phosphoric acid, and hot aqueous sodium hydroxide converts it into phosphine and sodium hypophosphite, the main reaction being

$$P_4 + 3NaOH + 3H_2O \rightarrow 3NaH_2PO_2 + PH_3$$

though hydrogen and a little of the spontaneously inflammable diphosphine, P_2H_4, are also produced.

Black phosphorus, the most stable allotrope, is obtained by heating white phosphorus at very high pressures. It resembles graphite in appearance and electrical conductivity, and has a double layer lattice in which each phosphorus atom is bonded to three others at 2.20 Å, the bond angles at the phosphorus atom being 99±3°. The shortest interatomic distance between layers is 3.9 Å. This allotrope is kinetically inert, and does not ignite in air even at 400 °C.

Red phosphorus is obtained by heating the white allotrope in an inert atmosphere at 250 °C, and is intermediate in reactivity between the black and white forms; it is not poisonous, is insoluble in organic solvents, and ignites in air only at 250 °C. Aqueous alkali does not attack it, and it reacts less vigorously than white phosphorus with halogens. Red phosphorus melts at 592 °C under pressure, but sublimes at 416 °C at 1 atm; like black phosphorus, it has a layer structure.

In view of the great difference in reactivity between red and white phosphorus it is interesting to note that ΔH° for the red phosphorus→white phosphorus transformation is only $+17.6\,\text{kJ mol}^{-1}$; the lower stability of the white form probably originates in the 60° bond angle in molecular P_4. Most of the chemical differences between the allotropes are differences in activation energies of their reactions.

Arsenic, antimony and bismuth may be extracted by roasting their sulphide ores in air to convert them into oxides, and reducing these with carbon. In the vapour phase all three metals exist as tetramers, and unstable yellow forms of solid arsenic and antimony probably also contain these units. The normal forms of the elements are grey solids which have corrugated layer lattices, each metal atom having three nearest neighbours and an interbond angle of about 96°. From arsenic to bismuth, however, although there is the expected substantial increase in interatomic distances within a layer, there is only a relatively small increase in the interlayer distance, and the structure of bismuth is in fact not very different from close packing. All three elements burn in air to give the trioxides and combine with the halogens; they are not attacked by non-oxidising acids, but react with concentrated nitric acid, forming As_2O_5, Sb_2O_5 and $Bi(NO_3)_3$ respectively. None of the elements reacts with aqueous alkali, but arsenic is attacked by fused sodium hydroxide with formation of sodium arsenite and hydrogen.

14.3 Hydrides

All of the Group V elements form trihydrides, and nitrogen and phosphorus also form N_2H_4 and P_2H_4; a third hydride of nitrogen, hydrogen azide (HN_3), will also be considered in this section. The variation in boiling-point among the trihydrides (NH_3, −33; PH_3, −88; AsH_3, −62; SbH_3, −18; BiH_3, estimated 17 °C) is one of the strongest pieces of evidence for hydrogen bond formation by nitrogen; ammonia also has a greater enthalpy of vaporisation and surface tension than its homologues. Thermal stability decreases down the group of hydrides, and bismuth hydride decomposes rapidly at room temperature; this behaviour is reflected in the bond energy terms for some of the hydrides given in Table 14.2. The formation of ammonia from the elements is slightly exothermic; all the other hydrides are endothermic compounds.

Ammonia is obtained by the action of water on the nitrides of lithium and magnesium, by heating ammonium salts with a base, or by reducing a nitrate or a nitrite in alkaline solution with zinc or aluminium. The other trihydrides are best made by acid hydrolysis of phosphides, arsenides etc., or by the action of lithium tetrahydridoaluminate on the corresponding chloride in ether. Since phosphonium iodide can be relatively easily made by dissolving phosphorus (in excess) and iodine in carbon disulphide, evaporating the solvent to leave a mixture of P_2I_4 and finely divided phosphorus, and dropping water on this in an inert atmosphere, the action of alkali on this salt provides a third method for the preparation of phosphine. On a large scale ammonia is always made from nitrogen and hydrogen (obtained as described in Section 9.4) by the Haber process, a classic application of physicochemical principles to an inorganic equilibrium.

For the reaction

$$N_2 + 3H_2 \rightleftharpoons 2NH_3$$

$\Delta H^\circ_{298} = -92\ \text{kJ mol}^{-1}$ and $\Delta G^\circ_{298} = -33\ \text{kJ mol}^{-1}$; the decrease in the number of gaseous molecules makes ΔS°_{298} negative. Increase in temperature reduces the yield of ammonia, but is necessary to increase the rate of the reaction; at a given temperature both the equilibrium yield and the reaction rate are increased by working at high pressures, and the rate can also be increased greatly by the use of a suitable catalyst: the rate-determining step is the dissociation of molecular nitrogen into atoms which are chemisorbed on the catalyst. The actual optimum conditions are a working temperature of 500 °C, a total pressure of at least 250 atm, and the presence of a finely divided iron catalyst containing promotors such as potassium and aluminium oxides. The ammonia formed is either liquefied or dissolved in water to form a saturated solution of specific gravity 0.880.

Ammonia is a colourless gas with a pungent odour. The molecule is pyramidal, but the barrier to inversion is very low (24 kJ mol^{-1}). Although liquid ammonia is not as strongly hydrogen-bonded as water, it nevertheless has a high enough enthalpy of vaporisation to be handled relatively easily in the laboratory despite its boiling-point of −33 °C, and to be a useful refrigerant. The properties of liquid ammonia as a solvent, and acid–base equilibria in liquid ammonia, were discussed at some length in Section 8.2.

Ammonia burns in oxygen to give nitrogen and water:

$$4NH_3 + 3O_2 \rightarrow 2N_2 + 6H_2O$$

In the presence of a platinum–rhodium catalyst at 750–850 °C, with a contact time of about a millisecond, however, the less exothermic reaction

$$4NH_3 + 5O_2 \rightarrow 4NO + 6H_2O$$

takes place. Since the nitric oxide so produced may be oxidised to nitric acid by the action of oxygen and water, the combination of this process with the Haber process provides a route of great importance for the conversion of atmospheric nitrogen into nitric acid.

Ammonia is more soluble in water than any other gas (doubtless because of hydrogen bond formation with the solvent), but nearly all of it dissolves unchanged; at 25 °C the equilibrium constant for the reaction

$$NH_3(aq) + H_2O \rightleftharpoons NH_4^+ + OH^-$$

is only 1.8×10^{-5}. This value is in accord with the fact that even a very dilute solution of ammonia has an odour, whereas dilute hydrochloric acid (in which hydrogen chloride is all converted into H_3O^+ and Cl^- ions) is odourless. Since K_w, the ionic product of water, is 10^{-14} at 25 °C, the hydrolysis constant (or as we may equally well describe it, the ionisation constant) of the ammonium ion, i.e. K for the reaction

$$NH_4^+ + H_2O \rightleftharpoons NH_3 + H_3O^+$$

must be 5.5×10^{-10} at this temperature. Thus solutions of ammonium salts of strong acids are slightly acidic. For the proton affinity of *gaseous* ammonia see Section 6.8.

The structures of ammonium salts are usually similar to those of the corresponding potassium, rubidium and caesium salts; the ionic radius of the NH_4^+ ion in a six-coordinated structure is 1.50 Å, almost exactly the same as that of Rb^+. Where hydrogen bonding is involved, however, ammonium salts have structures quite different from alkali metal salts, e.g. ammonium fluoride has the wurtzite structure. Nearly all ammonium salts are soluble in water, the hexachloroplatinate being one of the very few exceptions to this generalisation.

Ammonium salts are easily made by neutralisation of acids with aqueous ammonia, but industrial methods for the preparation of some of them merit brief comment. Thus ammonia may be converted into the chloride by means of the ammonia–soda process (Section 10.5) or into the sulphate by absorption in a suspension of calcium sulphate (anhydrite) into which carbon dioxide is passed:

$$CaSO_4 + 2NH_3 + CO_2 + H_2O \rightarrow CaCO_3 + (NH_4)_2SO_4$$

Passage of ammonia into nitric acid (itself made by catalytic oxidation of ammonia) is the usual preparation of ammonium nitrate. Both the sulphate and the nitrate are extensively used in fertilisers. The nitrate is also a constituent of some explosives: when heated at 200 °C it decomposes quickly into nitrous oxide and water, but above 250 °C decomposition is violent and at higher temperatures, or when the reaction

$$2NH_4NO_3 \rightarrow 2N_2 + O_2 + 4H_2O$$

is initiated by another explosive, detonation may result. Ammonium perchlorate, which is similarly metastable with respect to oxidation of the cation by the anion, is used as an oxidising agent in solid rocket propellants. Technical ammonium carbonate, used in 'smelling salts', is actually a mixture of the bicarbonate and carbamate, $NH_4HCO_3 \cdot NH_2COONH_4$, which smells strongly of ammonia because carbamic acid is an extremely weak acid; it is obtained by passing ammonia, carbon dioxide and steam into a lead chamber. At higher temperatures the interaction of carbon dioxide and ammonia in the presence of a little water gives urea:

$$2NH_3 + CO_2 \rightleftharpoons NH_2COONH_4 \rightleftharpoons NH_2CONH_2 + H_2O$$

Many complexes of transition metal ions and ammonia are mentioned in Chapters 19–21 and 24.

Phosphine is much less soluble than ammonia (the P—H bond is not polar enough to form hydrogen bonds to water), and the solution is neutral; the gaseous hydride has been shown in Section 6.8 to have a less negative enthalpy change than ammonia for the reaction with a proton, and (because the phosphonium ion is larger than the ammonium ion) the hydration enthalpy of PH_4^+ would be expected to be less negative than that of either NH_4^+ or H_3O^+. Phosphonium halides are thus decomposed by water even though the proton affinity of gaseous PH_3 is greater than that of gaseous H_2O (Section 6.8). It has long been known that phosphonium iodide is the most stable halide with respect to thermal dissociation into phosphine and hydrogen halide. If we consider the cycle shown in Fig. 14.1, we can see that the thermodynamic tendency of PH_3 to react with HX depends upon (i) the H—X bond energy (ii) the electron affinity of X and (iii) the lattice energy of PH_4X. Factor (i) has been discussed in Section 5.4; the bond which has to be broken is weakest for X = I. Factor (ii), discussed in Section 3.5, implies that X = Cl would be most likely to lead to PH_4X formation. Factor (iii) favours X = F, since the fluoride has the largest lattice energy of a series of halides. However, for a cation as large as PH_4^+ (which is about the same size as Cs^+) the difference between the lattice energies of the halides is smaller than the difference between the bond energies in the hydrogen halides; and since the variation in electron affinities among the halogens is much smaller than the variation in either of the other factors, the formation of PH_4X is most favoured by making X = I. Phosphonium chloride can be obtained at low temperatures, but is unstable above −30 °C; the bromide decomposes at 0 °C.

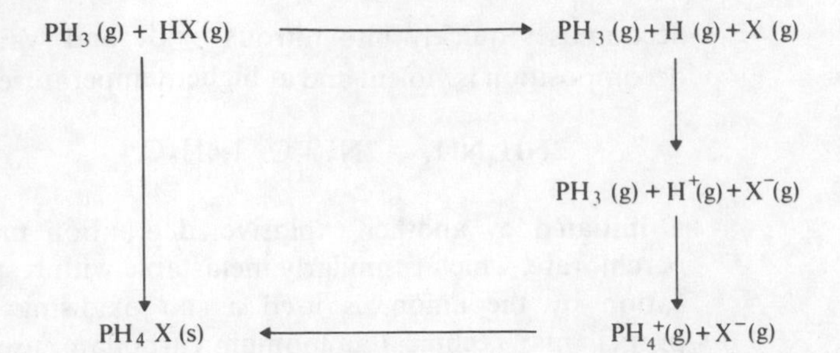

Figure 14.1
Cycle for the formation of a phosphonium halide from phosphine and a hydrogen halide.

Arsine and stibine closely resemble phosphine, but are less basic and less stable with respect to decomposition into their elements. Nevertheless, AsH_4I and SbH_4I can be shown to be formed from the hydrides and hydrogen iodide at 100 K.

Hydrazine, N_2H_4, is obtained by partial oxidation of ammonia with sodium hypochlorite. The reaction involves two stages:

$$NH_3 + NaOCl \rightarrow NH_2Cl + NaOH \text{ (fast)}$$

$$NH_3 + NH_2Cl + NaOH \rightarrow N_2H_4 + NaCl + H_2O \text{ (slow)}$$

Glue or gelatin is added to inhibit the side-reaction

$$2NH_2Cl + N_2H_4 \rightarrow N_2 + 2NH_4Cl$$

which otherwise consumes the hydrazine as it is formed. Part of the purpose of the inhibitor is to remove by complex formation traces of transition metal ions that catalyse the last reaction, but its function is probably more complicated than this simple comment suggests. The dilute solution of hydrazine obtained in this way can be concentrated by distillation to a composition $N_2H_4.H_2O$, from which pure hydrazine can be made by distillation over solid sodium hydroxide. The free base, being weaker than ammonia, can also be made by the action of ammonia on hydrazinium sulphate.

Pure hydrazine melts at 2 °C and boils at 114 °C; both the pure liquid and its aqueous solutions are powerful reducing agents, being oxidised successively to diimide, N_2H_2, and nitrogen. The pure liquid explodes when heated, and alkyl derivatives of hydrazine have been used as rocket fuels. In aqueous solution hydrazine usually forms $N_2H_5^+$ salts, but some salts of the doubly charged ion $N_2H_6^{2+}$ can be isolated, e.g. $N_2H_6SO_4$, which is sparingly soluble in dilute sulphuric acid. Hydrazine is used industrially for the removal of oxygen from boiler water.

Three possible conformations for N_2H_4 and the related molecules P_2H_4, N_2F_4 and P_2F_4, are shown in Fig. 14.2; the lone pairs of electrons on the nitrogen or phosphorus atoms are not shown, but may be presumed to occupy symmetrical positions with respect to the bonding pairs. Hydrazine has the *gauche* conformation, and so has P_2H_4 in the gas phase; P_2H_4 in the solid state, however, has the staggered conformation, and N_2F_4 exists in

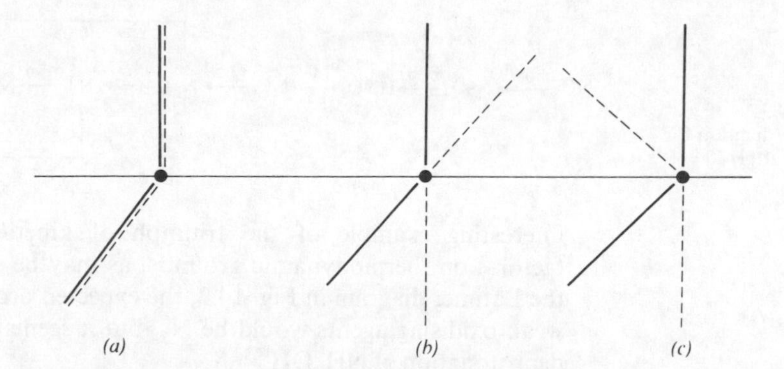

Figure 14.2
Possible conformations for the molecule N_2H_4 viewed along the N—N axis: (*a*) eclipsed; (*b*) staggered; (*c*) *gauche*.

both forms. The eclipsed conformation, which would maximise lone pair–lone pair repulsions, is not found in nitrogen or phosphorus compounds of this formula type. The N—N bond length in hydrazine is 1.45 Å, and the bond energy (derived from ΔH_f° on the basis of the assumption that the N—H bond energy is the same as in ammonia) is 160 kJ mol^{-1}.

Diphosphine, P_2H_4, boils at 51 °C, is spontaneously inflammable in air and has no basic properties; it is formed as a minor product in several reactions for the preparation of phosphine, e.g. the action of aqueous sodium hydroxide on white phosphorus or of water on calcium phosphide, and may be separated from phosphine by condensation in a freezing mixture.

Chloramine, NH_2Cl, is made by the interaction of gaseous ammonia and chlorine diluted with nitrogen, and is the cause of the odour of water containing nitrogenous matter that has been sterilised with chlorine. It is highly reactive; with dimethylamine it yields the rocket fuel dimethylhydrazine, Me_2NNH_2. Hydroxylamine, NH_2OH, is obtained by reduction of dinitrogen tetroxide in hydrochloric acid solution by hydrogen in the presence of a platinised charcoal catalyst, or by the action of concentrated sulphuric acid on nitromethane at 120 °C, a curious reaction of unknown mechanism:

$$CH_3NO_2 + H_2SO_4 \rightarrow [NH_3OH]HSO_4 + CO$$

The free base may be obtained from its salts by the action of sodium methoxide in methanol. Pure hydroxylamine melts at 33 °C and explodes at higher temperatures; it is a weaker base than ammonia or hydrazine. In aqueous solution it enters into a great variety of redox reactions. For example, it reduces Fe(III) in acidic solution, but oxidises Fe(II) in the presence of alkali:

$$2NH_2OH + 4Fe^{3+} \rightarrow N_2O + 4Fe^{2+} + H_2O + 4H^+$$

$$NH_2OH + 2Fe(OH)_2 + H_2O \rightarrow 2Fe(OH)_3 + NH_3$$

More powerful oxidising agents (e.g. bromate) oxidise it to nitric acid. The formation of N_2O in most oxidations of hydroxylamine is an

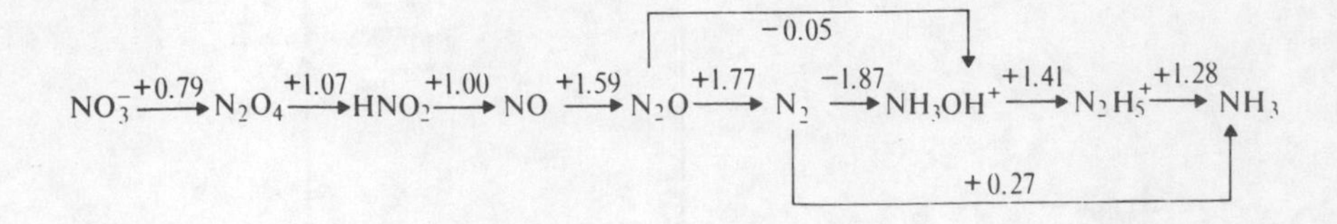

Figure 14.3
Potential diagram for nitrogen at $[H^+] = 1$.

interesting example of the triumph of kinetic over thermodynamic factors: on thermodynamic grounds, as may be seen from inspection of the Latimer diagram in Fig. 14.3, the expected product from the action of weak oxidising agents would be N_2, but it seems that reaction occurs by deprotonation of NH_2OH

$$NH_2OH \rightarrow NOH + 2H^+ + 2e$$

followed by combination of two NOH radicals to give HON=NOH (hyponitrous acid – mentioned again in Section 14.7) and decomposition of this to N_2O and water. Figure 14.3 also shows that at $[H^+] = 1$ hydroxylamine is unstable with respect to disproportionation into nitrogen and hydrazine or ammonia; a slow decomposition into nitrogen and ammonia does in fact take place.

Hydrogen azide, HN_3, is quite unlike ammonia and hydrazine, and in aqueous solution is a weak acid, pK_a 4.75. Sodium azide is prepared by passing nitrous oxide over molten sodium amide at 190 °C:

$$2NaNH_2 + N_2O \rightarrow NaN_3 + NaOH + NH_3$$

It can be separated by crystallisation from water. The acid may be obtained in dilute solution by ion-exchange; in the anhydrous form, obtained from the sodium salt by distillation with sulphuric acid, it is dangerously explosive: its standard enthalpy of formation is +290 kJ mol^{-1}. Silver and lead azides, which are insoluble in water, are also explosive; the latter is used as an initiator for less sensitive explosives. Azides of the alkali metals, on the other hand, decompose quietly when heated: lithium azide gives the nitride and nitrogen, but the others give the pure metal and nitrogen.

The azide group, like cyanide (though to a smaller extent) shows similarities to a halogen, and is another example of a pseudo-halogen. Azide complexes of transition metals and halogen compounds such as ClN_3 and BrN_3, all of which are explosive, are known, but N_6, analogous to the halogen molecules, has not yet been prepared.

Alkali metal azides contain the symmetrical linear anion $^-N{=}N^+{=}N^-$, which is isoelectronic with CO_2 and has a bond length of 1.15 Å. Protonation necessarily destroys the symmetry of the ion, and hydrogen azide has the structure

1.24 Å 1.13 Å
N—N—N
1.01 Å
H 120°

On the basis of a comparison of nitrogen–nitrogen bond lengths with those in diimide (1.23 Å) and nitrogen (1.10 Å) this may reasonably be written as a resonance hybrid of

$$\underset{\mathrm{H}}{\diagup}\overset{-}{\mathrm{N}}-\overset{+}{\mathrm{N}}\equiv\mathrm{N} \quad \text{and} \quad \underset{\mathrm{H}}{\diagup}\mathrm{N}=\overset{+}{\mathrm{N}}=\overset{-}{\mathrm{N}}$$

or as

$$\underset{\mathrm{H}}{\diagup}\mathrm{N}-\mathrm{N}=\mathrm{N}$$

with lone pairs on each nitrogen and a three-centre two-electron bonding molecular orbital extending over all three of them.

14.4 Nitrides, phosphides and arsenides

The classification of the nitrides poses a problem something like that encountered in dealing with carbides, though since there are fewer nitrides than carbides the overall picture is a clearer one, and nearly all nitrides fall into one of three groups: saline nitrides of the alkali and alkaline earth metals, beryllium, magnesium and aluminium; covalently bonded nitrides of the non-metallic elements; and macromolecular electron-deficient 'interstitial' nitrides of transition metals.

As we said in Section 14.1, whether the saline nitrides actually contain a monatomic ion carrying three negative charges must be considered doubtful; but the naïve formulation of Li_3N, Be_3N_2, Mg_3N_2, Ca_3N_2, Sr_3N_2, Ba_3N_2 and AlN as containing the N^{3-} ion is customary, and we shall adopt it here. The hydrolysis of these nitrides to ammonia and metal hydroxide is certainly compatible with, though it does not require, such a formulation. Lithium is the only alkali metal to combine with nitrogen, but all the Group II metals do so; some saline nitrides may also be obtained by thermal decomposition of metal amides, a process analogous to the thermal decomposition of metal hydroxides to oxides.

Nitrides of the non-metallic elements are described under the element concerned; the most important of them are boron nitride (Section 12.9), cyanogen (Section 13.12) and the nitrides of silicon (Section 13.12) and sulphur (Section 15.7).

Transition metal nitrides (e.g. TiN, VN) closely resemble the corresponding carbides in being hard, inert solids of very high melting-point; they look like metals and are good conductors of electricity. Nearly all of them have structures in which the nitrogen atoms occupy octahedral holes in a close-packed metal lattice; when all these holes are filled the stoichiometry corresponds to the formula MN, and many such substances have the sodium chloride structure. Nitrides of this kind are obtained by heating the metals in nitrogen or ammonia at about 1200 °C.

Many elements combine with phosphorus at high temperatures, and the products show the same wide range of properties and structures as the borides and silicides. Alkali and alkaline earth metals form phosphides such as Na_3P and Ca_3P_2 which are hydrolysed by water and may be considered as ionic. Alkaline earth metal phosphides of formula M_3P_{14} contain P_7^{3-} ions having the same structure as the isoelectronic sulphide P_4S_3 (Section 14.11). A few non-metals (e.g. sulphur) form discrete molecular species: the phosphorus sulphides, as they are usually called, are described briefly in Section 14.11. Transition metal phosphides are usually inert metallic-looking compounds which have high melting-points and conductivities; their formulae bear no relationship to ordinary oxidation states of the metals, and their structures may contain not only single phosphorus atoms, but P_2 groups, P_4 rings, or chains or layers of phosphorus atoms.

Metal arsenides resemble the corresponding phosphides. We shall not discuss most of them in this book, but since nickel arsenide, NiAs, gives its name to a well-known structure, it is appropriate to mention it briefly. If we describe the sodium chloride structure as cubic close-packed chloride ions with sodium ions in the octahedral holes, the nickel arsenide structure may be described as hexagonal close-packed arsenic atoms (the structure is certainly not a purely ionic one) with nickel atoms in the octahedral holes. Each nickel atom thus has six arsenic atoms as nearest neighbours (at 2.43 Å); however, because of the overall geometry of the structure it also has two nickel atoms only slightly further away (at 2.52 Å), and there is almost certainly nickel–nickel bonding running through the structure – hence the electrical conductivity of nickel arsenide. Each arsenic atom has six nickel atoms as nearest neighbours at the corners of a trigonal prism. This structure, a very common one for transition metal arsenides, antimonides, sulphides, selenides and tellurides, is illustrated in Fig. 14.4 (in which, however, it has been drawn so as to emphasise the

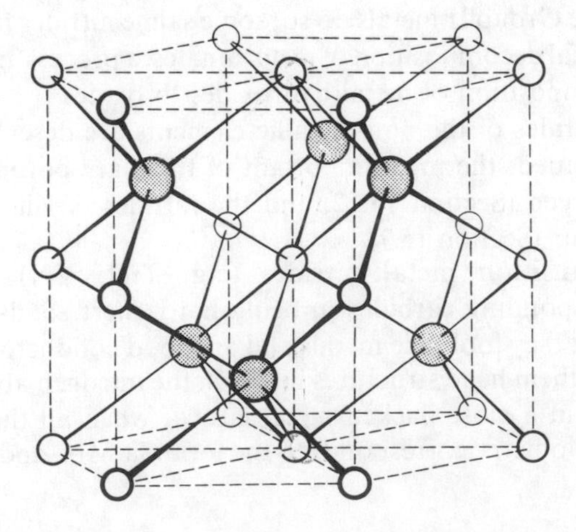

Figure 14.4
The structure of nickel arsenide (shaded circles are arsenic atoms) drawn so as to bring out the coordination of the central nickel atom.

coordination of the nickel atoms rather than the hexagonal close packing of the arsenic atoms).

14.5 Halides, oxohalides and complex halides

Among the halides of the Group IV elements, we noted an increasingly saline character and a decrease in the stability of the higher oxidation states of the elements as we progressed from carbon to lead. Since nitrogen is restricted to an octet of electrons in its valence shell, there are no pentahalides of this element, but otherwise the general trends among the Group IV halides are found in those of the Group V elements also, e.g. bismuth trifluoride is a salt, and bismuth pentafluoride (the only pentahalide of bismuth) an infinite chain polymer of BiF_6 octahedra sharing two opposite corners, whilst the fluorides of all the other elements are gases or volatile liquids or solids. Nitrogen halides, like carbon halides, are kinetically resistant to hydrolysis (though see below for the slow hydrolysis of nitrogen trichloride) and do not form complex anions. Halogen-containing cations of nitrogen, phosphorus and arsenic (e.g. NF_4^+, PCl_4^+, and AsF_2^+) are, however, well established. Bismuth differs from the other elements in the group by forming a subhalide containing a cluster of metal atoms: when bismuth trichloride is heated with molten bismuth a black solid $Bi_{24}Cl_{28}$ is obtained. X-ray diffraction shows that it contains two Bi_9^{5+} ions (each of which is a trigonal prism of six bismuth atoms with extra atoms outside the mid-points of the vertical faces), four square pyramidal $[BiCl_5]^{2-}$ ions, and a $[Bi_2Cl_8]^{2-}$ ion, the last consisting of two $BiCl_5$ square pyramids having a basal edge in common; the lone pairs are on opposite sides of the basal plane.

Another cluster cation, the trigonal bipyramidal Bi_5^{3+}, is known as the hexafluoroarsenate. The monatomic cation Bi^+, which also has no counterpart among the species known for the other elements in the group, occurs in the chlorohafnate $Bi_{10}Hf_3Cl_{18}$, which is really $Bi^+Bi_9^{5+}[HfCl_6^{2-}]_3$.

Nitrogen trifluoride, made by the controlled fluorination of ammonia in a copper-packed reactor or the electrochemical fluorination of amines in anhydrous hydrogen fluoride, is the only exothermic nitrogen trihalide ($\Delta H_f^\circ = -109$ kJ mol^{-1}). It is a colourless gas (b.p. -129 °C) which is unaffected by acids or alkalis but is decomposed by sparking with hydrogen:

$$2NF_3 + 3H_2 \rightarrow N_2 + 6HF$$

The structure of NF_3 and the fact that it has only a very small dipole moment (0.2 debye) have already been discussed (Section 5.7); the compound shows no donor properties. Nitrogen trichloride, an oily yellow liquid (b.p. 71 °C), is obtained by the action of chlorine on a concentrated aqueous solution of ammonium chloride, followed by extraction with carbon tetrachloride, or (diluted with air) for use in the bleaching of flour, by electrolysis of ammonium chloride solution at pH 4. It is strongly endothermic ($\Delta H_f^\circ = +230$ kJ mol^{-1}) and dangerously explosive. The difference in ΔH_f° between the trifluoride and the trichloride arises in part from the fact that the N—F bond is much stronger than the N—Cl bond, and in part from the greater bond strength in Cl_2 than in F_2. The trichloride is hydrolysed by alkali with formation of nitrogen and hypochlorite:

$$2NCl_3 + 6OH^- \rightarrow N_2 + 3OCl^- + 3Cl^- + 3H_2O$$

The reaction proceeds via formation of the ion Cl_2NClOH^- and decomposition of this to $NHCl_2$ and OCl^-. Nitrogen tribromide and triiodide resemble the chloride in both preparation and properties.

Other nitrogen fluorides are obtained by the reactions:

$$2NF_3 \xrightarrow[\text{Cu}]{400\,^\circ\text{C}} N_2F_4 + CuF_2$$

$$2N_2F_4 + 2AlCl_3 \xrightarrow{-70\,^\circ\text{C}} \textit{trans}\text{-}N_2F_2 + 3Cl_2 + N_2 + 2AlF_3$$

$$\downarrow 100\,^\circ\text{C}$$

$$\textit{cis}\text{-}N_2F_2$$

Tetrafluorohydrazine (b.p. −73 °C) is obtained as a colourless mixture of *gauche* and *trans* forms; in the gas phase it is reversibly dissociated into blue NF_2 radicals (cf. N_2O_4), which undergo many interesting reactions, e.g.

$$2NF_2 + S_2F_{10} \rightarrow 2F_2NSF_5$$

$$2NF_2 + Cl_2 \rightarrow 2NF_2Cl$$

$$NF_2 + NO \rightarrow F_2NNO$$

Salts of nitrogen–fluorine cations are obtained by the reactions:

$$NF_3 + F_2 + SbF_5 \rightarrow NF_4^+[SbF_6]^-$$

$$N_2F_4 + 2SbF_5 \rightarrow N_2F_3^+[Sb_2F_{11}]^-$$

$$N_2F_2 + AsF_5 \rightarrow N_2F^+[AsF_6]^-$$

Several oxofluorides and oxochlorides of nitrogen, all unstable gases or volatile liquids which are rapidly hydrolysed, are known. Nitrosyl fluoride, FNO, and nitrosyl chloride, ClNO, are obtained by combination of nitric oxide and the appropriate halogen. Nitryl fluoride, FNO_2, is obtained by the action of N_2O_4 on cobalt(III) fluoride at 300 °C:

$$N_2O_4 + 2CoF_3 \rightarrow 2FNO_2 + 2CoF_2$$

All these molecular oxohalides combine with suitable fluorides or chlorides to give salts of the NO^+ and NO_2^+ cations, e.g.

$$FNO + AsF_5 \rightarrow NO^+[AsF_6]^-$$

$$ClNO + SbCl_5 \rightarrow NO^+[SbCl_6]^-$$

$$FNO_2 + BF_3 \rightarrow NO_2^+[BF_4]^-$$

The complex fluorides may also be made conveniently in bromine trifluoride as solvent (Section 8.5). The principal factor involved in the change from a covalent simple halide to an ionic complex halide is believed to be the enthalpy of halide ion capture by boron trifluoride, arsenic pentafluoride or antimony pentachloride (cf. Section 6.8). By the action of the powerful fluorinating agent iridium hexafluoride on nitrosyl fluoride the unstable nitrogen(v) compound F_3NO (b.p. −85 °C) is obtained:

$$3FNO + 2IrF_6 \rightarrow 2NOIrF_6 + F_3NO$$

To preserve the octet restriction for nitrogen we may write the electronic structure of the product as $F_3N \rightarrow O$, but the shortness of the N—O bond (1.16 Å) suggests appreciable contributions from structures

```
F
 \
F—N⁺=O

F⁻
```

etc (cf. the structure of O_2F_2 in Section 15.4).

Phosphorus forms two series of halides: PF_3 (b.p. −95 °C), PCl_3 (b.p. 76 °C), PBr_3 (b.p. 173 °C) and PI_3 (m.p. 61 °C); PF_5 (b.p. −84 °C), PCl_5 (sublimes 163 °C) and PBr_5 (m.p. *c.* 100 °C). Most are made by direct combination, the product depending upon which element is present in excess; phosphorus trifluoride, however, cannot be prepared in this way and has to be obtained by a halogen exchange reaction such as

$$PCl_3 + AsF_3 \rightarrow PF_3 + AsCl_3$$

The trifluoride is much less rapidly hydrolysed than the other halides, and appears not to combine with metal fluorides to give salts of the $[PF_4]^-$ ion, which has not yet been characterised. It does, however, behave as an electron donor towards transition metals, forming complexes such as $Ni(PF_3)_4$ and $Cr(PF_3)_6$, in which, in addition to σ donation of the lone pair of electrons from the phosphorus atom to the metal atom being involved, there is back π donation from a filled orbital of the metal atom into an empty *d* orbital of the phosphorus atom. Such compounds, which are also formed by phosphorus trichloride, are closely related to the metal carbonyls, and are discussed further in Chapter 22.

Phosphorus trichloride is very rapidly hydrolysed to phosphorous and hydrochloric acids by water, and forms a complex with trimethylamine. Like the fluoride, it does not form complexes with alkali metal halides. It combines with oxygen to form the oxochloride $OPCl_3$, a liquid of b.p. 107 °C, which may also be obtained by heating the pentachloride with the pentoxide; the oxochloride is an important reagent for the preparation of phosphate esters.

Phosphorus pentafluoride has a trigonal bipyramidal structure with P—F (axial) = 1.577 Å and P—F (equatorial) = 1.534 Å; as we pointed out in Section 5.2, there is no way in which all positions of X atoms in a trigonal bipyramidal molecule PX_5 can be equivalent, and the observed difference in bond lengths can be rationalised on the basis of the bonding orbitals used by the phosphorus atom being divided into two groups, three hybrid sp_xp_y orbitals and two hybrid p_zd_z orbitals, the two sets of bonds formed having slightly different strengths. The ^{19}F nuclear magnetic resonance spectrum of PF_5 contains only a single line, showing that on the time-scale of this technique all the fluorine atoms are equivalent. This situation arises because of the ready availability of a simple intramolecular route for 'exchange' between axial and equatorial positions – see Fig. 1.4 in Section 1.5. Gaseous phosphorus pentachloride, if thermal dissociation is prevented by the presence of an excess of chlorine, has a similar structure; the solid pentachloride, however, has a structure closely related to that of caesium chloride, and contains tetrahedral $[PCl_4]^+$ and octahedral $[PCl_6]^-$ ions. The solid pentabromide is $[PBr_4]^+Br^-$. The mixed halide PF_3Cl_2 (in which the chlorines are in equatorial positions) is obtained as a gas (b.p. 7 °C) by addition of chlorine to the trifluoride, but as a solid (m.p. *c.* 130 °C) of structure $[Cl_4]^+[PF_6]^-$ by treatment of phosphorus pentachloride with arsenic trifluoride in arsenic trichloride solution.

Phosphorus pentafluoride, unlike the trifluoride, is a strong Lewis acid, and forms stable complexes with amines and ethers. Hexafluorophosphate ion, $[PF_6]^-$, may be made in aqueous medium by the action of concentrated hydrofluoric acid on phosphoric acid, oxofluoro complexes being intermediates; it is often used for the precipitation of salts of large organic cations. If potassium hydrogen fluoride, KHF_2, is heated with phosphorus pentachloride, solid KPF_6 is obtained, and its thermal decomposition affords a convenient preparation of phosphorus pentafluoride. The $[PF_6]^-$ ion is, of course, a regular octahedron.

The arsenic trihalides are similar to the corresponding phosphorus compounds in most respects, but a few differences between the trifluorides call for comment. Arsenic trifluoride, though hydrolysed by water, can be made by a reaction analogous to those used for the preparation of boron trifluoride and silicon tetrafluoride, the action of concentrated sulphuric acid on a mixture of calcium fluoride and arsenic(III) oxide:

$$As_2O_3 + 3CaF_2 + 3H_2SO_4 \rightarrow 2AsF_3 + 3CaSO_4 + 3H_2O$$

It forms complexes with both potassium fluoride and antimony pentafluoride and the structures of these complexes are $K^+[AsF_4]^-$ and $[AsF_2]^+[SbF_6]^-$ respectively; both increase the otherwise low conductivity of arsenic trifluoride, behaviour which is analogous to that of adducts of bromine trifluoride (cf. Section 8.5). The pentafluoride is the only stable pentahalide of arsenic, though the pentachloride can be obtained by the action of chlorine on the trichloride at −100 °C under the influence of ultraviolet radiation. Many complexes containing the $[AsF_6]^-$ ion are known. Arsenic trichloride may be made from the elements, but it is also formed, and distils out, when arsenic(III) oxide is heated with concentrated hydrochloric acid; water reverses the latter reaction.

Antimony trihalides are solids of low melting-point having structures which contain pyramidal molecules, though in each case the antimony atom has three more halogen atoms near enough to it to suggest some interaction between molecules. The trifluoride, usually made from antimony(III) oxide and concentrated hydrofluoric acid, is an important halogen-exchange reagent, converting many chlorides into the corresponding fluorides. It combines with alkali metal fluorides to form complexes containing a wide range of anions, though $[SbF_6]^{3-}$ is not among them: K_2SbF_5 contains the square pyramidal $[SbF_5]^{2-}$ ion, KSb_2F_7 the discrete $[SbF_4]^-$ anion (which is roughly a trigonal bipyramid with one equatorial position occupied by an unshared pair of electrons) and SbF_3 molecules, $KSbF_4$ the cyclic anion $[Sb_4F_{16}]^{4-}$ in which four SbF_5 groups each share two fluorine atoms, and $CsSb_2F_7$ the bridged anion $[F_3SbFSbF_3]^-$. Antimony trichloride is hydrolysed by water to an oxochloride SbOCl.

Antimony pentachloride (m.p. 3 °C) is obtained from the elements; the pentafluoride (m.p. 7 °C) may be made similarly or by the action of anhydrous hydrogen fluoride on the pentachloride:

$$SbCl_5 + 5HF \rightarrow SbF_5 + 5HCl$$

The solid pentafluoride contains tetramers in which each SbF_6 unit shares two adjacent corners; similar fluorine bridging in the liquid gives a polymeric structure which accounts for the very high viscosity of antimony pentafluoride. Both solid and liquid antimony pentachloride, on the other hand, contain individual trigonal bipyramidal molecules.

Both the pentafluoride and pentachloride form complexes with alkali metal halides; the latter gives rise only to $[SbCl_6]^-$, but the former forms $[SbF_6]^-$, $[Sb_2F_{11}]^-$ and $[Sb_3F_{16}]^-$, the last two species having fluorine bridging between SbF_6 units. Antimony pentafluoride is the most powerful fluoride ion acceptor known, and a very large number of salts of the $[SbF_6]^-$, $[Sb_2F_{11}]^-$ or $[Sb_3F_{16}]^-$ ions containing unusual cations (e.g. O_2^+, XeF^+, Br_2^+, ClF_2^+ and NF_4^+) are known. Antimony pentafluoride and hydrogen fluoride or fluorosulphuric acid form very powerful protonating mixtures which were mentioned in Section 8.4.

When antimony trichloride is partially oxidised by chlorine in the presence of caesium chloride, the dark blue compound Cs_2SbCl_6 is precipitated; the black salt $(NH_4)_2SbBr_6$ may be made by an analogous reaction. Both are isomorphous with $K_2[PtCl_6]$. Since these compounds are diamagnetic, they cannot contain Sb(IV); they are mixed oxidation state compounds like $GaCl_2$ ($Ga^+[GaCl_4]^-$ – Section 12.7) and contain Sb(III) and Sb(V), the dark colour arising from absorption of light associated with electron-transfer between the two anions present. In $(NH_4)_2SbBr_6$ these are $[SbBr_6]^{3-}$ (with Sb—Br 2.79 Å) and $[SbBr_6]^-$ (with Sb—Br 2.56 Å); both ions, it should be noted, are octahedral, despite the presence of a pair of unshared electrons on Sb(III). As for the isoelectric $[TeBr_6]^{2-}$ ion (Section 5.2), either a stereochemically inert 5*s* pair or six three-centre two-electron bonds may be involved.

Bismuth fluorides were mentioned at the beginning of this section. The trifluoride, a salt of m.p. 800 °C, and the molecular trichloride (m.p. 232 °C) are both hydrolysed by water to oxohalides containing layers of Bi^{3+}, O^{2-}, and F^- or Cl^- ions. The pentafluoride is the only pentahalide of bismuth; made by heating the trifluoride in fluorine, it is a powerful fluorinating agent that is rapidly and completely hydrolysed by water.

14.6 Oxides of nitrogen

There is so little in common between the oxides of nitrogen and those of later elements in Group V that it is desirable to follow the pattern set in the previous chapter and consider the oxides of the first element of the group (in the structures of which p_π–p_π bonding plays a dominant role) in a separate section. The oxo-acids of nitrogen are treated similarly. The known oxides are N_2O, NO, N_2O_3, NO_2 (in equilibrium with N_2O_4) and N_2O_5. In addition, it is useful at this point to discuss the cations NO^+ and NO_2^+ and their salts.

Before describing individual species in some detail it is helpful to recall the simple molecular orbital treatment of the N_2 and NO molecules (Sections 4.2 and 4.3 respectively). The bond order in N_2 is 3 (one σ and two π bonds), and all the bonding orbitals are fully occupied; in NO there is an extra electron, which has to be accommodated in an antibonding orbital, thereby reducing the bond order to 2.5. When NO is oxidised to the NO^+ ion, this extra electron is lost, and the bond length decreases from 1.15 Å to 1.06 Å; at the same time the NO vibration frequency rises from 1 876 cm^{-1} to about 2 300 cm^{-1}. Similar considerations apply when NO_2 (which has one more electron than CO_2) is oxidised to NO_2^+. It is also instructive to note that N_2O is isoelectronic with CO_2 and N_3^- and, like them, is linear; other examples of the isoelectronic principle that are helpful in remembering structures of nitrogen compounds include NO_2^-

and O_3 (both bent), NO_3^- and CO_3^{2-} (both planar), and N_2O_4 and $C_2O_4^{2-}$ (both planar). All the oxides of nitrogen are gaseous at the ordinary temperature, and all are slightly endothermic; redox potential data for the oxidation or reduction of some of them (or the products they form with water) have been given in Fig. 14.3, but it should be said again here that many systems involving nitrogen compounds are subject to kinetic control.

Dinitrogen monoxide or nitrogen(I) oxide, N_2O, commonly called nitrous oxide, is a colourless gas (b.p. −89 °C) which is nearly always prepared by decomposition of ammonium nitrate at 180–250 °C:

$$NH_4NO_3 \rightarrow N_2O + 2H_2O$$

An alternative method useful for preparation of a purer product is based on the aqueous reaction

$$NH_2OH + HNO_2 \rightarrow N_2O + 2H_2O$$

Nitrous oxide has a faint sweet odour and taste, and is soluble in about an equal volume of water at room temperature; although the oxide is formally the anhydride of hyponitrous acid, $H_2N_2O_2$, the solution is neutral, and in fact the position of the equilibrium

$$N_2O + H_2O \rightleftharpoons H_2N_2O_2$$

lies far to the left. The gas is a good supporter of combustion once reaction has started and decomposition of the oxide is under way, but at room temperature it is rather unreactive. At 190 °C it reacts with sodium amide to give sodium azide (Section 14.3). It is used as an anaesthetic and in the preparation of whipped cream. Bond lengths in the linear molecule are shown below:

$$N \overset{1.13\,Å}{\text{———}} N \overset{1.19\,Å}{\text{———}} O$$

Comparison with those in the N_3^- ion, NO and NO^+ indicates that the structure $N^-{=}N^+{=}O$ with some contribution from $N{\equiv}N^+{-}O^-$ is a good valence-bond representation of the molecule; alternatively we may write the structure as N=N—O with four lone pairs of electrons and a three-centre two-electron bond extending over all three atoms.

Nitrogen monoxide or nitrogen(II) oxide, NO, usually called nitric oxide, is made on an industrial scale by the catalytic oxidation of ammonia; in the laboratory it may be prepared by reduction of nitric acid by iron(II) salts or mercury in the presence of sulphuric acid:

$$NO_3^- + 3Fe^{2+} + 4H^+ \rightarrow NO + 3Fe^{3+} + 2H_2O$$

$$2NO_3^- + 6Hg + 8H^+ \rightarrow 3Hg_2^{2+} + 2NO + 4H_2O$$

The former reaction provides the basis of the 'brown ring' test for a nitrate; the NO formed yields a brown complex $[Fe(H_2O)_5NO]^{2+}$ with the excess of Fe(II) present; this decomposes when heated. Many complexes of nitric oxide with transition metals are known; they are described further in Chapter 22. Nitric oxide is a colourless gas (b.p. -152 °C) which is paramagnetic owing to the presence of the unpaired electron; unlike nitrogen dioxide, it does not dimerise to any appreciable extent at ordinary temperatures, though a dimer is probably an intermediate in the reactions

$$2NO + Cl_2 \rightarrow 2ClNO$$

$$2NO + O_2 \rightarrow 2NO_2$$

for which the rates are proportional to $p^2_{NO} \cdot p_{Cl_2}$ and $p^2_{NO} \cdot p_{O_2}$ respectively and decrease with rise in temperature. In the diamagnetic solid a loosely bonded dimer with a long N—N bond (2.18 Å) is present.

The reaction with oxygen is important in the manufacture of nitric acid. Ammonia is first oxidised to nitric oxide (Section 14.3); this is cooled and mixed with more air, and the resulting gas is absorbed in a countercurrent of water; the reactions involved include

$$2NO + O_2 \rightleftharpoons 2NO_2$$

$$2NO_2 \rightleftharpoons N_2O_4$$

$$N_2O_4 + H_2O \rightarrow HNO_3 + HNO_2$$

$$2HNO_2 \rightarrow NO + NO_2 + H_2O$$

$$3NO_2 + H_2O \rightarrow 2HNO_3 + NO$$

This gives nitric acid of concentration about 60 per cent by weight; it may be concentrated to 68 per cent by distillation. The 98 per cent acid may be obtained by oxidation of N_2O_4 with oxygen or air under pressure and in the presence of the quantity of water required for the reaction

$$2N_2O_4 + 2H_2O + O_2 \rightarrow 4HNO_3$$

Nitric oxide may be oxidised directly to nitric acid by acidified permanganate; sulphur dioxide reduces it to nitrous oxide, tin and acid reduce it to hydroxylamine. Although it is thermodynamically unstable with respect to decomposition into its elements, it does not decompose at an appreciable rate below about 1000 °C, and it is therefore usually a poor supporter of combustion. Since nitric oxide is an endothermic compound high temperatures favour its formation, and it is present in exhausts from cars and aeroplanes. Its oxidation products are important constituents of urban 'smog' and 'acid rain'. Its presence in the upper atmosphere has been considered to constitute an environmental hazard, since it decomposes the ozone layer (Section 15.2) that protects the earth from the sun's ultraviolet radiation:

$$NO + O_3 \rightarrow NO_2 + O_2$$

It now seems probable, however, that NO is less significant than atomic chlorine in this respect.

The formation of nitrosyl halides and their conversion into nitrosonium salts of complex halogeno acids were mentioned in Section 14.5. Many other nitrosonium salts are known; $NO^+HSO_4^-$, for example, is an intermediate in the lead-chamber process for the manufacture of sulphuric acid, and $NO^+ClO_4^-$ (which is isomorphous with ammonium perchlorate) is obtained when a mixture of NO and NO_2 is passed into pure perchloric acid. All are decomposed by water:

$$NO^+ + H_2O \rightarrow HNO_2 + H^+$$

Dinitrogen trioxide, N_2O_3, is obtained as a blue liquid (f.p. about −100 °C) by the interaction of 2NO and N_2O_4 at low temperatures; even at −78 °C extensive dissociation into nitric oxide and dinitrogen tetroxide takes place.

Dinitrogen tetroxide, N_2O_4, and nitrogen dioxide, NO_2, coexist in a rapidly established equilibrium

$$N_2O_4 \rightleftharpoons 2NO_2$$

The solid is diamagnetic and colourless, and consists of the pure dimer; it melts at −9 °C to a yellow liquid which contains a little of the paramagnetic monomer; at the boiling-point (21 °C), the brown vapour contains 15 per cent of NO_2. The colour of the vapour darkens up to about 140 °C, at which temperature dissociation is practically complete, and after that becomes lighter owing to a further reversible dissociation into nitric oxide and oxygen. Dinitrogen tetroxide is usually made in the laboratory by thermal decomposition of a heavy metal nitrate, e.g.

$$2Pb(NO_3)_2 \rightarrow 2PbO + 2N_2O_4 + O_2$$

It is a powerful oxidising agent which attacks many metals, including mercury, at room temperatures; it has been used, in conjunction with hydrazine as fuel, in rocket propellants. Water hydrolyses it to nitrous and nitric acids:

$$N_2O_4 + H_2O \rightarrow HNO_2 + HNO_3$$

In concentrated sulphuric acid it yields a mixture of nitrosonium and nitronium salts:

$$N_2O_4 + 3H_2SO_4 \rightarrow NO^+ + NO_2^+ + H_3O^+ + 3HSO_4^-$$

Other nitronium salts have been mentioned earlier (Sections 8.5 and 14.5). The N—O bond length in them is 1.15 Å, shorter than that in NO_2 because of the loss of an electron in an antibonding orbital.

In the solid state N_2O_4 has the planar structure

O O
N—1.75 Å—N) 135°
1.18 Å
O O

All the N—O distances are equal, and if the N—N bond is taken as a single bond, the N—O bond order is 1.5, there being two N—O σ bonds and a three-centre two-electron π bond per NO_2 group; alternatively, the NO_2 groups may be regarded as resonance hybrids with equally weighted contributing structures

$$-\overset{+}{N}(=O)-O^- \quad \text{and} \quad -\overset{+}{N}(-O^-)=O$$

The N—N bond is, however, much longer than that in hydrazine (1.45 Å); repulsion between positively charged nitrogen atoms has been suggested as the reason for this, but such an explanation is not entirely satisfactory, since in hydrazinium salts containing the $H_3N^+—N^+H_3$ ion there is no appreciable lengthening of the bond relative to that in hydrazine itself. The NO_2 molecule is bent, with N—O 1.20 Å and ONO bond angle 134°.

Liquid dinitrogen tetroxide is of considerable interest as a non-aqueous solvent; that aspect of its chemistry has already been described in Section 8.6.

Dinitrogen pentoxide, N_2O_5 (nitrogen pentoxide), is obtained by dehydration of nitric acid with P_2O_5 or by the oxidation of N_2O_4 with ozone. It is a colourless solid, stable below 0 °C, but slowly decomposing above this temperature into dinitrogen tetroxide and oxygen; it reacts vigorously with water to form nitric acid, and is a powerful oxidising agent. In the gaseous state and in solution in carbon tetrachloride it consists of N_2O_5 molecules, presumably with the structure O_2NONO_2, and the molecular form is obtained by sudden cooling of the vapour to −180 °C; the normal form of the solid, however, consists of linear NO_2^+ and planar NO_3^- ions.

14.7 *Oxo-acids of nitrogen*

The important oxo-acids of nitrogen are nitrous acid, HNO_2, and nitric acid, HNO_3, sometimes known as nitric(III) and nitric(V) acids respectively. There is also, however, a well-defined hyponitrous acid, $H_2N_2O_2$, obtained in ethereal solution by the action of anhydrous hydrogen chloride in dry ether on silver hyponitrite, which is precipitated from the aqueous sodium salt; this is made by the action of an organic nitrite on hydroxylamine in the presence of ethanolic sodium hydroxide:

$$RONO + NH_2OH + 2EtONa \rightarrow Na_2N_2O_2 + ROH + 2EtOH$$

The salt may also be obtained by reduction of sodium nitrite with sodium amalgam. Free hyponitrous acid is a weak acid and decomposes spontaneously into nitrous oxide and water; spectroscopic evidence indicates that it is *trans* HON=NOH.

Nitrous acid is known only in solution and in the vapour phase, in which it has the structure

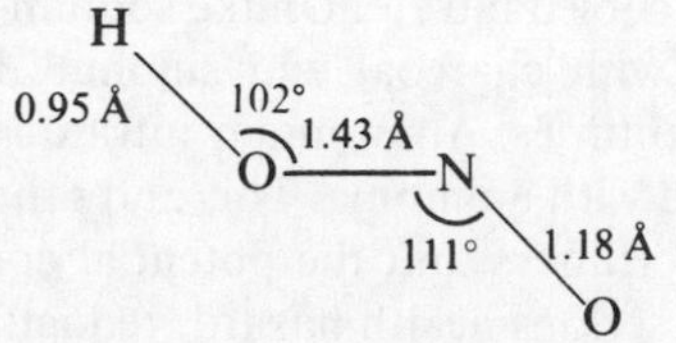

H
0.95 Å 102°
O 1.43 Å N
111° 1.18 Å
O

from which the electronic structure is clearly HON=O. The large difference in the O—N and N=O bond lengths should be noted. In the NO_2^- anion the ONO bond angle is 115° and the bond length 1.24 Å, corresponding to a bond order of 1.5. Nitrous acid in aqueous solution is a fairly weak acid ($pK_a = 3.5$); it may be obtained from a solution of barium nitrite and dilute sulphuric acid, or by dissolving an equimolar mixture of NO and NO_2 in ice-water. Alkali metal nitrates yield the nitrites when heated alone or, better, with metallic lead:

$$NaNO_3 + Pb \rightarrow NaNO_2 + PbO$$

Barium and silver salts may be obtained by precipitation, but most nitrites are freely soluble in water. Sodium nitrite, usually obtained as a very pale yellow solid, is an important reagent for the preparation of diazonium compounds and is used as a preservative for meat.

Nitrous acid may be oxidised to nitrate by powerful oxidising agents such as acidified permanganate. Reduction yields a variety of products: iodide and iron(II) salts give nitric oxide, tin(II) salts give nitrous oxide, sulphur dioxide gives hydroxylamine, and zinc and alkali give ammonia. An interesting manifestation of the importance of kinetic factors in nitrogen chemistry is the distinction between nitrate and nitrite based on the fact that in dilute solution only the latter liberates iodine from acidified iodide solution: it is easily shown from Fig. 14.3 that E° for the half-reaction

$$NO_3^- + 3H^+ + 2e \rightleftharpoons HNO_2 + H_2O$$

is +0.94 V, almost exactly the same as that for the half-reaction

$$HNO_2 + H^+ + e \rightleftharpoons NO + H_2O \quad E^\circ = +1.00\ V$$

and much higher than that for the half-reaction

$$\tfrac{1}{2}I_2 + e \rightleftharpoons I^- \quad E^\circ = +0.54\ V$$

So nitrous acid is a faster, rather than a more powerful, oxidising agent than dilute nitric acid.

We have already mentioned naturally occurring sodium nitrate (Section 14.1) and the manufacture of nitric acid from ammonia (Sections 14.3 and 14.6). Potassium nitrate is usually obtained from the sodium salt by adding a hot concentrated solution of potassium chloride to precipitate sodium chloride, and then crystallising potassium nitrate (which is much more soluble than sodium chloride at 70 °C but much less soluble at 0 °C) from the cooled filtrate. Unlike sodium nitrate, it is not deliquescent, and, together with charcoal and sulphur, has been used in gunpowder for several centuries. Ammonium nitrate is made by direct neutralisation of nitric acid with ammonia. This and other nitrates are used on an immense scale as fertilisers, but the potential contamination of drinking water by them is a serious health hazard: reduction to nitrite can lead to formation of carcinogenic organic nitroso compounds.

Pure nitric acid can be made in the laboratory by the action of sulphuric acid on potassium nitrate and distillation *in vacuo*; it is a colourless liquid which must be stored below 0 °C to prevent slight decomposition according to the equation

$$4HNO_3 \rightarrow 4NO_2 + 2H_2O + O_2$$

Ordinary concentrated nitric acid is the azeotrope containing 68 per cent by weight of HNO_3 and boiling at 120 °C; its yellow colour arises from photochemical decomposition according to the above equation; fuming nitric acid is orange owing to the presence of an excess of nitrogen dioxide.

In aqueous solution nitric acid behaves as an extremely strong acid, but in sulphuric acid (cf. Section 8.4) it functions as a base:

$$HNO_3 + H_2SO_4 \rightarrow H_2NO_3^+ + HSO_4^-$$

$$H_2NO_3^+ \rightarrow NO_2^+ + H_2O$$

$$H_2O + H_2SO_4 \rightarrow H_3O^+ + HSO_4^-$$

or, in sum,

$$HNO_3 + 2H_2SO_4 \rightarrow NO_2^+ + H_3O^+ + 2HSO_4^-$$

Nitrates of all metallic cations are soluble in water. Those of the alkali metals decompose when strongly heated to form nitrite and oxygen; ammonium nitrate gives nitrous oxide and water; and most other metal nitrates give the oxide, dinitrogen tetroxide and oxygen.

Most metals other than gold and the platinum metals are attacked by nitric acid (often more rapidly if a trace of nitrous acid is present), though chromium and iron are rendered passive by the concentrated acid, a very thin surface layer of oxide being formed. Tin, arsenic, antimony and a few transition metals are converted into oxides; other metals form nitrates. Metals which are more powerful reducing agents than hydrogen reduce the acid to N_2, NH_3, NH_2OH or N_2O; others liberate NO or NO_2. Only magnesium, manganese and zinc liberate hydrogen, and then only from the very dilute acid. Many organic and inorganic compounds are oxidised by concentrated nitric acid, but (as we have seen already) nitrate ion in aqueous solution is usually only a very slow oxidising agent. A mixture of concentrated nitric and hydrochloric acids (*aqua regia*) contains free chlorine and nitrosyl chloride, and attacks gold and platinum with formation of chloro complexes. Concentrated nitric acid oxidises iodine, phosphorus and sulphur to iodic acid, phosphoric acid and sulphuric acid respectively.

The molecular structure of nitric acid is:

H 0.96 Å 102° O—N 1.40 Å 130° 1.21 Å O O

The nitrogen and the three oxygen atoms are coplanar. Both N—O bonds in the NO_2 group are equivalent, just as they are in N_2O_4; the other N—O bond is much longer and corresponds to a single bond. The nitrate anion is planar and symmetrical, with a N—O bond length of 1.22 Å, reflecting a slightly lower bond order (one π bonding orbital distributed over four atoms) than in the NO_2 group of the acid (one π bonding orbital distributed over three atoms). Nitrate ion often acts as a bidentate ligand when it is bonded to a metal atom, e.g. in $Be_4O(NO_3)_6$ and the anions $[Fe(NO_3)_4]^-$ and $[Ce(NO_3)_6]^{2-}$.

The orthonitrate ion NO_4^{3-} has recently been obtained as the sodium salt by the interaction of $NaNO_3$ and Na_2O at 300 °C; the anion has the expected tetrahedral structure with a N—O bond length of 1.39 Å, corresponding to a bond order of unity, no π bonding being possible in this ion.

Before leaving the subject of nitric acid and its derivatives we should mention fluorine nitrate, $FONO_2$ (b.p. −46 °C), obtained by the action of fluorine on dilute nitric acid or potassium nitrate; it is an explosive gas which reacts slowly with water but rapidly with aqueous alkali according to the equation

$$2FONO_2 + 4OH^- \rightarrow 2NO_3^- + 2F^- + 2H_2O + O_2$$

Its structure is that of nitric acid with fluorine replacing the hydrogen in the molecule.

14.8 *Oxides of phosphorus, arsenic, antimony and bismuth*

Because of the complexity of the oxo-acids of phosphorus and the lack of structural relationships between many of them and the oxides, it is convenient to consider the oxides of the Group V elements apart from the acids.

Each of the elements forms two oxides in which the characteristic oxidation states of III and V are displayed; the higher oxidation state becomes less stable as we go from phosphorus to bismuth, but the pentoxides are all acidic; the lower oxides become more basic, P_4O_6 being exclusively acidic, As_4O_6 and Sb_4O_6 amphoteric, and Bi_2O_3 exclusively basic.

Phosphorus trioxide, diphosphorus trioxide, or phosphorus(III) oxide as it is best called, since the only known form is molecular P_4O_6, is obtained when white phosphorus is burned in a restricted supply of oxygen. It is a colourless volatile solid (m.p. 24 °C) which is soluble in ether or benzene, and is irreversibly hydrolysed by cold water to the acid H_3PO_3. Its structure, shown in Fig. 14.5(*a*), consists of a tetrahedron of phosphorus atoms with the oxygen atoms off the edges of the tetrahedron; the P—O distance is 1.65 Å, the POP bond angle 128°, and the OPO bond angle 99°. The corresponding oxides of arsenic and antimony are the normal products of combustion of these elements; although arsenic(III) oxide, which sublimes at 150 °C, is much more volatile than the antimony compound, the vapours of both substances contain molecules isostructural with those of P_4O_6. The higher-temperature forms of both solid oxides also contain such molecules, but lower-temperature forms are layer polymers in which each arsenic or antimony atom is bonded pyramidally to three oxygen atoms. Arsenic(III) oxide may also be obtained as a glass.

Arsenic(III) oxide may be oxidised to the pentoxide by hydrogen peroxide or nitric acid; it is reduced to the element by heating it with charcoal. It forms a very weakly acidic aqueous solution, but only the oxide separates on crystallisation. The oxide is, however, readily soluble in warm aqueous alkali with formation of arsenites, and in aqueous hydrochloric acid with formation of arsenic trichloride. Antimony(III) oxide shows similar properties.

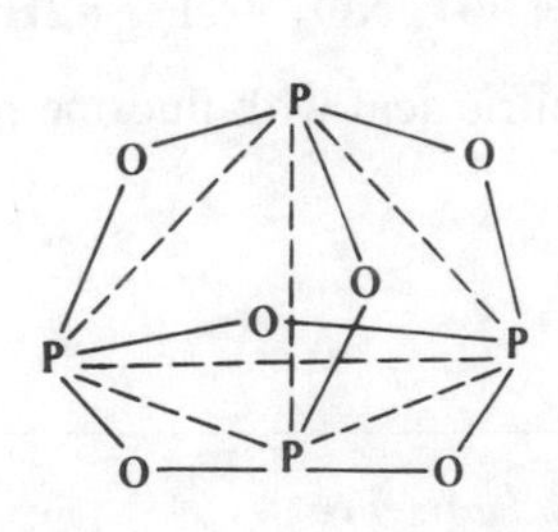

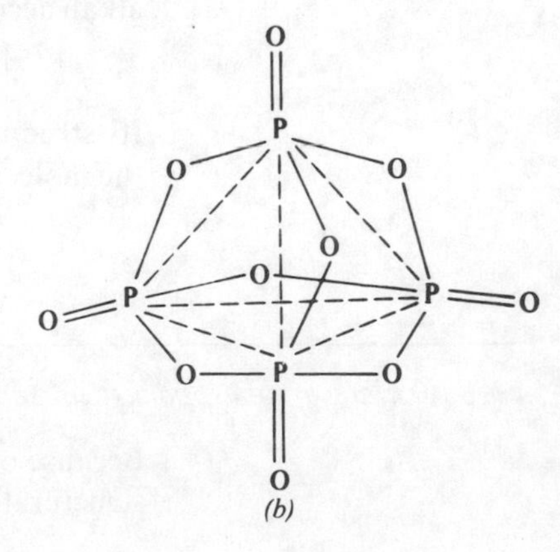

Figure 14.5
The structures of (*a*) P_4O_6 and (*b*) P_4O_{10}.

Bismuth(III) oxide may be made from the elements or by heating the hydroxide which is precipitated by alkali from solutions of bismuth salts; it dissolves in aqueous solutions of most strong acids but not in aqueous alkali. The oxide has a complicated crystal structure (probably because of the influence of the $6s^2$ lone pair), but there is an irregular three-dimensional coordination of the bismuth atoms by oxygen atoms, and the lattice is much more like that of a typical metal oxide than those of arsenic(III) and antimony(III) oxides.

The acidic pentoxides, M_2O_5, of arsenic, antimony and bismuth are all obtained by oxidation of the trioxides; nitric acid is a sufficiently powerful oxidant for the arsenic and antimony compounds, but the preparation of the ill-characterised bismuth pentoxide necessitates the use of alkaline hypochlorite or peroxodisulphate. The structures of the pentoxides are unknown. Phosphorus pentoxide or phosphorus(V) oxide is, of course, the normal product of combustion of phosphorus, and the anhydride of the wide range of acids described in the following section; it has a great affinity for water, and is much used as a drying and dehydrating agent, e.g. in the preparation of N_2O_5, Cl_2O_7 and C_3O_2.

Phosphorus pentoxide exhibits an interesting polymorphism, existing in three forms at ordinary pressure and temperature. The structures of these polymorphs, and indeed of oxo compounds of phosphorus(V) in general, resemble those of silica and the silicates; but whereas the latter are based on SiO_4 tetrahedra with sharing of up to all four oxygen atoms, those of phosphorus(V) oxide and the phosphates are based on PO_4 tetrahedra with one oxygen doubly bonded to phosphorus and hence not available for sharing, the fundamental unit being

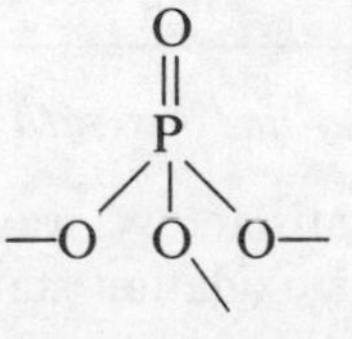

```
      O
      ‖
      P
     /|\
  —O  O  O—
       \
```

The vapour of phosphorus(V) oxide contains P_4O_{10} molecules whose structure (Fig. 14.5(*b*)) is essentially that of P_4O_6 with the addition of a doubly bonded oxygen atom attached to each phosphorus; P—O (terminal) is 1.40 Å, P—O (bridge) 1.60 Å. When the vapour is condensed rapidly the product, which is volatile and extremely hygroscopic, also contains P_4O_{10} molecules. If, however, this volatile polymorph is heated for several hours in a closed vessel or melted and the melt, after being maintained at a high temperature for some time, is allowed to cool, the other polymorphs are obtained. These are less volatile and less rapidly attacked by water. One has a layer structure built up of PO_4 tetrahedra sharing three oxygen atoms (Fig. 14.6), the other a three-dimensional structure built up of the same units; bond lengths and angles are approximately the same as in the molecular form, the difference in

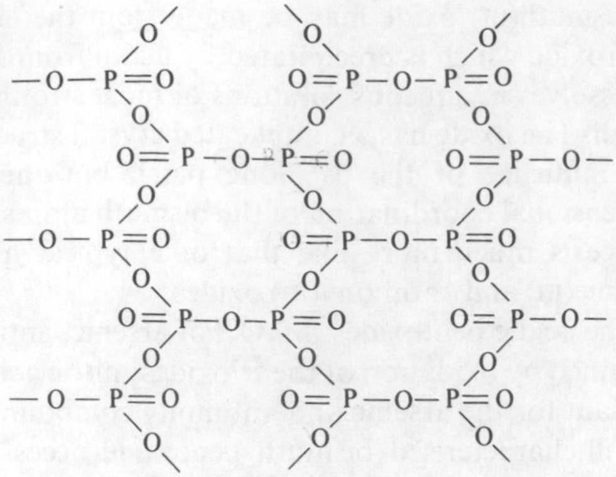

```
        O          O         O
  —O— P=O        O=P—O—P=O
        O          O         O
    O= P—O—P=O          O=P—O—
        O          O         O
  —O— P=O        O=P—O—P=O
        O          O         O
    O= P—O—P=O          O=P—O—
        O          O         O
  —O— P=O        O=P—O—P=O
        O          O         O
```

Figure 14.6
Plan of the structure of one of the polymorphic macromolecular forms of P_2O_5.

properties arising from the way in which the PO_4 units are joined up to give molecular or macromolecular species.

When phosphorus(III) oxide is heated in a sealed tube at 440 °C it decomposes to form an oxide of empirical formula PO_2 (which is also produced by the slow combustion of white phosphorus) and red phosphorus; this oxide, which yields H_3PO_3 and H_3PO_4 on hydrolysis, is actually P_4O_8, and has a structure which is that of P_4O_6 with addition of oxygen atoms to only two of the phosphorus atoms, i.e. it is $P^{III}_2P^{V}_2O_8$. Antimony forms a corresponding oxide SbO_2 when Sb_2O_3 is heated in air at 900 °C, but this contains Sb(III) and Sb(V) in an infinite solid structure $Sb^{III}Sb^{V}O_4$.

14.9 *Oxo-acids of phosphorus, arsenic, antimony and bismuth*

There is no entirely satisfactory way of classifying the oxo-acids of phosphorus in its lower oxidation states, and the variety of systems of nomenclature used for these compounds is confusing; in the following account, we have dealt with them in succession according to the formal oxidation states of the phosphorus atoms present. By comparison the chemistry of the oxo-acids and oxo-anions of arsenic, antimony and bismuth is relatively limited and straightforward; we shall turn to these species after giving a fairly detailed (though still considerably simplified) account of the phosphorus compounds.

When white phosphorus is heated with aqueous alkali, the main products are phosphine and a hypophosphite:

$$P_4 + 3OH^- + 3H_2O \rightarrow PH_3 + 3H_2PO_2^-$$

By using barium hydroxide as the alkali, precipitating the metal as barium sulphate by dilute sulphuric acid, and evaporating the solution, white deliquescent crystals of hypophosphorous acid, H_3PO_2, formally a derivative of phosphorus(I), are obtained. This is a fairly strong monobasic acid, and its infrared spectrum shows the presence of P—H bonds; it has the tetrahedral structure

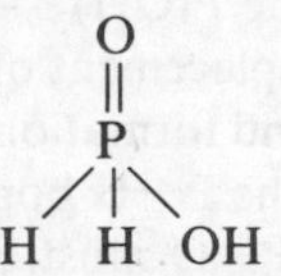

for which the recommended systematic name is phosphinic acid. Hypophosphorous acid and its salts are reducing agents, converting copper(II) salts into copper hydride, for example; they are used in electroless nickel plating. When heated, they disproportionate into phosphine and, according to the temperature, phosphite or phosphate:

$$3H_3PO_2 \rightarrow PH_3 + 2H_3PO_3$$

or

$$2H_3PO_2 \rightarrow PH_3 + H_3PO_4$$

Since the formation of phosphine and hypophosphite by the action of alkali is a disproportionation of phosphorus(0) into phosphorus(−III) and phosphorus(I), it is apparent that different oxidation states of phosphorus differ in stability towards oxidation or reduction by a much smaller margin than different oxidation states of nitrogen; this generalisation is also apparent from the potential diagram (calculated from thermal data) shown in Fig. 14.7. As in the case of nitrogen, the potential diagram is limited in usefulness because of kinetic factors.

Phosphorous acid, H_3PO_3, may be crystallised from the solution obtained by adding ice-cold water to phosphorus(III) oxide or phosphorus trichloride. It is dibasic and a good reducing agent; again the presence of a P—H bond is established spectroscopically, and the tetrahedral structure is

O
P
HO H OH

for which the approved systematic name is phosphonic acid. The P=O and P—O(H) bond lengths are 1.47 and 1.54 Å respectively. The ester $P(OMe)_3$, obtained from methanol and phosphorus trichloride, rearranges spontaneously to give $MeP(O)(OMe)_2$. Whilst we cannot give an

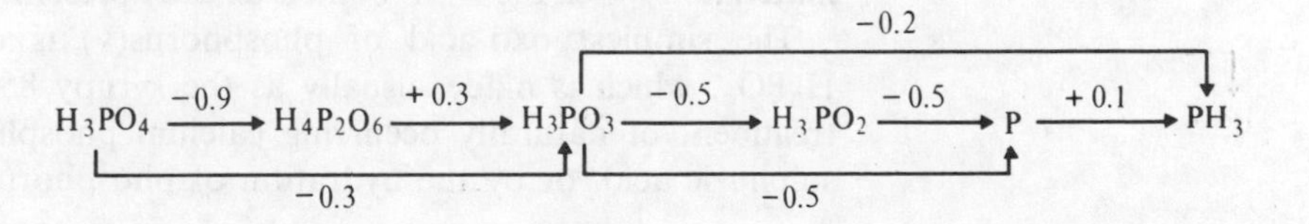

Figure 14.7
Potential diagram for phosphorus at $[H^+] = 1$.

exact discussion of why $P(OH)_3$ (which has never been isolated) and $P(OMe)_3$ are unstable with respect to isomerisation to $HP(O)(OH)_2$ and $MeP(O)(OMe)_2$ respectively, we can make some comments on this problem. If we consider the $P(OH)_3 \rightarrow HP(O)(OH)_2$ process, it involves fission of a O—H bond, replacement of P—O(H) by P=O (a shorter and therefore stronger bond) and formation of a P—H bond. The O—H bond is certainly stronger than the P—H bond, so the increased strength of the P=O bond compared to P—O must be great enough to overcome both this factor and the promotion energy for the change in the valence state of the phosphorus; similar considerations must apply to the isomerisation of the methyl ester. Like hypophosphorous acid, phosphorous acid disproportionates when heated:

$$4H_3PO_3 \xrightarrow{200\,^\circ\text{C}} PH_3 + 3H_3PO_4$$

The third oxo-acid of phosphorus in a formal oxidation state lower than v is hypophosphoric acid, $H_4P_2O_6$. The sodium salt, $Na_2H_2P_2O_6$, is obtained by oxidation of red phosphorus with sodium hypochlorite; it may be converted into a solution of the acid by ion-exchange, and evaporation *in vacuo* yields the free acid. The formula $H_4P_2O_6$ rather than H_2PO_3 was first indicated by the diamagnetism of the acid (H_2PO_3 would have an odd number of electrons) but has now been confirmed by an X-ray diffraction study of the ammonium salt $(NH_4)_2H_2P_2O_6$, which establishes the structure of the anion in this compound as

```
HO          OH
   \       /
  O=P—P=O
   /       \
 ⁻O          O⁻
```

with the negative charges delocalised so that all P—O (terminal) bonds become equivalent. We see from Fig. 14.7 that hypophosphoric acid is thermodynamically unstable with respect to disproportionation into phosphorous and phosphoric acids, and the reaction

$$H_4P_2O_6 + H_2O \rightarrow H_3PO_3 + H_3PO_4$$

occurs slowly in aqueous solution. This instability with respect to disproportionation explains why hypophosphoric acid cannot be made by reduction of phosphoric acid or by oxidation of phosphorous acid in aqueous media; hence the need for a special preparation starting with a material in which the P—P bond is already present.

The simplest oxo-acid of phosphorus(v) is orthophosphoric acid, H_3PO_4, which is made, usually as the syrupy 85 per cent acid, by the treatment of naturally occurring calcium phosphate with concentrated sulphuric acid, or by the hydration of phosphorus pentoxide. The pure acid is a colourless solid of melting-point 42 °C; its structure contains tetrahedral hydrogen-bonded $O{=}P(OH)_3$ molecules. There are, as expected, two different phosphorus–oxygen bond lengths, 1.52 (P=O) and 1.57 Å, but it should be noted that the difference between them is much less than in P_4O_{10}. Hydrogen bonding (like that in sulphuric acid) is responsible for the syrupy nature of the concentrated acid; in dilute aqueous solution the acid molecules or their anions are hydrogen-bonded to water rather than to one another, and the viscosity is much lower. Orthophosphoric acid is very stable and has no oxidising properties except at very high temperatures. At 25 °C, pK_a values for the three ionisations in aqueous solution are respectively 2.1, 7.1 and 12.4.

Naturally occurring phosphates are all orthophosphates. *Superphosphate* fertiliser is a mixture of $Ca(H_2PO_4)_2$, H_3PO_4 and $CaSO_4$ made by the action of a limited amount of sulphuric acid on powdered phosphate rock; *triple superphosphate*, mainly $Ca(H_2PO_4)_2$, comes from the treatment of phosphate rock with phosphoric acid. Three sodium salts are obtained by suitable neutralisation; ordinary sodium phosphate is $Na_2HPO_4.12H_2O$, the common potassium salt, KH_2PO_4. Sodium phosphates are extensively used for buffering aqueous solutions, and tri-n-butyl phosphate (made by the action of $POCl_3$ on the alcohol) is a very good solvent for the extraction of metals, commonly as their nitrates, from aqueous solution (Section 1.8). Most orthophosphates other than the alkali metal and ammonium compounds are sparingly soluble in water; those of certain tetrapositive heavy transition metal ions (e.g. Ce^{4+}, Th^{4+}, Zr^{4+}, U^{4+}) are insoluble even in 4 molar hydrochloric or nitric acid. The genetic substances DNA and RNA (nucleic acids) are orthophosphate esters.

As we remarked in dealing with the structures of the various forms of phosphorus pentoxide, all the oxygen chemistry of phosphorus(v) is based on the tetrahedral PO_4 unit, which may share one, two or three (but, except in BPO_4 and $AlPO_4$, not four) oxygen atoms. Species containing P—O—P bonds are commonly known as condensed phosphates; both they and the acids from which they are derived are usually obtained by heating substances containing P—OH groups:

$$\geq P{-}OH + HO{-}P\leq \;\rightarrow\; \geq P{-}O{-}P\leq + H_2O$$

Controlled hydrolysis of phosphorus pentoxide is also sometimes useful. In principle the condensation of orthophosphate ions, e.g.

$$2PO_4^{3-} + 2H^+ \rightleftharpoons P_2O_7^{2-} + H_2O$$

should be favoured by low pH, and vice versa, but in practice such reactions involving phosphates are usually slow.

In condensed phosphate anion formation, chain-terminating end groups are formed from HPO_4^{2-}, chain members from $H_2PO_4^-$ and cross-linking groups from H_3PO_4; these are respectively

$$^{-}O{-}\overset{\displaystyle O^{-}}{\underset{\displaystyle O}{\overset{|}{\underset{\|}{P}}}}{-}O{-} \qquad {-}O{-}\overset{\displaystyle O^{-}}{\underset{\displaystyle O}{\overset{|}{\underset{\|}{P}}}}{-}O{-} \quad \text{and} \quad {-}O{-}\overset{\displaystyle \overset{|}{O}}{\underset{\displaystyle O}{\overset{|}{\underset{\|}{P}}}}{-}O{-}$$

In the free acids a proton is attached at O^-, of course. These units can be distinguished either by ^{31}P nuclear magnetic resonance spectroscopy or chemically. In the same way as orthophosphoric acid has very different values for successive ionisation constants, so do condensed acids containing two hydroxyl groups on the same phosphorus atom. Thus terminal phosphorus atoms carry one strongly and one weakly acidic group, whilst those in the body of a chain carry only a single strongly acidic group; titration of a condensed phosphate with acid therefore enables distinctions to be drawn between different anions. Further, it is found that the cross-linking units are hydrolysed by water much faster than the others.

The simplest condensed phosphoric acid, pyrophosphoric or diphosphoric acid, $H_4P_2O_7$, may be obtained by heating orthophosphoric acid at 240 °C:

$$2H_3PO_4 \rightarrow H_4P_2O_7 + H_2O$$

A purer product results from heating orthophosphoric acid with $POCl_3$:

$$5H_3PO_4 + POCl_3 \rightarrow 3H_4P_2O_7 + 3HCl$$

It melts at 61 °C and is a stronger acid than orthophosphoric acid, its pK_a values at 25 °C being 0.8, 1.9, 6.5 and 9.6. Sodium pyrophosphate is obtained by heating Na_2HPO_4 at 240 °C. In aqueous solution the $P_2O_7^{4-}$ ion is only very slowly hydrolysed to orthophosphate, and it can be distinguished from the latter ion by simple chemical tests, e.g. the sparingly soluble $Ag_4P_2O_7$ is white, whereas Ag_3PO_4 is pale yellow. The structure of the anion is $O_3POPO_3^{4-}$, analogous to that of the disilicate ion $Si_2O_7^{6-}$ (Section 13.9).

What used to be known as metaphosphoric acid and given the empirical formula HPO_3 is a sticky mixture of polymeric acids obtained by heating ortho- or pyrophosphoric acid at 320 °C. Salts of the acids are much better characterised than the acids themselves. The sodium salt $Na_3P_3O_9$ (trimetaphosphate), formed when NaH_2PO_4 is heated to 600–640 °C and the melt is then maintained at 500 °C to allow water vapour to escape, contains the cyclic $P_3O_9^{3-}$ anion analogous to $Si_3O_9^{6-}$; in alkaline solution it hydrolyses to give $Na_5P_3O_{10}$, which contains the chain anion

$$^{-}O{-}\overset{\displaystyle O}{\underset{\displaystyle O^{-}}{\overset{\|}{\underset{|}{P}}}}{-}O{-}\overset{\displaystyle O}{\underset{\displaystyle O^{-}}{\overset{\|}{\underset{|}{P}}}}{-}O{-}\overset{\displaystyle O}{\underset{\displaystyle O^{-}}{\overset{\|}{\underset{|}{P}}}}{-}O^{-}$$

This is analogous to the silicate anion $Si_3O_{10}^{8-}$. Sodium triphosphate (or tripolyphosphate as it is often called) may also be made by evaporating a solution containing Na_2HPO_4 and NaH_2PO_4 in suitable proportions, and heating the residue at 300–400 °C:

$$2Na_2HPO_4 + NaH_2PO_4 \rightarrow Na_5P_3O_{10} + 2H_2O$$

It forms stable soluble complexes with alkaline earth metal cations and is much used in water-softening. Unfortunately high concentrations of phosphate lead to extensive algal growth which, by depleting the water of oxygen, causes extinction of animal life. Controlled precipitation of aluminium phosphate is an effective method of phosphate removal.

Sodium tetrametaphosphate, $Na_4P_4O_{12}$, may be prepared by heating NaH_2PO_4 with phosphoric acid at 400 °C and slowly cooling the melt, or by treating the volatile form of phosphorus(V) oxide with ice-cold aqueous sodium hydroxide and sodium bicarbonate, a process which involves partial opening of the P_4O_{10} cage. It contains the cyclic $P_4O_{12}^{4-}$ anion, isostructural with $Si_4O_{12}^{8-}$. Both the $P_3O_9^{3-}$ and $P_4O_{12}^{4-}$ ions exist in the chair forms. A hexametaphosphate containing the cyclic anion $P_6O_{18}^{6-}$, isostructural with $Si_6O_{18}^{12-}$, is also known.

Long-chain polyphosphates are produced by heating NaH_2PO_4 at 250 °C or above under special conditions; they contain chains like those in the silicates such as β-$CaSiO_3$ and, according to the orientations of the PO_4 tetrahedra, several different modifications are known. Finally, cross-linked phosphates, or ultraphosphates, are obtained by heating NaH_2PO_4 with P_2O_5 at above 350 °C; they have Na:P ratios of less than unity. At the present time, however, their structures have not been much investigated, and some are glasses.

We have referred to sodium salts throughout the foregoing description of condensed phosphates, but salts of many other cations are known. So also are esters; the reversible enzyme-catalysed hydrolysis of adenosine triphosphate (ATP) to the diphosphate (ADP) is involved in biosynthesis, transport of solutes in the body against concentration gradients and muscle contraction. The hydrolytic reaction may be written as

$$\underset{\text{(ATP)}}{R{-}O{-}\overset{\overset{\displaystyle O}{\|}}{\underset{\underset{\displaystyle O^-}{|}}{P}}{-}O{-}\overset{\overset{\displaystyle O}{\|}}{\underset{\underset{\displaystyle O^-}{|}}{P}}{-}O{-}\overset{\overset{\displaystyle O}{\|}}{\underset{\underset{\displaystyle O^-}{|}}{P}}{-}O^-} + 2H_2O \rightarrow \underset{\text{(ADP)}}{R{-}O{-}\overset{\overset{\displaystyle O}{\|}}{\underset{\underset{\displaystyle O^-}{|}}{P}}{-}O{-}\overset{\overset{\displaystyle O}{\|}}{\underset{\underset{\displaystyle O^-}{|}}{P}}{-}O} + HPO_4^{2-} + H_3O^+$$

where R stands for the adenosine parts of the molecules. At the standard state usually employed in the discussion of biochemical processes, pH = 7 and $[CO_2] = 10^{-5}$ molar, this reaction is accompanied by a decrease in free energy of 30.5 kJ mol^{-1} which can be used to do work or bring about other reactions. Conversely, energy released by, for example, oxidation of carbohydrate, can be used to reverse the reaction; the energy is thus stored in the form of ATP. Instead of simple hydrolysis, esterification of another organic molecule to a phosphate ester is often involved; both hydrolysis and phosphate transfer are catalysed by metal- (especially magnesium-) containing enzymes.

Although no solid arsenious acid has been isolated, $As(OH)_3$ is present in an aqueous solution of arsenic(III) oxide. Many solid arsenites, e.g. $NaAsO_2$ and Ag_3AsO_3, are known; the structure of the latter is not known, but the former contains Na^+ ions and an infinite chain ion in which each arsenic atom is linked pyramidally to three oxygen atoms, two of which are shared with other arsenic atoms. A point of similarity to phosphorous acid is shown by the existence of the isomeric methyl esters $As(OMe)_3$ and $MeAs(O)(OMe)_2$. Arsenic acid, H_3AsO_4, a somewhat weaker acid than phosphoric acid, is obtained by oxidation of arsenic(III) oxide with concentrated nitric acid; its salts are easily obtained by oxidation of arsenites in alkaline solution, but in acidic solution arsenic acid acts as an oxidising agent; the dependence of the ease of oxidation or reduction on the pH of the solution is easily understood on the basis of the half-reaction:

$$H_3AsO_4 + 2H^+ + 2e \rightleftharpoons H_3AsO_3 + H_2O \quad E^\circ = +0.56 \text{ V}$$

Condensed arsenate ions are kinetically much less stable with respect to hydrolysis than condensated phosphate ions, and only AsO_4^{3-} exists in aqueous solution. In solid salts, however, the existence of infinite chain anions built of AsO_4 tetrahedra sharing two oxygen atoms, i.e. analogous to the infinite chain anions in $NaPO_3$, is well established.

Very little is known about oxo-acids of antimony or their salts, but well-defined antimonates are obtained by dissolving the pentoxide in aqueous alkali and crystallising. These (e.g. the compound formerly written $Na_2H_2Sb_2O_7.5H_2O$) contain the $[Sb(OH)_6]^-$ anion.

Bismuth is acidic only in its unstable higher oxidation state. Sodium bismuthate, $NaBiO_3$, may be obtained by fusing Bi_2O_3 with NaOH in air or with Na_2O_2; it is an insoluble orange compound which is a very powerful oxidant. In the presence of acid, for example, it oxidises Mn(II) to permanganate, and it liberates chlorine from hydrochloric acid. The Bi(III)–Bi(V) compound $K_{0.4}Ba_{0.6}BiO_{3-x}$ (where x is about 0.02) is of great interest as a copper-free superconductor at 30 K (cf. Section 24.10). It has the perovskite structure.

14.10 *Phosphazenes*

Phosphazene is the unknown molecule $N{\equiv}PH_2$, which is unstable with respect to polymerisation; the compounds to be described in this section are derivatives of cyclic and linear polymers of phosphazene, particularly cyclic polymers of dichlorophosphazene, sometimes also known as phosphonitrilic chlorides.

When phosphorus pentachloride is heated with ammonium chloride in tetrachloroethane or chlorobenzene, a mixture of colourless solids of formula $(NPCl_2)_n$, with n mostly 3 or 4, is produced. The trimeric and tetrameric compounds are easily separated by distillation under reduced pressure. The stoichiometry of the reaction may be represented by the equation

$$nPCl_5 + nNH_4Cl \rightarrow (NPCl_2)_n + 4nHCl$$

Its mechanism is complicated; it is believed that $NH_4[PCl_6]$ is first formed and loses hydrogen chloride to give the unstable species $HN{=}PCl_3$, which then reacts as follows in the formation of the trimer:

$$[PCl_4]^+[PCl_6]^- + HN{=}PCl_3 \rightarrow [Cl_3P{=}N{=}PCl_3]^+[PCl_6]^- + HCl$$

$$\downarrow HN{=}PCl_3$$

$$(NPCl_2)_3 \text{ (ring: } Cl_2P{=}N,\ N,\ PCl_2,\ P{-}N,\ Cl_2) \xleftarrow{NH_3} [Cl_3PNPCl_2NPCl_3]^+[PCl_6]^-$$

$$+ PCl_5 + 3HCl$$

The corresponding bromo and methyl compounds are obtained by replacing phosphorus pentachloride by PBr_5 and by Me_2PCl_3 respectively; the difluoro compounds are not obtained if ammonium fluoride is used instead of ammonium chloride, the only product then being the stable hexafluorophosphate $NH_4[PF_6]$. Chlorine may, however, be replaced by fluorine by heating the chlorides with sodium fluoride suspended in acetonitrile or nitrobenzene.

The chlorine atoms in hexachlorocyclotriphosphazene (or the corresponding tetramer) are reactive, and substitution by $-NH_2$, $-NMe_2$, $-N_3$, $-OH$ and $-Ph$ takes place on treatment with liquid NH_3, Me_2NH, LiN_3, H_2O and LiPh respectively. Two types of substitution occur. If the entering group decreases the electron density at the phosphorus atom (e.g. if it is F), the second substitution takes place at the same phosphorus atom; if the first entering group results in an increase in electron density at the phosphorus atom (e.g. if it is NMe_2), the second substitution takes place elsewhere.

Linear polymers, formed in small amounts in the ordinary preparation of the cyclophosphazenes or in larger amounts if excess of phosphorus pentachloride is used, have the structure

$$Cl_3P{=}N{-}\left[\begin{matrix} Cl \\ | \\ P{=}N \\ | \\ Cl \end{matrix}\right]_n{-}PCl_4$$

They are also formed when $(NPCl_2)_3$ is heated to 250–350 °C in carbon tetrachloride in the presence of oxygen; in this way polymers with molecular masses of 20 000 or more are obtained. These polymers still contain reactive chlorine atoms; by the action of sodium alkoxides, however, water-repellent polymers of composition $[PN(OR)_2]_n$ may be obtained. When $R = CH_2CF_3$, the products are so inert that they can be used for the construction of artificial blood vessels and organs.

In both $(NPF_2)_3$ and $(NPCl_2)_3$ the P_3N_3 skeletons are planar almost regular hexagons and the P—N bond lengths (1.56 and 1.60 Å respectively) are substantially shorter than that in the anion of the salt $Na^+[H_3NPO_3]^-$ (1.77 Å), which may be taken to contain a P—N single bond. The trimers thus appear to show marked resemblances to benzene; however, in phosphazenes the double bonding arises not from p_π–p_π but p_π–d_π interactions, which cannot give rise to a strongly bonding orbital extending all the way round the ring; the analogy must therefore not be carried too far. This is at once seen to be the case when we study the structure of the tetramers: the P_4N_4 skeleton in $(NPF_2)_4$ is planar, but in $(NPCl_2)_4$ it is not, both boat and chair forms being known; in each compound, however, all the P—N bond lengths are equal, being 1.51 and 1.57 Å respectively; the bond angles at phosphorus are 120° in both compounds, but at nitrogen they are 147° in the fluoride and 131° in the chloride. The fact that the bond is shorter in the planar compound has been held to show that the lone pairs of electrons on the nitrogen atoms are also involved in bonding, i.e. the bond order is greater than 1.5: if the plane of the ring is taken as the xy plane and the single bond system is derived from sp^3 hybridised phosphorus and sp_xp_y hybridised nitrogen atoms, two π systems are possible, one based on d_{xz} phosphorus and p_z nitrogen orbitals, the other on $d_{x^2-y^2}$ phosphorus and the sp^2 nitrogen orbitals, these systems being perpendicular to and in the plane of the ring respectively. Maximum π bonding would be possible with a bond angle of 180° at nitrogen, so the shorter bond in the fluoride correlates with the larger bond angle at nitrogen in this compound. Further, thermochemical evidence indicates that, if the P—Cl bond energy term is taken as constant, the P—N bond energy term increases slightly as the bond length decreases and the bond angle at nitrogen increases. For a fuller discussion of the bonding in the cyclophosphazenes, however, the references given at the end of the chapter should be consulted.

14.11 Sulphides

Binary compounds of nitrogen and sulphur are usually called sulphur nitrides on the grounds that nitrogen is a more negative element than sulphur. Partly for this reason, partly because they are very different from the sulphides of phosphorus and arsenic, and partly because in many of the ternary compounds derived from them the sulphur is the element in the higher oxidation state (e.g. NSF and NSF_3 are derivatives of S(IV) and S(VI) rather than sulphur analogues of ONF and ONF_3), we have postponed the treatment of all nitrogen–sulphur compounds to the following chapter.

The principal sulphides of phosphorus and its homologues are

P_4S_3		P_4S_5		P_4S_7	P_4S_{10}
As_4S_3	As_4S_4		As_2S_3		As_2S_5
			Sb_2S_3		
			Bi_2S_3		

It should be noted that there is no P_4S_6, the hypothetical analogue of P_4O_6, though the latter compound combines with sulphur to form $P_4O_6S_4$, and P_4S_{10}, made by heating red phosphorus with excess of sulphur, has the same structure as molecular P_4O_{10}. The other phosphorus sulphides, also formed by union of the elements, all have cage structures (shown in Fig. 14.8), though there is no P_4 unit common to them. Bond lengths indicate that all the P—P and P—S bonds within the cages are single bonds; P—S bonds external to the cages are shorter and may be regarded as P=S bonds. The phosphorus sulphides are very easily ignited (P_4S_3 is used with $KClO_3$ in 'strike anywhere' matches) and are slowly attacked by water; P_4S_{10} gives orthophosphoric acid and hydrogen sulphide; the others give phosphine, various oxo-acids of phosphorus and hydrogen sulphide.

The arsenic sulphides As_2S_3 and As_2S_5 are usually precipitated from aqueous solutions of arsenite and arsenate respectively; As_4S_3 (*dimorphite*), As_4S_4 (*realgar*) and As_2S_3 (*orpiment*) all occur naturally; the last

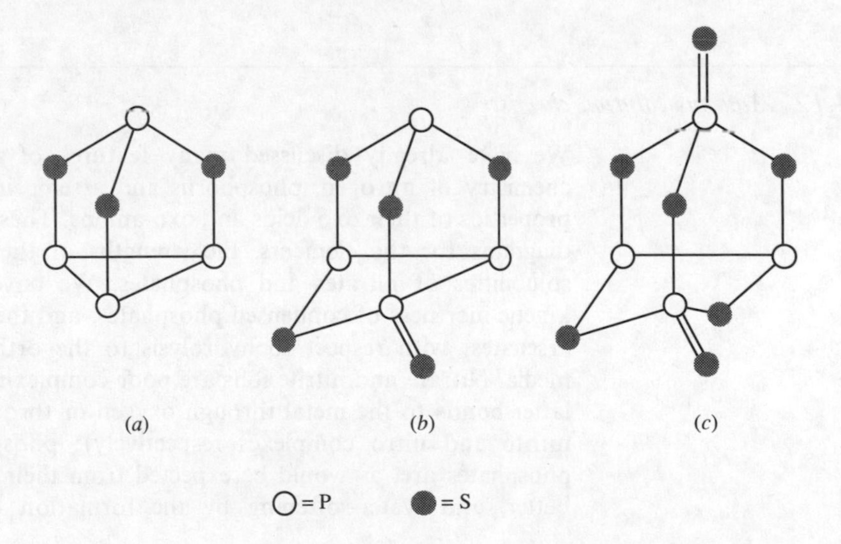

Figure 14.8
The structures of (*a*) P_4S_3 (*b*) P_4S_5 (*c*) P_4S_7.

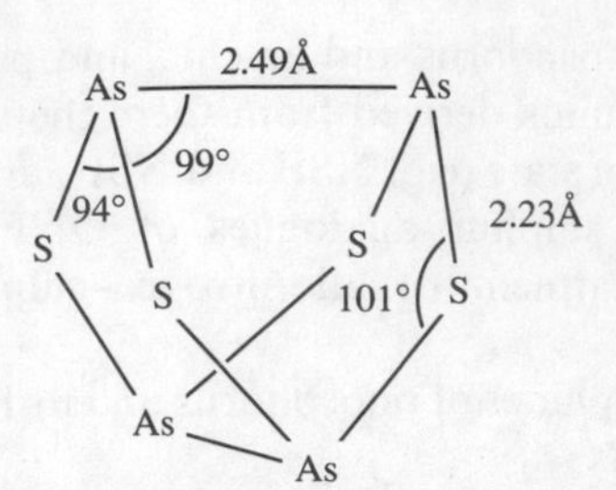

Figure 14.9
The structure of As_4S_4.

two, which are red and golden-yellow respectively, were used as pigments in the ancient world. The sulphide As_4S_3 is isostructural with P_4S_3. Realgar has a lattice of As_4S_4 molecules with the structure shown in Fig. 14.9. We shall see later (Section 15.7) that N_4S_4 has a rather similar structure, though with the sulphur atoms occupying the positions of the arsenic atoms in As_4S_4; in view of uncertainties about the electronic structure of N_4S_4 we should make it plain that As_4S_4 poses no problems: the As—As distances are the same as in As_4, and the As—S distances the same as in As_2S_3. Solid As_2S_3 has the same polymeric structure as low-temperature As_2O_3, but it vaporises to give As_4S_6 molecules isostructural with those of As_4O_6 and P_4O_6. The structure of As_2S_5 is not known. Both As_2S_3 and As_2S_5 are readily soluble in alkali metal sulphide solutions with formation of thioarsenites and thioarsenates; acids decompose these salts with reprecipitation of the sulphides.

Antimony forms only one well-authenticated sulphide, Sb_2S_3. A metastable red form (precipitated from aqueous solution) when heated slowly changes into the black form, which occurs naturally as *stibnite*. Antimony sulphide dissolves in alkali metal sulphide solutions, but bismuth sulphide does not.

14.12 *Aqueous solution chemistry*

We have already discussed many features of the aqueous solution chemistry of nitrogen, phosphorus and arsenic in connection with the properties of their oxo-acids and oxo-anions. These include the potential diagrams for the elements, the strengths of their oxo-acids, and the solubilities of nitrates and phosphates. We have also mentioned the kinetic inertness of condensed phosphates, and the lability of condensed arsenates, with respect to hydrolysis to the ortho anions in aqueous media. Nitrate and nitrite ions are poor complexing agents, whether the latter bonds to the metal through oxygen or through nitrogen (forming nitrito and nitro complexes respectively); phosphate and condensed phosphates are, as would be expected from their higher charges, much better, and water-softening by the formation of soluble condensed phosphate complexes of calcium and magnesium is of considerable practical importance. The role of hydrogen bonding to the solvent has been indicated in connection with the relative solubilities of ammonia and phosphine. It remains only to consider the solution chemistry of cationic species, and these are restricted to antimony(III) and bismuth.

Solutions of antimony(III) oxide in aqueous acids contain either hydrolysis products of Sb^{3+} or complexes. The former are commonly referred to as antimonyl ions and given the formula SbO^+, but analogies with bismuth (see below) indicate that this is almost certainly an oversimplification. Complexes of antimony(III) that may be isolated from aqueous solution include numerous halides (Section 14.5) and many containing organic anions; among these $K_3[Sb(C_2O_4)_3]$ is of interest as containing a Sb(III) atom having a coordination which may be described as pentagonal with one equatorial position occupied by a lone pair of electrons (c.f. thc octahedral $[SbCl_6]^{3-}$ ion).

There is no evidence for the presence of the simple aquo cation $[Bi(H_2O)_n]^{3+}$ in solutions of bismuth salts, the main species in the absence of complexing anions being $Bi_6O_6^{6+}$ in its hydrated form $[Bi_6(OH)_{12}]^{6+}$; in more alkaline media $[Bi_6O_6(OH)_3]^{3+}$ is formed and ultimately $Bi(OH)_3$ is precipitated. The structure of the $[Bi_6(OH)_{12}]^{6+}$ ion in solution is that of an octahedron of Bi^{3+} ions held together by bridging OH^- ions situated just off the edges of the octahedron.

14.13 *Organic derivatives*

The organic chemistry of nitrogen and phosphorus is so extensive that we shall not attempt to deal with it in this book beyond making a few points that are important in inorganic chemistry. The first of these concerns amines and phosphines as ligands. As we have already seen, ammonia is much more basic than phosphine, and towards most Lewis acids trimethylamine is a stronger base than trimethylphosphine. For some metal ions (class (b) ions), however, and for a few other species such as borane, this order is reversed; in such circumstances, it is believed that the conventional σ bonding from nitrogen or phosphorus to the acceptor species is reinforced in the case of phosphorus by π back-donation from the filled orbitals of the acceptor. Further reference will be made to this subject in Chapter 22.

The formation of polymeric species by phosphorus or arsenic, where the corresponding nitrogen compounds are monomers, has been illustrated several times earlier in this chapter; it is also a feature of organic derivatives of the elements. Thus there is no phosphorus or arsenic analogue of PhN=NPh or $CF_3N{=}NCF_3$; instead there are cyclic polymers such as P_6Ph_6, $P_4(CF_3)_4$, $As_4(CF_3)_4$ and $P_5(CF_3)_5$ containing puckered rings of phosphorus or arsenic atoms each bonded to a single organic group. When $R = 2,4,6\text{-}(Me_3C)_3C_6H_2$ or $(Me_3Si)_3C$, however, steric hindrance prohibits the dimerisation of RP=PR; the phosphorus—phosphorus distance in these compounds is 2.02 Å, compared with 2.21 Å in P_2I_4 and other derivatives of P_2H_4. Arsenic forms analogous compounds. The absence of penta-alkyl or -aryl derivatives of nitrogen is, of course, analogous to the non-existence of pentahalides. Pentaalkyl derivatives of phosphorus and antimony have not been reported, but the pentaaryl compounds are reasonably stable; $SbPh_5$ and even $BiPh_5$ are also known. Whereas the configuration around the phosphorus and arsenic atoms in the pentaphenyl is trigonal bipyramidal, that around the antimony atom in antimony pentaphenyl is square pyramidal with the antimony about 0.5 Å above the base of the pyramid. Bismuth pentaaryls are square pyramidal.

The organic derivatives of phosphorus and its homologues are usually made by the action of Grignard reagents or active organometallic compounds on halides, e.g.

$$PX_3 + 3RMgX \rightarrow PR_3 + 3MgX_2$$

$$PX_3 + 3LiR \rightarrow PR_3 + 3LiX$$

Mixed halo organo species can be obtained by using a deficiency of the organometallic compound, or by the use of a weaker alkylating or arylating agent such as a mercury compound, e.g.

$$PCl_3 + LiR \rightarrow RPCl_2 + LiCl$$

$$PCl_3 + 2HgR_2 \rightarrow R_2PCl + 2RHgCl$$

Such mixed derivatives are invaluable for synthesising other alkyl or alkyl substituted compounds. Direct synthesis is also sometimes possible, e.g.

$$CF_3I + P \xrightarrow{\text{heat}} P(CF_3)_3 + IP(CF_3)_2 + I_2P(CF_3)$$

$$IP(CF_3)_2 \xrightarrow{\text{heat + Hg}} (CF_3)_2P\text{—}P(CF_3)_2 \qquad I_2P(CF_3) \xrightarrow{\text{heat + Hg}} (PCF_3)_n + HgI_2$$

Pentaaryl antimony and bismuth compounds are obtained by the reaction sequence:

$$MX_3 \xrightarrow{RMgX} MR_3 \xrightarrow{Cl_2} MR_3Cl_2 \xrightarrow{RMgX} MR_5$$

Trimethyl derivatives of phosphorus, arsenic, antimony and bismuth are all attacked by air, but aryl derivatives, including those of bismuth, are relatively inert. Trimethylbismuth shows no basic properties. It is interesting to note that whilst triaryl compounds become progressively less stable from phosphorus to bismuth, the reverse is true of the pentaphenyls of these elements; but it is not clear how far these observations relate to kinetic, and how far to thermodynamic, factors.

PROBLEMS

1 Suggest syntheses for
(a) $Na^{15}NH_2$, $^{15}N_2$ and $^{15}NOAlCl_4$ from $K^{15}NO_3$;
(b) $^{32}PH_3$, $H_3{}^{32}PO_3$ and $Na_3{}^{32}PS_4$ from $Ca_3(^{32}PO_4)_2$.

2 Predict the structures of the following species
(a) NF_4^+, (b) $N_2F_3^+$, (c) NH_2OH, (d) $SPCl_3$, (e) PF_3Cl_2.

3 How may nuclear magnetic resonance spectroscopy be used
(a) to distinguish between solutions of $Na_5P_3O_{10}$ and $Na_6P_4O_{13}$;
(b) to determine whether fluorine atoms exchange rapidly between the axial and equatorial positions in AsF_5;
(c) to determine the positions of the NMe_2 groups in $P_3N_3Cl_3(NMe_2)_3$?

4 Deduce what you can about the nature of the following reactions:
(a) One mole of NH_2OH reacts with two moles of Ti(III) in the presence of excess of alkali, and the Ti(III) is converted into Ti(IV).
(b) Silver phosphite, Ag_2HPO_3, is warmed with water, and all the silver is precipitated as the metal.
(c) When one mole of hypophosphorous acid is treated with excess of iodine in acidic solution, one mole of iodine is reduced; on making the solution alkaline, another mole of iodine is consumed.

5 (a) 25 cm^3 of a 0.0500 molar solution of sodium oxalate ($Na_2C_2O_4$) reacted with 24.8 cm^3 of a solution of potassium permanganate **A** in the presence of excess of sulphuric acid. 25 cm^3 of a 0.0494 molar solution of hydroxylamine (NH_2OH) in sulphuric acid was boiled with excess of iron(III) sulphate solution, and when the reaction was complete the iron(II) which was produced was found to be equivalent to 24.65 cm^3 of solution **A**.

The product **B** formed from hydroxylamine in this reaction may be assumed not to interfere with the determination of iron(II); what can you deduce about its identity?

(b) 25 cm^3 of a 0.0500 molar solution of sodium oxalate reacted with 24.7 cm^3 of a solution of potassium permanganate **C** in the presence of excess of sulphuric acid.

25.00 cm^3 of a 0.0250 molar solution of hydrazine (N_2H_4) when treated with excess of alkaline hexacyanoferrate(III) ($[Fe(CN)_6]^{3-}$) solution gave hexacyanoferrate(II) ($[Fe(CN)_6]^{4-}$) and a product **D**. The hexacyanoferrate(II) formed was oxidised to hexacyanoferrate(III) by 24.80 cm^3 of solution **C**; **D** did not interfere with the titration. What can you deduce concerning the identity of **D**?

REFERENCES FOR FURTHER READING

Chatt, J. (1978) *Chem. Reviews*, **78**, 589. An advanced account of work on nitrogen fixation.

Corbridge, D.E.C. (1990) *Phosphorus*, 4th edition, Elsevier, Amsterdam. An up-to-date review of all aspects of phosphorus chemistry.

Cotton, F.A. and Wilkinson, G. (1988) *Advanced Inorganic Chemistry*, John Wiley, New York. Chapters 10 and 11 give a detailed account of the chemistry of nitrogen and phosphorus with special emphasis on coordination compounds.

Emsley, J. and Hall, D. (1976) *The Chemistry of Phosphorus*, Harper and Row, London. A thorough advanced treatment of most aspects of phosphorus chemistry.

Greenwood, N.N. and Earnshaw, A. (1984) *Chemistry of the Elements*, Pergamon Press, Oxford. Chapters 11–13 give a full account of the chemistry of the Group V elements.

Powell, P. and Timms, P. (1974) *The Chemistry of the Non-Metals*, Chapman and Hall, London. Covers most subjects discussed in this chapter at a slightly higher level and in a refreshingly different manner.

Wells, A.F. (1984) *Structural Inorganic Chemistry*, 5th edition, Oxford University Press, Chapters 18–20. Structural chemistry of Group V elements.

15 *Oxygen, sulphur, selenium, tellurium and polonium*

Introduction • The elements • Hydrides • Halides, oxohalides and complex halides • Oxides • Oxo-acids and their salts • Sulphur–nitrogen compounds • Aqueous solution chemistry of sulphur, selenium and tellurium • Organic derivatives

15.1 Introduction

Oxygen occupies so central a position in any treatment of inorganic chemistry that most of its compounds are dealt with under other elements, though in this chapter we have included hydrogen peroxide and the oxygen fluorides; water was of course considered in some detail in Chapter 7. Further, knowledge of the chemistry of polonium is limited by the absence of a stable isotope of this element and the difficulty of working with ^{210}Po, the most readily available form of the element. This, which is made by a (n,γ) reaction on ^{209}Bi, followed by β^--decay of the product, is an intense α emitter of half-life 138 days; it evolves 520 kJ per hour per gram (leading to decomposition of many of its compounds) and is a useful source of energy in space satellites; in addition, α-emission in aqueous solution leads to decomposition of water and enormously complicates the study of the solution chemistry of the element. It is, however, known that polonium crystallises in a simple cubic structure and forms volatile readily hydrolysed halides $PoCl_2$, $PoCl_4$, $PoBr_2$, $PoBr_4$ and PoI_4, complex halide ions of formula $[PoX_6]^{2-}$ (where X = Cl, Br or I), and a dioxide which is only sparingly soluble in aqueous alkali. These observations agree with expectations for the homologue of tellurium; we shall add little more about polonium in the remainder of this chapter.

The range of physical properties of oxygen, sulphur, selenium, tellurium and polonium given in Table 15.1 is somewhat more restricted than that given at the beginning of chapters on elements of Groups I–V.

Table 15.1
Some properties of oxygen, sulphur, selenium, tellurium and polonium.

Element	O	S	Se	Te	Po
Atomic number	8	16	34	52	84
Electronic configuration	$[He]2s^2 2p^4$	$[Ne]3s^2 3p^4$	$[Ar]3d^{10} 4s^2 4p^4$	$[Kr]4d^{10} 5s^2 5p^4$	$[Xe]4f^{14} 5d^{10} 6s^2 6p^4$
ΔH° atomisation[a]/kJ mol^{-1}	247	278	207	192	145
M.p./°C	−229	114	221	452	254
B.p./°C	−183	445	685	1087	962
First ionisation energy/kJ mol^{-1}	1314	999	941	869	813
Covalent radius[b]/Å	0.73	1.04	1.17	1.37	—
Ionic radius of X^{2-}/Å	1.40	1.85	1.95	2.20	—
Pauling electronegativity	3.4	2.6	2.6	2.0	—

[a] Note this refers to production of one mole of gaseous atoms from the element in its standard state; thus for oxygen it is half the dissociation energy of O_2.
[b] For two-coordinated element.

Table 15.2
Some covalent bond energy terms for compounds of oxygen, sulphur, selenium and tellurium ($kJ\ mol^{-1}$).

O—O	O—H	O—C	O—F (OF_2)	O—Cl (OCl_2)
146	467	359	190	205
S—S	S—H	S—C	S—F (SF_6)	S—Cl (S_2Cl_2)
266	374	272	326	255
Se—Se	Se—H		Se—F (SeF_6)	Se—Cl ($SeCl_2$)
192	276		285	243
	Te—H		Te—F (TeF_6)	
	238		335	

With the possible exception of polonium dioxide, one form of which has been reported to have the fluorite structure, there is no evidence for the formation of compounds containing monatomic cations by any of the Group VI elements; we have therefore given only the first ionisation energies to show that there is the usual decrease with increasing atomic number. Polyatomic cations such as O_2^+, S_8^{2+}, Se_8^{2+}, and Te_4^{2+}, on the other hand, are well known. Some covalent bond energy terms are given in Table 15.2; in addition we should mention the bond energies in O_2 and the unstable species S_2 (which, like O_2, is paramagnetic and hence has an electronic configuration analogous to that described for O_2 in Section 4.2). The bond energy in O_2 is 493 $kJ\ mol^{-1}$; that for S_2 is 430 $kJ\ mol^{-1}$. Thus O_2 is stable with respect to polymerisation to a single bonded species; S_2 is not, and forms S_8.

In discussing the elements of Groups IV and V we laid stress on the significance in the chemistry of the first period element of the greater importance of p_π–p_π bonding and of the effect of the restriction to an octet of electrons in the valence shell. These factors are also responsible for some of the differences between oxygen and its homologues. Thus there are no stable sulphur compounds corresponding to CO and NO (though the sulphur analogue of CO_2 is, of course, a stable species), and the highest fluoride of oxygen is OF_2, whereas sulphur forms SF_6. Again we may note the relative weakness of the bond when an atom of a first period element is linked to another atom having unshared electrons on it: thus O—O and O—F are much weaker bonds than S—S and S—F, though O—H and O—C are much stronger than S—H and S—C.

The fact that water is strongly hydrogen-bonded and hydrogen sulphide is not requires no further comment here; but we may point out that there is thermochemical evidence for weak O—H . . . S hydrogen bonding between, e.g. phenol and n-butyl sulphide in carbon tetrachloride solution.

The estimation of the sum of the first two electron affinities of the oxygen atom was discussed in Section 6.8; application of the Born cycle to the oxides of magnesium and the alkaline earth metals leads to a value of $\Delta H^\circ = +630\ kJ\ mol^{-1}$ for the gas-phase reaction

$$O + 2e \rightarrow O^{2-}$$

and it was concluded that oxide ions exist in lattices only because of the high lattice energies of metal oxides. Most (though by no means all) metal oxides have symmetrical three-dimensional structures: MgO, CaO, SrO, BaO, CdO, VO, MnO and CoO, for example, have the rock salt structure; SnO_2, PbO_2, VO_2 and MnO_2 have the rutile structure; CeO_2, ThO_2 and UO_2 have the fluorite structure; Li_2O, Na_2O, K_2O and Rb_2O have the antifluorite structure; and Al_2O_3, Cr_2O_3 and Fe_2O_3 have the corundum structure (Section 12.8). A Born cycle evaluation of ΔH° for the gas-phase reaction

$$S + 2e \rightarrow S^{2-}$$

gives a value of $+330$ kJ mol^{-1} for the sum of the first two electron affinities of sulphur: repulsion between electrons is less in a larger anion. On the other hand, lattice energies of sulphides are much lower than those of the corresponding oxides, and high oxidation-state oxides (e.g. MnO_2 and PbO_2) often have no sulphide analogues; further, it is frequently found that sulphides are not isostructural with oxides; for example VS and CoS have the nickel arsenide structure, SnS_2 the cadmium iodide structure. Agreement between cycle and calculated lattice energies for many transition metal sulphides is much poorer than it is for oxides, indicating that non-electrostatic interactions contribute substantially to the bonding in the sulphides. Similar considerations apply to selenides and tellurides.

Selenium and tellurium have a relatively simple oxo-acid chemistry, but that of sulphur resembles the complicated system of phosphorus oxo-acids and anions. There are structural analogies between sulphates and phosphates (though far fewer condensed sulphates are known), and oxidation and reduction processes involving sulphur oxo-anions, like those involving the phosphorus-containing species, are often slow. A special feature of the chemistry of sulphur is the existence of polysulphides and polythionates containing chains of sulphur atoms.

The isotope ^{18}O, present to the extent of 0.2 per cent in ordinary oxygen, is commonly used as a tracer for oxygen, but ^{17}O (abundance 0.04 per cent), which has a nuclear spin quantum number of $\frac{5}{2}$ and has been used in studies of hydrated ions in solution (Section 7.3), is also important. Both are obtained by fractional electrolysis or fractional distillation of water or by thermal diffusion of oxygen. The usual tracer for sulphur is ^{35}S, made by a (n, p) reaction on ^{35}Cl in the form of potassium chloride; it is a β^--emitter of half-life 87 days.

Oxygen is always obtained on an industrial scale by the fractional distillation of liquid air, and is stored and transported as a liquid. Smaller amounts of the pure gas are made by electrolysis of aqueous alkali using nickel electrodes, or by the platinum-catalysed decomposition of aqueous hydrogen peroxide. Oxygen is, of course, also produced in the thermal decomposition of many oxo salts (e.g. KNO_3, $KMnO_4$, $KClO_3$, $K_2S_2O_8$). Its chief uses are as a fuel, e.g. for oxyacetylene and hydrogen flames, as a supporter of respiration under special conditions (e.g. in aircraft), and in steel manufacture.

Sulphur occurs native in several countries and is mined by a special method (the Frasch process), conventional mining being made impossible because of quicksands above deposits. In this process, three large concentric pipes are sunk into the bed of sulphur. Superheated steam is forced down the outside pipe and melts the sulphur, which is forced up the next pipe by compressed air pumped down the middle pipe. Other sources of sulphur compounds are *anhydrite* (calcium sulphate), hydrogen sulphide (which occurs in natural gas and in various industrial waste gases), and sulphur dioxide (obtained as a by-product in the extraction of metals from sulphide ores). Sulphur may be obtained from hydrogen sulphide by controlled catalytic oxidation on active carbon or alumina:

$$2H_2S + O_2 \rightarrow 2S + 2H_2O$$

Most sulphur (about 85 per cent of total production) is used for the manufacture of sulphuric acid, smaller quantities for the preparations of sulphites for paper manufacture, gunpowder, matches, vulcanised rubber and drugs. Selenium and tellurium are usually obtained from sulphide ores, in which they occur as impurities; they are, for example, present in the flue dust from roasting such ores in air, or in the anode sludge obtained in the electrolytic refining of copper. Selenium is widely used in photoelectric cells (see the next section) and photocopiers. Tellurium is used to harden lead, but has otherwise found few applications, partly, no doubt, owing to the ease with which tellurium compounds are absorbed by the body and excreted in breath and perspiration as foul-smelling organic derivatives.

15.2 The elements

The decrease in non-metallic character with increase in atomic number is clearly seen in the properties and structures of the Group VI elements: oxygen exists only as two gaseous allotropes; sulphur is known in a variety of forms, but all are insulators; the stable forms of selenium and tellurium, which have infinite chain structures, are semiconductors; and polonium is a metallic conductor.

Oxygen is a colourless gas, but it condenses to a pale blue liquid or solid; it is paramagnetic in all three phases. Its ground state electronic configuration (Section 4.2) has an unpaired electron with the same spin in each of two antibonding orbitals, i.e. a triplet state. Oxygen in its ground state is, of course, a powerful oxidising agent (as we showed in Section 7.8, at $[H^+] = 1$, $E^\circ = +1.23$ V for the half-reaction

$$\tfrac{1}{2}O_2 + 2H^+ + 2e \rightleftharpoons H_2O$$

so that it is about as powerful as dichromate), but it is often a slow one; if it were not, virtually all organic chemistry would have to be investigated in closed systems. There is, however, an excited state of oxygen, in which the two antibonding electrons, though occupying different orbitals, have opposed spins (a singlet state), only 95 kJ mol^{-1} above the ground state. Singlet oxygen may be generated photochemically by irradiation of ordinary oxygen in the presence of an organic dye as sensitiser or in alcohol solution by means of the reaction

$$H_2O_2 + OCl^- \rightarrow Cl^- + H_2O + O_2$$

It is very reactive and combines with many organic compounds, a typical example being the Diels–Alder-like addition to butadiene:

singlet O_2

At higher temperatures oxygen combines with most elements (though not, it may be noted, with the noble gases or the halogens, or with nitrogen except under special conditions). With alkali metals, oxides, peroxides or superoxides (i.e. species containing O^{2-}, O_2^{2-} and O_2^- ions respectively) may be formed according to the size of the cation (cf. Section 10.4).

Oxygen is readily soluble in organic solvents and forms weak charge-transfer complexes with many organic molecules; reversible combination to form more stable adducts takes place with certain transition metal complexes, e.g. Vaska's compound (Section 22.5) and haemoglobin (Section 24.7).

The first ionisation energy of O_2 is 1168 kJ mol^{-1}, and it can be oxidised to the O_2^+ ion by very powerful oxidants such as platinum hexafluoride, which gives the orange salt $O_2^+[PtF_6]^-$; the bond length in the cation is, as expected, shorter than in O_2. Other dioxygenyl salts (as they are called) include $O_2^+[SbF_6]^-$ (made by irradiation of oxygen and fluorine in the presence of antimony pentafluoride) and $O_2^+[BF_4]^-$ (made from O_2F_2 and BF_3, elemental fluorine being formed as a by-product); $O_2^+[BF_4]^-$ is isomorphous with $NO^+[BF_4]^-$.

Ozone, O_3, is usually prepared in up to 10 per cent concentration by the action of a silent electrical discharge between two concentric metallised tubes in an apparatus called an ozoniser, but it is also formed by the action of ultraviolet radiation on oxygen, or by heating oxygen above 2500 °C followed by rapid quenching; in all these processes oxygen atoms are produced and combine with oxygen molecules. Pure ozone, a deep blue liquid which boils at −112 °C forming a perceptibly blue gas, may be separated by fractional liquefaction. Both the liquid and the gas are diamagnetic. The presence of ozone, which absorbs strongly in the ultraviolet, in the upper atmosphere of the earth (its maximum concentration occurs at a height of about 25 km) protects the earth's surface from over-exposure to ultraviolet radiation from the sun. (Radiation in the far-ultraviolet is absorbed by oxygen or nitrogen.) The formation and destruction of ozone are controlled by the reactions

$$O_2 + h\nu\ (<240\ \text{nm}) \rightarrow O + O$$

$$O + O_2 + M \rightarrow O_3 + M$$

(M is a third body, N_2 or O_2, which removes the energy liberated in this reaction as heat)

$$O_3 + h\nu\ (<300\ \text{nm}) \rightarrow O + O_2$$

$$O + O_3 \rightarrow 2O_2$$

These normally result in a low steady-state concentration of ozone. The last reaction, however, is relatively slow, but is catalysed by atomic chlorine produced, for example, by photochemical decomposition of chlorofluorocarbons:

$$Cl + O_3 \rightarrow ClO + O_2$$

$$ClO + O \rightarrow Cl + O_2$$

or, in sum

$$O + O_3 \rightarrow 2O_2$$

This is the reason for the campaign to phase out the use of chlorofluorocarbons as propellants and refrigerants. Nitric oxide has a similar but smaller effect.

Ozone is a strongly endothermic molecule: ΔH° for the reaction

$$\tfrac{3}{2}O_2 \rightarrow O_3$$

is $+142\ \text{kJ mol}^{-1}$. The pure liquid is dangerously explosive, and the gas is a very powerful oxidising agent; for the half-reaction

$$O_3 + 2H^+ + 2e \rightleftharpoons O_2 + H_2O$$

at $[H^+]=1$ in aqueous medium, $E^\circ = +2.07$ V. At lower $[H^+]$, E diminishes, the value at pH = 7 being +1.65 V; at pH = 14, +1.24 V (cf. Section 7.8). In the presence of high concentrations of alkali, ozone is kinetically as well as thermodynamically more stable. Ozone is much more reactive than oxygen (hence its use in water purification); for example, it oxidises moist sulphur to sulphuric acid, and (unlike oxygen) liberates iodine from neutral aqueous iodide:

$$O_3 + 2I^- + H_2O \rightarrow O_2 + I_2 + 2OH^-$$

Alkaline aqueous iodide is oxidised to iodate and periodate, and lead sulphide to lead sulphate. Alkenes react at ordinary temperatures to form ozonides. Ozone reacts with solid potassium hydroxide to form unstable red potassium ozonide, KO_3, containing the paramagnetic O_3^- ion:

$$2KOH + 5O_3 \rightarrow 2KO_3 + 5O_2 + H_2O$$

The molecule of ozone is bent, with OOO bond angle = 117° and bond length 1.28 Å, almost the same as that in the superoxide anion. We may describe the electronic structure of the molecule as a resonance hybrid of

$$\overset{\times\times}{\mathrm{O}}{}^{+}\ (=\mathrm{O},\ -\mathrm{O}^{-}) \quad \text{and} \quad \overset{\times\times}{\mathrm{O}}{}^{+}\ (-\mathrm{O}^{-},\ =\mathrm{O})$$

or, perhaps more neatly, as a sigma-bonded system

$$\overset{\times\times}{\mathrm{O}}\ (-\mathrm{O},\ -\mathrm{O})$$

with the four remaining electrons assigned to those multicentre orbitals formed from a *p* orbital of each atom perpendicular to the molecular plane. These three orbitals combine to form a bonding, a non-bonding and an antibonding orbital, of which only the first two are filled, giving a resulting π bond order of 0.5 between each pair of atoms. It may be noted that ozone is effectively isoelectronic with NO_2^-.

The allotropy of sulphur has been much investigated and is very complicated; we shall deal here only with the best-established species. Allotropes of known structure include those containing six- to twenty-membered puckered rings and fibrous sulphur, which contains infinite chains. In all of these the S—S distance is 2.06 ± 0.01 Å and the SSS bond angle is 102–108°; the dihedral angle (the angle between the planes containing any bond and the bond next but one to it, discussed for hydrogen peroxide in the next section) varies from 74° in S_6 to 98° in S_8. Thus the only differences between the allotropes are those which arise from variation in the amount of repulsion between the lone pairs of electrons on different atoms, and the thermodynamic differences between the different forms are very small.

The most stable form is S_α or orthorhombic sulphur, which occurs naturally as large yellow crystals in volcanic areas; it contains crown-shaped S_8 rings (Fig. 15.1). At 94.2 °C S_α is transformed reversibly into monoclinic S_β, which contains the same structural units as S_α but differently packed. When, however, S_α is heated rapidly, it melts at 112 °C before transformation into S_β occurs; S_β melts at about 128 °C, but this is not a true melting-point, since at this temperature some breakdown of the S_8 rings is occurring and the melting-point is being depressed by the decomposition products. Rhombohedral S_ρ is obtained by means of the reaction

$$S_2Cl_2 + H_2S_4 \rightarrow S_6 + 2HCl$$

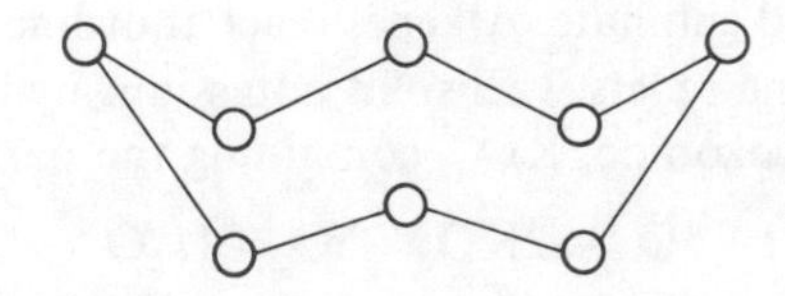

Figure 15.1
The structure of S_8.

in dry ether. It contains six-membered puckered rings in the chair form, and it is less stable than other forms, decomposing in light into S_8 and S_{12}. Other ring forms include S_9, S_{10}, S_{11}, S_{12}, S_{18} and S_{20}; they are mostly obtained by reactions between hydrogen polysulphides and polysulphur dichlorides, the general reaction being

$$H_2S_x + S_yCl_2 \rightarrow S_{x+y} + 2HCl$$

All of these finite ring polymers are soluble in carbon disulphide. By rapidly quenching molten sulphur at 300 °C in ice-water, fibrous sulphur, which is insoluble in this solvent, is obtained. Fibrous sulphur can be stretched under water and contains helical chains of sulphur atoms; it slowly reverts to S_α.

Sulphur melts to a mobile yellow liquid which darkens in colour as the temperature is raised; at 160 °C the viscosity increases enormously as S_8 rings break to give diradical species that polymerise to form chains containing up to 10^6 atoms. The viscosity reaches a maximum at about 200 °C and then decreases up to the boiling-point (444 °C), at which temperature the liquid is a complex mixture of rings and shorter chains.

The vapour above liquid sulphur at 200 °C consists mainly of cyclic S_8 molecules; at higher temperatures species containing fewer atoms predominate, and above 600 °C paramagnetic blue S_2 becomes the principal species. Dissociation into atoms occurs only above 2200 °C.

Sulphur reacts when heated with oxygen and the halogens (other than iodine); hot aqueous alkali gives a complex mixture of polysulphides and polythionates; oxidising acids convert it into sulphuric acid. Saturated hydrocarbons are dehydrogenated when heated with sulphur, and the element reacts with alkenes: the vulcanisation of rubber by hot sulphur involves the formation of sulphide bridges between carbon chains. Cyanides and sulphites combine with sulphur to form thiocyanates and thiosulphates respectively.

Cationic sulphur compounds are formed by reactions such as

$$S_8 + 3SbF_5 \rightarrow S_8^{2+} + 2SbF_6^- + SbF_3$$

$$S_8 + 2S_2O_6F_2 \rightarrow 2S_4^{2+} + 4SO_3F^-$$

An X-ray diffraction study of dark blue $S_8(AsF_6)_2$ (made by a reaction analogous to that given above for the fluoroantimonate) shows that the cation contains a puckered S_8^{2+} ring, but the conformation of the ring (Fig. 15.2) differs at one end from that of molecular S_8; S_4^{2+} is planar. The S—S distances round the ring in S_8^{2+} are 2.04 Å, slightly less than in S_8 rings, but the cross-ring distance S_3–S_7 (2.87 Å) is much less than in S_8 (4.7 Å) and shows that there is some transannular bonding. In this respect the system shows some similarity to S_4N_4 (Section 15.7).

The stable forms of selenium and tellurium, which are respectively grey and silvery-white metallic-looking solids, are infinite spiral chain polymers with bond angles of 103°; each atom has, however, four more

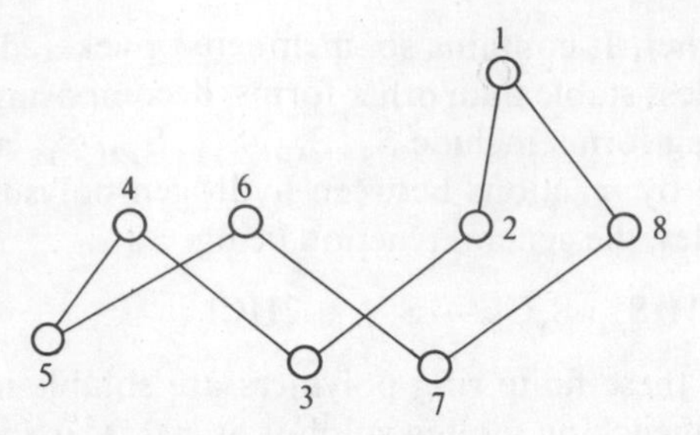

Figure 15.2
The structure of the S_8^{2+} cation.

next nearest neighbours belonging to other chains, giving it a distorted octahedral environment. Interatomic distances are Se–2Se 2.37 Å, Se–4Se 3.44 Å and Te–2Te 2.84 Å, Te–4Te 3.47 Å, showing that tellurium comes appreciably nearer than selenium to having a symmetrical three-dimensional structure. By rapid cooling of molten selenium and extraction with carbon disulphide, two red allotropes containing Se_8 rings (similar in structure to S_8 rings) may be obtained; there is also a black form consisting of large polymeric rings. No molecular form of tellurium is known, but there is a high-pressure modification in which, as in polonium, each atom has six nearest neighbours distributed around it in the form of a regular octahedron. Grey selenium, and to a smaller degree tellurium, are semiconductors which exhibit marked photoconductivity; selenium is widely used in photoelectric cells, photocopying machines and solar batteries. Its photoconductivity arises because the energy gap between the band of filled bonding orbitals and the band of empty antibonding orbitals in the solid (160 kJ mol^{-1}) is small enough for electrons to be promoted from the former to the latter under the influence of visible light (cf. Section 4.9).

In their chemical properties selenium and tellurium closely resemble sulphur; this resemblance extends to the formation of salts containing cyclic Se_4^{2+}, Te_4^{2+} and Se_8^{2+} ions by methods analogous to those used for the preparation of the sulphur-containing compounds. Tellurium also forms a Te_6^{4+} ion having a trigonal prismatic structure consisting of two Te_3 units (with Te—Te 2.68 Å) loosely held together by three weak bonds of length 3.13 Å.

15.3 Hydrides

We have already discussed the structure and properties of water in considerable detail (Chapter 7), but the preparation of pure water has not yet been described. The simplest method for the removal of all solid solutes is distillation, but because of the high boiling-point and enthalpy of vaporisation this method is rather expensive, and for the removal of electrolytes ion-exchange is cheaper. This involves the passage of water down a column of an organic resin containing acidic groups (e.g. $—SO_3H$) and then down a similar column containing basic groups (e.g. $—NR_3OH$). The former removes cations by means of the reaction

$$\text{Resin—}SO_3H + M^+X^- \rightarrow \text{Resin—}SO_3M + H^+ + X^-$$

and the latter the hydrogen ions which are produced by the reaction

$$\text{Resin—}NR_3OH + H^+ + X^- \rightarrow \text{Resin—}NR_3X + H_2O$$

The product is known as deionised water; the resins are reactivated as necessary by treatment with dilute sulphuric acid and sodium carbonate solution respectively. Reverse osmosis at high pressures is also an important process for water purification; cellulose acetate is the usual membrane.

The other hydride of oxygen is hydrogen peroxide, H_2O_2. Salts of this compound (e.g. Na_2O_2, BaO_2) are obtained by heating alkali and alkaline earth metals in air, and the oldest preparation of hydrogen peroxide is by the action of dilute sulphuric acid on barium peroxide:

$$BaO_2 + H_2SO_4 \rightarrow BaSO_4 + H_2O_2$$

Salts of peroxo-acids are often obtained by electrolytic oxidation in cold aqueous solutions at high current densities using platinum electrodes, e.g.

$$2NH_4HSO_4 \rightarrow H_2 + (NH_4)_2S_2O_8$$

The solution of ammonium peroxodisulphate so obtained is heated to bring about hydrolysis of the peroxodisulphate, and a solution of hydrogen peroxide may be obtained by distillation under reduced pressure:

$$(NH_4)_2S_2O_8 + 2H_2O \rightarrow 2NH_4HSO_4 + H_2O_2$$

This is concentrated to 98 per cent by low-pressure fractionation.

Modern methods for the manufacture of hydrogen peroxide avoid electrolytic processes (which are costly in terms of power and capital investment) and make use of the fact that in the partial oxidation of organic compounds by air, hydrogen peroxide is often produced. Oxidation of 2-ethylanthraquinol or one of its homologues, for example, can be carried out in a mixture of alcohols or esters (solvents for the quinol) and alkylbenzenes (solvents for the quinone):

OH
Et
OH
$\xrightarrow{O_2}$
O
Et
O
$+ H_2O_2$

The hydrogen peroxide formed is extracted into water, and the alkylanthraquinone is reduced back to the starting material by means of hydrogen in the presence of a palladium catalyst. An alternative process is based on the partial oxidation of 2-propanol:

$$\mathrm{Me_2CHOH + O_2 \rightarrow Me_2C\langle^{OH}_{OOH}}$$

$$\mathrm{Me_2C\langle^{OH}_{OOH} \rightarrow Me_2CO + H_2O_2}$$

Pure hydrogen peroxide freezes at −0.4 °C and boils with decomposition at 150 °C; like water, it is strongly hydrogen-bonded. The pure compound and its concentrated aqueous solutions are very readily decomposed according to the equation

$$\mathrm{H_2O_2(l) \rightarrow H_2O(l) + \tfrac{1}{2}O_2(g)} \qquad \Delta H^\circ = -98\ \mathrm{kJ\ mol^{-1}}$$

in alkaline solutions, or in the presence of heavy metal ions or of heterogeneous catalysts such as platinum and manganese dioxide; they are best stored in polythene bottles in a refrigerator. Traces of complexing agents (e.g. 8-hydroxyquinoline) or adsorbing materials (e.g. sodium stannate) are often added as stabilisers. Mixtures of hydrogen peroxide and organic or other easily oxidised materials are dangerously explosive; that of hydrogen peroxide and hydrazine has been used for rocket propulsion. Most of the hydrogen peroxide now produced is used for bleaching, control of water pollution, or the preparation of sodium peroxoborate and peroxocarbonate.

The ionisation constant of hydrogen peroxide in aqueous solution at 20 °C is given by

$$K = \frac{[\mathrm{H^+}][\mathrm{HO_2^-}]}{[\mathrm{H_2O_2}]} = 2 \times 10^{-12}$$

In alkaline media, therefore, it is present as the HO_2^- ion. For the half-reaction

$$\mathrm{H_2O_2 + 2H^+ + 2e \rightleftharpoons 2H_2O}$$

at unit hydrogen ion concentration, $E^\circ = +1.77$ V, so that hydrogen peroxide is a powerful oxidising agent; it will, for example, oxidise iodide to iodine, sulphur dioxide to sulphuric acid, and (in alkaline medium) chromium(III) to chromate(VI). With some powerful oxidising agents, however, it is itself oxidised to oxygen; the relevant half-reaction is

$$\mathrm{O_2 + 2H^+ + 2e \rightleftharpoons H_2O_2} \qquad E^\circ = +0.68\ \mathrm{V}$$

Permanganate or chlorine reacts with hydrogen peroxide for example:

$$\mathrm{2MnO_4^- + 5H_2O_2 + 6H^+ \rightarrow 2Mn^{2+} + 8H_2O + 5O_2}$$

$$\mathrm{Cl_2 + H_2O_2 \rightarrow 2HCl + O_2}$$

Tracer studies using ^{18}O show that in these reactions all the oxygen that is formed comes from the hydrogen peroxide, none from the solvent, so the O—O bond is not broken. Many reactions of hydrogen peroxide are radical reactions, e.g. the oxidation of Fe(II) at low hydrogen ion concentrations proceeds according to the equations

$$Fe^{2+} + H_2O_2 \rightarrow FeOH^{2+} + OH$$

$$Fe^{2+} + OH \rightarrow FeOH^{2+}$$

and a mixture of hydrogen peroxide and an iron(II) salt is sometimes used as a source of hydroxyl radicals for organic reactions. Unstable peroxo compounds (e.g. CrO_5) are formed in many reactions between transition metal compounds and hydrogen peroxide; many of these, like peroxo-acids of non-metallic elements, are mentioned elsewhere in this book.

The structure of gaseous hydrogen peroxide was shown in Fig. 5.4; the O—O bond length is 1.47 Å; the dihedral angle (the angle between the planes containing the O—H bonds) of 111° may be considered as arising from repulsions, between the pairs of unshared electrons on the oxygen atoms. However, the magnitude of the dihedral angle is very sensitive to the surroundings of the molecule: thus it is 90° in crystalline H_2O_2 (where there is intermolecular hydrogen bonding) and 180° (i.e. the molecule has the planar *trans* configuration) in the adduct $Na_2C_2O_4.H_2O_2$, in which the lone pairs appear to interact with the sodium ions. Values of the dihedral angle in organic peroxides also show wide variations, in this case from 81° to 146°.

Hydrogen sulphide, selenide and telluride are all poisonous foul-smelling gases (b.p. −61, −42 and −2 °C respectively); hydrogen sulphide is indeed more toxic than hydrogen cyanide, but because of its offensive odour of rotten eggs is in practice much less dangerous than the latter compound. Hydrogen sulphide occurs in some natural gas deposits; it may be removed by reversible absorption in a solution of an organic base, and converted into sulphur by controlled oxidation. In the laboratory it is usually made by the action of acids on metal sulphides. Hydrogen selenide and hydrogen telluride are made similarly. The standard enthalpies of formation of the three hydrides are: H_2S, −20; H_2Se, +86; H_2Te, +154 kJ mol^{-1}. In accordance with these values, hydrogen sulphide can be made by the action of hydrogen on boiling sulphur and is somewhat more stable with respect to decomposition into its elements than the other two compounds; all three, however, are readily oxidised.

In aqueous solution the hydrides behave as weak acids, their first ionisation constants being: H_2S, 10^{-7}; H_2Se, 10^{-4}; H_2Te, 10^{-3}. This increase in acidity is parallel to that found among the hydrides of the Group V and Group VII elements and doubtless originates in the decrease in bond energy in the hydrides with increase in atomic number (cf. Section 7.4). The second ionisation constant of hydrogen sulphide is approximately 10^{-19}; thus metal sulphides are hydrolysed in aqueous solution, and many can be isolated by the action of hydrogen sulphide on solutions of their salts only because they are extremely insoluble. Classical methods of qualitative analysis usually involve a distinction between sulphides which have solubility products less than about 10^{-30} (e.g. CuS, PbS, HgS, CdS, Bi_2S_3, As_2S_3, Sb_2S_3, SnS) and those which have solubility products of about 10^{-15} to 10^{-30} (e.g. ZnS, MnS, NiS, CoS). The former can be precipitated by hydrogen sulphide in the presence of dilute hydrochloric acid (which, of course, diminishes the available sulphide ion concentration by repressing the ionisation of hydrogen sulphide); the latter are precipitated only in neutral or weakly alkaline media. Protonation of hydrogen sulphide to form the H_3S^+ ion can be achieved by the use of HF/SbF_5 mixtures.

As would be expected, the molecules of H_2S and H_2Se are bent, but the bond angles (92° and 91° respectively) are considerably less than the bond angle in water (105°) and suggest that only *p* bonding is involved. Sulphur dissolves in aqueous solutions of alkali and alkaline earth metal sulphides to form polysulphides containing chain ions such as S_4^{2-} and S_5^{2-}; acidification of such solutions yields a mixture of hydrogen polysulphides as a yellow oil, which can be cracked and fractionated to give H_2S_2, H_2S_3, H_2S_4, H_2S_5 and H_2S_6. These and higher members of the series (*sulphanes*) can also be obtained by suitable condensation reactions at low temperatures, e.g.

$$2H_2S+S_2Cl_2 \rightarrow H_2S_4+2HCl$$

All are thermodynamically unstable with respect to decomposition into hydrogen sulphide and sulphur.

We have mentioned the structures of several metal sulphides in earlier chapters, but it may be useful to summarise them here. The alkali and alkaline earth metal sulphides have the antifluorite and sodium chloride structures respectively and appear to be typical salts. Lead sulphide (Section 13.11) also has the sodium chloride structure, though it is certainly not a simple ionic crystal, and the same is true of some other monosulphides (e.g. MnS). Most transition metal monosulphides, however, have the nickel arsenide structure (e.g. FeS, CoS and NiS) or the zinc blende or wurtzite structure (e.g. ZnS, CdS, HgS). Some disulphides (e.g. TiS_2, SnS_2) have the cadmium iodide structure, but others (e.g. CoS_2, NiS_2) have the *pyrites* (FeS_2) structure and contain S_2^{2-} ions, i.e. they are formally analogous to the peroxides and may be considered as salts of the hydride H_2S_2. The blue paramagnetic S_2^- ion (analogous to O_2^-) has been detected in solutions of alkali metal polysulphides in acetone or dimethyl sulphoxide, and is probably present in the silicate mineral ultramarine, but simple salts containing it are not known. Some transition metal sulphides exhibit non-stoichiometry (Section 6.10); samples of iron(II) sulphide, for example, are usually deficient in iron. All polysulphides contain unbranched zig-zag chains of sulphur atoms.

Many sulphides dissolve in aqueous solutions of alkali metal sulphides to form thio-anions analogous to oxo-anions (just as carbon disulphide dissolves in aqueous sodium sulphide to form the CS_3^{2-} ion). Complexes containing polysulphide ions as bidentate ligands are also known.

15.4 Halides, oxohalides and complex halides

The only halides of oxygen that will be considered in this section are the fluorides; chlorine, bromine and iodine oxides are discussed in the next chapter.

Oxygen difluoride, OF_2, a pale yellow gas (b.p. -145 °C), is obtained by passing fluorine rapidly through dilute aqueous sodium hydroxide. Although it is formally the anhydride of hypofluorous acid, HOF, the only reaction with water which occurs at ordinary temperature is slow hydrolysis to oxygen and hydrofluoric acid; with concentrated alkali hydrolysis is much faster, and with steam it is explosive. When pure, the compound may be heated to 200 °C without decomposition, but it reacts with many elements (to form fluorides and oxides) at, or slightly above, ordinary temperature.

The molecule OF_2 is bent, the bond angle being 103° and the bond length 1.41 Å. When subjected to ultraviolet radiation in an argon matrix at 4 K, the radical OF is produced:

$$OF_2 \rightarrow OF + F$$

On warming, the OF radicals combine to form dioxygen difluoride, O_2F_2, which may also be made by the action of a high voltage discharge on a mixture of oxygen and fluorine at 77–90 K and 10–20 mm pressure. Dioxygen difluoride is a yellow gas (m.p. -154 °C) which decomposes into the elements above -50 °C, the O_2F radical being formed as an intermediate. It is an extremely powerful fluorinating agent, even at low temperatures; for example, it inflames with sulphur at -180 °C. Boron trifluoride and antimony pentafluoride react with O_2F_2 forming dioxygenyl salts $O_2^+[BF_4]^-$ and $O_2^+[SbF_6]^-$ and fluorine, e.g.

$$O_2F_2 + BF_3 \rightarrow O_2^+[BF_4]^- + \tfrac{1}{2}F_2$$

The O_2F_2 molecule has the same shape as that of H_2O_2 (though the dihedral angle, 87°, is rather smaller). The bond lengths in this compound are remarkable. The O—O distance is only 1.22 Å (compared with 1.21 Å in O_2 and 1.48 Å in H_2O_2), and the O—F bond length (1.58 Å) is 0.17 Å greater than that in OF_2. The very long O—F bond probably accounts for the ease of dissociation into O_2F and F. In valence bond terminology the structure of O_2F_2 may be described as a resonance hybrid of

```
          F⁻                F
     +                    +/
      O=O       and     O=O
     /
    F                     F⁻
```

It is noteworthy that S_2F_2 contains an analogous short bond.

In contrast to the behaviour found in earlier groups, the stability of the lowest (dipositive) oxidation state of sulphur, selenium and tellurium in halides *decreases* with increase in atomic number. Thus the lowest fluorides of selenium and tellurium yet obtained are SeF_4 and TeF_4, and it may be noted that in hexafluorides tellurium forms a stronger bond to fluorine than either sulphur or selenium. In the following account we shall confine attention to fluorides and chlorides; where known, the bromides and iodides are closely similar to the chlorides.

The stable fluorides are S_2F_2 (two compounds), S_2F_4, SF_4, SF_6 and S_2F_{10}; all are gases or volatile liquids. The action of silver(I) fluoride on molten sulphur gives (in addition to SF_4) disulphur difluoride, S_2F_2 (analogous to O_2F_2) and thiothionyl fluoride, SSF_2; the former of these has the hydrogen peroxide structure with a very short sulphur–sulphur bond (1.89 Å); the latter is pyramidal, with a sulphur–sulphur bond length of 1.86 Å. These bond lengths, which are much shorter than that in S_8, provide evidence for high bond orders between the atoms of the Group VI elements in S_2F_2 and, by implication, O_2F_2. Disulphur difluoride slowly isomerises into thiothionyl fluoride, but both are also unstable with respect to disproportionation into sulphur tetrafluoride and sulphur; they are also extremely reactive, attacking glass and being rapidly hydrolysed by water. The fluoride S_2F_4 is $FSSF_3$.

Sulphur tetrafluoride (b.p. −38 °C) is best made by the reaction

$$3SCl_2 + 4NaF \xrightarrow[75\,^{\circ}C]{MeCN} SF_4 + S_2Cl_2 + 4NaCl$$

It is hydrolysed by water via SOF_2 to SO_2 and HF, and it combines with caesium fluoride to form the complex $CsSF_5$; this contains a square pyramidal $[SF_5]^-$ ion. Sulphur tetrafluoride is a useful fluorinating agent; it will, for example, convert carbonyl groups into CF_2 groups without destroying unsaturation in the molecule. Fluorine converts it into SF_6, chlorine monofluoride (or chlorine in the presence of caesium fluoride) into SF_5Cl, which is described later.

The structure of sulphur tetrafluoride is that of a trigonal bipyramid with one of the equatorial positions occupied by an unshared pair of electrons; the equatorial FSF angle is 103°. As we have mentioned earlier, however, in trigonal bipyramidal molecules the axial and equatorial bond lengths are never equal; in this instance they are, in fact, quite different, being 1.64 and 1.54 Å respectively, and the description of the molecule in terms of the simple VSEPR theory is a considerable oversimplification. Selenium and tellurium tetrafluorides resemble the sulphur compound, but solid TeF_4 (m.p. 130 °C) has a polymeric chain structure built up by sharing *cis* fluorine atoms between square pyramidal TeF_5 units.

Sulphur hexafluoride, a colourless, odourless, non-toxic gas (b.p. −64 °C) is widely used as an electrical insulator, and is made by burning sulphur in fluorine. A little S_2F_{10} is also formed and may be removed by pyrolysis at 400 °C, when it is converted into SF_6 and SF_4, and absorption of the latter in dilute alkali. Sulphur hexafluoride, which has a regular octahedral structure, is very inert chemically, being unaffected by steam at 500 °C or molten alkalis. This lack of reactivity is kinetic rather than thermodynamic in character: ΔG° for the gas phase reaction

$$SF_6 + 3H_2O \rightarrow SO_3 + 6HF$$

is -200 kJ mol^{-1}. The S—F bond in SF_6 is undoubtedly a strong one (326 kJ mol^{-1}), but the question of whether *d* orbitals are involved has been the subject of much controversy (cf. Section 4.6). A recent *ab initio* molecular orbital calculation indicates their contribution is only about one-sixth of that for sp^3d^2 hybridisation.

Selenium and tellurium hexafluorides, both gases made by burning the elements in excess of fluorine, are only slightly more reactive than sulphur hexafluoride, though the tellurium compound is slowly hydrolysed by water.

Many compounds containing the SF_5 group are now known, among them the gases SF_5Cl and SF_5NF_2, the latter being obtained by the action of S_2F_{10} on N_2F_4 in the presence of light. In most reactions of SF_5Cl only the S—Cl bond is broken, e.g.

$$2SF_5Cl + H_2 \xrightarrow{h\nu} F_5SSF_5 + 2HCl$$

$$2SF_5Cl + O_2 \xrightarrow{h\nu} F_5SOOSF_5 + Cl_2$$

The compound is, however, rapidly and completely hydrolysed by aqueous alkali. Disulphur decafluoride (b.p. 29 °C), in contrast to SF_6, is extremely poisonous, though it is similar to the hexafluoride in its lack of reactivity; the S—S bond length in this compound (2.21 Å) is substantially greater than that in elemental sulphur.

Sulphur also forms several oxofluorides. Thionyl or sulphinyl fluoride, SOF_2, is a colourless gas (b.p. -34 °C) made by the action of antimony trifluoride on thionyl chloride; the molcule is pyramidal, the sulphur atom having one unshared pair of electrons. It is only slowly hydrolysed by water. Sulphuryl or sulphonyl fluoride, SO_2F_2, b.p. -52 °C, is made from SO_2Cl_2 and NaF or by heating the barium salt of fluorosulphuric acid:

$$Ba(SO_3F)_2 \rightarrow BaSO_4 + SO_2F_2$$

It is unaffected by water, but is hydrolysed by concentrated aqueous alkali; its molecule is tetrahedral. Peroxodisulphuryl fluoride (b.p. 67 °C), $S_2O_6F_2$, is obtained by fluorination of sulphur trioxide in the presence of AgF_2 at 180 °C, SO_3F_2 (fluorine fluorosulphate) being formed as an intermediate:

$$SO_3 + F_2 \longrightarrow FSO_2OF \xrightarrow{SO_3} FSO_2OOSO_2F$$

This compound dissociates at 120 °C to give the brown paramagnetic radical FSO_2O, and among its many remarkable reactions are:

$$FSO_2OOSO_2F + C_2F_4 \rightarrow FSO_2OCF_2CF_2OSO_2F$$

$$2FSO_2OOSO_2F + KI \rightarrow K[I(OSO_2F)_4]$$

$$FSO_2OOSO_2F + Cl_2 \rightarrow 2ClOSO_2F$$

The ion SO_4F^- (the fluoride analogue of the peroxomonosulphate ion) is obtained by passing fluorine into a solution of a sulphate, and may be isolated as the caesium salt. It is a very powerful oxidant, $E°$ for

$$SO_4F^- + 2H^+ + 2e \rightleftharpoons HSO_4^- + HF$$

being about +2.5 V.

The range of chlorides and oxochlorides formed by sulphur is much more restricted than that of the corresponding fluoro compounds, and all of them are hydrolysed by water. Disulphur dichloride, S_2Cl_2, a fuming yellow liquid (b.p. 138 °C) is made from the elements. Decomposition by water yields a complex mixture containing sulphur, SO_2, $H_2S_5O_6$ and HCl. The compound is isostructural with disulphur difluoride, but the sulphur–sulphur bond length (1.97 Å), though shorter than that in S_8 or H_2S_2, is appreciably longer than in the difluoride. Condensation of S_2Cl_2 with hydrogen polysulphides gives a series of chlorosulphanes:

$$ClS_2Cl + HS_nH + ClS_2Cl \rightarrow ClS_{n+4}Cl + 2HCl$$

The action of excess of chlorine on S_2Cl_2 yields successively SCl_2 (b.p. 59 °C) and the very unstable SCl_4. Selenium tetrachloride is much more stable than the sulphur compound, and tellurium tetrachloride is the only stable chloride of tellurium; both of these compounds are solids. The molecular unit present in tellurium tetrachloride is Te_4Cl_{16}, which consists of a cube of composition Te_4Cl_4 (tellurium and chlorine atoms being located at opposite corners) with three further chlorine atoms bonded to each tellurium atom. Since there are two very different Te—Cl distances, Te–3Cl (cube corners) 2.93 Å and Te–3Cl (terminal) 2.31 Å, the structure may also be described in terms of pyramidal $TeCl_3^+$ and Cl^- ions, and it is worth noting that fused tellurium tetrachloride has quite a high electrical conductivity. Selenium tetrachloride is structurally similar. Both compounds are hydrolysed by water, but with alkali metal chlorides in the presence of concentrated hydrochloric acid, yellow complexes such as $K_2[SeCl_6]$ and $K_2[TeCl_6]$ are obtained. If we represent the anions in these compounds as complexes formed by donation of six electron pairs from Cl^- ions to Se^{4+} or Te^{4+} ions, the valence shells of the central atoms would contain 14 electrons, and the structures would be expected to be based on pentagonal bipyramids with one position occupied by a pair of unshared electrons, i.e. to be distorted octahedra. The anions are in fact regular octahedra and two explanations have been offered. For the $[TeCl_6]^{2-}$ ion, for example, the electronic configuration of the central atom could be $5s^25p^65d^46s^2$, with the 5s electrons occupying an orbital not involved in the hybridisation, only the 5p, 5d, and 6s orbitals counting as the effective valence orbitals. Alternatively, only the 5p orbitals of the tellurium atom may be involved in the bonding; combination of any one of the three 5p orbitals with two p orbitals of chloride ions leads to a bonding, a non-bonding, and an antibonding three-centre molecular orbital as shown in Fig. 16.3 in Section 16.5; twelve electrons occupy all the bonding and non-bonding orbitals, and the resulting Te—Cl bond order is 0.5. These two explanations thus imply different bond orders and hence different bond lengths and bond strengths. In this case there are no available experimental data on the basis of which we can decide between the explanations; in the related problems of the bonding in the I_3^- ion and XeF_2, however, the evidence strongly supports the idea of a low bond order (see Sections 16.5 and 17.2).

Thionyl chloride, $SOCl_2$, and sulphuryl chloride, SO_2Cl_2, are colourless fuming liquids (b.p. 78 and 69 °C respectively) which are usually made by the reactions:

$$SO_2 + PCl_5 \rightarrow SOCl_2 + POCl_3$$

$$SO_2 + Cl_2 \xrightarrow{C} SO_2Cl_2$$

They are isostructural with the analogous fluorine compounds. Thionyl chloride is used for the preparation of carboxylic acid chlorides and anhydrous metal chlorides, sulphuryl chloride as a chlorinating agent.

15.5 Oxides

The only important oxides of these elements are the dioxides and trioxides. A number of other unstable oxides of sulphur are also known, however; among them are S_2O (S=S=O), obtained by passing thionyl chloride over silver sulphide at 160 °C, and S_8O, obtained by oxidation of S_8 with trifluoroperacetic acid or from H_2S_7 and $SOCl_2$:

$$SOCl_2 + Ag_2S \rightarrow S_2O + 2AgCl$$

```
  H  Cl           S—S—S
 /     \         /     \
S7  +   S=O → S         S=O + 2HCl
 \     /         \     /
  H  Cl           S—S—S
```

The structure of S_8O is like that of S_8, but the two S—S(=O) bonds (2.20 Å) are 0.15 Å longer than the other S—S bonds.

Sulphur dioxide, a colourless gas with a pungent smell, is made on the large scale by burning sulphur or hydrogen sulphide, by roasting sulphide ores, or by reducing anhydrite ($CaSO_4$) with coke above 1350 °C:

$$CaSO_4 + C \rightarrow CaO + SO_2 + CO$$

Burning of sulphur is the most important of these processes. Atmospheric pollution by sulphur dioxide arises from combustion of sulphur-containing fuels and causes serious damage to respiratory organs, buildings and plants and aquatic life as a consequence of 'acid rain', the dilute solution of sulphuric acid produced by the NO- or OH-catalysed oxidation of the dioxide in the presence of water vapour. Recovery of the sulphur by absorption of the dioxide in a slurry of calcium or magnesium hydroxide is, unfortunately, by no means an economic process. In the laboratory, the gas is obtained from the action of dilute acid on a sulphite or from cylinders of the liquid, which boils at −10 °C. The liquid is a good solvent, though, as we mentioned in Section 8.1, the hypothesis that it undergoes self-ionisation according to the equilibrium

$$2SO_2 \rightleftharpoons SO^{2+} + SO_3^{2-}$$

has now been discarded.

Sulphur dioxide combines with oxygen (see below) and chlorine; in aqueous solution it is converted to only a slight extent into the weak acid sulphurous acid (cf. CO_2). The solution is oxidised to sulphate by many oxidising agents (e.g. I_2, MnO_4^-, $Cr_2O_7^{2-}$, Fe^{III}):

$$SO_2 + 2H_2O \rightarrow SO_4^{2-} + 4H^+ + 2e$$

As this equation implies, however, if the hydrogen ion availability is very high, sulphate can be reduced to sulphur dioxide: this happens, for example, when concentrated sulphuric acid is heated with zinc or copper. In the presence of concentrated hydrochloric acid, sulphur dioxide will itself act as an oxidising agent: iron(II) chloride, for example, is converted into a chloro complex of iron(III) with formation of elemental sulphur according to the equations

$$SO_2 + 4H^+ + 4Fe^{2+} \rightarrow S + 4Fe^{3+} + 2H_2O$$

$$Fe^{3+} + 4Cl^- \rightarrow FeCl_4^-$$

The molecule of sulphur dioxide is bent with OSO bond angle = 119°; the bond length is 1.43 Å. Unlike oxygen, sulphur is not restricted to an octet of electrons is its valence shell, and the sulphur–oxygen bonds are commonly taken as being double bonds rather than being analogous to the bonds of order 1.5 in ozone. In addition to its use for the manufacture of sulphuric acid, sulphur dioxide is widely used as a disinfectant and bleaching agent.

Selenium and tellurium dioxides are white solids obtained when the elements are oxidised by nitric acid. Selenium dioxide is readily soluble in water to form the acid H_2SeO_3, which is a good oxidising agent; tellurium dioxide, which is also a good oxidising agent, is only sparingly soluble, though it dissolves in aqueous alkali. Solid selenium dioxide has a non-planar infinite chain structure in which each selenium atom is bonded to three oxygen atoms, two of which are shared with adjacent selenium atoms. Tellurium dioxide has an unsymmetrical three-dimensional structure in which each tellurium atom is surrounded by four oxygen atoms, all shared with another tellurium atom, occupying four corners of a trigonal bipyramid. The trend among the Group IV elements towards metallic character is illustrated by the fact that polonium dioxide, which is only sparingly soluble in aqueous alkali, has the fluorite structure.

The oxidation of sulphur dioxide to the trioxide by atmospheric oxygen is very slow, but is catalysed by vanadium pentoxide or potassium vanadate in a cycle involving vanadium(IV). For the gas-phase reaction

$$SO_2 + \tfrac{1}{2}O_2 \rightarrow SO_3$$

$\Delta H^\circ = -96$ kJ mol^{-1}, so that although the reaction is accelerated the yield is diminished by the use of high temperatures; the yield is increased somewhat by the use of high pressures of air. In practice, modern contact processes usually employ temperatures of about 500 °C and achieve conversion factors above 98 per cent. Since gaseous sulphur trioxide is not readily absorbed by water (with which it forms a thick mist), it is removed by passage through concentrated sulphuric acid, in which it dissolves to form *oleum*, a mixture of polysulphuric acids (see the following section); subsequent dilution with water gives sulphuric acid. On a small scale sulphur trioxide is conveniently prepared by heating oleum.

Gaseous sulphur trioxide is an equilibrium mixture of monomer and trimer. The monomer is planar and symmetrical with a bond length of 1.43 Å, the same as in sulphur dioxide. Solid sulphur trioxide is polymorphic and shows considerable similarity to phosphorus pentoxide; all its forms are built up of SO_4 tetrahedra sharing two oxygen atoms with other tetrahedra, and the different forms react at different rates with water. As would be expected from the fact that the structural unit is the same in each of them, the difference in thermodynamic properties is, however, very small. Condensation of the vapour at low temperatures yields α-SO_3, m.p. 17 °C, which contains puckered rings of composition S_3O_9 (cf. the cyclic polyphosphates and silicates) as shown in Fig. 15.3(*a*); the justification for writing the sulphur–oxygen bridge and terminal bonds as single and double comes from their lengths, 1.60 and 1.40 Å respectively. The other two forms β-SO_3, m.p. 32 °C, and γ-SO_3, m.p. about 62 °C, both have infinite chain molecules as shown in Fig. 15.3(*b*); bond lengths are similar to those in α-SO_3.

Selenium trioxide (prepared as described above) and tellurium trioxide (obtained by heating telluric acid) are both readily decomposed into the dioxides and oxygen by the action of heat. Only the former compound is soluble in water, but tellurium trioxide dissolves in aqueous alkali.

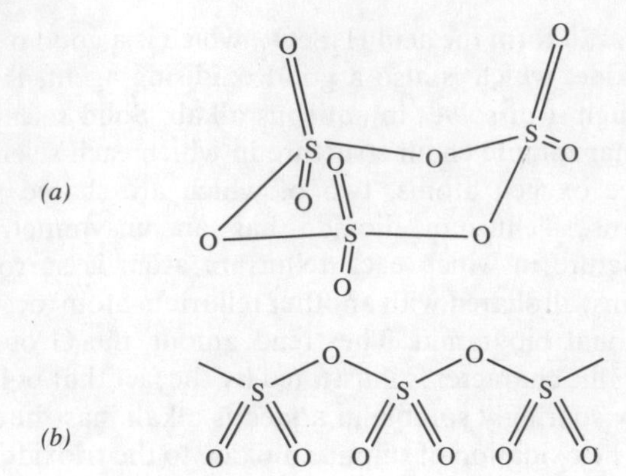

Figure 15.3
The structures of (*a*) α-SO_3 (*b*) β- and γ-SO_3

Selenium triodixe consists of tetrameric Se_4O_{12} moleculs with a structure like that of the $P_4O_{12}^{4-}$ ion. Tellurium trioxide has a three-dimensional structure (that of ReO_3 – Fig. 24.1) in which octahedral TeO_6 units share each oxygen atom with another tellurium atom.

15.6 Oxo-acids and their salts

By far the most numerous and important of the compounds to be described in this section are those of sulphur. This element forms some acids (or anions) in which the oxidation state of all the sulphur is four (e.g. H_2SO_3 and $H_2S_2O_5$) or is six (e.g. H_2SO_4 and $H_2S_2O_7$), but there are also a few species in which all the sulphur is in another oxidation state (e.g. $H_2S_2O_4$ and $H_2S_2O_6$) and a large number in which there are obviously two quite different kinds of sulphur atom. The last group includes, for example, thiosulphuric and tetrathionic acids, which have the respective structures

$(S)(O)S(OH)_2$ or $(O)_2S(SH)(OH)$ and $HO-S(O_2)-S-S-S(O_2)-OH$

As in the case of the oxo-acids and oxo-anions of nitrogen and phosphorus, redox reactions between sulphur-containing species are often very slow, and thermodynamic data alone do not give a very good picture of their chemistry.

Dithionous acid, $H_2S_2O_4$ (sometimes called hydrosulphurous acid) is also known only as salts which are powerful reducing agents. Dithionites are obtained by reduction of sulphites in aqueous solution by zinc or sodium amalgam:

$$2SO_3^{2-} + 2H_2O + 2e \rightarrow 4OH^- + S_2O_4^{2-}$$

Sodium dithionite has been shown to contain an anion having the eclipsed configuration

```
  S——S
 / \  / \
O  O O  O
```

with a very long S—S bond (2.39 Å); the S—O bond length is 1.51 Å. The presence of this very long, and therefore very weak, S—S bond is consistent with the very fast exchange of ^{35}S between $S_2O_4^{2-}$ and SO_2 in neutral or acidic solution; the presence of a low concentration of the SO_2^- ion in solutions of $Na_2S_2O_4$ has been demonstrated by electron spin resonance spectroscopy. Dithionite in solution is oxidised by air to sulphite; in the absence of air it slowly decomposes to form sulphite and thiosulphate:

$$2S_2O_4^{2-} + H_2O \rightarrow S_2O_3^{2-} + 2HSO_3^-$$

Sulphur dioxide in aqueous solution is nearly all present in molecular form. Although the HSO_3^- ion exists in solution and salts such as $NaHSO_3$ have been isolated, evaporation of a solution of sodium sulphite which has been saturated with sulphur dioxide gives the disulphite or pyrosulphite $Na_2S_2O_5$; unlike the disulphate anion, which contains a S—O—S bridge, this anion has the unsymmetrical structure $O_3SSO_2^{2-}$; the S—S bond is again long (2.17 Å), though not nearly so long as in the dithionite anion. The sulphite ion, SO_3^{2-}, in which the sulphur atom has two more electrons than in planar SO_3, is pyramidal.

Selenous acid, H_2SeO_3, which may be crystallised from an aqueous solution of selenium dioxide, gives rise to the expected series of salts containing the $HSeO_3^-$ and SeO_3^{2-} ions. When the former are heated they yield diselenites, but these, in contrast to the disulphites, contain an oxygen-bridged anion $O_2SeOSeO_2^{2-}$. Most tellurites contain the TeO_3^{2-} ion.

Dithionic acid, $H_2S_2O_6$, is yet another sulphur oxo-acid known only in aqueous solution (in which it is a strong acid) or in the form of salts. Since it is a derivative of sulphur (v), it might be expected to be formed by controlled oxidation of a sulphite or reduction of a sulphate; in practice only the former process is useful, being achieved either by anodic oxidation or by the use of manganese dioxide as oxidising agent:

$$MnO_2 + 2SO_3^{2-} + 4H^+ \rightarrow Mn^{2+} + S_2O_6{}^{2-} + 2H_2O$$

Dithionate ion may be isolated as the soluble barium salt, which is easily converted into salts containing other cations. The ion is not easily oxidised or reduced, but in acidic solution it slowly decomposes into sulphur dioxide and sulphate:

$$S_2O_6^{2-} \rightarrow SO_2 + SO_4^{2-}$$

Dithionate ion has a centrosymmetric structure $O_3SSO_3^{2-}$ with the sulphur–sulphur and sulphur–oxygen bond lengths 2.15 and 1.43 Å respectively.

We have already mentioned the manufacture of sulphuric acid by means of the contact process. An interesting alternative method for the oxidation of sulphur dioxide is by nitrogen dioxide in the presence of water, the nitric oxide produced being reconverted into the dioxide by air:

$$NO_2 + SO_2 + H_2O \rightarrow NO + H_2SO_4$$

$$2NO + O_2 \rightarrow 2NO_2$$

This is the essential basis of the now obsolete lead chamber process, the great disadvantage of which is that it is not suitable for the direct production of concentrated H_2SO_4.

Pure sulphuric acid is a viscous liquid which is strongly hydrogen-bonded; it has a specific gravity of 1.84 and melts at 10.5 °C. In the absence of water it will not react with metals to form hydrogen, though it is reduced to sulphur dioxide by many of them at moderate temperatures. It reacts violently with water, and is a powerful dehydrating agent. In aqueous solution it is almost completely ionised to H_3O^+ and HSO_4^- ions, but the second ionisation is appreciable only at high dilution. When pure sulphuric acid is heated, a little sulphur trioxide is produced, and an azeotrope containing 98.3 per cent H_2SO_4 distils at 338 °C.

The properties of pure sulphuric acid as a non-aqueous solvent have been described in some detail in Section 8.4. Its principal uses are in the manufacture of superphosphate and ammonium sulphate for fertilisers, in the production of other acids and of titanium dioxide, in petrol refining, and in the synthetic fibres, detergents, explosives and dyestuffs branches of the organic chemical industry.

Sulphuric acid gives rise to two series of salts, represented by K_2SO_4 and $KHSO_4$. In the molecule of H_2SO_4, sulphur is tetrahedrally surrounded by two OH at 1.54 Å and two O at 1.43 Å; in the HSO_4^- anion by one OH at 1.56 Å and three O at 1.47 Å; and in the SO_4^{2-} ion by four O at 1.49 Å, the increase in the sulphur–oxygen bond length corresponding to some decrease in bond order as the π bonding is distributed between a greater number of atoms.

Selenic acid, H_2SeO_4, may be crystallised from the solution obtained by oxidising selenous acid with 30 per cent hydrogen peroxide. It resembles sulphuric acid in most respects, but is a much more powerful oxidising agent, e.g. it liberates chlorine from concentrated hydrochloric acid. Telluric acid, although resembling selenic acid in preparation and oxidising properties, is structurally quite different from it; its molecular formula is H_6TeO_6, or $Te(OH)_6$, and it is a weak acid. Typical salts include those containing $[TeO(OH_5)]^-$ and $[TeO_2(OH)_4]^{2-}$ ions, and the existence of a simple TeO_4^{2-} ion has not yet been proved.

When sulphur trioxide dissolves in sulphuric acid, $H_2S_2O_7$ and $H_2S_3O_{10}$ are formed; potassium salts of these acids have been obtained by the action of gaseous sulphur trioxide on potassium sulphate, and have been shown to contain chain anions of formulae

$$^{-}OSO_2.O.SO_2O^{-} \text{ and } ^{-}OSO_2.O.SO_2.O.SO_2O^{-}.$$

Fluoro- and chloro-sulphuric acids, FSO_2OH and $ClSO_2OH$, are obtained by addition of the appropriate hydrogen halide to sulphur trioxide; the former is important in super-acid media (Section 8.4). Like many fluorine derivatives of sulphur(VI), it is only slowly hydrolysed by water, but chlorosulphuric acid reacts with water with great violence.

By the action of chlorosulphuric acid on cold anhydrous hydrogen peroxide, the peroxo-acids $HOSO_2OOH$ and $HOSO_2OOSO_2OH$ are obtained; the former of these may also be obtained by hydrolysis of the latter at 0 °C:

$$H_2O_2 \xrightarrow[-HCl]{ClSO_2OH} HOSO_2OOH \xrightarrow[-HCl]{ClSO_2OH} HOSO_2OOSO_2OH$$

$$HOSO_2OOSO_2OH + H_2O \rightarrow HOSO_2OOH + H_2SO_4$$

Few salts of peroxomonosulphuric acid are known, but those of peroxodisulphuric acid are easily made by anodic oxidation of the sulphates in acidic aqueous solution at low temperatures and high current densities. Peroxodisulphate oxidations are often catalysed by silver ion, Ag(II) species being formed as intermediates; E° for the half-reaction

$$S_2O_8^{2-} + 2e \rightleftharpoons 2SO_4^{2-}$$

has been estimated as +2.0 V, and among many changes peroxodisulphate effects are $Mn(II) \rightarrow MnO_4^-$ and $Cr(III) \rightarrow CrO_4^{2-}$; when an aqueous solution of $K_2S_2O_8$ is heated, ozonised oxygen is evolved. The fluoride analogue of the HSO_5^- ion was mentioned in Section 15.4.

The simplest acid containing two entirely different sulphur atoms is thiosulphuric acid, $H_2S_2O_3$, obtained from hydrogen sulphide and chlorosulphuric acid at low temperatures in absence of a solvent.

$$H_2S + ClSO_2OH \rightarrow HSSO_2OH + HCl$$

These reaction conditions suggest the formulation of the product as $HSSO_2OH$, but the formulation as the isomer $SSO_2(OH)_2$ is also possible.

The free acid is very unstable, but the sodium salt is readily prepared by heating aqueous sodium sulphite with sulphur and crystallising, when $Na_2S_2O_3.5H_2O$ separates. The thiosulphate ion, which is tetrahedral with one oxygen atom of the sulphate ion replaced by sulphur, is a very good complexing agent for silver ion, and sodium thiosulphate is widely used in photography for removing unchanged silver bromide from exposed photographic film; coordination to silver is through the sulphur atom. Most oxidising agents (including chlorine and bromine in aqueous solution) slowly oxidise thiosulphate to sulphate, but the ion is very rapidly oxidised by iodine to tetrathionate:

$$2S_2O_3^{2-} + I_2 \rightarrow S_4O_6^{2-} + 2I^-$$

This reaction is of immense importance in titrimetric analysis, since nearly all oxidising agents will liberate iodine quantitatively from acidified iodide solution.

Polythionates in general contain the ions $S_nO_6^{2-}$ and may be obtained by condensation between SCl_2 or S_2Cl_2 and sulphites or thiosulphates, e.g.

$$SCl_2 + 2HSO_3^- \rightarrow S_3O_6^{2-} + 2HCl$$

$$S_2Cl_2 + 2HSO_3^- \rightarrow S_4O_6^{2-} + 2HCl$$

Various special preparative methods are also available. All polythionates contain unbranched chains of sulphur atoms, but considerable variety is found in the configurations of the chains. Structures are known for anions containing up to six sulphur atoms; others containing up to twenty sulphur atoms have been prepared. In aqueous solution they slowly decompose with formation of sulphuric acid, sulphur dioxide and sulphur. A few compounds are known in which some of the sulphur atoms in polythionates are replaced by selenium or tellurium, e.g. $Ba[Se(SSO_3)_2]$ and $Ba[Te(SSO_3)_2]$; the other two elements cannot replace sulphur in the $—SO_3^-$ parts of the ion, presumably because in their highest oxidation states they are too powerfully oxidising and attack the remainder of the chain.

15.7 Sulphur–nitrogen compounds

Much the best known of the sulphur–nitrogen compounds is tetrasulphur tetranitride, S_4N_4, obtained as a diamagnetic orange solid by passing ammonia into a warm solution of disulphur dichloride in dry carbon tetrachloride:

$$6S_2Cl_2 + 16NH_3 \rightarrow S_4N_4 + 12NH_4Cl + S_8$$

Tetrasulphur tetranitride melts at 178 °C, explodes on percussion, and is hydrolysed slowly by water (in which it is insoluble) and rapidly by warm alkali:

$$S_4N_4 + 6OH^- + 3H_2O \rightarrow S_2O_3^{2-} + 2SO_3^{2-} + 4NH_3$$

Its structure is that of a cradle-like eight-membered ring (Fig. 15.4); all the sulphur–nitrogen bond lengths are 1.62 Å, a value between those for the single S—N bond in S_7NH (1.73 Å) and the double S═N bond in $S_4N_4F_4$ (1.54 Å) (see below for these compounds). To a first approxi-

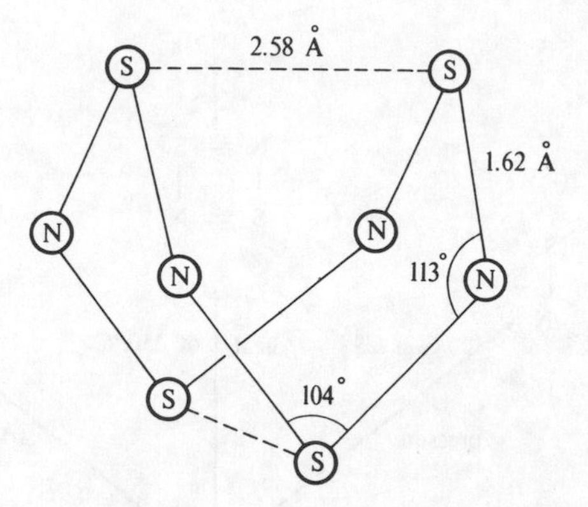

Figure 15.4
The structure of S_4N_4.

mation, therefore, the structure of S_4N_4 may be written as a hybrid of

```
N=S—N          N—S=N
|   ‖          ‖   |
S   S   and    S   S
‖   |          |   ‖
N—S=N          N=S—N
```

with a bond order of 1.5. This leaves a lone pair on every atom and an unpaired electron on each sulphur atom. However, the two pairs of sulphur atoms at the top and bottom of the cradle are near enough (2.58 Å) to suggest that there are weak bonds between them (cf. the S_8^{2+} cation mentioned in Section 15.2). Thus S_4N_4 is believed to be a cage compound, though the bonds closing the cage are very weak.

Some reactions of S_4N_4 are shown in Fig. 15.5, in which the transformations into rings of different sizes should be noted. Some sulphur–nitrogen bond lengths have already been mentioned; we may note in addition those in NSF (1.45 Å) and NSF_3 (1.42 Å) which contain sulphur–nitrogen triple bonds. In the puckered ring of $S_4N_4F_4$ there are two quite different S—N distances (1.54 and 1.66 Å), and electron delocalisation here appears to be only slight. The $S_4N_4^{2+}$ cation is planar in most of its salts, with S—N 1.55 Å.

Tetrasulphur tetraimide, $S_4(NH)_4$, is one of a number of compounds in which sulphur atoms in S_8 are replaced by imido groups; S_7NH, three isomers of $S_6(NH)_2$, and two isomers of $S_5(NH)_3$ are all obtained by the action of ammonia on S_2Cl_2, together with S_4N_4 and sulphur (compounds having adjacent NH groups are not known). These imides all have structures based on that of S_8, though in $S_4(NH)_4$ the ring is more nearly planar than in elemental sulphur; all the sulphur–nitrogen bond lengths in the compound are 1.66 Å. By the reaction between S_2Cl_2 and S_7NH, S_7N rings can be coupled via —S—S— bonds. Interaction of S_7NH and $SbCl_5$ in ammonia gives $NS_2[SbCl_6]$, the cation of which is the sulphur analogue of the nitronium ion.

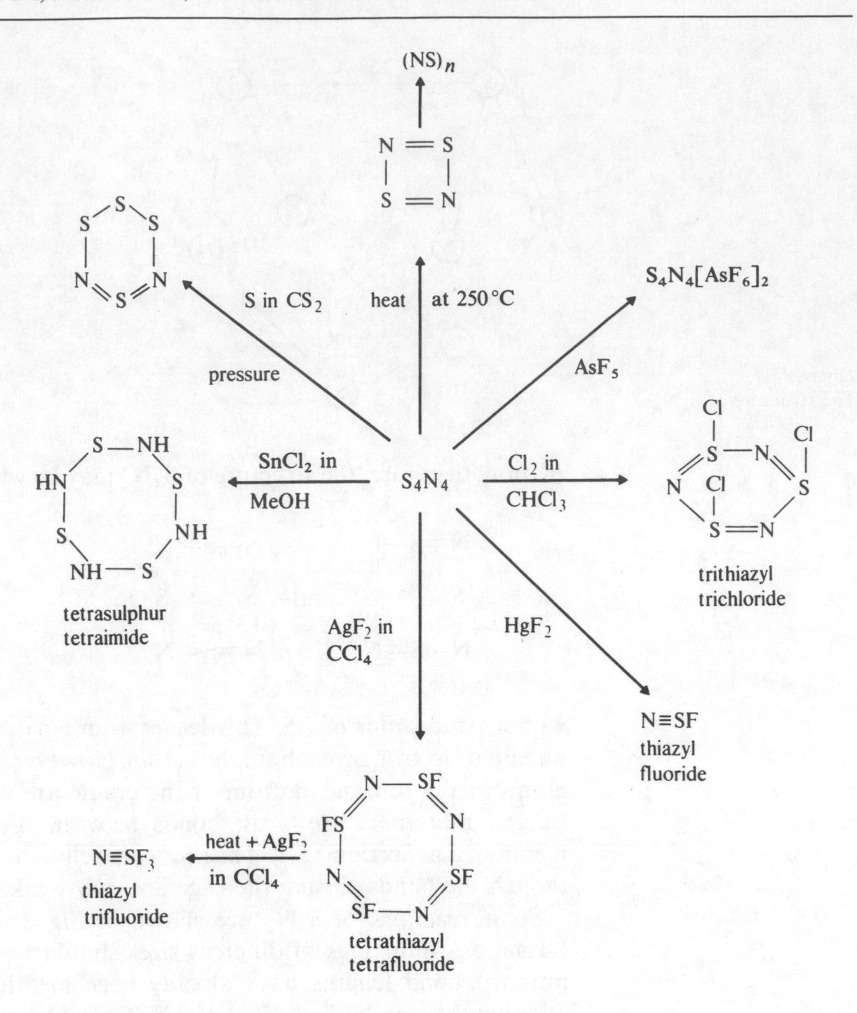

Figure 15.5
Some reactions of S_4N_4.

If S_4N_4 vapour is passed over silver wool at 250°C, rapid cooling permits the isolation of S_2N_2, a planar molecule in which all S—N distances are the same as in S_4N_4. At the ordinary temperature this slowly changes into a fibrous polymer $(SN)_n$ which is a lustrous golden yellow colour on the faces of the fibres and blue-black at their ends. The polymer decomposes explosively at 250°C but can be sublimed *in vacuo* at about 140°C. This remarkable substance is a covalently bonded polymer that shows metallic properties – a one-dimensional pseudometal. It has an electrical conductance about one-quarter of that of mercury along the polymer chains, and at 0.3 K it becomes a superconductor. According to the most recent X-ray diffraction study, there are two S—N bond lengths (1.59 and 1.63 Å), and the NSN and SNS angles are 106° and 119° respectively; the closest approaches of the chains are represented by sulphur–sulphur contacts at 3.5 Å. To a fair degree of approximation, therefore, the structure may be represented as an opened-up version of those of S_4N_4 and S_2N_2, i.e. a conjugated system:

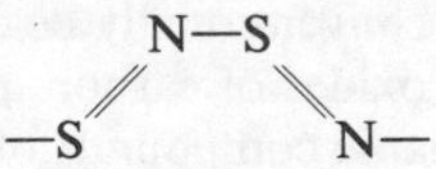

$$-S-N=S-N-$$

with the unpaired electrons on the sulphur atoms half-filling a conduction band of orbitals contributed on the basis of one per S—N unit.

15.8 Aqueous solution chemistry of sulphur, selenium and tellurium

As we mentioned earlier, redox reactions between compounds of sulphur in different oxidation states are often slow, so redox potential data are almost invariably obtained from thermochemical information or estimated on the basis of observed chemistry. We shall not attempt to deal here with transformations involving most of the sulphur oxo-acids and oxo-anions, but for purposes of comparison with selenium and tellurium we give some simplified potential diagrams for $[H^+] = 1$ in Fig. 15.6. These illustrate the much greater oxidising powers of selenate and tellurate than of sulphate, the smaller differences between selenite, tellurite and sulphite, and the instability of hydrogen selenide and telluride. It will be further observed that for sulphur there is not really very much difference in free energy between the various oxidation states, a fact which is doubtless involved in the complicated oxo-acid and oxo-anion chemistry of the element. Other aspects of the aqueous chemistry of sulphur, selenium and tellurium that we have already mentioned include the oxidising powers of peroxodisulphate, the ionisation of the hydrides and various oxo-acids, and the complexing of metal ions by sulphide and thiosulphate. Relatively little is known about complexing by other oxo-anions of sulphur and its homologues.

There is no cation chemistry in aqueous media for any of these elements.

$$SO_4^{2-} \xrightarrow{+0.17} H_2SO_3 \xrightarrow{+0.45} S \xrightarrow{+0.14} H_2S$$

$$SeO_4^{2-} \xrightarrow{+1.15} H_2SeO_3 \xrightarrow{+0.74} Se \xrightarrow{-0.40} H_2Se$$

$$H_6TeO_6 \xrightarrow{+1.02} TeO_2 \xrightarrow{+0.53} Te \xrightarrow{-0.72} H_2Te$$

Figure 15.6
Potential diagrams for sulphur, selenium and tellurium at $[H^+] = 1$.

15.9 Organic derivatives

Organic compounds of oxygen and sulphur form a large part of organic chemistry, and are conventionally dealt with there. We discussed carbon disulphide and the oxides of carbon in Chapter 13, and the only subject concerned with organic compounds of oxygen and sulphur that we shall mention here is complexing by dialkyl oxides (ethers), dialkyl sulphides and related compounds. In discussing complexing of silver ion by thiosulphate we noted that coordination to the metal occurs through a sulphur atom. This preference for cordination by sulphur rather than by oxygen is a feature of the chemistry of several metal ions (cf. Section 7.7); and although the evidence is mostly qualitative there is a clear distinction between those which form more stable complexes with ethers than with dialkyl sulphides and vice versa. Coordination by the thiocyanate ion, SCN^-, usually follows a similar pattern: class (*a*) metal ions, as defined in Section 7.7, bond through the nitrogen atom, class (*b*) metal ions through the sulphur atom.

The organic chemistry of selenium and tellurium is generally like that of sulphur, though much more restricted in scope. One interesting group of compounds that have no sulphur analogues is the alkyl and aryl selenium and tellurium chlorides, bromides and iodides, derivatives of the tetrahalides, which, as we have seen, are much more stable for selenium and tellurium than for sulphur. Dimethyl selenide and telluride, made from halides and mercury dimethyl, combine with chlorine, bromine and iodine to form Me_2SeX_2 and Me_2TeX_2; aryl compounds may be made by similar methods. In these compounds, the selenium or tellurium atom is at the centre of a trigonal bipyramid, one of the equatorial positions being occupied by a lone pair of electrons; halogen atoms always occupy axial positions. What was once believed to be a geometrical isomer of Me_2TeI_2 is in fact $Me_3Te^+[TeMeI_4]^-$ and contains a pyramidal cation and a square pyramidal anion.

PROBLEMS

1 Use the values of E° for the half-reactions

$$H_2O_2 + 2H^+ + 2e \rightleftharpoons 2H_2O \qquad E^\circ = +1.77\ \mathrm{V}$$
$$O_2 + 2H^+ + 2e \rightleftharpoons H_2O_2 \qquad E^\circ = +0.68\ \mathrm{V}$$

to show that hydrogen peroxide is thermodynamically unstable with respect to decomposition into oxygen and water. '20 Volume' aqueous hydrogen peroxide is so-called because 1 volume of the solution liberates 20 volumes of oxygen when it is decomposed. If the volumes are measured at 0 °C and 1 atm pressure, what is the concentration of the solution expressed in grams of H_2O_2 dm^{-3}?

2 Discuss the interpretation of each of the following observations:
(a) At low temperatures the ^{19}F nmr spectrum of SF_4 consists of two triplets of equal intensity; at higher temperatures only a single peak is observed.
(b) When metallic copper is heated with concentrated sulphur acid, in addition to copper(II) sulphate and sulphur dioxide some copper(II) sulphide is also formed.
(c) Silver nitrate gives a white precipitate with aqueous sodium thiosulphate; the precipitate dissolves in excess of thiosulphate. If the precipitate is heated with water it turns black, and the supernatant liquid then gives a white precipitate with acidified barium nitrate solution.
(d) The $[TeF_5]^-$ ion is square pyramidal.

3 (a) Sodium dithoinite, $Na_2S_2O_4$ (0.261 g) was added to excess of ammoniacal silver nitrate solution; the precipitated silver was removed by filtration, and dissolved in nitric acid; the resulting solution was found to be equivalent to 30.0 cm^3 of 0.1 molar thiocyanate solution.
(b) A solution containing 0.0725 g of sodium dithionite was treated with 50 cm^3 of 0.0500 molar iodine solution and acetic acid. After completion of the reaction the residual iodine was equivalent to 23.75 cm^3 of 0.1050 molar thiosulphate.

Interpret these results.

$$[Na_2S_2O_4 = 174]$$

4 The action of concentrated sulphuric acid on urea results in the production of a white crystalline solid **X** of formula H_3NO_3S. This is a monobasic acid. On treatment with sodium nitrite and dilute hydrochloric acid at 0 °C it liberates one mole of N_2 per mole of **X**, and on addition of aqueous barium chloride the resulting solution yields one mole of $BaSO_4$ per mole of **X** taken.

Deduce the structure of **X**.

REFERENCES FOR FURTHER READING

Gillespie, R.J. and Passmore, J. (1975) *Adv. Inorg. Chem. Radiochem.*, **17**, 49. A review of polyatomic cations, especially those of oxygen, sulphur, selenium and tellurium.

Greenwood, N.N. and Earnshaw, A. (1984) *Chemistry of the Elements*, Pergamon Press, Oxford. Chapters 14–16 cover the chemistry of the Group VI elements in detail.

Meyer, B. (1976) *Adv. Inorg. Chem. Radiochem.*, **18**, 287. A review of the structures of elemental sulphur.

Nickless, G., editor (1968) *Inorganic Sulphur Chemistry*, Elsevier, Amsterdam. A full account of most aspects of the inorganic chemistry of sulphur.

Powell, P. and Timms, P. (1974) *The Chemistry of the Non-Metals*, Chapman and Hall, London. Covers much of the same ground as the present chapter, but in a quite different manner.

Wells, A.F. (1984) *Structural Inorganic Chemistry*, 5th edition, Oxford University Press. Chapters 11–17 cover the structures of a large number of compounds of Group VI elements.

16 The halogens

16.1 Introduction

The chemistry of fluorine, chlorine, bromine and iodine is probably better understood than that of any other group of elements except the alkali metals. This is partly because much of the chemistry of the halogens is that of singly bonded atoms or singly charged anions, and partly because of the wealth of structural and physicochemical data for most of their compounds. The fundamentals of inorganic chemistry are very often best illustrated by the discussion of the properties of these elements and of the halides, and among the topics treated earlier in this book are the bond energies in the halogens and hydrogen halides (Section 5.4), force constants for these molecules (Section 5.5), the electronegativities of the halogens (Section 5.7), the structures of many metal halides (Section 6.3), lattice energies and standard enthalpies of formation of metal halides (Sections 6.8 and 10.3), the properties of halide ions in solution (Sections 7.3 and 8.1), the ionisation of the hydrogen halides (Section 7.4) and their relative efficiencies for the protonation of weak bases (Section 14.3), complex halide formation in solution (Section 7.7), hydrogen fluoride and bromine trifluoride as non-aqueous solvents (Sections 8.3 and 8.5) and hydrogen bonding in HF, NH_4F, and the HF_2^- and HCl_2^- ions (Section 9.5). Further, each of Chapters 10–15 has included a section on halides of the elements under consideration. The pseudo-halide ions CN^- and N_3^- and their derivatives were described in Sections 13.12 and 14.3 respectively. Most of this chapter will therefore be taken up by the discussion of the halogens themselves, their oxides and oxo-acids, and interhalogen compounds and polyhalogen ions, but we shall also take the opportunity to remind the reader (often by means of cross-references) of some general aspects of the chemistry of halogen-containing compounds. Halides of the noble gases and of the transition elements are, of course, covered in later chapters.

As in the last chapter, we shall confine the systematic treatment almost entirely to the stable elements. Astatine, the homologue of iodine, is known exclusively in the form of radioactive isotopes, the longest-lived of which have half-lives of only about eight hours; one of these, ^{211}At, an α-emitter, is obtained by the nuclear reaction $^{209}_{83}Bi(\alpha,2n)^{211}_{85}At$ and may be separated by vacuum distillation at 500 °C. Tracer studies (which are the only sources of information about the element) show that astatine is less volatile than iodine, is soluble in organic solvents, and is reduced by sulphur dioxide to an anion which is coprecipitated with AgI or TlI, i.e. At^-. Hypochlorite or peroxodisulphate oxidises astatine to an anion that is carried by IO_3^- (e.g. in precipitation of $AgIO_3$) and is therefore probably AtO_3^-; less powerful oxidising agents such as bromine also oxidise it, probably to AtO^- or AtO_2^-.

Some important properties of the halogens and related species are assembled in Table 16.1. Most of the differences in inorganic chemistry between fluorine and the other halogens can be attributed to the restriction of the valence shell of fluorine to an octet of electrons, the relatively small sizes of the fluorine atom and the fluoride ion, and the low dissociation energy of the F_2 molecule; but it is sometimes also necessary (as it very often is in organic chemistry) to invoke the less rigidly defined power of attracting an electron within a stable molecule (the electronegativity) of fluorine. Manifestations of the octet restriction are that

Table 16.1
Some properties of fluorine, chlorine, bromine and iodine.

Element	F	Cl	Br	I
Atomic number	9	17	35	53
Electronic configuration	$[He]2s^22p^5$	$[Ne]3s^23p^5$	$[Ar]3d^{10}4s^24p^5$	$[Kr]4d^{10}5s^25p^5$
M.p./°C	−220	−101	−7	113
B.p./°C	−188	−34	59	183
$\Delta H_f^\circ X_{(g)}$/kJ mol^{-1}	79	121	112	107
Dissociation enthalpy of X_2/kJ mol^{-1}	159	243	193	151
Electron affinity[a] of X/kJ mol^{-1}	−338	−355	−331	−302
$\Delta H_f^\circ X^-_{(g)}$/kJ mol^{-1}	−259	−234	−217	−197
First ionisation energy of X/kJ mol^{-1}	1680	1255	1140	1010
ΔH° hydration X^-/kJ mol^{-1}	−513	−370	−339	−274
ΔS° hydration X^-/J K^{-1} mol^{-1}	−150	−90	−70	−50
ΔG° hydration X^-/kJ mol^{-1}	−468	−344	−317	−280
$E^\circ\ \frac{1}{2}X_2 + e \rightleftharpoons X^-$/V	+2.9	+1.36	+1.07	+0.54
Pauling electronegativity	4.0	3.2	3.0	2.7
Bond energy in HX/kJ mol^{-1}	566	431	366	299
Covalent radius of X/Å	0.71	0.99	1.14	1.33
Ionic radius[b] of X^-/Å	1.33	1.81	1.96	2.20

[a]That is ΔH° for the gas-phase process $X + e \rightarrow X^-$.
[b]For six-coordination.

fluorine forms no compounds analogous to ClF_3, $HClO_3$ or Cl_2O_7, and where it is simultaneously bonded to two atoms (e.g. in the fluorine bridges of the polymeric forms of many transition metal pentafluorides), the M—F—M bridge bonds invariably have greater M—F distances than the corresponding terminal M—F bonds, indicating that they are of order lower than unity, probably three-centre two-electron bonds. The weakness of the bond in F_2 is generally attributed to repulsion between the pairs of unshared electrons on each atom (it is noteworthy that the O—O and N—N bonds in H_2O_2 and N_2H_4 are also weak). When fluorine is bonded to other elements, the bond in the fluoride is always stronger than that in the corresponding chloride, and where there are no unshared pairs of electrons on the combined atom of the other element (or where the other element has a large atom so that repulsion is minimised) the bond in the fluoride is much stronger than that in the chloride; for example the bond energy terms in NF_3 and NCl_3 are 272 and 193 kJ mol^{-1}, in PF_3 and PCl_3 they are 490 and 319 kJ mol^{-1}, and in CF_4 and CCl_4 they are 485 and 327 kJ mol^{-1} respectively. The small size of the fluorine atom not only permits high coordination numbers in molecular fluorides, but it also leads to better overlap of atomic orbitals and hence to shorter and stronger bonds. The volatility of molecular fluorine compounds (e.g. of fluorocarbons, which often have boiling-points little different from those of the corresponding hydrocarbons) originates in the weakness of the van der Waals' or London dispersion forces in these compounds, which may in turn be correlated with the low polarisability and small size of the fluorine atom. Consequences of the smallness of fluoride ion include high coordination numbers in saline fluorides, high lattice energies and very negative standard enthalpies of formation of these compounds, and a large negative standard enthalpy and entropy of hydration of the ion. The high electronegativity of fluorine (Section 5.7) is involved in the strong hydrogen bonding in hydrogen fluoride, the strength of fluorine-substituted carboxylic acids, the deactivating effect of the CF_3 group in electrophilic aromatic substitution, and the non-basic character of NF_3 and $(CF_3)_3N$.

In some ways the relationship between iodine, which has the lowest ionisation energy of the elements in Table 16.1, and the other halogens is like that between tellurium and oxygen, sulphur and selenium. Although none of the halogens has yet been shown to form a stable monatomic cation, complexed or solvated I^+ is well established in species such as $I(py)_2^+$ and $I(cryptand)^+$ salts, and appears to be present in solutions obtained by the interaction of iodine and silver salts, e.g.

$$I_2 + AgClO_4 \xrightarrow{Et_2O} AgI + IClO_4$$

The corresponding bromine- and chlorine-containing species are less stable, though they are probably involved in aromatic bromination and chlorination in aqueous solution. Further, iodine displays higher oxidation states in the halides ICl_3 and IF_7 than bromine or chlorine in their corresponding halides. Hydrogen iodide in solution is the strongest acid among the hydrogen halides, and solid periodic acid is not HIO_4, analogous to $HClO_4$ and $HBrO_4$, but H_5IO_6 or $OI(OH)_5$. Polyatomic cations X_2^+ and X_3^+ are known for chlorine, bromine and iodine, those of iodine being most stable; these species are discussed in Section 16.5.

The stable isotope of fluorine, ^{19}F, has a nuclear spin magnetic quantum number of $\frac{1}{2}$, and ^{19}F nuclear magnetic resonance spectroscopy is a valuable tool in the elucidation of the structures and reaction mechanisms of fluorine compounds. There is, unfortunately, no convenient radioactive tracer for fluorine (^{18}F, the most suitable isotope, has a half-life of only 109 min), but neutron activation analysis (Section 1.9) based on the formation and estimation of ^{20}F (half-life 12 s) can be used for the determination of the element, e.g. in fluorine dating of bones and teeth. These originally contain *apatite*, $Ca_5(OH)(PO_4)_3$; when they are embedded in the soil this mineral is slowly converted into the less soluble *fluorapatite*, $Ca_5F(PO_4)_3$; the fluorine content can therefore be used to estimate the age of remains, preferably in conjunction with ^{14}C dating of the organic matter present. The lower solubility of fluorapatite, incidentally, constitutes the thermodynamic basis for the argument in favour of fluoridisation of drinking water: teeth having a higher fluoride ion content are more resistant to attack and decay. Concentrations of F^- above 1 ppm, however, are harmful.

Fluorine occurs chiefly as *fluorspar* or *fluorite* (CaF_2), *cryolite* (Na_3AlF_6) and *fluorapatite* (above). Cryolite is used as such in the aluminium industry, but most fluorine compounds are obtained from fluorspar via hydrogen fluoride, which is manufactured by the action of concentrated sulphuric acid on the mineral. Chlorine is made almost entirely from sodium chloride, which occurs as solid deposits in many places, by electrolysis of the aqueous solution; sodium hydroxide (Section 10.4) is a by-product of the reaction. When chlorine is used for the chlorination of organic compounds, the hydrogen chloride which is also formed may be dissolved and electrolysed to regenerate further chlorine. Bromine is obtained from sea-water or from the residual liquid after the extraction of carnallite (Section 10.1). In either case, the solution is acidified to pH 3.5 and a mixture of chlorine and air is then swept through it to liberate and remove the bromine. Iodine is obtained mostly from natural brines occurring in the USA and Japan, or from sodium iodate, present in small amount in Chile saltpetre (sodium nitrate). In the former process the iodide present is oxidised by the requisite quantity of chlorine water; in the latter, controlled reduction by sulphur dioxide produces elemental iodine, complete reduction yields sodium iodide. The most important fluorine compounds on an industrial scale are hydrogen fluoride, boron trifluoride, calcium fluoride (as a flux in metallurgy), synthetic cryolite and chlorofluorocarbons. Chlorine, in addition to its major use for the preparation of organic compounds, is employed as a bleaching and sterilising agent. Bromine and iodine are used mainly for the preparation of organic compounds; additional uses include the production of silver bromide for photography. Iodine is an essential constituent of the thyroid hormones; a deficiency in it causes goitre.

16.2 The elements

The only important method for the preparation of fluorine is electrolysis of a molten mixture of potassium fluoride and hydrogen fluoride. The melting-point of such a mixture decreases as the hydrogen fluoride content increases; industrial cells use an electrolyte of composition KF.2–3HF, which is molten at 70–100 °C, and have a steel or copper cathode and an ungraphitised carbon anode; the cells themselves are made of steel, and contain a monel metal (Cu/Ni) diaphragm, perforated below the surface of the liquid but not above it, to prevent mixing of the hydrogen and fluorine which are produced. As electrolysis proceeds the hydrogen fluoride content of the melt is renewed by addition of the dry gas from a cylinder. Small 10 A cells are now available for laboratory use, or the compressed gas may be obtained. Fluorine is usually handled in metal apparatus, but it can be manipulated in glass if it is freed from hydrogen fluoride by passage through sodium fluoride, which removes the latter by forming $NaHF_2$.

Fluorine cannot be prepared in aqueous media, since it decomposes water with liberation of ozonised oxygen, E° for the $\frac{1}{2}F_2/F^-$ electrode being estimated from thermochemical data as +2.9 V. It is formed in the decomposition of a few fluorides of elements in high oxidation states, but the preparation of these compounds nearly always involves the use of elemental fluorine at some stage. The only alternative to the electrolytic method, first used by Moissan in 1886, yet found is the action of SbF_5 on $K_2[MnF_6]$ at 150 °C:

$$K_2[MnF_6] + 2SbF_5 \rightarrow 2K[SbF_6] + MnF_2 + F_2$$

Fluorine is a very pale yellow gas with a characteristic odour similar to that of ozone or chlorine. It is easily the most reactive element known, and combines directly, often with great violence at the ordinary temperature, with all other elements except oxygen, nitrogen and the lighter noble gases. Combustion in compressed fluorine (fluorine bomb calorimetry) is a suitable method for the determination of the standard enthalpies of formation of many binary fluorides. Many metals, however, become coated with a coherent layer of a non-volatile fluoride and are thereby made passive. Silica is thermodynamically unstable with respect to the reaction

$$SiO_2 + 2F_2 \rightarrow SiF_4 + O_2$$

but in the absence of hydrogen fluoride, which sets up the chain reaction

$$SiO_2 + 4HF \rightarrow SiF_4 + 2H_2O$$

$$2H_2O + 2F_2 \rightarrow 4HF + O_2$$

the reaction is slow at room temperature; powdered silica, however, reacts rapidly. Polytetrafluoroethylene may be used where fluorine is present, but most organic compounds inflame in the gas. The reactivity of fluorine arises partly from its own low dissociation energy and partly from the strengths of the bonds, whether covalent or ionic, formed with other elements; it should be noted that the electron affinity of the fluorine atom is out of line with values for the other halogens, ΔH° for the gas-phase reaction

$$X + e \rightarrow X^-$$

being -338, -355, -331 and -302 kJ mol^{-1} for X = F, Cl, Br and I respectively; addition of an electron to a small atom is accompanied by more repulsion from electrons already present than is the case for a large atom, and this probably explains why the process is less exothermic than might have been expected on chemical grounds.

The preparation of chlorine, bromine and iodine was mentioned in the last section; liquid chlorine is widely available in steel cylinders (which are not attacked by the element if moisture is absent), but small quantities of the gas may be made by the action of powerful oxidising agents (e.g. potassium permanganate, manganese dioxide) on concentrated hydrochloric acid; it is washed with water to remove hydrogen chloride and dried by passage through sulphuric acid.

The greenish-yellow, dark red and violet colours of chlorine, bromine and iodine in the gas phase (like the pale yellow colour of fluorine) all arise from the transition of an electron from the highest energy occupied molecular π^* orbital to the lowest energy unoccupied orbital, a σ^* orbital. The difference in energy between these orbitals decreases with increase in atomic number, leading to a progressive shift in the absorption maximum from the near-ultraviolet to the red region of the visible spectrum. Chlorine, bromine and iodine dissolve unchanged in many organic solvents (e.g. saturated hydrocarbons, carbon tetrachloride), but with those containing oxygen or nitrogen atoms (e.g. ethers, alcohols, ketones, pyridine) bromine and iodine (and to a smaller extent chlorine) form weak charge-transfer complexes. Solutions of iodine in these donor solvents are brown or yellow; even a solution of iodine in benzene can be seen to have a colour somewhat different from one of iodine in, say, cyclohexane, and a weak charge-transfer complex is formed here also. The spectra of such complexes contain an intense band in the ultraviolet, its frequency (and therefore energy) decreasing as the ionisation energy of the donor molecule decreases. Many charge-transfer complexes can be isolated in the solid state at low temperatures. Some of their structures are shown in Fig. 16.1; in complexes of poor electron donors such as benzene and

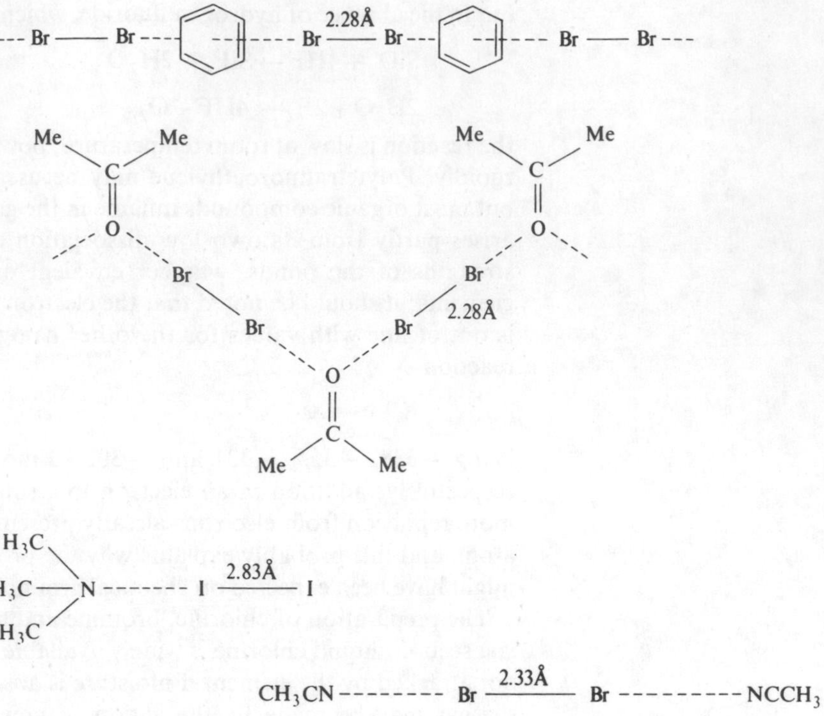

---- Br —— Br ---- (benzene) ---- Br —2.28Å— Br ---- (benzene) ---- Br —— Br ----

Me Me C O Br Br Me Me C O Br 2.28Å Br O C Me Me

H_3C H_3C—N ------ I —2.83Å— I H_3C

CH_3CN -------- Br —2.33Å— Br -------- $NCCH_3$

Figure 16.1
Structures of some charge-transfer complexes of bromine and iodine.

acetone, the bond length in the halogen molecule is unchanged or almost unchanged by complex formation; in amine or nitrile complexes of bromine or iodine, however, there is a substantial lengthening of the bond. Enthalpies of interaction of iodine with benzene and ethylamine are -5 and $-31\ \text{kJ mol}^{-1}$ respectively. Even for benzene complexes there is a significant decrease in the stretching frequency of the Br—Br or I—I bond, e.g. from $215\ \text{cm}^{-1}$ in I_2 to $204\ \text{cm}^{-1}$ in $C_6H_6.I_2$.

Chlorine, bromine and iodine are sparingly soluble in water, a fact which is by itself sufficient to show that interaction with the solvent according to the equation

$$X_2 + H_2O \rightleftharpoons HX + HOX$$

or

$$3X_2 + 3H_2O \rightleftharpoons 5HX + HXO_3$$

takes place only slightly, since all the acids formed are extremely soluble. By freezing the solutions, solid hydrates of chlorine and bromine of approximate composition $X_2.8H_2O$ may be obtained; in these, halogen molecules occupy the gaps in the open hydrogen-bonded network of water molecules that is the structure of ice; no chemical bonding is involved. Such hydrates, in which molecules are imprisoned by the hydrogen bonding, are sometimes known as *clathrate compounds*; other examples are the hydrates of the noble gases and several species formed by the inclusion of small gaseous molecules (e.g. SO_2) in the hydrogen-bonded structure of quinol (*o*-dihydroxybenzene).

In the solid state, chlorine and bromine have molecular lattices in which the intramolecular distance is the same as in the vapour. Iodine, however, exhibits signs of weak intermolecular covalent bonding (something like that of a metal) in the solid state. The intramolecular distance increases from 2.66 Å in the vapour to 2.72 Å in the solid; further, the shortest intermolecular distance in solid iodine (3.50 Å) is only slightly greater than that in solid chlorine (3.34 Å). In addition we may note the metallic lustre of solid iodine and its appreciable conductivity at higher temperatures; under very high pressure it becomes a metallic conductor.

Chemical reactivity decreases steadily from chlorine to iodine, notably in the action of the halogens on hydrogen, phosphorus, sulphur and most metallic elements. In two respects, however, iodine is more reactive than the other halogens. It is more easily oxidised to high oxidation states, e.g. to iodic acid (by nitric acid), and it is more easily converted into stable salts containing a monatomic cation coordinated by pyridine, e.g.

$$I_2 + AgNO_3 + 2py \rightarrow [I(py)_2]NO_3 + AgI$$

The increasing ease of formation of such compounds may be correlated with the decrease in ionisation energy as the atomic size of the halogen increases. Other compounds of halogen cations, containing X_2^+ and X_3^+ ions, are described in Section 16.5.

16.3 Hydrogen halides

All of the halogens react with hydrogen (as we described in Section 9.4), but bromine and iodine do so only at elevated temperatures, and in the case of iodine the reaction does not go to completion. Direct combination is therefore a potential preparative method only for hydrogen chloride and hydrogen bromide, in the case of the latter at 200–300 °C and in the presence of charcoal as a catalyst (cf. Section 13.3). Hydrogen fluoride and hydrogen chloride are obtained by the action of concentrated sulphuric acid on fluorides and chlorides, but analogous reactions with bromides and iodides result in partial oxidation of the hydrogen halide to the free halogen, and for these substances the action of water on the phosphorus trihalides prepared *in situ* is used.

All of the hydrogen halides are colourless gases with sharp acid smells; some of their physical properties are given in Table 16.2. The reaction of hydrogen fluoride with glass has already been mentioned; the anhydrous compound may be stored in steel cylinders, but the aqueous acid is best kept in polythene vessels. The variation in boiling-point among the hydrogen halides is one of the classic pieces of evidence for hydrogen

Table 16.2
Some properties of the hydrogen halides.

	HF	HCl	HBr	HI
B.p./°C	+20	−85	−67	−35
ΔH_f°/kJ mol^{-1}	−269	−92	−36	+26
ΔG_f/kJ mol^{-1}	−271	−95	−53	+1
Bond energy/kJ mol^{-1}	566	431	366	299
Bond length/Å	0.92	1.28	1.41	1.60
Dipole moment/debyes	1.8	1.1	0.8	0.4

bonding in hydrogen fluoride. In the solid state, this compound has the structure shown in Fig. 9.4 (Section 9.5); much of the hydrogen bonding persists in the liquid and in the vapour, which appears to contain a mixture of chain and ring polymers. The solvent properties of liquid hydrogen fluoride were discussed in Section 8.3. Both in the liquid and in the gaseous state it is an important reagent for the introduction of fluorine into organic and other compounds; the SbF_3-catalysed reaction

$$\rangle C{-}Cl + HF \rightarrow \rangle C{-}F + HCl$$

is the basis of the production of chlorofluorocarbons (Section 13.6).

In aqueous solution hydrogen fluoride behaves as a weak acid (though it is stronger at high concentrations owing to stabilisation of F^- by $[HF_2]^-$ formation), the other three hydrogen halides as very strong ones. We have earlier discussed the factors underlying the ionisation of the hydrogen halides and the related subject of their combination with bases (Sections 7.4 and 14.3), in which the variation in bond energy plays an important part. Hydrogen fluoride differs from the other hydrogen halides also in that it combines with alkali metal fluorides to form acid fluorides of formula MHF_2 (and others) that are stable at ordinary temperatures. Analogous compounds are, however, formed by hydrogen chloride, bromide and iodide at low temperatures, and all contain symmetrical linear anions. The estimation of the interaction energy for the gas-phase process

$$HX + X^- \rightarrow HX_2^-$$

has been discussed in Section 9.5; not surprisingly, ΔH° is most negative for X = F. Some mixed anions such as $ClHI^-$ are also known.

16.4 *The halides: some general considerations*

So much of inorganic chemistry is concerned with binary compounds of the halogens that it is appropriate at this stage to devote a section to the discussion of some general features of their structures and their thermochemistry.

All the halides of the alkali metals have three-dimensional ionic (NaCl or CsCl) structures, and their formation may be considered entirely adequately in terms of the Born–Haber cycle (Section 6.7). We have shown earlier that for a given formula type, the Madelung constant is almost independent of the structure, and the lattice energies of these compounds are therefore approximately inversely proportional to $(r_+ + r_-)$, where r_+ and r_- are the radii of the ions present. That this is so may be confirmed by considering the lattice energies given in Table 10.3 (Section 10.3). If we think of a halogen exchange reaction such as

$$\geq C{-}Cl + MF \rightarrow \geq C{-}F + MCl$$

in the absence of a solvent, it is easily seen that the energy change attending this reaction involves the difference between the C—Cl and C—F bond energy terms (a constant), the difference between the electron affinities of fluorine and chlorine (a constant) and the difference in lattice energy between MF and MCl, which is approximately proportional to

$$\frac{1}{r_+ + r_{Cl^-}} - \frac{1}{r_+ + r_{F^-}}$$

This is always negative since fluoride is a smaller ion than chloride, but it approaches zero as r_+ increases; thus caesium fluoride should be the best alkali metal fluoride for effecting this change, a prediction confirmed by experiment.

A few other monohalides have the NaCl or the CsCl structure; these include AgF, AgCl, AgBr, TlCl and TlBr. For these compounds, however, the agreement between lattice energies calculated from observed interatomic distances and those obtained via the Born cycle is less close than it is for the alkali metal halides, and it is generally accepted that some degree of covalent interaction is involved. The same is true for CuCl, CuBr, CuI and AgI, which have the wurtzite structure.

Most metal difluorides have the CaF_2 or the TiO_2 (rutile) structure, and for many of these the simple ionic model holds without elaboration (e.g. for CaF_2, SrF_2, BaF_2, MgF_2, MnF_2 and ZnF_2); we shall see later that, with slight modification to take into account the distribution of *d* electrons between different orbitals, it holds also for other transition metal difluorides. A few dichlorides have one of these symmetrical three-dimensional structures, but the great majority, and nearly all dibromides and diiodides, have the cadmium chloride or the cadmium iodide structure; both of these are layer structures in which all the nearest neighbours of the anions are on the same side of the ion, indicating that a purely electrostatic model cannot be satisfactory. On the other hand, these are obviously not simple molecular structures, so a treatment in terms of covalent bonding is not satisfactory either.

Metal trifluorides are crystallographically more complex than the difluorides, but again symmetrical three-dimensional structures are commonly found; for the other halides, layer structures predominate. A few metal tetrafluorides (e.g. ZrF_4) have what appear to be ionic structures, but most metal tetrahalides are either volatile molecular species (e.g. $SnCl_4$, $TiCl_4$) or contain rings or chains of MX_6 octahedra sharing four corners (e.g. SnF_4); as we remarked in Section 16.1, metal-bridging halogen distances in such compounds are always greater than metal–terminal halogen distances, and it is very likely that the metal–halogen–metal bonding involves multicentre orbitals and low bond orders. Metal pentahalides usually have chain or ring structures based on MX_6 octahedra sharing two corners (e.g. SbF_5, BiF_5) or have molecular structures (e.g. $SbCl_5$); metal hexahalides (e.g. MoF_6, UF_6, WCl_6) all have molecular structures. In general, therefore, increase in oxidation state results in a structure change in the direction three-dimensional ionic structure → layer or polymer structure → molecular structure.

If we are considering a change of oxidation state involving two ionic structures, e.g.

$$MF + \tfrac{1}{2}F_2 \rightarrow MF_2$$

we can discuss the thermochemistry of the reaction in terms of two lattice energies (which will be different because of three factors – differences in ionic charges, in ionic radii, and in Madelung constants) and the dissociation energy and electron affinity of fluorine. In extending this treatment to metal halides in general, i.e.

$$MX + \tfrac{1}{2}X_2 \rightarrow MX_2$$

there is some loss in reliability, since for X = Br or I, MX_2 is not very likely to be a purely ionic compound. Nevertheless, if we introduce the simplifying assumptions that (i) all the monohalides of a given metal M have the same Madelung constant A_1 and all of its dihalides the same Madelung constant A_2 (ii) MX and MX_2 are purely ionic for X = F, Cl, Br or I (iii) we can neglect the dissociation energies and electron affinities of the halogens (which partly cancel one another), we can see that the energy change for the reaction is then given by the difference between the second ionisation energy of M (energy absorbed) and the increase in lattice energy on going from MX to MX_2 (energy liberated). The second ionisation energy of M is constant, so whether the reaction goes is determined by the increase in lattice energy, which is proportional to

$$\frac{A_2}{r(M^{2+})+r(X^-)} - \frac{A_1}{r(M^+)+r(X^-)}$$

Since $A_2 > A_1$ (Section 6.6), the reaction is most likely to go when $r(X^-)$ is as small as possible, i.e. when X = F. Conversely, if we want to make a monohalide by decomposition of a dihalide, the reaction is most likely to go when X = I. These situations are represented in practice by

$$AgX + \tfrac{1}{2}X_2 \rightarrow AgX_2$$

(which goes only for X = F) and

$$CuX_2 \rightarrow CuX + \tfrac{1}{2}X_2$$

(which at room temperature goes only for X = I). It is, in fact, quite easy to avoid the need for invoking assumption (iii) above by introducing data for the halogens; doing so only reinforces the conclusions we have just reached. Where, in such a system, the increase in lattice energy is about equal to the ionisation energy involved, the two ionic halides will be of about equal stability; this is what commonly happens among compounds of the transition metals, as we shall see in later chapters.

Before we say anything about covalently bonded metal halides it is desirable first to consider some simpler cases of covalent bonding, such as nearly all the halides of the non-metallic elements. The formation of the hydrogen halides, of oxygen dihalides, or of nitrogen trihalides can be represented by union of ground state atoms formed by dissociation of diatomic molecules, e.g.

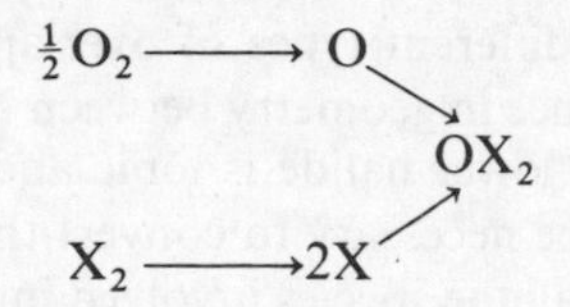

Thus the standard enthalpy of formation of OX_2 can in principle be evaluated from the bond energy terms in O_2, X_2 and OX_2, which in turn ought to be calculable theoretically (it will, of course, be necessary for the values to be very accurate, since only the small difference between them is involved in determining whether OX_2 exists). For other molecules (such as CX_4 or PX_5) however, the process is more complicated, as in these cases the carbon or phosphorus atom that is forming the bonds is not in its ground state but in its valence state (Section 4.6), and we then have to consider a sequence such as

$$\begin{array}{l} C(s) \rightarrow C(g)(\text{ground}) \rightarrow C(g)(\text{valence}) \searrow \\ \qquad\qquad\qquad\qquad\qquad\qquad\qquad\quad CX_4 \\ 2X_2 \longrightarrow\longrightarrow\longrightarrow\longrightarrow\longrightarrow 4X \nearrow \end{array}$$

Clearly, in a situation like this, the bigger the difference in energy between the ground state and the valence state, the less likely it is that the overlapping of orbitals will lead to interaction strong enough to result in compound formation; but since (i) many valence states are not observable spectroscopically (ii) we cannot yet calculate accurately the bonding energy term resulting from overlapping of the wave functions of the halogen atom and the valence state non-metal atom, we cannot give a very satisfactory quantitative discussion of this problem. We can only say that the weakness of the bond in X_2 for X = F and the likelihood of best overlap when X = F (the smallest atom among the halogens) give us some ideas of why maximum covalent oxidation states commonly occur in fluorides (e.g. SF_6, IF_7). If we extend the discussion to X = H we may note that any increase in bonding because of the smaller size of the H atom would be offset by the much greater bond energy in H_2, and neither SH_6 nor IH_7 is likely to exist. Before we leave this subject we should perhaps remind the reader once more that because of the inaccessibility of many valence states to spectroscopic investigation, bond energy terms as conventionally used throughout chemistry refer to formation from ground state atoms, and (except when the ground state is the valence state) are not to be equated with the bonding energies arising from overlapping of valence state wave functions that we need here.

Let us now consider a change of oxidation state involving two covalent halides, e.g.

$$MX_4 + X_2 \rightarrow MX_6$$

If MX_4 and MX_6 are both molecular species (e.g. if M = S), two valence states and two different types of overlapping of wave functions (arising from the difference in geometry between MX_4 and MX_6) will in general be involved; if the lower halide is ionic and the higher one covalent (e.g. if M = U), it will be necessary to convert the lower halide into ground state atoms, convert all the species involved into the appropriate valence states, and then proceed to consider the overlapping of wave functions. If one of the halides has a polymeric molecular structure, the presence of two different types of bonds to halogen (bridging and terminal) will further increase the complexity of the problem. In considering cases such as these, it soon becomes apparent that, once we have left behind those situations for which the ionic model is adequate, experimental information has outstripped our capacity to give a really satisfactory interpretation of it, let alone predict it. The realisation that this is so is essential to the assessment of the present value and limitations of valence theory in relation to inorganic chemistry.

16.5 Interhalogen compounds and polyhalogen ions

The stable interhalogen compounds characterised at the present time are shown in Table 16.3. All are prepared by direct combination; where there is more than one possible product, which compound is formed depends upon the relative proportions of the reactants and the temperature. Fluorine reacts with the other halogens at normal temperature and pressure to produce ClF, BrF_3 and IF_5, but under more vigorous

Table 16.3
Known stable interhalogen compounds.

Type		
XX′	ClF	colourless gas b.p. – 100 °C
	BrF	pale brown gas b.p. approx. 20 °C
	BrCl	exists only in equilibrium with dissociation products
	ICl	red solid m.p. 27 °C (α), 14 °C (β)
	IBr	black solid m.p. approx. 40 °C
XX'_3	ClF_3	colourless gas b.p. 12 °C
	BrF_3	yellow liquid b.p. 126 °C
	IF_3	yellow solid m.p. – 28 °C (decomp.)
	ICl_3	orange solid subl. 64 °C
XX'_5	ClF_5	colourless gas b.p. – 14 °C
	BrF_5	colourless liquid b.p. 41 °C
	IF_5	colourless liquid b.p. 100 °C
XX'_7	IF_7	colourless gas, solid sublimes 5 °C

conditions ClF_3, ClF_5, BrF_5 and IF_7 are formed. Much, though by no means all, of the interest in these compounds lies in their structures and those of the species they form by losing or combining with a halide ion. As is usual in such series of compounds, the bond energy terms decrease with increase in oxidation state (Section 5.4): for ClF, ClF_3 and ClF_5, for example, they are 257, 172 and 153 kJ mol^{-1} respectively. This decrease may reasonably be ascribed to the need for promotion of electrons to give the higher valence states.

The most stable of the diatomic compounds are ClF and ICl; BrCl is substantially, and IBr detectably, dissociated into the constituent elements at 25 °C, and BrF is unstable with respect to disproportionation into Br_2 and BrF_3. In general it may be said that all five compounds are intermediate between their parent halogens in physical properties and reactivity; ICl in polar solvents (but not in the vapour phase) acts as an iodinating agent for aromatic compounds, however, thus behaving under these conditions as a source of positive iodine. Bond lengths in the gaseous molecules are: ClF, 1.63; BrF, 1.76; BrCl, 2.14; ICl, 2.30 Å. Those in the halogen fluorides, in which there is a considerable difference in the electronegativities of the atoms, are shorter than the sum of the covalent radii; thermochemical data show that the bonds are also stronger than the mean of those in the parent halogens (cf. Section 5.7). In solid α-ICl there is a chain structure in which ICl molecules, here having bond lengths of either 2.37 or 2.44 Å, are only 3.00–3.08 Å away from other molecules, the iodine atom of one molecule being next to the chlorine atom of another molecule, and vice versa. Solid IBr has an essentially similar structure. These structures should be compared with that of solid iodine (Section 16.2), in which there is also evidence for considerable intermolecular interaction.

The higher fluorides of chlorine and bromine are extremely reactive substances which explode with water and many organic compounds;

the trifluorides even set fire to asbestos, converting it into a mixture of fluorides. The iodine fluorides are less reactive, being quietly hydrolysed by water; iodine pentafluoride is occasionally used as a mild fluorinating agent for organic compounds. Iodine trichloride decomposes readily into the monochloride and chlorine. Bromine trifluoride undergoes appreciable self-ionisation according to the equation

$$2BrF_3 \rightleftharpoons BrF_2^+ + BrF_4^-$$

This aspect of its chemistry and its use for the preparation of complex fluorides were discussed in Section 8.5. There is also limited evidence for the self-ionisation of chlorine trifluoride and iodine pentafluoride:

$$2ClF_3 \rightleftharpoons ClF_2^+ + ClF_4^-$$
$$2IF_5 \rightleftharpoons IF_4^+ + IF_6^-$$

The structures of ClF_3 and BrF_3 were discussed at length in Section 5.2, where we first introduced the VSEPR theory. Although they can be roughly described as trigonal bipyramids with two equatorial positions occupied by lone pairs of electrons, this description does not give sufficient weight to the differences in length observed in both molecules between the axial and equatorial bonds (as shown in Fig. 16.2(*a*) and (*b*)), and to this extent a description in which the orbitals involved fall into two distinct groups (sp^2 and pd) is preferable. A third view, that the Cl and

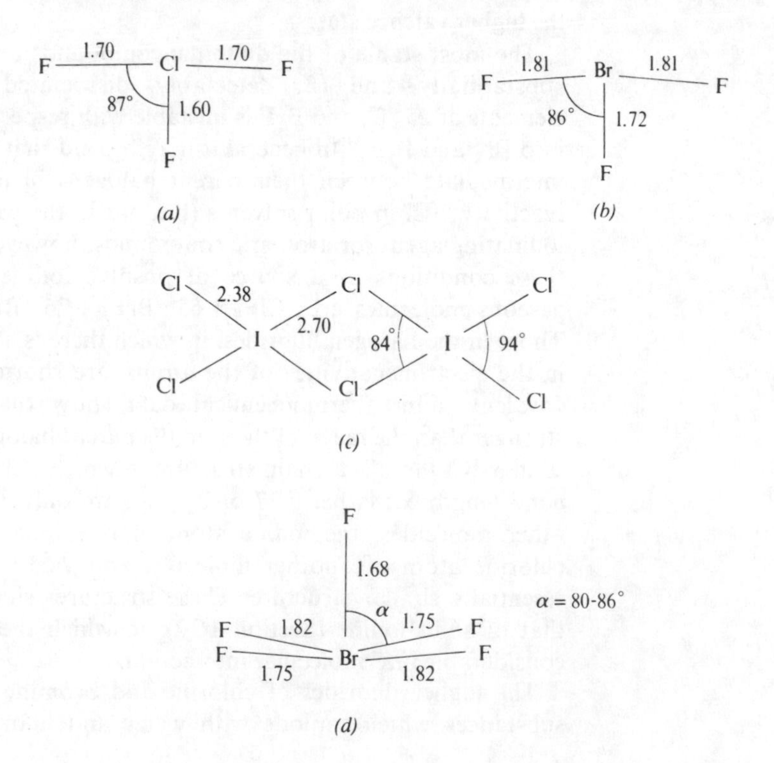

Figure 16.2
The structures of (*a*) ClF_3 (*b*) BrF_3 (*c*) I_2Cl_6 (*d*) BrF_5 (bond lengths in ångströms).

terminal F atoms form a three-centre two-electron system with bond order one-half (cf I_3^-, described later), whilst the middle bond is an ordinary single bond, has also been put forward; whether a bond length difference of 0.10 Å justifies such a description is, however, open to argument. Iodine trichloride in the solid state is planar I_2Cl_6 (Fig. 16.2(*c*)), the bridge bonds being much longer than the terminal bonds, indicating that the bond order in the bridges is lower. In solid BrF_5, the molecule is approximately a square pyramid with the bromine atom a little below the base; again an approximate description can be given on the basis of the VSEPR theory, this time as an octahedron with one position occupied by an unshared pair; once more, however, this simple description conceals the significant variation in bond lengths which is shown in Fig. 16.2(*d*). Iodine pentafluoride is known to have the same shape as BrF_5, but neither for this molecule nor for IF_7 (a pentagonal bipyramid) are accurate interatomic distances available.

All of the interhalogen compounds mentioned above (except chlorine pentafluoride) combine with halide ions (nearly always in the form of alkali metal halides) to give polyhalogen anions; other such anions are formed in the combination of the halogens with metal halides. For the less stable polyhalogen anions, the use of as large a cation as possible is advisable, so as to minimise the decrease in lattice energy when the polyhalide is formed (cf. Section 6.8); all known salts containing the Cl_3^-, I_5^- and I_7^- ions, for example, are those of quaternary ammonium or similar cations, but Br_3^- and I_3^- can be isolated in the form of caesium salts. For polyhalides containing two different halogens (e.g. $CsICl_2$, $KBrF_4$) it is found that thermal decomposition always yields the halide of highest lattice energy (in these examples, CsCl and KF), this factor outweighing all others when we consider possible modes of thermal decomposition.

Among the triatomic polyhalogen anions, ClF_2^- and BrF_2^- are known from vibrational spectroscopy to be linear, but no bond length data are available. The ICl_2^- anion has been shown to be linear and symmetrical in several salts, the bond lengths being 2.55 Å, considerably greater than that in either gaseous or solid ICl. In $KIBr_2.2H_2O$ the anion is isostructural with ICl_2^-, I—Br being 2.71 Å. In Me_4NBr_3 and Me_4NI_3 (and in most salts of other very large cations) the Br_3^- and I_3^- ions are also linear and symmetrical, with Br—Br 2.54 Å and I—I 2.92 Å; but in the analogous caesium salts the anions, though still linear, are unsymmetrical, with Br—Br 2.44 and 2.70 Å in Br_3^-, and I—I 2.83 and 3.03 Å in I_3^-. Two forms of Et_4NI_3 are known, one containing symmetrical, and the other unsymmetrical, linear anions; only the latter occur in RbI_3 and TlI_3. It seems that it is the crystal environment of the anions, rather than simply the size of the cation, that determines the anion structure.

We can describe the structures of the symmetrical trihalide ions in terms of a 10-electron valence shell of the central atom with three lone pairs of electrons occupying the equatorial positions of a trigonal bipyramid, but the lengths of the bonds in the ICl_2^-, IBr_2^-, Br_3^- and I_3^- ions raise serious doubts as to whether they really are ordinary single (i.e. two-centre two-electron) bonds. These doubts are reinforced when we compare the stretching frequencies in the ICl_2^- and I_3^- ions with those in ICl and I_2. In ICl_2^-, the symmetric and asymmetric stretching frequencies are at 267 and 222 cm^{-1} respectively, compared with a stretching frequency of 384 cm^{-1} in ICl; in I_3^-, the corresponding frequencies are 113 and 135 cm^{-1}, and that in I_2 is 215 cm^{-1}. It seems that if we take the bonds in ICl and I_2 as single bonds, we must conclude that those in ICl_2^- and I_3^- are much weaker. The presently accepted view of the electronic structures of these anions is that if we write the electronic structure of a halide ion as $s^2p_x{}^2p_y{}^2p_z{}^2$ and of a halogen atom as $s^2p_x{}^2p_y{}^2p_z{}^1$, the p_z orbitals of an ion and two atoms combine to form three orbitals (bonding, non-bonding and antibonding) as shown in Fig. 16.3, and that the four electrons to be accommodated occupy the bonding and non-bonding orbitals, giving a net bond order of one-half between each pair of atoms. For the ClF_2^- ion, only the asymmetric stretching frequency is available; this is 635 cm^{-1}, somewhat less than in ClF, where it is 772 cm^{-1}, but the difference is proportionately much smaller than in ICl_2^- and I_3^-. Whether the structure of ClF_2^- is not best described in terms of an electronic configuration $3s^23p^63d^2$ for the chlorine atom and occupation of three equatorial positions of a trigonal bipyramid by lone pairs, is therefore still an open question.

We shall not discuss the structures of other polyhalogen anions in great detail, though bond lengths are known for many of them. The ions IF_4^-, BrF_4^- and ICl_4^- are all planar, and are usually (though not necessarily correctly) believed to be based on a 12-electron configuration round the central atom, with two *trans* positions of an octahedron occupied by lone pairs. The I_5^- ion, however, is a V-shaped assembly of a central iodide ion and two stretched iodine molecules, successive I—I distances from the

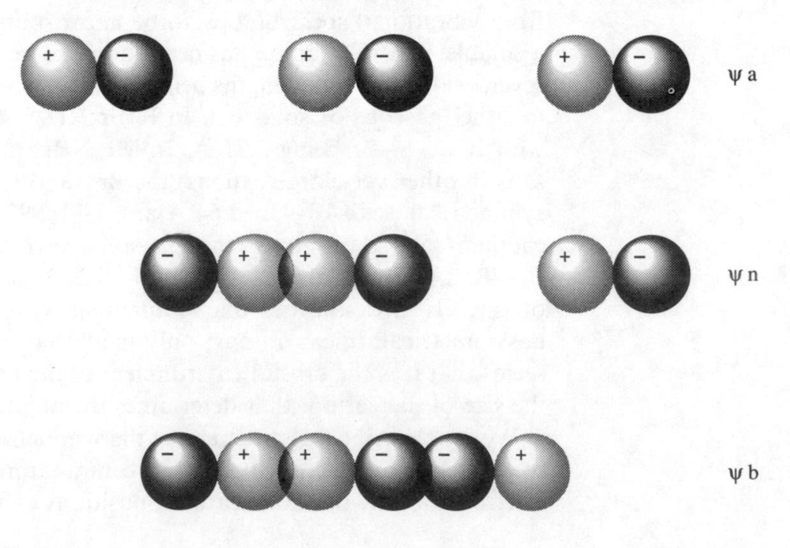

Figure 16.3
Interaction of three *p* orbitals in a trihalide ion.

centre being 3.17 and 2.81 Å; the structures of the I_7^-, I_8^{2-} and I_9^- ions are built up along rather similar lines. Great interest is attached to the structures of the BrF_6^- and IF_6^- ions (which on the VSEPR theory should be pentagonal bipyramids with one position occupied by a lone pair – cf. $SbCl_6^{3-}$, $[Sb(C_2O_4)_3]^{3-}$ and $TeCl_6^{2-}$); the vibrational spectrum of the IF_6^- ion shows it is certainly not a regular octahedron, and on the ^{19}F nmr timescale it is fluxional; the BrF_6^- ion is regular octahedral in both the solid caesium salt and in acetonitrile solution. There is evidently some subtle difference in the electronic structures of the two ions. The structure of the ion IF_8^- in $Me_4N[IF_8]$ has recently been shown to be a square antiprism.

Polyhalogen cations include Br_2^+ and I_2^+; I_3^+, ClF_2^+, BrF_2^+ and ICl_2^+; I_4^+; I_5^+, ClF_4^+, BrF_4^+ and IF_4^+; and ClF_6^+, BrF_6^+ and IF_6^+. Obviously, cations like Br_2^+ and I_2^+ are obtained only in oxidising conditions; cations derived from interhalogen compounds are typically obtained by halide ion transfer to a very strong halide ion acceptor such as SbF_5, AsF_5 or $SbCl_5$; since ClF_7 and BrF_7 are not known, the production of the ClF_6^+ and BrF_6^+ cations requires both very powerfully oxidising conditions and the presence of a halide ion acceptor. The following examples illustrate the various methods used:

$$Br_2 + SbF_5 \xrightarrow{BrF_5} Br_2^+[Sb_3F_{16}]^-$$

$$2I_2 + S_2O_6F_2 \xrightarrow{HSO_3F} 2I_2^+ + 2SO_3F^- \xrightarrow{-80\,°C} I_4^{2+} + 2SO_3F^-$$

$$5I_2 + 3AsF_5 \rightarrow 2I_5^+[AsF_6]^- + AsF_3$$

$$BrF_3 + SbF_5 \rightarrow [BrF_2]^+[SbF_6]^-$$
$$ICl_3 + AlCl_3 \rightarrow [ICl_2]^+[AlCl_4]^-$$
$$IF_5 + SbF_5 \rightarrow [IF_4]^+[SbF_6]^-$$
$$IF_7 + AsF_5 \rightarrow [IF_6]^+[AsF_6]^-$$
$$Kr_2F_3^+AsF_6^- + BrF_5 \rightarrow [BrF_6]^+[AsF_6]^- + 2Kr + F_2$$
$$ClF_5 + AsF_5 + KrF_2 \rightarrow [ClF_6]^+[AsF_6]^- + Kr$$

In $Br_2^+[Sb_3F_{16}]^-$ the cation has a Br—Br distance of 2.15 Å and a stretching frequency of 368 cm^{-1}; these, when compared with the corresponding data for bromine (2.27 Å and 320 cm^{-1}) show that the bond in the cation is stronger, obviously because the electron lost has come from an antibonding orbital. The bond in the cation of $I_2^+[Sb_2F_{11}]^-$ (length 2.58 Å) is similarly stronger than that in I_2. The blue I_2^+ cation, which is necessarily paramagnetic, may be prepared in several solvents; on cooling to −80 °C it forms the weakly bonded dimer I_4^{2+}.

The cations I_3^+, ClF_2^+, BrF_2^+ and ICl_2^+ are all bent; the bond lengths in ClF_2^+ and I_3^+ are about the same as those in ClF and I_2 respectively. It appears, therefore, that these cations contain ordinary single bonds, and their structures may be described as based on tetrahedra with two positions

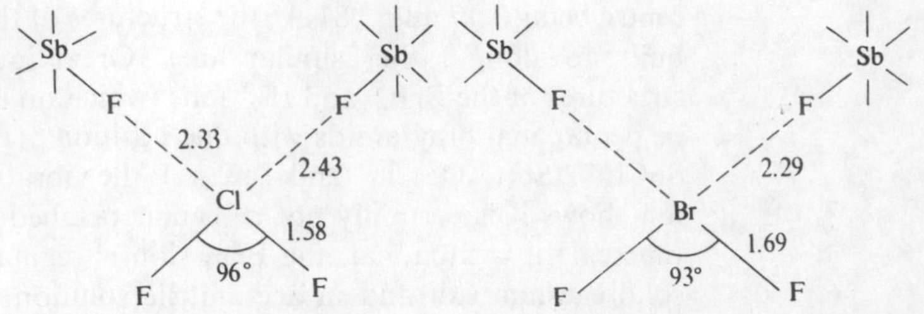

Figure 16.4
The environment of the Cl and Br atoms in ClF_2SbF_6 and BrF_2SbF_6 (bond lengths in ångströms).

occupied by lone pairs. There is, however, some interaction between anion and cation in their salts, as the bond length data shown in Fig. 16.4 show: although bromine is a larger atom than chlorine, the Br—FSb distance in BrF_2SbF_6 is less than the Cl—FSb distance in ClF_2SbF_6. In $[BrF_4]^+[Sb_2F_{11}]^-$, the BrF_4^+ ion has a structure like that of SF_4, i.e. a trigonal bipyramid with a lone pair in one of the equatorial positions, but again there is evidence from the interatomic distances of some interaction with the anion. The I_5^+, however, has the structure

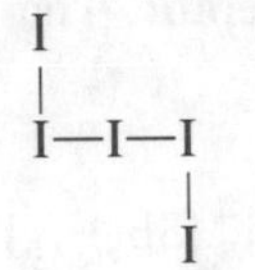

```
I
|
I—I—I
    |
    I
```

with the terminal and non-terminal bonds of length 2.64 and 2.89 Å respectively. The vibrational spectra of ClF_6^+, BrF_6^+ and IF_6^+ are all compatible with the regular octahedral structures expected for systems involving a central atom with twelve electrons in its valence shell and a coordination number of six: ClF_6^+, for example, is isoelectronic with SF_6.

16.6 Oxides and oxofluorides of chlorine, bromine and iodine

Iodine is the only one of the halogens to form an oxide (I_2O_5) which is thermodynamically stable with respect to decomposition into its elements; chlorine oxides, though not difficult to prepare, are all liable to decompose explosively, and bromine oxides are both very inaccessible and very unstable. Among these compounds only chlorine dioxide and iodine pentoxide have any practical importance. Oxofluorides of chlorine, however, have attracted great interest in recent years.

Dichlorine monoxide or simply chlorine monoxide, Cl_2O, is obtained as a yellow–brown gas by passing chlorine over freshly-prepared mercury(II) oxide or moist sodium carbonate:

$$2Cl_2 + 3HgO \rightarrow Cl_2O + Hg_3O_2Cl_2$$

$$2Cl_2 + 2Na_2CO_3 + H_2O \rightarrow 2NaHCO_3 + 2NaCl + Cl_2O$$

It hydrolyses to hypochlorous acid, of which it is formally the anhydride:

$$Cl_2O + H_2O \rightarrow 2HOCl$$

Chlorine monoxide liquefies at about 4 °C, and explodes on warming. The analogous bromine compound, prepared by a similar reaction, decomposes above − 50 °C. Chlorine dioxide, ClO_2, a yellow gas (b.p. 10 °C), is formed in the complex (and highly dangerous) reaction between potassium chlorate and concentrated sulphuric acid; a safer preparation is from the chlorate and moist oxalic acid:

$$2KClO_3 + 2H_2C_2O_4 \rightarrow K_2C_2O_4 + 2ClO_2 + 2CO_2 + 2H_2O$$

On an industrial scale it is used for the bleaching of flour and wood pulp and for water treatment, and is made by the action of dilute sulphuric acid and sulphur dioxide on sodium chlorate:

$$2NaClO_3 + SO_2 + H_2SO_4 \rightarrow 2NaHSO_4 + 2ClO_2$$

The molecule of ClO_2 is bent, with OClO bond angle 117° and Cl—O 1.47 Å; although it contains an unpaired electron, it shows no tendency to dimerise. Chlorine dioxide dissolves unchanged in water, but is slowly hydrolysed to chloric and hydrochloric acids. In alkali chlorate and chlorite are formed:

$$2ClO_2 + 2OH^- \rightarrow ClO_3^- + ClO_2^- + H_2O$$

Chlorine dioxide reacts with ozone at 0 °C to form Cl_2O_6, a dark red liquid which is also obtained by the reaction

$$ClO_2F + HClO_4 \rightarrow Cl_2O_6 + HF$$

This reaction and the hydrolysis of the compound to chlorate and perchlorate suggests the structure $O_2ClOClO_3$, i.e. chloryl perchlorate, the mixed anhydride of $HClO_3$ and $HClO_4$. The solid is $ClO_2^+ClO_4^-$. There is a similar mixed anhydride of HOCl and $HClO_4$, the compound $ClOClO_3$ (chlorine perchlorate), which is obtained by interaction of caesium perchlorate and chlorine fluorosulphate (Section 15.4):

$$CsClO_4 + ClSO_3F \rightarrow CsSO_3F + ClOClO_3$$

The anhydride of perchloric acid, Cl_2O_7, an oily explosive colourless liquid (b.p. approximately 80 °C), is obtained by dehydration of the acid with phosphorus pentoxide at low temperatures; it has the expected structure $O_3ClOClO_3$ with a bond angle of 119° at the central oxygen atom and Cl—O bond lengths of 1.71 Å (bridging) and 1.41 Å (terminal), indicating single and double bonds respectively.

Iodine pentoxide, a white solid obtained by dehydration of iodic acid at 250 °C, is stable to 300 °C. Its reaction with carbon monoxide at 90 °C (to form iodine and carbon dioxide) is used for the determination of the latter compound. Solid I_2O_5 consists of O_2IOIO_2 molecules linked by weak intermolecular bonds.

In addition to compounds like $ClOClO_3$ and $O_2ClOClO_3$ which are mixed anhydrides of two oxo-acids, there also exist compounds containing halogen–fluorine bonds, the halogenyl fluorides FXO_2 and the perhalogenyl FXO_3, for each of which X = Cl, Br or I. Chloryl fluoride, $FClO_2$, is a colourless gas (b.p. −6 °C) which is formed by the action of fluorine on chlorine dioxide. It is hydrolysed to chloric and hydrofluoric acids, and combines with antimony pentafluoride to form the chloryl salt $ClO_2[SbF_6]$ and with caesium fluoride to form the difluorochlorate $Cs[ClO_2F_2]$. The iodine-containing analogue of the last compound may be made by the action of aqueous hydrofluoric acid on caesium iodate. It has been shown to contain an anion similar in structure to SF_4, i.e. a trigonal bipyramid with one equatorial position occupied by a lone pair of electrons; the other two are occupied by oxygen atoms. Perchloryl fluoride, $FClO_3$ (b.p. −47 °C), is made by the action of fluorine on potassium perchlorate (fluorine perchlorate, $FOClO_3$, being also formed) or, better, by the action of a mixture of HF and SbF_5 on the same compound:

$$KClO_4 + 2HF + SbF_5 \rightarrow FClO_3 + KSbF_6 + H_2O$$

This compound, which has a tetrahedral structure, is surprisingly stable, for it does not decompose until heated above 400 C, and is only slowly attacked by alkali. It is, however, a powerful oxidising agent at higher temperatures. Perchloryl fluoride is also a mild fluorinating agent which has been used for the preparation of fluorinated steroids; it reacts with phenyl lithium to form the unstable perchlorylbenzene, $C_6H_5ClO_3$, and lithium fluoride. There is, however, no evidence for the formation of ClO_3^+ and $ClO_3F_2^-$ ions: perchloryl fluoride does not react with either antimony pentafluoride or caesium fluoride.

The more highly fluorinated oxofluoride F_3ClO is formed by the action of fluorine on Cl_2O at low temperatures; it reacts with SbF_5 and with CsF to yield compounds containing the F_2ClO^+ and F_4ClO^- ions respectively.

16.7 Oxo-acids of the halogens and their salts

Fluorine stands apart from the other halogens in forming no species in which it has a formal oxidation state other than −I, and the only known oxo-acid of fluorine is the unstable species HOF, commonly known as hypofluorous acid although it does not ionise in water and no salts are known. It is obtained by passing fluorine over ice at −40 °C and condensing the resulting gas; it decomposes rapidly at room temperature and reacts with water to form hydrogen peroxide and hydrogen fluoride:

$$HOF + H_2O \rightarrow HOOH + HF$$

A number of compounds which may be described as covalent hypofluorites, e.g. O_2NOF (more commonly known as fluorine nitrate, and referred to as such in Section 14.7) and $FOSO_2F$ (fluorine fluorosulphate – Section 15.4) are known.

The principal oxo-acids of the other halogens known in the free state, in aqueous solution, or in the form of salts, are:

HOCl	HOBr	HOI
HOClO		
$HOClO_2$	$HOBrO_2$	$HOIO_2$
$HOClO_3$	$HOBrO_3$	$HOIO_3$, $(HO)_5IO$

The compounds HOX, HXO_2, HXO_3 and HXO_4 are almost universally known as hypohalous, halous, halic, and perhalic acids respectively; halic(I), halic(III), halic(V) and halic(VII) acids are alternative names. Some of the redox reactions of these species are slow (notably those involving ClO_4^- or BrO_4^-), but enough are fast for a consideration of the potential diagrams of the halogens in aqueous solution to be a useful exercise. This, however, is postponed until Section 16.8; for the present we shall be concerned with other aspects of their chemistry.

The hypohalous acids are obtained in aqueous solution by interaction of the halogen and mercury(II) oxide, a reaction analogous to that used for the preparation of chlorine monoxide:

$$2X_2 + 3HgO + H_2O \rightarrow Hg_3O_2X_2 + 2HOX$$

All are weak acids (pK_a about 8) and are unknown in the free state. Sodium hypochlorite may be crystallised from a solution obtained by electrolysing aqueous sodium chloride in such a way that the chlorine liberated at the anode mixes with the alkali produced at the cathode; bleaching powder, a non-deliquescent mixture of calcium chloride, hydroxide and hypochlorite, is produced by the action of chlorine on solid calcium hydroxide. All the hypohalite ions are unstable with respect to the disproportionation

$$3OX^- \rightarrow XO_3^- + 2X^-$$

but whereas the reaction is slow at the ordinary temperature for X = Cl, it is fast for X = Br and very fast for X = I. Sodium hypochlorite disproportionates in hot aqueous solution

$$3NaOCl \rightarrow NaClO_3 + 2NaCl$$

and passage of chlorine into hot aqueous alkali gives chlorate and chloride instead of hypochlorite and chloride. Hypochlorite solutions decompose to form chloride and oxygen in the presence of cobalt(II) compounds as catalysts; they are very powerfully oxidising and in the presence of alkali convert IO_3^- into IO_4^-, Cr(III) into CrO_4^{2-} and even Fe(III) into FeO_4^{2-}. Aqueous sodium hypochlorite is extensively used as a cheap bleaching and sterilising agent.

Chlorous acid, $HClO_2$, is obtained in solution by the action of dilute sulphuric acid on aqueous barium chlorite; it is a stronger acid than HOCl, and has a pK_a of 2 (an increase in acid strength with increase in oxygen content is usual among oxo-acids, as we saw in Section 7.4). Sodium chlorite, which is used as a bleaching agent, is obtained by the action of chlorine dioxide on sodium peroxide:

$$2ClO_2 + Na_2O_2 \rightarrow 2NaClO_2 + O_2$$

In the ClO_2^- anion the bond angle is 111° and the bond length 1.57 Å; this is longer than in ClO_2 (1.47 Å) because the extra electron has had to go into an antibonding orbital. Alkaline solutions of chlorite persist unchanged over long periods, but in the presence of acid a complex decomposition takes place with formation of chlorine dioxide and chloride:

$$5HClO_2 \rightarrow 4ClO_2 + H^+ + Cl^- + 2H_2O$$

Chlorates, as we have already seen, are formed by the action of chlorine on hot aqueous alkali, but the main product of this reaction is chloride, and for preparative purposes it is better to electrolyse an aqueous solution of a chloride (with mixing of products) at 70 °C, and to crystallise out the chlorate. Bromates are made by alkaline hypochlorite oxidation of bromide, e.g.

$$KBr + 3KOCl \rightarrow KBrO_3 + 3KCl$$

Iodic acid may be obtained as a white solid by oxidation of iodine with nitric acid, and potassium iodate is conveniently prepared by heating iodine with concentrated potassium chlorate solution and a little nitric acid:

$$2KClO_3 + I_2 \rightarrow 2KIO_3 + Cl_2$$

Chloric and bromic acids are obtainable only in the form of aqueous solutions. All three halic acids are strong acids and powerful oxidising agents. Sodium chlorate is extensively used as a weedkiller, potassium chlorate in some matches and in fireworks, and potassium bromate and iodate in titrimetric analysis. The reaction

$$IO_3^- + 5I^- + 6H^+ \rightarrow 3I_2 + 3H_2O$$

may, under suitable conditions, be used for the determination of any of the reactants; but (since potassium iodate is easily obtained very pure) its most important application is as a source of iodine for the standardisation of thiosulphate solutions. All the halate ions are pyramidal; in ClO_3^- the OClO bond angle is 106° and the Cl—O bond length 1.48 Å. Some iodates (e.g. KIO_3) decompose when heated into the iodide and oxygen, others (e.g. $Ca(IO_3)_2$) into the oxide, iodine and oxygen; bromates behave similarly, and the interpretation of these observations is an interesting problem in energetics and kinetics. Acid iodates, e.g. $KH(IO_3)_2$, are formed from iodates and iodic acid; they presumably contain hydrogen-bonded anions, but their structures are not known.

Perchloric acid is the only oxo-acid of chlorine that can be obtained pure; it is an oily colourless liquid (b.p. 90 °C with some decomposition) prepared by heating the potassium salt with concentrated sulphuric acid under reduced pressure. The pure acid is liable to explode when heated or in the presence of organic matter, but in dilute aqueous solution the perchlorate ion is very difficult to reduce despite the standard potentials at $[H^+]=1$ for the half-reactions

$$ClO_4^- + 2H^+ + 2e \rightleftharpoons ClO_3^- + H_2O \qquad E^\circ = +1.19\ V$$

$$ClO_4^- + 8H^+ + 8e \rightleftharpoons Cl^- + 4H_2O \qquad E^\circ = +1.24\ V$$

Zinc, for example, merely liberates hydrogen, and iodide ion has no action. Reduction to chloride may, however, be brought about by titanium(III) in acidic solution or by iron(II) in the presence of alkali. Perchloric acid is an extremely strong acid in aqueous solution, and the anion shows less tendency in aqueous media than other common anions to complex cations; sodium perchlorate solution is therefore the standard medium for the investigation of ionic equilibria in aqueous systems (cf. Section 7.2). Alkali metal perchlorates may be obtained by the disproportionation of chlorates under carefully controlled conditions (traces of foreign matter cause complete decomposition to chloride and oxygen to take place). Thus potassium chlorate when heated at 500 °C in a clean silica vessel gives perchlorate and chloride

$$4KClO_3 \rightarrow 3KClO_4 + KCl$$

and the sparingly soluble perchlorate is easily separated. Alternatively, electrolytic oxidation of an aqueous solution of a chlorate at high current density may be employed. When heated, potassium perchlorate gives the chloride and oxygen, apparently without intermediate formation of chlorate. Silver perchlorate, like the silver salts of some other very strong acids (e.g. $AgBF_4$, $AgSbF_6$, CF_3COOAg) is soluble in many organic solvents, including benzene and diethyl ether, owing to complex formation between the organic molecule and the silver ion. Both it and perchlorate esters are dangerously explosive. A mixture of ammonium perchlorate and aluminium powder is a standard rocket propellant. The perchlorate anion is a regular tetrahedron with Cl—O = 1.44 Å, and many perchlorates are isomorphous with salts of other tetrahedral anions, e.g. $KClO_4$ with $BaSO_4$. The monohydrate of perchloric acid, a solid of m.p. 50 °C, is actually $H_3O^+ClO_4^-$ and has been mentioned earlier (Section 6.8) in connection with the estimation of the proton affinity of water.

Perbromic acid and the perbromates were first prepared as recently as 1968. The best method for the oxidation of bromate is by fluorine in dilute aqueous sodium hydroxide:

$$BrO_3^- + F_2 + 2OH^- \rightarrow BrO_4^- + 2F^- + H_2O$$

Electrolytic oxidation, however, also yields the anion, which may be precipitated as the salt of one of the larger alkali metal cations. A solution of the acid may be made by ion-exchange, but the anhydrous acid has not been isolated. Thermochemical data show that for the half-reaction

$$BrO_4^- + 2H^+ + 2e \rightleftharpoons BrO_3^- + H_2O$$

$E^\circ = +1.76$ V at $[H^+]=1$, making perbromate a slightly stronger oxidising agent than perchlorate or periodate under the same conditions. Like perchlorate, it is usually a slow oxidising agent in dilute neutral solution, but a much more rapid one at high acidities.

Several different periodic acids and types of periodate are known. Oxidation of potassium iodate by hot alkaline hypochlorite yields $K_2H_3IO_6$ (paraperiodate), which is converted by nitric acid into KIO_4 (metaperiodate) and by concentrated alkali into $K_4H_2I_2O_{10}$; the last compound is dehydrated at 80 °C to $K_4I_2O_9$ (mesoperiodate). The simple tetrahedral IO_4^- ion is present in KIO_4, but all the other compounds contain octahedrally coordinated iodine, the correct formulae of the anions present being

$$[IO_3(OH)_3]^{2-},\ [(HO)O_3I\overset{O}{\underset{O}{\diamond}}IO_3(OH)]^{4-}$$

and

$$[O_3I\overset{O}{\underset{O}{\diamond}}\!\!-O-\!\!IO_3]^{4-}$$

respectively. The relations between these ions may be expressed by the equilibria:

$$[IO_3(OH)_3]^{2-} + H^+ \rightleftharpoons IO_4^- + 2H_2O$$

$$2[IO_3(OH)_3]^{2-} \rightleftharpoons 2HIO_5^{2-} + 2H_2O$$

$$\updownarrow$$

$$H_2I_2O_{10}^{4-} \rightleftharpoons I_2O_9^{4-} + H_2O$$

Periodic acid, obtained by electrolytic oxidation of iodic acid or by the action of concentrated nitric acid on the barium salt, is H_5IO_6 or $OI(OH)_5$, a rather weak acid having pK_a 3.3; it is successively dehydrated by the action of heat to $H_4I_2O_9$ and HIO_4. Periodate oxidises iodide rapidly even in neutral solution:

$$IO_4^- + 2I^- + H_2O \rightarrow IO_3^- + I_2 + 2OH^-$$

In hot acidic solution ozonised oxygen is liberated from water, and Mn(II) is oxidised to MnO_4^-.

Reactions of the oxo-acids and oxo-anions of the halogens in aqueous solution are discussed further in the next section.

16.8 *Aqueous solution chemistry*

Since the halide ions are the simplest anions stable in aqueous solution, it was inevitable that much of Chapter 7 was concerned with their properties and those of systems which involve them. Topics discussed there included the thermodynamics of hydration of the gaseous ions and the solubilities of some metal halides (Section 7.3), the ionisation of the hydrogen halides, the oxo-acids of chlorine and chloroacetic acids (Section 7.4), and the complexing of metal ions by halide ions and the distinction between class (a) and class (b) cations (Section 7.7). In this section we shall therefore be considering for the most part oxidation and reduction processes.

The standard potentials for the half-reactions

$$\tfrac{1}{2}X_2 + e \rightleftharpoons X^-$$

may be measured directly for X = Cl, Br and I, the values being +1.36, +1.07 and +0.54 V respectively. The magnitudes of these quantities are determined by the bond energies in the halogen molecules, the electron affinities of the halogen atoms and the standard free energies of hydration of the halide ions, as may be seen by writing the half-reaction as a series of stages:

$$\tfrac{1}{2}X_2 \rightarrow X(g) \rightarrow X^-(g) \rightarrow X^-(aq)$$

(in the case of bromine and iodine the halogen must be got into the gas phase before dissociation). We must, of course, bear in mind that standard electrode potentials are relative to the standard hydrogen electrode, and consider the sum of the energy changes for the stages shown relative to that for the sequence

$$H^+(aq) \rightarrow H^+(g) \rightarrow H(g) \rightarrow \tfrac{1}{2}H_2$$

under standard conditions. Inspection of the values for the appropriate quantities in Table 16.1 shows that chlorine is a more powerful oxidant in aqueous media than bromine or iodine partly because of a more negative enthalpy of formation of the anion, but more because the chloride ion, being smaller, interacts more strongly with the solvent molecules. (In solid salt formation, the lattice energy factor similarly provides the main reason why chlorides are more exothermic than bromides or iodides.) Since fluorine liberates ozonised oxygen from water, E° for the $\tfrac{1}{2}F_2/F^-$ system has no physical reality, but it is interesting to estimate its value. This is most simply done by comparing the energy changes attending the various stages in the reduction of fluorine and chlorine, and hence deriving the difference in E° between the $\tfrac{1}{2}F_2/F^-$ and $\tfrac{1}{2}Cl_2/Cl^-$ systems; in this way E° for the half-reaction

$$\tfrac{1}{2}F_2 + e \rightleftharpoons F^-$$

is obtained as +2.9 V, most of the difference between the systems originating in the much more negative standard free energy of hydration of the smaller fluoride ion.

It is well known that iodine is much more soluble in aqueous solutions of iodides than in pure water; at low relative concentrations of iodine, the system can be described in terms of the single equilibrium

$$I_2 + I^- \rightleftharpoons I_3^-$$

and by partition of the molecular iodine between the aqueous layer and a solvent immiscible with water (e.g. carbon tetrachloride), the equilibrium constant of the reaction can be determined as approximately 10^2 at 25 °C. If K is measured at different temperatures ΔH° for the formation of the triiodide ion is about -15 kJ mol^{-1}, i.e. nearly zero. However, we must not conclude from this information that there is negligible bonding in the I_3^- ion: the hydration energy of I_3^- will be much less than that of I^-, and both these quantities, together with the enthalpy of solution of solid iodine (+21 kJ mol^{-1}), are involved in the overall reaction. Evidently it is much easier to determine ΔH° for such a reaction than to give it an unambiguous interpretation.

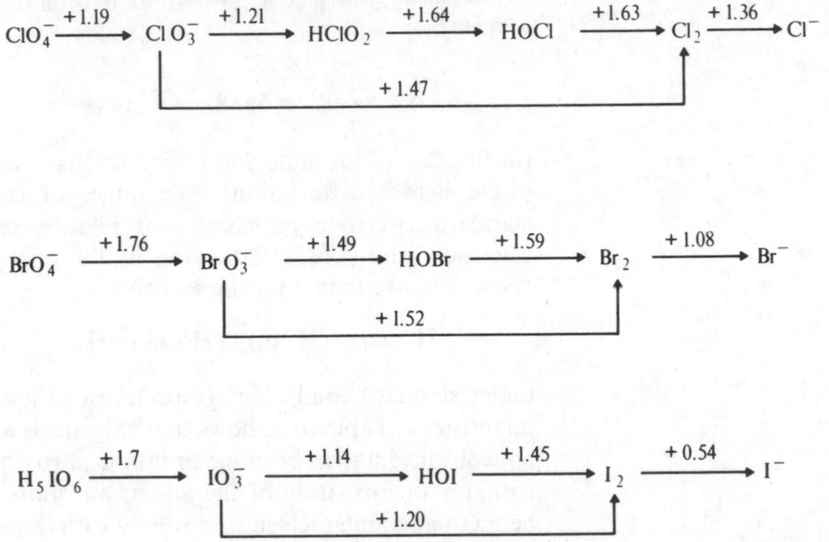

$$ClO_4^- \xrightarrow{+1.19} ClO_3^- \xrightarrow{+1.21} HClO_2 \xrightarrow{+1.64} HOCl \xrightarrow{+1.63} Cl_2 \xrightarrow{+1.36} Cl^-$$
$$ClO_3^- \xrightarrow{+1.47} Cl_2$$

$$BrO_4^- \xrightarrow{+1.76} BrO_3^- \xrightarrow{+1.49} HOBr \xrightarrow{+1.59} Br_2 \xrightarrow{+1.08} Br^-$$
$$BrO_3^- \xrightarrow{+1.52} Br_2$$

$$H_5IO_6 \xrightarrow{+1.7} IO_3^- \xrightarrow{+1.14} HOI \xrightarrow{+1.45} I_2 \xrightarrow{+0.54} I^-$$
$$IO_3^- \xrightarrow{+1.20} I_2$$

Figure 16.5
Potential diagrams for chlorine, bromine and iodine at $[H^+] = 1$.

Potential diagrams at $[H^+] = 1$ (partly calculated from thermochemical data) are given for chlorine, bromine and iodine in Fig. 16.5, in which, however, only a few of the possible disproportionations are indicated. Because of the weakness of several of the oxo-acids, the effects of hydrogen-ion concentration on the relative values of potentials are quite complicated: although, for example, the disproportionation of hypochlorite to chlorate and chloride can be written

$$3OCl^- \rightleftharpoons ClO_3^- + 2Cl^-$$

without involving protons, the fact that HOCl is a very weak acid whilst $HClO_3$ and HCl are strong ones means that in the presence of hydrogen ions OCl^- is protonated and the position of equilibrium is altered: hypochlorous acid is more stable with respect to disproportionation than hypochlorite ion. On the other hand, the disproportionation of chlorate into perchlorate and chloride is faithfully represented as

$$4ClO_3^- \rightleftharpoons 3ClO_4^- + Cl^-$$

This reaction is easily shown from the data in Fig. 16.5 to be thermodynamically favourable; nevertheless, the reaction does not occur in aqueous solution, some undetermined kinetic factor being also involved. Even more striking is the inference that oxygen at $[H^+] = 1$ should oxidise both bromide and iodide. Finally, the fact that chlorine rather than oxygen is evolved when hydrochloric acid is electrolysed is the consequence of the high overpotential for oxygen evolution at most surfaces. We see, therefore, that Fig. 16.5 has only a limited value in predicting the chemistry of the halogens, and we shall not give it a detailed discussion. Nevertheless, we should draw attention to some features shown or implied in Fig. 16.5, notably the more powerfully oxidising properties of periodate and perbromate than of perchlorate when all three are being reduced to halate ions, and the more weakly oxidising powers of iodate and iodine than of the other halates and halogens respectively. Although HOI is unstable with respect to disproportionation into iodate and iodine, and hence is not formed when iodate acts as an oxidant in aqueous solution, it is interesting to note that in hydrochloric acid solution HOI is converted into the fairly stable ICl_2^- ion:

$$HOI + 2HCl \rightarrow ICl_2^- + H^+ + H_2O$$

The potential diagram for reduction of IO_3^- then becomes

$$IO_3^- \xrightarrow{+1.23} ICl_2^- \xrightarrow{+1.06} I_2$$

and iodine(I) is now stable with respect to disproportionation. The half-reaction

$$IO_3^- + 2Cl^- + 6H^+ + 4e \rightleftharpoons ICl_2^- + 3H_2O$$

forms the basis of a very useful group of titrimetric determinations (e.g. of iodine, iodide, hydrazine and sulphite) in which iodate in the presence of hydrochloric acid is used as the oxidant. Carbon tetrachloride is added to assist in the location of the end-point; on the first addition of iodate to the reductant, iodine is formed and can be seen by its colour in the organic solvent. At the end-point, all of it has been oxidised to $H_3O^+ICl_2^-$ and is now in the aqueous solution, so that the organic solvent is colourless.

16.9 *Organic derivatives*

Some account of the carbon tetrahalides and of the preparation and properties of the fluorocarbons was given in Section 13.6. We shall not attempt to deal with alkyl and aryl halides in this book, but to complete our survey of the chemistry of the halogens we should mention briefly some compounds of iodine(III) and iodine(V) that have no counterparts in the organic derivatives of fluorine, chlorine and bromine.

Iodobenzene reacts with chlorine to form the compound $PhICl_2$, (iodobenzene dichloride or phenyliodine dichloride), a lemon-yellow solid which decomposes to iodobenzene and chlorine at about 100 °C and is hydrolysed by water to PhIO, usually known as iodosobenzene; the reaction may be reversed by the action of concentrated hydrochloric acid. When iodosobenzene is heated, it disproportionates into iodobenzene and iodoxybenzene, $PhIO_2$, which in turn disproportionates on treatment with hot aqueous sodium hydroxide, giving benzene and sodium iodate. Iodobenzene difluoride, $PhIF_2$, is obtained from iodosobenzene and concentrated aqueous hydrofluoric acid. By the action of silver hydroxide on a mixture of iodosobenzene and iodoxybenzene, diphenyliodonium hydroxide, Ph_2IOH, is formed:

$$PhIO + PhIO_2 + AgOH \rightarrow Ph_2IOH + AgIO_3$$

Iodobenzene dichloride has a structure based on a trigonal bipyramidally coordinated iodine atom formally containing ten electrons in its valence shell: the phenyl group (its plane perpendicular to the Cl—I—Cl axis) and two lone pairs occupy equatorial positions. The Ph_2I^+ ion has the expected angular shape, but the oxygen-containing compounds have not been examined for many years and nothing is known about their structures.

PROBLEMS

1 Discuss the interpretation of each of the following observations:
(a) 0.01 molar solutions of iodine in n-hexane, benzene, ethanol and pyridine are violet, purple, brown and yellow respectively. When 0.001 mol of pyridine is added to 100 cm^3 of each of the solutions of iodine in n-hexane, benzene and ethanol, all of these become yellow.
(b) Silver chloride and silver iodide are soluble in saturated aqueous potassium iodide, but insoluble in saturated aqueous potassium chloride.
(c) Thermal decomposition of $Bu_4N[ClHI]$ yields tetrabutylammonium iodide and hydrogen chloride.
(d) Although the hydrogen bonding in hydrogen fluoride is stronger than that in water, water has much the higher boiling-point.

2 Describe in outline how you would attempt:

(a) to determine the equilibrium constant and standard enthalpy change for the reaction

$$Cl_2 + H_2O \rightleftharpoons HCl + HOCl$$

in aqueous solution;

(b) to show that the oxide I_4O_9, reported to be formed by the action of ozone on iodine, reacts with water according to the equation

$$5I_4O_9 + 9H_2O \rightarrow 18HIO_3 + I_2$$

(c) to show that when alkali metal atoms and chlorine molecules interact in a solidified inert gas matrix at very low temperatures the ion Cl_2^- is formed.

Is the product formed in (c) likely to exist at ordinary temperature?

3 (a) In strongly alkaline solution containing excess of barium ions, a solution containing 0.01587 g of I^- was treated with 0.1 molar MnO_4^- until a pink colour persisted in the solution: 10.0 cm^3 of the MnO_4^- solution were required. Under these conditions MnO_4^- is converted into the sparingly soluble $BaMnO_4$. What is the product of oxidation of the iodide?

[I = 127]

(b) In neutral solution one mole periodate ion (IO_4^-) reacts with excess of iodide to produce one mole of iodine; on acidification of the resulting solution, a further three moles of iodine are liberated. Derive equations for the reactions which occur under these conditions.

REFERENCES FOR FURTHER READING

Christe, K.O. and Schack, C.J. (1976) *Adv. Inorg. Chem. Radiochem.*, **18**, 319. A review of the chemistry of chlorine oxofluorides which gives a good account of research in inorganic chemistry in a limited field.

Greenwood, N.N. and Earnshaw, A. (1984) *Chemistry of the Elements*, Pergamon, Oxford. Chapter 17 gives a very good account of the chemistry of all the halogens in more detail than that in this book.

Sharpe, A.G. (1990) *J. Chem. Education*, **67**, 309. A short review of the solvation of halide ions and its chemical significance.

Thompson, R. (1977) editor, *The Modern Inorganic Chemicals Industry*, The Chemical Society, London. Contains a very good account of the industrial chemistry of the halogens (and also of several other elements).

Wells, A.F. (1984) *Structural Inorganic Chemistry*, 5th edition, Oxford University Press, Chapters 9 and 10. The structural chemistry of the halogens and halides considered in detail.

Woolf, A.A. (1981) *Adv. Inorg. Chem. Radiochem.*, **24**, 1. A review of the thermochemistry of fluorine compounds.

17 The noble gases

17.1 Introduction

Until 1962 the chemistry of the noble gases was restricted to a few very unstable species such as HHe^+, He_2^+, ArH^+, Ar_2^+ and $HeLi^+$ formed by the union of an ion and an atom under highly energetic conditions, and detected spectroscopically. Molecular orbital theory (Section 4.2), according to which a non-bonding orbital is always destabilised more than the corresponding bonding orbital is stabilised, provides a simple explanation of why the diatomic neutral species He_2, Ne_2, Ar_2 etc. are not known; when, however, there is only a single electron in an antibonding orbital and there are two in the corresponding bonding orbital, there is an appreciable net bonding interaction; thus the bond energies in He_2^+, Ne_2^+ and Ar_2^+ are 126, 67 and 104 kJ mol^{-1} respectively. No stable compounds containing these ions have been isolated, but Xe_2^+ has recently been shown to be produced when xenon reacts with $O_2^+[SbF_6]^-$ or by the action of lead or mercury on the salt $XeF^+[Sb_2F_{11}]^-$ (see later) in SbF_5.

When water is frozen in the presence of argon, krypton or xenon at high pressures, hydrates of limiting composition $Ar.6H_2O$, $Kr.6H_2O$ and $Xe.6H_2O$ are obtained. These, however, are only clathrate compounds like the halogen hydrates, and no chemical combination is involved.

The reasoning which led to the study of the interaction of xenon and platinum hexafluoride (which actually gives a mixture of products) has already been outlined (Section 6.8); with the realisation that the noble gases are not all chemically inert, a very detailed search for compounds of these elements took place, and there is now quite an extensive chemistry of xenon, though so far as stable species are concerned it is very largely restricted to compounds in which the element is bonded to fluorine or oxygen. Compounds of krypton are at present confined to KrF_2 and its derivatives; in principle there should be many more compounds of radon, but since the radon isotope of longest half-life has $t_{\frac{1}{2}} = 3.8$ days and is intensely α-active, leading to decomposition of its compounds, in practice information is very limited. In this chapter we shall therefore review the chemistry of xenon first and follow it by brief accounts of what is known about krypton and radon compounds.

Before doing so, however, we shall discuss the isolation, properties and uses of all the noble gases, some properties of which are given in Table 17.1. Argon is present to the extent of 0.93 per cent by volume in the atmosphere; since it has almost the same boiling-point as oxygen (−183 C), it accompanies the latter in the fractionation of liquid air. The oxygen–argon mixture may be partially separated by further fractionation; the crude argon so obtained is mixed with hydrogen and sparked to remove the oxygen as water, the excess of hydrogen being then destroyed by passage over hot copper(II) oxide. The other gases are present in the atmosphere in much smaller quantities. Neon remains as a gas when the other constituents of air are liquefied, and is separated in that way; krypton and xenon are usually separated from oxygen by selective absorption on charcoal. Helium occurs to the extent of up to 7 per cent by volume in natural gas from some sources in the USA and Canada; it is separated by liquefaction of the other gases present and pumping it away. The origin of such helium was doubtless radioactive decay of heavier elements, in which α-particles (helium nuclei) are often formed; the gas is also found in various minerals containing α-emitting unstable isotopes, and, as mentioned in Section 1.9, its concentration may be used to estimate the age of these minerals. Helium was first detected spectroscopically in the atmosphere of the sun, where it is formed by nuclear fusion (Section 1.4). Radon is the α-decay product of radium and is usually collected from an aqueous solution of radium chloride; by passing it successively over heated copper and copper(II) oxide (to remove ozone, oxygen and hydrogen formed by the action of α-particles on water) and drying it with phosphorus pentoxide, the gas can be obtained pure.

The noble gases have the highest ionisation energies of the elements in their respective periods, but among them there is a steady decrease in ionisation energy with increase in atomic number. Their enthalpies of vaporisation increase appreciably as the interactions between the monatomic molecules of the elements become greater with increasing size and polarisability (cf. Section 6.1); helium is liquefied only with difficulty and can be solidified only under pressure. It will diffuse through rubber and even most glasses. In the solid state all the elements have close-packed structures. Below 2.18 K ordinary liquid ^4He (though not ^3He) is transformed into liquid helium(II), which has the remarkable properties of a thermal conductivity 600 times that of copper and a viscosity approaching zero, so that it forms films only a few hundred atoms thick which flow up and over the side of the containing vessel.

Table 17.1
Some properties of helium, neon, argon, krypton, xenon and radon.

Element	He	Ne	Ar	Kr	Xe	Rn
Atomic number	2	10	18	36	54	86
Electronic configuration	$1s^2$	$[He]2s^22p^6$	$[Ne]3s^23p^6$	$[Ar]3d^{10}4s^24p^6$	$[Kr]4d^{10}5s^25p^6$	$[Xe]4f^{14}5d^{10}6s^26p^6$
B.p./°C	−269	−246	−186	−152	−107	−62
ΔH vaporisation/kJ mol^{-1}	0.08	1.77	6.5	9.7	13.7	18.0
First ionisation energy/kJ mol^{-1}	2370	2080	1515	1350	1170	1040

Helium, being very light and non-inflammable, is used to inflate the tyres of large aircraft, thereby increasing their payload. An oxygen–helium mixture is used in place of oxygen–nitrogen by deep-sea divers: helium is much less soluble in blood than nitrogen, and so does not cause sickness by bubbling out when the pressure is released. It is also used to provide an inert atmosphere during certain welding operations, and as a heat-transfer agent in gas-cooled nuclear reactors, for which it has the advantage of being non-corrosive and of not becoming radioactive under irradiation. The superconductivity of metals cooled to the temperature of liquid helium suggests that the element may also become important in power transmission. Neon, krypton and xenon are used in electric discharge signs, and argon as a filling for metal filament bulbs to reduce evaporation from the filament. Argon is also used in welding for the same purpose as helium, and both gases are employed to provide inert atmospheres for zone-refining of silicon and germanium. Radon finds a limited use in cancer treatment.

17.2 *Compounds of xenon*

The most stable xenon compounds are the three colourless fluorides, XeF_2 (m.p. 140 °C), XeF_4 (m.p. 117 °C) and XeF_6 (m.p. 49 °C), all obtained by combination of the elements. Xenon reacts with fluorine at the ordinary temperature under the influence of ultraviolet light, forming XeF_2; combination to form this fluoride is also effected by the action of an electrical discharge on a mixture of the elements or by streaming them through a short nickel tube at 400 °C. The latter method gives a mixture of XeF_2 and XeF_4, though the yield of the latter may be increased by using excess of fluorine. Combination using a large excess of fluorine and a high pressure (50 atm) at 300 °C gives a good yield of the hexafluoride, which can be separated completely from XeF_2 and XeF_4 by combination with sodium fluoride to form $Na_2[XeF_8]$ and thermal decomposition of the complex. All of the fluorides sublime *in vacuo*, and all are decomposed by water, XeF_2 very slowly and the other two fluorides rapidly; the approximate stoichiometries of the reactions are represented by the equations

$$2XeF_2 + 2H_2O \rightarrow 2Xe + 4HF + O_2$$

$$6XeF_4 + 12H_2O \rightarrow 2XeO_3 + 4Xe + 24HF + 3O_2$$

$$XeF_6 + 3H_2O \rightarrow XeO_3 + 6HF$$

In the last reaction $XeOF_4$, a colourless liquid that freezes at −46 °C, can be isolated as an intermediate; another oxofluoride, XeO_2F_2 (m.p. 31 °C), can be obtained by interaction of $XeOF_4$ and XeO_3. All three fluorides are powerful oxidising agents, the reactivity being, as expected, least for

XeF_2 and greatest for XeF_6. Thus XeF_6 reacts with silica and with hydrogen at the ordinary temperature, the other fluorides only when heated.

All three fluorides are exothermic; approximate standard enthalpies of formation are XeF_2, −109; XeF_4, −216; XeF_6, −294 kJ mol^{-1}. Combination with enthalpies of sublimation leads to mean bond energy terms of 133, 131 and 126 kJ mol^{-1} in XeF_2, XeF_4 and XeF_6 respectively.

The molecule of XeF_2 is linear, with Xe—F 2.00 Å; XeF_4 is planar, with Xe—F 1.95 Å; both have molecular lattices. In the vapour phase the vibrational spectrum of XeF_6 indicates C_{3v} symmetry, i.e. an octahedron distorted by the lone pair in the centre of one face, but the molecule is very easily converted into other structures. Solid XeF_6 is polymorphic, but except at very low temperatures it consists essentially of rings of four square pyramidal XeF_5^+ ions each bonded to two similar ions by means of two bridging F^- ions; the Xe—F distance within the XeF_5^+ ion is 1.84 Å, and Xe—F^- (bridging) distances are 2.23 and 2.60 Å: a loose aggregate of four $XeF_5^+F^-$ ion-pairs is probably the best description of the molecular unit present. The structures of XeF_2 and XeF_4 may be described in terms of conventional two-centre two-electron bonds as involving xenon atoms having valence shells containing ten and twelve electrons, lone pairs of electrons occupying the three equatorial positions of a trigonal bipyramid and two axial positions of an octahedron respectively, in accordance with the VSEPR theory. Alternatively, the view can be taken that three-centre two-electron bonds are involved, as seems certainly to be the case in the related ICl_2^- and I_3^- ions (Section 16.5). In the case of XeF_2, for example, three p_z orbitals combine to form a bonding, a non-bonding and an antibonding orbital, of which only the first two are occupied, giving a net bond order of 0.5. If the XeF^+ ion (see below) is taken as containing a two-centre two-electron bond (by analogy with the diatomic halogen molecules), its observed length of 1.84–1.90 Å lends support to the case for believing there is three-centre two-electron bonding in XeF_2. The low bond energy term (which is, of course, relative to ground state atoms) may be interpreted as arising either from a low bond order or from a high promotion energy to a valence state in which *d* orbitals are involved. The structures of XeF_4 and XeF_6 can be explained similarly. The near-constancy of the bond energies in XeF_2, XeF_4 and XeF_6 (which is in marked contrast to the variation along the series ClF, ClF_3 and ClF_5) provides support for the former interpretation, since it implies a common valence state.

Xenon difluoride and hexafluoride form large numbers of complexes with fluoride ion acceptors such as SbF_5, AsF_5 and PtF_5; the tetrafluoride is much less reactive in this respect. From the difluoride three types of complex are obtained: $2XeF_2.MF_5$, $XeF_2.MF_5$ and $XeF_2.2MF_5$. These have the respective structures $Xe_2F_3^+[MF_6]^-$, $XeF^+[MF_6]^-$ and $XeF^+[M_2F_{11}]^-$, though in each case there is evidence for fluorine bridging between anion and cation as in $BrF_2^+[SbF_6]^-$ (Section 16.5). The

$Xe_2F_3^+$ ion is planar with a V-shaped structure like that of the I_5^- ion:

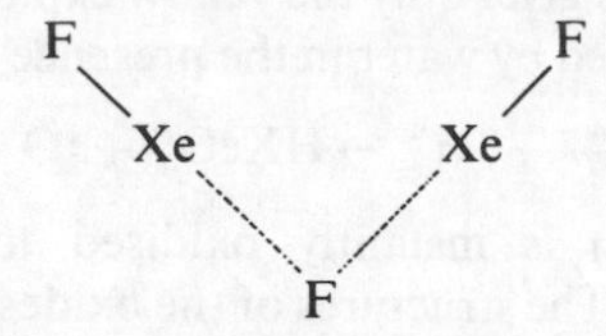

The Xe—F (terminal) and Xe—F (bridging) distances are 1.90 and 2.14 Å respectively, the bond angle 151°; the $[M_2F_{11}]^-$ ions are octahedra with one fluorine atom in common. Among the few complexes of the tetrafluoride is $XeF_3^+[Sb_2F_{11}]^-$, in which the cation is isostructural with ClF_3, Xe—F bond lengths being 1.83 and 1.89 Å. Xenon hexafluoride forms mostly 1:1 complexes with the pentafluorides; these contain the XeF_5^+ ion, a square pyramid with the xenon atom below the base (as in BrF_5) and an average bond length of 1.84 Å; again there is evidence for bridging to the anions. In $2XeF_6.AuF_5$ the cation $Xe_2F_{11}^+$ (or $F_5Xe^+F^-XeF_5^+$) is present; the Xe—F (bridging) distance of 2.24 Å is much greater than the Xe—F (terminal) distance (1.84 Å) and the xenon atoms are also involved in weaker bridging to the $[AuF_6]^-$ ion, the overall environment of the XeF_5^+ ions being very similar to that in solid XeF_6. Xenon hexafluoride will also act as a fluoride ion acceptor and combines with alkali metal fluorides; with RbF and CsF complexes $MXeF_7$ are formed at or a little above the ordinary temperature; these lose XeF_6 when heated to give $M_2[XeF_8]$, in which the anion is a square antiprism. The complexes Rb_2XeF_8 and Cs_2XeF_8 are the most stable compounds of xenon yet made, and decompose only when heated above 400 °C.

There is evidence for the existence of $XeCl_2$ (obtained on condensing the products of a microwave discharge in a mixture of chlorine and a large excess of xenon at 20 K) and $XeCl_4$ (detected by ^{129}Xe Mössbauer spectroscopy as a decay product of ^{129}I-labelled $KICl_4.2H_2O$), but both are extremely unstable. Attempts to make xenon chlorides by halogen-exchange reactions have all been unsuccessful.

We have already mentioned the formation of xenon trioxide in the hydrolysis of XeF_4 and XeF_6; it may be obtained as colourless, dangerously explosive crystals by evaporation of the solution. The trioxide is only weakly acidic and its aqueous solution is almost a non-conductor; xenates of formula $MHXeO_4$ (where M = K, Rb or Cs) are, however, obtained from the trioxide and metal hydroxides. These compounds slowly disproportionate in solution, forming xenon and perxenate (or xenate(VIII)) ion:

$$2HXeO_4^- + 2OH^- \rightarrow XeO_6^{4-} + Xe + O_2 + 2H_2O$$

Aqueous perxenate is also formed when ozone is passed through a dilute solution of XeO_3 in alkali. Insoluble salts such as $Na_4XeO_6.8H_2O$ and Ba_2XeO_6 may be precipitated. Perxenic acid, a weak acid in aqueous solution, has not been isolated; the action of concentrated sulphuric acid on perxenates gives only the yellow explosive oxide XeO_4. Perxenate ion is rapidly reduced by water in the presence of acid:

$$XeO_6^{4-} + 3H^+ \rightarrow HXeO_4^- + \tfrac{1}{2}O_2 + H_2O$$

Manganese(II) is instantly oxidised to permanganate even at room temperature. The structures of the oxides and oxo-anions of xenon are like those of the isoelectronic iodine-containing species; thus XeO_3 is pyramidal, XeO_4 tetrahedral and XeO_6^{4-} octahedral.

Among the remaining xenon compounds that have been characterised are the perchlorate and fluorosulphate obtained by the action of the acids on XeF_2 at low temperatures, the reaction being driven by the highly exothermic formation of hydrogen fluoride:

$$XeF_2 + HOClO_3 \xrightarrow{-HF} FXeOClO_3 \xrightarrow[-HF]{+HClO_4} Xe(OClO_3)_2$$

$$XeF_2 + HOSO_2F \xrightarrow{-HF} FXeOSO_2F \xrightarrow[-HF]{+HSO_3F} Xe(OSO_2F)_2$$

Stable oxofluoro complexes $KXeO_3F$ and $CsXeO_3F$ are obtained from the fluorides and solutions of xenon trioxide; they contain infinite chain anions with F^- ions bridging XeO_3 groups. Similar complexes are obtained from RbCl and CsCl; in these, infinite anions of composition $XeO_3Cl_2^{2-}$ contain XeO_3Cl groups linked by two more Cl atoms. A compound containing a Xe—N bond, $FXeN(SO_3F)_2$, made from XeF_2 and $HN(SO_3F)_2$ in a freon solvent at low temperatures, decomposes at 70 °C. During the reaction between xenon and $O_2^+[BF_4]^-$, $FXeBF_2$ is formed and has been characterised spectroscopically; above −30 °C, however, it breaks down into xenon and boron trifluoride. The salt $C_6F_5Xe^+[B(C_6F_5)_3F]^-$, made from XeF_2 and $B(C_6F_5)_3$, is more stable. There is also evidence for the existence of unstable $Xe(CF_3)_2$ as a product of the interaction of XeF_2 and CF_3 radicals.

17.3 *Compounds of krypton and radon*

The only binary compound of krypton yet known is the difluoride, a colourless solid which is obtained by passing an electric discharge through a mixture of krypton and fluorine at −196 °C and low pressure or by the action of oxygen difluoride on krypton in sunlight. It is much less stable than XeF_2, being rapidly decomposed by water and dissociating into krypton and fluorine at room temperature. Its standard enthalpy of formation is +60 kJ mol^{-1}, corresponding to a Kr—F bond energy term of 49 kJ mol^{-1}. Like XeF_2, it has a linear molecule; Kr—F is 1.89 Å.

Krypton difluoride combines with antimony pentafluoride, forming $KrF^+[Sb_2F_{11}]^-$ and $Kr_2F_3^+[SbF_6]^-$, and by the action of the compound on gold, the salt $KrF^+[AuF_6]^-$ is obtained.

The only known compound containing krypton bonded to an element other than fluorine is the salt $HCNKrF^+[AsF_6]^-$, made from KrF_2 and $HCN^+[AsF_6]^-$ in HF solution.

Radon is oxidised to a non-volatile compound, probably RnF_2, by fluorine or halogen fluorides; this fluoride is reduced by hydrogen at 500 °C. Because of the short half-life of radon and the α-activity of its compounds it has not been possible to study the compound in any detail.

PROBLEMS

1 With what other species are XeF_2, XeF_4, XeF_6, XeF_5^+ and $Xe_2F_3^+$ isoelectronic so far as valence shell electrons are concerned? To what extent is VSEPR theory a reliable guide to the structures of these species and of $[XeF_8]^{2-}$?

2 How would you attempt to determine the standard enthalpy of formation of XeF_2 and the Xe—F bond energy in this compound?

Why is $XeCl_2$ likely to be much less stable than XeF_2?

How may the standard enthalpy of formation of the unknown salt Xe^+F^- be estimated?

REFERENCES FOR FURTHER READING

Bartlett, N. and Sladky, F.O. (1973) *Noble Gas Chemistry*, Pergamon, London. Particularly good on comparisons between corresponding compounds of the noble gases and of other elements.

Ebsworth, E.A.V., Rankin, D.W.H and Cradock, S. (1987) *Structural Methods in Inorganic Chemistry*, Blackwell, Oxford. Chapter 10 contains a full account of the XeF_6 structure problem.

Seppelt, K. and Lentz, D. (1982) *Progress in Inorganic Chemistry*, **29**, 167. A review of developments in noble gas chemistry.

18 *The transition elements*

Introduction • Physical properties • Chemical properties • Coordination numbers and geometries in transition metal complexes • Isomerism in transition metal complexes • Structural isomerism • Geometrical isomerism • Optical isomerism

18.1 Introduction

In this book we have based our classification of the elements on the combinations of permitted quantum numbers for electrons in atoms (2, 6, 10 and 14), and, as we explained in Section 3.1, we have therefore defined the transition elements as the series of ten metals from scandium to zinc, yttrium to cadmium and lanthanum to mercury (omitting cerium to lutecium) inclusive. The observed ground state electronic configurations for these metals were given in Table 3.1; to a first approximation they correspond to the progressive filling of the $3d$, $4d$ and $5d$ orbitals respectively, though there are minor deviations from this pattern. In the first transition series, for example, the ground state of chromium is $[Ar]3d^5 4s^1$ rather than $[Ar]3d^4 4s^2$. Why this is so is an exceedingly difficult question beyond the scope of this book; to answer it we should need to know both the energy difference between the $3d$ and $4s$ orbitals when the nuclear charge is 24 (the atomic number of chromium) and the interelectronic interaction energies for each of the two electronic configurations. Fortunately, M^{2+} and M^{3+} ions of metals of the first transition series all have electron configurations $[Ar]3d^n$, and so the comparative chemistry of these metals is largely concerned with the consequences of the successive filling of the $3d$ orbitals. For metals of the second and third series (frequently referred to as the heavy transition metals) the picture is more complicated, and a really satisfactory systematic treatment of their chemistry cannot yet be given. Most of this and the next few chapters, therefore, will be chiefly concerned with the first transition series, but we shall include some material on the later transition elements and draw attention to the ways in which they show general differences from the elements scandium to zinc.

It will be clear from earlier chapters that the theoretical treatment of systems containing as many electrons as the atoms of the transition metals is still at a highly approximate stage. When we refer to radial distribution functions, for example, we usually mean those of one-electron species. A scientifically honest account of the chemistry of the transition elements would therefore consist of a long statement of the experimental facts, followed by a retrospective but still incomplete explanation of them. For the assimilation of the experimental material, however, such a treatment is not very acceptable, and a compromise has been adopted here. After a general, and for the most part non-theoretical, discussion in this chapter of the chemistry of the transition elements, we turn in the following one to theories of bonding in transition metal compounds, and then go on to use these theories as the bases for the treatment of spectroscopic, magnetic and thermodynamic properties. Two other major topics, kinetic aspects of transition metal chemistry and organometallic compounds, are next dealt with, and the book ends with four chapters on the systematic chemistry of metals of the first transition series, metals of the second and third transition series, and the two series of inner transition metals (the lanthanides and the actinides).

18.2 Physical properties

Nearly all the transition metals are hard, ductile and malleable, and have high electrical and thermal conductivities. With the exceptions of manganese, zinc, cadmium and mercury, at normal temperatures they have one or more of the typical metal structures described in Section 6.9 – cubic close-packed, hexagonal close-packed, and body-centred cubic. Their atomic radii for twelve-coordination, given in Table 6.6 in Section 6.9, are much smaller than those of alkali and alkaline earth metals of comparable atomic number. Transition metals are also (again with the exceptions of zinc, cadmium and mercury) much harder and less volatile than the Group I and Group II metals. Their enthalpies of atomisation were shown in Fig. 6.16 (Section 6.9); the maxima at about the middle of each series indicate that one unpaired electron per *d* orbital is a particularly favourable distribution for strong interatomic interaction. This corresponds to occupation of the band of bonding *d* orbitals on the simple theory of the metallic state (Section 4.9), the band of antibonding *d* orbitals being empty; but this interpretation is an oversimplification, as is shown by the complex magnetic properties of iron and other metals of the first transition series. An important generalisation that may be drawn from Fig. 6.16 is that metals of the second and third series have greater enthalpies of atomisation than the corresponding elements of the first series; this is a substantial factor in accounting for the much more frequent incidence of metal–metal bonding in compounds of the heavy transition elements. Because of the similar radii and other characteristics of the transition metals, alloys are readily formed, the ferrous alloys which form the many varieties of steel being particularly important. Several of the metals, especially later members of each series, are valuable heterogeneous catalysts.

The first ionisation energies of the transition metals (shown in Fig. 3.3 in Section 3.4) are higher than those of the alkali and alkaline earth metals, but they vary along a series much less than those of the typical elements from, say, potassium to krypton or from rubidium to xenon. Within each series, the overall trend is for the ionisation energy to increase with increasing atomic number, but many small variations occur. Chemical comparisons between the transition metals and the metals of Group I and II are, however, complicated by the number of factors involved. Thus all elements of the first transition series have first and second ionisation energies greater than those of calcium, and all except zinc also have a larger enthalpy of atomisation. These factors make the metals less reactive than calcium; on the other hand, since all known dipositive ions of elements of the first transition series are smaller than Ca^{2+}, lattice and solvation energy effects operate in the reverse direction. In practice, it turns out that, in the formation of species containing M^{2+} ions, all the metals of the first series are thermodynamically less reactive than calcium; because of the formation of a coherent surface film of oxide, or for other reasons, they are, indeed, often much less reactive than expected on the basis of quoted E° values. Nevertheless, a few transition elements at the beginnings of the series are very powerful reducing agents: E° for the Sc^{3+}/Sc couple (-2.08 V) is more negative than that for Al^{3+}/Al (-1.66 V), and E° for the Zr^{4+}/Zr couple (-1.5 V) is only a little different. Later in the series (particularly the second and third series), however, we find the platinum metals, silver, gold and mercury; these are, even in the thermodynamic sense, the least reactive metals known.

18.3 Chemical properties

The transition metals are, in general, moderately reactive and combine with oxygen, sulphur and halogens when heated with these elements. Many of them also combine with hydrogen, carbon and nitrogen, forming the so-called interstitial compounds that we discussed in Sections 9.6, 13.5 and 14.4 respectively; in these, hydrogen atoms occupy tetrahedral holes, and carbon or nitrogen atoms octahedral holes, in a close-packed structure (not necessarily that of the parent metal). Thermodynamic considerations given earlier lead to the conclusion that these compounds are macromolecular electron-deficient species involving multicentre covalent or metallic bonding. Most transition metals should, on thermodynamic considerations, liberate hydrogen from acids, but many do not do so, being rendered passive by a thin surface coating of oxide, or having a high hydrogen overpotential, or both.

Compounds of the elements in low oxidation states are generally more or less saline in character, e.g. difluorides have the rutile structure. Along a given series, compounds of the same formula type (e.g. MF_2, MF_3, $KM(SO_4)_2.12H_2O$) are commonly isomorphous. Such compounds, and the hydrated ions, are usually coloured except when the metal ion has the electronic configuration d^0 or d^{10}; e.g. $Cr(H_2O)_6^{2+}$ is sky-blue, $Mn(H_2O)_6^{2+}$ very pale pink, $Fe(H_2O)_6^{2+}$ pale green, $Co(H_2O)_6^{2+}$ pink, $Ni(H_2O)_6^{2+}$ green and $Cu(H_2O)_4^{2+}$ pale blue; $Zn(H_2O)_6^{2+}$, on the other hand, is colourless. This suggests that the colour originates in transitions of electrons between different *d* orbitals of the same principal quantum number. If we were dealing with an isolated gaseous ion, such transitions would be forbidden (cf. Section 2.4); the pale colours indicate that even in compounds they are relatively infrequent. Transition metal ions readily form complexes, and complex formation is often accompanied by a change of colour and sometimes also by a change in the intensity of the colour. When concentrated hydrochloric acid is added to a solution containing the $[Co(H_2O)_6]^{2+}$ ion, for example, the very rapid formation of the $[CoCl_4]^{2-}$ ion is marked by a colour change from pink to deep blue; we shall discuss this observation at some length in the next chapter. There are, however, many strongly-coloured transition metal compounds in which the colour originates in a different way, e.g. the metal oxides and sulphides, potassium permanganate and Prussian blue; in these, electron transfer between one ion and another is involved; such *charge-transfer absorptions* or *emissions* are not subject to a selection rule and are always much more intense than transitions between different *d* orbitals.

Many compounds of transition metals are paramagnetic, i.e. are attracted by a magnetic field. Paramagnetism arises from the presence of unpaired electrons, each such electron having a magnetic moment associated with its spin angular momentum and, unless it is an electron for which the second quantum number, *l*, is zero, another moment associated with its orbital angular momentum (Section 3.3). For compounds of elements of the first transition series we can, to a first approximation, usually ignore the contribution to the magnetic moment of the orbital angular momentum and regard the moment as determined by the number of unpaired electrons and given by

$$\mu = 2\sqrt{S(S+1)}$$

where S is the sum of the spin quantum numbers for the individual electrons ($S = \frac{1}{2}, 1, \frac{3}{2}, 2, \frac{5}{2}$ for 1, 2, 3, 4 and 5 unpaired electrons respectively) and μ is the magnetic moment in Bohr magneton units ($1\mu_B = eh/4\pi m = 9.27 \times 10^{-24}$ A m^2 or J T^{-1}). Now μ can be obtained from the measured molar magnetic susceptibility χ_M of the compound at absolute temperature T by means of the relationship

$$\mu = \sqrt{\frac{3k\chi_M T}{N_A \mu_0 \mu_B^2}}$$

(k = Boltzmann's constant; N_A = Avogadro's number; μ_0 = vacuum permeability). Hence we can derive S for the metal ion in the compound under investigation.

For many compounds of first transition series metals, S is the same as for the isolated gaseous ion, as may be seen from Table 18.1 by comparing calculated and observed values for μ (we shall refer to the occasional discrepancies between these values in the next chapter). Such compounds, which include all the fluorides of the ions listed in the table, and salts containing all the hexa-aquo ions except $[Co(H_2O)_6]^{3+}$, are known as *high-spin complexes*. In other species, notably the cyano complexes of all the ions, the value of the magnetic moment indicates that the electrons are completely or partly paired; $K_4[Fe(CN)_6]$, for example, is diamagnetic ($\mu = 0$, so $S = 0$), and $K_3[Fe(CN)_6]$ is only weakly paramagnetic ($\mu = 2\ \mu_B$ so $S = \frac{1}{2}$, corresponding to the presence of only one unpaired electron compared with five in FeF_3 or $[Fe(H_2O)_6]^{3+}$ salts). Species in which the electrons are paired (usually, though not invariably, as far as possible) are known as *low-spin complexes*, and the high-spin/low-spin distinction is, as we shall see later, one of the most important concepts in transition metal chemistry. Its value is, however, largely restricted to the first series; for the heavy transition metals the simplifying assumption that magnetic moments arise entirely from spin angular momentum is no longer reliable. For d^0, d^1, d^9 and d^{10} ions there can, of course, be no distinction between high-spin and low-spin complexes.

It will be noticed that in the foregoing paragraph we suddenly started writing about complexes rather than compounds or ions. This use of the

Table 18.1
Calculated (from the spin-only formula) and observed magnetic moments (in μ_B units) for transition metal ions of the first series in high-spin species.

Ion	S	$2\sqrt{S(S+1)}$	Observed
Sc^{3+}, Ti^{4+}	0	0	0
Ti^{3+}	$\frac{1}{2}$	1.73	1.7–1.8
V^{3+}	1	2.83	2.8–3.1
V^{2+}, Cr^{3+}	$\frac{3}{2}$	3.87	3.7–3.9
Cr^{2+}, Mn^{3+}	2	4.90	4.8–4.9
Mn^{2+}, Fe^{3+}	$\frac{5}{2}$	5.92	5.7–6.0
Fe^{2+}, Co^{3+}	2	4.90	5.0–5.6
Co^{2+}	$\frac{3}{2}$	3.87	4.3–5.2
Ni^{2+}	1	2.83	2.9–3.9
Cu^{2+}	$\frac{1}{2}$	1.73	1.9–2.1
Zn^{2+}	0	0	0

word complex, which is now general in the literature of transition metal chemistry, originated in the realisation that although in elementary inorganic chemistry we describe FeF_3 as a simple salt and $K_3[FeF_6]$ as a complex, the immediate environment of the Fe^{3+} ion is apparently the same in both, i.e. an octahedron of F^- ions; for many purposes it is convenient to use one word for Fe^{3+} in FeF_3, $[FeF_6]^{3-}$ or $K_3[FeF_6]$, and *complex* is then the preferred term. By analogy we might refer to $[Fe(H_2O)_6]^{3+}$ as the hexa-aquo complex, and $[FeCl_4]^-$ as the tetrachloro complex, of iron(III).

We have left until the end of this section the subject of the range of oxidation states exhibited by the transition metals – probably the most important feature of their chemistry, and certainly the most difficult one to discuss adequately. We touched on this problem in Section 16.4, where we referred to the thermodynamics of halide formation by reference to such processes as:

$$M + \tfrac{1}{2}X_2 \rightarrow MX \text{ (ionic)}$$

$$MX \text{ (ionic)} + \tfrac{1}{2}X_2 \rightarrow MX_2 \text{ (ionic)}$$

$$MX_4 \text{ (covalent)} + X_2 \rightarrow MX_6 \text{ (covalent)}$$

and

$$MX_2 \text{ (ionic)} + X_2 \rightarrow MX_4 \text{ (covalent)}$$

Where X = F the discussion in Section 16.4 at least defines all the types of reaction we at present need consider when M is a transition metal. For X = Cl, Br or I, however, two more possibilities arise quite frequently. First, the apparent oxidation state of M in MX_n may be misleading: LaI_2, for example, is a metallic conductor best formulated $La^{3+}(I^-)_2(e^-)$. Second, a simple empirical formula may conceal the fact that metal clusters are present: in $MoCl_2$, for example, only one-third of the chlorine is present as Cl^-, and an X-ray diffraction study shows that the correct formula is $[Mo_6Cl_8]Cl_4$, the cation present consisting of an octahedron of molybdenum atoms with a chlorine atom situated off the middle of each face of the octahedron. The presence of such metal clusters is common in halides (other than fluorides) of metals of the second and third transition series and, as we mentioned earlier, is doubtless associated with the very high enthalpies of atomisation of these metals.

There are many transition metal compounds in which it is impossible to assign oxidation states unambiguously. These include complexes formed with CO, bipyridyl, NO^+ and a few other ligands which can not only donate electrons to metal atoms or ions but also accept them in relatively low energy antibonding orbitals. We shall consider these complexes further in Chapter 22; for the present it will suffice to note that in species such as $[Fe(CO)_4]^{2-}$ and $V(bipy)_3$, which formally contain Fe(–II) and V(0) and uncharged ligands, there is strong evidence to suggest that considerable negative charge is delocalised onto the ligands.

Table 18.2
Oxidation states of the transition metals (the most stable ones are in bold type).

Sc	Ti	V	Cr	Mn	Fe	Co	Ni	Cu	Zn
		0	0	0	0	0	0		
				1	1	1	1	**1**	
	2	2	**2**	**2**	**2**	**2**	**2**	**2**	**2**
3	**3**	**3**	**3**	**3**	**3**	**3**	3	3	
	4	**4**	4	4	4	4	4	4	
		5	5	5					
			6	6	6				
				7					
Y	**Zr**	**Nb**	**Mo**	**Tc**	**Ru**	**Rh**	**Pd**	**Ag**	**Cd**
			0	0	0	0	0		
				1		1		**1**	
			2	2	2	2	**2**	2	**2**
3	3	3	3	3	**3**	**3**		3	
	4	4	**4**	**4**	**4**	4	**4**	4	
		5	**5**	5	5	5			
			6	6	**6**	6	6		
				7	7				
					8				
La	**Hf**	**Ta**	**W**	**Re**	**Os**	**Ir**	**Pt**	**Au**	**Hg**
			0	0	0	0	0		
				1		1		1	**1**
			2	2	2		**2**		**2**
3	3		3	3	3	**3**		**3**	
	4	4	**4**	**4**	**4**	**4**	**4**		
		5	**5**	**5**	5	5	5	5	
			6	**6**	**6**	6	6		
				7	7				
					8				

In compiling Table 18.2, which gives an overall view of the established oxidation states of the transition metals, we have included states known only in neutral metal carbonyls and in cluster compounds, but have excluded those for which there is strong evidence to suggest that extensive delocalisation of formal charge from the metal has taken place, such as La(II) in LaI_2 and Fe(−II) in $[Fe(CO)_4]^{2-}$; we have also excluded oxidation states found only in very unstable species. In a few instances a case could doubtless be made for including other oxidation states or deleting some shown. In general, however, Table 18.2 certainly gives a fair idea of the distribution of the known and of the most stable oxidation states of the elements of all three transition series.

As would be expected, the elements which display the greatest number of different oxidation states occur in or near the middle of each series. The first member of each series is, in fact, found to form only compounds in which it is in the tripositive state; for example, simple scandium(I) and scandium(II) compounds are unstable with respect to disproportionation into the metal and scandium(III) compounds. A similar state of affairs occurs at the other ends of the first and second series; but it should be noted that Zn_2^{2+} and Cd_2^{2+} ions are now well-established entities in molten salt media, though they are much less stable than the well-known Hg_2^{2+} ion. A general feature of Table 18.2 is that higher oxidation states are more stable for metals of the second and third series than for those of the first series. These higher oxidation states involve the formation of covalent molecular or macromolecular species rather than of simple ionic salts, which in any case are rarely formed by the heavy transition elements: dihalides and trihalides, for example, are very often metal cluster compounds. On the whole we may say that the chemistry of metals of the first series is predominantly that of the dipositive and tripositive ions and a few molecular species and oxo-anions in which higher oxidation states are exhibited; an extensive aqueous solution chemistry of the kind dealt with in Chapter 7 and involving redox equilibria, precipitation and complex formation is well documented and well understood. Metals of the second and third series form few simple ions, and much of the chemistry of their lower oxidation states is dominated by metal–metal interactions; in their higher oxidation states they form molecular or macromolecular species involving covalent bonding; their aqueous solution chemistry is more complicated and scarcely ever involves simple monatomic ions. Other differences will become apparent later, but it will already be clear that there are good grounds for discussing the metals of the first series separately from those of the other two.

18.4 *Coordination numbers and geometries in transition metal complexes*

We shall now attempt a survey of the structures of transition metal complexes analogous to that given in Section 5.2 for compounds of the typical elements. We shall not be much concerned with ionic lattices in which the transition metal has a symmetrical environment (e.g. AgF, MnF_2, ScF_3, $MgFe_2O_4$) but we shall mention some examples in which the environment in an essentially ionic structure is irregular. For the most part, however, we shall be discussing molecules or ions in which it is generally accepted that there is covalent bonding between the metal and the ligand. For the present we shall assume that such covalent bonding involves only electron-pair donation from the ligand to the metal atom or ion, and we shall draw particular attention to the electronic configuration of the metal atom or ion, e.g. Fe(0), d^8, in $Fe(CO)_5$; Ni(II), d^8, in $[Ni(H_2O)_6]^{2+}$; Cr(III), d^3, in $[Cr(C_2O_4)_3]^{3-}$. We shall see later that this simple picture of the bonding in transition metal complexes as involving only coordinate links from ligand to metal is often an oversimplification, but for a first look at transition metal stereochemistry it is very satisfactory.

For many years after the classic work of Werner, which laid the foundations for the correct formulation of transition metal complexes, it was assumed that a metal in a given oxidation state would have a fixed coordination number and geometry rather like carbon in saturated organic compounds. In the light of the widespread (though not universal) success of the VSEPR theory in accounting for the structures of compounds of non-transition elements, we might reasonably expect the ions $[V(H_2O)_6]^{3+}$ (d^2), $[Mn(H_2O)_6]^{3+}$ (d^4), $[Co(H_2O)_6]^{3+}$ (d^6), $[Ni(H_2O)_6]^{2+}$ (d^8) and $[Zn(H_2O)_6]^{2+}$ (d^{10}) to vary in structure as the electronic configuration of the central ion changes, but this is not so: all are octahedral. Thus it is clear that the VSEPR theory is not going to be very helpful in transition metal chemistry. We are, in fact, not yet at a stage at which the structures of transition metal complexes can often be predicted, though it is clear that among the factors which determine them are:

(a) the electronic configuration, oxidation state and energetically accessible orbitals of the metal;
(b) high-spin or low-spin character, which in turn depends on
(c) the nature of the ligand;
(d) size and steric effects.

Steric effects are especially important for polydentate ligands, but since the structures of complexes containing polydentate ligands tend to be complicated in any case, we shall not be much concerned with them here.

We saw that, in the case of several non-transition elements and their compounds, different structures are sometimes of very nearly equal energy, and we shall find that this is also true for derivatives of the transition metals. It is even found that for the neutral complex $[Ni(PBzPh_2)_2Br_2]$ (Bz = benzyl) and the ion $[Ni(CN)_5]^{3-}$ two different structures occur in the same crystal.

In the following systematic outline of the occurrence of different coordination numbers and geometries in solid transition metal complexes, we have limited examples to those for which structures have been determined by X-ray diffraction. Other physical methods can often be used to assign structures with a fair degree of certainty, but ultimately the X-ray method is the only absolute one for species containing as many atoms as those discussed in this section, and in a summary of this kind it is prudent to depend upon it alone.

Coordination number two

This is a very uncommon coordination number among transition metal compounds, and the few certain examples of it known at present are all complexes of d^{10} ions, such as the species $[Ag(NH_3)_2]^+$, $[Au(CN)_2]^-$ and $Hg(CN)_2$. In all these, the two bonds formed by the transition metal are collinear.

Coordination number three

This is also uncommon, but it is not restricted to d^{10} species, though the metal ions in $[HgI_3]^-$ and $[Cu(SPMe_3)_3]^+$ have this configuration. Other examples are $Fe\{N(SiMe_3)_2\}_3$ (made from $FeCl_3$ and $LiN(SiMe_3)_2$) and $Co\{N(SiMe_3)_2\}_2(PPh_3)$, derivatives of high-spin d^5 and high-spin d^7 ions respectively. The coordination of the metal is planar in each of these species; a pyramidal structure and a T-shaped structure like that of chlorine trifluoride have not yet been found to occur in transition metal complexes.

Coordination number four

After six, this is the commonest coordination number. Four-coordination is most commonly symmetrical tetrahedral, but in a few instances the tetrahedron is somewhat flattened; planar complexes are rather rare.

Tetrahedral complexes have not yet been reported for d^3 or d^4 ions, and they are relatively uncommon for metals of the second and third transition series. Where a distinction between high-spin and low-spin complexes is meaningful, tetrahedral complexes are always of the former type. Typical examples include: VO_4^{3-}, CrO_4^{2-}, MnO_4^- and OsO_4 (d^0); MnO_4^{2-} and RuO_4^- (d^1); FeO_4^{2-} and RuO_4^{2-} (d^2); $[FeCl_4]^-$ (d^5); $[FeBr_4]^{2-}$ (d^6); $[CoCl_4]^{2-}$ (d^7); $[NiCl_4]^{2-}$ (d^8); $[CuCl_4]^{2-}$ (a flattened tetrahedron) (d^9); and $Ni(CO)_4$, $[Cu(CN)_4]^{3-}$ and $[ZnCl_4]^{2-}$ (d^{10}). It is a curious fact that tetrahedrally four-coordinated metal atoms or ions are very much more common in neutral molecules or anions than in cations.

Most planar four-coordinated species are derivatives of ions of the low-spin d^8 configuration, e.g. $[Ni(CN)_4]^{2-}$, $[PdCl_4]^{2-}$, $[Pt(NH_3)_4]^{2+}$, $[PtCl_4]^{2-}$ and $[AuF_4]^-$. However, the *trans*-dimesityl bis-diethylphenylphosphine Co(II) complex

PEt_2Ph
Me Me
Me —Co— Me
Me Me
PEt_2Ph

(low-spin d^7) is planar, though the ligands may play substantial roles in determining the geometry. One form of $[Ni(PBzPh_2)_2Br_2]$ contains both planar and tetrahedral molecules, and the bulk magnetic moment shows that these have the low-spin and high-spin d^8 configurations respectively.

Coordination number five

The limiting structures for five-coordination are the trigonal bipyramid and the square pyramid; in practice, there are many structures which lie between these extremes. Even the limiting structures are not as different as they may at first sight appear, for a simple vibrational motion is sufficient for their interconversion, a point made in Section 1.5 (Fig. 1.4) in discussing the ^{19}F nuclear magnetic resonance spectrum of phosphorus pentafluoride. Many compounds containing five-coordinated metal atoms involve polydentate amine or arsine and other ligands, and fall outside our self-imposed restriction of this discussion to relatively simple species; among the latter, $[CuCl_5]^{3-}$ and $[CdCl_5]^{3-}$ (d^9 and d^{10} respectively) have the trigonal bipyramidal structure, whilst $[Ni(CN)_5]^{3-}$ (low-spin d^8) normally has the square pyramidal structure. In $[Cr(en)_3]$ $[Ni(CN)_5]1.5H_2O$, however, approximately trigonal bipyramidal and approximately square pyramidal ions (both low-spin) are present in the same lattice, and the energy difference between the two forms must be very small.

Coordination number six

For many years after Werner's proof from stereochemical studies that many six-coordination complexes of chromium and cobalt had octahedral geometry, it was believed that no other form of six-coordination occurred, and a vast amount of evidence from X-ray structure determinations supported this belief. The regular or very nearly regular octahedron is found for all electronic configurations from d^0 to d^{10} inclusive, some of the simplest examples being $[TiF_6]^{2-}$ (d^0), $[Ti(H_2O)_6]^{3+}$ (d^1), $[V(H_2O)_6]^{3+}$ (d^2), $[Cr(H_2O)_6]^{3+}$ (d^3), $[Mn(H_2O)_6]^{3+}$ (d^4), $[Fe(H_2O)_6]^{3+}$ (d^5), $[Fe(H_2O)_6]^{2+}$ (d^6), $[Co(H_2O)_6]^{2+}$ (d^7), $[Ni(H_2O)_6]^{2+}$ (d^8), $[Cu(NO_2)_6]^{4-}$ (d^9) and $[Zn(H_2O)_6]^{2+}$ (d^{10}). Where the distinction is meaningful, all of these are high-spin species, but many low-spin species having octahedral geometry are also known, among them $[Mn(CN)_6]^{3-}$ (d^4), $[Fe(CN)_6]^{3-}$ (d^5) and $[Co(H_2O)_6]^{3+}$ and $[Co(CN)_6]^{3-}$ (d^6). There is, however, no X-ray structural evidence for the formation of six-coordinated low-spin d^7 or d^8 metal ion complexes, a fact which we shall discuss further in the following chapter. Another fact we shall also discuss there is that the octahedral environment of high-spin d^4 and of d^9 ions is often distorted so that there are four nearest neighbours in a plane and two more rather further away; this is found for CrF_2, CuF_2 and $CrCl_2$ (distorted rutile structures), and $CuCl_2$, $CrBr_2$ and CrI_2 (distorted CdI_2 structures).

In addition to octahedral six-coordination, however, trigonal prismatic coordination occurs. This was first recognised for MoS_2, the structure of which may be described as superimposed layers of sulphur atoms with

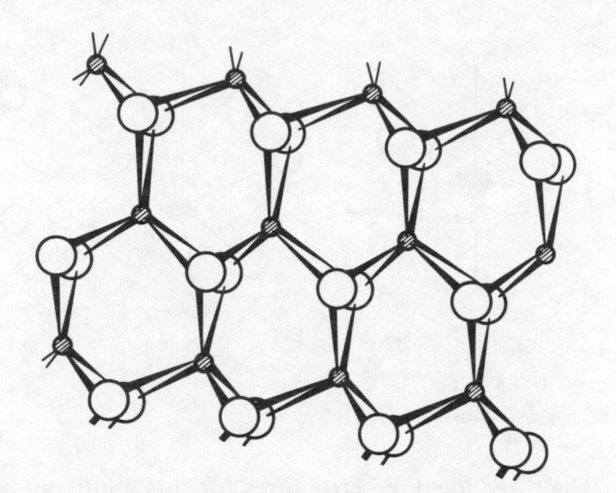

Figure 18.1
The structure of molybdenum disulphide, showing the trigonal prismatic coordination of the molybdenum atoms (shaded circles) by sulphur.

molybdenum atoms in the trigonal pyramidal holes (see Fig. 18.1). This is a layer structure with the sequence SMoSSMoS (compare the structure of CdI_2 described in Section 6.3), and MoS_2 is a good lubricant used in engine oils. The electronic configuration of Mo(IV) is d^2; NbS_2 (d^1) and ReS_2 (d^3) have structures very similar to that of MoS_2.

Trigonal prismatic coordination of the metal atom is also found for $W(CH_3)_6$ and a group of complexes of the dianion of diphenylethylene-dithiol, (HS)PhC=CPh(SH). These include the compounds $[Mo(S_2C_2Ph_2)_3]$ and $[Re(S_2C_2Ph_2)_3]$; if we write the ligand as

$$\begin{array}{l} Ph\text{—}C\text{—}S^- \\ \quad\;\; \| \\ Ph\text{—}C\text{—}S^- \end{array}$$

they contain Mo(VI) (d^0) and Re(VI) (d^1) respectively, but these are not necessarily correct formulations; it would, for example, be possible to write the molybdenum complex with Mo(0) and the neutral ligand

$$\begin{array}{l} Ph\text{—}C\text{=}S \\ \quad\;\; | \\ Ph\text{—}C\text{=}S \end{array}$$

or with some intermediate formulation. The structure of the rhenium compound is shown in Fig. 18.2.

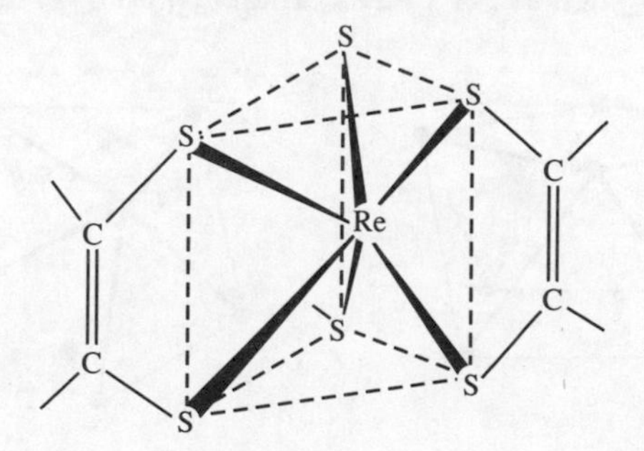

Figure 18.2
Structure of the complex $[Re(S_2C_2Ph_2)_3]$ (one pair of carbon atoms is omitted).

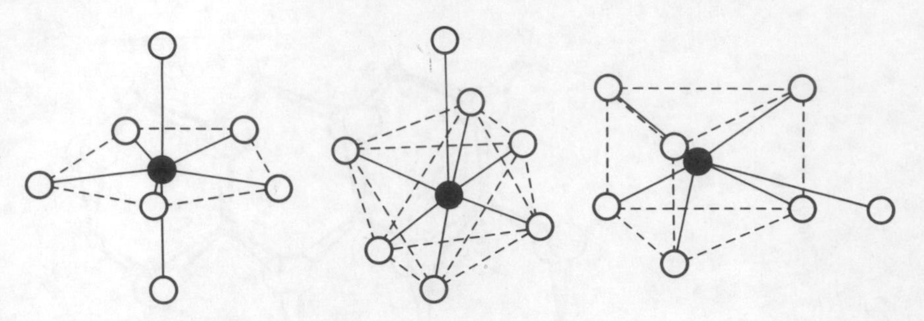

Figure 18.3
The pentagonal bipyramid, the capped octahedron, and the capped trigonal prism; broken lines define the equatorial plane of the bipyramid and the edges of the octahedron and the trigonal prism respectively.

Coordination number seven

The three idealised structures for this relatively uncommon coordination are the pentagonal bipyramid, the capped octahedron (an octahedron plus an extra atom off one face), and the capped trigonal prism (a trigonal prism plus an extra atom off one rectangular face) shown in Fig. 18.3. Not only are the differences between them small, but real complexes often have structures that deviate somewhat from the idealised ones; when this happens different authors may describe the same structure in different ways. The capped octahedron has not yet been shown to occur in a simple structure, but the pentagonal bipyramid occurs in $[V(CN)_7]^{4-}$ (d^2) and $[ZrF_7]^{3-}$ (d^0); in $[NbF_7]^{2-}$ and $[TaF_7]^{2-}$, however, which formally also contain a d^0 metal ion, the anions are capped trigonal prisms.

Coordination number eight

As the number of vertices in a polyhedron increases, so does the number of possible structures. For eight-coordination we shall limit the discussion to those which actually occur in complexes containing only simple ligands. These are the cube, the square antiprism and the dodecahedron, the last two of which are shown in Fig. 18.4 (note that in the dodecahedron there are two different types of position, marked *a* and *b* respectively). Cubic coordination, which should be least favoured because of inter-ligand repulsion, has so far been found only in fluoro complexes of the actinide elements, such as $Na_3[PaF_8]$ and $Na_3[UF_8]$. The square antiprism occurs

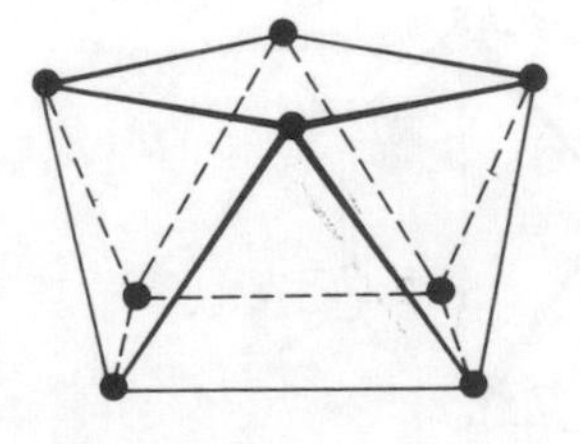

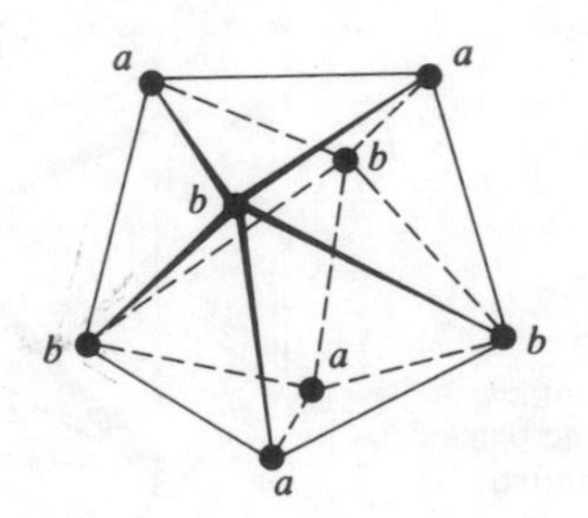

Figure 18.4
The square antiprism and the dodecahedron.

in $[Zr(acac)_4]$ and $Na_3[TaF_8]$ (d^0), $K_2[ReF_8]$ and $Na_3[W(CN)_8].4H_2O$ (d^1) and $H_4[W(CN)_8].6H_2O$ (d^2), whilst the dodecahedron is found in $Na_4[Zr(C_2O_4)_4].3H_2O$ (d^0), $(Bu_4N)_3[Mo(CN)_8]$ (d^1) and $K_4[Mo(CN)_8].2H_2O$ (d^2). The occurrence of both the square antiprism and the dodecahedron in cyano complexes of molybdenum and tungsten in a given oxidation state shows how small the differences between the energies of these structures must be. All the examples of eight-coordination given so far have involved heavy transition metals or inner transition metals; for metals of the first transition series only the dodecahedron has yet been found, and then only in complexes involving bidentate ligands, e.g. $[Ti(NO_3)_4]$ (d^0), $[Cr(O_2)_4]^{3-}$ (d^1), and $[Mn(NO_3)_4]^{2-}$ and $[Fe(NO_3)_4]^-$ (d^5). The donor atoms would necessarily be further apart in a square antiprismatic structure, which would thus be more probable with larger ligands.

Higher coordination numbers

The only structure for nine-coordination yet identified is the tricapped trigonal prism (a trigonal prism with an extra atom off each vertical face) found in $[ReH_9]^{2-}$ and $[TcH_9]^{2-}$, both d^0 systems, and shown in Fig. 18.5. Coordination numbers of ten, eleven and twelve occur only in complexes of the lanthanides and actinides; we shall not describe ten- and eleven-coordination, but in twelve-coordination we meet again a structure familiar from boron chemistry, the icosahedron (Section 12.2, Fig. 12.1). In the ions $[Ce(NO_3)_6]^{2-}$ and $[Th(NO_3)_6]^{2-}$ the oxygen atoms of six bidentate nitrate ions span adjacent positions in an icosahedron; the former ion is present in ceric ammonium nitrate or, as it should now be called, ammonium hexanitratocerate(IV).

It is always dangerous to draw conclusions on the basis of the non-existence of compounds, but it seems from this survey that higher coordination numbers occur mainly for ions of metals of the second and

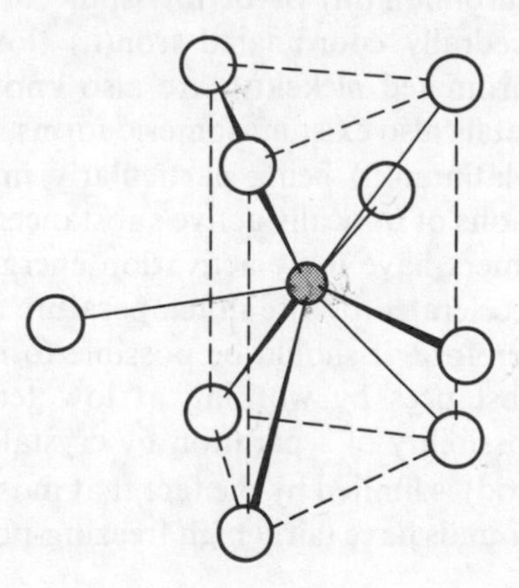

Figure 18.5
Structure of the $[ReH_9]^{2-}$ ion.

third series, and particularly for ions which, formally at least, have few or no *d* electrons. Even octahedral complexes, we note, are apparently not formed by ions having the low-spin d^7 and d^8 configurations. Although, therefore, the VSEPR theory is no guide to the structures of transition metal complexes, electronic configuration does have some influence, and in the next chapter we shall see how this comes about.

18.5 Isomerism in transition metal complexes

Most readers of this section will be acquainted with the subject of isomerism from their study of elementary organic chemistry, and we shall take for granted some familiarity with simple examples of geometrical and optical isomerism. In this book so far we have not mentioned isomerism very often, for the obvious reason that relatively few inorganic compounds exhibit it. This is mainly because, on the whole, reactions of inorganic compounds are relatively rapid, and most of inorganic chemistry is concerned with equilibrium conditions. As we shall demonstrate shortly, the complex ion $[Co(C_2O_4)_3]^{3-}$ lacks a plane or centre of symmetry, and it can be resolved into optically active forms by crystallisation of the salt of an optically active base. We now know from X-ray diffraction studies that the complex ions $[Al(C_2O_4)_3]^{3-}$ and $[Fe(C_2O_4)_3]^{3-}$ are isostructural with the cobalt complex, yet they cannot be resolved; evidence from other sources indicates that substitution reactions at aluminium and high-spin iron(III) are always very fast, whilst those at low-spin cobalt(III) are nearly always very slow. In the case of the iron and aluminium complexes, therefore, failure to isolate optically active forms is telling us something about kinetic lability rather than about structure.

Among complexes of metals of the first transition series, it is found that the overwhelming majority of those that exhibit isomerism are octahedral complexes of chromium(III) or of low-spin cobalt(III), but a few cases involving octahedrally coordinated iron(II) (low-spin) or nickel(II), or planar four-coordinated nickel(II), are also known. Many complexes of the platinum metals also exist in isomeric forms, planar four-coordinated complexes of platinum(II) being particularly important in this respect. Since racemisations of optically active substances and interconversions of geometrical isomers have finite activation energies, it follows that these processes are accelerated by rise in temperature and retarded by cooling. In principle, therefore, it should be possible to isolate isomeric forms of many more substances by working at low temperatures; in practice, however, the possibility of separation by crystallisation (by far the most important method) is limited by the fact that most solvents which dissolve inorganic compounds have fairly high freezing-points.

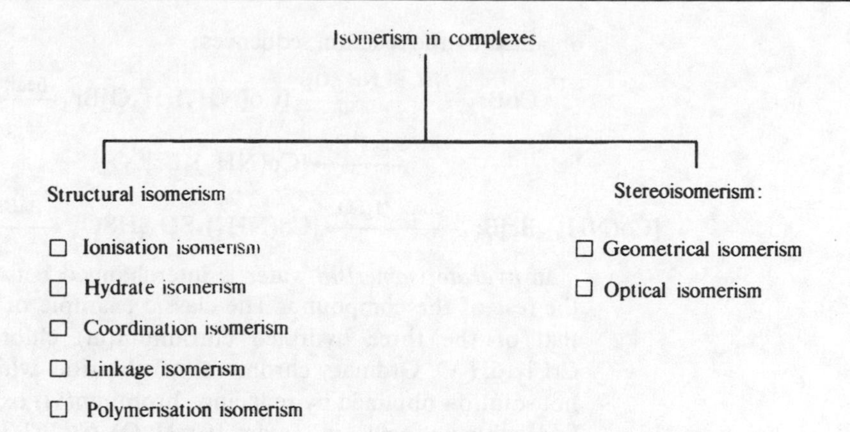

Figure 18.6
Classification of isomerism in metal complexes.

In discussing the principal types of isomerism that occur in metal complexes we shall use a classification going back to the work of Werner and shown in Fig. 18.6, considering in turn structural isomerism and the two forms of stereoisomerism (geometrical isomerism and optical isomerism). We should, however, mention that coordination chemists have recently begun to follow the usage of organic chemistry and to refer to stereoisomers as *enantiomers* (stereoisomers not superimposable on their mirror images) and *diastereoisomers* (all stereoisomers that are not enantiomers). A molecule is described as *asymmetric* if it is totally lacking in symmetry, and *disymmetric* if it lacks an S_n axis, i.e. a rotation–reflection axis (Section 5.3). A molecule that is asymmetric or disymmetric is *chiral, chirality* being the property of 'handedness'; molecules of opposite chirality (enantiomers) are related to one another in the same way as the left hand to the right hand. Chirality is usually detected by optical activity, the effect of a chiral molecule or ion on the plane of polarisation of plane-polarised light. Enantiomers of some transition metals complexes are described under the heading optical isomerism in Section 18.8.

18.6 Structural isomerism

Ionisation isomers are related by interchange of a ligand anion within the complex and an external anion associated with the complex, e.g. violet $[Co(NH_3)_5Br]SO_4$ and red $[Co(NH_3)_5SO_4]Br$, which give the usual reactions of ionic sulphate and bromide, but not of ionic bromide and sulphate, respectively. The isomers are also easily distinguished by infrared spectroscopy because of the different symmetries of the free and coordinated sulphate ions, which give rise to one and three infrared-active S—O stretching vibrations respectively. These compounds may be obtained by the reaction sequences:

$$CoBr_2 \xrightarrow{NH_4Br, NH_3, O_2} [Co(NH_3)_5H_2O]Br_3 \xrightarrow{heat} [Co(NH_3)_5Br]Br$$

$$\xrightarrow{Ag_2SO_4} [Co(NH_3)_5Br]SO_4$$

$$[Co(NH_3)_5Br]Br_2 \xrightarrow{conc.\ H_2SO_4} [Co(NH_3)_5SO_4]HSO_4 \xrightarrow{BaBr_2} [Co(NH_3)_5SO_4]Br$$

In *hydrate isomerism* water is interchanged between the complex and the rest of the compound. The classic example of hydrate isomerism is that of the three hydrated chromium(III) chlorides of composition $CrCl_3.6H_2O$. Ordinary chromium(III) chloride, which crystallises from a hot solution obtained by reducing chromium(VI) oxide with concentrated hydrochloric acid, is green $[Cr(H_2O)_4Cl_2]Cl.2H_2O$. When this is dissolved in water, the chloride ions in the complex are slowly replaced, and blue-green $[Cr(H_2O)_5Cl]Cl_2.H_2O$ and finally violet $[Cr(H_2O)_6]Cl_3$ may be isolated. On treatment with cold aqueous silver nitrate the compounds give one, two and three moles of silver chloride respectively.

Coordination isomerism is possible only for salts in which both cation and anion contain a metal ion, and arises from a different distribution of ligands between the two metal ions. Examples are provided by:

$[Co(NH_3)_6][Cr(CN)_6]$ and $[Cr(NH_3)_6][Co(CN)_6]$

$[Co(NH_3)_6][Co(NO_2)_6]$ and

$[Co(NH_3)_4(NO_2)_2][Co(NH_3)_2(NO_2)_4]$

$[Pt^{II}(NH_3)_4][Pt^{IV}Cl_6]$ and $[Pt^{IV}(NH_3)_4Cl_2][Pt^{II}Cl_4]$

Polymerisation isomerism is a rather unfortunate term (since polymers are not really isomers) used to denote compounds of the same empirical formula but different molecular formula: its use arose because of the difficulty before X-ray structure determinations became available of determining formula weights of insoluble and non-volatile compounds. There are, for example, two geometrical isomers of $[Pt(NH_3)_2Cl_2]$ and these have a polymerisation isomer $[Pt(NH_3)_4][PtCl_4]$; another example is provided by the non-electrolyte $[Co(NH_3)_3(NO_2)_3]$ and the salt $[Co(NH_3)_6][Co(NO_2)_6]$.

Finally, *linkage isomerism* arises because some ligands (e.g. NO_2^-, SCN^-) can coordinate to metals in more than one way. Isomeric nitrito and nitro complexes, for example, contain —ONO and —NO_2 as ligands, and may be obtained by methods such as those indicated in the scheme:

$$[Co(NH_3)_5Cl]Cl_2 \xrightarrow[NH_3]{dil} [Co(NH_3)_5H_2O]Cl_3$$

$[Co(NH_3)_5Cl]Cl_2$ ↓ $NaNO_2$ → $[Co(NH_3)_5ONO]Cl_2$ (red)

$[Co(NH_3)_5H_2O]Cl_3$ ↓ $NaNO_2$, conc.HCl → $[Co(NH_3)_5NO_2]Cl_2$ (yellow)

$$\underset{\text{red}}{[Co(NH_3)_5ONO]Cl_2} \underset{U.v.}{\overset{\text{Warm HCl or spont.}}{\rightleftharpoons}} \underset{\text{yellow}}{[Co(NH_3)_5NO_2]Cl_2}$$

In this case the number of M—O stretching vibrations cannot be used to distinguish the isomers, since in both complexes all the possible stretching vibrational modes of the ligand are infrared-active, but the values of the stretching frequencies are quite different for the two compounds. For the nitro complex the symmetric and asymmetric stretching frequencies (1310 and 1430 cm^{-1} respectively) are not very different; for the nitrito complex the corresponding frequencies are 1065 and 1470 cm^{-1}. Thiocyanate, which may coordinate through the nitrogen or through the sulphur atom, bonds preferentially to class (a) metal ions through the former and to class (b) metal ions through the latter. Ligands which can coordinate in two different ways are sometimes known as *ambidentate* ligands.

18.7 Geometrical isomerism

Geometrical isomers have the same structural framework, but differ in the spatial arrangement of the atoms, the arrangement in a particular isomer being called its configuration. Configurational isomers differ somewhat in physical properties and can be isolated under ordinary conditions, being thus distinguished from the different conformations of a molecule. Whether a molecule exhibits isomerism or is merely known to exist in different conformations depends on the activation energy of the interconversion process: *cis* and *trans* forms of alkenes, for example, persist unchanged for long periods at ordinary temperatures, since rotation about a C═C bond has a high activation energy; the staggered and eclipsed conformations of ethane, on the other hand, are very rapidly interconverted even at low temperatures.

A regular tetrahedral species Ma_4, Ma_2b_2 or Mabcd can exist in only one geometrical form, but a planar species Ma_2b_2 can exist in *cis*(*a*) and *trans* (*b*) forms:

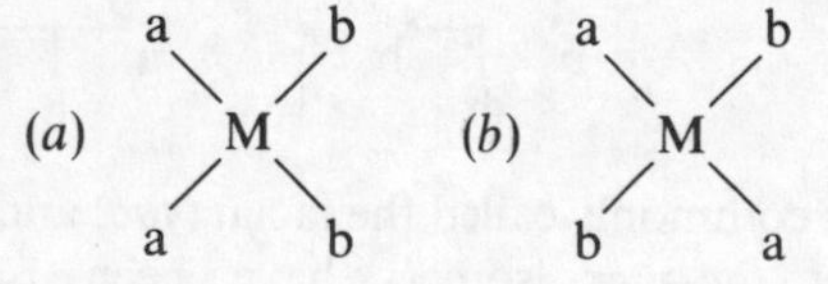

A planar species Mabcd can exist in three forms:

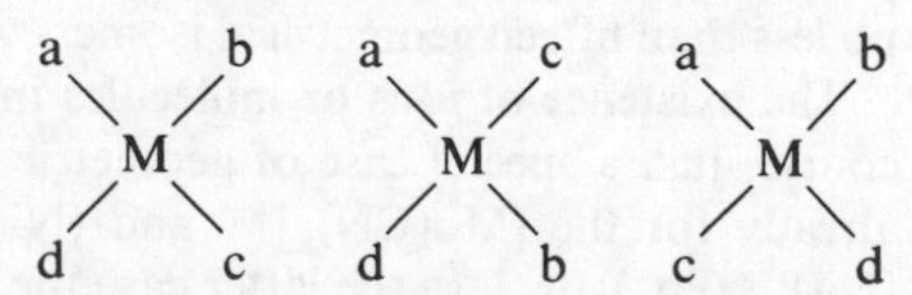

The long-established facts that there are two forms of the platinum complex $[Pt(NH_3)_2Cl_2]$ and three of the complex $[Pt(NH_3)(NH_2OH)$

$(NO_2)py]NO_2$ show that the coordination of the platinum in these compounds is not regular tetrahedral, but they do not conclusively establish that it is planar, since if the platinum were above the plane of the ligands at the apex of a square pyramid the same number of isomers would be possible. Although, therefore, much of the earlier knowledge of the stereochemistry of metals was based on the numbers of isomers that could be isolated for a particular compound, modern work relies almost entirely on X-ray diffraction determinations of structure, supplemented occasionally by measurements of dipole moments and by vibrational spectroscopy. In the case of $[Pt(NH_3)_2Cl_2]$, for example, the *trans* isomer has no dipole moment and has a simpler vibrational spectrum than the *cis* isomer.

For octahedral complexes there are, naturally, more possible geometrical isomers. An octahedral complex of formula Ma_6 or Ma_5b can exist in only one form, but Ma_4b_2 complexes can exist as *cis*(*a*) and *trans* (*b*) isomers:

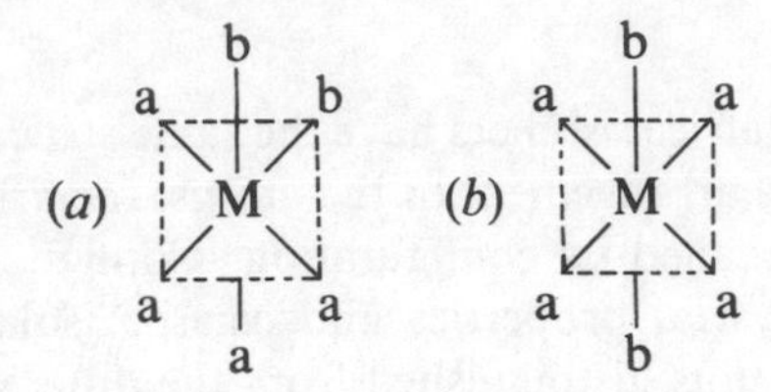

The ions $[Cr(NH_3)_4Cl_2]^+$ and $[Co(NH_3)_4Cl_2]^+$, for example, can exist in *cis* and *trans* forms which have different colours, and whose salts have different solubilities. For a formula Ma_3b_3, the three a groups may all be adjacent, i.e. at the corners of one face of the octahedron, or they may occupy three positions such that two are *trans* to each other: (*a*) and (*b*) below respectively, which

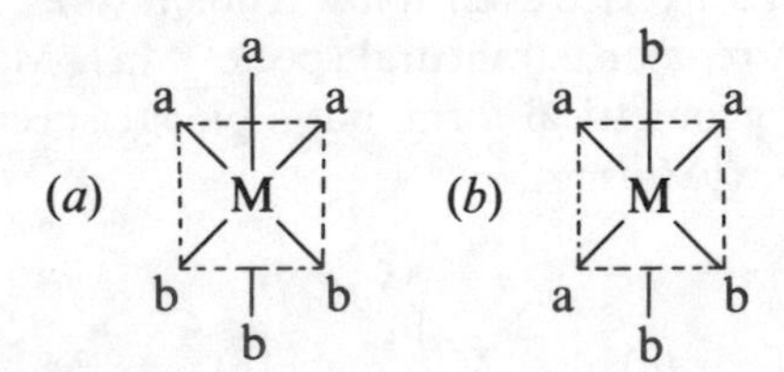

are commonly called the facial (*fac*) and meridional (*mer*) isomers. Only a few *fac–mer* isomers have been isolated, among them those of $[Co(NH_3)_3(NO_2)_3]$ and $[Rhpy_3Cl_3]$. For a complex of formula Mabcdef, no less than fifteen geometrical isomers would be possible.

The existence of ions or molecules in totally different structures is, of course, just a special case of geometrical isomerism which we have seen already for the $[Mo(CN)_8]^{4-}$ and $[Ni(CN)_5]^{3-}$ ions and the molecule $[Ni(PBzPh_2)_2Br_2]$. In the latter case the tetrahedral and planar molecules have different magnetic properties, so we shall discuss this particular case of isomerism again in the following chapter.

18.8 Optical isomerism

The differences between optical isomers or enantiomers can be detected only by using measuring devices that are themselves chiral, the commonest of these being polarised light. Ordinary light is vibrating in all directions perpendicular to its direction of travel, but by passage through polaroid or a Nicol prism only vibrations in one direction are transmitted; on passage through an optically active substance or solution, the plane of polarisation (i.e. the plane in which the light is vibrating) is altered by an amount depending on the wavelength of the light, the concentration of optically active species per unit length of light path, the temperature, the solvent, and the nature of the optically active compound. If all these factors are constant the rotations effected by two enantiomers are equal. The enantiomer which rotates the plane to the right is given a (+) sign, the other one a (−) sign. Since both the magnitude and the sign of the angle of rotation depend on the wavelength of the light used for the measurement of optical activity (an effect known as *optical rotatory dispersion*), it is always desirable to state this wavelength. As we have already said, the formal condition for a molecule to be chiral and (provided it is kinetically inert with respect to racemisation) to exhibit optical activity is the absence of an S_n rotation–reflection axis of symmetry. In practice, however, a slightly less complete statement, that a molecule having a plane or centre of symmetry (and therefore a S_1 or S_2 rotation–reflection axis) is not chiral, is nearly always satisfactory, and, being more easily applied, will be used here.

The simplest case of optical isomerism among transition metal complexes is that of salts containing an ion such as $[Co(en)_3]^{3+}$, the enantiomers of which we may represent as shown in Fig. 18.7. If we look at the octahedra along a line perpendicular to a pair of opposite faces (Fig. 18.8) we see the molecules as having the appearance of helices, rather like a ship's propeller, the twists of the helices being to the left and to the right respectively. These twists have nothing to do with the wavelength of the light used to investigate optical activity; they are absolute configurations, and are given the symbols Λ (capital lambda, for left) and Δ (capital delta, for right). Now although Λ and Δ $[Co(en)_3]^{3+}$ in isolation differ only in their action on polarised light, if we form salts of them with

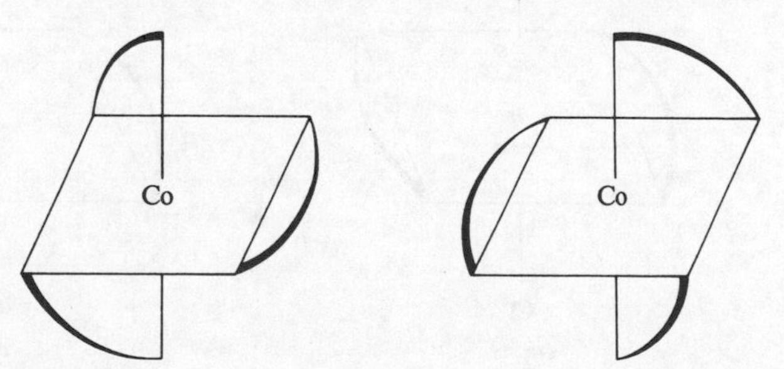

Figure 18.7
Enantiomeric forms of the $[Co(en)_3]^{3+}$ ion.

Λ Δ

Figure 18.8
The absolute configurations of the enantiomers of the $[Co(en)_3]^{3+}$ ion.

one enantiomer of an optically active anion, the salts of the Λ and Δ cations will differ in the packing of the ions and will not have the same interatomic distances, i.e. they will be geometrical isomers differing in physical properties and separable by, for example, fractional crystallisation. By methods which we cannot discuss here, it has been shown that the isomer which is (+) $[Co(en)_3]^{3+}$ when sodium light of $\lambda = 589$ nm is used to measure the rotation of the plane of polarisation is, in fact, the Δ isomer; and by means of optical rotatory dispersion measurements the structures of other ions can be correlated with this one.

As we showed in the last section, the ion $[Co(NH_3)_4Cl_2]^+$ can exist in *cis* and *trans* forms; but each of these possesses at least one plane of symmetry, and so neither can display optical activity. For the $[Co(en)_2Cl_2]^+$ ion this is no longer the case, as we may see from Fig. 18.9: the *trans* form has a centre and several planes of symmetry, but the *cis* form has neither, and (+) and (−) forms of *cis* $[Co(en)_2Cl_2]Cl$ have been separated.

Many other bidentate ligands have been shown to give rise to optical activity in octahedral metal complexes, among them $acac^-$ in $[Co(acac)_3]$, $C_2O_4^{2-}$ in $[Co(C_2O_4)_2]^{3-}$ and S_5^{2-} in $[Pt(S_5)_3]^{2-}$. The first purely inorganic compound to be resolved was a salt of the cation

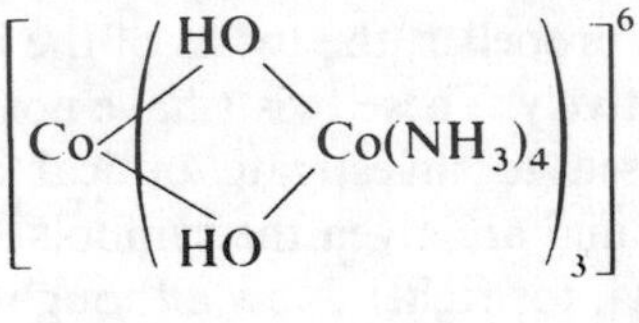

$$\left[Co\left(\begin{matrix} HO \\ HO \end{matrix} Co(NH_3)_4 \right)_3 \right]^{6+}$$

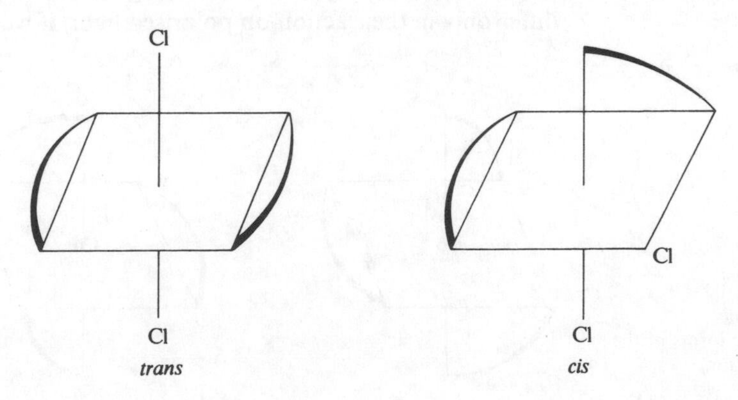

Figure 18.9
Isomeric forms of the $[Co(en)_2Cl_2]^+$ ion.

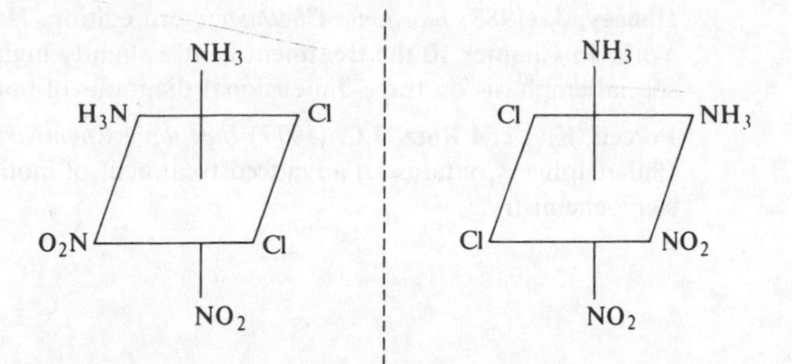

Figure 18.10
Optical isomers of one geometrical isomer of $[Pt(NH_3)_2(NO_2)_2Cl_2]$.

in which three bidentate $[Co(NH_3)_4(OH)_2]^+$ ions are distributed octahedrally around the central Co^{3+} ion. The presence of chelating ligands is not essential for the occurrence of enantiomorphism, however, as may be seen by inspection of Fig. 18.10, in which the optical isomers of one of the geometrical isomers of $[Pt(NH_3)_2(NO_2)_2Cl_2]$ are shown.

For many purposes it is permissible to regard the chelate rings in an ion such as $[Co(en)_3]^{3+}$ as planar (in conjugated systems such as oxalate and acetylacetonate complexes they are actually planar); but in reality the carbon and nitrogen atoms in saturated rings have their usual bond angles, and the rings are puckered. When we have three rings in one ion, their relative orientation slightly affects the energy of the system. For proton-containing ligands, nuclear magnetic resonance studies of complexes in solution can provide valuable evidence about the detailed structures of the ions, and show which orientation of the rings is most stable. For discussion of this topic and other aspects of stereoselectivity, however, reference must be made to more advanced works such as those cited in the references for further reading.

PROBLEMS

1 Show that the trigonal bipyramid, square pyramid, square antiprism and dodecahedron belong to the point groups D_{3h}, C_{4v}, D_{4d} and D_{2d} respectively.

2 What isomers would you expect to exist for the compounds
(a) $[Pt(H_2NCH_2CHMeNH_2)_2]Cl_2$
(b) $[Pt(H_2NCH_2CPh_2NH_2)(H_2NCH_2CMe_2NH_2)]Cl_2$?

3 Sketch all the isomeric forms of each of the following ions:

$$[Co(en)_2(C_2O_4)]^+, [Co(en)_2(NH_3)Cl]^{2+}, [Co(en)(NH_3)_2Cl_2]^+$$

REFERENCES FOR FURTHER READING

Douglas, B., McDaniel, D.H. and Alexander, J.J. (1982) *Concepts and Models of Inorganic Chemistry*, 2nd edition, Wiley, New York. A more formal presentation of stereochemistry making use of considerations of symmetry.

Huheey, J. (1983) *Inorganic Chemistry*, 3rd edition, Harper and Row, New York. In Chapter 10 the treatment is at a slightly higher level than here, with special emphasis on three-dimensional diagrams of optically active species.

Purcell, K.F. and Kotz, J.C. (1977) *Inorganic Chemistry*, W.B. Saunders, Philadelphia. Contains an advanced treatment of modern inorganic stereochemistry.

19 *Electronic configurations, electronic spectra, and magnetic properties of transition metal compounds*

19.1 Introduction

The bonding in transition metal complexes is not fundamentally different from that in other compounds. In this chapter we shall have reason to discuss the applications of the electrostatic model, valence bond theory and molecular orbital theory, just as we did for compounds of the typical elements; but throughout this discussion we shall return repeatedly to a key fact first mentioned at the end of Section 2.7 and illustrated in Fig. 2.9, that three of the five *d* orbitals for a given principal quantum number have their lobes between the axes of the p_x, p_y and p_z orbitals, and the other two are directed along the same axes as these *p* orbitals. As a consequence of this difference, the *d* orbitals in the presence of ligands are split into groups of different energies, the type of splitting and the magnitude of the energy differences depending on the geometrical distribution and nature of the ligands. Except when all the orbitals are equally populated (i.e. for d^0, high-spin d^5, and d^{10} configurations), the *d* electron distribution affects the energy of the complex. Much of this chapter will be about the splitting of the *d* orbitals and its effect on electronic spectra and magnetic properties; most of it will be about compounds of first transition series metals, for which the theories of bonding have been developed most successfully. The reader may wonder why, if splitting of the *d* orbitals is so important, nothing has been said about the splitting of the *p* orbitals into groups of different energies. This is because it is very rare for the *p* orbitals to be unequally populated in stable species, so the possibility of their having different energies in different geometrical environments is of little practical significance. In Chapter 26 we shall see that the *f* orbitals are very often unequally populated; on the other hand, the energies of the different groups of them in different coordinations are almost the same, and in this case also the splitting has no important chemical consequences.

We shall begin this chapter with a brief and partly historical account of the three theories of valence that have so far been the ones principally applied to transition metal complexes: the valence bond theory, the crystal field theory and the molecular orbital theory. In this book we shall follow the common practice of using the term ligand field theory to denote anything intermediate between crystal field theory (based on an electrostatic model) and molecular orbital theory (based on a covalent model). In recent years a modified molecular orbital approach known as the angular overlap method has attracted considerable interest. This is concerned, as the name implies, with the dependence of orbital overlapping between metal *d* orbitals and ligand orbitals on the angular positions of the ligands, the metal–ligand distance being fixed. A detailed examination of these different methods of treating transition metal complexes shows that they have more in common than may at first appear, but since our main concern in this book is with the experimental material, we shall not undertake such an examination, and we shall generally be satisfied with the simplest method that is reasonably adequate for our particular purpose. *All* theories of valence are only approximations when we are dealing with species as complicated as $[Fe(CN)_6]^{4-}$ and $[Ni(PBzPh_2)_2Br_2]$.

19.2 *Bonding in transition metal complexes: valence bond theory*

Although the valence bond theory in the form developed by Pauling in the 1930s is now not much used in the discussion of transition metal complexes, so much of the terminology and so many of the ideas associated with it have been retained in one form or another that some knowledge of it remains essential. It is, of course, only an extension of the application of the valence bond method to polyatomic molecules that we discussed in Section 4.6. In that section we showed how collinear *sp*, trigonal sp^2 and tetrahedral sp^3 hybrid orbitals may be obtained by linear combinations of atomic orbitals (s and p_z; s, p_x and p_y; and s, p_x, p_y and p_z respectively), and went on to mention combinations involving *d* orbitals such as octahedral sp^3d^2 hybrid orbitals obtained from s, p_x, p_y, p_z, $d_{x^2-y^2}$

and d_{z^2} orbitals. The central idea of valence bond theory as applied to transition metal complexes is that empty hybrid orbitals of the metal atom or ion having the required orientation in space accept pairs of electrons from the ligands to form σ bonds. If, for example, we want to describe an octahedral complex we must be able to write down an electronic configuration for the metal atom or ion that shows empty s, p, $d_{x^2-y^2}$ and d_{z^2} orbitals of comparable energies.

Some combinations of orbitals needed for hybridisation to produce various geometrical configurations are given in Table 19.1; the list there is by no means exhaustive. We have included the cube only to point out that a f orbital is needed to produce this configuration; hence its restriction to complexes of the inner transition elements.

We can illustrate the applications of valence bond theory by considering the octahedral complexes of chromium(III) and iron(III). The Cr^{3+} ion has the electronic configuration $3d^3$ with three unpaired electrons. These are placed in the $3d_{xy}$, $3d_{yz}$ and $3d_{xz}$ orbitals, which have their lobes between the ligands; then six pairs of electrons from the ligands occupy the $3d_{x^2-y^2}$, $3d_{z^2}$, $4s$, $4p_x$, $4p_y$ and $4p_z$ orbitals as shown below:

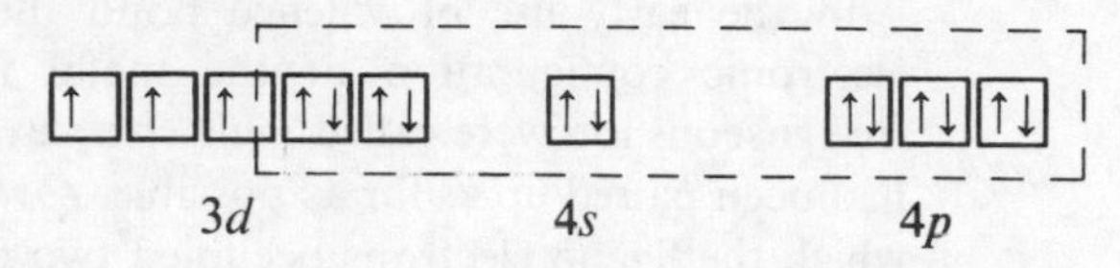

This diagram represents all octahedral Cr(III) complexes, since the three $3d$ electrons always occupy different orbitals. For octahedral Fe(III) (d^5) complexes, however, as we have already seen, the five $3d$ electrons may be paired as completely as possible, giving a low-spin complex, or be all

Table 19.1
Sigma bonding hybrid orbitals for some geometrical configurations.

Coordination number	Geometry	Orbitals hybridised	Example
2	linear	s, p_z	$Ag[(NH_3)_2]^+$
3	trigonal planar	s, p_x, p_y	$[HgI_3]^-$
4	tetrahedral	sp^3	$Ni(CO)_4$
		$d_{xy}, d_{yz}, d_{xz}, s$	MnO_4^-
	square planar	$d_{x^2-y^2}, s, p_x, p_y$	$[Ni(CN)_4]^{2-}$
5	trigonal bipyramidal	d_{z^2}, s, p^3	$[CuCl_5]^{3-}$
	square pyramidal	$d_{x^2-y^2}, s, p^3$	$[Ni(CN)_5]^{3-}$
6	octahedral	$d_{z^2}, d_{x^2-y^2}, s, p^3$	$[Co(NH_3)_6]^{3+}$
	trigonal prismatic	d^5, s or d_{xz}, d_{yz}, s, p^3	$Mo(S_2C_2Ph_2)_3$
7	pentagonal bipyramidal	$d_{xy}, d_{x^2-y^2}, d_{z^2}, s, p^3$	$[V(CN)_7]^{4-}$
	capped trigonal prism	$d_{xy}, d_{xz}, d_{z^2}, s, p^3$	$[NbF_7]^{2-}$
8	cube	$s, p^3, d_{xy}, d_{xz}, d_{yz}, f_{xyz}$	$[PaF_8]^{3-}$
	dodecahedron	$d_{z^2}, d_{xy}, d_{xz}, d_{yz}, s, p^3$	$[Mo(CN)_8]^{4-}$
	square antiprism	$d_{x^2-y^2}, d_{xy}, d_{yz}, d_{xz}, s, p^3$	$[TaF_8]^{3-}$
9	face-centred trigonal prism	d^5, s, p^3	$[ReH_9]^{2-}$

unpaired, giving a high-spin complex. For a low-spin octahedral complex such as $[Fe(CN)_6]^{3-}$ we can represent the electronic configuration by means of the diagram:

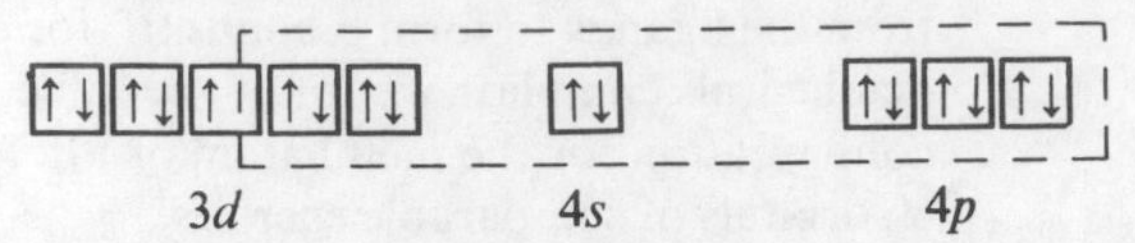

For an octahedral high-spin Fe(III) complex, e.g. $[Fe(H_2O)_6]^{3+}$ or $[FeF_6]^{3-}$, the five unpaired 3*d* electrons occupy all the 3*d* orbitals singly, and in order to maintain octahedral geometry we have to suggest that the pairs of electrons from the ligands go into the 4*s*, $4p_x$, $4p_y$, $4p_z$, $4d_{x^2-y^2}$ and $4d_{z^2}$ orbitals:

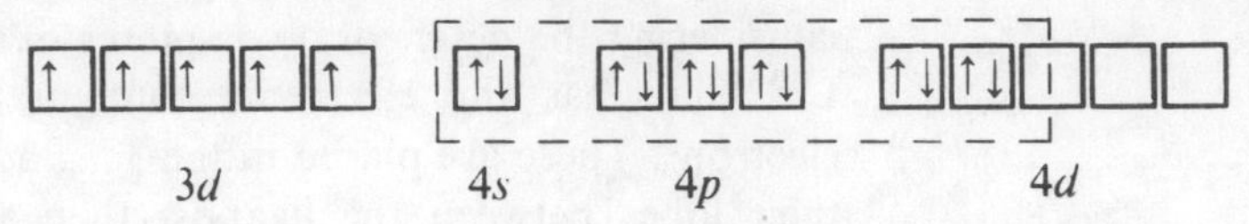

This suggestion, however, is unattractive inasmuch as the 4*d* orbitals are much higher in energy than the 3*d* orbitals.

In the early use of valence bond theory, complexes in which the electronic configuration of the metal ion was the same as in the free gaseous ion were called *ionic complexes*, those in which the electrons had been paired up as far as possible, *covalent complexes*. Later, those in which the ligand electrons occupied two of the 3*d* orbitals became *inner orbital complexes* and those in which the 4*d* orbitals were occupied, *outer orbital complexes*. Both of these terminologies are still used, so we should note:

high-spin ≡ ionic ≡ outer orbital
low-spin ≡ covalent ≡ inner orbital.

Valence bond theory can cope with other geometries, and its application to nickel(II) complexes is worth discussing here. Tetrahedral and octahedral Ni^{2+} (d^8) complexes are all paramagnetic; the configuration of the Ni^{2+} ion may be represented as

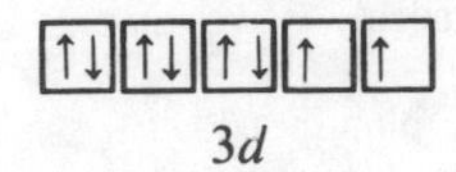

and four and six pairs of electrons from the ligands occupy the $4s4p^3$ and $4s4p^34d^2$ orbitals respectively. Planar complexes of nickel(II), e.g. $[Ni(CN)_4]^{2-}$, are diamagnetic, however, and the electronic configuration is then shown as

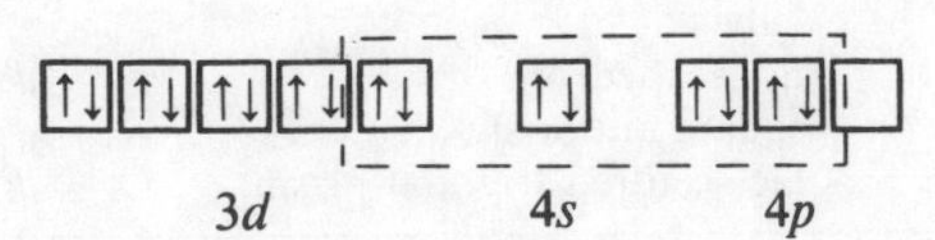

with ligand electrons occupying $3d4s4p^2$ hybrid orbitals. Nickel carbonyl, $Ni(CO)_4$, is tetrahedral, but it is a derivative of Ni(0), for which the valence state is $3d^{10}$; the four pairs of electrons from the carbon monoxide molecules then occupy the four $4s4p^3$ hybrid orbitals.

It seems improbable, however, that in a compound such as nickel carbonyl a very large negative charge will be accumulated on the nickel atom. Pauling suggested that this charge accumulation as a result of σ bonding is substantially offset by π bonding from the filled d_{xy}, d_{yz} and d_{yz} orbitals of the nickel atom into the relatively low energy π^* antibonding orbitals of the carbon monoxide molecules, which they overlap to a considerable extent; for a ligand such as chloride ion or phosphorus trifluoride, empty d orbitals of the chlorine or phosphorus atom may be used for a similar purpose. (With fluoride ion, water or ammonia as ligand such back-bonding is impossible, there being no d orbitals in the valence shell of the donor atom, but the electronegativity of the latter operates in the same direction.) This idea of charge redistribution is the basis of the so-called *electroneutrality principle*: charge tends to get distributed so that no atom in a complex has a resultant charge greater than about ± 1 electron. It embodies an important idea that we shall meet again in discussing other theories of bonding.

We see, then, that valence bond theory is able to deal satisfactorily with many stereochemical and magnetic properties (at least at the level at which we have sc far considered them). But in its simple form as presented here, it has nothing to say about electronic spectra or the reasons for the kinetic inertness of chromium(III) and low-spin cobalt(III) octahedral complexes. And inasmuch as the inner orbital–outer orbital distinction implies a substantial difference between the two types of complex in the cation–ligand interaction energy (because the $4d$ orbitals are much higher in energy than the $3d$ orbitals, so bonding involving them should be weaker), it is actually misleading. Nor does it it tell us why water and halide ions commonly form high-spin complexes whilst cyanide ion forms low-spin complexes. To understand this and many other features of transition metal chemistry we must turn to other theories.

19.3 *Bonding in transition metal complexes: crystal field theory*

Although crystal field theory was developed by Bethe and van Vleck at about the same time as valence bond theory by Pauling, its application to simple chemical problems came much later. It is a purely electrostatic theory, and it takes no account of ligand electrons except in so far as they create an electric field in which the d orbitals of the metal atom or ion may have different energies. It is thus a great oversimplification, but in a quite different way from valence bond theory, and its advantages and disadvantages are naturally quite different from those of the latter theory, as we shall now see.

Let us consider first the case of a cation surrounded octahedrally by six ligands placed on the x, y and z axes. There is a large net electrostatic attraction and the energy of the system is substantially lowered. If, however, there is an electron in each of the d orbitals, there is a local *increase* in the energy of all these orbitals as the anions (or negative ends of polar molecules) are brought up to the cation. Since the $d_{x^2-y^2}$ and d_{z^2} orbitals, which are directed between the x, y and z axes. These three (if they are occupied) are raised more than those of the d_{xy}, d_{xz} and d_{yz} orbitals, which are directed between the x, y and z axes. These three orbitals of lower energy are collectively known as the t_{2g} orbitals, and the two orbitals of higher energy as the e_g orbitals; these symbols will be discussed later, but in remembering them it is helpful to know that t means there are three, and e that there are two, of equal energy (i.e. that are degenerate); g (*gerade*) means, as we explained in Section 2.8, that the orbitals are centrosymmetric (unlike, for example, p orbitals).

We can represent the consequences for the energies of the d orbitals of bringing up the ligands by Fig. 19.1. If the field created by the ligands were spherically symmetrical, the energies of all the d orbitals would be equal, just as they are in an isolated metal ion. Now it can be shown (though we shall not prove it here) that if, instead of being distributed spherically, the charges of the ligands are concentrated at the six corners of a regular octahedron, the total energy of the d orbitals is unchanged: relative to their energies in a spherical field, therefore, the combined energies of the three t_{2g} orbitals must be lowered by the same amount as the combined energies of the e_g orbitals are raised. The dashed line at the right of Fig. 19.1 represents the barycentre or centre of gravity of the energies of the orbitals; and if Δ_0 (the subscript denoting octahedral geometry) is the energy separation between the two sub-sets of orbitals, each t_{2g} orbital is lowered in energy by $0.4\Delta_0$ and each e_g orbital raised in energy by $0.6\Delta_0$ relative to the barycentre. Other symbols, we should add, are sometimes used, Δ being replaced by $10Dq$, t_{2g} by d_ε, and e_g by d_γ.

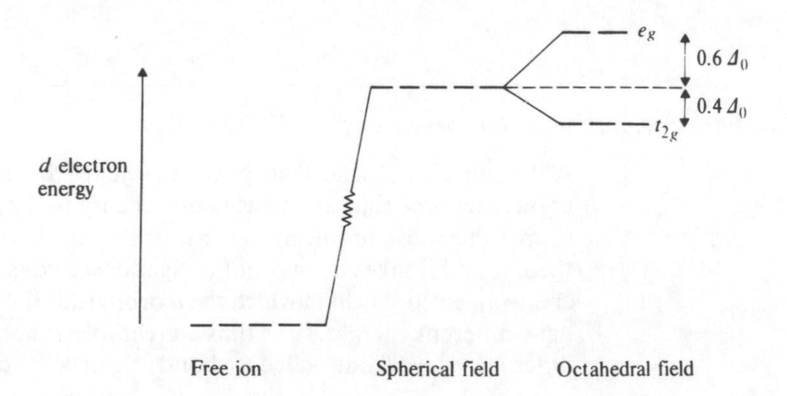

Figure 19.1
The effects on the energies of d electrons of a spherical and an octahedral field.

So far we have been considering an ion with one electron in each d orbital, i.e. a high-spin d^5 ion. Clearly the total energy of the d orbitals for this species will be the same as it would in a spherical field. For a d^1 system, however, with the electron in a t_{2g} orbital, there will be a crystal field stabilisation energy (CFSE) for the ion of $0.4\Delta_0$; for d^2 and d^3 ions the crystal field stabilisation energies will be 0.8 and $1.2\Delta_0$ respectively. For the electronic configuration d^4 there are two possibilities. Either the fourth electron can go into a e_g orbital (the high-spin d^4 configuration) with a total crystal field stabilisation energy of $1.2\Delta_0 - 0.6\Delta_0 = 0.6\Delta_0$, or it can pair with one of the electrons in a t_{2g} orbital (the low-spin d^4 configuration for octahedral geometry: it would require much more energy to pair all the electrons to form a diamagnetic complex). Which it does will depend upon whether Δ_0 is greater or smaller than the pairing energy P, which we may define as the energy required to transform two electrons with parallel spin in different degenerate orbitals into two paired electrons in the same orbital. Where we are dealing with the pairing of more than one electron P becomes the mean pairing energy; it may be obtained from the analysis of electronic spectra, though in general its value in complexes (derived from analysis of their spectra) is found to be about 20 per cent less than in the corresponding free gaseous ion. What we need next, therefore, is a method of evaluating Δ_0.

It is the great merit of crystal field theory that, in principle at least, there is a way of doing this. If we consider the absorption spectrum of the simplest case, the d^1 complex $[Ti(H_2O)_6]^{3+}$, shown in Fig. 19.2, there is only a single weak band, with a maximum at 20 300 cm^{-1} corresponding to an energy change of 243 kJ mol^{-1}, and this absorption results from the promotion of the electron from a t_{2g} to a e_g orbital; the ion is purple because the absorption maximum is in the green region of the visible spectrum. For systems containing more than one d electron the evaluation

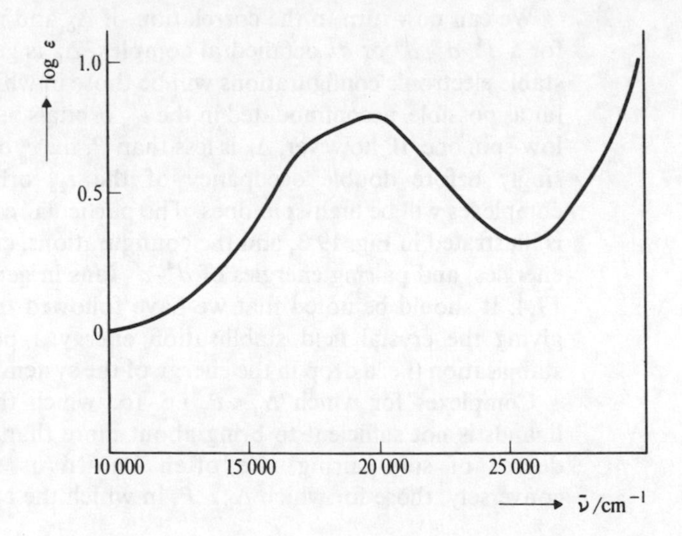

Figure 19.2
The electronic spectrum of the $[Ti(H_2O)_6]^{3+}$ ion in aqueous solution.

of Δ_0 is often considerably more complicated and is not always very accurate, but Δ_0 remains an experimental quantity. It is found to depend upon a number of factors: the identity and oxidation state of the metal, the geometry of the complex, and the nature of the ligand. If we keep all these factors except the last constant, and consider the octahedral complexes of a given cation, it is found that Δ_0 increases along the following series for ligands:

$$I^- < Br^- < Cl^- < F^- < OH^- < C_2O_4{}^{2-} < H_2O < NH_3 < en < bipy < phen < CN^- \sim CO.$$

For $[CrF_6]^{3-}$, $[Cr(H_2O)_6]^{3+}$, $[Cr(NH_3)_6]^{3+}$ and $[Cr(CN)_6]^{3-}$, for example, the Δ_0 values are estimated as 15 000, 17 400, 21 600 and 26 600 cm^{-1} respectively. This series (the *spectrochemical series*) is valid, with remarkably few exceptions, for any given ion in complexes of a given geometry. It may be noted that ligands which bond through a given donor atom (e.g. OH, $C_2O_4^{2-}$ and H_2O; NH_3, en, bipy, phen) are close together in the series. Thus ammonia and ethylenediamine complexes of a given cation are usually very similar in colour, but rather different from aquo and oxalato complexes of the same cation. For a given ligand and a given oxidation state of the metal ion, Δ_0 is found to vary irregularly along the first transition series (e.g. over the range 8 000 cm^{-1} to 14 000 cm^{-1} for $[M(H_2O)_6]^{2+}$ ions). For a given ligand and a given metal, it increases with increase in oxidation state (e.g. from 9 400 cm^{-1} for $[Fe(H_2O)_6]^{2+}$ to 13 700 cm^{-1} for $[Fe(H_2O)_6]^{3+}$. In the few cases for which analogous species exist for metals of all three transition series, Δ_0 increases by 30–50 per cent on going from one series to the next higher one (e.g. for $[Co(NH_3)_6]^{3+}$, $[Rh(NH_3)_6{}^{3+}]$) and $[Ir(NH_3)_6]^{3+}$ Δ_0 is 23 000, 34 000 and 41 000 cm^{-1} respectively). We should emphasise that the spectrochemical series is an empirical generalisation; its validity is in no way dependent upon the validity of explanations of the observed sequence.

We can now turn to the correlation of Δ_0 and magnetic properties. If, for a d^4, d^5, d^6 or d^7 octahedral complex, Δ_0 is greater than P, the most stable electronic configurations will be those in which the electrons are, so far as possible, accommodated in the t_{2g} orbitals, i.e. the complex will be a low-spin one. If, however, Δ_0 is less than P, the e_g orbitals will be occupied singly before double occupancy of the t_{2g} orbitals occurs, and the complexes will be high-spin ones. The particular case of Fe(III) complexes is illustrated in Fig. 19.3, and the configurations, crystal field stabilisation energies, and pairing energies of d^4–d^7 ions in general are shown in Fig. 19.4. It should be noted that we have followed the common practice of giving the crystal field stabilisation energy a positive sign for actual stabilisation (i.e. a drop in the energy of the system).

Complexes for which $\Delta_0 < P$, i.e. for which the field created by the ligands is not sufficient to bring about more than the minimum possible degree of spin-pairing, are often known as *weak-field* complexes; conversely, those for which $\Delta_0 > P$, in which the t_{2g} orbitals are occupied

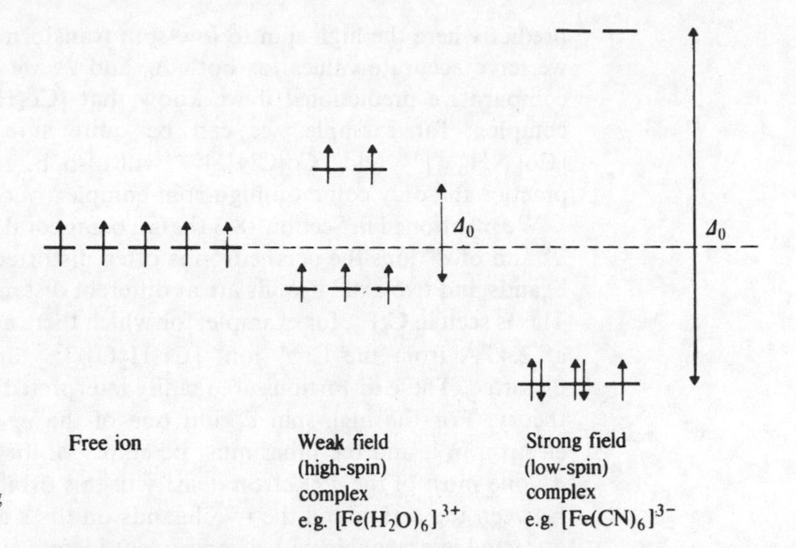

Figure 19.3
The occupation of the 3d orbitals in weak and strong field $Fe^{3+}(3d^5)$ complexes.

completely before any electrons go into the e_g orbitals, are known as *strong-field* complexes. These terms are the equivalents in crystal field (or ligand field) theory terminology of the experimental designations high-spin and low-spin respectively.

Since cyanide ion lies at the high-field end of the spectrochemical series we can now see why many cyano complexes are low-spin species whilst the corresponding aquo complexes are high-spin in character. Similarly, we can see that if any complexes of a given ion are to be high-spin species, they are most likely to be halide complexes. Thus although we cannot

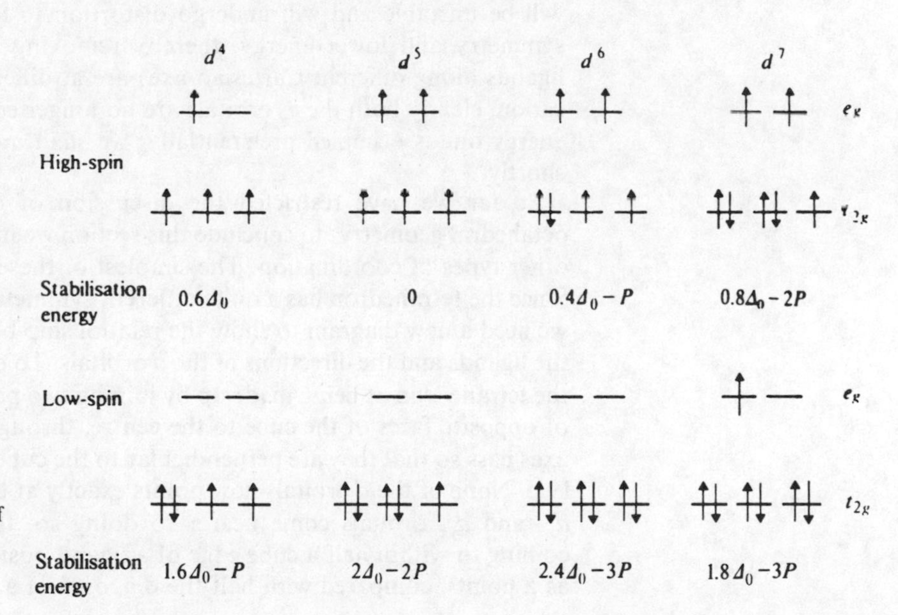

Figure 19.4
Electronic configurations and stabilisation energies of high-spin and low-spin d^4, d^5, d^6 and d^7 ions in octahedral complexes.

predict where the high-spin to low-spin transformation will occur unless we have accurate values for both Δ_0 and P, we can often make useful comparative predictions: if we know that $[Co(H_2O)_6]^{3+}$ is a low-spin complex, for example, we can be quite sure that $[Co(C_2O_4)_3]^{3-}$, $[Co[NH_3)_6]^{3+}$ and $[Co(CN)_6]^{3-}$ will also be low-spin complexes. In practice the only common high-spin complex of cobalt(III) is $[CoF_6]^{3-}$.

We mentioned in Section 18.3 that in octahedral complexes of high-spin d^4 and of d^9 ions the octahedron is often distorted so that four coplanar ligands and two axial ligands are at different distances from the metal ion. This is seen in CrF_2, for example, for which there are $4F^-$ at 2.00 and $2F^-$ at 2.43 Å from the Cr^{2+} ion; $[Cr(H_2O)_6]^{2+}$ and CuF_2 are similarly distorted. These distortions are readily interpreted in terms of crystal field theory. For the high-spin d^4 ion one of the e_g orbitals must have an electron in it and the other must be empty; if the occupied orbital is the d_{z^2} one most of the d electron density in this orbital will be concentrated between the cation and the two ligands on the z axis, so these should be attracted less strongly with a consequent increase in interionic separation. Conversely, if the fourth electron were in the $d_{x^2-y^2}$ orbital, two ligands should be closer than the other four. Accurate electron-density measurements have recently confirmed that the electronic configuration of the Cr^{2+} ion in $[Cr(H_2O)_6]^{2+}$ is *approximately* $d^1_{xy}d^1_{xz}d^1_{yz}d^1_{z^2}$. The corresponding effect when the t_{2g} orbitals are unequally occupied would be expected to be very much smaller, since these orbitals are directed between the ligands in an octahedral complex. This expectation is usually, though not invariably, fulfilled. Distortions of this kind are often called *Jahn–Teller* distortions, being named after the so-called *Jahn–Teller theorem*: any non-linear molecular system in a degenerate electronic state will be unstable and will undergo distortion to form a system of lower symmetry and lower energy, thereby removing the degeneracy. When ligands along different Cartesian axes are at different distances from the cation, clearly both the e_g orbitals are no longer equivalent and the lower energy one is occupied preferentially; we shall meet this situation again shortly.

So far we have restricted the discussion of crystal field theory to octahedral geometry; to conclude this section we must now consider some other types of coordination. The simplest of these is regular tetrahedral. Since the tetrahedron has a quite different symmetry from the octahedron we need a new diagram to show the relationship between the positions of the ligands and the directions of the d orbitals. To obtain this, we visualise the tetrahedron as being made up by joining two pairs of opposite corners of opposite faces of the cube to the centre, through which the x, y and z axes pass so that they are perpendicular to the cube faces as shown in Fig. 19.5. None of the d orbitals now points exactly at the ligands, but the d_{xy}, d_{xz} and d_{yz} orbitals come nearer to doing so, their axes of symmetry coming to within half a cube edge of a ligand position (treating the latter as a point), compared with half the diagonal of a cube face for the $d_{x^2-y^2}$

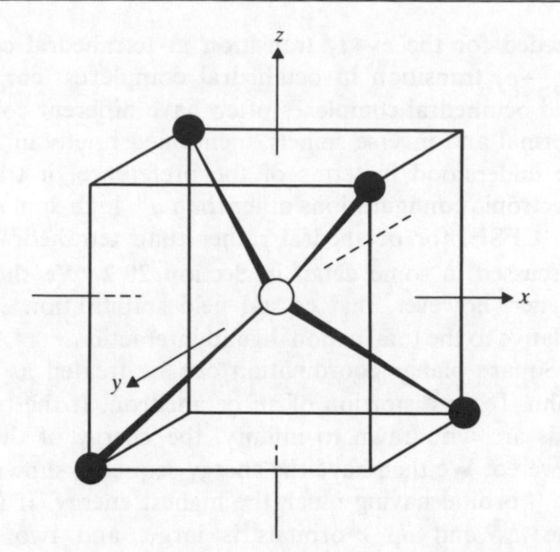

Figure 19.5
The relationship between tetrahedral coordination, octahedral coordination (with ligands on the x, y and z axes) and a cube.

and d_{z^2} orbitals. Thus for a regular tetrahedral structure the splitting is inverted compared with that for a regular octahedral structure, and the energy difference between the two types of orbital is smaller; it can be shown from solid geometry that, if all other things are equal (in reality, of course, they never are), the splitting in a tetrahedral field, Δ_t is given by

$$\Delta_t = \tfrac{4}{9}\Delta_0$$

Thus we have the situation shown in Fig. 19.6. Note that the subscript g has now been dropped from the orbital designations; this is because a tetrahedron has no centre of symmetry about which the symmetry properties can be specified.

One of the consequences of Δ_t being so much smaller than Δ_0 is that in practice all tetrahedral complexes of d^3, d^4, d^5, d^6, d^7 and d^8 ions are high-spin complexes. Another is that since smaller amounts of energy are

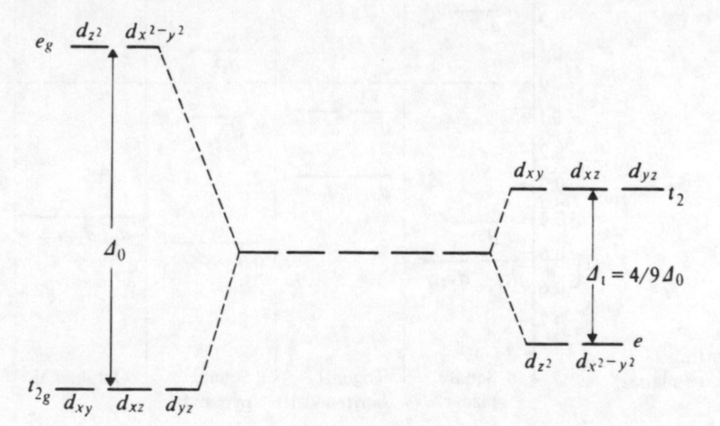

Figure 19.6
Crystal field splitting diagram for octahedral and tetrahedral fields.

needed for the $e \rightarrow t_2$ transition in tetrahedral complexes than for the $t_{2g} \rightarrow e_g$ transition in octahedral complexes, corresponding tetrahedral and octahedral complexes often have different colours. The existence of normal and inverse spinels, mentioned briefly in Section 12.8, can often be understood in terms of the preference of transition metal ions of electronic configurations other than d^0, high-spin d^5 and d^{10} (which have no CFSE) for octahedral rather than tetrahedral coordination, and is discussed in some detail in Section 20.2. We should again remind the reader, however, that crystal field stabilisation energies are very small relative to the total cation–ligand interaction.

Square planar coordination can be treated as the limiting case of a Jahn–Teller distortion of an octahedron. If the two ligands along the z axis are withdrawn to infinity, the energy of the d_{z^2} orbital is greatly lowered. We then have the energy sequence shown in Fig. 19.7, with the $d_{x^2-y^2}$ orbital having much the highest energy. If the separation between the d_{xy} and $d_{x^2-y^2}$ orbitals is large, and two electrons have to be accommodated in them, both will go into the d_{xy} orbital, leaving the $d_{x^2-y^2}$ orbital empty. This is what happens for planar Ni(II) complexes, which are derivatives of a d^8 ion and are diamagnetic, e.g. $[Ni(CN)_4]^{2-}$. For Pd(II) and Pt(II), which are homologues of Ni(II), the energy gap between

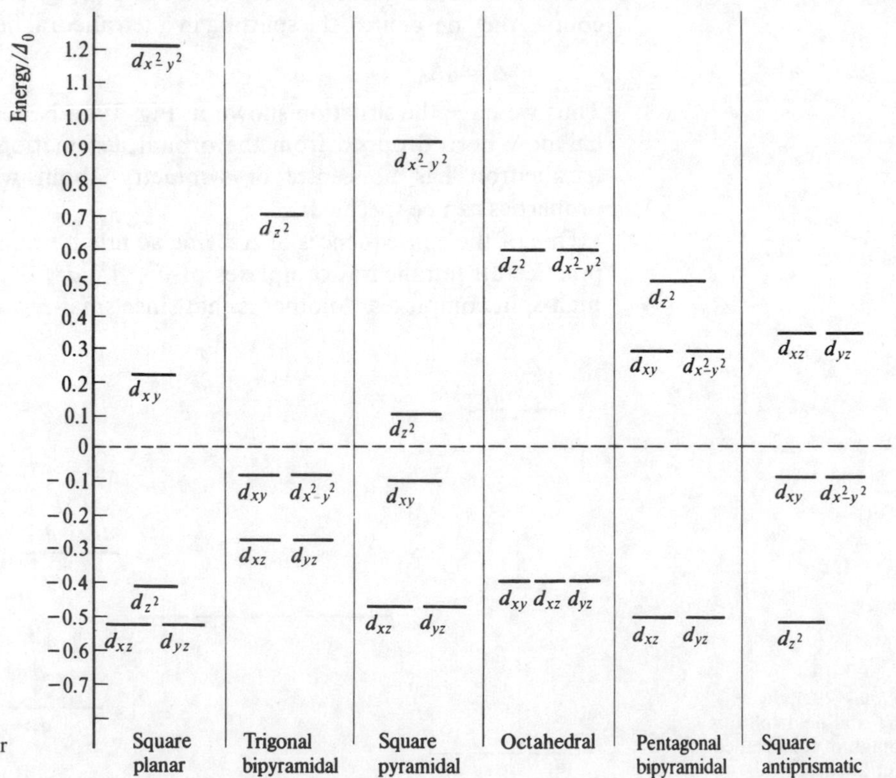

Figure 19.7
Crystal field splitting diagram for square planar and other fields.

the d_{xy} and $d_{x^2-y^2}$ orbitals is large enough even with chloride as ligand for the complexes $[PdCl_4]^{2-}$ and $[PtCl_4]^{2-}$ to be planar and diamagnetic, although $[NiCl_4]^{2-}$ is tetrahedral and paramagnetic. Qualitative crystal field splitting diagrams for trigonal bipyramidal, square pyramidal and other geometries are also given in Fig. 19.7 and may be discussed along similar lines.

Crystal field theory can therefore be seen to bring together structures, magnetic properties and electronic spectra, and we shall see in Chapters 20 and 21 that it also helps us to understand some thermodynamic and kinetic aspects of transition metal chemistry. It is, in fact, surprisingly useful when we consider the simplicity (and sometimes the obvious inaccuracy) of its starting point of a purely ionic structure. Thus although $Cr(CO)_6$, for example, may be treated as an octahedral low-spin complex of Cr(0) (d^6), it is inconceivable that this volatile compound is faithfully described as an ionic product of the interaction of an uncharged atom and a neutral molecule. Further, although we can interpret the contrasting magnetic properties of halo and cyano complexes in terms of the positions of halide and cyanide ions in the spectrochemical series, the theory provides no suggestion as to *why* these ions occupy the places in the series that they do. Nor has π bonding any place in crystal field theory. To deal with these and related matters some kind of molecular orbital theory is required.

19.4 Bonding in transition metal complexes: molecular orbital theory

The existence of stable complexes in which both metal atom and ligand are uncharged (such as the metal carbonyls) would of itself be sufficient evidence that any general treatment of complex formation must incorporate provision for covalent bonding. Even in conventional complex ions, however, there is often evidence for delocalisation of electrons from metal to ligand. Thus the fact that pairing energies are less in complexes than in free gaseous ions indicates that interelectronic repulsion is less in complexes and that the effective size of the metal orbitals containing the electrons has increased (this is the so-called *nephelauxetic* or (electron) cloud expanding effect). More direct proof of electron sharing comes from electron spin resonance spectroscopy. Unpaired electrons behave as magnets as a result of their spin, and (in the simplest possible case) may align themselves parallel or antiparallel to a magnetic field. The former configuration has a very slightly lower energy, and transitions from one configuration to the other can be observed by applying the necessary energy in the form of radiofrequency radiation. An electron in an isolated metal ion shows a single sharp absorption for such a transition, but if the metal ion carrying the unpaired electron is bonded to a ligand having a nuclear magnetic moment a multiline absorption results, showing that the electron has been partly transferred to ligand orbitals or, more accurately, that the orbital occupied by the electron is a molecular orbital obtained from atomic orbitals of both metal and ligand. This has been observed for a solid solution of $Na_2[IrCl_6]$ in $Na_2[PtCl_6]$, the former being a low-spin d^5 complex and the latter being diamagnetic. Accurate electron density maps of $K_2Na[Co(NO_2)_6]$ and other compounds also provide evidence of electron delocalisation.

We shall illustrate the application of molecular orbital theory to transition metal complexes first by considering an octahedral complex in which π bonding may be neglected, such as $[Co(NH_3)_6]^{3+}$ or $[CoF_6]^{3-}$. In the construction of an energy level diagram for such a species, many approximations are involved, and the result is usually only qualitatively accurate; even so, it has a notable contribution to make to our study of metal–ligand bonding. The orbitals we use to construct the molecular orbitals are six metal orbitals (s, p_x, p_y, p_z, $d_{x^2-y^2}$ and d_{z^2}) and six ligand group orbitals. The construction of these ligand group orbitals from atomic orbitals is itself a considerable problem, which we shall not discuss here; we may take it for granted that for any complex that does exist, it will be possible to obtain a set of ligand group orbitals of the required symmetries to overlap the six metal orbitals. The appropriate set for octahedral coordination is shown in Fig. 19.8. We have already subdivided the d orbitals into t_{2g} and e_g orbitals and commented on the designations t, e, g and u; for the designations of other orbitals we need also the label a, denoting a singly degenerate orbital; and we should also now add that the subscripts 1 and 2 mean respectively that the orbital or orbitals do not change sign on rotation about the Cartesian axes and that they do not change sign on rotation about axes diagonal to the Cartesian axes. Thus the metal s and p orbitals are designated a_{1g} and t_{1u}.

The molecular orbital diagram shown in Fig. 19.9 may be regarded as built up of a set of simpler molecular orbital diagrams such as we encountered in Chapter 4. The overlap of the metal s and p orbitals with ligand group orbitals is extensive, so there are large differences in energy between the resulting a_{1g} and a^*_{1g}, and t_{1u} and t^*_{1u}, bonding and antibonding orbitals; thus the a_{1g} and t_{1u} molecular bonding orbitals are lowest, and the a^*_{1g} and t^*_{1u} molecular antibonding orbitals highest, in energy. The e_g and e^*_g orbitals are less widely separated because of the poorer overlap between the $d_{x^2-y^2}$ and d_{z^2} orbitals and ligand group orbitals; even so, the better the overlap (or, as we might say, the more covalent bonding there is) the bigger the separation will be. In a system without π bonding the t_{2g} orbitals are non-bonding and therefore have the same energy as in the metal ion. The energy difference between the t_{2g} and e^*_g molecular orbitals corresponds to Δ_0 in crystal field theory.

The ion $[Co(NH_3)_6]^{3+}$ is, as we have seen earlier, a low-spin complex. With eighteen electrons (six from Co^{3+} and twelve from the ligands) to be accommodated in molecular orbitals, the a_{1g}, t_{1u}, e_g and t_{2g} molecular

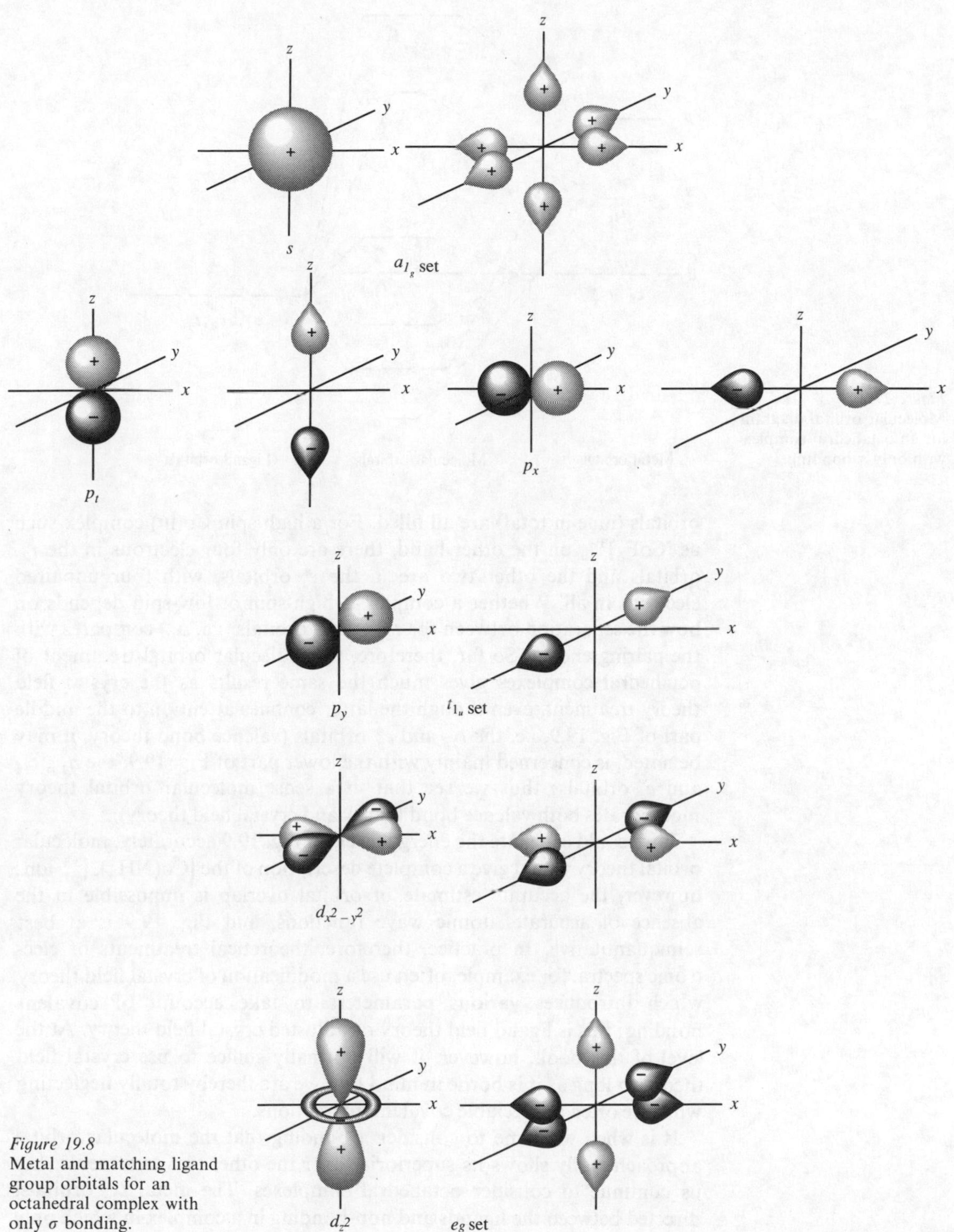

Figure 19.8
Metal and matching ligand group orbitals for an octahedral complex with only σ bonding.

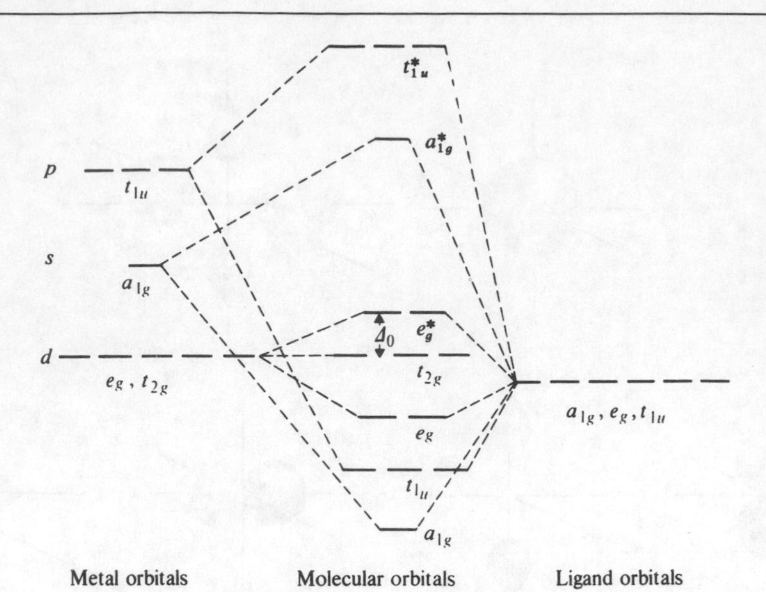

Figure 19.9
Molecular orbital diagram for an octahedral complex with only σ bonding.

orbitals (nine in total) are all filled. For a high-spin Co(III) complex such as $[CoF_6]^{3-}$, on the other hand, there are only four electrons in the t_{2g} orbitals and the other two are in the e_g^* orbitals, with four unpaired electrons in all. Whether a complex is high-spin or low-spin depends on how the separation between the t_{2g} and e_g^* orbitals (i.e. Δ_0) compares with the pairing energy. So far, therefore, the molecular orbital treatment of octahedral complexes gives much the same results as the crystal field theory treatment, even though the latter confines attention to the middle part of Fig. 19.9, i.e. the t_{2g} and e_g^* orbitals (valence bond theory, it may be noted, is concerned mainly with the lower part of Fig. 19.9, the a_{1g}, t_{1u} and e_g orbitals; thus we see that in a sense molecular orbital theory incorporates both valence bond theory and crystal field theory).

If we could calculate the energy levels in Fig. 19.9 accurately, molecular orbital theory would give a complete description of the $[Co(NH_3)_6]^{3+}$ ion; however, the accurate estimate of orbital overlap is impossible in the absence of accurate atomic wave functions, and Fig. 19.9 is at best semiquantitative. In practice, therefore, theoretical treatments of electronic spectra, for example, often use a modification of crystal field theory which introduces various parameters to take account of covalent bonding; this is ligand field theory or adjusted crystal field theory. At the level of this book, however, it will normally suffice to use crystal field theory so long as it is borne in mind that we are thereby totally neglecting what are often appreciable covalent interactions.

It is when we come to consider π bonding that the molecular orbital approach really shows its superiority over the other two treatments. Let us continue to consider octahedral complexes. The metal t_{2g} orbitals, directed between the ligands and non-bonding in a complex in which only

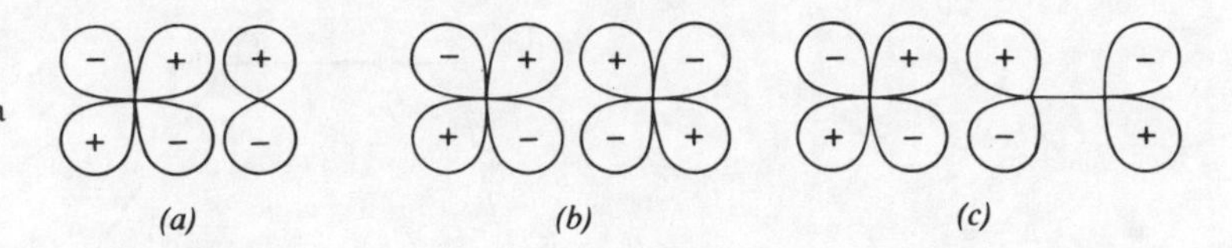

Figure 19.10
π bond formation between a metal t_{2g} orbital and a ligand (*a*) *p* orbital (*b*) *d* orbital (*c*) π^* orbital.

σ bonding is present, can readily overlap ligand group orbitals having the same symmetry, and these can be formed from *p* orbitals perpendicular to the σ bond axes, *d* orbitals lying in a plane which includes the metal atom, or π^* orbitals lying in a plane that includes the metal atom. The formation of π bonds in these three ways is represented in Fig. 19.10; in (*a*), (*b*) and (*c*) the ligand might be a halide ion, a phosphine and carbon monoxide respectively. If the ligand group orbitals lie at a lower energy than the metal t_{2g} orbitals (e.g. if the ligand is a halide ion using filled *p* orbitals for π bonding), π bonding affects the energy level diagram as shown in Fig. 19.11, with the π bonding orbitals resembling the ligand orbitals more than the metal orbitals. The e_g^* orbitals are not affected by π bonding though their energy does, of course, depend upon the nature of the ligand (even where there is no π bonding, Δ_0 varies because of the magnitude of σ bonding). Except for a d^0 complex, the π^* orbitals must be occupied to some extent, and Δ_0, now the energy difference between the π^* orbitals and the e_g^* orbitals (which will not be full except for a d^{10} complex) will be less than it would be in the absence of π bonding. Conversely, if the ligand group orbitals lie at a higher energy than the metal t_{2g} orbitals (e.g. if the ligand is a phosphine or carbon monoxide using empty *d* or π^* orbitals for bonding), the energy level diagram becomes as shown in Fig. 19.12. In these circumstances Δ_0 is larger than it would be in the absence of π bonding. Further, the e_g^* orbitals will be more strongly antibonding (being further above the highest energy bonding orbital), and will therefore be much less likely to be occupied. Thus octahedral complexes involving π bonding ligands of this kind (π acceptors) are very unlikely to be high-spin

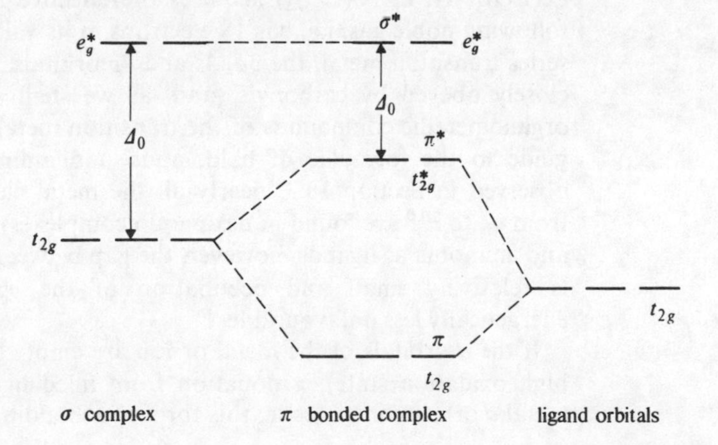

Figure 19.11
Energy level diagram showing the interaction of the t_{2g} orbitals with lower-energy filled orbitals of the ligands, leading to a decrease in Δ_0.

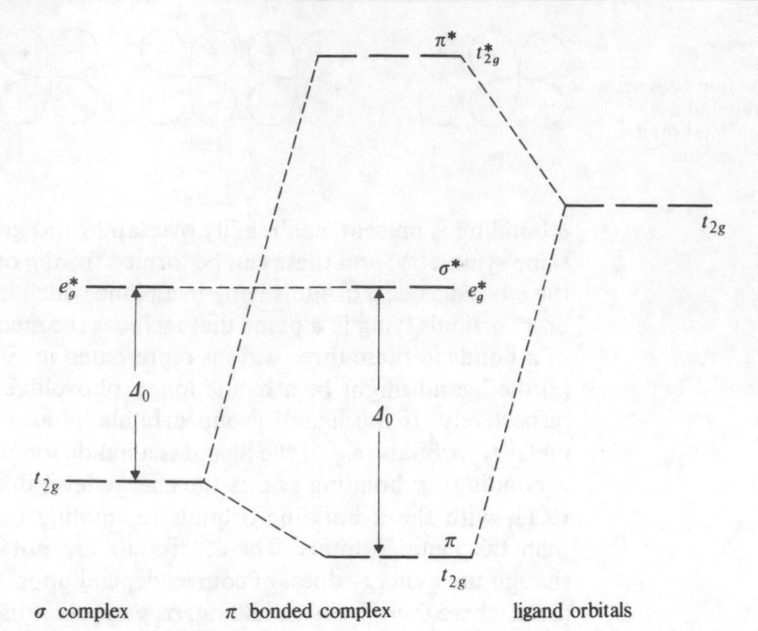

Figure 19.12
Energy level diagram showing the interaction of the t_{2g} orbitals with higher-energy empty orbitals of the ligands, leading to an increase in Δ_0.

complexes. What is more, since octahedral complexes involving π-accepting ligands must involve occupation of the e_g^* orbitals if the metal or ion has more than six d electrons, octahedral carbonyl, cyanide or phosphine complexes of d^7, d^8, d^9 and d^{10} configurations are very unlikely to exist.

This last point brings us straight back to some fundamental observations in experimental inorganic chemistry. It was noticed many years ago that transition metal carbonyls, in particular, nearly always obey the *effective atomic number rule* or 18-*electron rule*: they have formulae such that if each carbon monoxide molecule is counted as contributing two electrons the metal concerned (e.g. Cr in $Cr(CO)_6$, Fe in $Fe(CO)_5$, Ni in $Ni(CO)_4$) acquires the effective atomic number of the following noble gas, i.e. has 18 electrons in its valence shell of, for a first series transition metal, the 3*d*, 4*s* and 4*p* orbitals. This rule, though very closely obeyed by carbonyls (and, as we shall see in Chapter 23, by organometallic compounds of the transition metals), is quite useless as a guide to the formulae of halo, aquo and ammine complexes (as we observed in Section 18.3, nearly all the metal electronic configurations from d^0 to d^{10} are found in hexa aquo complexes); for halide ions, water and ammonia as ligands, however, the gap between the t_{2g} and e_g^* orbitals is relatively small and occupation of the e_g^* orbital is therefore energetically less unfavourable.

If the d orbitals of the metal or ion are empty (i.e. if the metal is in a high oxidation state), π donation from filled ligand orbitals to empty metal d orbitals may occur; this form of π bonding occurs in CrO_4^{2-} and

MnO_4^-. But here we are no longer dealing with octahedral geometry, and a molecular orbital diagram corresponding to Fig. 19.9 is required for tetrahedral bonding in order to discuss this subject. Energy level diagrams are available for tetrahedral and other geometries, but their discussion would not add very much to the extra insight that we have gained from considering the octahedral case, and so we shall not discuss them in this book. We shall, however, often make use of descriptions which consist essentially of an electrostatic (crystal field theory) treatment reinforced as necessary by taking account of π bonding.

To conclude this section we return briefly to the spectrochemical series. This, as we said earlier, is an experimental series which incorporates the effects of both σ bonding and the various possible types of π interaction. We cannot predict the series, partly because even the σ component cannot yet be calculated accurately, and partly because some ligands are potentially capable of π interactions of more than one kind: halide ions other than fluoride, for example, have empty *d* orbitals in their valence shells and could conceivably act as π acceptors using these rather than as π donors using filled *p* orbitals. Given the experimental evidence, however, we can at least offer the retrospective comment that the series can be reasonably well accounted for on the basis of increasing σ interaction along the series halide donor < O donor < N donor < C donor, π donation from halide ions, and π acceptance by many unsaturated N donors and C donors. The discussion of the last type of interaction will be resumed in Chapter 22.

19.5 Electronic spectra

Before reading this section it may be desirable to re-read Section 3.3, which deals with Hund's rules and state symbols for free ions. As we have mentioned earlier, not all bands in the visible and ultraviolet spectra of transition metal complexes originate in *d–d* transitions; charge-transfer absorptions (in which electrons are transferred from ligand to metal or vice versa, and for which there are no selection rules) also occur in this region (especially at the high energy end), and, being much more intense than *d–d* absorptions, frequently mask the latter. For a free ion, transitions which involve a change in multiplicity and transitions between states having the same values for the quantum numbers *n* and *l* (e.g. between one *d* orbital and another of the same principal quantum number) are forbidden. For a complex which has a centre of symmetry *d–d* transitions are forbidden by both these selection rules, and are described as both spin- (or mutiplicity-) forbidden and Laporte-forbidden; they are accordingly extremely rare and the corresponding absorptions are extremely weak. For a complex which has no centre of symmetry, however, the Laporte selection rule does not apply. This is why tetrahedral complexes (e.g. $[CoCl_4]^{2-}$) are more strongly coloured than octahedral complexes (e.g. $[Co(H_2O)_6]^{2+}$). For a d^7 ion like $[Co(H_2O)_6]^{2+}$, electronic configuration $t_{2g}^5e_g^2$, an electron can be promoted from a t_{2g} orbital to a e_g orbital without breaking the spin selection rule, but this is impossible for an octahedral high-spin d^5 ion like $[Mn(H_2O)_6]^{2+}$. Thus although a solution containing the $[Co(H_2O)_6]^{2+}$ ion is not very strongly coloured, one containing the $[Mn(H_2O)_6]^{2+}$ ion is practically colourless. Such absorption as there is arises from *vibronic coupling*: vibrations within the ion distort its symmetry slightly and then the Laporte selection rule does not hold so completely.

The subject of electronic spectra is a large and complicated one, and in so far as we consider at all species having 4–7 d electrons, we shall restrict attention here to a few high-spin (weak field) complexes. The spectrum of the $[Ti(H_2O)_6]^{3+}$ ion consists, as we have already seen from Fig. 19.2, of a single band (close inspection suggests it is, in fact, two closely-spaced bands arising from a Jahn–Teller effect in the excited state, but this point is of minor importance). The state symbol for the ground state of the Ti^{3+} ion (d^1; one electron for which $m_l = 2$, so $L = 2$, $S = \frac{1}{2}$) is ^2D; in an octahedral field, as we have seen, this is split into a $^2T_{2g}$ and a ^2E_g state (capital letters are used for states, small letters for electrons) separated by energy Δ_0, the magnitude of which increases with increase in ligand field strength. We can represent this state of affairs by the simple diagram shown in Fig. 19.13.

For the d^9 electronic configuration (e.g. of Cu^{2+}) in an octahedral field (actually a rare occurrence because of Jahn–Teller effects) the ground state of the free ion (^2D) is again split and the value of Δ_0 again corresponds to promotion of an electron from a t_{2g} to a e_g orbital; but in this case the state of lower energy is ^2E_g and that of higher energy is $^2T_{2g}$. This is so because the ground state $(t_{2g})^6(e_g)^3$ is doubly degenerate (it can be $(t_{2g})^6(d_{x^2-y^2})^2\,(d_{z^2})^1$ or $(t_{2g})^6(d_{x^2-y^2})^1(d_{z^2})^2$) whilst the excited state is triply degenerate (it can be $(d_{xy})^2(d_{yz})^2(d_{xz})^1(e_g)^4$ or $(d_{xy})^2(d_{yz})^1\,(d_{xz})^2(e_g)^4$ or $(d_{xy})^1(d_{yz})^2(d_{xz})^2(e_g)^4$). Thus for a d^9 ion in an octahedral field the diagram for a d^1 ion is inverted (alternatively, we can say that for the d^9 ion there is a hole in an e_g orbital in the ground state and a hole in a t_{2g} orbital in the excited state). The diagram for a d^1 or a d^9 ion is also inverted by a change from an octahedral to a tetrahedral field, although for a given metal–ligand distance Δ_t is only $\frac{4}{9}\Delta_0$. Further, since high-spin d^6 differs from d^1 only in that an electron has been added to each orbital, the diagram for this configuration is the same as for d^1; by analogy, high-spin d^4 has the same diagram as d^9. These relations are shown by means of an *Orgel diagram* in Fig. 19.14. Analogous reasoning shows that d^2 and d^7, and d^3 and d^8, also have the same diagrams; so if we discuss one of these configurations we shall in principle have covered all high-spin configurations from d^1 to d^9 inclusive except d^5. All of the observed d–d bands in the spectra of ions of this last configuration involve transitions

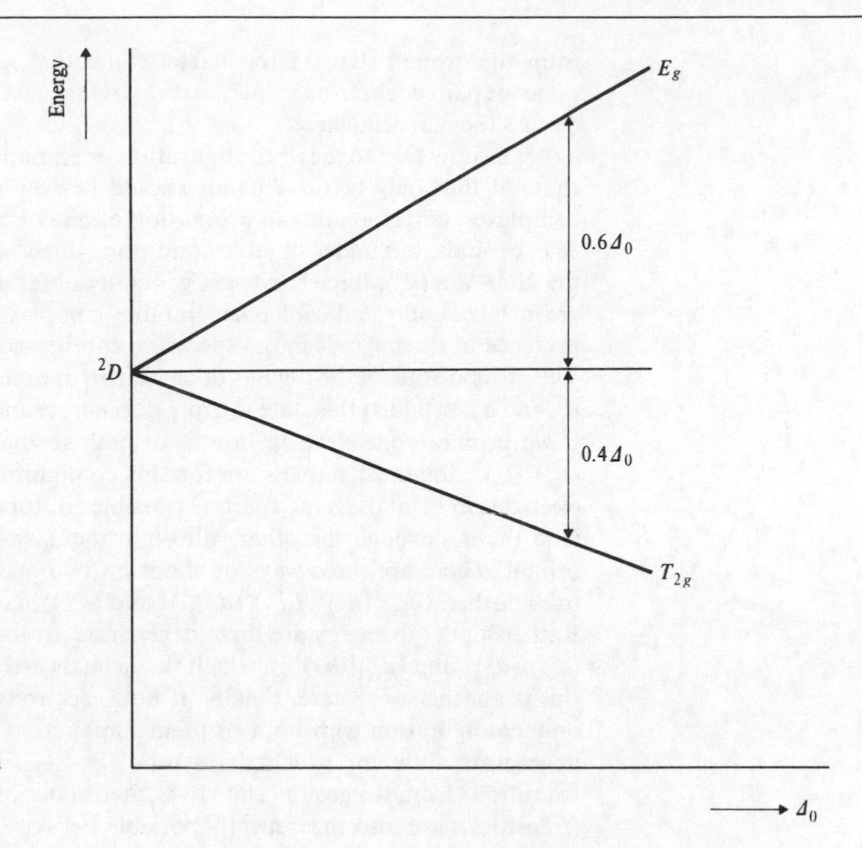

Figure 19.13
Energy level diagram for a d^1 ion in an octahedral field.

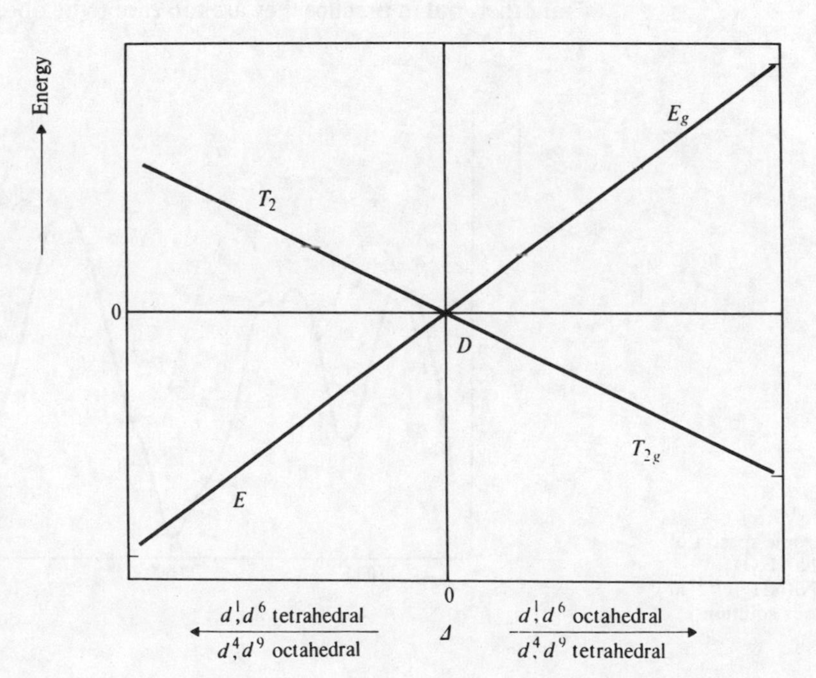

Figure 19.14
Orgel diagram for d^1, d^4 (high-spin), d^6 (high-spin) and d^9 ions in an octahedral or a tetrahedral field (for d^4 and d^9 ions the g subscripts should be transferred to the left-hand side of the diagram).

from the ground state ^6S to quartet states (i.e. states which have only three unpaired electrons); they are extremely weak and we shall not discuss them in detail here.

Let us now turn to the d^2 configuration, exemplified by V^{3+}. It might be thought that only two *d–d* bands should be seen in the spectra of V^{3+} complexes, corresponding to promotion of one or both electrons from t_{2g} to e_g orbitals, but under suitable conditions three bands are observed; for the Ni^{2+} ion (d^8), three bands are generally observed, as is shown for the green hexa-aquo and violet hexammine complexes in Fig. 19.15. The presence of three bands in the spectrum can be accounted for as follows. The ground state of the ion has one electron in each of any two of the d_{xy}, d_{yz} and d_{xz} orbitals; this state is triply degenerate and has the symbol $^3T_{1g}$. If we promote one electron to a e_g orbital, so that the configuration is $(t_{2g})^1(e_g)^1$, the most stable state for this configuration will have the two electrons in orbitals as far apart as possible: if, for example, one electron is in the d_{xy} orbital, the other will be in the d_{z^2} rather than in the $d_{x^2-y^2}$ orbital. There are three ways of choosing two orbitals perpendicular to one another, $(d_{xy})^1(d_{z^2})^1$, $(d_{xz})^1(d_{x^2-y^2})^1$ and $(d_{yz})^1(d_{x^2-y^2})^1$; this state is $^3T_{2g}$. Rather higher in energy are three degenerate arrangements $(d_{xy})^1(d_{x^2-y^2})^1$, $(d_{xz})^1(d_{z^2})^1$ and $(d_{yz})^1(d_{z^2})^1$ in which the orbitals are at 45° to one another; this is another $^3T_{1g}$ state. Finally, if both electrons are in e_g orbitals the only configuration with both of them unpaired is $(d_{x^2-y^2})^1(d_{z^2})^1$, a singly degenerate state for which the symbol is $^3A_{2g}$. There are thus three transitions from the ground state to excited states of the same multiplicity (transitions are also theoretically possible between one excited state and another, but in practice they are too rare to be observed).

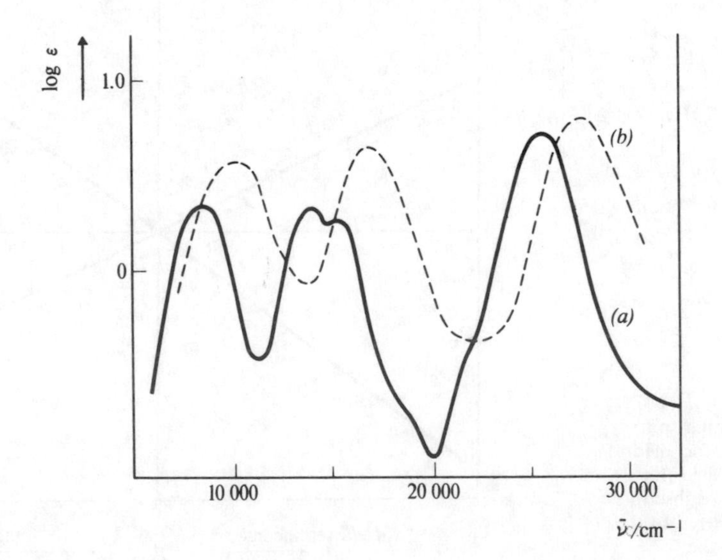

Figure 19.15
Electronic spectra of
(*a*) $[Ni(H_2O)_6]^{2+}$
(*b*) $[Ni(NH_3)_6]^{2+}$ in aqueous solution.

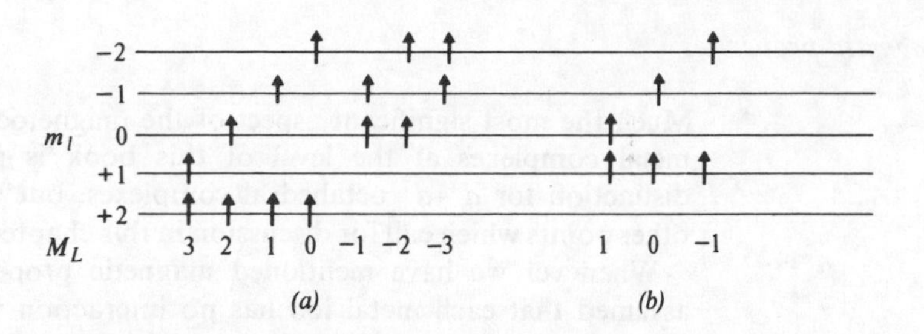

Figure 19.16
Possible combinations of values of m_l for $l = 2$, each electron having $m_s = +\frac{1}{2}$, grouped so as to show values of M_L corresponding to (*a*) $L = 3$ (^3F term) (*b*) $L = 1$ (^3P term).

The discussion in the previous paragraph is tantamount to saying that for a species involving two electrons, interelectronic repulsion must be taken into account and parameters additional to Δ_0 are required to give a quantitative description of the spectrum. These are the *Racah parameters* to which reference is made in more advanced works. It will, however, be clear without further elaboration that the evaluation of Δ_0 is now a more difficult matter than for a d^1 complex, and this is even more so for low-spin species. Some uncertainty is, in fact, often attached to reported values of Δ_0 for other than d^1 (or d^6 etc.) species.

In discussing the spectra of octahedral d^2 complexes we have deliberately sought to emphasise the physical principles underlying the number of bands in the spectrum. An outline of a more formal treatment of this and related problems is now given. If we write down, as is done in Fig. 19.16, all possible combinations of values of m_l for two unpaired d electrons with parallel spins, we see that in addition to the ground state of the d^2 ion (^3F), the only other triplet state is ^3P. It can be shown from group theory that in octahedral or tetrahedral fields D and F, but not S and P, states are split: a S state gives rise only to an A_1 state, a P state only to a T_1 state, a D state to an E state and a T_2 state, and a F state to an A_2 state, a T_1 state, and a T_2 state (all are g states if the use of the designation g or u is appropriate). For the d^2 ion in an octahedral field the possible triplet states are therefore $^3A_2$, $^3T_1$ and $^3T_2$ derived from the free ion ground state ^3F, and $^3T_1$ derived from the free ion excited state ^3P; we add the symbol (F) or (P) after $^3T_1$ to avoid ambiguity. The ground state of the ion in an octahedral field is $^3T_1$ (F), and the possible transitions from this state are to the $^3T_2$, $^3T_1$ (P) and $^3A_2$ states.

It may have been noticed that the bands in Fig. 19.15 are all rather broad. This is because a transition in which an electron moves from a t_{2g} to a e_g orbital or vice versa should, if equilibrium were obtained, lengthen or shorten the metal–ligand bond; however, since electronic transitions are much faster than atomic vibrations (the Franck–Condon principle) the excited state is not one of equilibrium interatomic distance but is a vibrationally excited level of the excited state. In solution, for example, it interacts to various extents with solvent molecules, and a broad band results.

19.6 Magnetic properties

Much the most significant aspect of the magnetochemistry of transition metal complexes at the level of this book is the high-spin/low-spin distinction for d^4–d^7 octahedral complexes, but there are a number of other points which call for discussion in this chapter.

Whenever we have mentioned magnetic properties so far, we have assumed that each metal ion has no interaction with other metal ions. This is true for substances in which the paramagnetic species are separated from one another by several diamagnetic species (e.g. by water molecules in hydrated cations); such substances are said to be magnetically dilute. When the paramagnetic species are very close together (as in metals) or are separated only by an atom or monatomic ion that can transmit magnetic interactions (as in many oxides, fluorides or chlorides), they may interact with one another. This interaction may give rise to ferromagnetism (in which large domains of magnetic dipoles are aligned in the same direction) or antiferromagnetism (in which neighbouring magnetic dipoles are aligned in opposite directions). *Ferromagnetism* leads to greatly enhanced paramagnetism as in iron at temperatures of up to 768 °C (the Curie temperature), above which thermal energy overcomes the alignment and normal paramagnetic behaviour occurs. *Antiferromagnetism* occurs below the Néel temperature; as less thermal energy is available with decrease in temperature the paramagnetic susceptibility falls rapidly. The classic example of antiferromagnetism is the oxide MnO, which has the NaCl structure and a Néel temperature of 118 K. Neutron diffraction, which is capable of distinguishing between sets of atoms having opposed moments, shows the side of the unit cell at 80 K to be twice that at 293 K, indicating that in the conventional unit cell shown in Fig. 6.1 metal atoms at adjacent corners have opposed moments at the lower temperature (and hence the unit cells cannot be stacked randomly to produce the overall crystal structure). More complex behaviour may occur if some moments are systematically aligned so as to oppose others, but the relative numbers or relative values of the moments are such as to lead to a finite resultant magnetic moment: this is *ferrimagnetism*. The relationship between paramagnetism, ferromagnetism, antiferromagnetism and ferrimagnetism is illustrated in Fig. 19.17. In Fe_3O_4, for example (an inverse spinel), the moments of Fe^{3+} in octahedral and in tetrahedral sites cancel, and the observed moment arises from the Fe^{2+} moments only.

When we discussed Russell–Saunders (LS) coupling in Section 3.3, and wrote ground states in the form $^{2S+1}L_J$, we mentioned that for the lighter elements variation in the value of J makes only a small difference to the energy of the atom or ion compared with variation in the values of S and L; and we implicitly assumed this in the last section, where we disregarded J completely. The energy difference between adjacent states of J values J' and $(J'+1)$ is in fact given by the expression $(J'+1)\lambda$, where λ is called the *spin–orbit coupling constant*. For the d^2 configuration, for

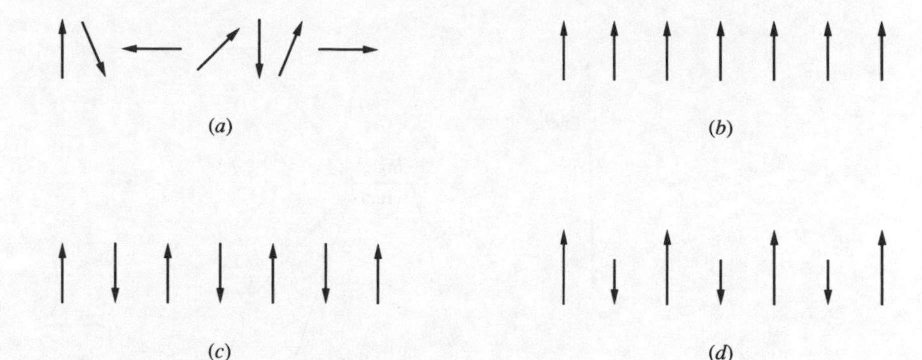

Figure 19.17
Representation of
(*a*) paramagnetism
(*b*) ferromagnetism
(*c*) antiferromagnetism
(*d*) ferrimagnetism.

example, the ^3F state in an octahedral field is split into states $^3F_2$, $^3F_3$ and $^3F_4$, the energy differences between successive pairs being 3λ and 4λ respectively. In a magnetic field these states having different J values are each split again to give $(2J+1)$ different levels separated by $g_J\mu_B B_0$, where g_J is a constant called the Landé splitting factor for the system and B_0 the magnetic field; it is with the very small differences between these levels that electron spin resonance spectroscopy is concerned. The overall splitting pattern for a d^2 ion is shown in Fig. 19.18.

The value of λ varies from a fraction of a cm^{-1} for the very lightest atoms to a few thousand cm^{-1} for the heaviest ones. Obviously the extents to which states of different J values are populated at the ordinary temperature depends upon how large their separation is compared with the thermal energy available, kT, which at 300 K is approximately 200 cm^{-1} or 2.6 kJ mol^{-1}. It can be shown theoretically that if the separation of the energy levels is large, the magnetic moment should be given by

$$\mu = g_J\sqrt{J(J+1)}$$

$$\text{where } g_J = 1 + \frac{S(S+1) - L(L+1) + J(J+1)}{2J(J+1)}$$

This formula gives values for the magnetic moments of the lanthanide ions (for which λ is usually about 1000 cm^{-1}) that are in good agreement with the observed values (see Section 26.4), but for ions of metals of the first transition series the agreement between calculated and observed values is very poor (see Table 19.2). For many (though not all) of these ions, λ is very small and the spin and orbital angular momenta of the electrons operate independently. For electron spin only, $L = 0$, $J = S$, and $g_J = 2$; for orbital motion only, $S = 0$, $J = L$, and $g_J = 1$. Then μ is given by

$$\mu = \sqrt{4S(S+1) + L(L+1)}$$

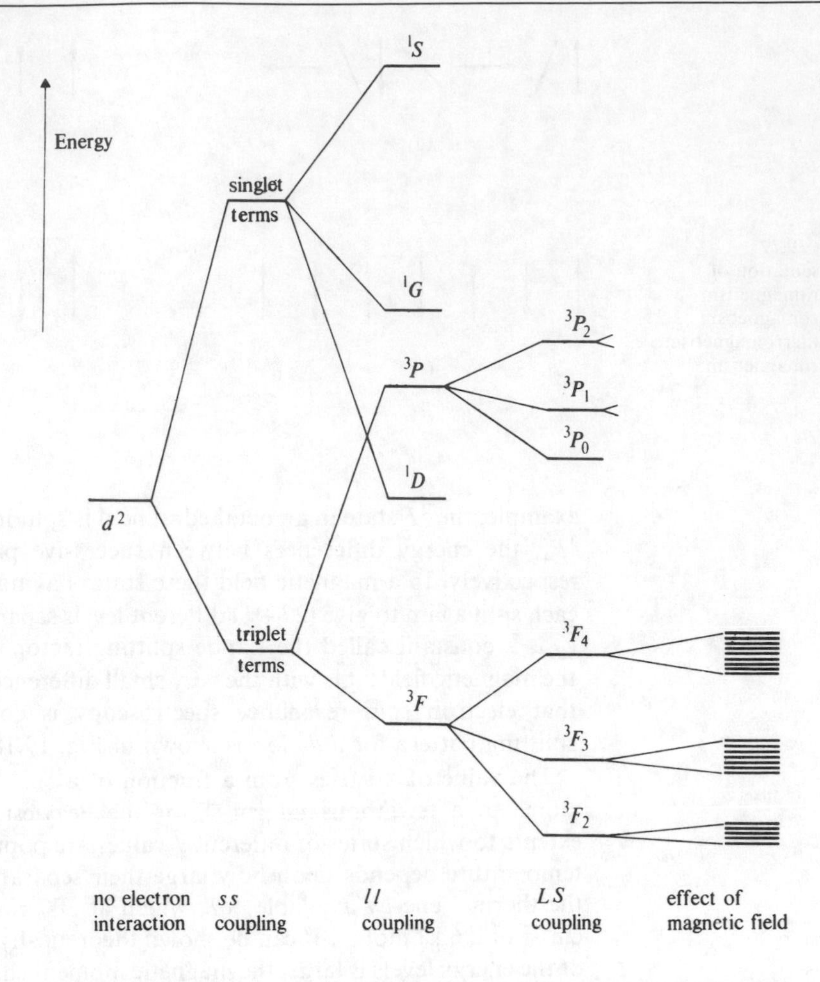

Figure 19.18
Splitting of the states of a d^2 ion (not to scale).

which, if there is no contribution from orbital motion, reduces to the spin-only formula

$$\mu = \sqrt{4S(S+1)} = 2\sqrt{S(S+1)}$$

that we met in Section 18.3.

Any ion for which $L=0$ (e.g. Mn^{2+} or Fe^{3+}, for which all the d orbitals are occupied singly by electrons for which $m_l = 2, 1, 0, -1$ and -2, giving $L=0$) should obviously obey the spin-only formula, but some others also do so, in which cases the orbital angular momentum is said to be *quenched*. In order for an electron to have orbital angular momentum, it must be possible to transform the orbital it occupies into an entirely equivalent and degenerate orbital by rotation; the electron is then effectively rotating about the axis used for the rotation of the orbital. In an octahedral complex, for example, the three t_{2g} orbitals can be interconverted by rotation through 90°, so an electron in a t_{2g} orbital has orbital angular momentum. The e_g orbitals, having different shapes,

Table 19.2
Magnetic moments of some first transition series ions in high-spin complexes at ordinary temperatures.

Ions	Ground term	μ/μ_B calculated from $g_J\sqrt{J(J+1)}$	$\sqrt{4S(S+1)+L(L+1)}$	$\sqrt{4S(S+1)}$	μ/μ_B Exptl.
Ti^{3+}	$^2D_{\frac{3}{2}}$	1.55	3.01	1.73	1.7–1.8
V^{3+}	$^3F_2$	1.63	4.49	2.83	2.8–3.1
V^{2+}, Cr^{3+}	$^4F_{\frac{3}{2}}$	0.70	5.21	3.87	2.7–3.9
Cr^{2+}, Mn^{3+}	$^5D_0$	0	5.50	4.90	4.8–4.9
Mn^{2+}, Fe^{3+}	$^6S_{\frac{5}{2}}$	5.92	5.92	5.92	5.7–6.0
Fe^{2+}, Co^{3+}	$^5D_4$	6.71	5.50	4.90	5.0–5.6
Co^{2+}	$^4F_{\frac{9}{2}}$	6.63	5.21	3.87	4.3–5.2
Ni^{2+}	$^3F_4$	5.59	4.49	2.83	2.9–3.9
Cu^{2+}	$^2D_{\frac{5}{2}}$	3.55	3.01	1.73	1.9–2.1

cannot be interconverted, so electrons in e_g orbitals never have orbital angular momentum. There is, however, a further factor that needs to be taken into consideration: if all the t_{2g} orbitals are singly occupied, an electron in, say, the d_{xz} orbital cannot be transferred into the d_{xy} or d_{yz} orbital because these already contain an electron having the same spin quantum number as the incoming electron; if all the t_{2g} orbitals are doubly occupied the transfer is also impossible. Thus only configurations which have a t_{2g} electron but not three or six t_{2g} electrons make orbital contributions to the magnetic moments of octahedral complexes: for high-spin complexes, this means only the configurations $(t_{2g})^1$, $(t_{2g})^2$, $(t_{2g})^4(e_g)^2$ and $(t_{2g})^5(e_g)^2$. For tetrahedral geometry it is easily shown that the configurations giving rise to an orbital contribution are $(e)^2(t_2)^1$, $(e)^2(t_2)^2$, $(e)^4(t_2)^4$ and $(e)^4(t_2)^5$. So an octahedral high-spin d^7 complex should have a magnetic moment greater than the spin-only value of 3.87 μ_B, but a tetrahedral d^7 complex should not. In practice $[Co(H_2O)_6]^{2+}$ and $[CoCl_4]^{2-}$ both have magnetic moments above the spin-only value, though the former certainly has the higher one, the values being 5.0 and 4.4 μ_B respectively. Another factor must therefore also be involved.

This is *spin–orbit coupling*, a subject too difficult to get more than a brief mention here. It corresponds to some population of a higher energy state and therefore leads to a temperature variation of the magnetic moment. Its effect is to modify μ_s, the value calculated on the spin-only formula, in the case of an octahedral complex according to the equation

$$\mu = \mu_s\left(1 - \frac{\alpha\lambda}{\Delta_0}\right)$$

where α is a constant depending on the ground state, and λ and Δ_0 have their usual meanings in this chapter. Since λ has a positive sign for d^1, d^2, d^3 and d^4 ions, and a negative sign for d^6, d^7, d^8 and d^9 ions, spin–orbit coupling leads to somewhat low magnetic moments for the first set of ions, and somewhat high magnetic moments for the second set.

All of the foregoing discussion in terms of crystal field theory can be elaborated on the basis of molecular orbital theory, but we shall not pursue the subject here. It will be clear that as soon as we proceed beyond the simple treatment of magnetochemistry, we enter a very difficult field. We can, however, take comfort in the fact that small differences between observed and predicted magnetic moments are seldom of much chemical significance, and we shall very rarely have occasion to refer to them again.

PROBLEMS

1 Discuss each of the following observations
 (a) The ion $[CoCl_4]^{2-}$ is a regular tetrahedron, but $[CuCl_4]^{2-}$ is a flattened tetrahedron.
 (b) The electronic spectrum of the $[CoF_6]^{3-}$ ion contains two bands with maxima at approximately 11 500 and 14 500 cm^{-1}.
 (c) Octahedral high-spin Ni(II) complexes have magnetic moments in the range 2.9–3.4 μ_B; tetrahedral Ni(II) complexes have moments of up to 4.1 μ_B; square planar Ni(II) complexes are diamagnetic.

2 Explain the forms of the *d* orbital splitting diagrams for trigonal bipyramidal and square pyramidal complexes of formula ML_5. What would you expect concerning the magnetic properties of such complexes of Ni(II)?

3 How many *d–d* bands would you expect to find in the electronic spectrum of an octahedral Cr(III) complex?

The colour of *trans* $[Co(en)_2F_2]^+$ is less intense than that of either *cis* $[Co(en)_2F_2]^+$ or *trans* $[Co(en)_2Cl_2]^+$. How do you account for this?

REFERENCES FOR FURTHER READING

Burdett, J.K. (1978) *Adv. Inorg. Chem. Radiochem.*, **21**, 113. An introduction to the Angular Overlap Method.

Earnshaw, A. (1968) *Introduction to Magnetochemistry*, Academic Press, London. A clear account of this difficult subject with due emphasis on experimental data.

Lever, A.B.P. (1984) *Inorganic Electronic Spectroscopy*, 2nd edition, Elsevier, Amsterdam. A full account of the subject.

Nicholls, D. (1974) *Complexes and First-Row Transition Elements*, Macmillan, London. Good discussion of basic theory with useful tables of experimental data.

Shriver, D.F., Atkins, P.W. and Langford, C.H. (1990) *Inorganic Chemistry*, Oxford University Press. Chapter 14 contains an advanced account of the material covered here and of some additional topics.

Sutton, D. (1968) *Electronic Spectra of Transition Metal Complexes*, McGraw-Hill, London. A good short account of this subject.

20 *Thermodynamic aspects of transition metal chemistry*

Introduction • Crystal field stabilisation energies of octahedral and tetrahedral complexes • Oxidation states in aqueous media • Ionisation energies

20.1 Introduction

We have already discussed at length the thermodynamics of the formation of ionic salts, covalent molecules and complexes in aqueous solution, and in this chapter we shall cover the same ground again only in so far as transition metal chemistry involves one more variable than the chemistry of the typical elements: the influence of crystal (or, in a broader sense, ligand) field stabilisation energy. In addition, however, we shall return to the topic of ionisation energies, discussed in Chapter 3, for a more detailed treatment. Some essentially thermodynamic matters will not be dealt with until later chapters, most of the material in this one being concerned with high-spin octahedral complexes of dipositive and tripositive cations in which π bonding plays only a minor part. More will be said about carbonyls and related compounds in Chapter 22. Bond energy terms in molecular compounds of the heavy transition metals will be mentioned in Chapter 25 (molecular compounds, other than those involving π bonding ligands, of the first series transition metals are very few in number and little is known about their thermochemistry). Finally, one interesting and potentially very important aspect of thermodynamics is discussed briefly in the following chapter, which is otherwise devoted to kinetics and mechanism. This is the attempt to calculate the contribution to the activation energy of a substitution at a transition metal atom of the loss or gain in crystal field stabilisation energy on going from the starting complex to a postulated transition state.

20.2 *Crystal field stabilisation energies of octahedral and tetrahedral complexes*

So far we have considered Δ_0 (or Δ_t) only as the quantity, derived from electronic spectroscopy, representing the energy required to transfer an electron from a t_{2g} to an e_g orbital (or from an e to a t_2 orbital). To the chemist, however, most of the interest in Δ_0 and Δ_t lies in the use that can be made of these quantities in explanations (or, if possible, predictions) in other fields of chemistry. We have already indicated that the crystal field stabilisation energies for high-spin octahedral systems are zero for d^0, d^5 and d^{10} electronic configurations, 0.4 Δ_0 for d^1 and d^6, 0.8 Δ_0 for d^2 and d^7, 1.2 Δ_0 for d^3 and d^8, and (if we ignore the consequences of the Jahn–Teller effect) 0.6 Δ_0 for d^4 and d^9. If we are considering only the extra stability that results from the splitting of the d orbitals in high-spin complexes, we can to a first approximation disregard the pairing energy term, since even in the free gaseous d^6–d^{10} ions pairing is inevitable, and so only the difference in pairing energy between that in the free ion and that in the complex is involved. For tetrahedral complexes, all of which are of the high-spin type, the crystal field stabilisation energies are zero for d^0, d^5 and d^{10} configurations, 0.6 Δ_t for d^1 and d^6, 1.2 Δ_t for d^2 and d^7, 0.8 Δ_t for d^3 and d^8, and 0.4 Δ_t for d^4 and d^9. Since, if all other factors are equal, $\Delta_t = \frac{4}{9}\,\Delta_0$, we can express tetrahedral crystal field stabilisation energies in terms of Δ_0; for a d^1 ion in a tetrahedral field, for example, the crystal field stabilisation energy is

$$0.6\,\Delta_t = 0.6 \times \tfrac{4}{9}\,\Delta_0 = 0.27\,\Delta_0$$

The corresponding calculations for other electronic configurations lead to the values given in Table 20.1, and a plot of the crystal field stabilisation energies for high-spin ions in both octahedral and tetrahedral complexes is shown in Fig. 20.1.

For d^4–d^7 ions in octahedral complexes we must also consider the low-spin electronic configurations, for which crystal field stabilisation energies and pairing energies (in terms of the average pairing energy per electron, P) are shown in Table 20.2; values for high-spin complexes are included for comparison. It will be clear that when spin-pairing takes place, there is a large increase in the crystal field stabilisation energy, though this is partly offset by pairing energy considerations. In practice it is often very difficult to interpret the electronic spectra of low-spin d^4–d^7 complexes unambiguously, and so we shall not say much about their crystal field stabilisation energies in this book; we should, however, draw attention to their relevance to the discussion of the complex cyanides of iron in Section

Table 20.1
Crystal field stabilisation energies for high-spin complexes (in terms of Δ_0).

Configuration	d^0	d^1	d^2	d^3	d^4	d^5	d^6	d^7	d^8	d^9	d^{10}
Octahedral	0	0.4	0.8	1.2	0.6	0	0.4	0.8	1.2	0.6	0
Tetrahedral	0	0.27	0.53	0.35	0.18	0	0.27	0.53	0.35	0.18	0

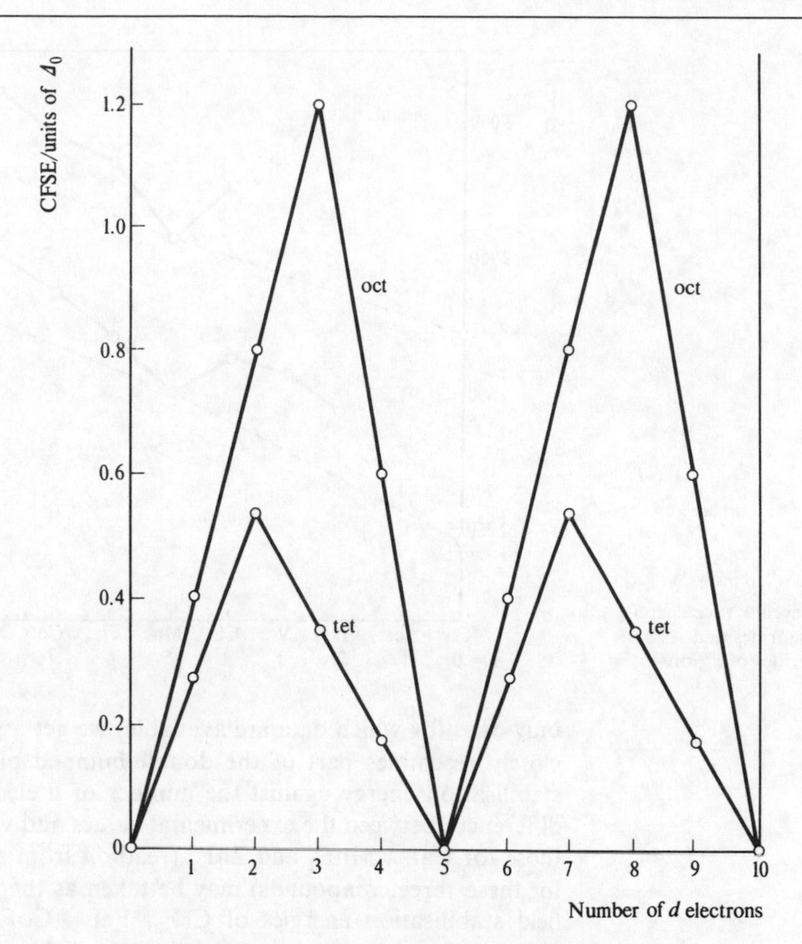

Figure 20.1
Crystal field stabilisation energies of high-spin ions in octahedral and tetrahedral fields.

24.7 and of the Co(III)/Co(II) standard potential in Section 24.8. For the present, however, we shall confine attention to simpler systems.

We shall begin by discussing the lattice energies of the difluorides of metals of the first transition series. As we saw in Chapter 6, lattice energies may be obtained either from the Born cycle (experimental values) or, by calculation, from X-ray and compressibility data (calculated values); for CaF_2, MnF_2 and ZnF_2 (d^0, high-spin d^5 and d^{10} systems), the values obtained by the two methods agree closely. If we plot the experimental lattice energies of CrF_2, MnF_2, FeF_2, CoF_2, NiF_2, CuF_2 and ZnF_2 (the

Table 20.2
Crystal field stabilisation energies for high-spin and low-spin d^4–d^7 ions in octahedral fields.

Configuration		d^4	d^5	d^6	d^7
Crystal field	High-spin	$0.6\Delta_0$	0	$0.4\Delta_0$	$0.8\Delta_0$
stabilisation energy	Low-spin	$1.6\Delta_0$	$2\Delta_0$	$2.4\Delta_0$	$1.8\Delta_0$
Pairing energy	High-spin	0	0	P	$2P$
	Low-spin	P	$2P$	$3P$	$3P$

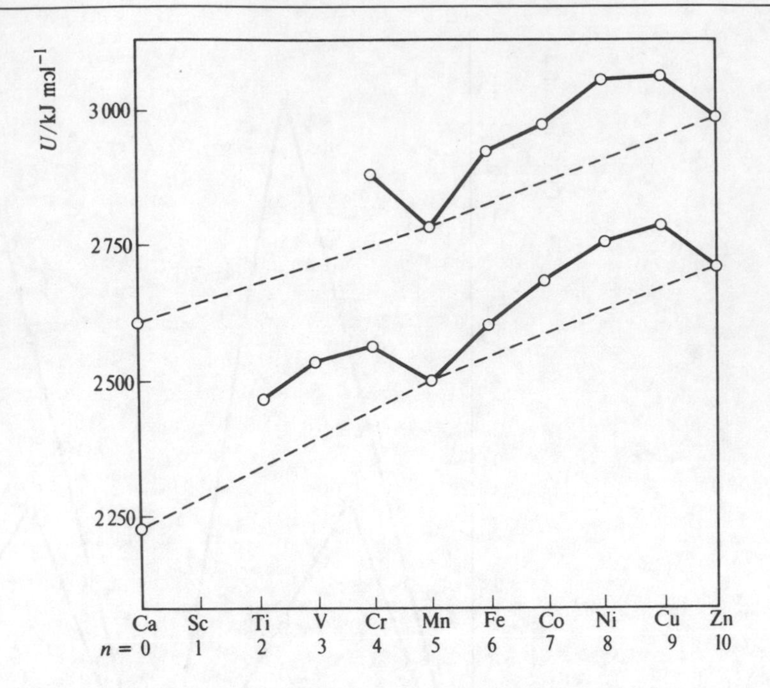

Figure 20.2
Born cycle lattice energies of difluorides and dichlorides of d^n ions.

only ones for which data are available) we get a graph (Fig. 20.2) which closely resembles part of the double-humped plot for the crystal field stabilisation energy against the number of *d* electrons in Fig. 20.1; the differences between the experimental values and values interpolated from those for CaF_2, MnF_2 and ZnF_2 (read off from a line joining the values for these three compounds) may be taken as the thermochemical crystal field stabilisation energies of CrF_2, FeF_2, CoF_2, NiF_2 and CuF_2. A similar treatment may be extended to the dichlorides (though since most of these have layer structures there is much less justification for working with the ionic model in this case), the monoxides (though these are often non-stoichiometric) and the trifluorides and their complexes of formula K_3MF_6 (in this case of the elements from scandium to cobalt). In Fig. 20.2 we show the variation in lattice energy against electronic configuration for the dichlorides as well as the difluorides.

As would be expected, the interionic distances in the difluorides vary in a manner which is approximately inversely related to that of the lattice energies. If we take the radius of the fluoride ion as 1.33 Å, the radii of the ions are: Ca^{2+}, 1.00; Cr^{2+}, 0.83; Mn^{2+}, 0.78; Fe^{2+}, 0.70; Co^{2+}, 0.71; Ni^{2+}, 0.68; Cu^{2+}, 0.71; Zn^{2+}, 0.71 Å. (Where the structures are distorted by Jahn–Teller effects, average values are given.)

As a second example of the relevance of crystal field stabilisation energy considerations to thermochemical data for transition metal complexes, we may consider the enthalpies of hydration of the dipositive ions. Like the difluorides, all hydrated dipositive ions of metals of the first transition series are high-spin complexes; and comparisons of the electronic spectra of the ions in solution and in solid salts (where the presence of $[M(H_2O)_6]^{2+}$ ions has been demonstrated by X-ray crystallography) show that the same regular (or Jahn–Teller distorted) octahedral ions are present in both. For this series of ions enthalpies of hydration may be obtained by the general methods discussed in Section 7.2; Fig. 20.3 again shows a portion of a double-humped curve, and again the departure from the interpolated value (using the values for Ca^{2+}, Mn^{2+} and Zn^{2+} as reference points) may be taken as a measure of the thermochemical crystal field stabilisation energy.

In general the agreement between thermochemical crystal field stabilisation energies obtained in this way and those calculated from the observed values of Δ_0 is fairly close; for the $[Ni(H_2O)_6]^{2+}$ ion, for example, the values are 120 kJ mol^{-1} and 126 kJ mol^{-1} (1.2 Δ_0, where $\Delta_0 = 8\,500$ cm^{-1}) respectively. Perhaps in this case the agreement is, in fact, better than it ought to be, firstly because only part of the enthalpy of hydration corresponds to interaction of the cation with its six nearest water molecules, and secondly because the two definitions of crystal field stabilisation energy are not really equivalent. The thermochemical one relates to a value for a species in equilibrium compared with the interpolated value for a species in equilibrium, the validity of the interpolation being taken for granted; the spectroscopic one, on the other hand, measures the energy difference between the ground state of the complex and an excited state in which an electron has been promoted to a higher energy orbital but the system has not had time to adjust itself to the new equilibrium interatomic distances (cf. Section 19.5). We make this point to emphasise that, interesting and useful as these discussions of the role of crystal field stabilisation energy in the thermodynamics of transition metal complexes are, they are never more than approximations. Further, crystal field stabilisation energy terms are only small parts of the

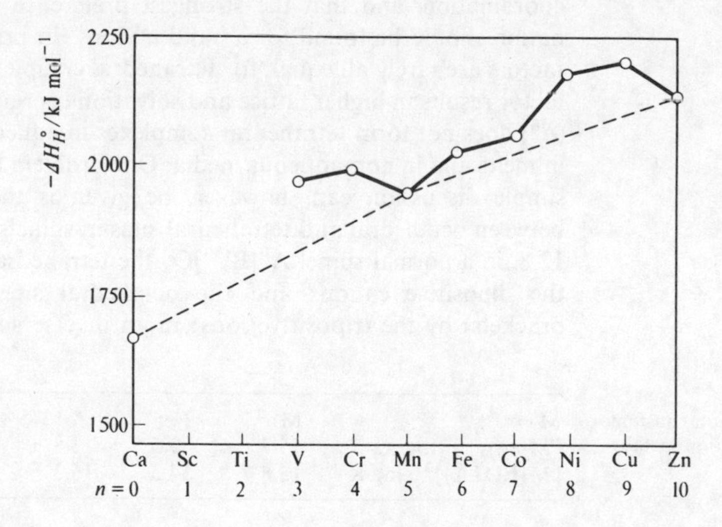

Figure 20.3
Absolute enthalpies of hydration of dipositve d^n ions.

total interaction energies, seldom being as great as 10 per cent. Nevertheless, it is very useful to be able to discuss the erratic variations in lattice energies or hydration enthalpies with *d* electron configuration as successfully as we have done on the basis of so simple a theory of the bonding in transition metal complexes.

Complex formation in aqueous solution, which we have chosen as a final illustration of the use of the crystal field stabilisation energy concept as applied to high-spin octahedral species, involves the displacement of water by ligands, so differences between two stabilisation energies are involved. Numerical data for the formation constants of tris(ethylenediamine) and $EDTA^{4-}$ complexes of the dipositive cations from Mn^{2+} to Zn^{2+} are given in Table 20.3; for a given ligand and a given cationic charge ΔS^0 should be nearly constant along this series, so the variation in log K should be parallel to that in $-\Delta H^0$. Again we see that the data would fit on roughly the same sort of double-humped curve that we have met before, the sequence being $Mn^{2+} < Fe^{2+} < Co^{2+} < Ni^{2+} \sim Cu^{2+} > Zn^{2+}$, indicating that water, ethylenediamine and the $EDTA^{4-}$ ion all interact with these cations in very much the same way; but since complexing in aqueous media is so much concerned with small differences between the interaction of the cation with water and with the ligand, and between the hydration of the resulting species, we shall not discuss the values given in Table 20.3 in detail.

There is little to be said about crystal field stabilisation energies of tetrahedral complexes, since very few quantitative data are available for them; further, many tetrahedral complexes certainly involve π bonding and so would not be accurately described by the simple ionic model. The question of octahedral versus tetrahedral coordination, nevertheless, is illuminated by crystal field stabilisation energy considerations. We see from Fig. 20.1 that, if all other factors are equal, d^0, high-spin d^5, and d^{10} ions should have no preference between octahedral and tetrahedral coordination, and that the strongest preference for octahedral coordination should be found for d^3 and d^8 ions. In practice, of course, other factors are rarely all equal, for tetrahedral complexes are smaller and this factor results in higher lattice and solvation energies; thus although Ni^{2+} (d^8) does not form tetrahedral complexes in aqueous solution, it does so in melts and in non-aqueous media. One problem for which a meaningful simple discussion can, however, be given is the distribution of ions between octahedral and tetrahedral sites in spinels. As we said in Section 12.8, in a normal spinel $A^{II}[B_2^{III}]O_4$ the tetrahedral sites are occupied by the dipositive cations and the octahedral sites (denoted by square brackets) by the tripositive ions; in an inverse spinel the distribution is

Table 20.3
Overall formation constants for some transition metal complexes.

M =	Mn^{2+}	Fe^{2+}	Co^{2+}	Ni^{2+}	Cu^{2+}	Zn^{2+}
$[M(en)_3]^{2+}$ log K	5.7	9.5	13.8	18.6	18.7	12.1
$[M(EDTA)]^{2-}$ log K	13.8	14.3	16.3	18.6	18.7	16.1

indicated by the formula $B^{III}[A^{II}B^{III}]O_4$. For spinel itself A = Mg, B = Al. If at least one of the cations is that of a transition metal, the inverse structure is frequently, though by no means always, found: $Zn^{II}Fe_2{}^{III}O_4$, $Fe^{II}Cr_2{}^{III}O_4$ and $Mn^{II}Mn_2{}^{III}O_4$ are normal spinels and $Ni^{II}Ga_2{}^{III}O_4$, $Co^{II}Fe_2{}^{III}O_4$ and $Fe^{II}Fe_2{}^{III}O_4$ are inverse spinels. In seeking to account for these observations we need to know that the Madelung constants for the two structures are usually very nearly equal, to assume that the actual charges on the ions are independent of their environments, and to recall that Δ_0 values for complexes of tripositive ions are substantially greater than those for the corresponding complexes of dipositive ions. Consider the compounds having the normal structure: in the first, neither ion has any crystal field stabilisation energy; in the second, Cr^{3+} has a much greater additional stabilisation energy in an octahedral site than Fe^{2+}; and in the third, only Mn^{3+} has any stabilisation energy, this being greater in an octahedral site. In the first of the compounds having the inverse structure, only Ni^{2+} has any crystal field stabilisation energy (and it has a strong preference for octahedral sites); in the second and third, only Co^{2+} and Fe^{2+} respectively have any stabilisation energy, and both have slight preferences for octahedral sites. Occasional exceptions to what would be expected on the basis of crystal field stabilisation energy calculations occur where the margin of preference for one structure is only small: $Fe^{II}Al_2{}^{III}O_4$, for example, is a normal spinel. In general, however, it can be said that this approach offers an interesting comment on what are, on ionic size considerations, some very puzzling observations.

20.3 Oxidation states in aqueous media

The factors which determine the values of standard electrode potentials, and the influence of hydrogen-ion concentration, complex formation and precipitation on the relative stabilities of oxidation states in aqueous solution, were discussed at length in Chapter 7. Further, we saw in the case of the alkali metals (Chapter 10) how, as the atomic and ionic sizes increase, the decreases in enthalpy of atomisation and ionisation energy are almost exactly offset by the decrease in hydration energy, so that E^0 for the $M^+(aq)/M$ system is nearly constant. Now there is an overall decrease in atomic and ionic size along a series of transition metals, and some of the irregularities in the variations of quantities that do not change smoothly with increase in atomic number (e.g. hydration energies) can be attributed to the effect of crystal field stabilisation energy; it is interesting, therefore, to see if we can account for the variation in E^0 for, say, the $M^{2+}(aq)/M$ system along the first transition series.

This turns out to be a difficult problem. Water is relatively easily oxidised or reduced, and the range of oxidation states on which measurements can be made under aqueous conditions is therefore somewhat restricted; scandium(II) and titanium(II), for example, would liberate hydrogen. The observed values for E^0 for M^{2+}(aq)/M are: V, −1.18; Cr, −0.91; Mn, −1.18; Fe, −0.44; Co, −0.28; Ni, −0.25; Cu, +0.34; Zn, −0.76 V. A more negative value denotes that M^{2+}(aq) is less easily reduced or, conversely, that M is more easily oxidised. There is (as we see from Fig. 20.3) an overall increase in the amount of energy liberated when the gaseous M^{2+} ion is hydrated as we move along the first transition series; there is also, we see from Table 20.4, a successive increase in the sum of the first two ionisation energies (from 1866 kJ mol^{-1} for scandium to 2640 kJ mol^{-1} for zinc) except for out-of-line values for manganese and copper. The enthalpies of atomisation (Fig. 6.16), however, vary not only erratically but also over a wide range (from 126 kJ mol^{-1} for zinc to 515 kJ mol^{-1} for vanadium). In so far as the variation in E^0 can be said to be reflected in that of any one of the three variables, it is probably in the sum of the ionisation energies (with the qualification that the very low enthalpy of atomisation of zinc puts it in a special position); it is clearly not worth while to discuss the relatively small variations in crystal field stabilisation energy in the context of the much larger variations in other relevant quantities.

If we turn to the variation in the M^{3+}/M^{2+} standard potential in aqueous solution, the enthalpy of atomisation of M is no longer relevant, and we are concerned only with the third ionisation energy of M and the hydration energies of M^{3+} and M^{2+}. Experimental values for this potential are restricted to elements in the middle of the series; as we mentioned earlier, scandium(II) and titanium(II) would reduce water, and we must now add that nickel(III), copper(III), and zinc(III) would oxidise it. But there is a very large variation in E^0 M^{3+}/M^{2+} even among the five elements for which experimental data are available, observed potentials being V, −0.26; Cr, −0.41; Mn, +1.51; Fe, +0.77; Co, +1.95 V. These values, together with the more negative but unknown ones for Sc and Ti, and the more positive but unknown ones for Ni, Cu and Zn, show a trend which is closely similar to that found in the third ionisation energies (see Table 20.4). This suggests that a steady overall increase in the difference between the hydration energies of M^{3+} and M^{2+} (which would become larger as the ions become smaller) is generally outweighed by the variation in third ionisation energy. The only pair of metals between which the change in E^0 is in the opposite direction to what would be expected on the

Table 20.4
Ionisation energies/kJ mol^{-1} of metals of the first transition series.

	Sc	Ti	V	Cr	Mn	Fe	Co	Ni	Cu	Zn
I_1	631	656	650	653	717	762	758	736	745	906
I_2	1235	1309	1414	1592	1509	1561	1644	1752	1958	1734
I_3	2393	2657	2833	2990	3260	2962	3243	3402	3556	3837

basis of the variation in the third ionisation energy is vanadium and chromium. The third ionisation energy of chromium is 157 kJ mol^{-1} greater than that of vanadium, so it is harder to oxidise Cr^{2+}(g) than V^{2+}(g); in aqueous solution, however, Cr^{2+} is a slightly more powerful reducing agent than V^{2+}. These two oxidations correspond to changes in electronic configuration of $d^3 \rightarrow d^2$ for V and $d^4 \rightarrow d^3$ for Cr. Since both cations of both elements in aqueous solution are present as hexa aquo complexes, all having the maximum possible number of unpaired electrons, there is a loss of crystal field stabilisation energy of $0.8\ \Delta_0^{III} - 1.2\ \Delta_0^{II}$ on oxidation of V^{2+} and a gain of crystal field stabilisation energy of $1.2\ \Delta_0^{III} - 0.6\ \Delta_0^{II}$ on oxidation of Cr^{2+} (the minor consequences of the Jahn–Teller effect being neglected in the latter case); Δ_0^{III} and Δ_0^{II} are the values of Δ_0 for the tri- and di-positive ions respectively. If we now make use of the spectroscopic values of Δ_0, we find that for the oxidation of V^{2+} ($\Delta_0^{II} = 12\,300$ cm^{-1}, $\Delta_0^{III} = 18\,600$ cm^{-1}), Δ(CFSE) is approximately zero, whereas for the oxidation of Cr^{2+}($\Delta_0^{II} = 14\,100$ cm^{-1}, $\Delta_0^{III} = 17\,000$ cm^{-1}), Δ(CFSE) is 11 900 cm^{-1}, which is equivalent to about 143 kJ mol^{-1}. Thus the difference in Δ (crystal field stabilisation energy) largely cancels that in ionisation energy, and the apparent anomaly in E^0 values is mostly accounted for – a considerable achievement in view of the simplicity of the theory used.

20.4 Ionisation energies

When we discussed ionisation energies in general in Section 3.4, we were content to stop at showing how they depended on the first and second quantum numbers of the electron being removed. An observant reader looking at Fig. 3.3 might have noticed that although the first ionisation energies of the elements from boron to neon (electronic configurations $2s^2 2p^1$ to $2s^2 2p^6$) for the most part increase between successive elements, this is not true of the values for nitrogen and oxygen; the fact that the latter has the lower ionisation energy means that the energy difference between the electronic configuration $2p^4$ and $2p^3$ is for some reason less than that between $2p^3$ and $2p^2$, i.e. there appears to be some sort of extra stability associated with the half-filled $2p$ shell. Similar observations may be made for other pairs of typical elements: the first ionisation energy of sulphur is less than that of phosphorus, and the second ionisation energy of fluorine is slightly less than that of oxygen. Among pairs of elements of higher atomic number, however, the reverse order may be found: the first ionisation energy of tellurium is greater than that of antimony, and the second ionisation energy of chlorine is greater than that of sulphur. Evidently other facts here override what may be called *the half-filled shell effect*.

Something of the same kind is found among the ground-state electronic configurations of transition metal atoms, inasmuch as that of chromium is $3d^5 4s^1$ rather than $3d^4 4s^2$, though in this case we are considering one electronic configuration compared with another containing the same number of electrons. Again, however, the effect is not general, for of the homologues of chromium in the second and third transition series, molybdenum has $4d^5 5s^1$ but tungsten $5d^4 6s^2$ as the ground state configuration.

When we consider doubly or more highly charged cations of the transition elements, however, the ground state is always of the form d^n until all the d electrons have been removed; most singly charged cations also have d^n ground states. Moreover, in other respects the variation in properties along a series of transition elements is much less than along a series of typical elements (a fact attributed to the relatively low screening power of occupied d orbitals, as a result of which one d electron does not shield another in a different orbital from the influence of the nucleus to any great extent). It is, therefore, not surprising to find that among these ions of the transition elements (and also, as we shall see in Chapters 26 and 27, among comparable f^n ions of the inner transition elements) the concept of there being some special stability associated with the d^5 (or f^7) electronic configuration is a generally valid and a very useful one. We may note in Table 20.4 that the steady increases in the second and third ionisation energies are broken at manganese and iron, in each case a d^5 ion (Mn^{2+} and Fe^{3+}) then being formed; similar breaks are found to occur at the corresponding elements in the later transition series.

Useful as this generalisation is, we must not carry it too far: it always requires very large amounts of energy to remove electrons from atoms, and the difference in ionisation energies between $nd^6 \rightarrow nd^5$ and $nd^5 \rightarrow nd^4$ processes for a given extent of ionisation, striking though it is, is much less than the difference between successive ionisation energies for any one atom. Thus we must not expect that because of the extra stability associated with, say, the $3d^5$ configuration that Co^{4+}, Ni^{5+} and Cu^{6+} will be able to be stable in ordinary chemical environments because they have this electronic configuration.

The interpretation of the variation in a given ionisation energy with electronic configuration is as follows for an electronic configuration d^n (a similar treatment may be given for f^n). First, we consider the attraction of each electron for the nucleus; if the energy of attraction is $-U$ per electron, the total attractive energy is $-nU$. Next, we take account of the repulsion between the n d electrons: any two electrons repel one another, and the total repulsive energy is approximately proportional to the number of ways of choosing two electrons out of a total of n, denoted by ${}^nC_2$. The total repulsive energy is then $+{}^nC_2 J$, where J is a proportionality constant. For a system that obeyed the classical laws of electrostatics, only these two terms would be involved. For an electronic system, however, there is a third term, the existence of which is responsible for the fact that when a number of electrons occupy a set of degenerate orbitals, the lowest energy state corresponds to the maximum possible extent of single occupation and parallel spins (Hund's first rule). This third term is a stabilisation term called the *exchange energy*, which is approximately proportional to the total number of possible pairs of parallel spins; if for the electronic configuration d^n this is m_n, the exchange energy is $-m_nK$, where K is another proportionality constant. Thus the total repulsive energy is given by

$$E(d^n) = -nU + {}^nC_2J - m_nK$$

On ionisation the d^n shell becomes d^{n-1} with

$$E(d^{n-1}) = -(n-1)U + {}^{n-1}C_2J - m_{n-1}K$$

The ionisation energy is then given by:

$$I = E(d^{n-1}) - E(d^n) = U + ({}^{n-1}C_2 - {}^nC_2)J - (m_{n-1} - m_n)K$$

If we have m identical objects and have to choose n of them, where $n < m$, the number of ways of choosing, ${}^mC_n$, is given by

$${}^mC_n = \frac{m!}{n!(m-n)!}$$

Here $n = 2$ and it is easily shown that

$${}^{n-1}C_2 - {}^nC_2 = -(n-1)$$

$$\text{Hence } I = U - (n-1)J + K\delta m$$

where $\delta m = m_n - m_{n-1}$, the decrease in the number of possible pairs of spin–spin interactions. The overall increase in any ionisation energy along a transition series shows that the rise in U with increase in n outweighs that in $(n-1)J$, i.e. the effect of increase in nuclear charge outweighs the repulsion between the d electrons. If we assume a smooth variation in each of these quantities with increase in n, $U - (n-1)J$ should increase smoothly with n, and the function $K\delta m$ should be superimposed on this increase. If a d^n configuration has x electrons with spin $+\frac{1}{2}$ it must have $n-x$ with spin $-\frac{1}{2}$, and the number of pairs of parallel spin interactions, m, is given by

$$m = {}^xC_2 + {}^{n-x}C_2$$

Table 20.5
The number, m, of pairs of parallel spins for d^n configurations and the decrease on ionisation.

n	m	δm
0	0	—
1	0	0
2	1	1
3	3	2
4	6	3
5	10	4
6	10	0
7	11	1
8	13	2
9	16	3
10	20	4

The number of possible pairs of parallel spins for d^n configurations and their decrease on ionisation is shown in Table 20.5. As we increase n from 1 to 5, δm increases from 0 to 4, but at $n = 6$, δm drops back to 0 before increasing to 4 again. Thus for $d^n \rightarrow d^{n-1} + e$ there is no loss of exchange energy at $n = 6$, and this is why Mn^+ and Fe^{2+} have lower ionisation energies than Cr^+ and Mn^{2+} respectively. These conclusions are represented graphically in Fig. 20.4, in which ABC is the pattern of

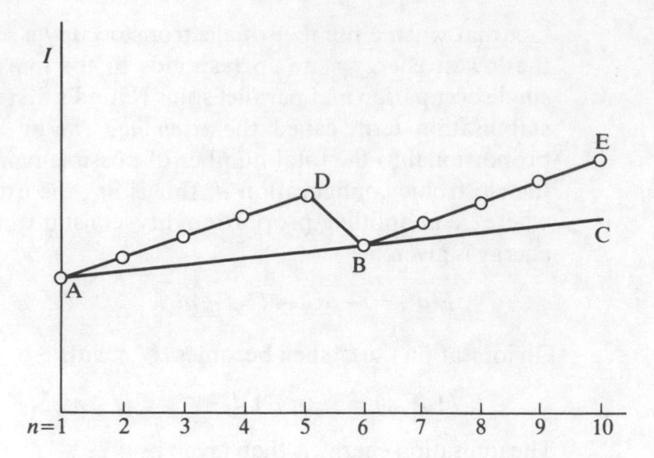

Figure 20.4
The predicted variation in ionisation energy with d^n electron configurations.

ionisation energies to be expected if exchange energy considerations are neglected, and ADBE is the pattern to be expected if they are taken into account. Inspection of the data in Table 20.4 shows that where ionisation corresponds to loss of a *d* electron (most second and all third ionisations), the latter pattern is always observed.

PROBLEMS

1 The overall formation constant of the $[Co(NH_3)_6]^{2+}$ ion in aqueous solution is 10^5 and the standard potentials for the reduction of Co^{3+}(aq) and $[Co(NH_3)_6]^{3+}$(aq) are as follows:

$Co^{3+}(aq) + e \rightleftharpoons Co^{2+}(aq)$ $\qquad E^\circ = +1.9$ V

$[Co(NH_3)_6]^{3+}(aq) + e \rightleftharpoons [Co(NH_3)_6]^{2+}(aq)$ $\qquad E = +0.1$ V

Calculate the overall formation constant of the $[Co(NH_3)_6]^{3+}$ ion.

2 Values of Δ_0 for the high-spin cations $[Ni(H_2O)_6]^{2+}$ and $[Mn(H_2O)_6]^{3+}$ have been evaluated spectroscopically as 8400 and 21 000 cm^{-1} respectively. On the assumption that these values hold also for the corresponding oxide lattices, predict whether $Ni^{II}Mn_2{}^{III}O_4$ should have the normal or inverse spinel structure. What factors might make your prediction unreliable?

3 Discuss each of the following observations:

(a) Although Co^{2+}(aq) forms a tetrahedral tetrachloro complex on treatment with concentrated hydrochloric acid, Ni^{2+}(aq) does not do so.

(b) E° for the half-reaction

$$[Fe(CN)_6]^{3-} + e \rightleftharpoons [Fe(CN)_6]^{4-}$$

depends on the pH of the solution in which the measurement is made, being most positive in strongly acidic solution.

(c) E° for the Mn^{3+}/Mn^{2+} couple is much more positive than that for Cr^{3+}/Cr^{2+} or Fe^{3+}/Fe^{2+}.

REFERENCES FOR FURTHER READING

Johnson, D.A. (1982) *Some Thermodynamic Aspects of Inorganic Chemistry*, 2nd edition, Cambridge University Press. Another version of substantially the same material as this chapter, though with a wider range of examples than we have given.

The Open University (1976) S304 Unit 15, *The Stabilities of Oxidation States*, Open University Press, Milton Keynes. Covers much the same ground as this chapter but with slightly different emphasis.

21 *Kinetic aspects of transition metal chemistry*

21.1 Introduction

Most of this chapter will be taken up with the discussion of reactions which are either *substitutions* in which change in the coordination of the transition metal is not accompanied by change in oxidation state, or are *oxidation–reduction reactions* in which electrons are transferred between two metal atoms without change in the coordination of either; as we shall see at the end of the chapter, however, there are some oxidation–reduction reactions that involve both processes. Further, we shall also have occasion to mention some intramolecular changes (which may be regarded as internal substitution processes) and some reactions which turn out to be really those of the coordinated ligands. In general, though, reactions of the last kind, particularly where the ligands are organic, will not be dealt with in this book.

In earlier chapters, kinetics and mechanisms have been discussed only under individual elements, since a classification and unified treatment of inorganic reactions is a very difficult task; with the transition metals, however, we reach a series of elements for which it is possible to study the effects of the electronic configuration of the metal ion in a series of complexes which are in most respects very closely similar. From such studies some generalisations emerge, and it is interesting to see how the simple crystal field theory once again throws light on the subject even though we recognise it to be an oversimplification.

Sixty years ago, the classic studies of the kinetics and mechanisms of substitution at a saturated carbon atom, some of the best understood processes in the whole of chemistry, were already under way. Partly because most inorganic reactions are too fast or too slow for investigation by conventional physical methods, and partly because substitution at an octahedrally coordinated metal atom (the commonest case in transition metal chemistry) offers many more stereochemical possibilities than substitution at a tetrahedrally coordinated carbon atom, progress in the elucidation of the mechanisms of substitution in metal complexes was for many years relatively slow. Much the same was true of electron-transfer reactions, especially those between two complexes of the same metal in different oxidation states, such as hexacyanoferrate(II) and hexacyanoferrate(III). With the development of relaxation methods for studying fast reactions, however, the scope of inorganic kinetics has been greatly increased. In this book we shall not be much concerned with the methods used (which are, of course, not specific to inorganic chemistry), but mainly with the results obtained and the interpretation of them; details of the experimental techniques may be found in the references for further reading given at the end of the chapter.

21.2 Substitution at a metal atom: some general considerations

The first point to be made in this section is that the true nature of a reaction is not always obvious from the stoichiometric equation. For example, the reaction

$$[(H_3N)_5Co(CO_3)]^+ + 2H_3O^+ \rightarrow [(H_3N)_5Co(H_2O)]^{3+} + CO_2 + 2H_2O$$

looks at first sight like the substitution of coordinated water for carbonate, but the use of ^{18}O-labelled water as solvent shows that all the oxygen in the aquo complex formed is derived from the carbonate. The reaction is therefore really one of substitution at the carbonate oxygen atom which is bonded to the metal:

$$[(H_3N)_5CoOCO_2]^+ + H_3O^+ \rightarrow \left[(H_3N)_5Co{-}O\begin{matrix} H \\ \\ CO_2^- \end{matrix}\right]^{2+} + H_2O$$

$$\downarrow$$

$$[(H_3N)_5Co(H_2O)]^{3+} + H_2O \xleftarrow{H_3O^+} [(H_3N)_5CoOH]^{2+} + CO_2$$

Another example is provided by the exchange of ^{59}Fe between the two cyanide complexes of iron:

$$[^{56}Fe(CN)_6]^{3-} + [^{59}Fe(CN)_6]^{4-} \rightarrow [^{59}Fe(CN)_6]^{3-} + [^{56}Fe(CN)_6]^{4-}$$

Both of these complexes are kinetically very inert towards substitution (as may be shown by the extreme slowness of their exchange of cyanide with aqueous $^{14}CN^-$), and what is described as exchange of iron is in fact only electron-transfer between the complexes.

There is no connection between the thermodynamic stability of a complex and its lability towards substitution. The standard free energies of hydration of the Cr^{3+} and Fe^{3+} ions, for example, are almost exactly equal, yet $[Cr(H_2O)_6]^{3+}$ (d^3) undergoes substitution very slowly and $[Fe(H_2O)_6]^{3+}$ (high-spin d^5) very rapidly; similarly, although the overall formation constant of $[Hg(CN)_4]^{2-}$ is greater than that of $[Fe(CN)_6]^{4-}$, the former rapidly exchanges cyanide with labelled aqueous cyanide and the latter does so extremely slowly. We can never satisfactorily establish the mechanism of a reaction that does not proceed at a rate fast enough for investigation of its kinetics, but, as we shall see, there is now good reason to ascribe the kinetic inertness of d^3 (and low-spin d^6) octahedral complexes to crystal field stabilisation energy factors. The kinetic lability of $[Fe(H_2O)_6]^{3+}$ and $[Hg(CN)_4]^{2-}$ appears to be due to rapid reversible dissociation into complexes in which the metals have lower coordination numbers.

The general practice in the use of the terms *labile* and *inert* (strictly, *kinetically inert*) as applied to metal complexes is to describe as labile those complexes which undergo reactions having half-lives of about one minute or less; others are described as inert. Occasionally this classification is an oversimplification, since reactivity may depend upon the reagent: some ammine complexes of cobalt(III), for example, are decomposed rapidly by alkali but very slowly by water or acid, even when all the possible reactions involved are thermodynamically favourable. And even reactions of ammine complexes with water may be rapid when more than two different ligands are present: $[Co(NH_3)_3(H_2O)Cl_2]^+$, for example, is converted into $[Co(NH_3)_3(H_2O)_2Cl]^{2+}$ immediately upon dissolution in water, but $[Co(NH_3)_5Cl]^{2+}$ and *cis* and *trans* $[Co(en)_2Cl_2]^+$ undergo only slow substitution of chloride by water. In order to simplify the kinetics, most studies of substitution in metal complexes have been made under conditions such that only one ligand is replaced, i.e. most work has been done on complexes of general formula ML_5X, where X is the ligand replaced. Some caution should perhaps be exercised in applying conclusions based on the study of unsymmetrical complexes to symmetrical ones.

For various reasons, though chiefly because of their rapidity, substitution reactions of tetrahedral transition metal complexes have been little studied. Square planar complexes studied have mostly been those of platinum(II), for which it has been found that the incoming ligand always replaces the outgoing one without any other change in the configuration at the metal atom. For coordination numbers greater than four, substitution possibilities are more numerous, and detailed studies often show that more than one process may be involved. Thus in the reaction between (−) *cis* $[Co(en)_2Cl_2]^+$ and hydroxide ion, the products are *trans* $[Co(en)_2Cl(OH)]^+$ (63 per cent), (−) *cis* $[Co(en)_2Cl(OH)]^+$ (21 per cent), and (+) *cis* $[Co(en)_2Cl(OH)]^+$ (16 per cent). These results can best be interpreted on the basis of the formation of a trigonal bipyramidal $[Co(en)_2Cl]^{2+}$ cation which is attacked by hydroxide ion in three different ways.

Since charged species are very often involved in reactions of complexes, water is frequently the only possible solvent for all the reactants and products; studies are then being carried out in the presence of a 55 molar concentration of a very good ligand whose intervention is not discernible from the rate equation. Suppose, for example, that for a reaction

$$L_nMX + Y \rightarrow L_nMY + X$$

the rate is proportional to $[L_nMX]^1[Y]^0$, where the square brackets denote concentrations. Then *either*

$$L_nMX \xrightarrow[k_1]{\text{slow}} L_nM + X$$

$$L_nM + Y \xrightarrow{\text{fast}} L_nMY$$

or

$$L_nMX + H_2O \xrightarrow[k_1]{\text{slow}} L_nM(H_2O) + X$$

$$L_nM(H_2O) + Y \xrightarrow{\text{fast}} L_nMY + H_2O$$

may represent the mechanism of the reaction, though k_1 in the former mechanism is greater than that in the latter mechanism by a factor of 55. Further, if ions are formed or disappear during the reaction, the consequent variation in the ionic strength of the solution may have some effect on the rate. This difficulty is usually circumvented by working in a fairly concentrated solution of an inert electrolyte, e.g. molar aqueous $NaClO_4$. Ion-pairing may complicate the interpretation of kinetic observations, especially in concentrated solutions or in solvents of lower dielectric constant than water; thus the use of such solvents in an attempt to avoid some of the problems that arise from working in aqueous media, or to permit studies to be made at low temperatures, often introduces fresh difficulties.

For substitution at a metal atom two terminologies are in current use. In the common case where the attacking entity carries a negative charge or a lone pair of electrons and would in organic chemistry be termed a nucleophile, reactions are classed as belonging to one of the following types:

Evidence for an intermediate of lower coordination number	Dissociative (D) ≡ limiting SN1
Evidence for an intermediate of higher coordination number	Associative (A) ≡ limiting SN2
No evidence for an intermediate	Interchange (I) ≡ SN1 or SN2 (I_d) (I_a)

The symbols D, A and I are often used by workers in the field of coordination chemistry. In elementary textbooks, however, the SN1 or SN2 (substitution, nucleophilic, uni- or bi-molecular) terminology is usually preferred in order to facilitate comparisons with organic chemistry; it is for this reason that from time to time we shall employ it here.

On the basis of simple electrostatic arguments and the idea that the important features of dissociative and associative processes are bond-breaking and bond-making respectively, we can make some predictions about the expected effects on the reaction rate of changing certain variables. Thus increasing the positive charge on the central atom should decrease the rate of a dissociative process but increase the rate of an associative one. Since the rate-determining step in a dissociative process does not involve the attacking ligand, increasing the negative charge on, or decreasing the size of, the latter should have no effect on the rate of a dissociative process, but should increase the rate of an associative one. Increasing the negative charge and the size of the non-labile ligands should both increase the rate of a dissociative process and decrease the rate of an associative one. Enthalpies and entropies of activation, obtained from temperature variations of rate constants, may shed light on reaction mechanisms, but for many reactions, accurate measurements have been made only at one temperature and no data for these activation parameters are available. In recent years there has been considerable interest in the pressure dependence of rate constants, leading to volumes of activation: a reaction in which the transition state has a greater volume than the initial state shows a positive $\Delta V^{\ddagger}$, and the reverse is true for a reaction in which the transition state is compressed relative to the reactants. After allowance for any change in volume of the solvent (which is important if solvated ions are involved), the sign of $\Delta V^{\ddagger}$ should then in principle distinguish between an associative and a dissociative mechanism. All these considerations are sometimes used in assigning mechanisms, as we shall see in later sections of this chapter.

21.3 Substitution in octahedral complexes

Most of the work in this field that has been carried out by classical methods involves complexes of chromium(III) (d^3) or of low-spin cobalt(III) (d^6), and it is with these that we shall mostly be concerned, since an appreciation of kinetic and mechanistic considerations is essential for even a basic understanding of the chemistry of the two metals. But before we turn to the detail of their reactions we should discuss one very important process that has been studied for a wide range of cations, the substitution of coordinated water, either by isotopically labelled water or by other ligands. For most cations the rate of substitution is nearly independent of the nature of the incoming ligand, a fact which suggests exchange is predominantly dissociative (SN1) in character. The first-order rate constants for the exchange of a coordinated water molecule in some hydrated cations are shown in Fig. 21.1 in which, for purposes of comparison, we have included some non-transition metal ions and also some ions in which the coordination is probably not octahedral (e.g. La^{3+}). The value for Cr^{3+} is about 10^{-6} s^{-1}. We see that within a group of typical elements, the rate constant increases with increasing cationic (crystal) radius; further, comparison of ions of similar radii (such as Li^+, Mg^{2+} and Ga^{3+}) shows that increase in ionic charge retards substitution, in line with the expectations outlined in the last section. Among dipositive cations of the first transition series, however, there is no correlation with ionic size, and we deduce that electronic configurations must play a decisive part. The limited data for tripositive transition metal ions support a similar conclusion for these species, too. We shall return to this point

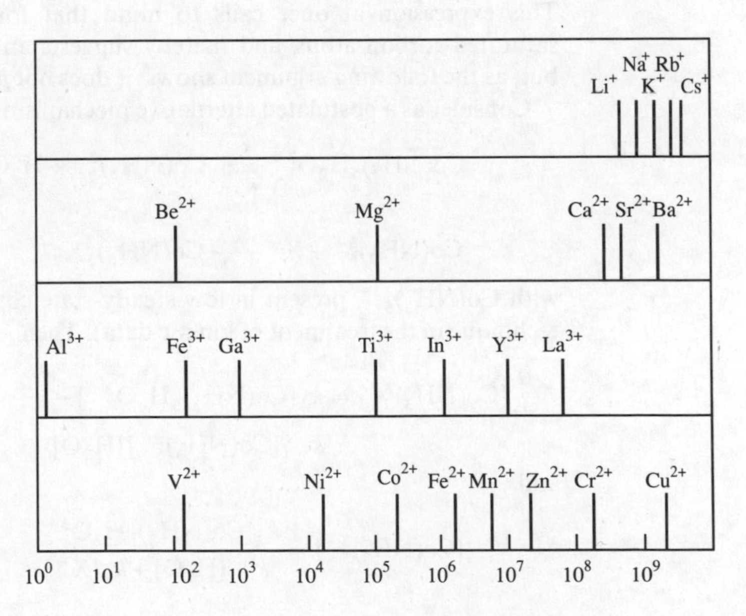

Figure 21.1
First-order rate constants for the exchange of coordinated water in hydrated metal ions (in s^{-1}).

later. Before we leave solvent exchange we should, however, mention that $\Delta V^{\ddagger}$ is slightly negative for V^{2+} and Mn^{2+} (and also for Fe^{3+}) but slightly positive for Fe^{2+}, Co^{2+}, Ni^{2+} and Cu^{2+}, implying a change from associative to dissociative mechanism. This implication has not met with universal approval, but at least it seems certain that bond-making becomes less, and bond-breaking more, important as we move along the series.

In the remainder of this section we shall be concerned with the displacement of water by a ligand (*anation*) or the reverse process *hydrolysis*), concentrating on reactions in which only one molecule of water or entering ligand is involved:

$$L_5M(H_2O)+X \underset{\text{hydrolysis}}{\overset{\text{anation}}{\rightleftharpoons}} L_5MX+H_2O$$

For clarity we shall for the present use square brackets only to denote concentrations.

Anation of cobalt(III) complexes

For the reaction

$$Co(NH_3)_5H_2O^{3+}+X^- \rightarrow Co(NH_3)_5X^{2+}+H_2O$$

where X is one of a number of singly charged anions, it is found that the rate law is

$$\frac{d}{dt}[Co(NH_3)_5X^{2+}]=k[Co(NH_3)_5H_2O^{3+}][X^-]$$

This expression at once calls to mind that for a SN2 reaction at a saturated carbon atom, and thereby suggests an associative mechanism but, as the following argument shows, it does not *prove* this.

Consider as a postulated alternative mechanism

$$Co(NH_3)_5H_2O^{3+} \underset{k_{-1}}{\overset{k_1}{\rightleftharpoons}} Co(NH_3)_5^{3+}+H_2O$$

$$Co(NH_3)_5^{3+}+X^- \xrightarrow{k_2} Co(NH_3)_5X^{2+}$$

with $Co(NH_3)_5{}^{3+}$ present in low steady-state concentration (a common technique in the treatment of kinetic data). Then

$$\frac{d}{dt}[Co(NH_3)_5^{3+}]=k_1[Co(NH_3)_5H_2O^{3+}]-$$
$$k_{-1}[Co(NH_3)_5^{3+}][H_2O]-k_2[Co(NH_3)_5^{3+}][X^-]=0$$

and

$$[Co(NH_3)_5^{3+}]=\frac{k_1[Co(NH_3)_5H_2O^{3+}]}{k_{-1}[H_2O]+k_2[X^-]}$$

Therefore

$$\frac{d}{dt}[Co(NH_3)_5X^{2+}] = k_2[Co(NH_3)_5^{3+}][X^-]$$
$$= \frac{k_1k_2[Co(NH_3)_5H_2O^{3+}][X^-]}{k_{-1}[H_2O]+k_2[X^-]}$$

If $k_{-1}[H_2O] \gg k_2[X^-]$, as it often will be in dilute aqueous solution, this reduces to

$$\frac{d}{dt}[Co(NH_3)_5X^{2+}] = \frac{k_1k_2[Co(NH_3)_5H_2O^{3+}][X^-]}{k_{-1}[H_2O]}$$

corresponding to second-order kinetics as observed. Thus the observed rate law does not tell us the mechanism of the reaction.

We can, however, obtain indirect evidence in support of our postulated dissociative mechanism. First, for $X^- = Cl^-$, Br^-, NCS^-, N_3^-, NO_3^- and NH_3, the value of k varies only by a factor of two along the series; if the mechanism was associative, a much bigger variation would be expected. Second, if $k_2[X^-] \gg k_{-1}[H_2O]$

$$\frac{d}{dt}[Co(NH_3)_5X^{2+}] = k_1[Co(NH_3)_5H_2O^{3+}]$$

corresponding to a reaction which is first order with respect to the concentration of the complex, and zero order with respect to $[X^-]$, the rate constant being the same as that for the loss of water from the complex. Such a reaction has not been found for a cationic complex (which would in any case form ion-pairs at very high anion concentrations), but it is found in the case of substitutions into the anionic complex $[Co(CN)_5H_2O]^{2-}$:

$$Co(CN)_5H_2O^{2-} + X^- \rightarrow Co(CN)_5X^{3-} + H_2O$$

In this reaction, in which ion-pairing is impossible because both ions are negatively charged, first-order kinetics are found and the rate constant is nearly the same as that of the $[Co(CN)_5H_2O]^{2-} - H_2{}^{18}O$ exchange; the dissociative (limiting SN1) mechanism is thus supported. Further, a positive value of $\Delta V^{\ddagger}$ for anation of $[(Co(NH_3)_5H_2O]^{3+}$ when $X = Cl^-$ or SO_4^{2-} also indicates a dissociative reaction. Overall, therefore, the balance of evidence is strongly in favour of a dissociative mechanism for the anation of the cobalt(III) complexes mentioned here. Extensive studies of complex formation from $[Ni(H_2O)_6]^{2+}$ support a similar conclusion for this species. We must not, however, rush to conclude that anation reactions involving other metal complexes are all dissociative. Negative volumes of activation for anation of $[Cr(NH_3)_5H_2O]^{3+}$ may indicate associative mechanisms in this case.

Hydrolysis of cobalt(III) complexes

Whether hydrolysis of a complex $[Co(NH_3)_5X]^{2+}$ yields $[Co(NH_3)_5H_2O]^{3+}$ or $[Co(NH_3)_5OH]^{2+}$ depends upon the hydrogen-ion concentration of the solution, the aquo complex being produced in acidic solution and the hydroxo complex in basic solution. The general rate law may be expressed by the equation

$$\frac{d}{dt}[Co(NH_3)_5X^{2+}] = k_A[Co(NH_3)_5X^{2+}] + k_B[Co(NH_3)_5X^{2+}][OH^-]$$

where $k_B = 10^4\text{–}10^8 k_A$, k_A and k_B being the rate constants for acidic and basic hydrolysis respectively. It is generally found that cobalt ammine complexes are much more reactive towards bases than towards acids.

At low pH values, the second term disappears. The rate law is then compatible with either dissociation into $Co(NH_3)_5^{3+}$ and X^- or replacement of X^- by H_2O (present in effectively constant high concentration) as the rate-determining step. Evidence for a dissociative mechanism in the case $X = H_2O$ is provided by the observation that the rate of exchange of water between $[Co(NH_3)_5H_2O]^{3+}$ and $H_2{}^{18}O$ decreases at high pressures. Further, a large increase in rate when NH_3 as ligand in $[Co(NH_3)_5Cl]^{2+}$ is replaced by butylamine also indicates a dissociative mechanism, and the same is true for another substantial increase on going from the doubly charged $[Co(NH_3)_5Cl]^{2+}$ to the singly charged species *trans*-$[Co(NH_3)_4Cl_2]^+$.

At pH values above about 5, the first term disappears, and the rate of the reaction is then given by

$$-\frac{d}{dt}[Co(NH_3)_5X^{2+}] = k_B[Co(NH_3)_5X^{2+}][OH^-]$$

Again this might appear to indicate an associative (SN2) mechanism, but two observations provide evidence against this. First, if ammonia is replaced as ligand by pyridine or another tertiary amine, the rate of basic hydrolysis is very much reduced; second, the exchange of hydrogen for deuterium in alkaline D_2O is much faster than the rate of basic hydrolysis. Both of these observations are compatible with (though of course they cannot rigidly establish) the mechanism as, for example, when $X = Cl$,

$$Co(NH_3)_5Cl^{2+} + OH^- \underset{k_{-1}}{\overset{\text{fast}\; k_1}{\rightleftharpoons}} Co(NH_3)_4(NH_2)Cl^+ + H_2O$$

$$Co(NH_3)_4(NH_2)Cl^+ \xrightarrow[k_2]{\text{slow}} Co(NH_3)_4(NH_2)^{2+} + Cl^-$$

$$Co(NH_3)_4(NH_2)^{2+} + H_2O \xrightarrow{\text{fast}} Co(NH_3)_5OH^{2+}$$

Application of the steady-state treatment to $Co(NH_3)_4(NH_2)Cl^+$ gives

$$[Co(NH_3)_4(NH_2)Cl^+] = \frac{k_1[Co(NH_3)_5Cl^{2+}][OH^-]}{k_{-1}[H_2O]+k_2}$$

and

$$\frac{d}{dt}[Co(NH_3)_5OH^{2+}] = k_2[Co(NH_3)_4(NH_2)Cl^+]$$

$$= \frac{k_1k_2[Co(NH_3)_5Cl^{2+}][OH^-]}{k_{-1}[H_2O]+k_2}$$

$$= k[Co(NH_3)_5Cl^{2+}][OH^-]$$

with

$$k = \frac{k_1k_2}{k_{-1}[H_2O]+k_2}$$

Thus rapid reversible ionisation of the complex $[Co(NH_3)_5Cl]^{2+}$, followed by a rate-determining slow loss of chloride from the amide complex (the conjugate base of the starting material) leads to second-order kinetics. This is the SN1 conjugate base (SN1 CB) dissociative mechanism. Why the conjugate base loses chloride ion much faster than the parent pentammine complex is not certain, but it may well be the consequence of the smaller positive charge on the complex.

It seems, therefore, that in both anation and hydrolysis, cobalt ammine complexes probably react by dissociative mechanisms, and the current position appears to be that no case of an associative (SN2) process at an octahedrally coordinated cobalt(III) atom has yet been established.

21.4 The racemisation of octahedral complexes

The particular interest of racemisation reactions lies in the fact that the variation of optical activity with time can be studied independently of the rate of exchange with a labelled ligand, thus providing extra information about the reaction pathway. The octahedron is stereochemically a rigid structure, and it is likely that for most optically active complexes involving only simple ligands it is easier to break a bond than to transform the octahedron into a symmetrical trigonal prism by a twisting mechanism: this state of affairs is rather like that in organic chemistry, where it is easier to break a bond in a carbon compound $CR_1R_2R_3R_4$ (in either a SN1 or a SN2 reaction) than to transform the molecule into an optically inactive planar form. By way of contrast, the racemisation of a pyramidal molecule requires only inversion (hence the failure to resolve tertiary amines $NR_1R_2R_3$), and for transition metal complexes there is often little difference in energy between planar and tetrahedral structures (as we saw in Section 18.3), so that molecular vibrations will suffice, as they will also do for a trigonal bipyramidal or a square pyramidal molecule, to bring about loss of optical activity. Whether there are any truly intramolecular racemisation processes for octahedral complexes that do not involve bond breaking is therefore an interesting question.

Often there can be no doubt that bond-breaking is involved. The rate of racemisation of (+) *cis* $[Co(en)_2Cl_2]^+Cl^-$ in methanol, for example, is equal to the rate of exchange with labelled chloride in the solution, and the product is a mixture of 70 per cent *trans* complex and 30 per cent racemic *cis* complex. Further, the rate of loss of optical activity is the same as the rates of substitution of one chloride ion in the complex by bromide, thiocyanate or nitrate. In this case a symmetrical five-coordinated intermediate is therefore involved, and in the first reaction it recombines with chloride ion at different positions at slightly different rates. For $[Ni(phen)_3]^{2+}$ and $[Ni(bipy)_3]^{2+}$ also, the rates of exchange with ^{14}C-labelled ligand have been found to be the same as the rates of racemisation; either the intermediate bis complex is symmetrical or it racemises faster than it recombines with the ligand.

For $[Cr(C_2O_4)_3]^{3-}$, $[Co(C_2O_4)_3]^{3-}$ and the low-spin complex $[Fe(bipy)_3]^{2+}$, on the other hand, the rate of racemisation is greater than that of ligand exchange. This still does not require racemisation without bond-breaking, however; if one end of a bidentate ligand becomes free with formation of a symmetrical five-coordinated intermediate, followed by recombination to give equal amounts of the (+) and (−) complexes, this would have the same effect. In aqueous solution, at least, this appears to be the route by which $[Cr(C_2O_4)_3]^{3-}$ racemises. This ion exchanges all its oxygen with $H_2{}^{18}O$ much faster than it exchanges its carbon with $^{14}C_2O_4^{2-}$ and at approximately the same rate as it racemises. But $K_3[Co(C_2O_4)_3].3H_2O$ also racemises in the solid state, and the rate of racemisation increases at high pressure (indicating that the formation of the transition state is accompanied by a decrease in volume). Even here, however, the possibility that the lattice water is somehow involved in the racemisation cannot be entirely discounted, and until it is shown that an anhydrous compound racemises under similar conditions it seems premature to accept that a totally intramolecular process involving only the twisting of opposite faces of an octahedron has been established for the racemisation of a regular octahedral complex.

21.5 *The correlation of rates of substitution in octahedral complexes with electronic configuration*

We may summarise the experimental information given so far about rates of reactions of octahedral complexes as follows. Reactions of complexes of low-spin cobalt(III) and of chromium(III) are particularly slow, and those of low-spin iron(II) and of some nickel(II) complexes are slow enough to be followed by conventional methods. By the use of fast reaction techniques it has been shown that substitution of water in octahedral aquo complexes (all high-spin where the high-spin or low-spin classification is relevant) is:

$$V^{2+} < Ni^{2+} < Co^{2+} < Fe^{2+} < Mn^{2+} < Zn^{2+} < Cr^{2+} < Cu^{2+}$$

and

$$Cr^{3+} \ll V^{3+} < Fe^{3+} < Ti^{3+}$$

Early in the attempts to correlate reactivity and electronic configuration it was believed that, for a first transition series ion, a vacant 3d orbital (to accept electrons from the substituting ligand) was a prerequisite for rapid substitution; labile complexes had to be derivatives of d^0, d^1 or d^2 cations, or to be of the outer orbital type. This idea, however, presupposed that substitution is an associative process involving an increase in coordination number and, as we have seen, there is now a large amount of evidence to suggest that this is not usually so. What we must do, therefore, is to consider also the implications of a decrease in coordination number on the likely reaction rate. For a series of ions of the same charge and about the same size undergoing the same reaction by the same mechanism, we may reasonably suppose that collision frequencies and entropies of activation will be nearly constant, and that the variations in rates will arise from variations among the enthalpies of activation. We assume that these are caused mainly by the loss or gain of crystal field stabilisation energy on going from the original complex to the transition state: a loss of crystal field stabilisation energy means an increase in the activation energy for the reaction and hence a decrease in its rate. Now we have already seen that for different geometries the splitting of the d orbitals occurs in different ways (Fig. 19.7) and we can, for example, calculate the change in crystal field stabilisation energy on formation of a square pyramidal or pentagonal bipyramidal complex from the original octahedral one if we assume that there is no change in the number of unpaired electrons, metal–ligand distance, or strength of the ligand field. The last two assumptions are unlikely to be valid, but they may well be equally invalid for a series of similar cations undergoing the same process, and this is the justification for their introduction.

The results of these calculations are shown in Table 21.1; similar calculations can be made for the formation of a trigonal bipyramidal transition state, but since the square pyramid never has a smaller crystal field stabilisation energy than the corresponding trigonal bipyramid, this is not really necessary. It is seen that the electronic configuration which loses most crystal field stabilisation energy on either model for the transition state is low-spin d^6; the next most affected are d^3 and d^8. The qualitative agreement with the list of the least labile species (complexes of

Table 21.1
Changes in crystal field stabilisation energy on converting an octahecral complex into a square pyramid (CN5) or pentagonal bipyramid (CN7), other factors remaining constant (in units of Δ_0).

Number of *d* electrons	High-spin complexes		Low-spin complexes	
	CN5	CN7	CN5	CN7
0	0	0	0	0
1	+0.06	+0.13	+0.06	+0.13
2	+0.11	+0.26	+0.11	+0.26
3	−0.20	−0.43	−0.20	−0.43
4	+0.31	−0.11	−0.14	−0.30
5	0	0	−0.09	−0.17
6	+0.06	+0.13	−0.40	−0.85
7	+0.11	+0.26	+0.11	−0.53
8	−0.20	−0.43	−0.20	−0.43
9	+0.31	−0.11	+0.31	−0.11
10	0	0	0	0

low-spin Co(III) and Fe(II) (d^6), Cr(III) (d^3), Ni(II) (d^8) and V(II) (d^3)) is very striking; and although the quantitative agreement is less satisfactory (Cr(III) complexes, for example, are far less labile than those of Ni(II)), we should remember that ions of different charges may be involved. The high rates for Cr^{2+} and Cu^{2+} may well arise from Jahn–Teller distortions. Once more, therefore, we find that despite the simplicity of crystal field theory, it has a useful contribution to make to the discussion of a fundamental chemical problem. An analogous treatment can, of course, be given on the basis of molecular orbital theory, but it is more difficult to follow.

21.6 *Substitution in planar complexes*

Nearly all kinetic work on planar complexes, which are almost all derivatives of d^8 configurations, has been carried out on platinum(II) compounds; the very limited data for nickel(II), palladium(II) and gold(III) complexes show their reactions to be similar to those of platinum(II) complexes but much faster. Substitution in a planar complex is much less subject to steric factors than substitution in an octahedral complex, and since large numbers of five-coordinated d^8 complexes are known there is obviously no reason against postulating a five-coordinated intermediate or transition state. Further, the rate constants for the replacement of a chloride ion in each of the species $[PtCl_4]^{2-}$, $[Pt(NH_3)Cl_3]^-$, $[Pt(NH_3)_2Cl_2]$ and $[Pt(NH_3)_3Cl]^+$ by water are almost identical; this strongly suggests an associative mechanism, since for a dissociative process dependence on the charge on the complex would be expected. Negative entropies and volumes of activation for many substitutions in planar complexes also support associative mechanisms.

For reactions of the general type

$$L_nPtCl_{4-n} + Y \rightarrow L_nPtCl_{3-n}Y + Cl^-$$

the usual form of the rate law is found to be

$$-\frac{d}{dt}[L_nPtCl_{4-n}] = k_1[L_nPtCl_{4-n}] + k_2[L_nPtCl_{4-n}][Y]$$

indicating that the reaction proceeds simultaneously by two independent routes. The first term in the rate equation appears only when the solvent is a potential ligand, e.g. water; it is believed that this term corresponds to a two-stage process in which the leaving ligand is slowly displaced by the solvent and the solvent is then rapidly displaced by the incoming ligand Y. The second term corresponds to an ordinary SN2 bimolecular displacement. In the overwhelming majority of reactions, substitution at a platinum(II) atom takes place with retention of geometrical isomerism, which is most simply interpreted as showing that the entering ligand attacks the complex from one side of the plane with formation of a trigonal bipyramidal intermediate or transition state as shown for a metal complex containing four different ligands T, A, B and X (the last, *trans* to T, being replaced by Y) in Fig. 21.2. For most substitutions at platinum(II), k_2 is found to increase in the sequence

$$H_2O < NH_3 \sim Cl^- < Br^- < py < I^- < CN^- < R_3P$$

This sequence is sometimes referred to as the *nucleophilicity sequence* for substitution at platinum(II); unfortunately attempts to correlate it with any one non-kinetic property of the entering ligand have met with little success, though it may be noted that for halide ions there is a parallel with formation constants for complex halide ions, platinum(II) being a class (*b*) cation (Section 7.7). The sequence of ease of replacement of an already present ligand is, as would be expected, the reverse of the nucleophilicity series.

A particular feature of substitution in platinum(II) complexes that attracts general interest is the so-called *trans effect*, the relationship between the reaction rate and the nature of the ligand *trans* with respect to the departing species. That there is such a relationship is illustrated by the reaction sequences:

$$[PtCl_4]^{2-} \xrightarrow{NH_3} [Pt(NH_3)Cl_3]^- \xrightarrow{NH_3} cis\text{-}[Pt(NH_3)_2Cl_2]$$

$$[Pt(NH_3)_4]^{2+} \xrightarrow{Cl^-} [Pt(NH_3)_3Cl]^+ \xrightarrow{Cl^-} trans\text{-}[Pt(NH_3)_2Cl_2]$$

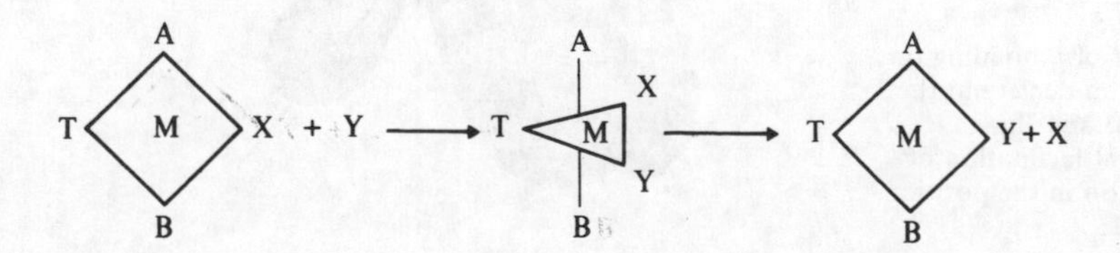

Figure 21.2
The mechanism of substitution in a planar complex.

These can be rationalised on the basis that a chloride ion in the complex directs the entering ligand into the position *trans* to itself more strongly than an ammonia molecule does. In a similar way, the study of the proportions of the products in the general reaction

$$\begin{array}{ccc} L & & X \\ & Pt & \\ L' & & X \end{array} + Y^- \longrightarrow \begin{array}{ccc} L & & X \\ & Pt & \\ L' & & Y \end{array} + \begin{array}{ccc} L & & Y \\ & Pt & \\ L' & & X \end{array}$$

can be used to assess the relative magnitudes of the *trans* effects of L and L′. The general order of *trans* effects spans a factor of about 10^6 in rates and is:

$$H_2O \sim OH^- \sim NH_3 \sim py < Cl^- < Br^- < I^- \sim NO_2^- <$$
$$C_6H_5^- < CH_3^- < PR_3 \sim H^- \ll CO \sim CN^- \sim C_2H_4$$

This is an empirical order, and it is concerned only with the rates of competing reactions; in recent years ground-state *trans* effects of some of these ligands have been established and are discussed in Section 25.10; it is therefore advisable that the effect described in this section should henceforth be known as the *kinetic trans effect*, the more so since it turns out that there is no close connection between the relative magnitudes of the ground-state and kinetic (i.e. transition state) effects. So far as the kinetic *trans* effect is concerned, it is generally accepted that the strong *trans* directing tendencies of CO, CN^- and C_2H_4 are associated with the π-accepting character of these ligands (discussed in the next two chapters); if we represent one of these ligands by T, the leaving ligand by X, and the entering ligand by Y, Fig. 21.3 indicates how overlap between the d_{xy} orbital of the metal and a π^* orbital of the ligand can lower the electron density at the *trans* position and facilitate a bimolecular substitution.

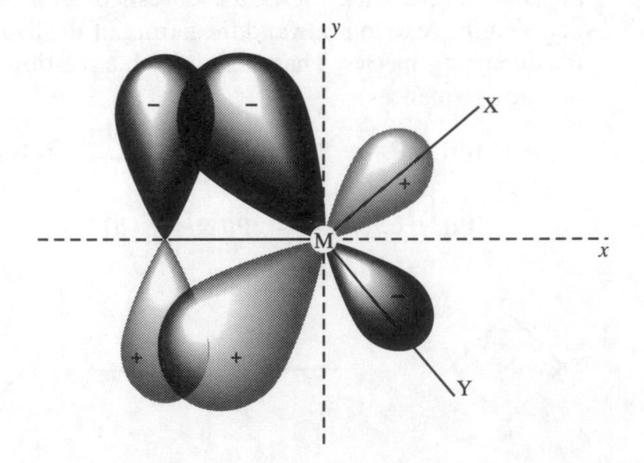

Figure 21.3
The effect of π bonding on the electron density in the d_{xy} orbital and the consequent facilitation of substitution in the position *trans* to T.

The *trans* effect is a useful aid in devising syntheses of platinum(II) complexes: thus for the preparation of *cis* and *trans* $[Pt(NH_3)(NO_2)Cl_2]^-$ from $[PtCl_4]^{2-}$, for example, all we have to do is to vary the order of the two substitutions:

$$[PtCl_4]^{2-} \xrightarrow{NH_3} [Pt(NH_3)Cl_3]^- \xrightarrow{NO_2^-} \left[\begin{array}{ccc} Cl & & NH_3 \\ & \diagdown \; \diagup & \\ & Pt & \\ & \diagup \; \diagdown & \\ Cl & & NO_2 \end{array}\right]^-$$

$$[PtCl_4]^{2-} \xrightarrow{NO_2^-} [Pt(NO_2)Cl_3]^{2-} \xrightarrow{NH_3} \left[\begin{array}{ccc} Cl & & NH_3 \\ & \diagdown \; \diagup & \\ & Pt & \\ & \diagup \; \diagdown & \\ O_2N & & Cl \end{array}\right]^-$$

Many other syntheses can be devised in this way.

21.7 *Redox processes: outer-sphere reactions*

The simplest of all redox reactions are those that involve only the transfer of electrons, e.g. between $[Fe(CN)_6]^{4-}$ and $[Fe(CN)_6]^{3-}$, MnO_4^{2-} and MnO_4^-, or $[IrCl_6]^{3}$ and $[IrCl_6]^{2-}$. Such reactions are most often examined by the use of an isotopic tracer. If, for example, MnO_4^- labelled with ^{54}Mn is mixed with inactive MnO_4^{2-}, it is found that however rapidly the MnO_4^{2-} is precipitated as $BaMnO_4$ exchange has taken place. In the case of the $[Os(bipy)_3]^{2+}$–$[Os(bipy)_3]^{3+}$ system, the rate of electron transfer is measured by studying the loss of optical activity during the reaction

$$(+)[Os(bipy)_3]^{2+} + (-)[Os(bipy)_3]^{3+} \rightleftharpoons (-)[Os(bipy)_3]^{2+} + (+)[Os(bipy)_3]^{3+}$$

In order that we may be sure that only electron transfer takes place during a redox reaction, it is desirable that both reactants should be known to be kinetically inert with respect to ligand exchange; the absence of a term $[L]^{-1}$ in the rate expression, where [L] is the concentration of free ligand in the solution in which the redox reaction is proceeding, is also indicative of a simple electron transfer reaction, since its presence implies that dissociation of a complex is involved.

Where both reactants are inert with respect to ligand exchange, e.g. $[Fe(CN)_6]^{4-}$ and $[Fe(CN)_6]^{3-}$, a close approach of the metal atoms is impossible, and electron transfer must take place via a tunnelling or *outer sphere* mechanism. For an isotopic exchange reaction ΔG° is of course very nearly zero, but activation energy is needed to overcome the

Table 21.2
Second-order rate constants for some outer-sphere redox reactions at 25 °C in aqueous solution.

Reactants	k_2/l mol^{-1} s^{-1}
(a) No net chemical reaction	
$[Fe(bipy)_3]^{2+}$, $[Fe(bipy)_3]^{3+}$	$>10^6$
$[Os(bipy)_3]^{2+}$, $[Os(bipy)_3]^{3+}$	$>10^6$
$[IrCl_6]^{3-}$, $[IrCl_6]^{2-}$	$>10^6$
$[Fe(CN)_6]^{4-}$, $[Fe(CN)_6]^{3-}$	10^5
$[Fe(H_2O)_6]^{2+}$, $[Fe(H_2O)_6]^{3+}$	3
$[Co(en)_3]^{2+}$, $[Co(en)_3]^{3+}$	10^{-4}
$[Co(NH_3)_6]^{2+}$, $[Co(NH_3)_6]^{3+}$	10^{-6}
(b) Net chemical reaction	
$[Os(bipy)_3]^{2+}$, $[Mo(CN)_8]^{3-}$	2×10^9
$[Fe(CN)_6]^{4-}$, $[Fe(phen)_3]^{3+}$	10^8
$[Fe(CN)_6]^{4-}$, $[IrCl_6]^{2-}$	4×10^5

electrostatic repulsion between ions of like charge, to change bond lengths so that they are the same in the transition state, and to alter the solvent structure round each complex. Rates of outer-sphere exchange reactions are found to vary very widely, as may be seen from the rate constants given in the first part of Table 21.2. The very fast reactions in this table are between two low-spin complexes which differ from one another only in the presence of an extra non-bonding electron in the t_{2g} orbitals of the complex containing the metal in the lower oxidation state. Low-spin complexes of this kind have nearly the same metal–ligand bond lengths even in their ground states: Fe—C, for example, is 1.92 Å in $[Fe(CN)_6]^{4-}$ and 1.95 Å in $[Fe(CN)_6]^{3-}$. Where one complex is high-spin and the other low-spin in type, the ground-state metal–ligand distances are widely different, and more activation energy is needed to make them the same in the transition state: Co—N, for example, is 2.11 Å in $[Co(NH_3)_6]^{2+}$ and 1.96 Å in $[Co(NH_3)_6]^{3+}$, the electronic configurations of the metal atoms in these complexes being $(t_{2g})^5(e_g)^2$ and $(t_{2g})^6$ respectively.

Outer-sphere reactions between complexes of different metals are usually very fast; for these reactions there is a decrease in standard free energy (there must be, or the reactions would not occur), and the energy thus available can be regarded as contributing to the activation energy needed for the purposes mentioned above. Values for the rate constants of some of these reactions are also given in Table 21.2. According to a theory developed by Marcus, for a reaction

$$Ox_1 + Red_2 \rightleftharpoons Ox_2 + Red_1$$

the rate constant k is given approximately by

$$k^2 = k_1 k_2 K$$

where k_1 and k_2 are rate constants for the Ox_1–Red_1 and Ox_2–Red_2 exchanges and K is the equilibrium constant of the net reaction.

21.8 Redox processes: inner-sphere reactions

In the outer-sphere mechanism described in the last section, oxidant and reductant have no common atom by means of which electron transfer may be facilitated. In the *inner-sphere* mechanism that we shall now describe, a common atom (or, sometimes, a larger ligand, particularly if it is a conjugated species) acts as a bridge between the two reactants. It is often transferred from one metal atom to the other as a result of the reaction, but this is not necessarily so.

The classic demonstration of an inner-sphere reaction was made by Taube in 1953 on a skilfully chosen system in which the reduced forms were substitution-labile and the oxidised forms substitution-inert:

$$[Co(NH_3)_5Cl]^{2+} + [Cr(H_2O)_6]^{2+} + 5H_3O^+ \rightarrow$$

low-spin Co(III) non-labile; high-spin Cr(II) labile

$$[Co(H_2O)_6]^{2+} + [Cr(H_2O)_5Cl]^{2+} + 5NH_4^+$$

high-spin Co(II) labile; Cr(III) non-labile

All the chromium(III) formed was shown to be in the form $[Cr(H_2O)_5Cl]^{2+}$, and tracer experiments in the presence of added chloride showed that all the chloride in this complex came from the original $[Co(NH_3)_5Cl]^{2+}$. Since the cobalt could not have lost its chloride before reduction and the chromium could not have acquired its chloride after oxidation, the transferred chloride must have been bonded to both metals during the reaction, the intermediate being

$$[(H_3N)_5CoClCr(OH_2)_5]^{4+}$$

In this instance the chloride ion is transferred; in the reaction between $[Fe(CN)_6]^{3-}$ and $[Co(CN)_5]^{3-}$, however, the intermediate $[(NC)_5FeCNCo(CN)_5]^{6-}$ (which is stable enough to be precipitated as its barium salt) is slowly hydrolysed to $[Fe(CN)_6]^{4-}$ and $[Co(CN)_5H_2O]^{2-}$ without transfer of the bridging ligand.

In the last example the rate-determining step in the overall reaction is the breaking of the bridge, but bridge formation or electron transfer within the bridge may also be the slowest stage. The former occurs when the substitution process required to form the bridge is less rapid than the rate of electron transfer; this is not so in the case of Taube's experiment, since substitution in $[Cr(H_2O)_6]^{2+}$ is, as we saw in Section 21.3, very rapid, but it is if $[V(H_2O)_6]^{2+}$ (electronic configuration d^3) is used as a reductant. It is then found that the rate constants for reduction are not very different from the rate constant for water exchange of the cation if $[Co(NH_3)_5Cl]^{2+}$, $[Co(NH_3)_5Br]^{2+}$ and $[Co(CN)_5N_3]^{3-}$ are the oxidants (if however $[Ru(NH_3)_5Cl]^{2+}$ and $[Ru(NH_3)_5Br]^{2+}$ are the oxidants, the rate constants are larger than the rate constant for water exchange, and outer-sphere mechanisms are suggested). For the $[Cr(H_2O)_6]^{2+}$–$[Co(NH_3)_5X]^{2+}$ reaction, where X = F^-, Cl^-, Br^-, I^- or PO_4^{3-}, electron transfer within the bridge is rate-determining.

Many species other than halide and cyanide ions may act as bridging ligands. When thiocyanate does so, as in the $[Co(NH_3)_5SCN]^{2+}$–$[Cr(H_2O)_6]^{2+}$ reaction, it can function in two ways and lead to the formation of two isomeric products. If the Cr atom becomes bonded to the N atom of the bridging ligand, the chromium-containing product of the reaction is $[Cr(H_2O)_5NCS]^{2+}$; if, however, the Cr atom becomes attached to the S atom to give an intermediate of structure

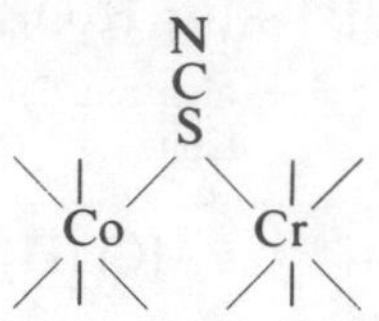

N
C
S
Co Cr

$[Cr(H_2O)_5SCN]^{2+}$ is formed. These processes take place to the extents of 70% and 30% respectively. Conjugated organic anions (e.g. oxalate and salicylate) lead to much faster inner-sphere reactions than non-conjugated anions (e.g. succinate), suggesting that there is an easy flow of electrons along the conjugated system. In a series of reactions between $[Fe(CN)_5H_2O]^{3-}$ and $[Co(NH_3)_5L]^{3+}$ where L is

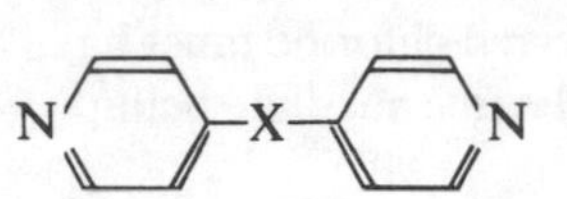

N⟨ ⟩–X–⟨ ⟩N (X = CH_2, C_2H_4, C_3H_6, C_2H_2 or CO)

the reaction is fast when either X is such as to lead to a conjugated bridging ligand or when it is long enough for the ligand to be sufficiently flexible to act by bringing the iron and chromium atoms close enough for an outer-sphere reaction to take place.

The simultaneous transfer of more than one electron in an outer sphere process has not yet been established; if, as is likely to be the case, it would involve equalising bond lengths in species of which one contains two more bonding electrons than the other, it would be improbable because of a high activation energy. That two electrons may be transferred in an inner-sphere reaction is, however, shown by the catalysis of the *trans* $[Pt(en)_2Cl_2]^{2+}$ – $^*Cl^-$ exchange by $[Pt(en)_2]^{2+}$. The rate of exchange is proportional to $[Pt^{II}][Pt^{IV}][^*Cl^-]$ where Pt^{II} stands for $[Pt(en)_2]^{2+}$, and Pt^{IV} for *trans* $[Pt(en)_2Cl_2]^{2+}$. The former ion is known to combine rapidly with chloride to form a five-coordinated species $[Pt(en)_2Cl]^+$, and the overall reaction may then be represented as a two-electron transfer accompanied by chloride ion transfer between this Pt(II) complex and the

Pt(IV) complex:

$$[Pt(en)_2]^{2+} + {}^*Cl^- \rightleftharpoons [Pt(en)_2{}^*Cl]^+$$

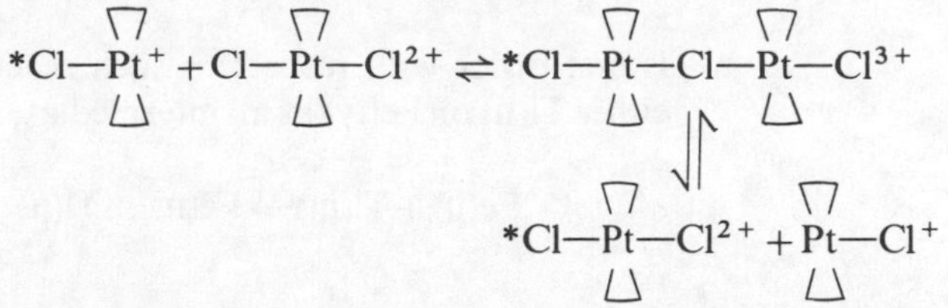

$$^*Cl{-}Pt^+ + Cl{-}Pt{-}Cl^{2+} \rightleftharpoons {}^*Cl{-}Pt{-}Cl{-}Pt{-}Cl^{3+}$$

$$\rightleftharpoons {}^*Cl{-}Pt{-}Cl^{2+} + Pt{-}Cl^+$$

Other examples of ligand substitution that involve catalysis by another compound of the same metal in a different oxidation state are given in Sections 24.5 and 24.9.

21.9 Complementary and non-complementary reactions

In the redox reactions we have so far discussed, oxidant and reductant change their oxidation states by an equal number of units; such reactions are sometimes known as *complementary reactions*. Only one molecule of each reactant is required to react. When oxidation states of the reactants change by a different number of units (in *non-complementary reactions*) different numbers of molecules of oxidant and reductant must be involved in the stoichiometric equation, for example:

$$2Fe(III) + Sn(II) \rightarrow 2Fe(II) + Sn(IV)$$

$$2Cr(II) + Tl(III) \rightarrow 2Cr(III) + Tl(I)$$

$$Mn(VII) + 5Fe(II) \rightarrow Mn(II) + 5Fe(III)$$

Since reactions of molecularity three or more are extremely improbable, non-complementary reactions must proceed in stages, and they often involve intermediate oxidation states which are not known in stable compounds. When, for example, Sn(II) reduces Cr(VI) or Mn(VII), a species (presumed to be Sn(III)) is formed which is capable of rapidly reducing the complex $[Co(C_2O_4)_3]^{3-}$, which is not reduced very quickly by Sn(II); on the other hand, if $[Co(C_2O_4)_3]^{3-}$ is present during the reduction of Tl(III) to Tl(I) or of Hg(II) to Hg(0) by Sn(II), no reduction of the cobalt(III) complex occurs, and oxidation of the Sn(II) then appears to take place in a two-electron transfer process. Acidic dichromate (Cr(VI)) does not oxidise Mn(II), but if Mn(II) is present during the oxidation of isopropanol by dichromate, MnO_2 is precipitated, indicating that Cr(V) or Cr(IV) is formed as an intermediate. This is probably also the case for the Fe(II) catalysed reaction between Cr(VI) and iodide ion in acidic solution, since Cr(VI) in the presence of Fe(II) oxidises iodide faster than either Cr(VI) or Fe(III) (the oxidation product of Fe(II)) alone.

The reaction

$$2Fe(II) + Tl(III) \rightarrow 2Fe(III) + Tl(I)$$

is first order with respect to each reactant, suggesting the formation of either Tl(II) or Fe(IV) as an intermediate:

$$Fe(II) + Tl(III) \rightarrow Fe(III) + Tl(II)$$

or

$$Fe(II) + Tl(III) \rightarrow Fe(IV) + Tl(I)$$

Since addition of Fe(III), but not of Tl(I), diminishes the reaction rate, the intermediate must be Tl(II).

Because many transition metals form compounds in which they exhibit a range of oxidation states, often differing only by unity, their ions are frequently good catalysts for non-complementary reactions. Thus the oxidation of Cr(III) to Cr(VI) by peroxodisulphate, for example, is catalysed by Ag^+, and the following mechanism is suggested:

$$Ag(I) + S_2O_8^{2-} \rightarrow Ag(III) + 2SO_4^{2-}$$

$$Ag(III) + Ag(I) \rightarrow 2Ag(II) \quad \text{(fast)}$$

$$Cr(III) + Ag(II) \rightarrow Cr(IV) + Ag(I)$$

$$Cr(IV) + Ag(II) \rightarrow Cr(V) + Ag(I)$$

$$Cr(V) + Ag(II) \rightarrow Cr(VI) + Ag(I)$$

Many other redox reactions are also catalysed by silver or copper compounds.

PROBLEMS

1 Discuss each of the following observations:

(a) When the reaction

$$[Co(NH_3)_4CO_3]^+ \xrightarrow[H_2O]{H_3O^+} [Co(NH_3)_4(H_2O)_2]^{3+} + CO_2$$

is carried out in $H_2{}^{18}O$, the water in the complex contains equal proportions of $H_2{}^{16}O$ and $H_2{}^{18}O$.

(b) The reaction

$$[Cr(NH_3)_5Cl]^{2+} + NH_3 \rightarrow [Cr(NH_3)_6]^{3+} + Cl^-$$

in liquid ammonia is catalysed by KNH_2.

(c) Anhydrous $CrCl_3$ dissolves much more rapidly in a dilute solution of $CrCl_2$ than in pure water.

2 Account for the relative values for the following rate constants for electron-transfer reactions in aqueous solution:

$[Ru(NH_3)_6]^{3+}$–$[Ru(NH_3)_6]^{2+}$ 10^8 l mol^{-1} s^{-1}
$[Co(NH_3)_6]^{3+}$–$[Ru(NH_3)_6]^{2+}$ 10^{-2} l mol^{-1} s^{-1}
$[Co(NH_3)_6]^{3+}$–$[Co(NH_3)_6]^{2+}$ 10^{-9} l mol^{-1} s^{-1}

3 Suggest syntheses from
(a) $K_2[PtCl_4]$ of

(i) 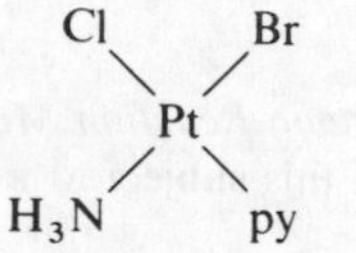 Pt with Cl, Br, H_3N, py (ii) 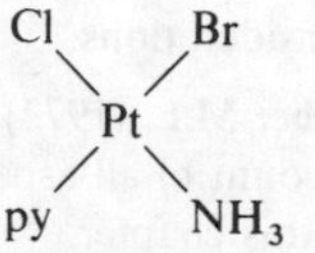 Pt with Cl, Br, py, NH_3

and from
(b) *trans* $[Pt(en)_2Cl_2]Cl_2$ of *trans* $[Pt(en)_2Br_2]Br_2$.

4 The rate of racemisation of the complex

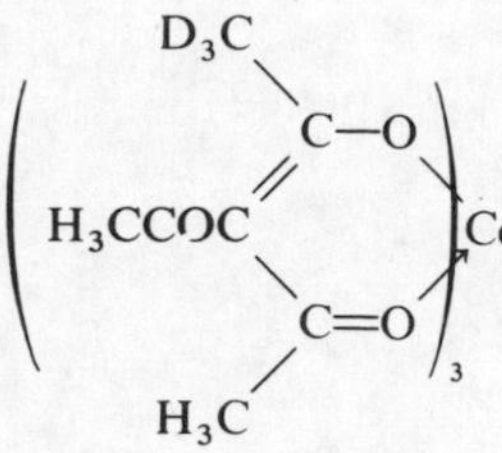

(D_3C–C–O, H_3CCOC, C=O–H_3C chelate → Co)$_3$

is approximately the same as its rate of isomerisation into the complex

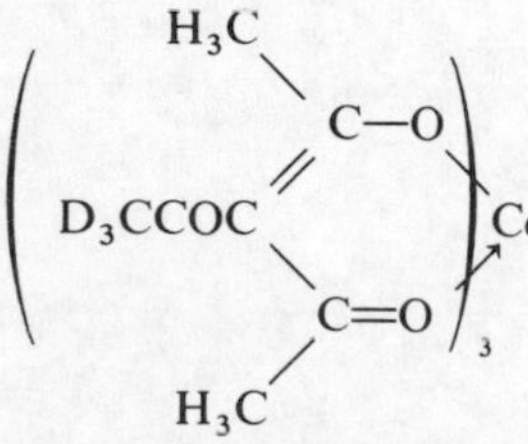

(H_3C–C–O, D_3CCOC, C=O–H_3C chelate → Co)$_3$

What can you deduce about the mechanisms of these reactions?

REFERENCES FOR FURTHER READING

Burgess, J. (1988) *Ions in Solution*, Ellis Horwood, Chichester. Chapters 8–12 give an alternative treatment of topics covered in this chapter, with emphasis on different systems.

Constable, E. (1990) *Metals and Ligand Reactivity*, Ellis Horwood, Chichester. A general account of metal complexes which emphasises reactions of ligands.

Douglas, B., McDaniel, D.H. and Alexander, J.J. (1983) *Concepts and Models of Inorganic Chemistry*, 2nd edition, John Wiley, New York. Chapter 9 gives a very clear and fuller account of the topics covered in this book.

Healy, M. de S. and Rest, A.J. (1978) *Adv. Inorg. Chem. Radiochem.*, **21**, 1. A review of template reactions, the syntheses of coordinated ligands in which reactions of organic molecules already present in complexes are involved.

Katakis, D. and Gordon, G. (1987) *Mechanisms of Inorganic Reactions*, John Wiley, New York. A general treatment of inorganic kinetics.

Purcell, K.F. and Kotz, J.C. (1977) *Inorganic Chemistry*, W.B. Saunders, Philadelphia. Chapters 12–14 contain detailed treatments of some of the reactions covered in this chapter, with more emphasis on electronic considerations.

Tobe, M.L. (1972) *Inorganic Reaction Mechanisms*, Nelson, London. A general account of all aspects of this subject at a rather higher level than the treatment in this chapter.

22 Transition metal carbonyls and related compounds

22.1 Introduction

The classic σ bonding ligands of inorganic chemistry (such as water, ammonia and halide ions) form stable complexes with both main group and transition metal ions. This is not generally true of the ligands with which we shall be concerned in this chapter, and in the case of the most important of them (carbon monoxide) it is not true at all: with the exception of borane carbonyl, H_3BCO (Section 12.3), the only known carbonyls (i.e. complexes at which carbon monoxide is the ligand) are those of the transition metals ($K_6(CO)_6$, obtained from potassium and carbon monoxide, is a derivative of hexahydroxybenzene). In transition metal carbonyls the σ bonding from carbon to the metal is reinforced by π bonding from filled *d* orbitals of suitable symmetry into the empty π^* orbitals of the carbon monoxide molecules, which are thus said to be π acceptors or π (Lewis) acids. Such bonding, in which the accumulation of negative charge on the metal as a result of σ bonding is largely cancelled by π bonding, is sometimes described as synergic. The most direct evidence for this bonding model comes from a study by X-ray and neutron diffraction of the electron density distribution in $Cr(CO)_6$ in which the charges on the atoms are $+0.15 \pm 0.12$ on Cr, $+0.09 \pm 0.05$ on C and -0.12 ± 0.05 on O. Further evidence is given in Sections 22.3 and 22.4; nevertheless, it should be remembered that we can only measure the properties of the bond (e.g. length, stretching frequency, force constant) as a whole, and that any attempt to sort out the σ and π components must necessarily involve the introduction of another variable. We discussed π bonding in an octahedral complex in Section 19.4, and that discussion would apply exactly to $Cr(CO)_6$; most transition metal carbonyls, however, are not octahedral species, and so it is customary to use the more vague representation of π bonding shown in Fig. 22.1, in which the orbitals involved are not specified. Nevertheless, in an accurate discussion of the total metal–carbon bond energies in a series of carbonyls of different structures (e.g. $Cr(CO)_6$, $Fe(CO)_5$ and $Ni(CO)_4$, which are octahedral, trigonal bipyramidal, and tetrahedral respectively) it would be necessary to consider the variation in the degree of overlapping of the metal t_{2g} orbitals with the π^* orbitals of the ligands.

As we explained in Section 19.4, the consequence of this type of π bonding is to increase Δ, the difference in energy between the highest energy filled molecular orbital and the lowest energy empty molecular orbital, i.e. to make the latter less likely to be occupied. Further, the possibilities of π bonding will be greatest when all the metal d orbitals used for π bonding are fully occupied. This corresponds to the presence of 18 electrons in the valence shell of the metal atom. With the exception of $V(CO)_6$, the stable mono- and di-nuclear metal carbonyls obey the 18-electron rule, which is thus an invaluable guide to their formulae (though not, as we shall see later, to their actual structures). Cations and anions derived from these carbonyls also obey the 18-electron rule.

An inevitable consequence of metal-to-ligand π bonding when ligand π^* orbitals are involved is a weakening of the bonding in the ligand (usually detected in the case of carbonyls and their derivatives by a decrease in the carbon–oxygen stretching frequency, since the carbon–oxygen bond length is not very sensitive to a change in bond order –it is, for example, 1.131 and 1.163 Å in CO and CO_2 respectively). So far as symmetry considerations are concerned, ligands which are isoelectronic with carbon monoxide (such as N_2, NO^+, CN^- and C_2^{2-}) would be just as suitable for π bond formation, but the charge on the ligand must also be taken into consideration. Thus NO^+, by virtue of its positive charge, is a stronger π acceptor than CO, whilst CN^- is a weaker one: the range of stretching frequencies found decreases along the series nitrosyl complexes

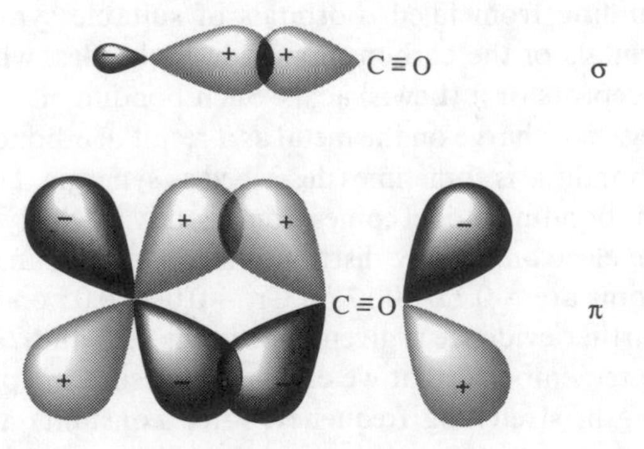

Figure 22.1
The components of a transition metal–carbonyl bond.

$>$ carbonyls $>$ cyano complexes. Further differences between the cyanide ion and carbon monoxide as ligands arise from the fact that the former combines with cations rather than uncharged atoms, and in many respects the chemistry of cyano complexes is therefore more like that of the classical complexes we mentioned at the beginning of the section. (The same should be true of complexes of the acetylide ion, but these compounds are dangerously explosive and have been little investigated, and we shall not discuss them here.) On balance, however, there is enough in common between complexes of CO, N_2, NO^+ and CN^- to justify treatment together in this chapter.

One other group of complexes in which the ligands act in a dual capacity will also be discussed briefly. Complexes of the halides and other compounds of Group V elements (e.g. PH_3, PF_3, PCl_3, PPh_3, $P(OEt)_3$, $AsCl_3$, $SbCl_3$) involve back-donation from filled metal *d* orbitals to empty ligand *d* orbitals. Nitrogen derivatives, of course, cannot form complexes of this kind.

The larger subject of metal-to-carbon bonded complexes of organic ligands is reserved for the next chapter.

22.2 *The preparation and properties of transition metal carbonyls*

A large number of stable metal carbonyls, many of them containing several metal atoms, are now known, but in this book we shall concentrate attention on some of the simpler species, the formulae and physical properties of which are given in Table 22.1. Many unstable carbonyls have been obtained by *matrix isolation,* the action of carbon monoxide on metal atoms in a noble gas matrix at very low temperatures or the photolysis of stable metal carbonyls under similar conditions. Among species made by these methods are $Ti(CO)_6$, $Pd(CO)_4$, $Pt(CO)_4$, $Cu_2(CO)_6$, $Ag_2(CO)_6$ and $Au(CO)_2$ (derivatives of metals not previously known to form carbonyls), and $Cr(CO)_4$, $Mn(CO)_5$, $Fe(CO)_4$, $Fe(CO)_3$ and $Ni(CO)_3$ (fragments produced by decomposition of stable carbonyls). The members of the former group are of particular interest in that they show there is only a small difference between metals forming stable carbonyls and those forming not quite stable carbonyls; members of both groups pose interesting problems in valence theory which lie outside the scope of this book but are discussed in the references given at the end of this chapter. In the remainder of this section we shall consider only carbonyls which are isolable at ordinary temperatures.

Only nickel carbonyl and iron pentacarbonyl are normally obtained by the action of carbon monoxide on the finely divided metal; the former is obtained at, or a little above, room temperature and at atmospheric pressure, but the preparation of the latter requires a temperature of

Table 22.1
Some stable binary carbonyls.

Electrons needed by metal to attain noble gas configuration	13	12	11	10	9	8
First transition series	$V(CO)_6$ black solid dec. 70 °C	$Cr(CO)_6$ white solid dec. 130 °C sublimes *in vacuo*	$Mn_2(CO)_{10}$ yellow solid m.p. 154 °C	$Fe(CO)_5$ yellow liquid b.p. 103 °C	$Co_2(CO)_8$ orange solid m.p. 51 °C	$Ni(CO)_4$ colourless liquid b.p. 43 °C
				$Fe_2(CO)_9$ yellow solid m.p. 100 °C (dec.)	$Co_4(CO)_{12}$ black solid	
				$Fe_3(CO)_{12}$ black solid dec. 140 °C	$Co_6(CO)_{16}$ black solid	
Second transition series		$Mo(CO)_6$ white solid sublimes *in vacuo*	$Tc_2(CO)_{10}$ white solid	$Ru(CO)_5$ colourless liquid m.p. −22 °C		
				$Ru_3(CO)_{12}$ orange solid m.p. 154 °C	$Rh_4(CO)_{12}$ orange solid	
					$Rh_6(CO)_{16}$ black solid	
Third transition series		$W(CO)_2$ white solid sublimes *in vacuo*	$Re_2(CO)_{10}$ white solid m.p. 177 °C	$Os(CO)_5$ colourless liquid m.p. 2 °C		
				$Os_3(CO)_{12}$ yellow solid m.p. 224 °C	$Ir_4(CO)_{12}$ yellow solid	
					$Ir_6(CO)_{16}$ red solid	

150–250 °C and 200 atm pressure. Most other simple metals carbonyls are obtained by *reductive carbonylation*, the action of carbon monoxide and a reducing agent (occasionally excess of carbon monoxide itself) on a metal oxide, halide or other compound, e.g.

$$OsO_4 + CO \xrightarrow[350\ \text{atm}]{250\ ^\circ\text{C}} Os(CO)_5 + CO_2$$

$$CrCl_3 + CO + LiAlH_4 \xrightarrow[\text{ether}]{115\ ^\circ\text{C},\ 70\ \text{atm}} Cr(CO)_6 + LiCl + AlCl_3$$

$$MoCl_5 + CO + AlEt_3 \xrightarrow[200\ atm]{100\ ^\circ C} Mo(CO)_6 + AlCl_3$$

$$CoCO_3 + CO + H_2 \xrightarrow[250\ atm]{150\ ^\circ C} Co_2(CO)_8 + CO_2 + H_2O$$

Yields are often poor, and we have therefore not attempted to write stoichiometric equations. In the last reaction a carbonyl hydride (see later) is an intermediate product. Other methods of preparations are illustrated by the following examples:

$$VCl_3 + Na + CO \xrightarrow[diglyme]{150\ ^\circ C,\ 150\ atm} [Na(diglyme)_2][V(CO)_6] \xrightarrow[ether]{HCl} V(CO)_6$$

$$WCl_6 + Fe(CO)_5 \xrightarrow{100\ ^\circ C} W(CO)_6 + FeCl_2$$

The carbonyls $Ni(CO)_4$, $Fe(CO)_5$, $Mo(CO)_6$ and $W(CO)_6$ are available commercially. Polynuclear carbonyls may be obtained by thermal or photochemical decomposition of simple carbonyls (e.g. $2Fe(CO)_5 \rightarrow Fe_2(CO)_9 + CO$), or by what are often very complicated decompositions of carbonylate anions in alkaline media (e.g. $Fe_3(CO)_{12}$ by the action of alkali on iron pentacarbonyl, which gives $Na[HFe(CO)_4]$, and oxidation of the latter to $Na[HFe_3(CO)_{11}]$, followed by treatment of this complex with acid). Carbonyls containing two different metals are known; typical preparations are:

$$Na[Co(CO)_4] + Mn(CO)_5Br \xrightarrow{THF} (OC)_4CoMn(CO)_5 + NaBr$$

$$3Fe(CO)_5 + Ru_3(CO)_{12} \xrightarrow{110\ ^\circ C} FeRu_2(CO)_{12} + Fe_2Ru(CO)_{12} + 3CO$$

Most carbonyls are volatile solids, but $Fe(CO)_5$, $Ru(CO)_5$, $Os(CO)_5$ and $Ni(CO)_4$ are liquids at ordinary temperature; nearly all of those discussed in this section are soluble in organic solvents. Some carbonyls (e.g. $Ni(CO)_4$) are insoluble in water, others (e.g. $Fe(CO)_5$) react with it; none dissolves unchanged. All carbonyls are, of course, thermodynamically unstable with respect to oxidation in air, but the speeds of the reactions vary widely: $Co_2(CO)_8$ reacts at ordinary temperatures; $Fe(CO)_5$ and $Ni(CO)_4$ are also very easily oxidised, their vapours forming explosive mixtures with air; but $Cr(CO)_6$, $Mo(CO)_6$ and $W(CO)_6$ do not react unless they are heated.

Among the most important reactions of metal carbonyls are:

(i) the formation of carbonylate anions by the action of alkali or by reduction, e.g.

$$Fe(CO)_5 + 3NaOH \xrightarrow{H_2O} Na[HFe(CO)_4] + Na_2CO_3 + H_2O$$

$$Co_2(CO)_8 + 2Na/Hg \xrightarrow{THF} 2Na[Co(CO)_4]$$

(ii) the formation of carbonyl halides by the action of halogens, e.g.

$$Mn_2(CO)_{10} + Br_2 \xrightarrow{40\,^{\circ}C} 2Mn(CO)_5Br$$

(iii) the displacement of carbon monoxide by other ligands, notably PF_3, PCl_3, PPh_3, $AsPh_3$ and unsaturated organic compounds, e.g.

$$Ni(CO)_4 + 2PF_3 \rightarrow Ni(CO)_2(PF_3)_2 + 2CO$$

$$Fe(CO)_5 + PPh_3 \rightarrow (Ph_3P)Fe(CO)_4 + CO$$

$$Cr(CO)_6 + C_6H_6 \rightarrow (C_6H_6)Cr(CO)_3 + 3CO$$

(iv) the formation of nitrosyls by the action of nitric oxide on dinuclear carbonyls, e.g.

$$Fe_2(CO)_9 + 4NO \rightarrow 2Fe(CO)_2(NO)_2 + 5CO$$

$$Co_2(CO)_8 + 2NO \rightarrow 2Co(CO)_3(NO) + 2CO$$

The classes of compounds formed in these reactions (not all of which are, however, given by all carbonyls) are discussed in later sections of this chapter or in the following chapter.

22.3 *The structures of transition metal carbonyls*

The mononuclear carbonyls all have symmetrical structures. Vanadium, chromium, molybdenum and tungsten hexacarbonyls are octahedral with metal–carbon distances: $V(CO)_6$, 2.00 Å; $Cr(CO)_6$, 1.92 Å; $Mo(CO)_6$, 2.06 Å; $W(CO)_6$, 2.07 Å. Since the carbonyl stretching frequencies are nearly the same in $Cr(CO)_6$, $Mo(CO)_6$ and $W(CO)_6$, the observed variation in the standard enthalpy of formation of these compounds may reasonably be ascribed to variations in the enthalpies of atomisation of the metals and in the metal–carbon bond strength; on this basis and on the necessary but much more questionable assumption that the bond energy in the carbonyl group is the same as that in carbon monoxide, the metal–carbon bond energy terms are evaluated as 107, 151 and 178 kJ mol^{-1} for $Cr(CO)_6$, $Mo(CO)_6$ and $W(CO)_6$ respectively. Iron pentacarbonyl has a trigonal bipyramidal structure with the customary small difference between axial and equatorial bond lengths (1.81 and 1.83 Å respectively); the mean iron–carbon bond energy term is 118 kJ mol^{-1} if the carbonyl group is again taken as the same as that in carbon monoxide. Nickel carbonyl is a regular tetrahedron with Ni—C 1.84 Å; the mean nickel–carbon bond energy term is 147 kJ mol^{-1}.

The fact that the bond energy increases along the series $Cr(CO)_6$, $Fe(CO)_5$, $Ni(CO)_4$ whilst the thermal stability decreases along the same series shows how difficult it is to interpret chemical properties in terms of thermochemical data for a series of complexes like the metal carbonyls, in which the properties of the coordinated ligand may vary from one compound to another according to the extent of π bonding. There is, for example, some change in the carbonyl stretching frequencies between $Cr(CO)_6$ and $Ni(CO)_4$, and the structure changes from octahedral to tetrahedral; further, bond energy terms relate to dissociation into ground-state gaseous atoms, whilst thermal decomposition produces excited state fragments, e.g. of $Ni(CO)_3$ and CO from $Ni(CO)_4$, so that bond dissociation energies (which are of more relevance to kinetics) may be quite different from bond energy terms.

Manganese carbonyl, $Mn_2(CO)_{10}$, has a structure composed of two staggered $Mn(CO)_5$ units linked only by a metal–metal bond. In $Fe_2(CO)_9$ there are three bridging carbonyl groups, and in solid $Co_2(CO)_8$ there are two. The overall symmetries of these three molecules, shown in Fig. 22.2, are D_{4d}, D_{3h} and C_{2v} respectively. Bridging and terminal carbonyls are found to have different stretching frequencies, typically in the ranges 1800–1900 cm^{-1} and 1950–2150 cm^{-1} respectively for neutral metal carbonyls; solid $Co_2(CO)_8$, for example, shows infrared carbonyl stretching bands at 1857, 1886, 2001, 2031, 2044, 2059, 2071 and 2112 cm^{-1}. Further, in the case of $Co_2(CO)_8$ (one of the few poly-nuclear carbonyls for which accurate data are available), there is a significant difference in bond length in bridging and terminal carbonyls, 1.21 and 1.17 Å respectively. There is also a difference between bridging and terminal cobalt–carbon distances, 1.92 and 1.79 Å respectively, showing that the bridge bonds are much weaker than the terminal ones. This is accounted for by postulating that each metal–carbon bridge bond involves only one electron from the carbon monoxide molecule, i.e. that if the σ bond order of the terminal bonds is unity, that of the bridge bonds is one-half (cf. the structure of Al_2Cl_6,

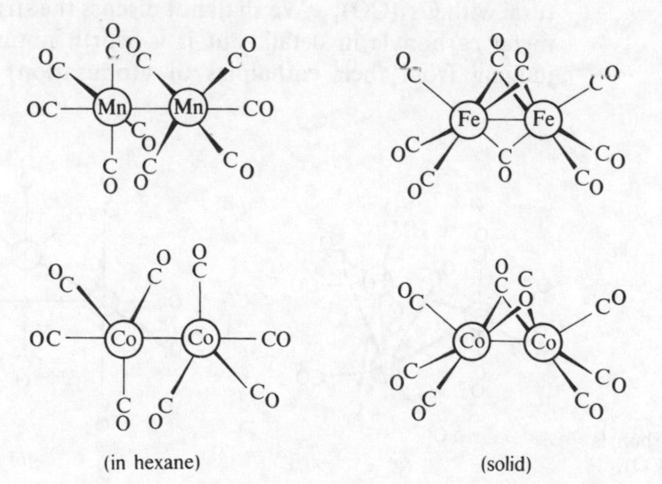

Figure 22.2
The structures of some dinuclear metal carbonyls: $Mn_2(CO)_{10}$, $Fe_2(CO)_9$ and $Co_2(CO)_8$.

Section 12.7). On the basis that each terminal ligand is a two-electron donor and each bridging ligand a one-electron donor, the effective atomic number of each cobalt atom in $Co_2(CO)_8$ would be $27+(3\times2)+(2\times1)$, i.e. 35, corresponding to a valence shell of 17 electrons, and the molecule would be paramagnetic owing to the presence of two unpaired electrons. Since it is actually diamagnetic it is deduced that there is a Co—Co bond; the distance between the cobalt atoms (2.52 Å) is certainly compatible with this deduction, since it is almost the same as that in metallic cobalt. (The Mn—Mn distance in $Mn_2(CO)_{10}$, 2.92 Å, which *must* contain a metal–metal bond, is actually greater than that in metallic manganese.) The diamagnetism of $Fe_2(CO)_9$ similarly indicates the presence of a metal–metal bond in this molecule.

When $Co_2(CO)_8$ is dissolved in hexane, the bridging carbonyl bands disappear from the infrared spectrum and the compound has the staggered structure of D_{3d} symmetry shown in Fig. 22.2, for which it is easily verified that the 18-electron rule is just as valid as for the solid state structure. The non-bridged structure is, in fact, easily derived from the bridged structure by breaking one of the two metal–carbon bonds formed by each bridging ligand and slightly changing the ligand positions. It is found that ΔH^0 for the conversion of the bridged into the non-bridged form is only 5.5 kJ mol^{-1}. Thus there is neither a substantial thermodynamic nor a substantial kinetic barrier to the interconversion of the two forms. Similar intramolecular rearrangements (usually investigated by ^{13}C nuclear magnetic resonance spectroscopy) have been established for many other metal carbonyls and related compounds; the case of $Fe(CO)_5$ was discussed in Section 1.5.

The iron carbonyl $Fe_3(CO)_{12}$ has a structure (shown in Fig. 22.3(*a*)) in which there are two carbonyl bridges between one pair of iron atoms; $Os_3(CO)_{12}$, on the other hand (Fig. 22.3(*b*)), contains only terminal carbonyl groups. It is interesting to note that $Fe_2Ru(CO)_{12}$ has the $Fe_3(CO)_{12}$ structure, whilst $FeRu_2(CO)_{12}$ and $Ru_3(CO)_{12}$ are isostructural with $Os_3(CO)_{12}$. We shall not discuss the structures of more complex metal carbonyls in detail, but it is worth noting that the larger (and, judging from their enthalpies of atomisation) more strongly bonded

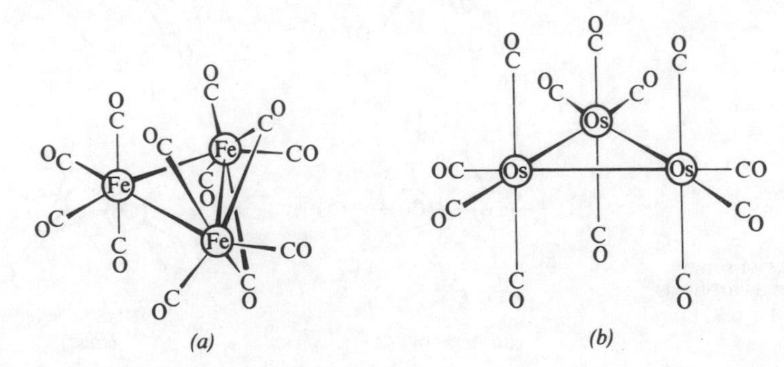

Figure 22.3
The structures of the polynuclear metal carbonyls $Fe_3(CO)_{12}$ and $Os_3(CO)_{12}$.

metals of the later transition series are in general somewhat less prone to form compounds containing bridging carbonyl groups than the metals of the first transition series. Carbonyl groups bridging three metal atoms are known, e.g. in $Rh_6(CO)_{16}$, which contains a Rh_6 octahedron with twelve terminal carbonyl groups and four others symmetrically placed above faces of the octahedron; the latter have carbonyl stretching frequencies in the range 1 600–1 750 cm^{-1}. For some of the larger polynuclear carbonyls including $Rh_6(CO)_{16}$ the 18-electron rule as we have discussed it in this section is no longer a reliable guide to formulae, and electron-deficient metal clusters are believed to be present; the discussion of the structures of these compounds, however, lies outside the scope of this book.

22.4 *Carbonyl hydrides and carbonylate anions and cations*

The formation of the carbonylate anions $[V(CO)_6]^-$, $[HFe(CO)_4]^-$ and $[Co(CO)_4]^-$ was mentioned in Section 22.2. Many other similar anions are obtained by the action of bases, alkali metal amalgams or sodium borohydride on carbonyls; the reactions which occur are often complicated and lead to polynuclear products, but some of the main results may be indicated as follows:

$$Fe(CO)_5 \xrightarrow{Na} Na_2[Fe(CO)_4]$$

$$Cr(CO)_6 \xrightarrow[NH_3]{NaBH_4} Na_2[Cr_2(CO)_{10}]$$

$$Mn_2(CO)_{10} \xrightarrow[THF]{Li} Li[Mn(CO)_5]$$

$$TaCl_5 + CO + Na \xrightarrow[\text{diglyme}]{100\,^\circ C\ 200\ atm} [Na(\text{diglyme})_2][Ta(CO)_6]$$

Salts containing some of these anions yield carbonyl hydrides on acidification of their solutions: $H_2Fe(CO)_4$, $HMn(CO)_5$ and $HCo(CO)_4$ are obtained in this way. Other preparations of these compounds include:

$$Fe(CO)_4I_2 \xrightarrow[THF]{NaBH_4} H_2Fe(CO)_4$$

$$Mn_2(CO)_{10} + H_2 \xrightarrow[200\ atm]{200\,^\circ C} HMn(CO)_5$$

$$Co + 4CO + \tfrac{1}{2}H_2 \xrightarrow[50\ atm]{150\,^\circ C} HCo(CO)_4$$

Carbonyl hydrides corresponding to many known carbonylate anions are, however, unstable and have not been isolated.

A few cationic carbonyl complexes are also known; the action of carbon monoxide and aluminium chloride on manganese carbonyl chloride, for example, gives a salt of the cation $[Mn(CO)_6]^+$:

$$Mn(CO)_5Cl + CO + AlCl_3 \rightarrow [Mn(CO)_6][AlCl_4]$$

Protonated metal carbonyls in which a proton is attached to the metal atom are produced when some metal carbonyls are dissolved in strongly acidic media such as sulphuric acid or a mixture of hydrogen fluoride and boron trifluoride:

$$Fe(CO)_5 + H^+ \rightarrow [HFe(CO)_5]^+$$

Large numbers of protonated organometallic carbonyl complexes have been characterised.

Manganese, iron and cobalt carbonyl hydrides (to which we shall confine attention) are colourless or yellow liquids, the manganese compound being the most stable; the other two compounds decompose above $-10\ ^\circ C$. All have proton magnetic resonances which indicate that the hydrogen is attached to the metal. This conclusion has been confirmed in the case of $HMn(CO)_5$ by an X-ray and neutron diffraction study, which shows the molecule to be octahedral with Mn—H = 1.60 Å. The pK values for $HMn(CO)_5$ and $H_2Fe(CO)_4$ in aqueous solution are 7 and 4.5 (pK_1) and 14 (pK_2) respectively; $HCo(CO)_4$ is almost insoluble in water but is a strong acid in methanol.

Few structural data are available for the carbonylate anions and cations, but there are striking variations in the corresponding infrared stretching frequencies in the series of isoelectronic species $[V(CO)_6]^-$, $Cr(CO)_6$ and $[Mn(CO)_6]^+$:

	$[V(CO)_6]^-$	$Cr(CO)_6$	$[Mn(CO)_6]^+$
νMC/cm^{-1}	460	441	416
νCO/cm^{-1}	1859	1981	2101

These show beyond doubt that as the negative charge on the metal increases from the manganese complex to the vanadium complex it is delocalised on to the ligands by the π bonding mechanism described in Sections 19.4 and 22.1, with an increase in the strength of the bonding between metal and carbon and a reduction in that in the carbonyl group.

Dicobalt octacarbonyl is an important catalyst for many organic reactions, among them the conversion of alkenes to aldehydes by the action of carbon monoxide and hydrogen (the 'oxo' reaction) and the reduction of aldehydes to alcohols. The rates of both these reactions are increased by increasing the pressure of hydrogen and reduced by the presence of excess of carbon monoxide. The carbonyl hydride $HCo(CO)_4$ is believed to be formed intermediately in the reactions, but the retardation by carbon monoxide indicates that $HCo(CO)_3$, rather than $HCo(CO)_4$ itself, is the active species.

Many covalently bonded compounds containing $Mn(CO)_5$, $Fe(CO)_4$ and $Co(CO)_4$ groups may be obtained from the carbonyl hydrides or their salts, typical preparations being indicated by the equations:

$$Hg + Co_2(CO)_8 \xrightarrow{150\,^\circ C} Hg[Co(CO_4)]_2$$

$$SiH_3I + [Co(CO_4)]^- \rightarrow H_3SiCo(CO)_4 + I^-$$

$$Ph_3PAuCl + [Mn(CO)_5]^- \rightarrow Ph_3PAuMn(CO)_5 + Cl^-$$

These species are much more stable than the carbonyl hydrides themselves.

22.5 *Carbonyl halides*

Carbonyl halides are known for most metals that form stable carbonyls, and also for palladium, platinum, copper, silver and gold. They are made by interaction of a metal halide and carbon monoxide at high pressures, or by the action of halogens on metal carbonyls:

$$Fe(CO)_5 \xrightarrow{I_2} Fe(CO)_4I_2$$

$$RhCl_3.3H_2O \xrightarrow[100\,^\circ C]{CO} [Rh(CO)_2Cl]_2$$

$$PtCl_2 \xrightarrow[200\,^\circ C]{CO} [Pt(CO)Cl_2]_2$$

Monomeric compounds analogous to the carbonyl hydrides have structures similar to those of the latter compounds; dimeric compounds are always bridged through the halogen atoms rather than through the carbonyl groups. The carbonyl halides are usually white or yellow solids that are soluble in organic compounds but are decomposed by water.

Most carbonyl halides and their substitution products obey the 18-electron rule, but this is not true of $[Rh(CO)_2Cl]_2$ or 'Vaska's compound', *trans* $(Ph_3P)_2Ir(CO)Cl$, which is obtained from sodium hexachloroiridate(IV) and triphenylphosphine in ethylene glycol under an atmosphere of carbon monoxide. This compound undergoes a wide variety of addition reactions, some of them reversible, as a result of which the valence shell of the metal atom acquires 18 electrons, e.g.

$$
\begin{array}{c}
\begin{array}{c} PPh_3 \\ OC\diagdown\;\;\diagup H \\ Ir \\ Cl\diagup\;\;\diagdown Cl \\ PPh_3 \end{array}
\xleftarrow{HCl}
\begin{array}{c} Ph_3P\diagdown\;\;\diagup Cl \\ Ir \\ OC\diagup\;\;\diagdown PPh_3 \end{array}
\underset{vacuum}{\overset{O_2}{\rightleftharpoons}}
\begin{array}{c} PPh_3 \\ OC\diagdown\;\;\diagup O \\ Ir \;\;| \\ Cl\diagup\;\;\diagdown O \\ PPh_3 \end{array}
\\
\downarrow H_2
\\
\begin{array}{c} PPh_3 \\ OC\diagdown\;\;\diagup H \\ Ir \\ Cl\diagup\;\;\diagdown H \\ PPh_3 \end{array}
\end{array}
$$

It is almost certain that in the addition of H_2 a three-centred intermediate

$$
\begin{array}{c} \diagdown\;|\;\diagup H \\ Ir \;\;\;\; | \\ \diagup\;|\;\diagdown H \end{array}
$$

is involved: several compounds in which the H_2 molecule remains intact whilst chelated to a metal atom are now known, e.g. $W(CO)_3(PPr_3)_2(H_2)$ and $Cr(CO)_5(H_2)$. Those reactions in which planar four-coordinated Ir(I) becomes octahedrally six-coordinated Ir(III) are often known as *oxidative additions*.

22.6 *Phosphine and phosphorus trihalide complexes*

We have mentioned already that carbon monoxide is displaced from many metal carbonyls by triphenylphosphine or by phosphorus trifluoride, with formation of species such as $Ni(CO)_2(PPh_3)_2$ (an important catalyst for the polymerisation of olefins) and $Ni(CO)_2(PF_3)_2$. Phosphine itself reacts similarly to triphenylphosphine, forming, for example $Ni(CO)_3(PH_3)$ on treatment with nickel carbonyl. Complete substitution of carbon monoxide, however, is rarely achieved, and for the preparation of complexes containing only phosphorus(III) compounds as π bonded ligands other methods are necessary, e.g.

$$Ni + 4PF_3 \xrightarrow[I_2\ \text{catalyst}]{150\,^\circ C,\ 400\ \text{atm}} Ni(PF_3)_4$$

$$PtCl_2 + 4PF_3 + Cu \xrightarrow{100\,^\circ C,\ 100\ \text{atm}} Pt(PF_3)_4 + CuCl_2$$

$$2Co + H_2 + 8PF_3 \xrightarrow{200\,^\circ C,\ 200\ atm} 2HCo(PF_3)_4$$

$$K_2PtCl_4 \xrightarrow{PPh_3,\ N_2H_4\ in\ EtOH} Pt(PPh_3)_4$$

The compound $Ni(PF_3)_4$ is a colourless liquid (b.p. 73 °C) which can be steam-distilled without decomposition; this appears to be the consequence of the drift of electron density to the phosphorus atoms. Similar compounds are formed by many other trivalent phosphorus compounds; infrared spectroscopic data indicate π acidity decreases along the series PF_3, PCl_3, $P(OEt)_3$, PEt_3. The stability of $Pt(PF_3)_4$ and $Pt(PPh_3)_4$ compared with $Pt(CO)_4$ should be noted. Triphenylphosphine is a very bulky ligand, and for steric reasons complexes such as $Pt(PPh_3)_4$ readily undergo partial dissociation, e.g.

$$Pt(PPh_3)_4 \rightleftharpoons Pt(PPh_3)_3 + PPh_3$$

22.7 *Dinitrogen complexes*

Much of the significance attached to complexes of molecular nitrogen and transition metals arises from the possibility of reducing the ligand to ammonia and regenerating the nitrogen-absorbing species, thus obtaining an ordinary-temperature alternative to the Haber process. Little success in this direction has yet been achieved, though a few complexes yield ammonia on decomposition, e.g. *cis* $[W(N_2)_2(PMe_2Ph)_4]$ (which is prepared by the interaction of WCl_4, PMe_2Ph, Mg and N_2) on treatment with methanolic sulphuric acid.

The simplest dinitrogen complex is $[Ru(NH_3)_5N_2]^{2+}$, which is obtained by the action of hydrazine on aqueous ruthenium trichloride, by the action of azide ion on the complex $[Ru(NH_3)_5H_2O]^{3+}$, or by the reaction sequence:

$$[Ru(NH_3)_5H_2O]^{3+} \xrightarrow{Zn/Hg} [Ru(NH_3)_5H_2O]^{2+} \xrightarrow[N_2]{100\ atm} [Ru(NH_3)_5N_2]^{2+} + H_2O$$

It may be isolated as its chloride. The cation $[(H_3N)_5RuN_2Ru(NH_3)_5]^{4+}$, in which the nitrogen molecule acts as a bridge, is obtained by the action of zinc amalgam on $[Ru(NH_3)_5Cl]^{2+}$(aq) in the presence of nitrogen, and may be precipitated with fluoroborate; it is also obtained as a by-product in the $[Ru(NH_3)_5H_2O]^{2+}$–N_2 reaction.

Other dinitrogen complexes are produced by direct uptake of gaseous nitrogen at atmospheric pressure in the presence of reducing agents and tertiary phosphines or arsines, typical examples being

$$Co(acac)_3 + 3Ph_3P + N_2 \xrightarrow{Al(isoC_4H_9)_3} CoH(N_2)(PPh_3)_3$$

$$FeCl_2 + 3PEtPh_2 + N_2 \xrightarrow[EtOH]{NaBH_4} FeH_2(N_2)(PEtPh_2)_3$$

$$OsCl_3(AsR_3)_3 + N_2 \xrightarrow[THF]{Zn/Hg} OsCl_2(N_2)(AsR_3)_3$$

Reactions of coordinated nitrogen-containing ligands may also serve, e.g.

$$[Os(NH_3)_5N_2]^{2+} + HNO_2 \rightarrow cis\,[Os(NH_3)_4(N_2)_2]^{2+} + 2H_2O$$

The most stable dinitrogen complexes are those of ruthenium, osmium and iridium, some of which can be heated to 100–200 °C without decomposition. Most dinitrogen complexes decompose when gently heated, and coordinated nitrogen is readily displaced by triphenylphosphine or carbon monoxide.

The bonding in mononuclear dinitrogen complexes is qualitatively similar to that in carbon monoxide complexes. In the ruthenium ammine dinitrogen complexes, for example, the Ru—N≡N group is linear, and the Ru—N_2 distance is about 0.2 Å shorter than the Ru—NH_3 distance, indicating metal–dinitrogen π bonding; the N≡N stretching frequency is lower, and the bond length slightly greater, than for molecular nitrogen. In the linear $[(H_3N)_5RuN_2Ru(NH_3)_5]^{4+}$ ion, however, the bridging N_2 molecule is bonded to ruthenium at each end, rather than like CO in bridged carbonyls.

22.8 *Nitric oxide complexes*

Nitric oxide has an unpaired electron in an antibonding orbital; as we saw in Chapter 14, when it loses this electron to form the diamagnetic NO^+ ion (which is isoelectronic with N_2 and CO) the bond length decreases from 1.15 to 1.06 Å and the stretching frequency increases from 1876 cm^{-1} to about 2300 cm^{-1}. For present purposes it is useful to regard most metal nitrosyls as derived from the NO^+ cation, which is considered as being produced by transfer of an electron from NO to the transition metal atom; a σ bond is then formed by donation of the lone pair of electrons on the nitrogen atom of the NO^+ ion; finally, a large part of the negative charge which would thus accumulate on the metal is counterbalanced by back-donation from filled metal *d* orbitals to π^* antibonding orbitals of NO^+, a process exactly analogous to back-donation in the carbonyls. This is a highly artificial picture of the bonding, of course, but at this level of treatment it has three advantages: it emphasises the relationships between carbonyl and nitrosyl compounds; by treating NO as a three-electron donor it enables us to rationalise the formulae of many nitrosyl compounds in terms of the 18-electron rule; and it correctly implies a linear M—N≡O configuration analogous to M—C≡O in species for which the NO stretching frequency is similar to, or higher than, that for nitric oxide itself. The few available accurate bond length data for nitric oxide complexes unfortunately all refer to rather complicated species and vary so widely that no conclusions can be drawn from them.

Only a few nitrosyl carbonyls and two neutral metal nitrosyls, the compounds $Cr(NO)_4$ and $Co(NO)_3$, are known. Nitrosyl carbonyls include: $Mn(NO)_3CO$, $Fe(NO)_2(CO)_2$, $Co(NO)(CO)_3$ (like $Co(NO)_3$, all isoelectronic with $Ni(CO)_4$); $Mn(NO)(CO)_4$ (isoelectronic with $Fe(CO)_5$); and $V(NO)(CO)_5$ (isoelectronic with $Cr(CO)_6$). All of these compounds are unstable volatile solids or liquids obtained by the action of nitric oxide on metal carbonyls; they are decomposed by air and by water. The NO stretching frequencies in them lie in the range 1700–1900 cm^{-1}, indicating extensive π bonding.

Probably the best-known nitric oxide complex is the nitrosopentacyanoferrate(II) or nitroprusside ion (Section 24.7), $[Fe(CN)_5NO]^{2-}$, in which the N—O distance is 1.13 Å and the NO stretching frequency 1939 cm^{-1}, thus supporting the formulation as a NO^+ complex of Fe(II) with strong π bonding. In the brown $[Fe(H_2O)_5NO]^{2+}$ complex the NO stretching frequency is only 1 745 cm^{-1} and the formulation in terms of Fe(I) and NO^+ is acceptable only if we postulate also very strong back-donation. There are several species (e.g. $[Co(NH_3)_5NO]^{2+}$, $[IrCl_2(NO)(PPh_3)_2]$) which have even lower values for ν (NO), down to 1 500 cm^{-1}. These can be regarded as complexes of NO^-, i.e. a diamagnetic anion (formed by giving an extra electron to nitric oxide) whose electronic structure may be written

$$\overset{\times}{\underset{\times}{{}_{\times}^{\times}N}}=\overset{\times}{\underset{\times}{O_{\times}^{\times}}}$$

This structure implies that if a σ bond from nitrogen to a metal atom is formed, the metal, nitrogen and oxygen will no longer be collinear; that this is so has been confirmed by X-ray diffraction.

22.9 Cyano complexes

Most complex cyanides are discussed in this book under individual transition metals, but some aspects of their chemistry, particularly structural studies, are highly relevant to any general treatment of π bonding by transition metals and are therefore dealt with here.

Cyano complexes do not obey the 18-electron rule, as is obvious when we consider the two series of complexes $[Cr(CN)_6]^{3-}$, $[Mn(CN)_6]^{3-}$,

$[Fe(CN)_6]^{3-}$, $[Co(CN)_6]^{3-}$ and $[V(CN)_6]^{4-}$, $[Cr(CN)_6]^{4-}$, $[Mn(CN)_6]^{4-}$, $[Fe(CN)_6]^{4-}$. Nevertheless, there are few if any stable cyano complexes in which the valence shell of the metal contains *more* than 18 electrons; in this connection the fact that the species formed by dissolving cobalt(II) cyanide in aqueous potassium cyanide is $[Co(CN)_5]^{3-}$ rather than $[Co(CN)_6]^{4-}$ is particularly significant, since Co^{2+} is a d^7 ion. Evidently the e_g^* orbitals are far enough above the t_{2g}^* orbitals in energy to prohibit their occupation; the energy gap here arises, as we showed in Section 19.4, from π bonding between the filled t_{2g} metal orbitals and the π^* orbitals of the ligands; cyanide ion and carbon monoxide are at the strong-field end of the spectrochemical series. There is also some evidence that cyanide ion stabilises low oxidation states, e.g. in the complex $[Ni(CN)_4]^{4-}$, though, as would be expected for a negatively charged ligand, it is certainly much less effective than carbon monoxide in this respect. The series of octahedral hexacyano complexes mentioned above, however, have no counterparts in carbonyl chemistry, and they provide us with a unique opportunity to study the effect of successively filling the metal t_{2g} orbitals along a series of isostructural complexes.

This effect is best seen by comparing the metal–carbon bond lengths, and the metal–carbon and carbon–nitrogen Raman-active (i.e. symmetrical) stretching frequencies; the carbon–nitrogen bond length (like the carbon–oxygen bond length) is sometimes not very sensitive to change in bond order as inferred from vibrational spectroscopy. Data for the potassium salts $K_3[M(CN)_6]$ and the $[M(CN)_6]^{3-}$ ions in aqueous solution are given in Table 22.2(*a*) for M = Cr, Mn, Fe and Co.

All of these cyano complexes are, of course, low-spin species. We should note first of all that the erratic variation in metal–ligand distance

Table 22.2
Some structural data for hexacyano complexes.

(*a*) Complexes of formula $[M(CN)_6]^{3-}$

M	M—C/Å	C—N/Å	νMC/cm^{-1}	νCN/cm^{-1}
Cr (d^3)	2.08	1.14	348	2132
Mn (d^4)	2.00	1.14	375	2129
Fe(d^5)	1.95	1.14	392	2135
Co (d^6)	1.89	1.15	400	2151

(*b*) Complexes of formula $[M(CN)_6]^{4-}$

M	M—C/Å	C—N/Å	νMC/cm^{-1}	νCN/cm^{-1}
V (d^3)	2.16	1.15		
Cr (d^4)	2.05	1.16		
Mn (d^5)	1.95	1.16		
Fe (d^6)	1.92	1.17	393	2095

with electronic configuration that we find for high-spin complexes (e.g. Cr—F, 1.90 Å; Mn—F 1.79–2.09 Å (Jahn–Teller distortion); Fe—F, 1.92 Å; Co—F, 1.89 Å in the trifluorides) is absent here; the metal ion steadily decreases in size as the nuclear charge increases. From the small variation in ν (CN), the increase in M—C bond strength appears to arise from stronger σ bonding as we go from chromium to cobalt. For the $[M(CN)_6]^{4-}$ ions (Table 22.2(*b*)) there is, however, a small but significant increase in the carbon–nitrogen bond length as the metal–carbon bond length decreases (unfortunately reliable vibrational spectra are not yet available for M = V, Cr or Mn). But much more striking than these trends are the relative values for $[Fe(CN)_6]^{3-}$ and $[Fe(CN)_6]^{4-}$, for which we see that the Fe(III) complex has the longer metal–carbon bond, the (slightly) lower metal–carbon stretching frequency, and the higher carbon–nitrogen stretching frequency. The same pattern of variations is also found for other salts containing the $[Fe(CN)_6]^{3-}$ and $[Fe(CN)_6]^{4-}$ ions, and for $[Mn(CN)_6]^{3-}$ and $[Mn(CN)_6]^{4-}$. In other Fe(III) and Fe(II) compounds (e.g. the aquo complexes or the halides) Fe^{3+} is, of course, smaller than Fe^{2+}, so that data for the cyano complexes establish beyond doubt that there is stronger π bonding in the lower oxidation state cyano complex.

This greater strength of the bond in the dipositive oxidation state complex at first sight appears to be at variance with the standard electrode potentials

$$Fe^{3+} + e \rightleftharpoons Fe^{2+} \qquad E^0 = +0.77 \text{ V}$$

$$[Fe(CN)_6]^{3-} + e \rightleftharpoons [Fe(CN)_6]^{4-} \qquad E^0 = +0.36 \text{ V}$$

which, as we showed in Section 7.9, indicate that $[Fe(CN)_6]^{3-}$ has an overall formation constant about 10^7 times that of $[Fe(CN)_6]^{4-}$, i.e. that CN^- stabilises Fe(III). These data, however, relate to formation from aquo complexes in aqueous solutions, and involve hydration enthalpies and entropies as well as metal–ligand bond energies, and a detailed examination of these systems shows that the lower value of E^0 for the $[Fe(CN)_6]^{3-}/[Fe(CN)_6]^{4-}$ couple arises mainly from a very negative entropy of hydration of the $[Fe(CN)_6]^{4-}$ ion: the high charge greatly restricts the freedom of motion of the nearby solvent molecules. Thus the often-quoted example of the cyano complexes of iron as an illustration of complex formation stabilising the higher oxidation state of a metal is much less simple than it appears to be when presented merely as a question of whether Fe^{3+} or Fe^{2+} (state not specified) will interact more strongly with CN^-. We shall return briefly to this subject when we discuss the systematic chemistry of iron in Section 24.7. For the present it will suffice to issue a warning that the connection between bond energy terms in gas-phase species and equilibria in aqueous solution is a very tenuous one except under strictly defined conditions such as we employed in Section 20.2.

PROBLEMS

1 Discuss each of the following observations.

(a) The infrared stretching frequencies in CO, $Mo(CO)_6$, $Mo(CO)_3(NH_3)_3$ and $Mo(CO)_3(PPh_3)_3$ are 2143, 2004, 1855 and 1950 cm^{-1} respectively.

(b) The symmetric CO stretching frequencies in $[Fe(CO)_4]^{2-}$, $[Co(CO)_4]^-$ and $[Ni(CO)_4]$ are 1788, 1918 and 2121 cm^{-1} respectively.

(c) The rate of exchange of CO in $Ni(CO)_4$ with ^{14}CO in toluene solution is proportional to the concentration of $Ni(CO)_4$ and independent of that of ^{14}CO.

(d) The V—C bond lengths in $[V(CO)_6]^-$ and $V(CO)_6$ are 1.93 and 2.00 Å respectively.

2 (a) The rate of exchange of CO in $Co_2(CO)_8$ with ^{14}CO in toluene solutions is independent of the ^{14}CO concentration, and all eight CO molecules exchange at the same rate. Suggest an explanation of this observation.

(b) How would you attempt to estimate ΔH^0 for the interconversion of the two forms of $Co_2(CO)_8$?

3 The compound $[Fe_2(NO)_6](PF_6)_2$ has recently been prepared. Suggest possible structures for the cation it contains. How would you attempt to distinguish between them?

4 The NO group in the complex $[Co(diarsine)_2NO]^{2+}$ has a length of 1.68 Å and the CoNO angle is 180°. When the ion reacts with thiocyanate to form $[Co(diarsine)_2(NCS)NO]^+$ the NO bond length increases to 1.85 Å and the CoNO angle decreases to 135°. Account for these observations.

REFERENCES FOR FURTHER READING

Cotton, F.A. and Wilkinson, G. (1988) *Advanced Inorganic Chemistry*, 5th edition, Interscience, New York. Chapter 22 gives a fuller account of the metal carbonyls.

Hughes, M.N. (1981) *The Inorganic Chemistry of Biological Processes*, 2nd edition, Wiley, Chichester and New York. Chapter 6 covers nitrogen fixation and the nitrogen cycle.

Purcell, K.F. and Kotz, J.C. (1977) *Inorganic Chemistry*, W.B. Saunders, Philadelphia. Chapters 16 and 18 give more advanced treatments of the spectra of metal carbonyls and the structures of polynuclear carbonyls than are given here.

Sharpe, A.G. (1976) *The Chemistry of Cyano Complexes of the Transition Metals*, Academic Press, London. Chapter I gives a general account of cyano complexes.

23 *Organometallic compounds of the transition metals*

Introduction • Alkyl and aryl complexes • Alkene complexes • Allyl and butadiene complexes • Complexes containing delocalised cyclic systems • Carbene (alkylidene) and carbyne (alkylidyne) complexes • Organometallic compounds in homogeneous catalysis

23.1 Introduction

Organic compounds of the non-transition elements have been discussed briefly in several sections of Chapters 10–16 inclusive. Where, as is usually the case, the bonding in such compounds is covalent, analogies with hydrides and halides are sufficient to give a clear idea of structures, the organic group being σ bonded by an electron-pair bond to a single atom of a non-transition element or involved in multicentre bonding to two or more such atoms. Transition elements also form simple alkyls or aryls, but except for those of zinc and mercury (which in many respects differ from most transition metals), they are usually unstable, and most organic compounds of transition metals are derivatives of alkenes, alkynes or unsaturated ring systems. The yellow alkene compound potassium ethylenetrichloroplatinate(II) or *Zeise's salt*, $K[PtCl_3(C_2H_4)]$, was discovered as long ago as 1830, but extensive progress in this field has been made only since the chance discovery of ferrocene, $(C_5H_5)_2Fe$, in 1951. In this chapter we can give only a brief introduction to the very large and complex subject of the organometallic chemistry of the transition metals, and we have chosen to discuss a small number of systems in some detail rather than to give only an outline of a wide range of compounds and reactions. In so doing, we have emphasised the significance of the transition metal, rather than that of the ligand, in the compounds discussed. Equally valid treatments which lay more weight on the organic chemistry involved, and which therefore provide fuller coverage of the role of organometallic compounds in organic reactions, will be found in the references cited at the end of the chapter.

Among the methods used for the preparation of organic compounds of the non-transition elements are the direct interaction of hydrocarbons containing acidic protons and metals (e.g. for the preparation of alkali metal derivatives), the action of alkyl or aryl halides on metals or metalloids (e.g. for the preparation of Grignard reagents and alkyl chlorosilanes), the use of Grignard reagents, mercury dialkyls or lithium alkyls for the introduction of alkyl groups (e.g. for the preparation of alkyl derivatives of boron, aluminium, silicon and phosphorus), and addition of an alkene or alkyne to a hydride (e.g. for the preparation of alkyl derivatives of boron and aluminium). All of these methods are used also in the preparation of organic derivatives of transition metals, but many other reactions are also available here.

Before we go on to preparations and properties, we should mention the nomenclature for compounds in which a ligand is attached to one metal atom through more than one of its carbon atoms, e.g. a benzene molecule in dibenzene chromium, which contains a chromium atom sandwiched symmetrically between two parallel benzene molecules. If two, three, four or more atoms of the ligand are within bonding distance of the same metal atom, this is indicated by prefixing the name of the ligand with dihapto-, trihapto-, tetrahapto- etc. (from the Greek for *fasten*) or the symbol η^2-, η^3-, η^4- etc. (h^2-, h^3-, h^4- etc. are also used). Thus the formula of dibenzene chromium is written $(\eta^6\text{-}C_6H_6)_2Cr$. Alkyl and aryl groups which form only σ bonds to metal atoms are monohapto groups. It should be noted that there are some differences in meaning between hapto- and -dentate as applied to ligands; hapticity refers to the number of ligand atoms bonded to the same metal atom (irrespective of the number of electrons involved); denticity refers to the number of donor bonds formed by the ligand (whether they are formed with the same or with different metal atoms). Several further examples of polyhapto ligands will be given later in this chapter.

In general, organometallic compounds of transition metals, other than σ bonded alkyl compounds containing no π bonding ligands, obey the 18-electron rule, though not so faithfully as the carbonyls and their derivatives. There are, for example, a considerable number of fairly stable square planar complexes of d^8 metal ions (e.g. the ion $[PtCl_3(C_2H_4)]^-$ that we mentioned earlier) that have only 16 electrons around the metal atom. Such complexes, in which the coordination number of the metal can be increased, are said to be coordinatively unsaturated; they are often useful homogeneous catalysts. A few compounds are known in which the metal has a valence shell of more than 18 electrons (e.g. $(\eta^5\text{-}C_5H_5)_2Co$), but these are readily oxidised to 18-electron systems. In counting electrons, ligands such as CO, PPh_3 and halide ions are considered as two-electron donors. The neutral molecules C_2H_4, C_4H_6 (butadiene) and C_6H_6 functioning as η^2-, η^4- and η^6- ligands contribute two, four and six electrons respectively, one pair from each double bond if we neglect electron delocalisation. For hydrogen and methyl groups two conventions are in current use and the reader should be prepared to meet both. Either we may count them as H^- and CH_3^- respectively, treating them as two-electron donors like halide ions, or we may count them as radicals having a single unpaired electron which they contribute to a two-electron bond, when we treat them as one-electron donors. The difference is often more apparent than real. If, for example, we have a compound MH we may write it as M^+H^- formed from a cation and a hydride anion, or as a covalently bonded species M—H formed from a metal atom and a hydrogen atom: if we treat hydrogen and methyl as two-electron donors (as we shall in this book) we must be sure to write the metal in MH in a positive oxidation state; if we treat them as one-electron donors, we must write the metal atom in MH as an uncharged entity in oxidation state zero. Other alkyl groups are treated in the same way. Similar considerations apply to the η^3-allyl and η^5-cyclopentadienyl species, which we treat here as four- and six-electron donating entities $CH_2{=}CH.CH_2^-$ and $C_5H_5^-$ respectively, thus describing ferrocene (Section 23.5), $Fe(C_5H_5)_2$, for example as the covalent product of the combination of a Fe^{2+} cation and two $C_5H_5{}^-$ anions rather than as a compound of Fe(0) and two C_5H_5 radicals, each obtained by loss of a hydrogen atom from cyclopentadiene (C_5H_6) and each having five π electrons. So our electron count for ferrocene is either $Fe^{2+}(d^6)+(2\times 6)=18$ or $Fe(0)(d^8)+(2\times 5)=18$. So far as hydrogen and alkyl groups are concerned, the practice adopted here has the advantage that it is in line with what is customary in some other areas of transition metal chemistry: in discussing the *trans* effect, for example, we wrote H^- and CH_3^- as ligands, and the complexes $[Co(CN)_5H]^{3-}$ and $[Co(CN)_5CH_3]^{3-}$ that we shall meet in Section 24.8 are certainly better represented as Co(III) complexes analogous to $[Co(CN)_6]^{3-}$ than as Co(II) complexes. Further, it is sometimes useful to be reminded of the aromaticity of ligands such as $C_5R_5^-$ and $C_8R_8^{2-}$, which have 6 and 10 π electrons respectively. On the other hand the consistent application of this practice leads to difficulty in a reaction such as the protonation of ferrocene in strongly acidic media:

$$(C_5H_5)_2Fe+H^+ \rightarrow [(C_5H_5)_2FeH]^+$$

This has to be represented as the oxidation of iron(II) to iron(IV), the cation produced being regarded as made up from Fe^{4+}, $2C_5H_5^-$ and H^-, obviously a highly artificial representation. There is no entirely satisfactory way out of these difficulties, which are inherent in the application of the oxidation state concept to compounds in which the distribution of electrons between metal and ligand is uncertain.

We shall now proceed to consider in succession a few of the principal classes of organic complexes of transition metals, starting with simple alkyl derivatives and gradually moving on to species containing cyclic polyhapto ligands.

23.2 Alkyl and aryl complexes

Simple σ bonded organic derivatives of transition metals are, in general, much more reactive than the analogous compounds of non-transition elements, though the origin of this observation is kinetic rather than thermodynamic: the Ti–Me bond, for example, is stronger than Pb–Me and about as strong as Sn—Me, and the reactivity of the titanium compound comes from the availability of vacant orbitals of the metal atom.

Transition metal alkyls are particularly prone to decomposition by alkene elimination according to the equation

$$MCH_2CH_2R \rightarrow MH + RCH = CH_2$$

The mechanism of the reaction may be represented as:

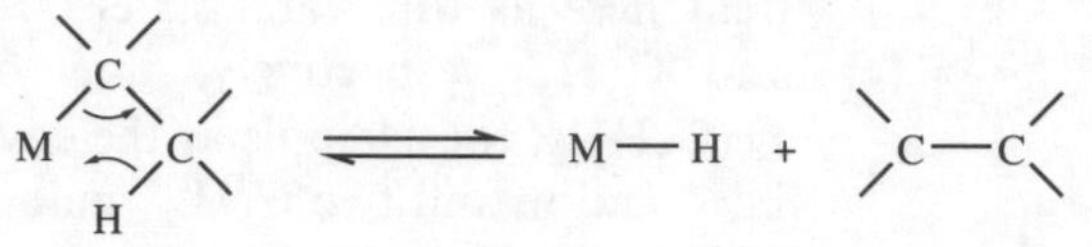

$$M\text{(C)(C)H} \rightleftharpoons M-H + \;>C=C<$$

Methyl derivatives cannot decompose by this route, and it is noteworthy that they are usually more stable than ethyl derivatives. Decomposition may also be hindered by substitution of all β hydrogen atoms (as in $Ti(CH_2Ph)_4$ or $Cr(CH_2SiMe_3)_4$).

Alkyl and aryl derivatives of transition metals may be made by the general reactions mentioned in the last section, e.g.

$$TiCl_4 + 4LiCH_3 \xrightarrow[\text{ether}]{-80\,°C} Ti(CH_3)_4 + 4LiCl$$

$$WCl_6 + 6LiCH_3 \xrightarrow[\text{ether}]{} W(CH_3)_6 + 6LiCl$$

$$HCo(CO)_4 + C_2H_4 \rightarrow C_2H_5Co(CO)_4$$

$$HCo(CO)_4 + C_2F_4 \rightarrow HCF_2CF_2Co(CO)_4$$

Other routes to such compounds when π donating groups are also present are illustrated by the reactions:

$$CH_3I + Na[Mn(CO)_5] \rightarrow CH_3Mn(CO)_5 + NaI$$

$$CF_3COCl + Na[Mn(CO)_5] \rightarrow CF_3COMn(CO)_5 + NaCl$$

$$CF_3COMn(CO)_5 \xrightarrow{\text{heat}} CF_3Mn(CO)_5 + CO$$

Alkyls containing no other groups are generally very unstable: $Ti(CH_3)_4$, for example, decomposes at $-50\,^\circ C$. The presence of electron-withdrawing groups (such as CF_3 or C_6H_5) or of π acceptor groups (such as CO, bipyridyl or PPh_3) results in a substantial increase in stability. The compound (bipy)$Ti(CH_3)_4$, for example, may be warmed to 30 °C before decomposition occurs, and $CF_3Co(CO)_4$ may be distilled unchanged at 90 °C although $CH_3Co(CO)_4$ decomposes above -20 °C and simple cobalt alkyls are not at present known. It is suggested that the drawing away of electrons by the fluorine atoms permits π donation from the filled metal orbitals to carbon atoms, thus strengthening the metal–carbon bond. The alkyl $W(CH_3)_6$ is remarkable as the first example of a discrete trigonal prismatic molecule.

23.3 Alkene complexes

In this section we shall restrict attention to complexes of monoalkenes; non-conjugated dialkenes behave like separate alkene molecules, but conjugated dialkenes such as butadiene undergo substantial changes in bond lengths on coordination to a metal atom and their complexes are discussed in the next section.

Alkene complexes are usually made by displacement of carbon monoxide or halide ions. Zeise's salt, $K[PtCl_3(C_2H_4)]$, has already been mentioned; it is prepared by the action of ethylene on potassium tetrachloroplatinate(II) in aqueous solution. The reaction is catalysed by $SnCl_2$, the ion $[PtCl_3(SnCl_3)]^{2-}$ being an intermediate. Silver(I) and copper(I) compounds in aqueous solutions combine with many alkenes; stable solid products can seldom be isolated but the solubility of alkenes in such solutions can be used for their separation from alkanes. Vaska's compound, *trans* $(Ph_3P)_2Ir(CO)Cl$, and the corresponding bromine compound both form readily dissociated complexes with ethylene but very stable complexes with tetracyanoethylene, $C_2(CN)_4$. Platinum(0) complexes such as $(Ph_3P)_2Pt(C_2H_4)$ are made by the action of ethylene and hydrazine hydrate on $(Ph_3P)_2PtCl_2$ in ethanolic solution.

The structure of the anion of Zeise's salt has been examined in detail by X-ray and neutron diffraction, and is shown in Fig. 23.1. Both carbon atoms of the ethylene molecule, it should be noted, are equidistant from the platinum atom, but the carbon–carbon bond length has increased from 1.337 Å in ethylene to 1.375 Å in the complex; this increase is accompanied by a decrease in the C=C stretching frequency from 1623 cm^{-1} in the Raman spectrum of ethylene to 1526 cm^{-1} in the infrared spectrum of Zeise's salt. The C=C bond in Zeise's salt is perpendicular to the $PtCl_3$ plane (thus minimising repulsions between hydrogen and chlorine atoms); further, the originally planar ethylene molecule has been

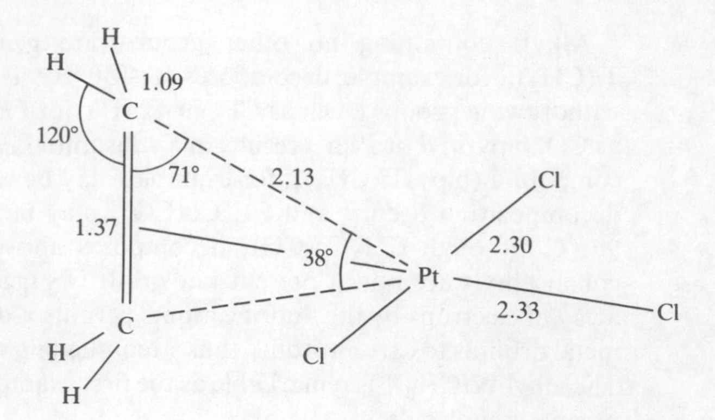

Figure 23.1
The structure of the anion in Zeise's salt (bond lengths in ångströms).

deformed slightly so that the hydrogen atoms are bent back from the rest of the ion. The general bonding scheme for such complexes involves donation from the filled π orbital of the alkene to an empty metal orbital of suitable symmetry and back-donation from a filled metal orbital of suitable symmetry to an empty π^* orbital of the alkene, an overall interaction similar to that which operates in transition metal carbonyls, though the configuration of the carbonyl group with respect to the metal atom is, of course, quite different. Figure 23.2 indicates the orbitals believed to be involved in the particular case of Zeise's salt.

Complexes of tetrafluoroethylene or tetracyanoethylene and transition metals would be expected to involve only weak ligand→ metal donor interactions but strong metal→ligand interactions. In

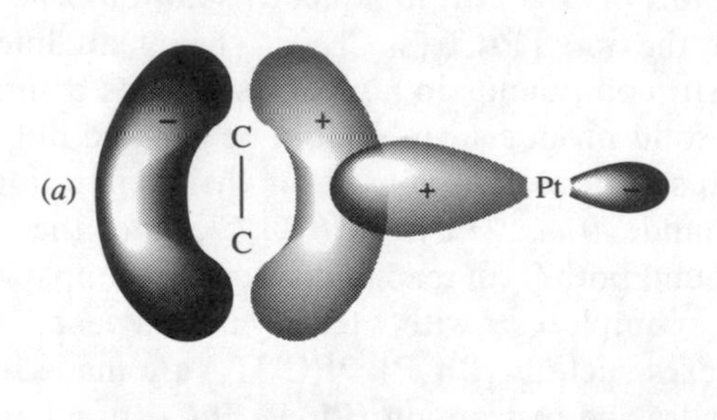

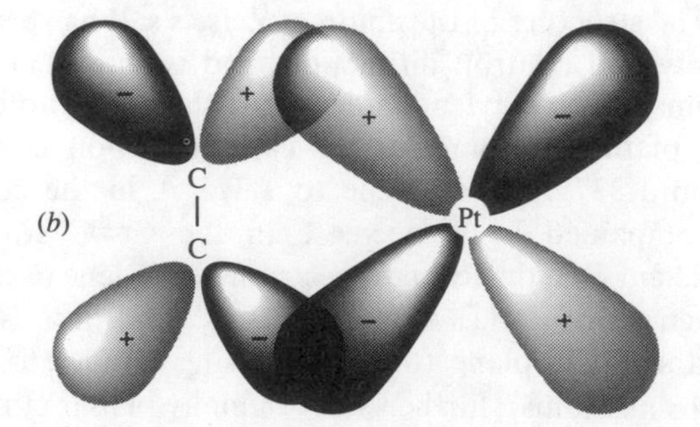

Figure 23.2
Platinum– ethylene bonding in the $[PtCl_3(C_2H_4)]^-$ anion showing the overlapping of (*a*) a $5d\ 6s\ 6p^2$ metal orbital with the ethylene π orbital (*b*) the d_{xz} metal orbital with the ethylene π^* orbital.

$(Ph_3P)_2Pt[C_2(CN)_4]$, in which the coordination of the metal atom is planar, the carbon—carbon bond length in the organic ligand is 1.52 Å, nearly the same as in ethane (compared with 1.34 Å in free tetracyano-ethylene), and the best formulation is:

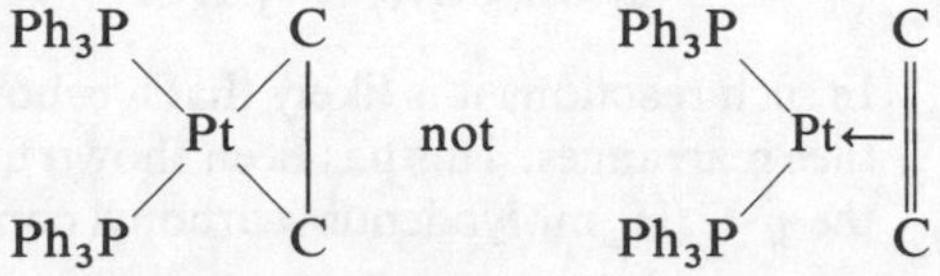

$$\begin{array}{c} Ph_3P \quad\quad C \\ \diagdown \;\; \diagup | \\ Pt \;\;\; | \\ \diagup \;\; \diagdown | \\ Ph_3P \quad\quad C \end{array} \quad \text{not} \quad \begin{array}{c} Ph_3P \quad\quad C \\ \diagdown \quad \| \\ Pt\leftarrow \| \\ \diagup \quad \| \\ Ph_3P \quad\quad C \end{array}$$

The stereochemistry of the molecule, it may be noted, is characteristic of Pt(II) rather than Pt(0).

In $Pt(C_2H_4)_2(C_2F_4)$, in which the platinum and all six carbon atoms are almost coplanar, carbon—carbon distances are 1.38 and 1.43 Å in the C_2H_4 and C_2F_4 ligands respectively; platinum—carbon (of C_2H_4) and platinum—carbon (of C_2F_4) distances are 2.22 and 2.03 Å respectively.

The orientation of the alkene with respect to the plane of the rest of the ion in Zeise's salt and similar species is not a matter of great significance; the σ component of the bond offers no resistance to rotation of the alkene about the z axis, and since the d_{xz} and d_{yz} orbitals are degenerate in the complex, metal–alkene π overlap is independent of the orientation of the alkene with respect to such rotation. It is, accordingly, not surprising that the ^1H nmr spectrum of coordinated alkenes is often temperature-dependent, showing that rotation of the alkene about the metal–alkene axis takes place at higher temperatures. In the case of $(\eta_5\text{-}C_5H_5)Rh(C_2H_4)_2$, for example, two types of alkene proton are distinguishable at 0°C, but all become equivalent at 60 °C. Such *fluxionality* is very common amongst transition metal organometallic compounds.

Unlike free alkenes, which are reactive towards electrophiles, the coordinated alkenes, because of electron donation to the metal, are more susceptible to attack by nucleophiles.

Many reactions of alkenes in solution are subject to homogeneous catalysis by transition metal complexes, and alkene complexes are frequently involved in such processes. A few examples are discussed in some detail in Section 23.6.

23.4 *Allyl and butadiene complexes*

The allyl group forms both monohapto (or σ-allyl) complexes analogous to those formed by methyl and ethyl groups and trihapto (or π-allyl) complexes in which it acts as a trihapto ligand contributing, if we follow the convention of regarding the ligand as the $CH_2{=}CH.CH_2^-$ anion, four electrons to the metal cation. Illustrative preparations are:

$$NiCl_2 + 2C_3H_5MgBr \xrightarrow[\text{ether}]{-10\,°C} Ni(\eta^3\text{-}C_3H_5)_2$$

$$Na[Mn(CO)_5] + C_3H_5Cl \rightarrow (\eta^3\text{-}C_3H_5)Mn(CO)_4 + CO + NaCl$$

In such reactions it is likely that a σ-bonded derivative is first formed and then rearranges. This has been shown to be the case for the preparation of the η^5-C_5H_5 molybdenum carbonyl complex:

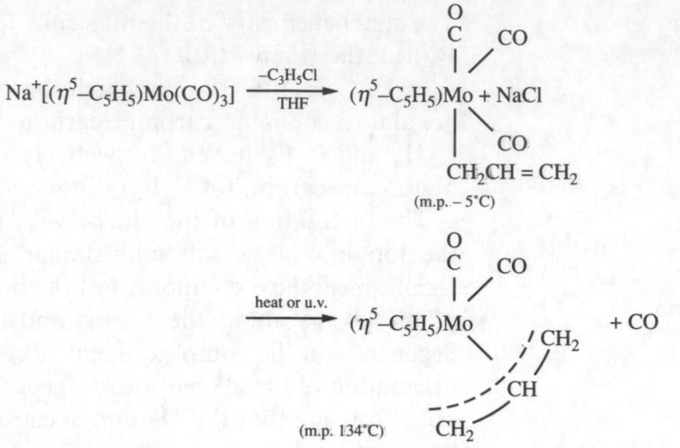

$$Na^+[(\eta^5\text{-}C_5H_5)Mo(CO)_3]^- \xrightarrow[\text{THF}]{-C_3H_5Cl} (\eta^5\text{-}C_5H_5)Mo(CO)_3(CH_2CH{=}CH_2) + NaCl$$

(m.p. – 5°C)

$$\xrightarrow{\text{heat or u.v.}} (\eta^5\text{-}C_5H_5)Mo(CO)_2(\eta^3\text{-}CH_2CHCH_2) + CO$$

(m.p. 134°C)

Note that this $\eta^1 \rightarrow \eta^3$ allylic conversion involves the simultaneous displacement of a molecule of CO as a two-electron donor is converted into a four-electron donor.

A typical structure for a η^3-allyl complex is that of the compound $[PdCl(\eta^3\text{-}C_3H_5)]_2$, obtained by the reaction sequence

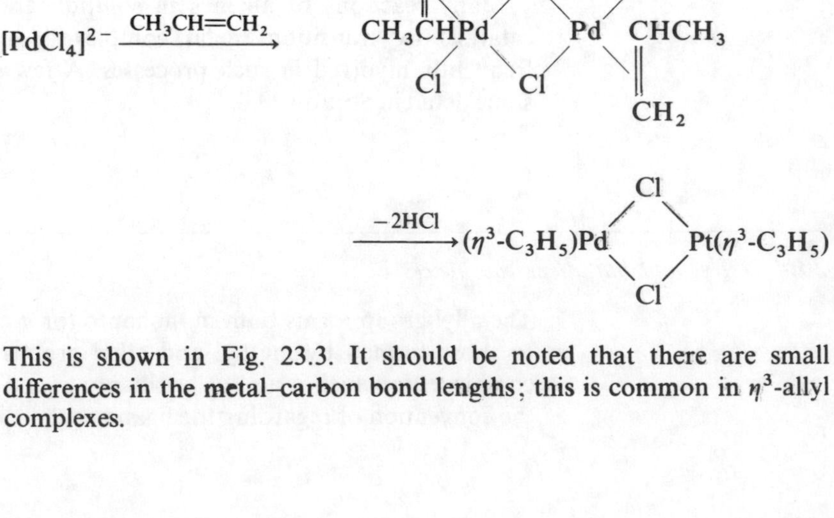

$$[PdCl_4]^{2-} \xrightarrow{CH_3CH{=}CH_2} [PdCl_2(CH_2{=}CHCH_3)]_2 \;(\text{Cl-bridged})$$

$$\xrightarrow{-2HCl} (\eta^3\text{-}C_3H_5)Pd(\mu\text{-}Cl)_2Pt(\eta^3\text{-}C_3H_5)$$

This is shown in Fig. 23.3. It should be noted that there are small differences in the metal–carbon bond lengths; this is common in η^3-allyl complexes.

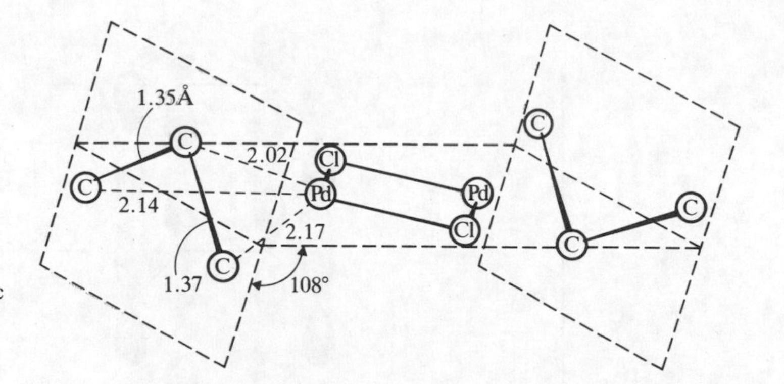

Figure 23.3
The structure of the dimeric allylpalladium(II) chloride (bond lengths in ångströms).

In addition to their formation in the reactions described above, η^3-allyl complexes are also formed by protonation of η^4-butadiene complexes, e.g.

CH—CH
H_2C CH_2
Fe
OC C CO
O

$\xrightarrow{HCl}$

CH CH_3
H_2C CH
Fe
OC C C Cl
O O

The starting material for this reaction, an air-stable yellow solid melting at 19 °C, is obtained by the action of 1,3-butadiene on iron pentacarbonyl under pressure; although butadiene exists in the *trans* form it isomerises to give better overlap with the metal orbitals. The complexed butadiene is difficult to hydrogenate and does not give the Diels–Alder reaction characteristic of conjugated dienes; further, an X-ray diffraction study shows that the iron atom is almost equidistant from all four carbon atoms and that all three carbon–carbon bond lengths are 1.45 Å (in butadiene itself the terminal bond lengths are 1.36 Å and the central bond length is 1.45 Å). The structure of the compound is therefore often written

HC—CH
H_2C CH_2
Fe
OC C CO
O

In complexes of substituted butadienes, however, it is not uncommon to find small differences in bond lengths, indicating that electron delocalisation is incomplete: where bonding to the metal is particularly strong, as in perfluoro complexes, the middle bond becomes the shortest, as in the first electronically excited state of the butadiene molecule itself.

Since the environment of the metal atoms in η^3-allyl complexes is often unsymmetrical (although C—C distances in the ligand are equal), we shall

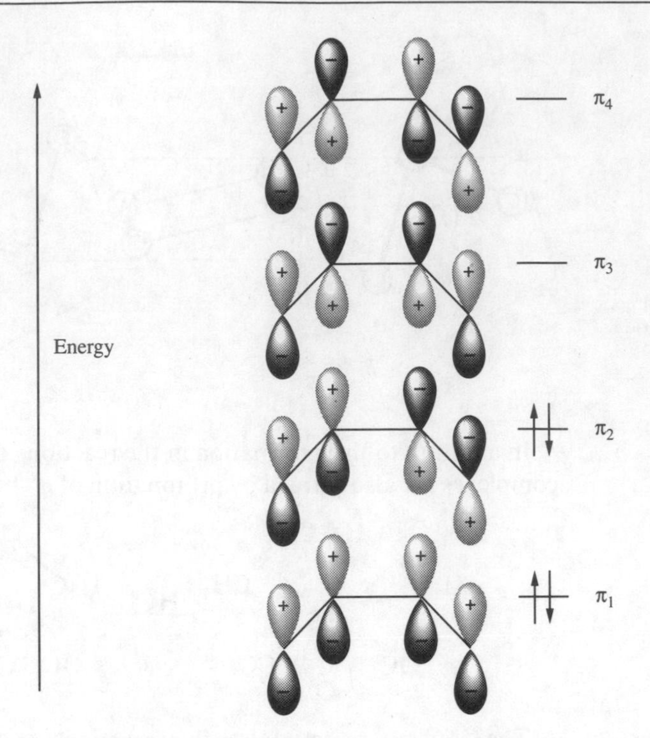

Figure 23.4
π Molecular orbitals in 1,3-butadiene.

not comment on the bonding in them, but it is instructive to consider η^4-butadiene complexes briefly. The molecular orbitals for butadiene may be represented as in Fig. 23.4. The difference in bond lengths arises because when only the two bonding orbitals (π_1 and π_2) are occupied the overall order is about 1.9 for the terminal bonds and 1.2 for the middle bond. When η^4-butadiene acts as a donor it loses bonding electron density from the middle bond into empty metal orbitals of suitable symmetry; when it acts as a π acceptor it gains antibonding electron density in the terminal bonds of π_3 and bonding electron density in the middle bond. The overall effect is thus to weaken the terminal bonds considerably and strengthen the middle bond.

The diminished reactivity of coordinated 1,3-dienes is the basis of the use of the $Fe(CO)_3$ moiety as a protecting group. By the action of an iron carbonyl most 1,3-dienes are easily converted into complexes, and reactions can then be carried out in other parts of the molecule, after which the $Fe(CO)_3$ group is removed by mild oxidants such as aqueous iron(III) or trimethylamine oxide in a hydrocarbon solvent.

23.5 Complexes containing delocalised cyclic systems

The best-known delocalised cyclic ligands are the $C_5H_5^-$ anion and the C_6H_6 molecule, each of which contains six π electrons and exhibits aromatic character. According to Hückel's rule for the occurrence of aromaticity (which is dealt with in textbooks of organic chemistry), only systems containing $4n+2$, where $n=0, 1, 2 \ldots$, π electrons are aromatic. Although cyclopropene, C_3H_4, is not known in the free state and cyclobutadiene, C_4H_6, is very readily polymerised, transition metal derivatives of the ions $C_3Ph_3^+$ and $C_4H_4^{2-}$ (two and six π electrons respectively) may be obtained, e.g. by the reactions:

$$Ph_3C_3^+Cl^- + Ni(CO)_4 \rightarrow [(\eta^3\text{-}Ph_3C_3)Ni(CO)_2]^+Cl^-$$

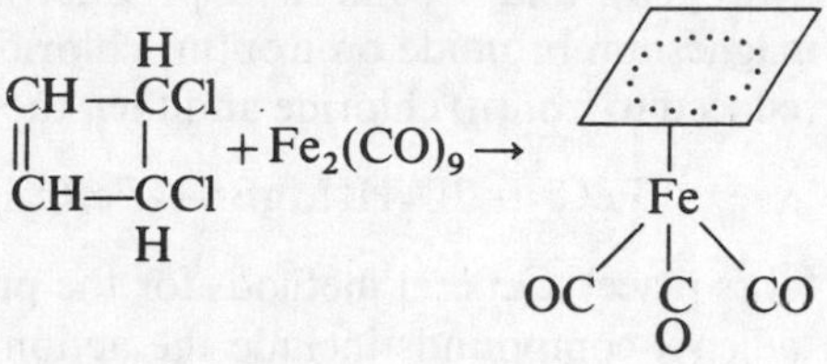

H
CH—CCl
‖ | + $Fe_2(CO)_9$ →
CH—CCl
H

Fe
OC C CO
O

Unlike cyclobutadiene, the ring in $C_4H_4Fe(CO)_3$ is square, and its aromatic nature is shown by the fact that it undergoes Friedel–Crafts acylation when treated with acetyl chloride and aluminium chloride.

The compounds $(\eta^6\text{-}C_6H_6)_2Cr$, $(\eta^5\text{-}C_5H_5)_2Fe$ and $(\eta^4\text{-}C_4H_4)_2Ni$ would form an isoelectronic series; the tetraphenyl derivative of the last of these is obtained by the reactions:

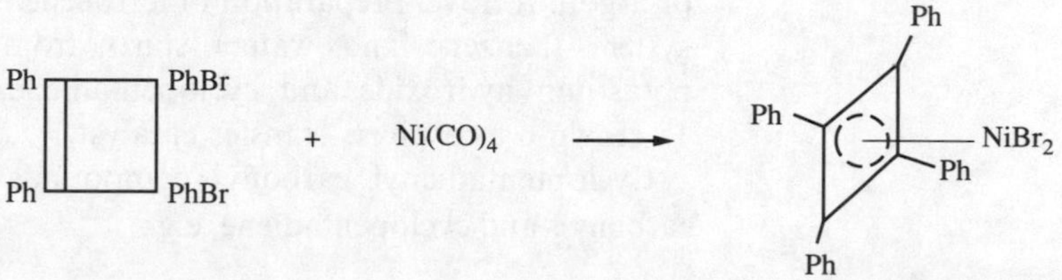

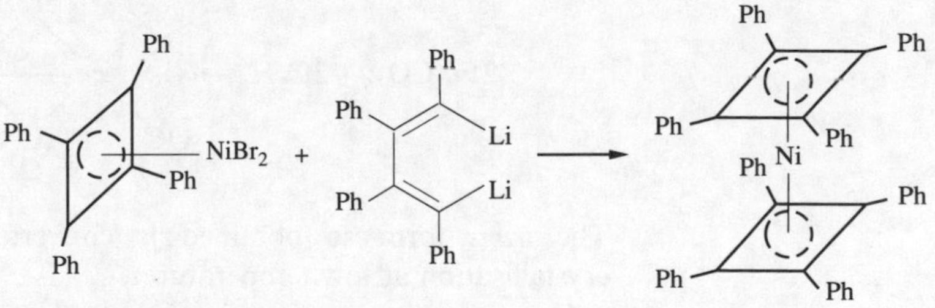

Ph PhBr
Ph PhBr + $Ni(CO)_4$ → Ph Ph Ph Ph $NiBr_2$

Ph Ph Ph Ph $NiBr_2$ + Ph Ph Ph Ph Li Li → Ph Ph Ph Ph Ni Ph Ph Ph Ph

The cyclopentadienyl anion, $C_5H_5^-$, may bond as η^1, η^3 or η^5, the mode of attachment being such as to maintain if possible an 18-electron system: thus we have, for example, $(\eta^5\text{-}C_5H_5)_2Ti(CO)_2$, $(\eta^3\text{-}C_5H_5)(\eta^5\text{-}C_5H_5)Cr(CO)_2$, and $(\eta^1\text{-}C_5H_5)(\eta^5\text{-}C_5H_5)Fe(CO)_2$, with $\eta^1\text{-}C_5H_5$ and $\eta^3\text{-}C_5H_5$ behaving like alkyl and allyl groups respectively. In this section we are concerned only with $\eta^5\text{-}C_5H_5$ derivatives; for the fluxionality of η^1- and η^5-systems see Section 24.3.

The complex $Fe(C_5H_5)_2$, which we are here regarding as the Fe^{2+} derivative of the $C_5H_5^-$ anion, is the best known of all transition metal organic compounds, and is usually referred to as *ferrocene*; the analogous compounds $Mn(C_5H_5)_2$, $Co(C_5H_5)_2$ and $Ni(C_5H_5)_2$ are commonly called *manganocene, cobaltocene* and *nickelocene* respectively. All contain a metal atom sandwiched between two planar C_5H_5 rings; details of their structures are given later. Ferrocene was discovered, like many other organometallic compounds, by accident, first in a study of the action of cyclopentadiene on reduced iron at 300 °C (which gives ferrocene and hydrogen), and second as a product of the action of cyclopentadienyl magnesium bromide on iron(III) chloride (in which the iron compound is reduced to iron(II) chloride and then the reaction

$$FeCl_2 + 2C_5H_5MgBr \rightarrow Fe(C_5H_5)_2 + 2MgBrCl$$

takes place). General methods for the preparation of pentahaptocyclopentadienyl compounds include the action of the sodium salt $Na^+C_5H_5^-$ or the thallium salt $Tl^+C_5H_5^-$ on a metal halide in ethereal solvents, especially tetrahydrofuran, and the interaction of cyclopentadiene itself and a metal halide in the presence of excess of diethylamine or piperidine:

$$2NaC_5H_5 + MnBr_2 \rightarrow Mn(C_5H_5)_2 + 2NaBr$$

$$2C_5H_6 + NiCl_2 + 2Et_2NH \rightarrow Ni(C_5H_5)_2 + 2Et_2NH_2Cl$$

For most complexes the preparation must be carried out under nitrogen or argon. A novel preparation of ferrocene in a rapidly stirred two-phase system (benzene and water) starts from aqueous iron(II) chloride, potassium hydroxide and cyclopentadiene, and uses the crown ether 18-crown-6 as a phase-transfer catalyst.

Cycloptentadienyl carbonyl compounds are usually obtained from carbonyls and cyclopentadiene, e.g.

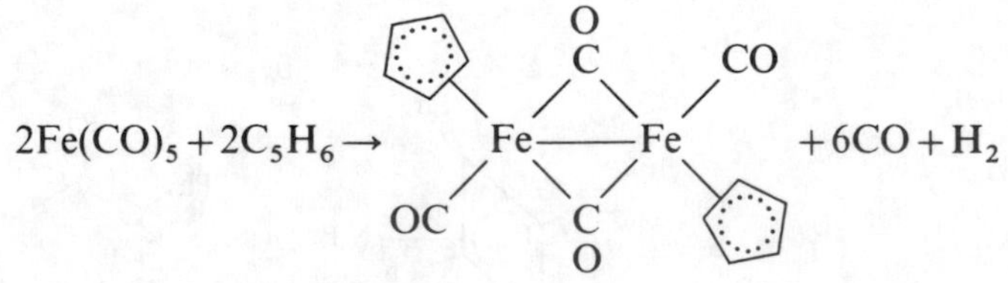

$$2Fe(CO)_5 + 2C_5H_6 \rightarrow \begin{array}{ccc} C_5H_5 & \overset{O}{C} & CO \\ Fe & \text{———} & Fe \\ OC & \underset{O}{C} & C_5H_5 \end{array} + 6CO + H_2$$

The *trans* form so obtained is converted into the *cis* isomer by crystallisation at low temperatures.

Ferrocene is a diamagnetic orange solid (m.p. 174 °C) that sublimes at 100 °C; it is unaffected by air and insoluble in water, but dissolves in most organic solvents. It is exceptionally stable for an organometallic compound, and in the absence of air may be heated to almost 500 °C without decomposition. It is, however, readily oxidised by aqueous silver ion or dilute nitric acid to the blue *ferricinium* ion $[Fe(C_5H_5)_2]^+$. Cobaltocene, which contains 19 electrons in the valence shell of the metal atom, is obtained only under reducing conditions; salts of the ion $[Co(C_5H_5)]^+$ are often produced in reactions expected to yield the neutral complex, and may be reduced with lithium tetrahydridoaluminate. Nickelocene, a 20-electron system, is very prone to undergo reactions which result in the formation of an 18-electron system; it is, for instance, easily converted by sodium amalgam into (η^5-cyclopentadienyl) (η^3-cyclopentenyl) nickel. Manganocene, unlike the other complexes of metals of the first transition series, is dimorphic: the brown ordinary-temperature form, which is apparently ionic, has an infinite chain structure, is a high-spin complex, and is vigorously decomposed by water; the high-temperature form is much more like ferrocene.

The aromatic character of the rings in ferrocene is well established; Friedel–Crafts acylation, metalation by butyl lithium and sulphonation all take place, and there is now an extensive organic chemistry of ferrocene.

If we represent ferrocene as the Fe^{2+} compound of the $C_5H_5^-$ ion and consider replacement of this anion by the neutral C_6H_6 molecule (also a 6 π electron system) it is obvious that for a neutral species the compound which would contain an 18-electron valence shell is $Cr(C_6H_6)_2$. This compound is best made by the sequence of reactions:

$$3CrCl_3 + 2Al + AlCl_3 + 6C_6H_6 \longrightarrow 3[(C_6H_6)_2Cr]AlCl_4$$

$$2[(C_6H_6)_2Cr]^+ + 4OH^- + S_2O_4^{2-} \longrightarrow 2Cr(C_6H_6)_2 + 2H_2O + 2SO_3^{2-}$$

More recently it has also been obtained by the action of benzene on atomic chromium produced by rapid condensation of the vapour at 77 K.

Dibenzene chromium, a brown solid (m.p. 284 °C), is surprisingly easily oxidised to the 17-electron system $[(C_6H_6)_2Cr]^+$, and does not withstand the experimental conditions of electrophilic aromatic substitution. Several other benzene or substituted benzene complexes of transition metals are known; they include $V(C_6H_6)_2$, $Mo(C_6H_6)_2$ and $Fe(C_6Me_6)_2$. The last compound is a very powerful reducing agent which is oxidised successively to the $[(C_6Me_6)_2Fe]^+$ and $[(C_6Me_6)_2Fe]^{2+}$ cations. One benzene ring in $Cr(C_6H_6)_2$ is readily displaced by carbon monoxide, giving the compound

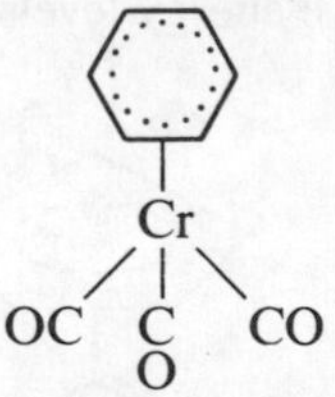

$(\eta^6\text{-}C_6H_6)Cr(CO)_3$: Cr bonded to a benzene ring and to three CO groups (OC, C≡O, CO)

which is much more stable than the dibenzene complex itself. The arene ring in such compounds is susceptible to nucleophilic attack; with sodium methoxide, for example, the chlorobenzene complex readily gives a methyl ether.

Other carbocyclic groups which exhibit aromatic character by being planar with all carbon–carbon bonds identical in length are the $C_7H_7^+$ (tropylium) cation and the $C_8H_8^{2-}$ anion. The former occurs in $(\eta^5\text{-}C_5H_5)(\eta^7\text{-}C_7H_7)Cr$, and the latter in $U(\eta^8\text{-}C_8H_8)_2$. For the preparation and properties of these compounds, however, more specialised texts should be consulted.

To conclude this section we must briefly discuss the structures of ferrocene and dibenzene chromium, and the nature of the bonding in them. Ferrocene in the solid state has the eclipsed configuration, unlike the corresponding compounds of cobalt and nickel (Fig. 23.5). This difference probably arises from packing considerations, since the barrier to rotation of the rings in ferrocene, whether in the gas phase or in solution, is very small (about 5 kJ mol^{-1}). In dibenzene chromium the configuration is eclipsed. In ferrocene the carbon–carbon distance in the rings is 1.39 Å and the iron–carbon distance 2.04 Å. In dibenzene chromium the carbon–carbon distance in the eclipsed rings is 1.42 Å (slightly greater than in benzene itself), and the chromium–carbon distance 2.15 Å.

We can do no more than indicate the general features of the approach to the electronic structures of ferrocene and dibenzene chromium, taking the latter as an example since we have described the π molecular orbitals of benzene in Section 4.7. Each carbon atom in benzene contributes one π orbital (the other three orbitals being used for σ bonding); two rings therefore contribute twelve orbitals and the transition metal another nine (the five $3d$, the $4s$ and the three $4p$ orbitals), giving a total of twenty-one. (For ferrocene there are correspondingly nineteen.) We then consider what metal orbitals have the correct symmetry to overlap the twelve ligand group π orbitals: if both rings have the positive lobes of their lowest energy π orbitals on the sides nearer to the metal, for example, the s and d_{z^2} orbitals would have the correct symmetry; if one ring has the positive lobe nearer the metal and the other the negative lobe nearer the metal, the p_z orbital has the correct symmetry. Some ligand group orbitals turn out to be non-bonding, there being no suitable metal orbital available to overlap them. Correct symmetry will not necessarily lead to appreciable bonding interaction, however; for this purpose it is also necessary that the orbitals involved should be of similar energies. The calculation of the relative energy levels in species as complicated as these

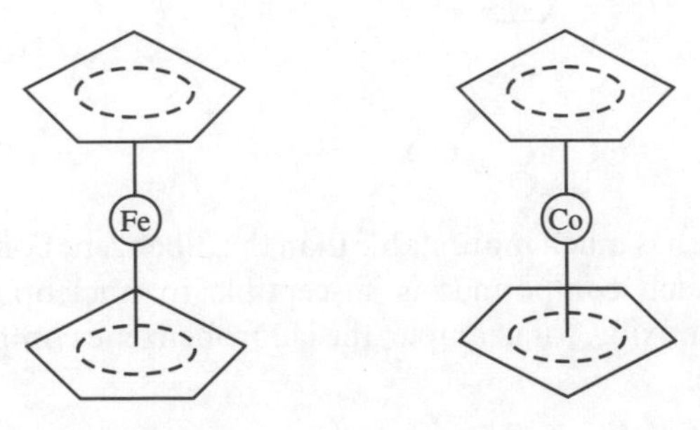

Figure 23.5
The structures of ferrocene and cobaltocene.

organometallic compounds is still subject to some uncertainty, and a number of different energy level diagrams have been suggested for each molecule. For both molecules, nevertheless, it transpires that there are six bonding and three non-bonding or relatively weakly bonding orbitals, much as for an octahedral complex; and since both ferrocene and dibenzene chromium are 18-electron systems, all these orbitals, and no others, are filled. For dibenzene chromium the strongest metal–ring bonding arises from the overlapping of the p_x, p_y, d_{xz} and d_{yz} orbitals and the two pairs of degenerate higher energy bonding orbitals (i.e. of e_{1g} symmetry) of the benzene molecules. Similar conclusions hold for the ferrocene molecule, though in this case the overlapping involving the d_{xz} and d_{yz} orbitals is decidedly greater. Two of the strongly bonding molecular orbitals of dibenzene chromium are shown in Fig. 23.6. For more detailed discussion, reference may be made to the works listed at the end of the chapter: it should, however, be made clear here that no exact discussion of the energetics of the molecules on the basis of valence theory is yet possible.

Standard enthalpies of formation of ferrocene, dibenzene chromium and a few other sandwich complexes have been obtained by combustion calorimetry, but the strength of the metal–ring bonds can be evaluated only if we make assumptions about the bonding in the rings themselves. If, for example, we assume that benzene in dibenzene chromium is identical with the free ligand (which is certainly not strictly true, since both the carbon–carbon bond length and the infrared spectrum are somewhat different from those of free benzene), the mean metal–ligand bond energy term is 170 kJ mol^{-1}. In the case of ferrocene the interpretation of the enthalpy of formation is further complicated by the absence of an accurate value for the enthalpy of formation of the $C_5H_5^-$ ion or C_5H_5 radical; the best estimate at present available is that the mean metal–ligand covalent bond energy (i.e. now considering ferrocene to be obtained from ground-state iron atoms and C_5H_5 radicals) is 290 kJ mol^{-1}. It is, however, interesting that on the basis of all the same assumptions the mean metal–ligand bond energy in nickelocene (which has two more electrons than ferrocene and which is much less stable) is

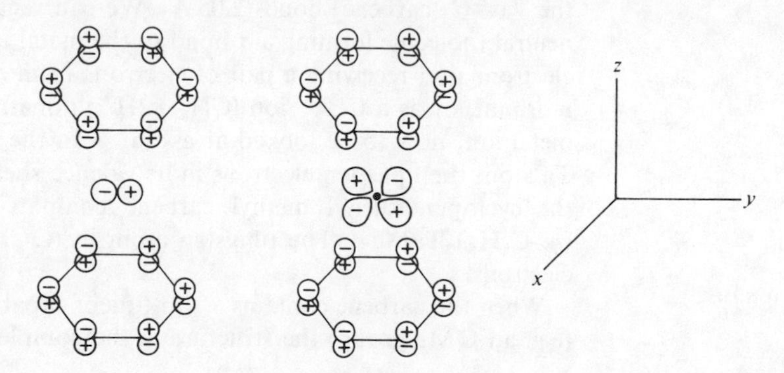

Figure 23.6
Bonding in dibenzene chromium. Overlapping of the metal p_y and d_{yz} orbitals and suitable π orbitals of the benzene molecules only is shown.

only 237 kJ mol^{-1}; further, the metal–carbon distance is slightly greater in the latter compound. These facts and the observation that ferrocene is diamagnetic whilst nickelocene has a magnetic moment of 2.86 μ_B are compatible with the presence of the two extra electrons in nickelocene in degenerate antibonding orbitals.

Silver(I) and copper(I), it may be mentioned, also form complexes with benzene, and salts $C_6H_6.AgClO_4$ and $C_6H_6.CuAlCl_4$ have been characterised. In these compounds, however, the metal ions appear to interact strongly only with one pair of carbon atoms per benzene molecule.

23.6 *Carbene (alkylidene) and carbyne (alkalidyne) complexes*

These complexes, the subjects of much recent work, contain (at least formally) M=C and M≡C bonds respectively. The simplest carbene derivatives contain only alkyl-substituted CH_2 groups and alkyl ligands, and are obtained by α-deprotonation of suitable alkyl compounds, e.g.

$$TaCl_5 \xrightarrow{Zn(CH_2CMe_3)_2} (Me_3CCH_2)_3TaCl_2$$

$$\xrightarrow{LiCH_2CMe_3} (Me_3CCH_2)_2Ta(Cl)=C(CMe_3)(H) + CMe_4$$

$$\xrightarrow{LiCH_2CMe_3} (Me_3CCH_2)_3Ta=C(CMe_3)(H)$$

In the last compound, no alternative bond distribution is possible, and in the analogous compound $(\eta^5\text{-}C_5H_5)_2(CH_3)Ta=CH_2$ it is found that, as expected, the Ta—C (methyl) bond (2.25 Å) is substantially longer than the Ta=C (carbene) bond (2.05 Å). We can regard the CH_2 either as a neutral molecule forming a σ bond to the metal using its lone pair of sp^2 electrons and receiving a pair of electrons from a Ta^{3+} ion in its empty p orbital, or as a CH_2^{2-} ion ($CH_4 - 2H^+$) donating four electrons to the metal ion, now to be looked at as Ta^{5+}. In the all-alkyl compound the Ta atom then has ten electrons in its valence shell (just as in $TaMe_5$); in the cyclopentadienyl methyl carbene complex it has 18 (just as in $(\eta^5\text{-}C_5H_5)_2TaMe_3$). The tungsten atom in $(OC)_5W=CPh_2$ also has 18 electrons.

When the carbene contains a substituent capable of donating electrons (e.g. an OMe group) the structure of the complex formed is less simple.

Consider the product of the following reactions:

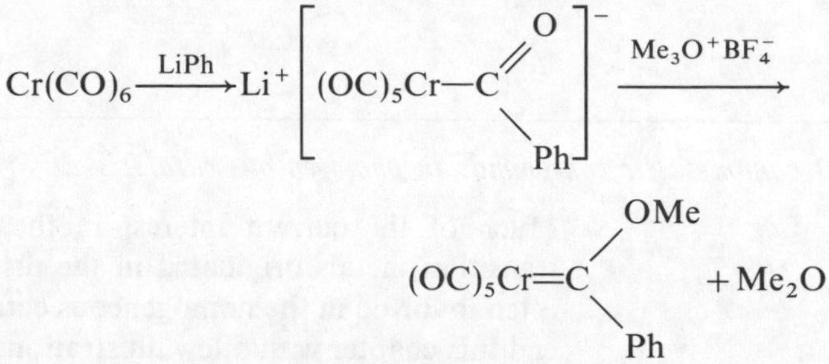

$$Cr(CO)_6 \xrightarrow{LiPh} Li^+ \left[(OC)_5Cr{-}C{\overset{O}{\underset{Ph}{\diagup\!\!\!\diagdown}}} \right]^- \xrightarrow{Me_3O^+BF_4^-}$$

$$(OC)_5Cr{=}C(OMe)(Ph) + Me_2O, LiBF_4$$

The Cr—CO distance of 1.88 Å is nearly the same as in $Cr(CO)_6$ (1.91 Å), but the Cr—C (carbene) distance of 2.04 Å implies a much weaker bond with relatively little metal–carbon π bonding. Further, the C—O (methyl) distance of 1.33 Å is intermediate between those in $(CH_3)_2O$ (1.43 Å) and $(CH_3)_2CO$ (1.23 Å). These data suggest that the molecule is best represented as:

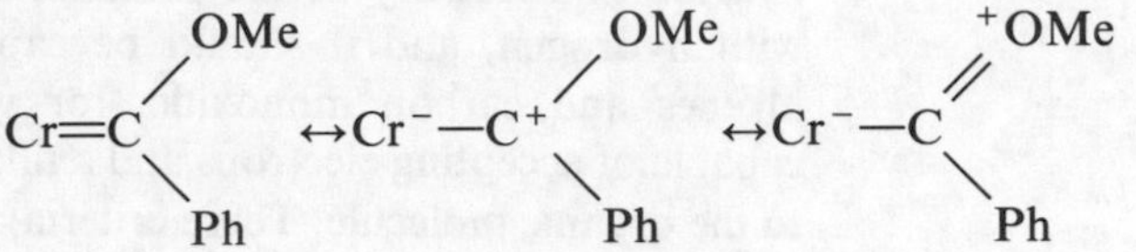

$$Cr{=}C(OMe)(Ph) \leftrightarrow Cr^-{-}C^+(OMe)(Ph) \leftrightarrow Cr^-{-}C({=}{}^+OMe)(Ph)$$

This formulation is equivalent to writing the structure

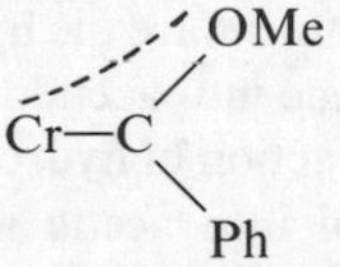

$$Cr{-}C(OMe)(Ph) \text{ (with a dashed delocalization line from Cr to OMe)}$$

The 18-electron rule is again obeyed.

The presence of electron-donating atoms or groups clearly affects the metal–carbene bond substantially, and differences in bond strength are reflected in chemical behaviour, the carbon atoms of the carbene groups in the chromium and tantalum compounds being susceptible to nucleophilic and electrophilic attack respectively.

Typical preparations of carbyne (alkylidyne) complexes, which contain M≡CR units, are:

$$WCl_6 + LiCH_2CMe_3 \rightarrow Me_3CC{\equiv}W(CH_2CMe_3)_3 + CMe_4 + LiCl$$

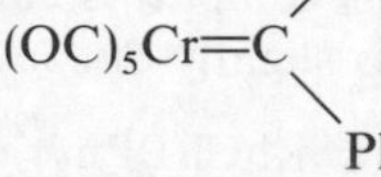

$$(OC)_5Cr{=}C(OMe)(Ph) + BX_3 \rightarrow (OC)_4XCr{\equiv}CPh + BX_2OMe + CO$$

Very short chromium—carbon (carbyne) bonds in such compounds (e.g. 1.69 Å in $(OC)_4ICr{\equiv}CCH_3$, in which the halogen atom and carbyne occupy *trans*-positions of an octahedron, compared with 1.95 Å for the chromium—carbonyl carbon distance) confirm the triply bonded structure.

23.7 Organometallic compounds in homogeneous catalysis

Much of the current interest in the organometallic chemistry of the transition metals originated in the discovery that these compounds are often involved in the homogeneous catalysis of organic reactions, and we end this chapter with a few illustrations of such processes. For a metal to be likely to form compounds that are homogeneous catalysts, it needs to have at least two oxidation states of comparable stability, and it needs to be capable of forming complexes with a wide range of coordination numbers and geometries. If, as is frequently the case, we are considering reactions of alkenes (since these are the most readily available reactive compounds obtainable from natural gas or petroleum), it should form a hydride (if necessary in the presence of other ligands) by combination with hydrogen, and it should be capable of complex formation with alkenes and carbon monoxide, for which it needs an empty orbital capable of accepting electrons and a filled one capable of π back-donation to the organic molecule. These criteria are best met by iron, cobalt, nickel and their homologues (the six platinum metals), and it is with derivatives of these elements that we shall be concerned here. The four processes that we have selected for comment are the hydrogenation of alkenes to alkanes, the conversion of ethylene into acetaldehyde, the conversion of an alkene into an aldehyde by the action of hydrogen and carbon monoxide, and the conversion of methanol into acetic acid. This choice has been based primarily on chemical interest.

The homogeneous hydrogenation of alkenes

The most widely used procedures for the hydrogenation of alkenes nearly all employ heterogeneous catalysts (commonly nickel, palladium or platinum in various forms), but for certain specialised purposes homogeneous catalysts find favour; one of these is chlorotris(triphenylphosphine)rhodium, $RhCl(PPh_3)_3$, a red-violet solid obtained when an ethanolic solution of rhodium trichloride is treated with an excess of triphenylphosphine, which serves as both reductant and ligand. The rhodium(I) complex is commonly used in benzene/ethanolic solution, in which it is slightly dissociated:

$$RhCl(PPh_3)_3 \rightleftharpoons RhCl(PPh_3)_2 + PPh_3$$

The following scheme (in which possible solvent participation is disregarded) shows a simplified mechanism for the catalytic hydrogenation of an alkene:

$$RhCl(PPh_3)_2 + H_2 \rightleftharpoons RhClH_2(PPh_3)_2$$

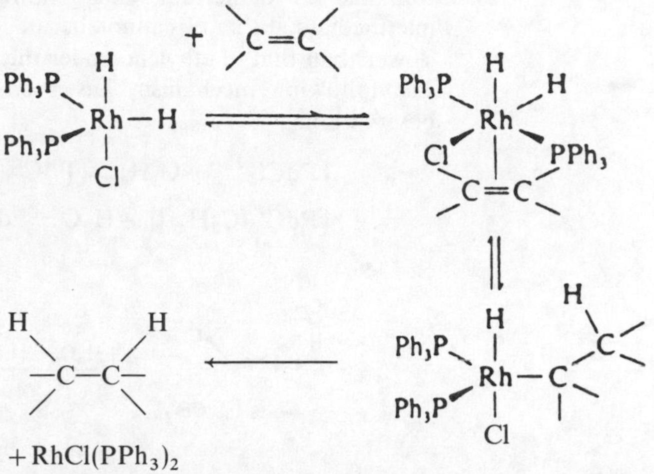

$$RhClH_2(PPh_3)_2 + \text{>C=C<} \rightleftharpoons RhClH_2(PPh_3)_2(\text{>C=C<}) \rightleftharpoons RhH(Cl)(PPh_3)_2(\text{>C(H)–C<}) \longrightarrow \text{H–C–C–H} + RhCl(PPh_3)_2$$

It will be noted that 14, 16, and 18-electron complexes of rhodium are all involved in the cycle.

The conversion of ethylene into acetaldehyde (the Wacker process)

This involves the less stable and more labile palladium analogue of the anion present in Zeise's salt (Section 23.3), i.e. $[PdCl_3(C_2H_4)]^-$. The organic reaction is

$$C_2H_4 + [PdCl_4]^{2-} + H_2O \rightarrow CH_3CHO + Pd + 2HCl + 2Cl^-$$

The palladium formed is oxidised back to the $[PdCl_4]^{2-}$ ion by copper(II) in weakly acidic chloride-containing solution, and the copper(I) produced is reoxidised to copper(II) by blowing in air:

$$Pd + 2Cu^{2+} + 8Cl^- \rightarrow [PdCl_4]^{2-} + 2[CuCl_2]^-$$

$$2[CuCl_2]^- + \tfrac{1}{2}O_2 + 2HCl \rightarrow 2CuCl_2 + 2Cl^- + H_2O$$

Overall, therefore, we have

$$C_2H_4 + \tfrac{1}{2}O_2 \xrightarrow[\text{dil HCl}]{[PdCl_4]^{2-} + Cu(II)} CH_3CHO$$

The rate law for the process is given by

$$-\frac{d}{dt}[C_2H_4] = \frac{k[PdCl_4^{2-}][C_2H_4]}{[H_3O^+][Cl^-]^2}$$

Further, the rates of oxidation of C_2H_4 and C_2D_4 are almost identical, showing that C—H bond breaking is not involved in the rate-determining step; and if the reaction is carried out in D_2O, the CH_3CHO produced contains no deuterium, thus eliminating free vinyl alcohol as an intermediate. Ethanol cannot be an intermediate since its oxidation is slower than that of ethylene under the conditions of the Wacker process. The following mechanism has been suggested to account for these observations.

$$[PdCl_4]^{2-} + C_2H_4 \rightleftharpoons [PdCl_3(C_2H_4)]^- + Cl^-$$

$$[PdCl_3(C_2H_4)]^- + H_2O \rightleftharpoons PdCl_2(H_2O)(C_2H_4) + Cl^-$$

$$\xrightarrow[\text{slow}]{+H_2O,\ -H^+}$$

$$\xrightarrow{-Cl^-} \longrightarrow$$

$$\longrightarrow CH_3CHO + H_2O + Pd + HCl$$

The oxidation of Pd(0) by Cu(II) and of the reoxidation of the chloro complex of Cu(I) by air are both relatively rapid processes.

The 'oxo' reaction (hydroformylation)

This is a general reaction of alkenes with hydrogen and carbon monoxide (or, formally, of a hydrogen atom and a formyl group HCO) to yield aldehydes:

$$RCH{=}CH_2 + H_2 + CO \rightarrow RCH_2CH_2CHO$$

Dicobalt octacarbonyl and the rhodium complex $Rh(H)(CO)(PPh_3)_3$ (obtained by the action of hydrogen and carbon monoxide under pressure on almost any rhodium compound in the presence of excess of triphenylphosphine) are the standard catalysts for this reaction; the rhodium compound has the advantage of being active under milder conditions but is, of course, more expensive. However, since we have already discussed one process catalysed by a rhodium complex, we shall choose the cobalt carbonyl catalysed reaction for examination here.

Under the conditions employed in this reaction (100–200 °C, 100–400 atm), $Co_2(CO)_8$ is converted into $HCo(CO)_4$, which is known to bring about hydroformylation of alkenes under mild conditions. The rate of the overall reaction is proportional to the concentrations of alkene and combined cobalt; it is increased by increasing the pressure of hydrogen and decreased by increasing the pressure of carbon monoxide. A suggested mechanism for the reaction is as follows:

$$HCo(CO)_4 \rightleftharpoons HCo(CO)_3 + CO$$

$$HCo(CO)_3 + RCH{=}CH_2 \rightleftharpoons HCo(CO)_3(CH_2{=}CHR)$$

$$HCo(CO)_3(CH_2{=}CHR) \rightarrow RCH_2CH_2Co(CO)_3$$

$$RCH_2CH_2Co(CO)_3 + CO \rightarrow RCH_2CH_2Co(CO)_4$$

$$RCH_2CH_2Co(CO)_4 \rightleftharpoons RCH_2CH_2COCo(CO)_3$$

$$RCH_2CH_2COCo(CO)_3 + H_2 \rightarrow RCH_2CH_2COCo(H)_2(CO)_3$$

$$RCH_2CH_2COCo(H)_2(CO)_3 \rightarrow RCH_2CH_2CHO + HCo(CO)_3$$

The third and fifth stages of this mechanism call for further comment. The third stage, which could be described as the insertion of the alkene into the Co—H bond, is better looked upon as a hydride ion migration:

$$(H_2C{=}CHR)\text{—}Co(CO)_3H \longrightarrow H\text{—}\overset{R}{\underset{H}{C}}\text{—}\overset{H}{\underset{H}{C}}\text{—}Co(CO)_3$$

Stage five is analogous to the reaction

$$CH_3Mn(CO)_5 + CO \rightarrow CH_3COMn(CO)_5$$

which, by using ^{14}CO for the free carbon monoxide and showing that none of it appears in the acetyl group, has been found to be a methyl group migration. Once again 16-electron species, in this case $HCo(CO)_3$ and its alkyl and acyl derivatives, play an essential part in the catalytic cycle, and the diminution of the rate by high pressures of carbon monoxide is simply accounted for by its reversal of the dissociation of $HCo(CO)_4$ leading to the production of this intermediate.

The conversion of methanol into acetic acid

The overall reaction

$$CH_3OH + CO \rightarrow CH_3CO_2H$$

is catalysed by several metal carbonyls in the presence of iodide, and the rhodium(I) dicarbonyl iodide anion $[Rh(CO)_2I_2]^-$ is particularly effective. This process (the Monsanto process) involves the conversion of methanol into methyl iodide and thence into acetyl iodide and acetic acid; it proceeds by the stages shown below:

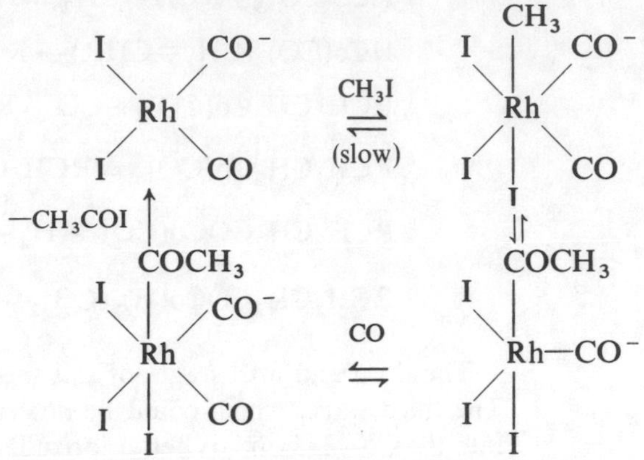

I CO⁻ Rh I CO

CH_3I ⇌ (slow)

CH_3 I CO⁻ Rh I CO I

$-CH_3COI$

$COCH_3$ I CO⁻ Rh I CO I

CO ⇌

$COCH_3$ I Rh—CO⁻ I I

$$CH_3COI + H_2O \rightarrow CH_3CO_2H + HI$$
$$CH_3OH + HI \rightleftharpoons CH_3I + H_2O$$

PROBLEMS

1 Show that the η^5-cyclopentadienyl carbonyl complexes $(\eta^5\text{-}C_5H_5)V(CO)_4$, $[\eta^5\text{-}C_5H_5Cr(CO)_3]_2$, $\eta^5\text{-}C_5H_5Mn(CO)_3$, $[(\eta^5\text{-}C_5H_5)Fe(CO)_2]_2$ and $(\eta^5\text{-}C_5H_5)Co(CO)_2$ all obey the 18-electron rule and suggest two possible structures for the chromium compound. How would you attempt to decide which structure is correct?

2 Cycloheptatriene displaces three molecules of carbon monoxide from molybdenum hexacarbonyl, and the proton magnetic resonance spectrum of the product (**A**) consists of four peaks of equal intensity. **A** reacts with triphenylmethyl fluoroborate to yield a compound **B**, $MoC_{10}H_7O_3BF_4$, and triphenylmethane. The proton magnetic resonance spectrum of **B** consists of a single peak.

Suggest a structure for each of the compounds **A** and **B**. How would you expect the infrared absorption spectrum in the region 1800–2100 cm^{-1} of **B** to differ from that of **A**?

3 Comment on each of the following observations:

(a) Nickelocene reacts with tetrafluoroethylene to form $(\eta^5\text{-}C_5H_5)(\eta^3\text{-}L)Ni$, where

L = H (structure: cyclopentenyl ring bearing H atoms, fused to a CF_2–CF_2 bridge)

(b) For free $CH_3CH{=}CH_2$, $\nu(C{=}C)$ is at 1652 cm^{-1}, but in the complex $K[PtCl_3(CH_3CH{=}CH_2)]$ it is at 1504 cm^{-1}.

(c) In $(C_6Me_6)_2Ru$ one hexamethylbenzene ligand is a non-planar η^4-donor.

(d) When methylpentacarbonylmanganese reacts with carbon monoxide according to the equation

$$(OC)_5MnCH_3 + CO \rightarrow (OC)_5MnCOCH_3$$

^{14}C tracer studies show that the CO molecule that becomes the acyl ligand is derived from $(OC)_5MnCH_3$.

REFERENCES FOR FURTHER READING

Collman, J.P., Hegedus, L.S., Norton, J.R. and Finke, R.G. (1987) *Principles and Applications of Organotransition Metal Chemistry*, 2nd edition, University Science Books, Mill Valley, California. A recent and comprehensive account of the organic reactions of organometallic compounds of transition metals.

Cotton, F.A. and Wilkinson, G. (1988) *Advanced Inorganic Chemistry*, 5th edition, Interscience, New York. Chapters 24 to 28 cover much the same ground as this chapter but at a much more advanced level.

Parkins, A.W. and Poller, R.C. (1986) *An Introduction to Organometallic Chemistry*, Macmillan, London. A good general account with emphasis on organic chemistry.

Powell, P. (1988) *Principles of Organometallic Chemistry*, 2nd edition, Chapman and Hall, London. The best all-round short treatment of the subject.

Yamamoto, A. (1986) *Organotransition Metal Chemistry*, Wiley, New York. A general survey incorporating much material of historical and industrial interest.

24 *Transition metals of the first series*

Introduction •	**Scandium** •	**Titanium** •	**Vanadium** •	**Chromium** •
Manganese •	**Iron** •	**Cobalt** •	**Nickel** •	**Copper** • **Zinc**

24.1 *Introduction*

So much of Chapters 18–23 was concerned with metals of the first transition series that only a brief reminder of their contents will be given here. All these metals other than scandium and zinc exhibit more than one stable oxidation state in monatomic species. In their compounds they display a wide range of coordination numbers and geometries, but the octahedron occupies a dominant position in their stereochemistry. Because of the kinetic inertness of octahedral complexes of ions having the electronic configurations d^3 and low-spin d^6, most studies of isomerism in complexes of first series transition metals involve derivatives of Cr(III) and Co(III), and the same is true of studies of substitution reaction mechanisms. For octahedral complexes at least, dissociative mechanisms appear to prevail over those involving an increase in coordination number in the rate-determining step.

The electronic spectra and magnetic properties of many complexes of first series transition metals are surprisingly well accounted for on the basis of simple electrostatic crystal field theory. The concept of crystal field stabilisation energy, which arises from this theory, provides a useful basis upon which to discuss variations in lattice energies, enthalpies of hydration and other properties of d^n ions, and also affords some useful insights into why some substitution and electron-exchange reactions are much slower than others. For complexes involving neutral metal atoms and uncharged ligands (e.g. the metal carbonyls), however, the simple electrostatic theory is inadequate and a molecular orbital theory must be used: in such complexes, and in most organometallic compounds of transition metals, π bonding plays an important part, and most complexes

Table 24.1
Electronic configurations and some properties of metals of the first transition series and their ions.

Element		Sc	Ti	V	Cr	Mn	Fe	Co	Ni	Cu	Zn
Atomic number		21	22	23	24	25	26	27	28	29	30
Ground states	atom	$3d^14s^2$	$3d^24s^2$	$3d^34s^2$	$3d^54s^1$	$3d^54s^2$	$3d^64s^2$	$3d^74s^2$	$3d^84s^2$	$3d^{10}4s^1$	$3d^{10}4s^2$
	M^+	$3d^14s^1$	$3d^24s^1$	$3d^4$	$3d^5$	$3d^54s^1$	$3d^64s^1$	$3d^8$	$3d^9$	$3d^{10}$	$3d^{10}4s^1$
	M^{2+}	$3d^1$	$3d^2$	$3d^3$	$3d^4$	$3d^5$	$3d^6$	$3d^7$	$3d^8$	$3d^9$	$3d^{10}$
	M^{3+}	[Ar]	$3d^1$	$3d^2$	$3d^3$	$3d^4$	$3d^5$	$3d^6$	$3d^7$	$3d^8$	$3d^9$
ΔH atomisation/kJ mol^{-1}		376	473	515	397	281	416	425	430	339	126
Ionisation	1st	631	656	650	653	717	762	758	736	745	906
energy/kJ mol^{-1}	2nd	1235	1309	1414	1592	1509	1561	1644	1752	1958	1734
	3rd	2393	2657	2833	2990	3260	2962	3243	3402	3556	3837
E°/V	M^{2+}/M			−1.18	−0.90	−1.18	−0.44	−0.28	−0.25	+0.34	−0.76
	M^{3+}/M^{2+}			−0.26	−0.41	+1.51	+0.77	+1.93			
r/Å	M	1.64	1.47	1.35	1.29	1.37	1.26	1.25	1.25	1.28	1.37
	M^{2+}			0.79	0.82	0.82	0.77	0.74	0.70	0.73	0.75
	M^{3+}	0.73	0.67	0.64	0.62	0.65	0.65	0.61	0.60		

containing strongly π bonding ligands have eighteen electrons in their valence shells (to maximise π bonding), but very few indeed have more.

The reader who has mastered the material of the preceding six chapters will already have quite a good grasp of transition metal chemistry, but any treatment of general principles naturally concentrates on the discussion of topics for which the general principles are useful. In this chapter, therefore, we give an outline treatment of the most important features of the chemistry of individual elements of the first transition series irrespective of whether these features lend themselves to discussion in terms of general principles; we shall, however, make frequent cross-references to Chapters 18–23 where it is appropriate to do so. It may be desirable at this point to draw attention to the importance of the subject matter of Chapters 6 and 7 in transition metal chemistry, since much of the chemistry of the elements involves the solid state or equilibria in aqueous solution. We shall meet numerous examples of the roles played by pH, precipitation and complex formation in stabilising oxidation states.

Many experimental data for the elements Sc–Zn have already been given, but for convenience most of these, together with some not previously provided, are assembled in Table 24.1. A list of oxidation states of these elements in stable compounds was given in Table 18.2 (Section 18.3). In this chapter we shall deal with the elements mainly on the basis of oxidation states, as is usual in inorganic chemistry, and we shall not pay much attention to compounds involving π accepting organic ligands, for which the oxidation state concept is less useful than the 18-electron rule; their chemistry was, of course, covered in Chapters 22 and 23.

24.2 *Scandium*

Scandium usually occurs with the later lanthanides (Chapter 26), from which it may be separated by ion-exchange or solvent extraction, but the silicate *thortveitite*, $Sc_2Si_2O_7$ (Section 13.9), is also a potential source. In its chemistry scandium is more like aluminium than its homologues yttrium and lanthanum: for example E° for the aqueous Sc^{3+}/Sc system is -1.9 V compared with -1.66 V for the Al^{3+}/Al system. Scandium metal is obtained by electrolysis of a melt of $ScCl_3$, KCl and LiCl; it dissolves in both acids and alkalis, combines with halogens and reacts with nitrogen at elevated temperatures to form a nitride, ScN, that is hydrolysed by water. Saline compounds are formed only by the element in the tripositive state, but lower halides, e.g. ScCl (a layer structure ClScScCl ... ClScScCl) and Sc_7Cl_{10} and Sc_7Cl_{12} (containing Sc_6 clusters), are obtained from $ScCl_3$ and the metal at high temperatures.

Scandium forms an insoluble hydrous oxide, ScO(OH), which is isostructural with AlO(OH) and, like it, is a weak base soluble in aqueous alkali. Dehydration yields Sc_2O_3. The insoluble fluoride forms complexes

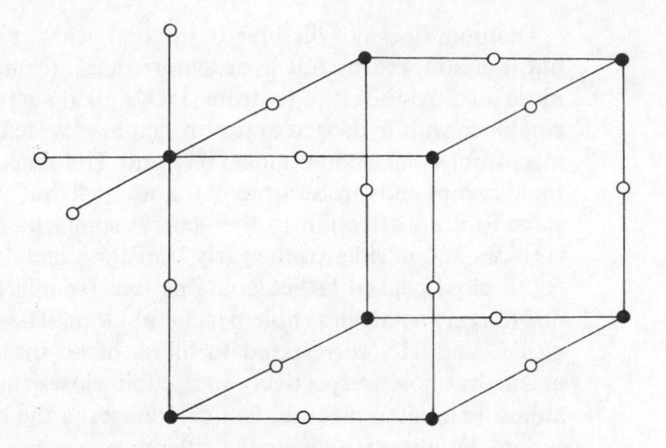

Figure 24.1
The ReO_3 structure of ScF_3.

containing the $[ScF_6]^{3-}$ ion on treatment with ammonium or alkali metal fluorides. Scandium trifluoride has the simplest AB_3 type of structure, that of rhenium trioxide (ReO_3), which is shown in Fig. 24.1; the metal ion coordination is octahedral, that of the fluoride ion linear. The trichloride $ScCl_3$ is isostructural with $AlCl_3$ and is hydrolysed by water, but is not a Friedel–Crafts catalyst.

At the present time scandium compounds have no technical applications and are very seldom encountered.

24.3 Titanium

The main sources of titanium are *rutile*, TiO_2, and *ilmenite*, $FeTiO_3$; the structure of the former was described in Section 6.3; the latter is a mixed oxide rather than a salt of an oxo-acid, and has a structure related to that of corundum, α-Al_2O_3, i.e. a hexagonal close-packed lattice of oxide ions with the cations (which are not very different in size) occupying two-thirds of the octahedral holes. Since titanium forms a very stable macromolecular carbide (Section 13.5), it is not feasible to produce the metal by reduction of the oxide with carbon. Either rutile or ilmenite, when heated with carbon in a stream of chlorine at 900 °C, gives the volatile tetrachloride $TiCl_4$ (b.p. 136 °C) which is purified by fractional distillation. This is reduced by magnesium at 800 °C in an atmosphere of argon, the magnesium chloride is sublimed away, and the silvery-white metal is fused under argon or helium and cast into ingots. The pure metal may be obtained by thermal decomposition of the vapour of the tetraiodide on a hot wire. Titanium is resistant to corrosion at ordinary temperatures, and its relative lightness combined with high mechanical strength makes it an excellent material (especially when alloyed with aluminium) for aircraft construction.

Titanium does not dissolve in mineral acids at ordinary temperatures, but it is attacked by hot hydrochloric acid, forming Ti(III) and H_2. Hot nitric acid oxidises it to hydrous TiO_2; alkalis have no action. The metal combines with hydrogen and nitrogen at elevated temperatures, forming macromolecular compounds TiH_2 and TiN respectively. We mentioned these compounds in Sections 9.6 and 14.4, but it is appropriate at this stage to draw attention to the general similarities between the hydrides, carbides and nitrides of the early transition metals. As we saw in Section 6.2, a close-packed lattice contains two (smaller) tetrahedral holes and one (larger) octahedral hole per metal atom. The structures of TiH_2 and of TiC and TiN correspond to filling of all the tetrahedral and all the octahedral holes respectively in a cubic close-packed lattice of titanium atoms. (Titanium metal in fact crystallises in the hexagonal close-packed system, but there is a negligible difference in energy between the two close-packed structures.) Since the compounds TiH_2, TiC and TiN are all inert solids of high melting-point which have negative standard enthalpies of formation, strong bonding must be involved in order to compensate for the energy needed to atomise hydrogen, carbon or mitrogen. It may be noted that cubic close-packed lattices with all tetrahedral and with all octahedral holes filled are geometrically equivalent to the fluorite and sodium chloride structures respectively, but such descriptions would be inappropriate here since the compounds are metallic conductors.

Titanium exhibits oxidation states of IV (by far the most stable), III, II and (very rarely) 0. Since all the most important compounds of titanium are those of titanium(IV), we begin with this oxidation state.

Titanium(IV)

The tetrachloride, prepared as already described, is a colourless fuming liquid which is rapidly hydrolysed to the oxide by water. On an industrial scale the oxide, an important white pigment, is made by passing $TiCl_4$ vapour mixed with air through a flame produced by combustion of a hydrocarbon, when the reaction

$$TiCl_4 + O_2 \rightarrow TiO_2 + 2Cl_2$$

takes place. Titanium dioxide, when made in the dry way, is difficult to dissolve in acids, but the hydrous material obtained by precipitation from a solution of $TiCl_4$ in concentrated hydrochloric acid dissolves in HF, HCl and H_2SO_4, forming fluoro, chloro, and sulphato complexes respectively. There is no simple aquo ion Ti^{4+}, and species sometimes formulated as titanyl compounds containing the ion TiO^{2+} really contain polymeric cations $(TiO^{2+})_n$, or hydrates of these. Although TiO_2 is usually formulated $Ti^{4+}(O^{2-})_2$, the very high value of the sum of the first four ionisation energies of the metal (8 800 kJ mol^{-1}) makes it doubtful whether this representation can be strictly correct; it may be noted that

TiF_4 (obtained from the metal and fluorine at 200 °C or by the action of anhydrous hydrogen fluoride on the tetrachloride) is a volatile hygroscopic powder rather than a typical salt.

Barium 'titanates' are mixed oxides, some of which are ferroelectrics, i.e. they have an electric dipole moment even in the absence of an external electric field. This arises because in the unit cell of, for example, $BaTiO_3$, which is related to that of perovskite, the Ti(IV) ion is relatively so small that it tends to be off the centre of the TiO_6 octahedron (or, as it is sometimes described, it can rattle around in its octahedral hole). Application of an electric field causes all such ions to be drawn to one side of the holes and leads to a great increase in specific permittivity; for this reason barium titanates are used in capacitors. Further, application of pressure to one side of a $BaTiO_3$ crystal, by causing the Ti(IV) ions to migrate, generates an electric current, and vice versa (the *piezoelectric effect*); use is made of this in microphones and gramophone pickups, for example.

The molecule of $TiCl_4$ is tetrahedral. It combines with tertiary amines and phosphines to form six-coordinated complexes such as $TiCl_4(NMe_3)_2$ and $TiCl_4(PEt_3)_2$. Salts containing the yellow octahedral anion $[TiCl_6]^{2-}$ are best made in thionyl chloride solution since they are hydrolysed by water, but salts containing the more stable colourless $[TiF_6]^{2-}$ ion can easily be prepared in aqueous acidic media. With the diarsine 1,2-$C_6H_4(AsMe_2)_2$, the dodecahedral complex $TiCl_4$ (diarsine)$_2$ is formed. The coordination of the titanium atom is also dodecahedral in $Ti(NO_3)_4$ (m.p. 58 °C), obtained from the tetrachloride and dinitrogen pentoxide; the four nitrate groups act as bidentate ligands. The titanium alkoxides, e.g. $Ti(OEt)_4$, which results from the interaction of the tetrachloride with ethanol in the presence of ammonia, are tetrameric, and contain molecular units, such as that shown in Fig. 24.2, in which each metal atom is six-coordinated. Titanium alkoxides are widely used in waterproofing fabrics and in heat-resisting paints; on exposure to the atmosphere eventual hydrolysis to titanium dioxide takes place.

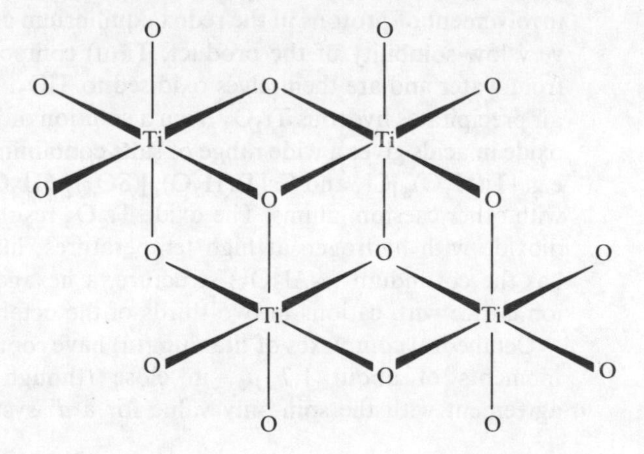

Figure 24.2
The structure of $[Ti(OEt)_4]_4$ (only Ti and O atoms are shown).

A red peroxo complex obtained from acidic titanium (IV)-containing solutions and hydrogen peroxide contains one peroxo group per titanium atom; various solid peroxo compounds have been isolated, but their structures are unknown. The bleaching of the colour of the complex by fluoride but not by chloride indicates that Ti(IV), like most ions of metals of the first transition series, is a class (a) cation in the sense of the discussion in Section 7.7.

Titanium tetraalkyls (Section 23.2) are unstable, but many cyclopentadienyl derivatives are known. X-ray diffraction studies show that $Ti(C_5H_5)_4$, obtained from $TiCl_4$ and NaC_5H_5, is $Ti(\eta^1\text{-}C_5H_5)_2(\eta^5\text{-}C_5H_5)_2$. The ^1H and ^{13}C nmr spectra of the compound at 20 °C, however, show all protons and all carbon atoms to be equivalent: the point of attachment of the metal to each ring is shifting rapidly over all possible positions. The kind of fluxionality, in which the metal atom is bonded in turn to each carbon atom of a $\eta^1\text{-}C_5H_5$ ring, is sometimes known as *ring whizzing*; in this example it is supplemented by fluxionality involving interchange between η^1 and η^5 ligands.

Titanium(III)

Violet $TiCl_3$ (obtained by reduction of the tetrachloride with hydrogen at 600 °C) and blue TiF_3 (made from TiH_2 and hydrogen fluoride at 700 °C) both contain octahedrally coordinated Ti(III), the former in a layer lattice and the latter in a three-dimensional structure closely related to that of ScF_3. The aquo ion, purple $[Ti(H_2O)_6]^{3+}$, is obtained when solutions containing Ti(IV) are reduced by metallic zinc. Since E° for the half-reaction usually represented as

$$TiO^{2+} + 2H^+ + e \rightleftharpoons Ti^{3+} + H_2O$$

is +0.1 V, Ti^{3+} is a powerful reductant, and is used in titrimetric analysis for the determination of Fe(III) and nitro groups; its solutions must be protected from oxidation by air. In alkaline media, partly because of the involvement of protons in the redox equilibrium and partly because of the very low solubility of the product, Ti(III) compounds liberate hydrogen from water and are themselves oxidised to TiO_2. Alkali in the absence of air precipitates hydrous Ti_2O_3 from a solution of $TiCl_3$; dissolution of the oxide in acids gives a wide range of salts containing the $[Ti(H_2O)_6]^{3+}$ ion, e.g. $[Ti(H_2O)_6]Cl_3$ and $Cs[Ti(H_2O)_6](SO_4)_2{\cdot}6H_2O$, which is isomorphous with other caesium alums. The oxide Ti_2O_3 results from reduction of the dioxide with hydrogen at high temperatures; like ilmenite ($FeTiO_3$), it has the corundum ($\alpha\text{-}Al_2O_3$) structure, a hexagonal close-packed oxide ion lattice with cations in two-thirds of the octahedral holes.

Octahedral complexes of titanium(III) have room-temperature magnetic moments of about 1.7 μ_B, in close (though somewhat fortuitous) agreement with the spin-only value for a d^1 system; they have a single band in the electronic spectrum arising from the transition $t_{2g} \rightarrow e_g$, though, as the spectrum shown in Fig. 19.2 (Section 19.3) illustrates, there is sometimes evidence from the asymmetry of the band of a Jahn–Teller distortion in the excited state.

A fibrous form of $TiCl_3$, obtained by the reduction of titanium tetrachloride with aluminium triethyl at 100 °C, contains $TiCl_6$ octahedra each sharing two faces; this material is the heterogeneous catalyst for the stereoregular polymerisation of alkenes in the presence of aluminium alkyls by the Ziegler–Natta process.

Lower oxidation states

A dichloride, dibromide and diiodide of titanium are obtained by the thermal disproportionation of the trihalides, or by reduction of the tetrahalides with titanium:

$$2TiX_3 \rightarrow TiX_2 + TiX_4$$

$$TiX_4 + Ti \rightarrow 2TiX_2$$

They are red or black solids having the cadmium iodide structure which react violently with water, liberating hydrogen. There is therefore no aqueous solution chemistry of titanium(II). The oxide TiO, made by heating the dioxide with the metal, is a metallic conductor having a sodium chloride structure with one-sixth of sites for both ions unoccupied. No stable carbonyl of titanium is known, but $Ti(CO)_6$ and the isoelectronic species $Ti(N_2)_6$ have been characterised in a noble gas matrix at very low temperatures. In the presence of π accepting groups, however, carbonyl complexes such as $(\eta^5\text{-}C_5H_5)_2Ti(CO)_2$ may be isolated. The compound $(Ti(bipy)_3$, made by the action of lithium on a solution of titanium tetrachloride and bipyridyl in tetrahydrofuran, formally contains Ti(0). but it is likely that delocalisation of electrons into the antibonding orbitals of the ligands occurs to a considerable extent.

24.4 *Vanadium*

Vanadium occurs in many deposits, but few of them are concentrated, and much of the metal is now obtained from Venezuelan petroleum, in which it is present in small amounts; other sources are the minerals *carnotite*, $K(UO_2)VO_4.H_2O$, and *vanadinite*, $Pb_5(VO_4)_3Cl$. Roasting of these ores (or of flue dusts from plants burning the appropriate oil) with sodium carbonate gives water-soluble $NaVO_3$, from a solution of which sparingly soluble NH_4VO_3 is precipitated and heated to give the oxide V_2O_5. The metal may be obtained by reduction of this oxide with calcium at high temperatures, but is not manufactured on a large scale; ferrovanadium, used for toughening steels, is made directly by reduction of a mixture of V_2O_5 and Fe_2O_3 with aluminium. On the technical scale the most important compound of vanadium is the pentoxide, which is used as a catalyst in the oxidation of sulphur dioxide to the trioxide and of naphthalene to phthalic acid.

Vanadium metal is in many respects similar to titanium. Although it is thermodynamically a powerful reductant (E° for the aqueous system V^{2+}/V is -1.2 V), it is easily made passive, and is insoluble in most non-oxidising acids and in alkalis. It is, however, readily attacked by nitric acid and by ammonium peroxodisulphate solution. It reacts with halogens at moderate temperatures, forming VF_5,VCl_4, VBr_3 and VI_3. The products of its reactions with hydrogen, carbon and nitrogen are analogous to the titanium compounds described in the last section.

The normal oxidation states of vanadium are V, IV, III and II; 0 occurs only in the carbonyl $V(CO)_6$ (Section 22.2) and a few organic compounds such as $V(C_6H_6)_2$, the analogue of dibenzene chromium. As in the last section, we shall consider the highest and most familiar oxidation state first.

Vanadium(V)

Pure vanadium pentoxide is an orange or red powder according to its state of division; it is sparingly soluble in water, but dissolves in alkalis to form a wide range of vanadates and in strong acids to form complexes of the VO_2^+ ion. The principal species in vanadium(V)-containing solutions depend upon the hydrogen-ion concentration; with increase in this quantity they are VO_4^{3-}, $[VO_3(OH)]^{2-}$ (in equilibrium with $V_2O_7^{4-}$), $V_3O_9^{3-}$, $V_{10}O_{28}^{6-}$, V_2O_5 and VO_2^+. It should be noted that the O:V ratio decreases along this series; the role of the hydrogen ion in removing O^{2-} is apparent from the equilibria involved, e.g.

$$2VO_4^{3-} + 2H^+ \rightleftharpoons 2[VO_3(OH)]^{2-} \rightleftharpoons V_2O_7^{4-} + H_2O$$

$$3[VO_3(OH)]^{2-} + 3H^+ \rightleftharpoons V_3O_9^{3-} + 3H_2O$$

Vanadate ions of other formulae are found in solid salts. The structural chemistry of V_2O_5 and the vanadates is exceedingly complicated, and we shall not attempt to discuss it in any detail here. Some analogies with phosphates can be discerned: the VO_4^{3-} ion, for example, is tetrahedral, and there are polyvanadates containing the infinite chain anion $(VO_3)_n^{n-}$ built up from VO_4 tetrahedra sharing two corners. Often, however, the metal atom has a coordination number higher than four: in V_2O_5 itself, for example, each V atom has one O at 1.59 Å (not shared), one at 1.78 Å (shared with one other V), and two O at 1.88 Å and one O at 2.02 Å (all shared with two other V) in an approximately square pyramidal configuration with the short V=O bond at the apex. Isopolyanion formation is a characteristic feature of the chemistry of several of the earlier transition metals, notably V, Nb, Ta, Cr (to a smaller extent), Mo and W; we shall meet it again in the next chapter.

The only halide of vanadium(V) is the fluoride, a volatile white solid obtained by fluorination of the metal at 300 °C; it is readily hydrolysed and is a vigorous fluorinating agent; it is a linear polymer of *cis*-VF_6 octahedra.

Reduction of vanadium(V) in acidic solution yields successively blue VO^{2+}, green V^{3+} and violet V^{2+}; from the oxidation state diagram given in Fig. 24.3 it will be clear that all oxidation states of vanadium in aqueous solution are stable with respect to disproportionation.

Vanadium(IV)

As will be apparent from Fig. 24.3, vanadium(V) is quite a powerful oxidant, and only mild reducing agents (e.g. SO_2) are needed to convert it into vanadium(IV); in aqueous media this is present as hydrated VO^{2+}, and many salts of this ion are known. In $VOSO_4.5H_2O$, for example, the vanadium atom has one O (that of the V═O group) at 1.59 Å, and five other O (belonging to water molecules or to the anion) at 1.98–2.22 Å. The V═O bond involves π bonding from a filled $2p$ oxygen orbital to an empty $3d$ vanadium orbital; in compounds for which interatomic distances are not available its presence may be inferred from that of a characteristic stretching frequency in the infrared spectrum at about 980 cm^{-1} (the corresponding frequency for a V—O single bond occurs at about 480 cm^{-1}).

The oxide from which these vanadyl(IV) salts are derived is blue VO_2, conveniently obtained by reduction of V_2O_5 with oxalic acid; the high-temperature form of this, a metallic conductor, is isostructural with TiO_2, but in the ordinary form, an insulator, the structure is distorted so that successive pairs of vanadium atoms in the crystal are alternately 2.62 Å and 3.17 Å apart, suggesting the occurrence of weak metal–metal interactions. Vanadium dioxide is amphoteric, and in alkali forms polyanions such as $V_{18}O_{42}^{12-}$.

Vanadium tetrachloride, the highest chloride of the element, is a dark red liquid which is hydrolysed by water and which when gently warmed decomposes into the trichloride and chlorine. Like titanium tetrachloride, it forms a diarsine adduct in which the metal atom is dodecahedrally coordinated. The green tetrafluoride, obtained by the action of hydrogen fluoride on the tetrachloride, resembles titanium tetrafluoride in properties.

Figure 24.3
Potential diagram for vanadium at $[H^+] = 1$.

$$VO_2^+ \xrightarrow{+1.0} VO^{2+} \xrightarrow{+0.4} V^{3+} \xrightarrow{-0.2} V^{2+} \xrightarrow{-1.2} V$$

Vanadium(III)

The oxide V_2O_3 is exclusively basic; the anhydrous compound (which is isostructural with Ti_2O_3) may be made by partial reduction of V_2O_5 with hydrogen, the hydrous oxide by precipitation from green solutions of vanadium(III) salts. These are obtained by reduction of vanadate with zinc or iodide ion, or electrolytically. The yellow-green trifluoride (obtained from the trichloride and hydrogen fluoride) is a sparingly soluble solid of high melting-point; the violet anhydrous trichloride is soluble in water without decomposition. Vanadium(III) complexes are derivatives of a d^2 ion and have magnetic moments of about 2.8 μ_B; most of them contain an octahedrally coordinated ion, and their spectra were discussed in detail in Section 19.3. The only cyano complex to have been characterised, however, is $[V(CN)_7]^{4-}$ in the salt $K_4[V(CN)_7].2H_2O$, made from vanadium(III) chloride and excess of aqueous potassium cyanide; the anion is a pentagonal bipyramid.

Vanadium(II)

This is obtained as the violet octahedral $[V(H_2O)_6]^{2+}$ ion by reduction of aqueous solutions of vanadium in higher oxidation states by means of zinc amalgam or electrolytically. The ion is strongly reducing and is rapidly oxidised on exposure to air, but compounds such as $[V(H_2O)_6]SO_4.H_2O$ and $K_4[V(CN)_6]$ may be isolated. The green chloride VCl_2 is formed by heating the trichloride in hydrogen at 500 °C; when heated in hydrogen fluoride and hydrogen it gives the blue difluoride, which has the rutile structure. The oxide VO can be obtained by reduction of higher oxides with hydrogen; like TiO, it has the sodium chloride structure but is markedly non-stoichiometric.

Substitution in vanadium(II) complexes is relatively slow, a fact which can be interpreted in terms of their being derived (like Cr(III) complexes) from a d^3 electronic configuration (cf. Section 21.5).

24.5 Chromium

Chromium is one of the more familiar transition metals owing to the uses of chromium(VI) oxide and dichromates as oxidising agents and of the metal as an electroplated protective coating for steel and in the form of stainless steels. Its principal ore is *chromite*, $FeCr_2O_4$, which has the normal spinel structure (cf. Section 20.2). For the production of ferrochromium for steel manufacture, this is reduced directly with carbon. If chromium is to be separated from iron, fusion with sodium carbonate in the presence of air yields sodium chromate, which is soluble in water, and iron(III) oxide, which is not:

$$4FeCr_2O_4 + 8Na_2CO_3 + 7O_2 \rightarrow 8Na_2CrO_4 + 2Fe_2O_3 + 8CO_2$$

Extraction with water, followed by acidification with sulphuric acid, gives a solution from which sodium dichromate, $Na_2Cr_2O_7$, can be crystallised. This may be reduced to chromium(III) oxide by heating it with carbon, and the oxide, after removal of the sodium carbonate which is also formed, can be reduced by aluminium:

$$Na_2Cr_2O_7 + 2C \rightarrow Cr_2O_3 + Na_2CO_3 + CO$$

$$Cr_2O_3 + 2Al \rightarrow Al_2O_3 + 2Cr$$

Chromium is also obtained by electrolysis of chromium(III) sulphate solution. It is a hard silvery-white metal which has the highest melting-point (1890 °C) of the first transition series elements. At ordinary temperatures it is very resistant to chemical attack (though it dissolves in dilute hydrochloric and sulphuric acids), but this inertness is kinetic rather than thermodynamic in origin, as may be seen from the values for the standard potentials of the following half-reactions:

$$Cr^{2+} + 2e \rightleftharpoons Cr \qquad E^\circ = -0.90 \text{ V}$$

$$Cr^{3+} + 3e \rightleftharpoons Cr \qquad E^\circ = -0.74 \text{ V}$$

Nitric acid renders it passive; alkalis have no action. At higher temperatures the metal is reactive: it decomposes steam and combines with oxygen, the halogens and most other non-metallic elements.

The chief oxidation states of chromium are VI, III and II. A few compounds of Cr(V) and Cr(IV) are known, but they are unstable with respect to disproportionation; Cr(0) occurs in $Cr(CO)_6$ and $Cr(C_6H_6)_2$, which have been described in some detail in Chapters 22 and 23, and still lower formal oxidation states occur in carbonylate anions; these species have also been described in Chapter 22, however, and in this section we shall therefore confine attention to Cr(VI), Cr(II) and intermediate oxidation states. The potential diagram for chromium at unit hydrogen-ion concentration is shown in Fig. 24.4.

Chromium (VI)

We have already described the preparation of the yellow sodium chromate (or chromate(VI), as it should strictly be called), Na_2CrO_4, and the red dichromate $Na_2Cr_2O_7$. The less soluble potassium dichromate, a

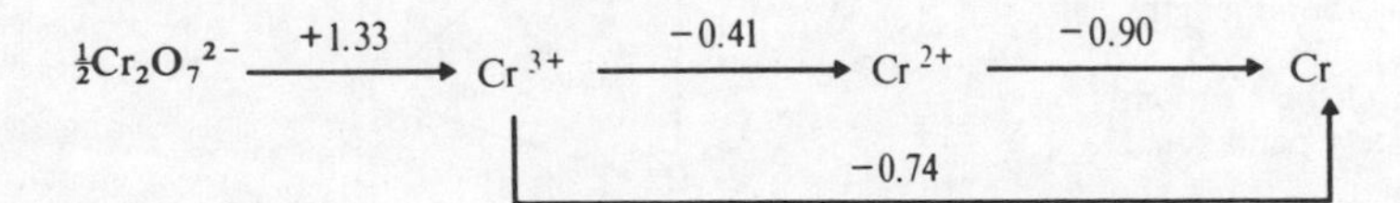

Figure 24.4
Potential diagram for chromium at $[H^+] = 1$.

familiar oxidant in titrimetric analysis (especially for the determination of iron(II)) is obtained by addition of potassium chloride to sodium dichromate solution. Chromium(VI) oxide (or trioxide, as it is still often called) separates as a purple-red solid when concentrated sulphuric acid is added to a solution of a dichromate; it melts at 198 °C and decomposes into Cr_2O_3 and oxygen slightly above this temperature; CrO_2 is formed as an intermediate. By the action of CrO_3 on dichromates, red polychromates such as $K_2Cr_3O_{10}$ and $K_2Cr_4O_{13}$ may be obtained. Chromium(VI) oxide and the various chromates all have structures based on CrO_4 tetrahedra: for example CrO_3 consists of infinite chains of such tetrahedra each sharing two corners (with Cr—O (terminal) 1.60 Å and Cr—O (bridging) 1.75 Å), and the $Cr_2O_7^{2-}$ ion of two such tetrahedra sharing one corner (with bond lengths 1.63 and 1.79 Å); in the tetrahedral CrO_4^{2-} ion the bond length is 1.66 Å. Their structures are thus much simpler than those of comparable vanadium compounds.

When hydrogen peroxide is added to an acidified solution of a chromate, the deep violet-blue peroxo compound $CrO(O_2)_2$ is formed:

$$CrO_4^{2-} + 2H^+ + 2H_2O_2 \rightarrow CrO(O_2)_2 + 3H_2O$$

In aqueous solution this rapidly decomposes, forming Cr(III) and oxygen, but an ethereal solution is more stable, and from it a pyridine adduct $pyCrO(O_2)_2$ may be isolated. This has an approximately pentagonal pyramidal structure with the O of the Cr=O bond occupying the apical position, as shown in Fig. 24.5. Like other Cr(VI) compounds, $pyCrO(O_2)_2$ has a very small paramagnetic susceptibility (arising from coupling of the diamagnetic ground state with excited states). By the action of hydrogen peroxide on neutral or slightly acidic solutions of dichromates, or by the action of alkalis on $CrO(O_2)_2$, diamagnetic violet salts believed to contain the $[CrO(O_2)_2OH]^-$ ion are obtained, but these are very dangerously explosive and their structures are not known.

The only reported binary halide of chromium(VI) is the yellow CrF_6, obtained by heating chromium in fluorine at high pressure and 400 °C and rapidly chilling the product; even at −100 °C it decomposes into the

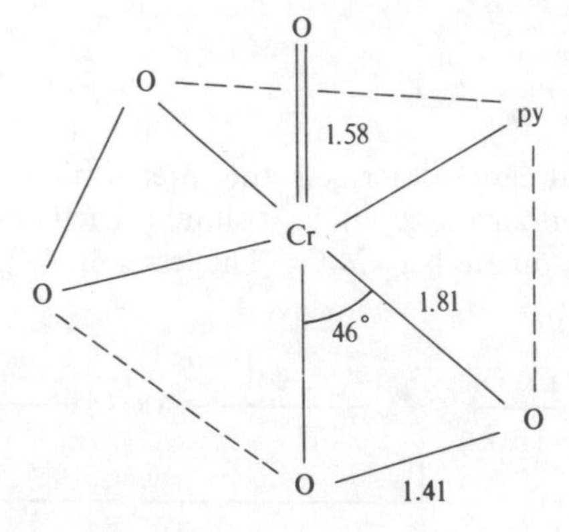

Figure 24.5
The structure of $(py)CrO(O_2)_2$ (bond lengths in ångströms). The coordination of the Cr atom is approximately pentagonal pyramidal.

pentafluoride and fluorine. The oxohalides CrO_2F_2 and CrO_2Cl_2 are much more stable; the last of these, chromyl chloride, is a red liquid (b.p. 117 °C) which is obtained by distillation of a chloride, a dichromate and concentrated sulphuric acid. Water hydrolyses it completely, but if it is added to a concentrated solution of potassium chloride, the monochlorochromate(VI), $KCrO_3Cl$, separates; this salt can also be obtained by crystallisation from a solution of $K_2Cr_2O_7$ in hydrochloric acid.

Chromium(VI) in acidic solution is a powerful oxidant, as we see from the value of E° for the half-reaction

$$\tfrac{1}{2}Cr_2O_7^{2-} + 7H^+ + 3e \rightleftharpoons Cr^{3+} + \tfrac{7}{2}H_2O \qquad E^\circ = +1.33\ V$$

It is, however, often a slow one, no doubt because of the mechanistic complexities involved in the overall three-electron change. In alkaline media it is much less powerfully oxidising, and under these conditions chromium(III) is fairly easily oxidised to chromium(VI), e.g. by heating with hydrogen peroxide. In acidic solution, on the other hand, the overall action of acidic hydrogen peroxide on chromium(VI) is to reduce it to chromium(III).

Chromium(V) and chromium(IV)

When an acidic solution in which dichromate is oxidising isopropanol is added to aqueous manganese sulphate, manganese dioxide is precipitated, although acidified dichromate alone does not effect this oxidation. This evidence for the participation of chromium(V) or (IV) in dichromate oxidations suggests that it might be possible to isolate compounds of chromium in these oxidation states under special conditions, and dark blue Sr_2CrO_4, for example, has been obtained by heating $SrCrO_4$, Cr_2O_3 and $Sr(OH)_2$ at 1000 °C. The volatile pentafluoride and tetrafluoride are obtained by the action of fluorine on the metal at 400 °C. All these compounds are rapidly hydrolysed to Cr(VI) and Cr(III).

Chromium dioxide is usually made by controlled decomposition of CrO_3; it has the rutile structure and, like the high-temperature form of VO_2, is a metallic conductor. It is also ferromagnetic, and is widely used in magnetic recording tapes.

Peroxo compounds of Cr(V) are obtained by the interaction of chromate and hydrogen peroxide in alkaline solution at 0 °C, oxygen also being formed (in acid, as we saw earlier, chromium(VI) compounds are produced). Red $K_3[Cr(O_2)_4]$, for example, is paramagnetic (with a moment corresponding to the presence of one unpaired electron); the oxygen atoms of the four peroxo groups are distributed round the metal atom in a somewhat distorted dodecahedral configuration. This salt is considerably less dangerous than the blue peroxochromates(VI), but explodes when heated. A peroxo compound of chromium(IV), $Cr(O_2)_2.3NH_3$, results from the interaction of dichromate, aqueous ammonia and hydrogen peroxide; this is also explosive. The redox relationships between Cr(VI) and H_2O_2 evidently involve kinetic as well as thermodynamic factors and are by no means well understood.

Chromium(III)

This is the most stable oxidation state of chromium and since, as we have seen in Chapters 18 and 21, Cr(III) complexes are kinetically inert to substitution, immense numbers of them are known. The hydrated ion $[Cr(H_2O)_6]^{3+}$, which is pale violet in colour, is obtained when dichromate is reduced by sulphur dioxide or by ethanol and sulphuric acid below 70 °C; above this temperature, the green sulphato complex $[Cr(H_2O)_5SO_4]^+$ is produced. The commonest salt containing the hexa aquo ion is chrome alum, $KCr(SO_4)_2.12H_2O$; ordinary chromium(III) chloride is dark green *trans* $[Cr(H_2O)_4Cl_2]Cl.2H_2O$. From solutions of Cr(III) salts, alkali precipitates hydrous Cr_2O_3, which dissolves in excess, forming the $[Cr(OH)_6]^{3-}$ ion. The aquo ion is quite acidic, and hydroxo-bridged polymers are present in solutions containing it, e.g.

$$2[Cr(H_2O)_6]^{3+} \underset{+2H^+}{\overset{-2H^+}{\rightleftharpoons}} 2[Cr(H_2O)_5OH]^{2+}$$

$$2[Cr(H_2O)_5OH]^{2+} \rightleftharpoons [(H_2O)_4Cr \begin{matrix} H \\ O \\ / \quad \backslash \\ \\ \backslash \quad / \\ O \\ H \end{matrix} Cr(OH_2)_4]^{4+} + 2H_2O$$

In the presence of ammonia, ammine complexes are slowly formed.

The oxide Cr_2O_3, a stable green pigment, is formed by burning the metal in air, by reduction of dichromate with carbon, or in the thermal decomposition of ammonium dichromate:

$$(NH_4)_2Cr_2O_7 \rightarrow Cr_2O_3 + N_2 + 4H_2O$$

It is isostructural with Ti_2O_3 and V_2O_3. When ignited it becomes inert towards both acids and bases. Chromium(III) fluoride and chloride, made from Cr_2O_3 and hydrogen fluoride, and from the metal and chlorine respectively, both contain octahedrally coordinated Cr(III) in the same structures as the trifluorides and trichlorides of scandium, titanium and vanadium. Chromium trifluoride is only sparingly soluble, and may be precipitated as a hexahydrate; the chloride, though at equilibrium very soluble in water, dissolves only very slowly. The dissolution process may, however, be accelerated by addition of a trace of chromium(II) chloride: the rapid redox reaction between Cr(III) in the lattice and Cr(II) in solution (which involves a chloride-bridged intermediate) is then followed by rapid substitution of chloride by water at the solid surface.

Complex halides of chromium(III) include salts containing $[CrF_6]^{3-}$, $[CrCl_6]^{3-}$ and $[Cr_2Cl_9]^{3-}$ ions. The last ion has a structure which consists of two $CrCl_6$ octahedra sharing a face. There is no interaction between the Cr^{3+} ions, each of which has a magnetic moment corresponding to the presence of the usual three unpaired electrons (in the analogous molybdenum and tungsten species, metal–metal bonds are formed).

The electronic spectra of Cr(III) complexes contain three bands due to *d–d* transitions; the reasons for this number of bands are exactly analogous to those given in the discussion of the spectra of V(III) complexes except that here we are dealing with two holes instead of two electrons, and the lowest energy state, corresponding to one electron in each of the t_{2g} orbitals, is the singly degenerate state $^4A_2$, the transitions being to the three other possible quartet states.

Chromium(II)

Solutions containing the sky-blue $[Cr(H_2O)_6]^{2+}$ ion are obtained by dissolving metallic chromium in acids or by reduction of chromium(III)-containing solutions by zinc amalgam or electrolytically. Hydrated salts such as $Cr(ClO_4)_2.6H_2O$, $CrCl_2.4H_2O$ and $CrSO_4.7H_2O$ may be isolated from solution, but cannot be dehydrated without decomposition. The standard potential for the half-reaction

$$Cr^{3+} + e \rightleftharpoons Cr^{2+}$$

is -0.41 V, and chromium(II) compounds slowly liberate hydrogen from water as well as undergo oxidation by atmospheric oxygen; they are, however, just stable with respect to disproportionation (Fig. 24.4). The study of the oxidation of Cr^{2+} species has played an important part in establishing the mechanisms of redox reactions (Section 21.8).

The anhydrous fluoride and chloride can be obtained from the metal and hydrogen halides; the iodide is formed merely by heating chromium(III) iodide, the consequence of lattice energy factors (cf. Section 16.4). All contain octahedrally coordinated Cr(II) in distorted rutile or cadmium iodide structures. The distortion is the consequence of the Jahn–Teller effect for a high-spin d^4 in an octahedral field (Section 19.3). Low-spin $K_4[Cr(CN)_6].3H_2O$ has a magnetic moment indicating two unpaired electrons.

One of the less unstable chromium(II) compounds is the dimeric red acetate, $Cr_2(OCOCH_3)_4.2H_2O$, which is precipitated when chromium(II) chloride solution is added to saturated aqueous sodium acetate. The structure of the dimer is shown in Fig. 24.6; the distorted octahedral coordination round the Cr^{2+} ion consists of four oxygen atoms of bidentate acetate groups, one oxygen atom of a water molecule, and the other Cr^{2+} ion. The diamagnetism of the compound and the rather short Cr—Cr distance (2.35 Å compared with 2.58 Å in the metal) indicate that

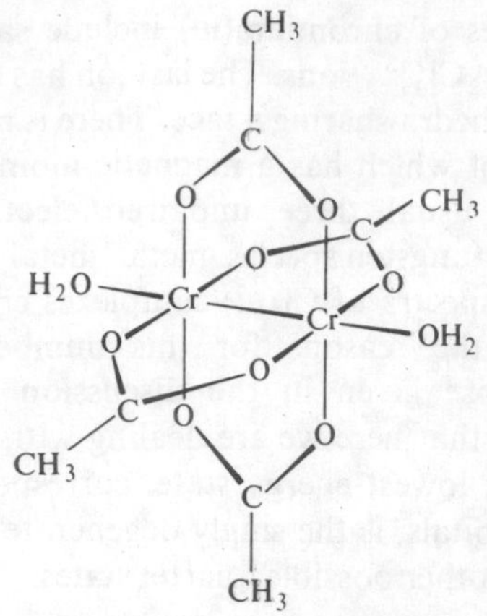

Figure 24.6
The structure of chromium(II) acetate.

the four unpaired electrons on each Cr^{2+} ion are all involved in metal–metal bonding: if the $d_{x^2-y^2}$ orbital of each Cr^{2+} ion (together with the s, p_x, p_y and p_z orbitals) is involved in metal–ligand bonding, the d_{z^2} orbitals can overlap to form a σ bond, the d_{xz} and d_{yz} orbitals to form π bonds, and the d_{xy} orbitals to form a δ bond. In $Li_4[Cr_2(CH_3)_8].4C_4H_8O$ and anhydrous $Cr_2(OCOCH_3)_4$, which have similar structures, the Cr—Cr distance is as short as 1.98 Å. This type of quadruple bond is found in several compounds of the heavy transition metals, and will be encountered again in the next chapter.

24.6 *Manganese*

Several oxides of manganese occur native, but by far the most important of them as a source of the metal is *pyrolusite*, MnO_2. In the principal use of the element, steel-making, this is mixed with Fe_2O_3 and reduced with coke to give ferromanganese (*ca.* 80 per cent Mn). Nearly all steels contain some manganese, which, by combining with dissolved oxygen and sulphur, prevents brittleness. The presence of larger quantities of the metal (up to 12 per cent) produces very hard steels suitable for crushing and grinding machinery. The metal is obtained by electrolysis of manganese sulphate solution. The most important compound of manganese is potassium permanganate or manganate(VII), $KMnO_4$, a versatile and powerful oxidising agent. Small quantities of manganese(II) sulphate are added to fertilisers: the metal is an essential trace element for plants.

Metallic manganese is rather like iron in its physical and chemical properties, but at ordinary temperatures it has a slightly less regular crystal structure containing metal atoms having 12, 13 or 16 nearest neighbours; this is probably the reason for its relative brittleness. At higher temperatures it adopts the body-centred cubic or the cubic close-packed structure. Manganese is slowly attacked by water and dissolves readily in acids. The finely divided metal is pyrophoric in air, but the massive metal is not attacked unless it is heated. Under the same conditions it combines with the halogens and most other non-metallic elements.

Manganese exhibits the widest range of oxidation states of any of the first series transition metals. Very low formal oxidation states occur in the carbonyl $Mn_2(CO)_{10}$, the carbonylate anion $[Mn(CO)_5]^-$ and the nitrosyl carbonyl $Mn(NO)_3CO$, all of which were discussed in Chapter 22. Manganese(I) occurs in various carbonyl derivatives such as $CH_3Mn(CO)_5$ (if we adopt the convention that alkyl groups in such species are counted as anions) and $Mn(CO)_5Cl$; the cyano complex $Na_5[Mn(CN)_6]$, obtained by dissolving manganese powder in air-free aqueous sodium cyanide, also contains manganese(I). For the most part, however, the inorganic chemistry of manganese is that of the oxidation states II–VII inclusive, and it is with these that the remainder of this section is concerned. There is an abrupt change in the stability of the dipositive state with respect to oxidation between chromium and manganese (E° values for the Cr^{3+}/Cr^{2+} and Mn^{3+}/Mn^{2+} systems are -0.41 and $+1.51$ V respectively, the difference arising mainly from the much higher third ionisation energy of manganese); all oxidation states of manganese above Mn(II) are powerful oxidising agents. We shall at this point therefore reverse the order in which we have discussed oxidation states in the first part of this chapter, and begin with manganese(II). In Section 7.10 we described potential diagrams and oxidation state diagrams with special reference to those for manganese, and we shall not repeat these discussions here; for convenience, however, a Latimer potential diagram at unit hydrogen-ion concentration is shown in Fig. 24.7.

Manganese(II)

Manganese(II) salts are obtained from the dioxide by a variety of methods. The soluble chloride and sulphate result from heating it with the appropriate concentrated acid:

$$MnO_2 + 4HCl \rightarrow MnCl_2 + Cl_2 + 2H_2O$$

$$2MnO_2 + 2H_2SO_4 \rightarrow 2MnSO_4 + O_2 + 2H_2O$$

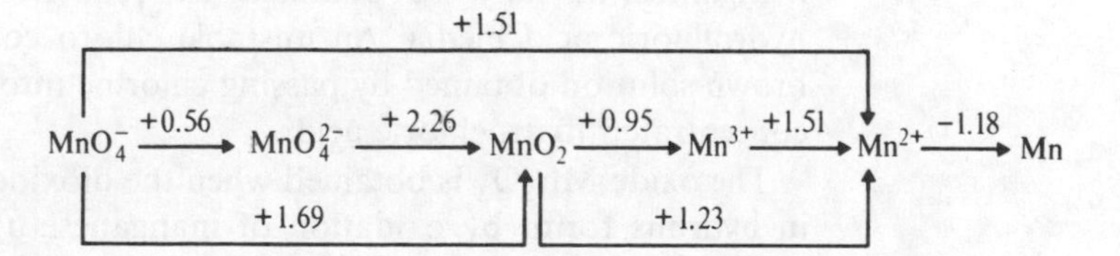

Figure 24.7
Potential diagram for manganese at $[H^+] = 1$.

From them the carbonate is obtained by precipitation, and may be converted into salts of other acids. The fluoride MnF_2 is sparingly soluble and has the rutile structure; anhydrous $MnCl_2$ has the cadmium chloride structure.

All these salts are pale pink or colourless; manganese(II) has the electronic configuration d^5 and for its octahedral high-spin complexes *d–d* transitions are both spin- and Laporte-forbidden (Section 19.5). The electronic spectrum of $[Mn(H_2O)_6]^{2+}$ does in fact contain several bands, but these are all weaker by a factor of about 10^2 than spin-allowed transitions for complexes of other first-series metal complexes; they arise from promotion of an electron to give various excited states containing only three unpaired electrons. Octahedrally coordinated Mn(II) is also present in $Me_4N[MnCl_3]$, which contains a chain anion, and in $K_4[MnCl_6]$; the rather more strongly coloured yellow-green tetrahedral anion $[MnCl_4]^{2-}$, for which the Laporte selection rule does not apply, has been characterised as the Et_4N^+ salt. The only common low-spin complex of manganese(II) is the blue salt $K_4[Mn(CN)_6].3H_2O$, which is isostructural with the hexacyanoferrate(II).

The oxidation of manganese(II) compounds in acidic solution requires the use of a powerful oxidising agent such as bismuthate, periodate or peroxodisulphate, but in alkaline media oxidation is much easier owing to the fact that hydrous Mn_2O_3 is very much less soluble than $Mn(OH)_2$; thus when alkali is added to a solution of a manganese(II) salt in the presence of air, the white precipitate of $Mn(OH)_2$ that is first formed rapidly darkens owing to atmospheric oxidation. The green oxide MnO (sodium chloride structure) has to be obtained by thermal decomposition of the oxalate or by heating $MnCO_3$ in hydrogen; for its magnetic properties see Section 19.6.

Manganese(III)

The only binary compounds of manganese in this oxidation state that are stable at ordinary temperatures are the trifluoride and the oxide Mn_2O_3. The trifluoride, the only stable trihalide of manganese(III), is obtained as a purple solid by action of fluorine on manganese halides at 250 °C; it is immediately hydrolysed by water. Manganese trifluoride has a structure related to those of TiF_3, VF_3, CrF_3, FeF_3 and CoF_3, but as a high-spin d^4 complex it exhibits the Jahn–Teller effect; there are, however, *three* pairs of Mn—F distances, of 1.79, 1.91 and 2.09 Å. Fluoro complexes of manganese(III) may be obtained by reduction of permanganate in hydrofluoric acid media. An unstable chloro complex is present in the brown solution obtained by passing chlorine into a solution of $MnCl_2$ in concentrated hydrochloric acid.

The oxide Mn_2O_3 is obtained when the dioxide is heated at 800 °C or, in hydrous form, by oxidation of manganese(II) in alkaline media. At higher temperatures it forms Mn_3O_4, a normal spinel. The red aquo ion $[Mn(H_2O)_6]^{3+}$ can be obtained by electrolytic oxidation of Mn^{2+}(aq), and may be isolated as the caesium alum $CsMn(SO_4)_2.12H_2O$; this, surprisingly, shows no Jahn–Teller distortion, at least down to −195 °C. The tris (acetylacetonate) complex is dimorphic, one form having two oxygen atoms 0.19 Å more distant, and the other having two 0.05 Å nearer, than other octahedral ligand atoms. The Mn^{3+} ion is appreciably hydrolysed to form H^+ and polymeric cations in aqueous solution, and is also unstable with respect to the disproportionation

$$2Mn^{3+} + 2H_2O \rightarrow Mn^{2+} + MnO_2 + 4H^+$$

as would be expected from the potentials in Fig. 24.7; it is less unstable in the presence of high concentrations of Mn^{2+} or H^+ ions. It is also stabilised by complexing with F^-, PO_4^{3-} or SO_4^{2-}; the pink colour sometimes seen before the end of a permanganate–oxalate titration is due to an oxalato complex of Mn(III).

The only well-known low-spin complex of Mn(III) is the dark red cyano complex $K_3[Mn(CN)_6]$, obtained by the action of potassium cyanide solution on either the fluoro complex or on $MnPO_4$, a dark green solid precipitated when a manganese(II) compound is heated with phosphoric and nitric acids. It contains the expected regular octahedral anion.

Manganese(IV)

The chemistry of this oxidation state is more limited than that of manganese(III). In addition to the dioxide (which has the rutile structure at high temperatures but is polymorphic and often markedly non-stoichiometric), there is an unstable blue tetrafluoride made by direct combination of the elements. Complexes include salts containing the yellow $[MnF_6]^{2-}$ ion, obtained either by fluorination of mixtures of chlorides or by reduction of permanganate with diethyl ether or hydrogen peroxide in aqueous hydrofluoric acid; they are hydrolysed by water. When $K_2[MnF_6]$ is heated with antimony pentafluoride, fluorine is evolved:

$$K_2[MnF_6] + 2SbF_5 \rightarrow 2K[SbF_6] + MnF_2 + F_2$$

This reaction represents the first really practicable non-electrolytic method for the production of fluorine.

Hydrous forms of MnO_2, which is extremely insoluble, are often obtained in oxidations by permanganate in acidic, but insufficiently acidic, media: it is very stable with respect to disproportionation (cf. Fig. 24.7). In the commonest form of dry battery, a zinc container is the negative electrode; it is filled with a paste of carbon black, ammonium chloride and manganese dioxide; a carbon rod is the positive electrode.

The cell reactions are:

$$Zn \rightarrow Zn^{2+} + 2e$$
$$Zn^{2+} + 2NH_4Cl + 2OH^- \rightarrow [ZnCl_2(NH_3)_2] + 2H_2O$$
$$2MnO_2 + 2H_2O + 2e \rightarrow 2MnO(OH) + 2OH^-$$

$$Zn + 2NH_4Cl + 2MnO_2 \rightarrow [ZnCl_2(NH_3)_2] + 2MnO(OH)$$

Manganese(V)

The only accessible Mn(V) compound is blue Na_3MnO_4, sodium manganese(V), obtained by formate reduction of permanganate in concentrated aqueous sodium hydroxide at 0 °C; it very readily disproportionates, to MnO_4^{2-} and MnO_2 in weakly alkaline solution, and to MnO_4^- and MnO_2 in acidic solution:

$$2MnO_4^{3-} + 4H^+ \rightarrow MnO_4^{2-} + MnO_2 + 2H_2O \text{ or}$$
$$2MnO_4^{3-} + 2H_2O \rightarrow MnO_4^{2-} + MnO_2 + 4OH^-$$
$$3MnO_4^{3-} + 8H^+ \rightarrow MnO_4^- + 2MnO_2 + 4H_2O$$

Manganese(VI)

The only compounds of manganese(VI) are the dark green manganates, or manganates(VI), obtained by fusion of manganese dioxide with sodium or potassium hydroxide in the presence of air or other oxidising agents, and also by the action of alkalis on permanganate:

$$4MnO_4^- + 4OH^- \rightarrow 4MnO_4^{2-} + 2H_2O + O_2$$

This change may be reversed by electrolytic oxidation or the action of chlorine:

$$2MnO_4^{2-} + Cl_2 \rightarrow 2MnO_4^- + 2Cl^-$$

Manganate is unstable to disproportionation in the presence of even weak acids such as carbonic acid:

$$3MnO_4^{2-} + 4H^+ \rightarrow 2MnO_4^- + MnO_2 + 2H_2O$$

It therefore is not formed in the reduction of acidified permanganate.

Manganese(VII)

We have already indicated the principal routes for the preparation of permanganates: on the industrial scale, conversion of the dioxide into manganate, followed by electrolytic oxidation; in analytical chemistry, for the determination of manganese, by oxidation of manganese(II) with bismuthate, periodate or peroxodisulphate in acidic solution. The commonest salt, potassium permanganate or manganate(VII), $KMnO_4$, forms dark purple (almost black) crystals which are isostructural with those of $KClO_4$. The free acid can be obtained by low-temperature evaporation of its aqueous solution, which is made by ion-exchange; it is a violent oxidising agent and is explosive above 0 °C. The oxide Mn_2O_7, obtained as a green liquid by the action of concentrated sulphuric acid on potassium permanganate, is also explosive.

Potassium permanganate has two physical characteristics of considerable interest: its intense colour and its weak temperature-independent paramagnetism. The former is believed to be due to charge-transfer from oxygen to manganese; the latter arises from the coupling of the diamagnetic ground state of the MnO_4^- ion with paramagnetic excited states under the influence of the magnetic field.

If we represent the reduction of permanganate to manganate(VI), manganese dioxide and manganese(II) by the half-reactions

$$MnO_4^- + e \rightarrow MnO_4^{2-}$$

$$MnO_4^- + 4H^+ + 3e \rightarrow MnO_2 + 2H_2O$$

$$MnO_4^- + 8H^+ + 5e \rightarrow Mn^{2+} + 4H_2O$$

we can see that the hydrogen-ion concentration of the solution plays an important part in influencing which reduction takes place, and many reactions of potassium permanganate can be better understood by consideration of redox potentials. Nevertheless, kinetic factors are also important. Permanganate at $[H^+] = 1$ should oxidise water, but in practice the reaction is extremely slow. It should also oxidise oxalate, but again the reaction is extremely slow unless either manganese(II) ions are added or the temperature is increased. Many interesting studies have been made of the mechanisms of such reactions and, as in oxidations by dichromate, it has been shown that intermediate oxidation state species are involved. In addition to its long-established uses as an oxidant for organic compounds and as a disinfectant, permanganate is now being used in water purification; it is preferable to chlorine not only in that it does not affect the taste of the water, but also in that the manganese dioxide produced is a coagulant for colloidal impurities.

24.7 Iron

Iron, the most important of all metals, is also one of the most abundant in the earth's crust, being second only to aluminium. The most important ores are *haematite*, Fe_2O_3, *magnetite*, Fe_3O_4, and *siderite*, $FeCO_3$. *Iron pyrites*, FeS_2, is also common, but owing to its high sulphur content it is not used for the manufacture of metallic iron. The earth's core is believed to consist mainly of iron (one of the elements for which the nuclear binding energy per nucleon – Section 1.2 – is a maximum), and it is also the principal constituent of metallic meteorites, a fact which suggests that it is abundant throughout the solar system. The element is of immense biological importance; the oxygen-carriers haemoglobin and myoglobin, the electron-transfer agents the cytochromes (which are involved in the conversion of adenosine diphosphate into adenosine triphosphate by energy liberated in the oxidation of glucose), and the enzyme system nitrogenase (used in nitrogen fixation in biological systems) all contain iron. The chemistry and biochemistry of these complex species is discussed in some of the references listed at the end of the chapter; in this section we refer only to the mechanism of oxygen uptake by deoxyhaemoglobin.

Pure iron is made by reduction of the oxides with hydrogen or by thermal decomposition of iron pentacarbonyl; it is a white, lustrous metal which is not particularly hard. Iron dissolves in dilute mineral acids to form iron(II) salts, but concentrated nitric acid and other powerful oxidants make it passive. It reacts with halogens at 200–300 °C to form FeF_3, $FeCl_3$, $FeBr_3$ and FeI_2; FeI_3 is unstable. Steam is decomposed at a red heat, Fe_3O_4 and hydrogen being formed. The finely divided metal is pyrophoric is air; the massive metal oxidises in dry air only when heated. In moist air iron rusts to hydrated Fe_2O_3. Rusting occurs only in the presence of oxygen, water and an electrolyte (which may be water, especially if it contains sulphur dioxide, but a dilute solution of sodium chloride in the form of sea spray or liquid splashed from frozen roads that have been treated with salt is also very effective). Rusting is an electrochemical process: iron exposed to a higher concentration of oxygen accepts electrons from iron where the concentration of oxygen is lower.

$$2Fe \rightarrow 2Fe^{2+} + 4e$$

$$O_2 + 2H_2O + 4e \rightarrow 4OH^-$$

Diffusion of the ions formed results in the production of $Fe(OH)_2$ at places between the points of attack, and this is further oxidised to hydrated Fe_2O_3:

$$4Fe(OH)_2 + O_2 \rightarrow 2Fe_2O_3 + 4H_2O$$

Rusting may be prevented by the application of a protective coating (e.g. of paint, zinc or tin – though the last is completely ineffective if the coating is broken), or by connection to a metal (e.g. magnesium or zinc), of more negative metal ion/metal standard electrode potential, which is thus attacked preferentially, leaving the iron or steel uncorroded.

We shall describe the manufacture of iron and steel only in outline. In the blast furnace, which is made of steel with a lining of firebricks, a charge of oxide ore, limestone and coke is fed into the top of the furnace, and preheated air at about 600 °C is blown in at the bottom. The burning of coke to carbon monoxide supplies most of the heat needed for the working temperature of the furnace, which ranges from about 1600 °C at the bottom to about 250 °C at the top. In the top part of the furnace the reducing agent is carbon monoxide, but at the bottom it is carbon itself: as we saw in Section 13.7, of the reactions

$$2CO + O_2 \rightarrow 2CO_2$$

$$C + O_2 \rightarrow CO_2$$

and

$$2C + O_2 \rightarrow 2CO$$

the first has the most negative value of ΔG° below about 700 °C and the last above this temperature. The limestone decomposes at about 800 °C to produce calcium oxide, which combines with silica from the ore to form a slag of molten calcium silicate; this floats on the liquid iron and they are continuously tapped off separately. The product (*pig iron*) contains about 4 per cent of carbon and many other impurities; mixed with scrap iron and cast, it forms *cast iron*, which contains about 3 per cent of carbon. Although pig iron is very hard, it is also very brittle, and in order to convert it into steel most of the carbon and all of the other non-metallic impurities must be removed by one of several methods. In the now obsolescent open-hearth process this is done by heating it with the oxide haematite (Fe_2O_3) in a furnace lined with a mixture of calcium and magnesium oxides; pure oxygen is injected to speed up the oxidation. Most of the carbon is converted into the monoxide, and phosphorus and silicon are converted into a slag containing phosphates and silicates. Alloys with other metals (e.g. Mn, Cr, Ni, W) may be added towards the end of the operation to produce steels of the desired characteristics. In modern plants the crude iron is melted and the impurities oxidised with a blast of oxygen blown through the converter. Oxidised impurities form a slag or are volatile, and the molten metal is run off and converted into alloy steels. Mild steel contains 0.1–0.4 per cent of carbon; hard steel 0.5–1.5 per cent, together with other metals. Thus stainless steel contains about 18 per cent of chromium, tungsten steel (which is very hard) about 5 per cent of tungsten and manganese steel (which is very tough) about 13 per cent of manganese. The role of carbon in steels is a specialised topic which we shall not discuss here beyond remarking that solid solutions, supersaturated solid solutions and the carbide Fe_3C (*cementite*) are all involved in different systems.

Most of the chemistry of iron is that of Fe(II) or Fe(III). Lower formal oxidation states occur in the carbonyls and carbonylate anions such as $Fe(CO)_5$ and $[Fe(CO)_4]^{2-}$ (Sections 22.1–4), and iron(I) in the NO^+ complex $[Fe(H_2O)_5NO]^{2+}$ (Section 22.8), which is formed by the action of nitrate and concentrated sulphuric acid on a Fe(II) salt. Iron(IV) and iron(VI) are also known in a small number of compounds. The extensive organometallic chemistry of iron, some aspects of which were mentioned in Chapter 23, will not be discussed again here.

Iron(II)

Anhydrous FeF_2 and $FeCl_2$ may be made by the action of hydrogen halides on the heated metal. The former is a sparingly soluble white solid which has the rutile structure, though the environment of the high-spin Fe^{2+} ion is surprisingly irregular with 4F at 2.12 Å and 2F at 1.98 Å; the chloride, which is readily soluble, has the cadmium chloride layer structure and forms a pale green hydrate $[Fe(H_2O)_6]Cl_2$. This aquo ion is present in most hydrated iron(II) salts such as the perchlorate, the sulphate and the double salt $(NH_4)_2[Fe(H_2O)_6](SO_4)_2$, which is kinetically more stable towards oxidation than most Fe(II) salts (since E° for the half-reaction

$$Fe^{3+} + e \rightleftharpoons Fe^{2+}$$

is 0.77 V, all Fe(II) compounds should, thermodynamically, be oxidised in air at $[H^+]=1$). The ion $[Fe(H_2O)_6]^{2+}$ shows the electronic spectrum expected for a high-spin d^6 species.

Iron(II) oxide, obtained by thermal decomposition of iron(II) oxalate *in vacuo*, is black and has the sodium chloride structure, but is deficient in iron (Section 6.10). Iron(II) sulphide is similar. The white hydroxide $Fe(OH)_2$ is precipitated by the action of alkali on solutions of iron(II) salts, but it rapidly absorbs oxygen, turning dark green and then brown, the products being a mixed Fe(II)–Fe(III) hydroxide and hydrous Fe_2O_3 respectively. This rapid conversion of $M(OH)_2$ into hydrous M_2O_3 originates in the much lower solubility of the latter and, as we have seen, is common to many transition metals. The hydroxide $Fe(OH)_2$ dissolves not only in acids, but also in concentrated aqueous sodium hydroxide, and from the solution the blue-green hydroxo complex $Na_4[Fe(OH)_6]$ may be crystallised. No peroxide of iron is known, but iron pyrites is $Fe^{2+}S_2^{2-}$ and contains low-spin Fe(II) in a distorted NaCl structure.

Iron(II) halides combine with gaseous ammonia, forming salts containing the $[Fe(NH_3)_6]^{2+}$ ion, but in aqueous solution this decomposes and $Fe(OH)_2$ is precipitated. With ethylenediamine, 2,2′-bipyridyl or 1,10-phenanthroline, however, complexes stable in aqueous media are formed. Oxidation of red $[Fe(phen)_3]^{2+}$ to blue $[Fe(phen)_3]^{3+}$ is more difficult than that of $[Fe(H_2O)_6]^{2+}$ to $[Fe(H_2O)_6]^{3+}$ in aqueous solution, as the following standard potentials imply:

$$[Fe(phen)_3]^{3+} + e \rightleftharpoons [Fe(phen)_3]^{2+} \qquad E^\circ = +1.12\ V$$

$$[Fe(H_2O)_6]^{3+} + e \rightleftharpoons [Fe(H_2O)_6]^{2+} \qquad E^\circ = +0.77\ V$$

Hence arises the use of $[Fe(phen)_3]SO_4$ as a redox indicator. This stabilisation of the Fe(II) state is ascribed to the existence of π bonding between the metal ion and the ligands, using relatively low-energy molecular π^* orbitals of the latter. Bipyridyl similarly stabilises iron(II). Both tris-complexes of iron(II) are among the few diamagnetic (low-spin) complexes of this ion.

The best known of such complexes, however, is undoubtedly yellow potassium hexacyanoferrate(II) or ferrocyanide, $K_4[Fe(CN)_6].3H_2O$, obtained by the action of excess of aqueous cyanide on any iron(II) compound. The anion in this salt is kinetically inert with respect to substitution, as is shown by the absence of exchange with labelled cyanide; further, the salt is not poisonous. The $[Fe(CN)_6]^{4-}$ ion is, of course, a low-spin d^6 system like the similarly unreactive cobalt(III) complexes. The blue precipitates obtained from Fe^{3+} and $[Fe(CN)_6]^{4-}$ or Fe^{2+} and $[Fe(CN)_6]^{3-}$ in aqueous media (formerly known as Prussian blue and Turnbull's blue) have both been shown by infrared and Mössbauer spectroscopy to be Fe(III) compounds containing the $[Fe(CN)_6]^{4-}$ anion, though alkali metal cations may also be present. The colour is the result of electron transfer between Fe(II) and Fe(III); it is noteworthy that $K_2Fe[Fe(CN)_6]$, which contains only Fe(II), is white.

Unlike $[Fe(phen)_3]^{2+}$, $[Fe(CN)_6]^{4-}$ is easier to oxidise than $[Fe(H_2O)_6]^{2+}$ in aqueous solution:

$$[Fe(CN)_6]^{3-} + e \rightleftharpoons [Fe(CN)_6]^{4-} \qquad E^\circ = +0.36\ V$$

This must mean that Fe(III) forms a more stable complex than Fe(II) which, since CN^- is also a good π bonding ligand, seems surprising. A detailed examination of the system based on thermochemical data and values for E° at different temperatures, however, reveals that ΔH° for the oxidation of $[Fe(CN)_6]^{4-}$, like that for the oxidation of $[Fe(phen)_3]^{2+}$, is more positive than that for the oxidation of $[Fe(H_2O)_6]^{2+}$, and that the difference in E° values for the systems $[Fe(phen)_3]^{3+}/[Fe(phen)_3]^{2+}$ and $[Fe(CN)_6]^{3-}/[Fe(CN)_6]^{4-}$ is mainly caused by a very large negative entropy of hydration of the highly charged $[Fe(CN)_6]^{4-}$ ion: as is so often the case, the solvent plays an important part in the system. Recent structural studies, in fact, establish beyond doubt that the Fe—C distance in $[Fe(CN)_6]^{4-}$ (1.92 Å) is shorter than that in $[Fe(CN)_6]^{3-}$ (1.95 Å), the reverse of what is found for Fe(II) and Fe(III) hexaquo and hexachloro complexes and is reflected in the smaller radius assigned to the Fe^{3+} ion. Since the C≡N bond lengths and stretching frequencies differ very little between $[Fe(CN)_6]^{4-}$ and $[Fe(CN)_6]^{3-}$, the Fe—C bond lengths provide convincing evidence for stronger π bonding in the lower oxidation state compound (cf. Section 19.4). In this connection it is interesting to note that whereas the Mössbauer chemical shifts for $FeCl_2$ and $FeCl_3$ are quite different (+1.3 and +0.5 mm s^{-1} respectively), those for $K_4[Fe(CN)_6]$ and $K_3[Fe(CN)_6]$ are both -0.1 mm s^{-1}.

There are many known monosubstitution products of the $[Fe(CN)_6]^{4-}$ ion, the most important being the red salt sodium nitrosopentacyanoferrate(II), $Na_2[Fe(CN)_5NO].2H_2O$, commonly called sodium nitroprusside and mentioned earlier in Section 22.8. This results from the action of nitric acid or sodium nitrite on the hexacyano complex:

$$[Fe(CN)_6]^{4-} + 4H^+ + NO_3^- \rightarrow [Fe(CN)_5NO]^{2-} + CO_2 + NH_4^+$$

$$[Fe(CN)_6]^{4-} + NO_2^- + H_2O \rightarrow [Fe(CN)_5NO]^{2-} + CN^- + 2OH^-$$

The anion of this compound, which is diamagnetic, contains coordinated NO with an interatomic distance of 1.13 Å and a stretching vibration at 1 947 cm^{-1}, considerably higher than that for NO itself; it is therefore formulated as a NO^+ complex of Fe(II). With sulphide ion it gives the purple anion $[Fe(CN)_5NOS]^{4-}$, a reaction which forms the basis of a sensitive test for sulphide; hydroxide ion (i.e. oxide + water) similarly gives $[Fe(CN)_5NO_2]^{4-}$:

$$[Fe(CN)_5NO]^{2-} + 2OH^- \rightarrow [Fe(CN)_5NO_2]^{4-} + H_2O$$

The most important function of iron in biological systems is as an oxygen carrier in the blood pigments haemoglobin and myoglobin. Both of these contain the haem unit shown in Fig. 24.8, the structure of which is closely related to that of chlorophyll (Section 11.6). From the inorganic chemical viewpoint the main interest in the uptake of oxygen by deoxyhaemoglobin (haem with a large protein molecule bonded to the iron atom through nitrogen on one side of the ring system) lies in the oxidation state and magnetic properties of the iron atom and the orientation of the absorbed oxygen molecule. In deoxyhaemoglobin the high-spin iron(II) ion lies about 0.70 Å from the plane of the porphyrin system and on the globin side, its coordination being completed by a weakly held water molecule *trans* to the globin moiety. When O_2 is coordinated, the haemoglobin formed is

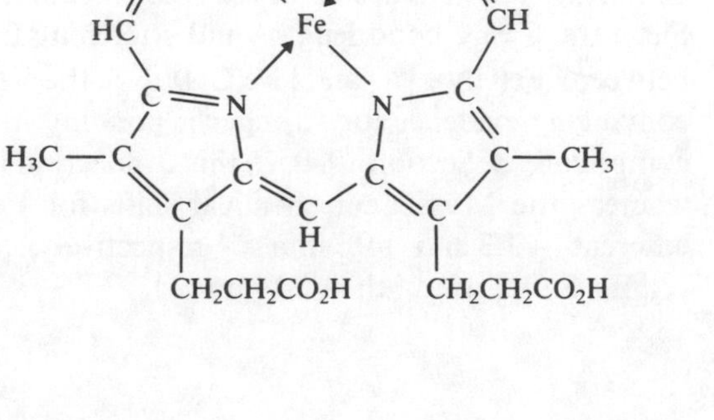

Figure 24.8
The structure of haem.

diamagnetic and the iron atom is in the porphyrin plane with the FeOO angle about 120°; the O—O stretching frequency is at 1106 cm^{-1}, about the same as in the superoxide (O_2^-) ion. These facts have been variously interpreted in terms of low-spin Fe(II) bonded to oxygen in a singlet state and of low-spin Fe(III) bonded to a superoxide ion with the unpaired spins of the two species paired as a result of antiferromagnetic interaction. Since high-spin iron(II) contains an electron in an e_g orbital it is larger than low-spin iron(II) or iron(III), and it has been suggested that it is too large to be accommodated in the porphyrin plane; on oxidation the smaller ion formed slips into the porphyrin ring. Carbon monoxide binds to haem like O_2, but more strongly; hence its toxicity.

Iron(III)

Iron(III) fluoride, a white involatile solid isostructural with ScF_3, and iron(III) chloride, an almost black solid which sublimes at 300 °C and has a layer structure like that of $ScCl_3$, are both made by heating the metal in the appropriate halogen. Iron(III) fluoride is only sparingly soluble, but combines with alkali metal fluorides to form a series of complexes; the $[FeF_6]^{3-}$ ion, salts of which are made by fluorinations of chlorides with elemental fluorine, is one of the classic examples of an octahedral high-spin d^5 complex, and was mentioned as such in Section 19.2. Fluoride complexes Fe(III) much more strongly than Fe(II) and hence reduces the oxidising powers of Fe(III); many other anions (notably phosphate) have similar effects. Iron(III) chloride is usually encountered as the very soluble yellow-brown hydrate $FeCl_3.6H_2O$, which is really *trans* $[Fe(H_2O)_4Cl_2]Cl.2H_2O$. In solutions containing high concentrations of chloride ion it forms the yellow tetrahedral anion $[FeCl_4]^-$; iron(III) may be extracted from hydrochloric acid solution into various ethers as oxonium salts of this anion.

Iron(III) oxide exists in a number of forms: α-Fe_2O_3 has the corundum structure based on a hexagonal close-packed oxide ion lattice; γ-Fe_2O_3, obtained by careful oxidation of Fe_3O_4, has the oxide ions in cubic close-packing, with Fe^{3+} randomly distributed between octahedral and tetrahedral holes; it is widely used in magnetic recording tapes. Iron(III) oxide does not combine with water to form $Fe(OH)_3$, but the hydrate FeO(OH) is known. The ignited oxide is difficult to dissolve in acids, but in the freshly precipitated hydrous form it dissolves in both acids and concentrated aqueous alkalis. In the former it gives the pale violet $[Fe(H_2O)_6]^{3+}$ ion (which has an electronic spectrum like that of the isoelectronic Mn(II) species); in the latter it forms the $[Fe(OH)_6]^{3-}$ ion. Solid compounds which could be formulated as ferrites or ferrates(III), however, are usually mixed oxides: $LiFeO_2$, for example, has the sodium chloride structure, Li^+ and Fe^{3+} being of about the same size and having an average charge of +2 to balance that of the oxide ion. The most familiar oxo-acid salt of iron(III) is the violet ammonium alum, $(NH_4)_2[Fe(H_2O)_6](SO_4)_2.6H_2O$. In aqueous media $[Fe(H_2O)_6]^{3+}$ ionises appreciably according to the primary equation

$$[Fe(H_2O)_6]^{3+} \rightleftharpoons [Fe(H_2O)_5OH]^{2+} + H^+$$

and the hydroxo species then polymerises to the yellow binuclear ion $[(H_2O)_5FeOFe(H_2O)_5]^{4+}$ in which the bridging oxygen and iron atoms are collinear, indicating O—Fe π bonding. Such behaviour is, of course, typical of highly charged aquo ions.

When Fe_2O_3 is heated to 1400 °C it is converted into black Fe_3O_4, a metallic conductor, which has the inverse spinel structure, i.e. has the Fe^{2+} and half the Fe^{3+} ions in octahedral sites. This oxide and other inverse ferrites, as they are commonly called, are ferrimagnetic: the ions occupying octahedral sites all have parallel spins, and so do those occupying tetrahedral sites, but the spins of the latter are antiparallel to those of the former, leading to greatly reduced magnetic moments: thus at low temperatures the inverse ferrite $Fe^{III}[Fe^{III}Ni^{II}]O_4$ has a moment corresponding to the presence of only two unpaired spins per formula unit. The spin alignment can be altered by application of an external magnetic field of sufficient magnitude, and these compounds are very important in electromagnetic devices concerned with the storage and retrieval of information.

We have already mentioned some Fe(III) complexes, notably blue $[Fe(phen)_3]^{3+}$ and red $[Fe(CN)_6]^{3-}$, both obtained by oxidation of the corresponding Fe(II) complexes, e.g. by MnO_4^- or electrolytically. (Iron(III) salts on treatment with aqueous cyanide precipitate the hydrous oxide: this dissolves in warm aqueous cyanide, but with formation of hexacyanoferrate(II) and cyanate.) Potassium hexacyanoferrate(III) or ferricyanide, $K_3[Fe(CN)_6]$, is somewhat more reactive than the iron(II) compound (it is, for example, poisonous), but exchange of labelled cyanide with the anion is slow.

Iron(IV) and iron(VI)

Oxidation of the low-spin *trans* $[Fe(diarsine)_2Cl_2]^+$ (diarsine = *o*-phenylenebis(dimethylarsine)) with nitric acid leads to the formation of the iron(IV) low-spin species *trans* $[Fe(diarsine)_2Cl_2]^{2+}$, characterised as its fluoroborate. Among the few other reported compounds of iron in the tetrapositive oxidation state are the mixed oxides $BaFeO_3$ and Ba_2FeO_4, obtained by heating hydrous iron(III) oxide and barium oxide in oxygen at 1100 °C; they disproportionate into Fe(III) and Fe(VI) on treatment with water. It is interesting that all attempts to make fluoro complexes of iron(IV) have been unsuccessful.

Purple or dark red ferrates(VI) are obtained by oxidation of a suspension of hydrous Fe_2O_3 in concentrated alkali with hypochlorite:

$$Fe_2O_3 + 3OCl^- + 4OH^- \rightarrow 2FeO_4^{2-} + 3Cl^- + 2H_2O$$

Alkali and alkaline earth metal salts may be so obtained; they contain the tetrahedral FeO_4^{2-} anion and have magnetic moments corresponding to the presence of the expected two unpaired electrons; K_2FeO_4 is isostructural with K_2SO_4 (and hence with K_2CrO_4 and K_2MnO_4), and $BaFeO_4$ with $BaSO_4$ (which it also resembles in being only sparingly soluble). The ferrate(VI) ion is fairly stable in concentrated alkali, but in neutral or acidic solution it rapidly decomposes:

$$2FeO_4^{2-} + 10H^+ \rightarrow 2Fe^{3+} + 5H_2O + \tfrac{3}{2}O_2$$

It is a powerful oxidising agent (for the half-reaction

$$FeO_4^{2-} + 8H^+ + 3e \rightleftharpoons Fe^{3+} + 4H_2O$$

E° at $[H^+] = 1$ has been estimated as approximately $+2.0$ V) and will, for example, convert chromium(III) into chromate.

24.8 *Cobalt*

Cobalt occurs mainly as arsenide and sulphide ores (e.g. *cobaltite*, CoAsS, which is probably a lattice aggregate of Co^{2+}, As_2^{2-} and S_2^{2-} ions), but the metal and its compounds are produced chiefly as by-products from the extraction of other metals, particularly nickel. The various processes in use end in the separation of cobalt as the oxide Co_3O_4, which is reduced by aluminium or carbon and refined electrolytically. The metal is used mainly in the form of hard non-corroding non-ferrous alloys (e.g. with chromium and tungsten), in special steels and (as an alloy with aluminium and nickel) in permanent magnets. Cobalt compounds are widely used as pigments and as catalysts (cf. Section 23.8). Vitamin B_{12}, which prevents pernicious anaemia, is a cobalt complex.

Cobalt is a bluish-white metal somewhat less reactive than iron; E° for the Co^{2+}/Co system is -0.28 V. It dissolves slowly in dilute mineral acids, but concentrated nitric acid makes it passive; alkalis have no action. It does not react with oxygen unless it is heated or is very finely divided, when it is pyrophoric. Fluorine reacts at 250 °C to give CoF_3, but the other halogens give only dihalides.

The decrease in the stability of high oxidation states that took place between manganese and iron is carried further between iron and cobalt. The occurrence of cobalt(IV) has been firmly established only in yellow $Cs_2[CoF_6]$, a low-spin d^5 complex obtained by fluorination of a mixture of CsCl and $CoCl_2$ at 300 °C (the corresponding Co(III) complex is a high-spin species – note the effect of increase in Δ_0 with increase in oxidation state); an ill-defined oxide CoO_2, made by alkaline hypochlorite oxidation of Co(II) salts, and an oxo salt Ba_2CoO_4 (made in an analogous way to Ba_2FeO_4) have, however, also been described. There are few binary cobalt(III) compounds, and the aquo ion $[Co(H_2O)_6]^{3+}$ is a powerful oxidant, E^0 for the half-reaction

$$Co^{3+} + e \rightleftharpoons Co^{2+}$$

being +1.93 V. Complex formation by ammonia, amines or cyanide, however, stabilises Co(III) greatly, and, as we have already seen, there is a very extensive chemistry of low-spin octahedral Co(III) (d^6) complexes, which are noteworthy for their kinetic inertness (cf. Section 21.5). Stable simple cobalt compounds are mostly those of Co(II), and many Co(II) complexes, both octahedral and tetrahedral, are also known. Cobalt(I) compounds and derivatives of the metal in lower oxidation states usually contain π accepting ligands (e.g. $HCo(CO)_4$, $CoBr(PPh_3)_3$, $Co_2(CO)_8$, $[Co(CO)_4]^-$) and some of them were described in Chapter 22. In the remainder of this section we shall, however, be concerned only with cobalt(II) and cobalt(III).

Cobalt(II)

Pink CoF_2, which has the rutile structure, is obtained by the action of hydrogen fluoride on the chloride at 300 °C; blue $CoCl_2$ (cadmium chloride structure) is made by combination of the elements. The red hexahydrate of $CoCl_2$ is actually *trans* $[Co(H_2O)_4Cl_2].2H_2O$, but the pink $[Co(H_2O)_6]^{2+}$ ion occurs in several solid salts (e.g. in the perchlorate) and in aqueous solution. By the action of concentrated hydrochloric acid on the aquo ion the intensely blue tetrahedral anion $[CoCl_4]^{2-}$ is formed; its electronic spectrum, it should be noted, is quite different from that of $CoCl_2$, in which Co^{2+} is octahedrally coordinated. Both $[Co(H_2O)_6]^{2+}$ and $[CoCl_4]^{2-}$, like nearly all other Co(II) complexes, are high-spin species. The more intense colour of the tetrahedral species (cf. $[MnCl_4]^{2-}$ and $[FeCl_4]^-$) originates in the fact that, since the tetrahedron has no centre of symmetry, *d–d* transitions, although subject to the spin selection rule, are not Laporte-forbidden. The relative ease of the octahedral $\rightleftharpoons$ tetrahedral coordination change, though dependent on a number of factors (including solvation energies), may be correlated with the small change in crystal field stabilisation energy (if all other factors are equal) between high-spin octahedral and tetrahedral d^7 complexes (Section 20.2); it may be remarked that cobalt(II) differs noticeably from nickel(II) in the ease with which tetrahedral complexes are formed. The magnetic properties of octahedral and tetrahedral cobalt(II) complexes were discussed in Section 19.6, where we explained why, although all cobalt(II) complexes have moments higher than the spin-only value of 3.87

μ_B, octahedral complexes have markedly higher moments than tetrahedral ones.

The olive-green oxide CoO is best obtained by thermal decomposition of the insoluble carbonate or of the nitrate; it has the sodium chloride structure. When heated in air at 500 °C it gives black Co_3O_4, a normal spinel containing high-spin Co^{2+} in tetrahedral holes and low-spin Co^{3+} in octahedral holes and therefore a much worse conductor than Fe_3O_4. The sparingly soluble hydroxide $Co(OH)_2$ is blue when freshly precipitated, but turns pink on standing, presumably because of a change in the coordination of the metal ion; in the presence of air oxidation to hydrous Co_2O_3 takes place. It dissolves in hot concentrated alkalis to form the $[Co(OH)_4]^{2-}$ ion.

As we have seen, $[Co(H_2O)_6]^{2+}$ and $[CoCl_4]^{2-}$ are both very stable with respect to oxidation; this is also true of the blue thiocyanato complex $[Co(NCS)_4]^{2-}$, the insoluble mercury(II) salt of which is the standard calibrant for use in magnetic susceptibility determinations. On the other hand the $[Co(NH_3)_6]^{2+}$ ion is very easily oxidised, and the same is true of amine complexes.

Special interest attaches to low-spin Co(II) complexes. The ligand par excellence for low-spin complex formation is CN^-, and since Co(II) has the electronic configuration d^7, an octahedral complex would have nineteen electrons in the valence shell of the metal ion. Now we saw in Section 19.4 that the effect of back-bonding from metal to a ligand like CN^- is to increase Δ_0 and thereby make the metal e_g orbitals strongly antibonding. It is therefore gratifying to find that when Co(II) salts are treated with excess of cyanide $[Co(CN)_6]^{4-}$ is not formed, but only green $[Co(CN)_5]^{3-}$ (which is square pyramidal) and its dimer, purple $[Co_2(CN)_{10}]^{6-}$ (which is isostructural with $Mn_2(CO)_{10}$, i.e. is $[(NC)_5CoCo(CN)_5]^{6-}$). These have been characterised in the form of $Et_2Pr'_2N^+$ and Ba^{2+} salts respectively; the former is paramagnetic (one unpaired electron per cobalt atom) and the latter diamagnetic. Some reactions of the $[Co(CN)_5]^{3-}$ ion are described later.

Cobalt(III)

Cobalt trifluoride is a light brown solid which is rapidly hydrolysed and is isostructural with FeF_3 etc.; it is a useful fluorinating agent (cf. Section 13.6). The blue complexes of formula $M_3[CoF_6]$, where M is an alkali metal cation, are obtained by fluorination of mixture of metal chlorides; they are among the very few high-spin complexes of cobalt(III). The anhydrous oxide Co_2O_3 does not appear to exist, but a hydrous oxide is precipitated when excess of alkali reacts (sometimes slowly) with most Co(III) compounds, or when $Co(OH)_2$ in aqueous suspension undergoes atmospheric oxidation. The nitrate $Co(NO_3)_3$, obtained from N_2O_5 and CoF_3 at -70 °C, has a molecular structure containing an octahedrally coordinated metal ion, the nitrate ions acting as bidentate ligands. Very few other simple Co(III) compounds are known.

By electrolytic oxidation of aqueous $CoSO_4$ in acidic solution at 0 °C, however, the blue diamagnetic cation $[Co(H_2O)_6]^{3+}$ is obtained; it is best isolated as the sparingly soluble caesium alum. Like $[Fe(H_2O)_6]^{3+}$, the cation is appreciably ionised, and except at high acidities polymeric oxo- or hydroxo-bridged partially hydrolysed cations are formed. It is also a very powerful oxidant, and in aqueous solution it decomposes with liberation of ozonised oxygen. The reported $E°$ for the half-reaction

$$[Co(H_2O)_6]^{3+} + e \rightleftharpoons [Co(H_2O)_6]^{2+}$$

of +1.93 V might therefore seem subject to some doubt, but the value has been confirmed by a thermochemical determination of $\Delta H°$ for the aqueous reaction

$$Co^{3+} + Fe^{2+} \rightarrow Co^{2+} + Fe^{3+}$$

(for which $\Delta S°$ is expected to be very small), accepting the standard value for $E°$ Fe^{3+}/Fe^{2+}. When water as ligand is replaced by ammonia, there is a dramatic change in $E°$:

$$[Co(NH_3)_6]^{3+} + e \rightleftharpoons [Co(NH_3)_6]^{2+} \qquad E° = +0.1 \text{ V}$$

This shows that the overall formation constant of $[Co(NH_3)_6]^{3+}$ is about 10^{30} times that of $[Co(NH_3)_6]^{2+}$. Much of this difference arises from the crystal field stabilisation energy, Δ_0, for the ammine complex being substantially greater than for the aquo complex in both oxidation states. Both Co(II) complexes, however, are high-spin d^7 species, so their crystal field stabilisation energies are only $0.8\Delta_0^{II} - 2P$, whilst the Co(III) complexes are both low-spin d^6 complexes, with crystal field stabilisation energies $2.4\Delta_0^{III} - 3P$, with $\Delta_0^{III} > \Delta_0^{II}$ in each case (Section 20.2).

Since Co(II) complexes are substitution-labile, whilst most Co(III) complexes are substitution-inert, a general method for the preparation of the latter is by oxidation of the appropriate Co(II) complex. Sometimes the identity of the product depends upon careful control of experimental conditions or the use of a catalyst: atmospheric oxidation of aqueous $CoCl_2$ in the presence of ammonia and ammonium chloride, for example, gives $[Co(NH_3)_5Cl]Cl_2$, but if charcoal is also added $[Co(NH_3)_6]Cl_3$ is obtained. Oxidation of Co^{2+} in the presence of excess of oxalate by PbO_2 gives $[Co(C_2O_4)_3]^{3-}$; in the formation of the nitro complex $[Co(NO_2)_6]^{3-}$ by the action of excess of nitrite and acid on a Co(II) salt, some of the nitrite acts as oxidant and is liberated as nitric oxide. As we discussed at length in Chapter 21, cobalt(III) complexes (with the exception of the high-spin fluoro complex $[CoF_6]^{3-}$) are generally kinetically inert to substitution, though much less so towards OH^- than towards other attacking entities. For ammine complexes this is satisfactorily accounted for in terms of a conjugate base mechanism for substitution.

In the atmospheric oxidation of the $[Co(CN)_5]^{3-}$ ion in aqueous cyanide solution, the first product is the brown peroxo complex $[(NC)_5CoOOCo(CN)_5]^{6-}$ (which may be precipitated as its potassium salt); this contains an O_2^{2-} unit with interatomic distance 1.45 Å (slightly shorter than in H_2O_2). Some red $[(NC)_5CoOOCo(CN)_5]^{5-}$ is also formed; the potassium salt containing this anion may be obtained if bromine is added to oxidise the peroxo complex. That the product is a superoxo (O_2^-) complex of Co(III) rather than a Co(III)–Co(IV) complex is established beyond doubt by a structure determination which shows that the O—O bond length in the red salt is only 1.26 Å, an antibonding electron having been lost from the O_2^{2-} ion. (Analogous pentammine complexes are also known.) When solutions containing these peroxo and superoxo complexes are boiled, hydrolysis and substitution by cyanide occur, with formation of yellow $K_3[Co(CN)_6]$.

The $[Co(CN)_6]^{3-}$ ion is so stable that if a solution of $K_3[Co(CN)_5]$ containing excess of KCN is heated, hydrogen is evolved and the Co(III) complex is formed. In this reaction the hydrido complex $[Co(CN)_5H]^{3-}$ is an intermediate. It is obtained in almost quantitative yield by the reversible reaction

$$2[Co(CN)_5]^{3-} + H_2 \rightleftharpoons 2[Co(CN)_5H]^{3-}$$

and may be precipitated as $Cs_2Na[Co(CN)_5H]$. The position of the first *d–d* transition in the electronic spectrum of this ion is at nearly the same position as that for $[Co(CN)_6]^{3-}$ and thus shows that H^- lies near CN^- in the spectrochemical series. The $[Co(CN)_5]^{3-}$ ion is an effective hydrogenation catalyst for olefins, which are inserted into the Co—H bond of the hydride complex formed as an intermediate:

$$[Co(CN)_5H]^{3-} + CH_2{=}CHX \rightarrow [Co(CN)_5CH_2CH_2X]^{3-}$$

$$[Co(CN)_5CH_2CH_2X]^{3-} + [Co(CN)_5H]^{3-} \rightarrow CH_3CH_2X + 2[Co(CN)_5]^{3-}$$

As a final illustration of the interesting chemistry of the complex $[Co(CN)_5]^3$ we may mention its role in the Co(II)-catalysed substitution of CN^- into $Co(NH_3)_5X$ complexes, where, for example, X = Cl or NH_3. For X = Cl, the rate is given by $k[Co(NH_3)_5Cl^{2+}][Co(CN)_5{}^-]$ and the product is $[Co(CN)_5Cl]^{3-}$ formed by the redox bridged process

$$\underset{\text{non-labile}}{[Co(NH_3)_5Cl]^{2+}} + [Co(CN)_5]^{3-} \rightarrow \underset{\text{non-labile}}{[Co(CN)_5Cl]^{3-}} + \underset{\text{labile}}{[Co(NH_3)_5]^{2+}}$$

$$\downarrow CN^-$$

$$[Co(CN)_5]^{3-}$$

For X = NH_3, on the other hand, the rate is given by $k[Co(NH_3)_6{}^{3+}][Co(CN)_5^{3-}][CN^-]$ where the last term represents CN^- in excess of that needed to convert all the Co(II) into $[Co(CN)_5]^{3-}$; in this case the product is $[Co(CN)_6]^{3-}$ believed to be formed by the mechanism

$$[Co(CN)_5]^{3-} + CN^- \rightleftharpoons [Co(CN)_6]^{4-}$$

followed by the outer-sphere electron-transfer

$$[Co(CN)_6]^{4-} + [Co(NH_3)_6]^{3+} \rightarrow [Co(CN)_6]^{3-} + [Co(NH_3)_6]^{2+}$$

$$\downarrow CN^-$$

$$[Co(CN)_5]^{3-}$$

24.9 Nickel

Like cobalt, nickel occurs mainly in the form of sulphide and arsenide ores, together with other metals, some of which are usually isolated from residues obtained in the extraction of nickel. Sulphide ores (e.g. *pentlandite*, (Fe,Ni)S) are roasted in air to convert them into the oxide, which is reduced to the crude metal by carbon. The metal is refined by electrolysis or by conversion into the volatile carbonyl $Ni(CO)_4$ by the action of carbon monoxide at 50 °C and decomposition at 150–300 °C (this is the Mond process, which is based on the fact that nickel forms a carbonyl much more readily than any other metal). Nickel is used as a protective coating for other metals and in coinage alloys (with copper) and heating elements (with iron and chromium); it is also a very important catalyst, e.g. for the hydrogenation of unsaturated organic compounds and for the cracking of methane in the presence of steam to produce carbon monoxide and hydrogen.

Nickel is a greyish-white metal, in reactivity rather like cobalt, E° for the Ni^{2+}/Ni system being −0.25 V. Thus it is attacked by dilute mineral acids, made passive by nitric acid, and, when compact, is oxidised by air or steam only at high temperatures. Fluorine produces a coherent coating of NiF_2 which prevents further attack, and nickel apparatus is commonly used in handling this element; the other halogens yield the dihalides.

By far the most important oxidation state of the metal is nickel(II). Apart from the carbonyl and the trifluorophosphine complex $Ni(PF_3)_4$ (Chapter 22), nickel(0) is represented by the yellow cyano complex $K_4[Ni(CN)_4]$, obtained by reduction of $K_2[Ni(CN)_4]$ in liquid ammonia with excess of potassium, which oxidises immediately on exposure to air. Nickel(I) is uncommon, but dark red $K_4[Ni_2(CN)_6]$ may be made by sodium amalgam reduction of aqueous $K_2[Ni(CN)_4]$. The complex has the structure $[(NC)_3NiNi(CN)_3]^{4-}$ with the two planar $Ni(CN)_3$ units perpendicular to one another; it liberates hydrogen from water with formation of the $[Ni(CN)_4]^{2-}$ ion. Higher oxidation states are nickel(III) and nickel(IV); these occur only in a hydrous oxide, probably NiO(OH), complex fluorides such as violet $K_3[NiF_6]$ and red $K_2[NiF_6]$ (both of which are low-spin complexes, the former showing the expected Jahn–Teller distortion), and a few other species, e.g. $NiBr_3(PEt_3)_2$ and $NaNiIO_6$. Impure NiF_3 is obtained by the action of arsenic pentafluoride on $K_2[NiF_6]$ in anhydrous hydrogen fluoride, but it has not been characterised fully. The black hydrous oxide is obtained by alkaline hypochlorite oxidation of an aqueous nickel(II) salt, and is a strong oxidising agent, liberating chlorine from hydrochloric acid. The Edison battery operates on the reaction

$$Fe + 2NiO(OH) + 2H_2O \underset{\text{charge}}{\overset{\text{discharge}}{\rightleftharpoons}} Fe(OH)_2 + 2Ni(OH)_2$$

in concentrated aqueous KOH as the electrolyte. The fluoro complexes result from high-temperature fluorination of mixtures of metal halides, $NiBr_3(PEt_3)_2$ and $NaNiIO_6$ from atmospheric or peroxodisulphate oxidations of nickel(II) compounds. Nickel(II) compounds exhibit a great variety of structures, sometimes in equilibrium with one another, and we shall devote the remainder of this section to them.

Nickel(II)

The only nickel halides that have been obtained pure are NiF_2 (best made by fluorination of the chloride at 350 °C) and the other dihalides mentioned already; NiF_2 and $NiCl_2$ are yellow solids, only the chloride being readily soluble in water. Both are isostructural with the dihalides of other first series transition elements. The compound $KNiF_3$ (made from the constituent fluorides) has the perovskite structure, but K_2NiF_4 has an infinite anion layer structure in which NiF_6 octahedra share all their equatorial corners; in $CsNiCl_3$, $NiCl_6$ octahedra share opposite faces to form an infinite anion chain; a strongly coloured blue tetrahedral $[NiCl_4]^{2-}$ ion, however, exists in melts and in nitromethane or ethanolic solution, and can be characterised as its Et_4N^+ salt. Crystal field stabilisation energy considerations provide some explanation for the relative rarity of tetrahedral Ni(II) complexes.

Green nickel oxide (sodium chloride structure) is obtained by thermal decomposition of the carbonate or nitrate. It is an insulator like CuO but unlike other monoxides of first transition series metals. This results from the d^8 configuration for Ni^{2+}; the two unpaired electrons are in e_g orbitals directed towards the O^{2-} ions and therefore unavailable for band formation (Section 6.11). As mentioned earlier, hypochlorite converts NiO into black NiO(OH); nickel sulphide oxidises similarly in air to form NiS(OH), a fact which explains why although NiS is not precipitated in acidic solution, after exposure to air it is insoluble in dilute acid. Nickel also form a green insoluble hydroxide, $Ni(OH)_2$, which is soluble in ammonia because of formation of the violet ion $[Ni(NH_3)_6]^{2+}$ but is insoluble in aqueous sodium hydroxide.

Hydrated nickel salts usually contain the ion $[Ni(H_2O)_6]^{2+}$, the electronic spectrum of which we mentioned in Section 19.5. We have also mentioned the magnetic properties of octahedral and tetrahedral nickel complexes earlier: both are derived from the d^8 ion having two unpaired electrons. For an octahedral complex there is no orbital contribution, and the magnetic moment is close to the spin-only value of 2.83 μ_B, but for a tetrahedral complex the orbital contribution (the origin of which was explained in Section 19.6) raises the value to about 4 μ_B; distorted tetrahedral complexes have intermediate values.

Planar four-coordinated nickel(II) complexes are diamagnetic, the $3d_{x^2-y^2}$ orbital (together with the 4*s* and two of the 4*p* orbitals) being involved in accepting electrons from the ligands (Section 19.3). Typical examples are the yellow $[Ni(CN)_4]^{2-}$ anion (obtained when aqueous cyanide is added in the calculated quantity to almost any nickel compound – it is a very stable complex anion with an overall formation constant of 10^{30}) and red nickel dimethylglyoximate. The latter is a familiar substance because of its use for the detection and determination of nickel; only palladium is precipitated with nickel when dimethylglyoxime in weakly ammoniacal solution is used as a reagent. This specificity originates not in an abnormally high formation constant for nickel dimethylglyoximate (the analogous complexes of other first series transition metal ions have similar values), but in its low solubility, which is in turn derived from its crystal structure. An individual molecule of the complex has the structure shown in Fig. 24.9. Strong hydrogen bonding, with an overall O . . . H . . . O distance of only 2.45 Å, links the two ligand anions and may play some part in making the structure planar; individual molecules are stacked parallel in the crystal so that their nickel atoms form a chain with a Ni—Ni distance of 3.25 Å. That weak metal–metal bonding is responsible for the insolubility of the compound is shown by the observation that nickel methylethylglyoximate, in which the metal atoms are forced further apart by the bigger ligand, is more soluble although the dioxime from which it is derived is less soluble than dimethylglyoxime.

For some nickel(II) complexes there appears to be only a small difference in energy between the tetrahedral structure (with two unpaired

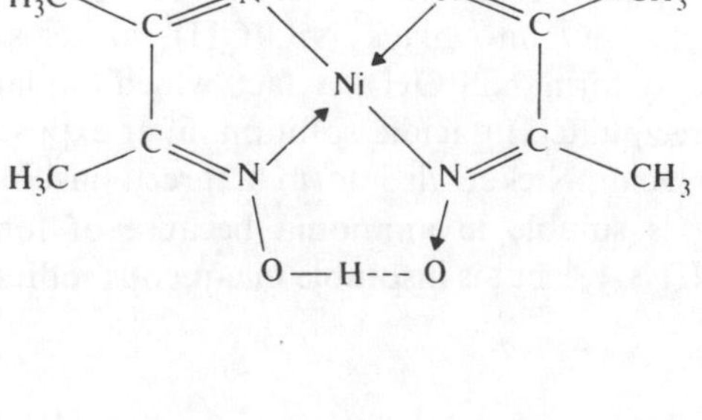

Figure 24.9
The molecular structure of nickel bis(dimethylglyoximate).

electrons) and the planar structure (with none). Thus for nickel halide complexes of formula $NiX_2.2L$ the molecules are generally planar when L is a trialkylphosphine, but tetrahedral when L is a triarylphosphine. When X = Br and L = $PBzPh_2$, both types of molecule occur in the same crystal, and the magnetic moment is intermediate between those for planar and tetrahedral complexes.

Planar ⇌ octahedral equilibria have also been established for nickel complexes: $Ni(py)_4(ClO_4)_2$ exists in a yellow diamagnetic form containing the planar cation $[Ni(py)_4]^{2+}$ and ClO_4^- ions, and in a blue paramagnetic form containing *trans* $[Ni(py)_4(ClO_4)_2]$ molecules in which oxygen atoms from the ClO_4^- ions complete the octahedral coordination of the metal ion (this is one of the few cases of complexing by perchlorate ion). Nickel salicylaldoximate (Fig. 24.10) forms colourless crystals and is diamagnetic, but it yields a green paramagnetic solution in pyridine, in which the addition of two solvent molecules as ligands transforms the planar molecule into an octahedral one.

Five coordinated nickel(II) also occurs. Where polydentate amines are used as ligands, as is often the case, they may impose a particular geometry: $[NiN(CH_2CH_2NMe_2)_3Cl]Cl$, for example, is trigonal bipyramidal with the unique nitrogen atom and the chlorine atom in the axial positions. This complex is paramagnetic, but $Ni(CN)_2(PMePh_2)_3$ is diamagnetic. The red diamagnetic ion $[Ni(CN)_5]^{3-}$, obtained when excess of cyanide is added to a nickel salt ($[Ni(CN)_6]^{4-}$ is not formed) exists as a square pyramid in solution and in some salts; in $[Cr(en)_3][Ni(CN)_5].\frac{3}{2}H_2O$, however, approximately trigonal bipyramidal and approximately square pyramidal ions both occur. In the square pyramidal ion, the axial and equatorial Ni—C bond lengths are 2.17 and 1.86 Å and the metal ion is 0.34 Å above the plane of the equatorial cyanides; in the trigonal bipyramidal species the average Ni—C axial bond length is 1.84 Å and the equatorial bond lengths are 1.91 Å (two) and 1.99 Å (one); the C≡N bond lengths are all very nearly identical. With so many variations in bond length it is clearly impossible to give any simple discussion of the trigonal bipyramidal⇌square pyramidal equilibrium; we quote the details of this structure to show how formidable a task it would be to try to predict it on theoretical grounds.

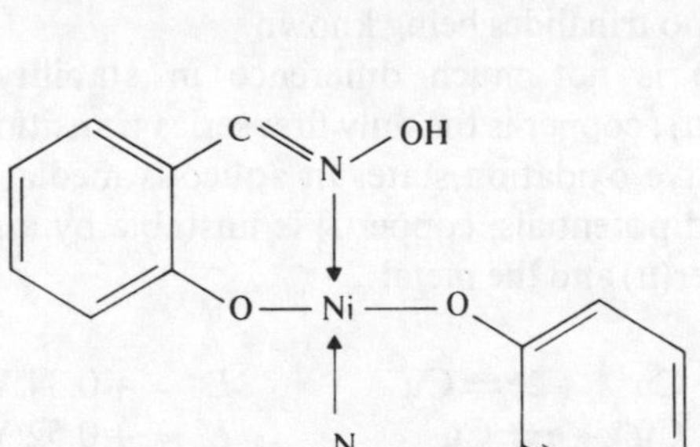

Figure 24.10
The molecular structure of nickel bis(salicylaldoximate).

24.10 Copper

Copper is by a considerable margin the most noble of the first transition series metals, and it occurs native in several places. Its commonest ore is *chalcopyrite* or *copper pyrites*, $CuFeS_2$, which has a zinc blende structure with the zinc atoms replaced half by copper and half by iron; others are *malachite*, $Cu_2(OH)_2(CO_3)$, *atacamite*, $Cu_2(OH)_3Cl$, and *cuprite*, Cu_2O. Sulphide ores are first roasted with limited access of air, when Cu_2S and FeO are formed; the latter is removed by combination with silica to form a slag, and the Cu_2S is converted into metallic copper by controlled oxidation:

$$Cu_2S + O_2 \rightarrow 2Cu + SO_2$$

Alternatively, prolonged exposure of copper pyrites to air and rain leads to formation of a dilute solution of copper(II) sulphate, from which the metal is precipitated by addition of scrap iron. It is always refined electrolytically.

Since copper, a soft reddish metal, is an excellent conductor of heat and electricity and is also resistant to corrosion, it is in great demand as an electrical conductor and for water and steam piping. It is also used in several alloys that are rather tougher than copper itself, e.g. brass (with zinc), bronze (with tin), and coinage metal (with nickel). Copper(II) sulphate is used extensively as a fungicide. Small amounts of copper occur in several enzymes concerned with biochemical redox reactions. Copper compounds are also catalysts for many inorganic and organic reactions in industry and in the laboratory.

Copper is not attacked by non-oxidising acids in the absence of air, but it liberates sulphur dioxide from hot concentrated sulphuric acid (forming a mixture of copper(II) sulphate and sulphide) and reacts readily with nitric acid of all concentrations. In the presence of air, copper reacts with many dilute acids (the green patina on copper roofs in towns is a basic sulphate), and also dissolves in aqueous ammonia and aqueous potassium cyanide, the complex ions $[Cu(NH_3)_4]^{2+}$ and $[Cu(CN)_4]^{3-}$ being formed. It reacts with oxygen at red heat to give CuO, which is decomposed to Cu_2O at higher temperatures. Fluorine and chlorine give CuF_2 and $CuCl_2$, no trihalides being known.

There is not much difference in stability between copper(I) and copper(II); copper is the only first series transition metal to exhibit a stable unipositive oxidation state. In aqueous media, as may be seen from the standard potentials, copper(I) is unstable by a small margin with respect to copper(II) and the metal:

$$Cu^{2+} + 2e \rightleftharpoons Cu \qquad E^\circ = +0.34\ V$$
$$Cu^{+} + e \rightleftharpoons Cu \qquad E^\circ = +0.52\ V$$
$$Cu^{2+} + e \rightleftharpoons Cu^{+} \qquad E^\circ = +0.15\ V$$

This disproportionation is usually fast, but in the case of Cu^+(aq) made by reduction of aqueous Cu^{2+} with V^{2+} or Cr^{2+}, decomposition in the absence of air requires several hours for completion. If copper(I) is protected by the formation of an insoluble compound (e.g. CuCl) or a complex (e.g. $[Cu(CN)_4]^{3-}$) it becomes the more stable oxidation state. In both non-aqueous media and the solid state, the environment is important. Thus when copper powder reacts with silver nitrate in aqueous solution, $1Cu \equiv 2Ag^+$, but in acetonitrile, $1Cu \equiv 1Ag^+$ and the ion $[Cu(CH_3CN)_4]^+$ is formed. There is no known copper(I) fluoride, i.e. CuF is unstable with respect to disproportionation into $Cu + CuF_2$; on the other hand CuI_2 spontaneously decomposes into CuI and iodine. Lattice energy considerations, as discussed in Section 16.4, are undoubtedly important in determining the positions of such equilibria.

The only reported copper(0) compounds are unstable carbonyls, e.g. $Cu_2(CO)_6$, isolated in a matrix at a very low temperature. Copper(III) and copper(IV) are known only in pale green $K_3[CuF_6]$ (a high-spin octahedral d^8 complex), orange $Cs[CuF_4]$ (a diamagnetic planar d^8 complex), and red $Cs_2[CuF_6]$ (a low-spin octahedral d^7 complex) both obtained by fluorination of mixtures of the appropriate metal chlorides, and in a few other species such as $KCuO_2$ and $K_7[Cu(IO_6)_2]$. There is no evidence for Cu^{3+} in aqueous media.

Certain oxides containing Cu(II) and Cu(III) are relatively high-temperature (90 K) superconductors. One of the most investigated of these is $YBa_2Cu_3O_{7-x}$, where x is about 0.1. This is made by heating together oxides or carbonates in oxygen under rigidly specific conditions. It may be noted that if copper is present only as Cu^{2+} or Cu^{3+}, stoichiometry requires that for $x = 0$ one third of the copper is Cu^{3+}.

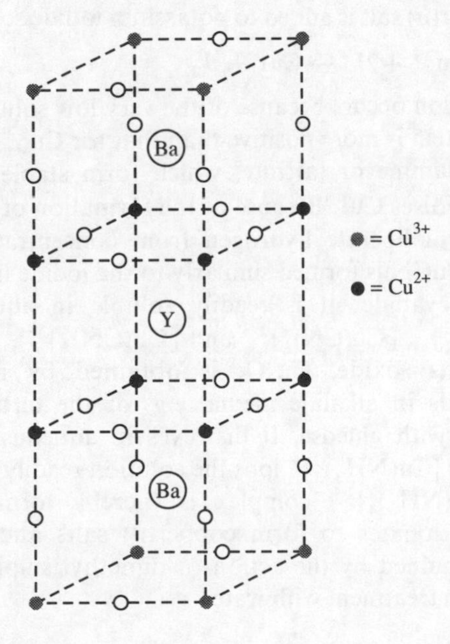

Figure 24.11
Structure of the $YBa_2Cu_3O_7$ phase.

The structure of the phase of ideal composition $YBa_2Cu_3O_7$ is shown in Fig. 24.11. The unit cell is three times that of perovskite (Fig. 6.12) with Y^{3+} and Ba^{2+} replacing Ca^{2+}, Cu^{2+} and Cu^{3+} replacing Ti^{4+}, and eight O^{2-} ions missing. Further O atoms (each shared by four unit cells) are lost when $x > 0$; loss of the four O^{2-} ions bonded only to Cu^{3+} gives a phase of composition $YBa_2Cu_3O_6$ ($YBa_2Cu^+Cu_2^{2+}O_6$ or $YBa_2Cu_2^+Cu^{3+}O_6$) which is a semiconductor. The coordination of Cu^{2+} and Cu^{3+} in $YBa_2Cu_3O_{7-x}$ is square pyramidal and approximately square planar respectively, the apical O atom of the square pyramid being shared. It was at first believed that the presence of copper in two oxidation states was essential for the occurrence of relatively high-temperature superconductivity, but this is not so: the perovskite $Ba_{1-x}K_xBiO_{3-y}$ (with x about 0.4 and y about 0.2) is also a superconductor up to 30 K. The technical value of these materials is limited by the loss of superconductivity if they have to carry large currents, but many specialised applications will doubtless be found.

Copper(I)

The ion Cu^+ has the d^{10} electronic configuration and is diamagnetic; its compounds are colourless except when the anion present is coloured or when charge-transfer absorptions take place in the visible region of the spectrum (e.g. for red Cu_2O). White CuCl and CuBr are easily made by reduction of a copper(II) salt in the presence of halide ions, e.g. by sulphur dioxide. When a solution of copper(II) chloride is boiled with concentrated hydrochloric acid and copper, the complex ion $[CuCl_2]^-$ is formed; on dilution CuCl is precipitated. The iodide CuI is obtained when any copper(II) salt is added to potassium iodide solution:

$$Cu^{2+} + 2I^- \rightarrow CuI + \tfrac{1}{2}I_2$$

This reaction occurs because of the very low solubility of CuI; E^0 for the $\frac{1}{2}I_2/I^-$ system is more positive than that for Cu^{2+}/Cu^+. In the presence of ethylenediamine or tartrate, which form stable complexes with Cu^{2+}, iodine oxidises CuI. Because of the formation of anionic iodo complexes, copper will liberate hydrogen from concentrated hydriodic acid. The cyanide CuCN is formed similarly to the iodide from a copper(II) salt and aqueous cyanide; it is readily soluble in aqueous cyanide, forming $[Cu(CN)_2]^-$, $[Cu(CN)_3]^{2-}$ and $[Cu(CN)_4]^{3-}$.

Copper(I) oxide, Cu_2O, is obtained by reduction of copper(II) compounds in alkaline media, e.g. of the tartrate complex (Fehling's solution) with glucose. It dissolves in aqueous ammonia, forming the colourless $[Cu(NH_3)_2]^+$ ion; the solution readily absorbs oxygen, and the blue $[Cu(NH_3)_4]^{2+}$ complex is thereby formed. With acids, Cu_2O disproportionates to form copper(II) salts and copper; copper(I) sulphate, obtained by the action of dimethyl sulphate on copper(I) oxide, does so on treatment with water.

Carbon monoxide and alkenes displace ammonia or chloride from aqueous $[Cu(NH_3)_2]^+$ or $[CuCl_2]^-$, forming unstable Cu(I) complexes; alkynes with $[Cu(NH_3)_2]^+$ yield insoluble explosive Cu(I) salts, but with $[CuCl_2]^-$ they undergo HCl addition or polymerisation.

In the copper(I) halides and the $[Cu(CN)_4]^{3-}$ ion the coordination of the metal atom is tetrahedral; the halides all have the zinc blende structure. In Cu_2O the structure is basically that of cristobalite, with linear and tetrahedral coordination of the metal and oxygen atoms respectively. In $K[Cu(CN)_2]$, however, there is a chain anion in which planar $Cu(CN)_3$ units are bonded by sharing of two cyanides.

Copper(I) hydride, CuH, which has the wurtzite structure, is obtained as a red solid by reduction of copper(II) salts with hypophosphorous acid; it is decomposed by acids with liberation of hydrogen.

Copper(II)

We have already mentioned the preparations of the copper(II) halides; their structures are interesting as manifestiations of the Jahn–Teller effect for d^9 complexes. The fluoride, which is white, has a distorted rutile structure with four Cu—F 1.93 Å and two Cu—F 2.27 Å; the brown chloride contains infinite chains of $CuCl_4$ squares sharing all corners (with Cu—Cl 2.30 Å) so stacked that each copper atom has two more Cl atoms from other chains at 2.95 Å, completing a distorted octahedron. A distorted octahedral environment is also found in green $CuCl_2.2H_2O$, which has a molecular lattice with two Cu—Cl 2.29 Å and two Cu—OH_2 1.96 Å in a plane and two Cu—Cl (belonging to other molecules) at 2.94 Å. It will be seen that the amount of distortion varies considerably; the Jahn–Teller theorem merely predicts a distortion, and says nothing about its direction or magnitude. Copper(II), however, also shows quite different structures in chloro complexes: thus orange $Cs_2[CuCl_4]$ contains a flattened tetrahedral anion, and $[Cr(NH_3)_6][CuCl_5]$ a trigonal bipyramidal one. The stabilisation of the $[CuCl_5]^{3-}$ anion in a lattice by means of the large cation should be noted.

In black copper(II) oxide each copper atom has four oxygen atoms in planar coordination and each oxygen atom four copper atoms in tetrahedral coordination; in this compound there are no other atoms within what might be considered a bonding distance from the copper atom. The insoluble blue hydroxide $Cu(OH)_2$ is readily dehydrated to the oxide; it is soluble not only in acids, but also in concentrated aqueous alkalis, in which ill-defined hydroxo complexes are formed. It may be presumed that the hydrated Cu^{2+} ion in solution is $[Cu(H_2O)_6]^{2+}$ with a distorted octahedral structure; this ion has been found in $[Cu(H_2O)_6](ClO_4)_2$.

In the familiar blue sulphate $CuSO_4.5H_2O$, planar $[Cu(H_2O)_4]^{2+}$ ions are present; the coordination of the copper is completed by two more distant oxygen atoms of sulphate ions, and the remaining water molecule is linked by hydrogen bonding to both cation and anion. Dehydration yields the almost colourless anhydrous salt in which the coordination of the cation remains distorted octahedral, though in this case all the oxygen atoms belong to anions.

Two other interesting copper(II) salts of oxo-acids are the nitrate and the acetate. Anhydrous copper(II) nitrate cannot be obtained by dehydration of the blue hexahydrate, a basic salt being formed, and it has to be made by the interaction of the metal and dinitrogen tetroxide (as described in Section 8.6) and decomposition of the compound $NO[Cu(NO_3)_3]$ first produced. Copper(II) nitrate volatilises *in vacuo* at 150 °C to give molecular $Cu(NO_3)_2$, in which two nitrate groups both act as bidentate ligands. Copper(II) acetate monohydrate is dimeric with a structure similar to that of chromium(II) acetate, but with a larger metal–metal separation of 2.64 Å, greater than that in the metal (2.56 Å); a magnetic moment of $1.4\mu_B$ per Cu^{2+} ion suggests that in this compound there is only weak coupling between the unpaired electrons and that little metal–metal interaction occurs.

When the Cu^{2+} ion interacts with ammonia in aqueous solution, no more than four NH_3 molecules replace coordinated water, but $[Cu(NH_3)_6]^{2+}$ salts containing the distorted octahedral cation can be isolated from liquid ammonia; $[Cu(en)_3]^{3+}$ is formed in very concentrated aqueous ethylenediamine. Large numbers of Cu(II) complexes involving multidentate oxygen- and nitrogen-containing ligands are known; their structures show that the discussion in this section has by no means covered all the stereochemical configurations found for copper(II). The deep blue aqueous solution of $[Cu(NH_3)_4](OH)_2$ has the remarkable property of dissolving cellulose, and if the solution is squirted into acid one variety of artificial silk is produced; the solution is also used for coating canvas with a water-tight coating of cellulose.

When a copper(II) salt is treated with excess of potassium cyanide at ordinary temperatures, cyanogen is evolved and a complex cyanide of copper(I) is formed, a reaction similar to that which occurs with potassium iodide. In aqueous methanol at low temperatures, however, violet $[Cu(CN)_4]^{2-}$ can be characterised; its electronic spectrum is compatible with the expected planar structure.

24.11 Zinc

The first and last elements of a transition series show a much more restricted range of oxidation states than the other elements, and the chemistry of zinc is confined to that of Zn(II) except in so far as the Zn_2^{2+} ion has been identified in the yellow diamagnetic glass obtained by cooling a solution of metallic zinc in molten $ZnCl_2$; this ion is the analogue of the Cd_2^{2+} and Hg_2^{2+} ions that we shall consider in the following chapter, but it is less stable than they are, and readily disproportionates to Zn^{2+} and zinc metal.

Zinc is usually isolated from *zinc blende*, ZnS, by roasting this in air and reducing the resulting oxide with carbon; the metal is more volatile than most other transition metals (it boils at 908 °C) and can be separated by rapid chilling (to avoid reversing the reaction), and purified by distillation or electrolytically.

Zinc is a silvery-white metal which is fairly soft, a property that is connected with its relatively low melting-point (419 °C) and may be attributed to the fact that the hexagonal close-packed structure of the metal is somewhat distorted, with the result that the interatomic distance is larger than in most of the metals of the first transition series. The metal is used mainly for galvanising iron (cf. Section 24.7) and in alloys (especially brass); its most important compound is the oxide, which is used as a pigment, as a filler for rubber and as an emollient in zinc ointment. For its use in dry batteries see Section 24.6.

Zinc is not attacked by air or water at ordinary temperatures, but the hot metal burns in air and decomposes steam to form the white oxide ZnO, which has the wurtzite structure; when this is heated in the absence of oxygen it turns yellow and becomes a semiconductor owing to the loss of oxygen and the production of some interstitial zinc atoms (Section 6.10). The metal liberates hydrogen from dilute mineral acids and from alkalis; zinc hydroxide is soluble in aqueous alkalis to form zincates containing the tetrahedral $[Zn(OH)_4]^{2-}$ ion. With the halogens the white dihalides are formed.

The sparingly soluble zinc fluoride has the rutile structure, and its lattice energy calculated from the interatomic distance is in quite good agreement with that derived from the Born–Haber cycle, indicating an ionic structure; the other halides, which are extremely soluble, have three-dimensional structures like that of cristobalite with tetrahedrally coordinated zinc. Complexes such as $[ZnCl_4]^{2-}$ and $[Zn(CN)_4]^{2-}$ are also tetrahedral, but the octahedral aquo complex $[Zn(H_2O)_6]^{2+}$ and ammine complex $[Zn(NH_3)_6]^{2+}$ occur in several salts. The commonest salt is the hydrated sulphate $ZnSO_4.7H_2O$, which is isostructural with the corresponding compounds of magnesium and of several other transition metals. Dodecahedral coordination occurs in the ion $[Zn(NO_3)_4]^{2-}$.

Two other zinc compounds worthy of mention are the basic acetate $Zn_4O(CH_3COO)_6$, which is isostructural with the beryllium compound (Section 11.5) and the acetylacetonate $Zn(acac)_2.H_2O$, in which the coordination of the zinc is approximately square pyramidal, with the metal atom 0.4 Å above the plane containing four ligand oxygen atoms.

Zinc is an important element biologically and occurs in a wide variety of enzymes involved in the catalysis of the CO_2–H_2O reaction (Section 13.7) and protein metabolism; but its role in these processes is complicated and for an account of this subject reference should be made to one of the books cited at the end of the chapter.

PROBLEMS

1 Comment on each of the following observations:
(a) Li_2TiO_3 forms a continuous range of solid solutions with MgO.
(b) When $TiCl_3$ is heated with concentrated aqueous sodium hydroxide, hydrogen is evolved.
(c) Electron transfer between $[Fe(CN)_6]^{3-}$ and $[Fe(CN)_6]^{4-}$ is much faster than between $[Co(NH_3)_6]^{3+}$ and $[Co(NH_3)_6]^{2+}$.
(d) $[Co(NH_3)_6]Cl_3$ is not affected by heating it with dilute sulphuric acid, but heating it with dilute sodium hydroxide results in the precipitation of hydrous cobalt(III) oxide.
(e) The freezing-point of an aqueous solution of KCN is raised after it has been shaken with CuCN, which is itself insoluble in water.

2 Copper(II) chloride is not completely reduced by sulphur dioxide in concentrated hydrochloric acid solution. Suggest an explanation of this observation, and state how you would try to establish its correctness.

3 How would you attempt to estimate:
(a) the crystal field stabilisation energy of FeF_2;
(b) the overall formation constant of the $[Co(NH_3)_6]^{3+}$ ion in aqueous solution;
(c) the standard enthalpy of formation of NiF_3?

4 Use the following data for aqueous solutions to predict qualitatively the outcome of the following experiments at 25 °C:
(a) Chromium metal is dissolved in excess of molar $HClO_4$ and the solution is shaken in air.
(b) Manganese is dissolved in molar $HClO_4$ and the solution is shaken in air, made strongly alkaline (pH 14), and shaken in air again.

$Cr^{2+} + 2e \rightleftharpoons Cr$ $\qquad E^\circ = -0.90$ V
$Cr^{3+} + e \rightleftharpoons Cr^{2+}$ $\qquad E^\circ = -0.41$ V
$\frac{1}{2}Cr_2O_7^{2-} + 7H^+ + 3e \rightleftharpoons Cr^{3+} + \frac{7}{2}H_2O$ $\qquad E^\circ = +1.33$ V
$Mn^{2+} + 2e \rightleftharpoons Mn$ $\qquad E^\circ = -1.18$ V
$Mn^{3+} + e \rightleftharpoons Mn^{2+}$ $\qquad E^\circ = +1.51$ V
$Mn(OH)_2 \rightleftharpoons Mn^{2+} + 2OH^-$ $\qquad K_{sp} = 10^{-13}$
$Mn(OH)_3 \rightleftharpoons Mn^{3+} + 3OH^-$ $\qquad K_{sp} = 10^{-36}$
$\frac{1}{2}O_2 + 2H^+ + 2e \rightleftharpoons H_2O$ $\qquad E^\circ = +1.23$ V

What are the limitations of such predictions?

5 What evidence would be required:
(a) to establish whether $[(H_3N)_5CoO_2Co(NH_3)_5]_2(SO_4)_5$ is a peroxo or a superoxo complex;
(b) to distinguish between the formulations $Cu(II)Fe(II)S_2$ and $Cu(I)Fe(III)S_2$ for the mineral *chalcopyrites*;
(c) to show that Fe^{3+} is a class (*a*) cation;
(d) to show that the blue compound precipitated when a solution of permanganate in concentrated aqueous potassium hydroxide is reduced by cyanide contains Mn(V)?

6 An acidified solution of 0.1000 molar ammonium vanadate (25 cm^3) was reduced by sulphur dioxide, and after boiling off excess of the reductant the

blue solution remaining was found to require addition of 25.0 cm^3 of 0.0200 molar permanganate to impart a pink colour to the solution.

Another 25 cm^3 portion of the vanadate solution was shaken with zinc amalgam and then immediately poured into excess of the ammonium vanadate solution; on titration of the resulting liquid with the permanganate solution 74.5 cm^3 of the latter were required.

Deduce what occurred during these experiments.

REFERENCES FOR FURTHER READING

Cotton, F.A. and Wilkinson, G. (1988) *Advanced Inorganic Chemistry*, 5th edition, Interscience, New York. The best detailed account of the chemistry of the transition metals available at the present time.

Greenwood, N.N. and Earnshaw, A. (1984) *Chemistry of the Elements*, Pergamon Press, Oxford. Very good on historical, technological and structural aspects of transition metal chemistry; the three transition series are treated together.

Hughes, M.N. (1981) *Inorganic Chemistry of Biological Processes*, Interscience, New York. A good introduction to bioinorganic chemistry.

Nicholls, D. (1974) *Complexes and First-Row Transition Elements*, Macmillan, London. Covers much the same ground as this chapter.

Parish, R.V. (1977) *The Metallic Elements*, Longman, London. A very good account of some aspects of transition metal chemistry (especially oxides, halides and aqueous solution chemistry) presented in an arrangement quite different from that in this book.

25 Transition metals of the second and third series

25.1 Introduction

In discussing the general principles underlying transition metal chemistry in Chapters 18–23 inclusive, we devoted most of our attention to metals of the first series. This was partly because these elements are more familiar, but mainly because their chemistry (much of which is that of octahedral or tetrahedral high-spin complexes) is much better understood than the chemistry of second and third series transition metals. In order to show why this is so, we begin this chapter with a general comparison between metals of the first and of later transition series. After this we shall go on to consider the chemistry of individual elements, though in somewhat less detail than that of elements of the first series. Actinium and rutherfordium, the only elements of the fourth ordinary transition series about whose chemistry anything is known, are discussed briefly in Chapter 27.

The electronic configurations, metallic radii (except for mercury, which does not crystallise in one of the typical metallic structures), and enthalpies of atomisation of the heavy transition metals have already been given elsewhere, but for convenience they are assembled, together with the corresponding data for the first series metals, in Table 25.1. Oxidation states of the heavy transition metals were shown in Table 18.2 (Section 18.3); since we shall give these again in separate sections, they are not reproduced now, but we must draw attention to one very important

Table 25.1
Atomic numbers, electronic configurations, metallic radii/Å (for twelve-coordination) and enthalpies of atomisation/kJ mol^{-1} for the transition metals.

Sc	Ti	V	Cr	Mn	Fe	Co	Ni	Cu	Zn
21	22	23	24	25	26	27	28	29	30
$3d^14s^2$	$3d^24s^2$	$3d^34s^2$	$3d^54s^1$	$3d^54s^2$	$3d^64s^2$	$3d^74s^2$	$3d^84s^2$	$3d^{10}4s^1$	$3d^{10}4s^2$
1.64	1.47	1.35	1.29	1.37	1.26	1.25	1.25	1.28	1.37
376	473	515	397	281	416	425	430	339	126
Y	Zr	Nb	Mo	Tc	Ru	Rh	Pd	Ag	Cd
39	40	41	42	43	44	45	46	47	48
$4d^15s^2$	$4d^25s^2$	$4d^45s^1$	$4d^55s^1$	$4d^55s^2$	$4d^75s^1$	$4d^85s^1$	$4d^{10}$	$4d^{10}5s^1$	$4d^{10}5s^2$
1.82	1.60	1.47	1.40	1.35	1.34	1.34	1.37	1.44	1.52
410	611	724	659	649	650	577	381	286	111
La	Hf	Ta	W	Re	Os	Ir	Pt	Au	Hg
57	72	73	74	75	76	77	78	79	80
$5d^16s^2$	$5d^26s^2$	$5d^36s^2$	$5d^46s^2$	$5d^56s^2$	$5d^66s^2$	$5d^76s^2$	$5d^94s^1$	$5d^{10}6s^1$	$5d^{10}6s^2$
1.87	1.59	1.47	1.41	1.37	1.35	1.36	1.39	1.44	—
435	618	781	837	791	790	690	566	368	61

general feature of Table 18.2 which was not emphasised earlier: the increasing stability of higher oxidation states, except for the first and last members of the series, as we go from the first to the second to the third series. We return to this subject later.

The first point to be noticed in Table 25.1 is that the electronic configurations (of ground state gaseous atoms, of course) of the heavy transition metals change rather irregularly with increase in atomic number: the *nd* and $(n+1)s$ orbitals are closer in energy for $n=4$ or 5 than for $n=3$. For ions of first transition series metals, we found that the general electronic configuration was of the form d^n and that this introduced a considerable degree of order into the discussion of the properties of di- and tri- positive ions; but for most heavy transition metals, simple monatomic cations are rarely encountered and hence their electronic configurations, even when known (which is not always the case), are of only minor importance.

Next we should note the similarity in atomic radii of zirconium and hafnium, niobium and tantalum, molybdenum and tungsten, and so on. This is the consequence of the so-called *lanthanide contraction* (the steady decrease in size along the fourteen inner transition elements between lanthanum and hafnium) which is discussed in the following chapter. The similarity extends to ionic radii (where these are meaningful) and covalent radii; for example, the internuclear distances in the high-temperature forms of ZrO_2 and HfO_2 (which have the fluorite structure), and in the octahedral molecules $Mo(CO)_6$ and $W(CO)_6$, differ by less than 0.01 Å in each pair of compounds. Atomic, ionic and covalent radii of the second and third series metals are all, however, considerably greater than the corresponding radii of first series metals. Those properties which depend mainly on size, e.g. lattice energies, solvation energies and complex formation constants, are nearly the same for corresponding species of second and of third transition series elements; this is doubtless why pairs of such elements (e.g. Zr and Hf, Nb and Ta) often occur together and are difficult to separate. Since the atomic numbers (and therefore nuclear charges) of a second series metal and its analogue in the third series differ by thirty-two units there is, nevertheless, an appreciable difference in electronic energy levels, and hence in electronic spectra and ionisation energies. In general it is found that first ionisation energies are higher for third than for first and second series metals, but the reverse is often true for removal of further electrons. As we have remarked earlier, however, the quantitative discussion of ionisation energies requires accurate consideration of both the species being ionised and the species formed, and we shall not pursue this subject here beyond remarking that even where two analogous transition elements and their compounds are isodimensional, there are often quite marked differences in stability towards oxidation or reduction. This is seen particularly clearly for palladium and platinum, and silver and gold (see Sections 25.10 and 25.11).

The larger sizes of the elements of the second and third series compared with those of the first series lead to a general tendency to exhibit higher coordination numbers, especially for elements near the beginnings of the series. Thus coordination numbers of seven, eight and nine are quite common in compounds of heavy transition metals; further, because the differences in energy between different possible structures involving high coordination numbers are usually small, many of these metals have a very involved crystal chemistry, as we shall see in later sections.

The third significant feature apparent in Table 25.1 is that the heavy transition metals usually have much greater enthalpies of atomisation than their first series analogues. This is the consequence of the increase in spatial extension of the *d* orbitals with increase in the principal quantum number. It is very important in understanding why metal–metal bonding is so much more common in compounds of second and third series metals than in compounds of first series metals (though it is not by any means unknown in the latter – cf. the structures of chromium(II) acetate, di-iron enneacarbonyl and nickel dimethylglyoximate). We can carry out an instructive calculation on the formation of, for example, $CrCl_2$ and WCl_2, assuming both to have ionic structures, by means of the Born–Haber cycle. The experimental value of ΔH_f° ($CrCl_2$) is -397 kJ mol^{-1}, and the cycle value for its lattice energy is 2510 kJ mol^{-1}. Thus the lattice energy of

WCl_2 should be approximately

$$2\,510 \times \frac{r(Cr^{2+}) + r(Cl^-)}{r(W^{2+}) + r(Cl^-)}$$

Since $r(Cl^-) \gg r(Cr^{2+})$ or $r(W^{2+})$, the difference between the radii of Cr^{2+} and W^{2+} will not affect the value of the lattice energy of WCl_2 very much; let us take it as 2 400 kJ mol^{-1}. This would be high enough to make WCl_2 a strongly exothermic compound if it were not for the higher ionisation energies of tungsten (770 and 1710 kJ mol^{-1}, compared with 652 and 1592 kJ mol^{-1} for chromium) and its much greater enthalpy of atomisation (837 kJ mol^{-1} for tungsten, 397 kJ mol^{-1} for chromium). As a consequence of these values, the estimated value of ΔH_f° for WCl_2 is +430 kJ mol^{-1}. Thus ionic $W^{2+}(Cl^-)_2$ is most unlikely to exist. There is, in fact, a compound of empirical formula WCl_2, but X-ray diffraction shows it to contain a $[W_6Cl_8]^{4+}$ ion in which six tungsten atoms form an octahedral unit with eight chlorine atoms situated off the faces of the W_6 octahedron. In forming this cation from elemental tungsten, relatively few metal–metal bonds have to be broken. Such *metal cluster compounds*, as they are called, are very common in the chemistry of the heavy transition metals (especially those of the third series, which have the highest atomisation energies) in low oxidation states. Metal–metal bonding, of course, often results in diamagnetism owing to the pairing of all electrons.

There are a few sets of analogous mononuclear complexes of metals of all three series, and we shall comment briefly on their electronic spectra and magnetic properties. It seems to be clear that for, say, a series of octahedral ions, Δ_o increases roughly in the ratio 1:1.5:2 as we go from the first to the third series; for the complexes $[M(NH_3)_6]^{3+}$, for example, Δ_o = 23 000, 34 000 and 41 000 cm^{-1} for M = Co, Rh and Ir respectively. Since there is less repulsion between electrons in the same $4d$ or $5d$ orbitals than in the smaller $3d$ orbitals, there are relatively few paramagnetic complexes of second and third series elements. These rarely have magnetic moments corresponding even approximately to the spin-only values: spin–orbit coupling constants are much larger for second and third series than for first series ions, with consequences that we have mentioned in Section 19.6. With so many complications (metal–metal interactions, a wide range of geometrically different environments, spin–orbit coupling, and difficulties of interpretation of spectra), most of the advantages of the simple crystal field theory are lost when we come to deal with the heavy transition metals and, regrettably, we shall be able to make very little use of it in this chapter. Nor shall we be able to make extensive use of electrode potentials, complex formation constants and solubility products: simple ions in solution are relatively rare, few electrodes involving heavy transition metals are reversible, and other thermochemical data are, at the present time, scanty.

To end this section we return briefly to the subject of oxidation states. As we saw in Section 16.4, it is rarely possible to discuss the relative stabilities of a range of oxidation states (even in simple compounds such as halides) very satisfactorily, since the lower ones tend to involve ionic species and the higher ones molecular or macromolecular species. For the second and third series, the problem is further complicated by the formation of metal cluster compounds by many of the elements in low oxidation states. Nevertheless, metals which form cluster compounds in low oxidation states often give rise to stable mononuclear molecular species when they are in high oxidation states: tungsten, for example, forms stable hexahalides WF_6 and WCl_6, whilst CrF_6 is extremely unstable and the highest chloride of chromium is $CrCl_3$. Further, although CrO_3 and chromates(VI) are powerful oxidising agents, WO_3 and tungstates(VI) are almost devoid of oxidising powers and MoO_3 and molybdates(VI) are not very different. In general the stability of high oxidation states increases in the sequence 1st series metals $\ll$ 2nd series metals $<$ 3rd series metals, as is shown by the existence of WCl_6, ReF_7, and AuF_5, none of which has an analogue among compounds of the second series elements. Two factors appear to be involved in the greater stability of high oxidation states among the third series elements: easier promotion of electrons to give the appropriate valence state, and better overlapping by 5*d* than by 4*d* orbitals, alone or in hybrid orbitals. Thus not only is the M—F bond energy term (which is relative to ground-state atoms) some 55 kJ mol^{-1} greater for WF_6 than for the almost isodimensional MoF_6, but the symmetric M—F stretching frequency and force constant are also higher; similar increases in stretching frequency are also found between TcF_6 and ReF_6, RuF_6 and OsF_6, and RhF_6 and IrF_6.

25.2 *Yttrium and lanthanum*

These two metals are so similar to the lanthanides, with which they occur in nature, that in this chapter we shall deal only briefly with their chemistry, which in any case is not extensive. Their only stable oxidation state under chemical conditions is the tripositive one, a few anomalous species such as LaI_2 and LaC_2 being metallic conductors best formulated as $La^{3+}(I^-)_2(e^-)$ and $La^{3+}(C_2)^{3-}$ respectively.

Some physicochemical data for scandium, yttrium and lanthanum are given in Table 25.2, from which it may be seen that the variation in properties with increase in atomic number is rather like that among the alkaline earth metals: there is a steady decrease in ionisation energies and a steady increase in atomic and ionic radius. Between yttrium and lanthanum there is little difference in enthalpy of atomisation, and E° M^{3+}/M is nearly constant. Whereas scandium hydroxide is amphoteric

Table 25.2
Some properties of scandium, yttrium and lanthanum.

Element	Sc	Y	La
Electronic configuration	$[Ar]3d^14s^2$	$[Kr]4d^15s^2$	$[Xe]5d^16s^2$
Metallic radius/Å	1.64	1.82	1.87
ΔH° atomisation/kJ mol^{-1}	376	410	435
M.p./°C	1540	1500	920
First ionisation energy/kJ mol^{-1}	631	616	540
Second ionisation energy/kJ mol^{-1}	1235	1180	1070
Third ionisation energy/kJ mol^{-1}	2390	1980	1850
Ionic radius of M^{3+}/Å	0.73	0.89	1.06
E° $M^{3+}(aq) + 3e \rightleftharpoons M$/V	−1.88	−2.30	−2.37

and scandium fluoride dissolves in alkali metal fluoride solutions to form complexes, the hydroxides of yttrium and lanthanum are exclusively basic ($La(OH)_3$ strongly so, absorbing atmospheric carbon dioxide), and their sparingly soluble fluorides form only double salts, e.g. $NaCaCdYF_8$ and $KLaF_4$, both of which have disordered fluorite structures. In these, as in other respects, the two elements are like the lanthanides; as expected from their ionic radii, lanthanum is more like the earlier, and yttrium more like the later, ones, from which they are separated by the methods described in the following chapter.

25.3 Zirconium and hafnium

Unlike yttrium and lanthanum, zirconium and hafnium are separated by a difference of thirty-two in atomic number, and their atomic, covalent and ionic radii are almost identical as a consequence of the lanthanide contraction. Metallic radii are 1.60 Å for Zr, and 1.59 Å for Hf, compared with 1.47 Å for Ti. Since zirconium and hafnium compounds usually crystallise in complicated structures, and few accurate data for hafnium compounds are available, comparisons of covalent and ionic radii necessarily rest on a few examples. Values of 1.20, 1.34 and 1.34 Å for covalent radii of Ti, Zr and Hf respectively rest on bond lengths in the tetrahedral tetrachlorides and a few complex ions; the ionic radii 0.61, 0.72 and 0.71 Å for Ti^{4+}, Zr^{4+} and Hf^{4+} are based on the assumption that the dioxides TiO_2 (rutile structure) and ZrO_2 and HfO_2 (fluorite structure at high temperatures) actually contain M^{4+} ions. Despite doubts whether these values are absolutely correct, it is clear that all known isostructural zirconium and hafnium compounds are isodimensional or very nearly so, and that interatomic distances are appreciably greater than in analogous titanium compounds. Ionisation energies for the three elements are given in Table 25.3.

Zirconium and hafnium occur together in *zircon*, $ZrSiO_4$, and *baddeleyite*, ZrO_2, the hafnium content in each case being only about 2

Table 25.3
Ionisation energies of titanium, zirconium and hafnium.

Ionisation energy/kJ mol^{-1}	1st	2nd	3rd	4th
Titanium	660	1310	2650	4180
Zirconium	660	1270	2220	3310
Hafnium	654	1440	2250	3200

per cent. The metals may be obtained by the reaction sequence

$$MO_2 \xrightarrow[500\,°C]{CCl_4} MCl_4 \xrightarrow[1150\,°C]{\text{Mg in Ar}} M$$

As an additive for strengthening steel, the mixture of metals so obtained is satisfactory. Zirconium, however, is also used for atomic pile construction, for which its resistance to corrosion and extremely low cross-section for neutron capture make it particularly suitable; for these purposes it must be free from hafnium (which is a very good neutron absorber). Since lattice energies and solubilities of zirconium and hafnium compounds, and the stabilities of their complexes, are almost the same, fractionation procedures similar to those used for the lanthanides (e.g. ion-exchange, solvent extraction) are necessary for the separation of the elements. Very pure metals may be obtained by zone-refining or by thermal decomposition of the iodides on a hot metal filament.

Because of its rarity and the difficulty of separating it from zirconium, the chemistry of hafnium has not been investigated in detail. Both metals are attacked by hot aqua regia or hydrofluoric acid (halo complexes being formed) and react when heated with halogens and, at high temperatures, with oxygen; in all these reactions only colourless compounds of the metals in the tetrapositive state are formed. Zirconium dioxide is very inert, and melts at 2700 °C; it is used as a refractory material and as an insulator. It is extremely insoluble in water, and is unaffected by aqueous alkali, but zirconates (probably mixed oxides) can be obtained from ZrO_2 and KOH at high temperatures. The fluorides ZrF_4 and HfF_4, white solids melting at about 900 °C, have three-dimensional lattices in which the metal atom is at the centre of a square antiprism of fluorine atoms; the chlorides, which melt at about 450°C, and are hydrolysed by water, have polymeric structures based on MCl_6 octahedra sharing edges. A typical illustration of the complicated stereochemistry of the early transition metals of the later series is provided by the fluoro complexes of zirconium: $[ZrF_8]^{4-}$ occurs as both dodecahedra and square antiprisms in different salts, $[ZrF_7]^{3-}$ as both pentagonal bipyramids and capped trigonal prisms, and species containing octahedral $[ZrF_6]^{2-}$ ions or polymeric anions built up from most of the discrete structures by sharing of edges are also known. The only known chloro complexes, on the other hand, contain $[ZrCl_6]^{2-}$ ions.

In aqueous acidic solution zirconium compounds are present as partially hydrolysed species such as $[Zr_3(OH)_4]^{8+}$ and $[Zr_4(OH)_8]^{8+}$. Zirconyl chloride, commonly formulated $ZrOCl_2.8H_2O$, which crystall-

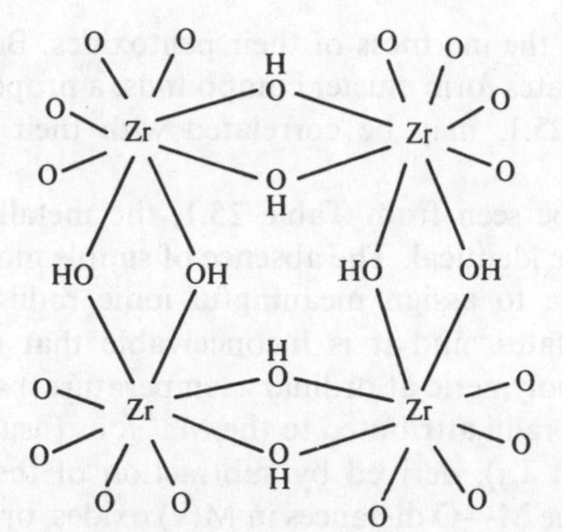

Figure 25.1
A representation of the $[Zr_4(OH)_8(H_2O)_{16}]^{8+}$ ion (H atoms of water molecules are not shown).

ises from a solution of the tetrachloride in dilute hydrochloric acid, contains the ion $[Zr_4(OH)_8(H_2O)_{16}]^{8+}$ whose structure is indicated in Fig. 25.1; the zirconium atoms lie roughly in a square, and each has distorted dodecahedral coordination. Many other zirconium-containing complexes contain eight-coordinated metal atoms, e.g. the molecular species $Zr(NO_3)_4$ and $Zr(acac)_4$ (dodecahedral and antiprismatic respectively) and the oxalato complex $[Zr(C_2O_4)_4]^{4-}$ (dodecahedral).

The lower oxidation states of zirconium and hafnium are much less stable with respect to oxidation than titanium(III), though quantitative data to illustrate this statement are not available. Blue or black halides ZrX_3, ZrX_2 and ZrX (where X = Cl, Br or I) are obtained by reduction of the tetrahalides with the metal or with hydrogen. The monohalides contain sheets of metal and halogen atoms in the sequence XMMX ... XMMX and are metallic conductors; the trihalides have chain structures consisting of ZrX_6 octahedra sharing faces and Zr—Zr bonds between alternate atoms. There is no aqueous solution chemistry of Zr(I), Zr(II) or Zr(III).

Up to the present time no carbonyl or cyano complex of zirconium or hafnium has been reported. A few unstable alkyl derivatives (e.g. $Li_2[ZrMe_6]$, made from LiMe and $ZrCl_4$ at low temperatures) are known, together with cyclopentadienyl derivatives, such as $Zr(C_5H_5)_4$ and $Zr(C_5H_5)_2Cl_2$; the latter is $Zr(\eta^5\text{-}C_5H_5)_2Cl_2$, the former $Zr(\eta^1\text{-}C_5H_5)_2(\eta^5\text{-}C_5H_5)_2$.

25.4 Niobium and tantalum

General relationships within the triad vanadium, niobium and tantalum are very like those within the preceding triad. The two heavy transition metals resemble one another very closely (the consequence of the lanthanide contraction again) and differ from vanadium in the relative instability of their lower oxidation states, the failure to form simple ionic species, and the inertness of their pentoxides. Both metals in their lower oxidation states form cluster compounds, a property which, as we showed in Section 25.1, may be correlated with their very high enthalpies of atomisation.

As may be seen from Table 25.1, the metallic radii of niobium and tantalum are identical. The absence of simple mononuclear species makes it impossible to assign meaningful ionic radii to the elements in low oxidation states, and it is inconceivable that in species such as MF_5 (which are polymeric at ordinary temperatures) and M_2O_5, M^{5+} ions are present; the radii attributed to the M^{5+} ions (usually quoted as 0.64 Å for M = Nb and Ta), derived by subtraction of the conventional radius of O^{2-} from the M—O distances in M(v) oxides, or of the radius of F^- from the M—F distances in complex fluorides, are therefore quite unreal quantities.

Niobium and tantalum occur together in *columbite*, a mixed oxide of formula $(Fe,Mn)(Nb,Ta)_2O_6$. Fusion of this with alkali gives polyniobates and -tantalates, which on treatment with dilute acid yield white Nb_2O_5 and Ta_2O_5. A separation based on the slightly more basic character of tantalum is possible: at a controlled concentration of HF and KF in aqueous solution, the pentoxides are converted into $K_2[NbOF_5]$ and $K_2[TaF_7]$ respectively; the former is much more soluble than the latter. Modern practice, however, is to separate the elements by fractional extraction into methyl isobutyl ketone from aqueous hydrofluoric acid.

The metals themselves can be isolated by reduction of the pentoxides with sodium or aluminium, or by electrolysis of complex fluorides. Niobium (m.p. 2470 °C) is a component of some tough steels; tantalum (m.p. 3000 °C) is used for the manufacture of corrosion-resistant equipment. Both are attacked by oxygen at high temperatures and by halogens, but are inert towards non-oxidising acids.

The principal oxides are the pentoxides obtained as described above. They are infinite polymeric solids of very complicated structure, insoluble in acids other than hydrofluoric acid, but soluble in molten alkalis; by dissolution of the product in water followed by slow concentration, salts such as $K_8[Nb_6O_{19}].16H_2O$, and the corresponding tantalum compound, may be isolated. The ion $Nb_6O_{19}^{8-}$ is a rather simple example of a heavy transition metal polyanion. Its structure, shown in Fig. 25.2, consists of an octahedron of NbO_6 octahedra each of which has four oxygen atoms each shared with one other octahedron and one oxygen atom shared with the other five octahedra, the remaining oxygen atom being unshared. Thus the overall composition is $Nb:O = 1:1+(4\times\frac{1}{2})+(1\times\frac{1}{6}) = 1:3\frac{1}{6}$ or 6:19. The insoluble compounds KMO_3 have the perovskite structure.

Niobium and tantalum pentafluorides are volatile white solids (m.p. 80 °C and 95 °C respectively) containing molecular M_4F_{20} units in which MF_6 octahedra are linked by linear M—F—M bridges as shown in Fig. 25.3. In Nb_4F_{20} the Nb—F bond lengths are 1.77 Å (terminal) and 2.06 Å (bridging), indicating much weaker bonding in the bridge. The fluorides

Figure 25.2
The structure of the $Nb_6O_{19}^{8-}$ ion.

are conveniently obtained by the reaction sequence

$$M_2O_5 \xrightarrow[300^\circ C]{CCl_4} M_2Cl_{10} \xrightarrow{HF} (MF_5)_4$$

The pale yellow pentachlorides, also volatile solids but much more readily hydrolysed than the pentafluorides, contain the dimeric unit shown for Nb_2Cl_{10} in Fig. 25.4; terminal and bridging Nb—Cl distances are 2.25 and 2.56 Å respectively.

Both pentafluorides form a range of fluoro complexes which may be isolated from aqueous solutions containing different concentrations of hydrofluoric acid, solubility considerations as well as complex formation constants being involved in determining the nature of the solid which separates from the solution; representative compounds are $Cs[NbF_6]$ and $K[TaF_6]$ (both containing discrete octahedral anions), $K_2[NbF_7]$ and $K_2[TaF_7]$ (containing capped trigonal prismatic anions), and $Na_3[TaF_8]$ (containing a square antiprismatic anion). A mixture of tantalum

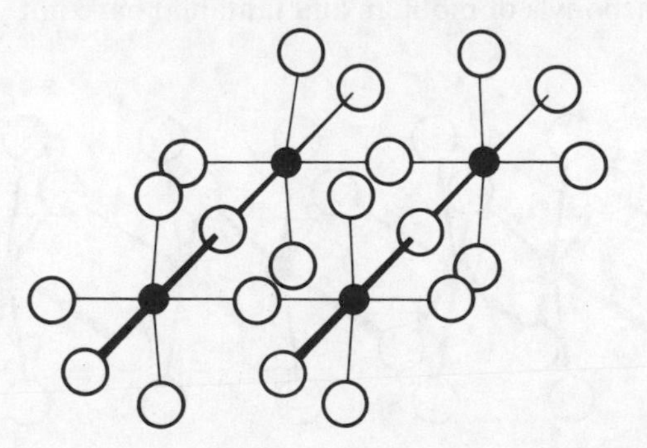

Figure 25.3
The $(NbF_5)_4$ structure.

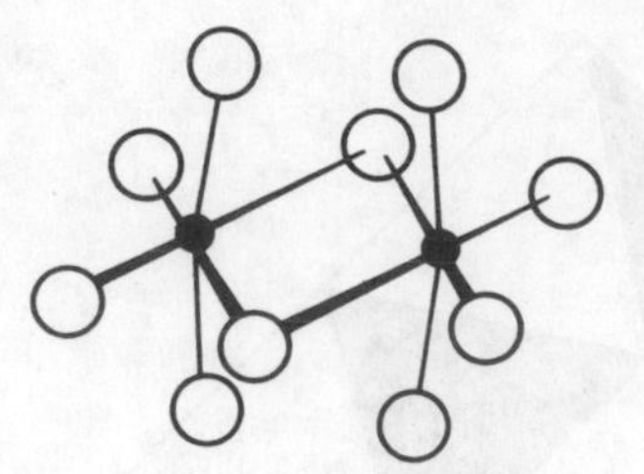

Figure 25.4
The Nb_2Cl_{10} structure.

pentafluoride and hydrogen fluoride is a very powerfully acidic solvent (cf. SbF_5/HF, Section 8.3) and has been used as a catalyst for rearrangement reactions of aromatic hydrocarbons.

Although the lower oxidation states of niobium and tantalum are of relatively little chemical importance, the structures of the lower halides provide a valuable simple introduction to metal cluster compounds. Black tetrahalides (other than TaF_4, which is not known) are obtained by reduction of pentahalides with the metals at 300–400 °C. Niobium tetrafluoride is paramagnetic, as expected for a Nb(IV) compound, and is isostructural with SnF_4, having an infinite layer lattice in which NbF_6 octahedra share four corners. The other tetrahalides are diamagnetic and have chain structures based on MX_6 octahedra sharing corners; there are, however, two quite different M—M distances, indicating that the metal atoms occur in pairs with metal–metal bonding. The structure of NbI_4 is shown in Fig. 25.5: alternate metal–metal distances are 3.31 and 4.36 Å; as usual, Nb—I bridging bonds are longer than terminal ones. (Similar pairs of Nb atoms occur in the distorted rutile structure of NbO_2.) Other lower halides obtained at higher temperatures include Nb_6I_{11} and Nb_6Cl_{14}, best formulated as $[Nb_6I_8]^{3+}I_3$ and $[Nb_6Cl_{12}]^{2+}Cl_2$. The cations in each of these substances contain octahedra of Nb atoms; in $[Nb_6I_8]^{3+}$ there is an iodine atom off the centre of every face, and in $[Nb_6Cl_{12}]^{2+}$ a chlorine atom off the centre of every edge, of the Nb_6 octahedron, as shown in Fig. 25.6. The $[Nb_6Cl_{12}]^{2+}$ ion can be oxidised electrolytically to $[Nb_6Cl_{12}]^{3+}$ and $[Nb_6Cl_{12}]^{4+}$ without change of structure. Tantalum cluster compounds have been less investigated but certainly exist in considerable numbers.

Carbonyls of niobium and tantalum have not yet been isolated, though

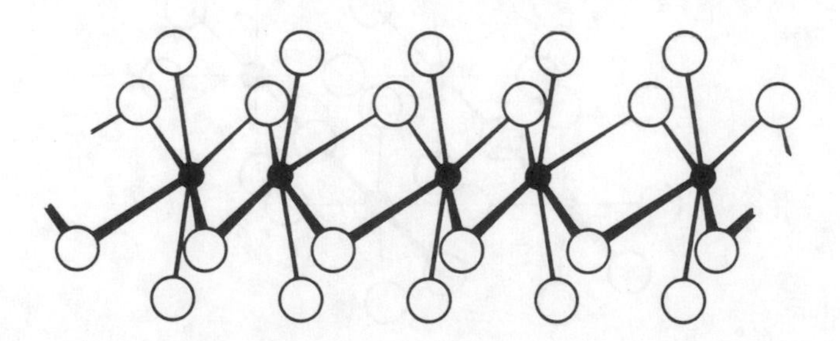

Figure 25.5
The NbI_4 structure.

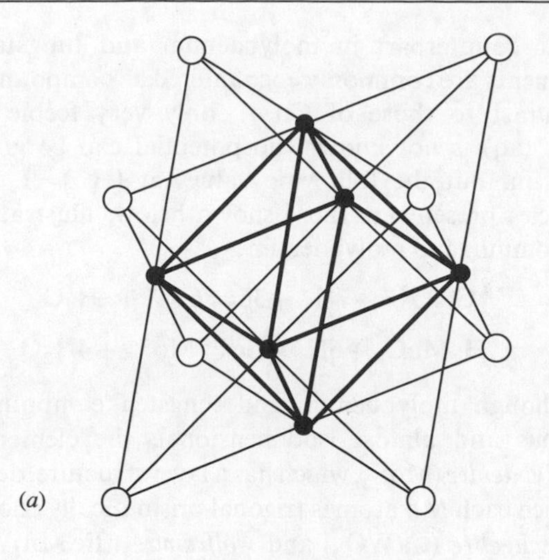

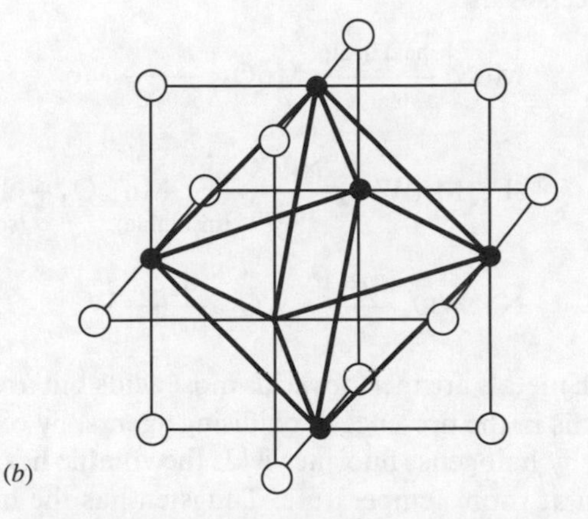

Figure 25.6
The structures of the (*a*) $Nb_6I_8^{3+}$ and (*b*) $Nb_6Cl_{12}^{2+}$ ions.

carbonylate anions $[M(CO_6]^-$ are well known. An unstable pentamethyl has recently been made by the reactions

$$TaCl_5 \xrightarrow{Zn(CH_3)_2} (CH_3)_3TaCl_2 \xrightarrow{LiCH_3} Ta(CH_3)_5$$

Some η^5-C_5H_5 compounds have been reported for both metals (e.g. $(\eta^5$-$C_5H_5)_2MH_3)$, and a cyano complex $K_4[Nb(CN)_8].2H_2O$, obtained by addition of potassium cyanide to a reduced methanolic solution of niobium pentachloride, is known.

25.5 Molybdenum and tungsten

Molybdenum and tungsten show a wide range of oxidation states (0, II, III, IV, V and VI) though simple mononuclear species are not known for all of these. The extensive chemistry of chromium(II) and chromium(III) has little counterpart in molybdenum and tungsten chemistry, and both elements are commonly encountered as compounds of M(VI) which are, in contrast to those of Cr(VI), only very feeble oxidising agents. Since W^{3+}(aq) is not known, no potential can be given for the W(VI)/W(III) system, but the following values at $[H^+]=1$, where the predominant species present are those shown below, illustrate the difference between chromium and molybdenum:

$$\tfrac{1}{2}Cr_2O_7^{2-} + 7H^+ + 3e \rightleftharpoons Cr^{3+} + \tfrac{7}{2}H_2O \qquad E^\circ = +1.33 \text{ V}$$

$$H_2MoO_4 + 6H^+ + 3e \rightleftharpoons Mo^{3+} + 4H_2O \qquad E^\circ = +0.1 \text{ V}$$

Although molybdenum and tungsten compounds are usually isomorphous and almost isodimensional, the elements occur separately, as *molybdenite* (MoS_2, which has a layer structure, described in Section 18.4, in which each Mo atom is trigonal prismatically coordinated by S atoms) and as *scheelite* ($CaWO_4$) and *wolframite* ((Fe,Mn)WO_4). Typical extraction processes are:

$$MoS_2 \xrightarrow[600\ ^\circ C]{\text{heat in air}} MoO_3 \xrightarrow[600\ ^\circ C]{H_2} Mo$$

$$(Fe, Mn)WO_4 \xrightarrow[\text{fusion}]{Na_2CO_3} \underset{\text{(insoluble)}}{(Fe, Mn)_2O_3} + \underset{\text{(soluble)}}{Na_2WO_4}$$

$$Na_2WO_4 \xrightarrow{HCl} WO_3 \xrightarrow[600\ ^\circ C]{H_2} W$$

Both metals are inert towards most acids but are rapidly attacked by fused alkalis in the presence of oxidising agents, by oxygen at high temperatures and by halogens; fluorine yields the volatile hexafluorides MoF_6 and WF_6 even at room temperature. Tungsten has the highest melting-point (3420 °C) and enthalpy of atomisation (837 kJ mol^{-1}) of all metals. Both metals are used in the manufacture of toughened steels (for which wolframite may be reduced directly with aluminium), tungsten also in lamp filaments and, as tungsten carbide, in cutting tools and abrasives. Some molybdenum compounds are useful catalysts; the key enzyme in nitrogen-fixing bacteria contains molybdenum (and also iron), and a model system containing $MoFe_3S_4$ clusters reproduces its characteristics to some extent. For dinitrogen complexes of tungsten see Section 22.7.

Molybdenum (VI) and tungsten(VI)

The most important compounds of the metals in this oxidation state are white MoO_3 (m.p. 800 °C) and yellow WO_3 (m.p. 1200 °C); in each compound the metal is octahedrally coordinated by oxygen, but whereas WO_3 has (approximately) the ReO_3 (or ScF_3) structure, MoO_3 has a complex layer structure. Neither oxide reacts with acids, but both form a range of salts with aqueous alkali. The best-characterised molybdate ions are MoO_4^{2-}, $Mo_7O_{24}^{6-}$ (which is present in ordinary 'ammonium molybdate') and $Mo_8O_{26}^{4-}$; as we saw for oxo-anions of vanadium, the oxygen:metal ratio decreases as the acidity of the solution increases, H^+ ions removing oxide ions from the anion initially present. For tungstates the best-characterised species are WO_4^{2-}, $H_2W_6O_{22}^{6-}$, $W_7O_{24}^{6-}$ and $H_2W_{12}O_{42}^{10-}$. The structures of all the poly-molybdates and -tungstates are based on sharing of oxygen atoms between MoO_6 or WO_6 octahedra (cf. chromates, whose structures are based on sharing of oxygen atoms between CrO_4 tetrahedra). Anions containing only one kind of metal atom are commonly known as isopolyanions to distinguish them from heteropolyanions such as $PMo_{12}O_{40}^{3-}$ (formed in the molybdate test for phosphate) and $TeMo_6O_{24}^{6-}$ which contain a PO_4 unit and a TeO_6 unit respectively sharing all their oxygen atoms with MoO_6 groups. The structures of most iso- and hetero- polyanions are very difficult to represent satisfactorily on paper, and the discussion of their characterisation and chemistry lies mostly beyond the scope of this book, but that of the simplest of them, $TeMo_6O_{24}^{6-}$, is indicated in Fig. 25.7.

Tungsten hexachloride and hexabromide are blue solids which are readily hydrolysed; there is some doubt about the existence of molybdenum hexachloride, and the hexabromide is unlikely to exist (cf. the comment in Section 25.1 on the bond energies in the hexafluorides). Oxohalides such as MoO_2F_2 (prepared by the action of HF on MoO_2Cl_2, itself obtained from MoO_2 and Cl_2) and WO_2Cl_2 (made from WO_3 and PCl_5) are molecular species or oxygen-bridged macromolecules; all are hydrolysed by water, just like CrO_2Cl_2. Tungsten hexachloride reacts with lithium methyl in ether to form unstable $W(CH_3)_6$.

Molybdenum (V) and tungsten (V)

Typical compounds of the metals in this oxidation state are the yellow pentafluorides, made by the reduction of the hexafluorides with the metal, and the black pentachlorides, the molybdenum compound made by the

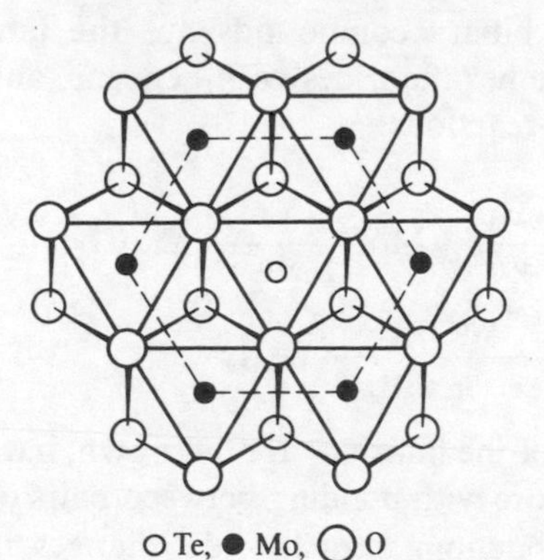

Figure 25.7
The structure of the $TeMo_6O_{24}^{6-}$ ion showing the mode of linking of the MoO_6 octahedra.

action of chlorine on the metal, and the tungsten compound by controlled thermal decomposition of the hexachloride; these compounds are paramagnetic and isostructural with Nb_4F_{20} and Nb_2Cl_{10} respectively. Complex fluorides KMF_6 are obtained by the interaction of KI, $M(CO)_6$, and IF_5; they contain octahedral MF_6^- ions.

Both metals form a series of oxides between MO_3 and MO_2 when the trioxide is heated with the metal. A large group of compounds containing tungsten(V) is known, however. These are the *tungsten bronzes*, inert metallic-looking solids of formula Na_xWO_3 (where x is between 0 and 1) which are obtained by heating sodium tungstate in hydrogen or, better, by heating mixtures of sodium tungstate, tungsten trioxide and tungsten at 850 °C:

$$\frac{x}{2}Na_2WO_4 + \frac{3-2x}{3}WO_3 + \frac{x}{6}W \rightarrow Na_xWO_3$$

For $x \sim 0.3$, the product is violet, for $x \sim 0.6$, red, and for $x \sim 0.9$, golden; tungsten bronzes with $x > 0.25$ exhibit metallic conductivity (i.e. their conductivity decreases with temperature), but for $x < 0.25$ the materials are semiconductors (i.e. their conductivity increases with temperature). The structures of the tungsten bronzes are closely related to those of perovskite and tungsten trioxide, which has a slightly deformed ReO_3 structure. The structure of perovskite, $CaTiO_3$, was discussed in Section 6.3 and was shown in Fig. 6.12. The tungsten bronze of limiting composition $NaWO_3$ would have the perovskite structure, but if all the Ca atoms are removed from perovskite and the Ti is replaced by Re (or W) the resulting structure is that of ReO_3 (or, very nearly, WO_3). A tungsten bronze of composition Na_xWO_3 thus has a fraction x of the Ca sites in the perovskite lattice occupied by Na and all the Ti sites by W. Somewhat similar compounds are formed by molybdenum, and also by titanium and vanadium, but these have been much less thoroughly investigated.

Molybdenum(IV) and tungsten(IV)

Representative binary compounds are the tetrafluorides, obtained by reduction of the hexafluorides with benzene, and the tetrachlorides, best obtained by the reactions:

$$MoO_3 \xrightarrow[450\,°C]{H_2} MoO_2 \xrightarrow[250\,°C]{CCl_4} MoCl_4$$

$$WCl_6 \xrightarrow[\text{reflux in } C_6H_5Cl]{W(CO)_6} WCl_4$$

The structures of the fluorides are unknown, but both tetrachlorides have the NbI_4 structure with bonding between pairs of metal atoms. At 250 °C, however, molybdenum tetrachloride changes into a cyclic hexamer in which the $MoCl_6$ octahedra are regular and there is no metal–metal bonding. The dioxides of both metals have the same metal–metal bonded distorted rutile structure as NbO_2. Neither dioxide dissolves in non-oxidising acids, but some complexes of the metals in the tetrapositive oxidation state are known. The most interesting of these are the cyano complexes $K_4[M(CN)_8]$, obtained by the interaction of K_2MO_4, KCN and KBH_4 in the presence of acetic acid. Both $[Mo(CN)_8]^{4-}$ and $[W(CN)_8]^{4-}$ are kinetically inert with respect to substitution, but they are rapidly oxidised (e.g. by Ce(IV) or MnO_4^-) to $[Mo(CN)_8]^{3-}$ and $[W(CN)_8]^{3-}$ respectively. Most of these octacyano anions exist in both dodecahedral and square antiprismatic forms according to their environments; clearly there is very little difference in energy between them.

Molybdenum(III) and tungsten(III)

The increasing tendency towards metal–metal bonding on going from the second to the third transition series is well illustrated by this oxidation state of molybdenum and tungsten, since mononuclear species are known only for the former element. A solution containing the red $[MoCl_6]^{3-}$ ion (which is paramagnetic, with a moment of 3.7 μ_B) is obtained by electrolytic reduction of molybdenum trioxide in concentrated hydrochloric acid; the ion is readily aquated to form yellow substitution-labile $[Mo(H_2O)_6]^{3+}$, one of the very few simple aquo ions of the heavy transition metals. Under somewhat different conditions, $[Mo_2Cl_9]^{3-}$ is formed in place of $[MoCl_6]^{3-}$; $[W_2Cl_9]^{3-}$ is always the product from the analogous reduction of tungstate. The ions $[Mo_2Cl_9]^{3-}$ and $[W_2Cl_9]^{3-}$, like $[Cr_2Cl_9]^{3-}$, consist of two MCl_6 octahedra sharing a face; the increasing interaction between the metal atoms in the two octahedra is shown not only by magnetic properties (the chromium-containing species is paramagnetic, the other two diamagnetic) but also by the M—M distances, which are 3.12, 2.67 and 2.45 Å in the chromium, molybdenum and tungsten complexes respectively.

The binary chlorides corresponding to these complexes are obtained by reduction of Mo_2Cl_{10} with hydrogen at 400 °C, and by the controlled chlorination of the lower halide of empirical formula WCl_2. Molybdenum trichloride is diamagnetic and has a layer structure in which pairs of molybdenum atoms are bonded; tungsten trichloride, however, is really $[W_6Cl_{12}]Cl_6$, the cation having the same structure as $[Nb_6Cl_{12}]^{2+}$.

Molybdenum(II) and tungsten(II)

Very few mononuclear species are known for the dipositive metals, though cyano complexes containing the pentagonal bipyramidal $[Mo(CN)_7]^{5-}$ ion may be obtained by reduction of molybdates with hydrogen sulphide in the presence of cyanide. The yellow chlorides of empirical formula MCl_2, obtained by decomposition of higher chlorides, are both cluster compounds $[M_6Cl_8]Cl_4$, the $[M_6Cl_8]^{4+}$ ions being isostructural with $[Nb_6Cl_8]^{3+}$. Molybdenum complexes containing only two metal atoms are obtained by the reactions

$$Mo(CO)_6 \xrightarrow{CH_3COOH} Mo_2(CH_3COO)_4 \xrightarrow{HCl} [Mo_2Cl_8]^{4-} \xrightarrow[H_2O]{CF_3SO_3H} Mo_2^{4+}(aq)$$

$$\downarrow LiCH_3,\ ether$$

$$Li_4[Mo_2(CH_3)_8].4(C_2H_5)_2O$$

All three species are diamagnetic; the $[Mo_2X_8]^{4-}$ ions have the structure shown in Fig. 25.8 with the MoX_4 groups in the eclipsed configuration. The adoption of this configuration, despite the repulsion between ligands, is attributed to the existence of a quadruple bond between the molybdenum atoms. If the $4d_{x^2-y^2}$ orbital of each molybdenum atom is used (together with the $5s$, $5p_x$ and $5p_y$ orbitals) in bonding the ligands, the d_{z^2} orbitals may be envisaged as forming the σ Mo—Mo bond, the d_{xz} and d_{yz} orbitals as forming π bonds, and the d_{xy} orbitals as forming a δ bond. This structure is, of course, like that of chromium(II) acetate dimer; metal–metal bond lengths in the molybdenum compounds are 2.09–2.18 Å. Species containing W≣W bonds are difficult to make, being more powerfully reducing than their molybdenum-containing analogues, but $[W_2Cl_8]^{4-}$ and $W_2(C_2H_5COO)_4$ have recently been characterised and found to be very like the corresponding Mo≣Mo species in structure; W≣W bond lengths are typically 0.10 Å longer than corresponding Mo≣Mo bond lengths, an increase attributed to greater internuclear repulsion in the tungsten compounds.

Lower oxidation states

These occur almost entirely in carbonyls and their derivatives and in organic compounds: $Mo(CO)_6$, $W(CO)_6$ and $Mo(C_6H_6)_2$, for example, were dealt with in Chapters 22 and 23.

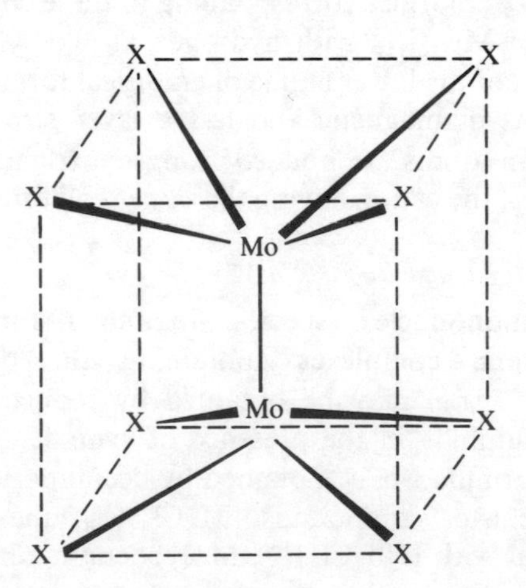

Figure 25.8
The structure of the $[Mo_2Cl_8]^{4-}$ and $[Mo_2Me_8]^{4-}$ ions (X = Cl or Me).

25.6 Technetium and rhenium

Although these elements have a complicated chemistry and exhibit all oxidation states between 0 and VII, we shall deal with them only by brief reference to a few aspects of particular interest.

Technetium has no stable isotope, and is available in considerable quantities only as ^{99}Tc, a β^--emitter of $t_{\frac{1}{2}} = 2.1 \times 10^5$ years, which is isolated from fission product wastes by oxidation to Tc_2O_7 and volatilisation. A γ-emitting metastable form of ^{99}Tc, made by a (n, β^-) reaction on ^{98}Mo, has a half-life of 6 hours and is widely used in medicine. The technetium is converted into a complex suitable for absorption by the organ or tissue to be examined, and an image made by γ-ray scanning. Rhenium occurs in small amounts in molybdenum ores, and volatilises as Re_2O_7 when these are roasted in air; this oxide is dissolved in water and rhenium precipitated as the colourless perrhenate $KReO_4$. (The ReO_4^- ion shows an absorption corresponding to the one responsible for the colour of MnO_4^-, but it occurs in the ultraviolet.) Technetium and rhenium differ from manganese in several respects; they have little cationic chemistry, their high oxidation states are much more stable with respect to reduction, and cluster compound formation is a prominent feature of their di-, tri- and tetra-positive oxidation states. The metals, obtained by reduction of NH_4MO_4 or $(NH_4)_2[MCl_6]$ in hydrogen, react readily with halogens and burn in oxygen above 400°C to form the oxides M_2O_7, which are the anhydrides of strong acids HMO_4. These acids react with hydrogen sulphide to precipitate the sulphides M_2S_7, a striking contrast to the reduction of permanganate to manganese(II) by hydrogen sulphide. Other oxides include TcO_2 and ReO_2 (isostructural with NbO_2) and ReO_3, the structure of which we have mentioned earlier; all result from controlled reduction of the oxides M_2O_7.

Largely, no doubt, because of the absence of detailed studies on technetium–halogen systems, the only halides yet reported for this metal are $TcCl_4$, TcF_5 and TcF_6. The first of these, unlike $ReCl_4$, shows no evidence of metal–metal bonding and is isostructural with $ZrCl_4$. For the remainder of this section we shall confine attention to rhenium.

Rhenium reacts with fluorine to form ReF_6 and ReF_7 according to the conditions employed; both are volatile solids which are hydrolysed by water. The heptafluoride is isostructural with IF_7; the hexafluoride molecule is octahedral. Reduction of the hexafluoride gives ReF_5, which has a chain structure built up of ReF_6 octahedra sharing *cis* fluorine atoms in contrast to the tetrameric molecular structures of other heavy transition metal pentafluorides. The highest chloride of rhenium is the volatile $ReCl_6$ obtained by halogen exchange between ReF_6 and BCl_3; chlorination of the metal yields Re_2Cl_{10} (isostructural with Nb_2Cl_{10}); this is reduced under different conditions to $ReCl_4$ (one form of which is isostructural with $NbCl_4$) and Re_3Cl_9, a cluster compound having the structure shown in Fig. 25.9(*a*), though there is also weak chlorine bridging

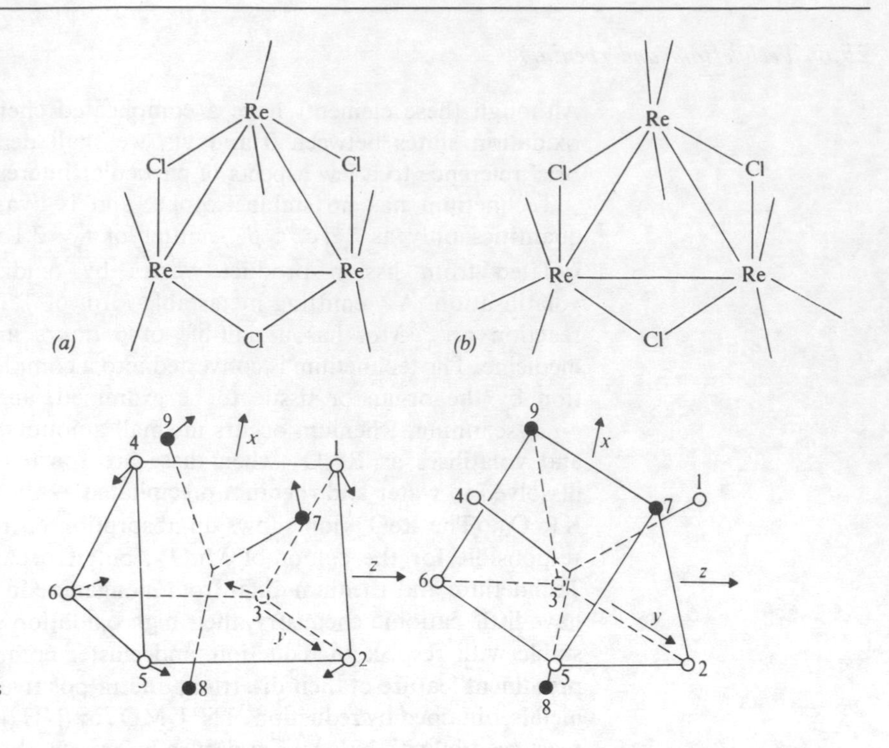

Figure 25.9
Partial structures of (*a*) Re_3Cl_9 and (*b*) the $Re_3Cl_{12}^{3-}$ ion (terminal Cl atoms are not shown).

Figure 25.10
The postulated mechanism for the exchange of ligands in the $[ReH_9]^{2-}$ ion.

between Re_3Cl_9 units which is not shown. When treated with HCl and CsCl, Re_3Cl_9 yields $Cs_3[Re_3Cl_{12}]$, the anion of which has the structure shown in Fig. 25.9(*b*). A quite different Re(III) chloro complex, $[Re_2Cl_8]^{2-}$, is obtained by reduction of perrhenate in hydrochloric acid with hydrogen or hypophosphite. This ion, characterised as its potassium salt, is isostructural with $[Mo_2Cl_8]^{4-}$. Metal–metal bond lengths in α-$ReCl_4$, Re_3Cl_9, and $[Re_2Cl_8]^{2-}$ are 2.73, 2.49 and 2.24 Å respectively, indicating single, double and quadruple bonds.

Other rhenium complexes worth mentioning are $[ReCl_6]^{2-}$ (obtained by reduction of $KReO_4$ with HCl and KI), the trigonal prismatic complex $Re(S_2C_2Ph_2)_3$ (Section 18.3), and the hydrido complex $K_2[ReH_9]$ (the structure of which was shown in Fig. 18.5). The last compound is prepared by the action of potassium on the perrhenate in ethylenediamine or ethanolic solution. The proton magnetic resonance spectrum of the $[ReH_9]^{2-}$ ion consists of a single sharp line, indicating that a rapid intramolecular site-exchange takes place; a possible mechanism for this process is shown in Fig. 25.10.

25.7 The platinum metals: general aspects

Ruthenium, osmium, rhodium, iridium, palladium and platinum are commonly known as the platinum metals. They occur together as the metals or in sulphide ores of copper and nickel; after the separation and electrolytic refinement of copper or nickel, the platinum metals, silver and gold remain in the anode sludge, from which they may be isolated by various complicated methods, most of which are based on differences in the relative stabilities of the oxidation states of the metals.

The platinum metals are relatively noble, and most of their compounds yield the metals when heated. Nevertheless, osmium can be oxidised by air to the volatile tetroxide and ruthenium to the non-volatile dioxide; and all the metals are attacked by mixtures of hydrochloric acid and oxidising agents, or by fluorine and chlorine at 300–600 °C. Palladium and platinum are also attacked by fused alkali. The metals have several features in common. Their chemistry is almost entirely that of complexes, simple cations being all but unknown. All of them exhibit several different oxidation states. Numerous complexes are formed with π acceptor ligands such as CO, NO, PMe_3 and PPh_3, and many stable organometallic compounds are known. The metals themselves, palladium and platinum especially, are very important heterogeneous catalysts: palladium for hydrogenation and dehydrogenation; platinum for ammonia oxidation, hydrocarbon reforming, and the complete combustion of noxious organic compounds in car exhausts. They are also used where a hard inert metal is required (e.g. for electrical contacts, pen nib tips, special crucibles) and as jewellery. Organic compounds of palladium and rhodium are versatile homogeneous catalysts. Compounds of the same formula type of all the platinum metals are nearly always isostructural, and compounds of ruthenium and osmium, of rhodium and iridium, and of palladium and platinum, are almost isodimensional; the metal–chlorine distances in $[PdCl_4]^{2-}$ and $[PtCl_4]^{2-}$ in the potassium salts, for example, are 2.318 and 2.316 Å respectively. Unlike metals discussed earlier in this chapter, the platinum metals rarely have coordination numbers greater than six.

These similarities notwithstanding, there are substantial differences between the metals, particularly in the relative stabilities of the different oxidation states. As we showed in Table 18.2 (Section 18.3), the most stable oxidation states are: Ru, 3, 4 and 6; Os, 4, 6 and 8; Rh, 3; Ir, 3 and 4; Pd, 2; Pt, 2 and 4. Further, the principal similarities to first series transition metals and the platinum metals undoubtedly occur between Fe, Ru and Os, between Co, Rh and Ir, and between Ni, Pd and Pt. It is therefore convenient to deal with the platinum metals in pairs, following the general method of treatment used so far in this chapter.

25.8 Ruthenium and osmium

Higher oxidation states

Both metals form volatile yellow oxides MO_4 (RuO_4 m.p. 25 °C, b.p. 40 °C; OsO_4 m.p. 40 °C, b.p. 101 °C) but, in accordance with the general pattern of behaviour among the heavy transition metals, the osmium compound is the more stable with respect to reduction. It is obtained when the finely divided metal is heated in oxygen or with nitric acid, whereas the preparation of ruthenium tetroxide requires the use of bromate or permanganate in the presence of sulphuric acid as the oxidant. Both tetroxides have penetrating ozone-like odours and are sparingly soluble in water, from which they may be extracted by carbon tetrachloride. Their molecules are regular tetrahedra.

Ruthenium tetroxide, which explodes above 180 °C, forming the dioxide and oxygen, is a very powerful oxidant which reacts violently with organic compounds. Osmium tetroxide is a useful oxidising agent in organic chemistry (e.g. for the conversion of alkenes into *cis* diols) and is also used as a biological stain; in both processes it is reduced to the metal. This ease of reduction, coupled with its volatility, makes it very dangerous to the eyes. By heating the tetroxide with osmium the dioxide OsO_2 is obtained.

When RuO_4 is dissolved in aqueous alkali, oxygen is evolved, RuO_4^- and RuO_4^{2-} (the latter analogous to FeO_4^{2-}, the only derivative of iron(VI)) being formed; K_2RuO_4 is obtained also by fusion of the metal with KNO_3 and KOH. Alkali and OsO_4, on the other hand, give the perosmate or osmate(VIII) anion *trans* $[OsO_4(OH)_2]^{2-}$, which is reduced by ethanol to osmate(VI), *trans* $[OsO_2(OH)_4]^{2-}$. By the action of ammonia on osmate(VIII), the nitrido complex OsO_3N^- is formed and may be isolated as potassium osmiamate, $K[OsO_3N]$. The infrared spectrum of the OsO_3N^- anion contains bands at 871 and 897 cm^{-1} attributed to Os=O stretching (the corresponding bands in the spectrum of OsO_4 are at 954 and 965 cm^{-1}) and a higher frequency band at 1 021 cm^{-1} attributed to Os≡N stretching.

The highest fluoride of ruthenium is RuF_6, obtained as an unstable solid by quenching the vapour formed when ruthenium is heated in fluorine under pressure. The normal product of fluorination is RuF_5, which has a tetrameric structure like that of Nb_4F_{20} but with non-linear Ru—F—Ru bridges; it is converted into RuF_4 and RuF_3 (which is isostructural with FeF_3) by the action of mild reducing agents. The highest chloride is $RuCl_3$, but the complex $[Ru_2^{IV}OCl_{10}]^{4-}$ is formed, together with $[RuCl_5(H_2O)]^{2-}$ and $[RuCl_6]^{3-}$, by the action of hydrochloric acid on RuO_4; the $[RuCl_6]^{2-}$ion is best obtained by oxidation of $[RuCl_6]^{3-}$ with chlorine. Osmium was for long believed to form an octafluoride, but it is now established that the ultimate product of fluorination of the metal is OsF_7, which in the absence of excess of fluorine decomposes to form OsF_6; this may be reduced to Os_4F_{20} and OsF_4. Osmium forms a stable tetrachloride, and OsO_4 is reduced by hydrochloric acid only to *trans* $[Os^{VI}O_2Cl_4]^{2-}$ (the analogue of the ion $[OsO_2(OH)_4]^{2-}$ mentioned earlier), $[Os_2^{IV}OCl_{10}]^{4-}$ and $[OsCl_6]^{2-}$.

The compound $K_2[RuCl_6]$ has a room-temperature magnetic moment of 2.8 μ_B, corresponding to the spin-only value for a low-spin d^4

octahedral complex; for $K_2[OsCl_6]$, the lower value (1.5 μ_B) arises from the much greater spin–orbit coupling constant for the heavier metal ion. The linear oxo complex $[Cl_5RuORuCl_5]^{4-}$ is diamagnetic, a fact which is explained by postulating that the electron configuration of the Ru^{IV} atoms in the pseudo-octahedral complex is $d_{xy}^2 d_{xz}^1 d_{yz}^1$, the d_{xz} orbital of each Ru atom and the p_z orbital of the O atom then combine to form three three-centre molecular orbitals, of which the bonding and non-bonding ones are occupied by one electron from each Ru atom and two electrons from the O atom; similar three-centre orbitals are formed from the d_{yz} orbital of each Ru atom and the p_y orbital of the O atom.

Lower oxidation states

We have already mentioned $RuCl_3$ and the complexes $[RuCl_6]^{3-}$ and $[RuCl_5(H_2O)]^{2-}$. Substitution in Ru(III) complexes is generally very slow (cf. low-spin Fe(III) complexes), and large numbers of aquo halo complexes have been characterised; they form the chief constituents of the material commonly described as '$RuCl_3.3H_2O$' and obtained by the action of concentrated hydrochloric acid on RuO_4. By prolonged aquation the aquo complex $[Ru(H_2O)_6]^{3+}$ can eventually be obtained in solution; it can be reduced to $[Ru(H_2O)_6]^{2+}$ electrolytically:

$$[Ru(H_2O)_6]^{3+} + e \rightleftharpoons [Ru(H_2O)_6]^{2+} \qquad E^\circ = +0.23 \text{ V}$$

$$\text{cf. } [Fe(H_2O)_6]^{3+} + e \rightleftharpoons [Fe(H_2O)_6]^{2+} \qquad E^\circ = +0.77 \text{ V}$$

By treatment of '$RuCl_3.3H_2O$' with zinc dust in strongly ammoniacal solution, $[Ru(NH_3)_6]^{2+}$ is obtained, and may be oxidised to $[Ru(NH_3)_6]^{3+}$ by air:

$$[Ru(NH_3)_6]^{3+} + e \rightleftharpoons [Ru(NH_3)_6]^{2+} \qquad E^\circ = +0.24 \text{ V}$$

Like $[Ru(NH_3)_6]^{3+}$, $[Ru(NH_3)_6]^{2+}$ is inert to substitution, but electron-exchange between the two low-spin species is very fast, with the second-order rate constant 10^8 l mol^{-1} s^{-1} (cf. Section 21.8). Reduction of '$RuCl_3.3H_2O$' with hydrazine leads to the formation of the dinitrogen complex $[Ru(NH_3)_5N_2]^{2+}$, the chemistry of which was described in Section 22.7. Ruthenium nitrosyl complexes, typified by $[Ru(NH_3)_5NO]^{3+}$, made by the action of nitrous acid on $[Ru(NH_3)_6]^{2+}$, are both numerous and stable; many are soluble in the organic solvents used in recovery of uranium and plutonium from nuclear waste and their removal is a difficult operation.

Bipyridyl complexes of ruthenium are of great photochemical interest, $[Ru(bipy)_3]^{2+}$ being extensively used as a photosensitiser. In the photochemically excited ion $*[Ru(bipy)_3]^{2+}$ an electron has been transferred from the metal to a ligand π^* orbital, so that the product may be described as a derivative of Ru(III), bipy, and bipy$^-$. The excited complex is both a better oxidant and a better reductant than the ground state complex, the relationship between the species involved being summarised by the redox

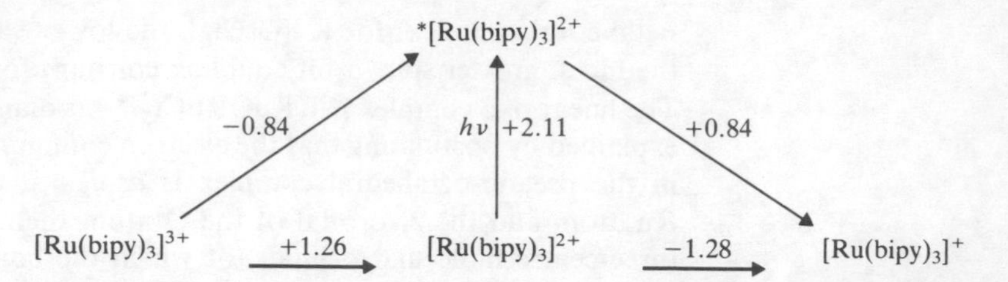

Figure 25.11
Potential diagram for $[Ru(bipy)_3]$ complexes.

diagram Fig. 25.11. In neutral solution, for example, water can be either oxidised or reduced by the excited complex. In order to make these reactions useful for the utilisation of solar energy for the hydrogen economy (Section 9.4) it is, of course, necessary for the two processes to take place at different locations, otherwise the oxygen and hydrogen recombine; unfortunately attempts to achieve this end have not yet been completely successful. Complexes containing two ruthenium atoms in different formal oxidation states sometimes show charge localisation and sometimes do not: in

$$\left[(H_3N)_5RuN\bigcirc NRu(NH_3)_5\right]^{5+}$$

for example, the metal atoms are indistinguishable, but in

$$\left[(bipy)_2ClRuN\bigcirc NRuCl(bipy)_2\right]^{3+}$$

an absorption band corresponding to internal electron transfer is observed. There is charge separation also in

$$\left[(H_3N)_5RuN\bigcirc NRuCl(bipy)_2\right]^{4+}$$

the Ru(III) being at the pentammine end of the complex. These complexes are, of course, analogous to the intermediates in redox reactions in solution (see Section 21.8).

There is relatively little chemistry of low oxidation states of osmium involving only σ bonding ligands, though ammine complexes of both Os(III) and Os(II) are known. The rate of the fast electron transfer

$$(+)[Os(bipy)_3]^{2+} + (-)[Os(bipy)_3]^{3+} \rightleftharpoons (-)[Os(bipy)_3]^{2+} + (+)[Os(bipy)_3]^{3+}$$

has been determined by measurement of the rate of loss of optical activity after the reactants are mixed. Large numbers of low oxidation state compounds of both ruthenium and osmium are known in which triphenylphosphine ligands are present: $RuCl_2(PPh_3)_3$, for example, made by the action of excess of methanolic triphenylphosphine on

'$RuCl_3.3H_2O$', is readily converted into $Ru(H)(Cl)(PPh_3)_3$ and is a versatile hydrogenation catalyst.

The trinuclear carbonyls of ruthenium and osmium, $M_3(CO)_{12}$, mentioned (together with the pentacarbonyls $M(CO)_5$) in Section 22.3, are the simplest metal cluster compounds of these elements. Many polynuclear complexes of osmium, especially, are now known.

25.9 *Rhodium and iridium*

The greater stability of higher oxidation states in the case of the third transition series element that we noted in the last section is also a feature of the chemistry of rhodium and iridium. For these elements, however, both the range of oxidation states found and the stability of the highest ones are much less than for ruthenium and osmium. All complexes of Rh(III) and Ir(III), including $[RhF_6]^{3-}$ ($[IrF_6]^{3-}$ is not known), are diamagnetic, and the kinetic inertness associated with the low-spin d^6 configuration is a feature that is common to the chemistry of Co(III), Rh(III) and Ir(III); thus large numbers of geometrical and optical isomers of Rh(III) and Ir(III) compounds are known, though we shall not discuss them in detail here.

Rhodium(VI) and iridium(VI) occur only in the unstable hexafluorides obtained by heating the metals in fluorine under pressure and rapidly quenching the volatile products. The ordinary products of fluorination of the metals at 300–400 °C are the tetrameric pentafluorides Rh_4F_{20} and Ir_4F_{20}, both isostructural with Ru_4F_{20} and Os_4F_{20}. Complexes such as $Cs[RhF_6]$ and $Cs[IrF_6]$ can be made in halogen fluoride solvents, but liberate oxygen on treatment with water, forming Rh(IV) and Ir(IV) compounds. Neither rhodium nor iridium forms a volatile oxide or an oxo-anion.

Rhodium(IV) is also an unstable oxidation state characterised only in RhO_2 (obtained by oxidation of Rh(III) in alkaline solution), RhF_4 (obtained by the action of BrF_3 on $RhCl_3$) and the complex anions $[RhF_6]^{2-}$ and $[RhCl_6]^{2-}$, obtained in the form of alkali metal salts by the reactions

$$2KCl + RhCl_3 \xrightarrow{BrF_3} K_2[RhF_6]$$

$$RhCl_6^{3-} + CsCl \xrightarrow[aq]{Cl_2} Cs_2[RhCl_6]$$

Both complexes are hydrolysed to RhO_2 by excess of water. Iridium(IV) is comparatively stable: IrO_2 is the normal product of oxidation of the metal, and $Na_2[IrCl_6]$, obtained by chlorination of a mixture of iridium powder and sodium chloride, is one of the commonest compounds of iridium. We mentioned it in Section 19.4 in connection with electron spin resonance spectroscopic evidence for delocalisation of the unpaired electron on the d^5 Ir(IV) ion. In alkaline solution $[IrCl_6]^{2-}$ decomposes with liberation of oxygen, but the reaction is reversed in strongly acidic media: $E°$ for the half-reaction

$$[IrCl_6]^{2-} + e \rightleftharpoons [IrCl_6]^{3-}$$

is +1.02 V, near enough to the value $E° = +1.23$ for the half-reaction

$$\tfrac{1}{2}O_2 + 2H^+ + 2e \rightleftharpoons H_2O$$

at $[H^+]=1$ for a change in the hydrogen-ion concentration of the solution to have a marked effect on the position of the equilibrium. Electron-exchange between $[IrCl_6]^{2-}$ and $[IrCl_6]^{3-}$ is, as expected, a rapid process.

Except for fluorides, the only binary halides of rhodium and iridium are those of the tripositive metals, obtained from the elements at elevated temperatures. Like iridium, finely-divided rhodium is attacked by chlorine in the presence of alkali metal chlorides, but in this case to form the red complex $Na_3[RhCl_6].12H_2O$, the commonest compound of rhodium. The reactions of the $[RhCl_6]^{3-}$ ion with alkali and water are indicated below:

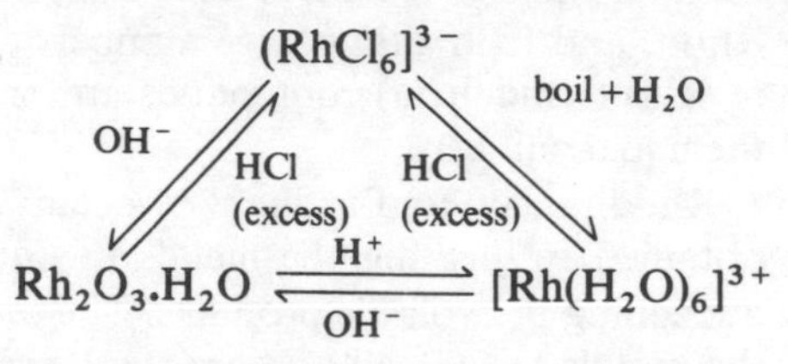

The yellow $[Rh(H_2O)_6]^{3+}$ ion forms salts (e.g. an alum) structurally similar to those of the $[Co(H_2O)_6]^{3+}$ ion but much more stable towards reduction. When $Rh_2O_3.H_2O$ is dissolved in a limited quantity of hydrochloric acid, $RhCl_3.3H_2O$ (probably a mixture of *fac* and *mer* $[RhCl_3(H_2O)_3]$, since it gives no precipitate with silver ion) is formed. Zinc and excess of ammonia form the hydrido complex $[RhH(NH_3)_5]^{2+}$, whilst ammonia alone gives the hexammine $[Rh(NH_3)_6]^{3+}$ or $[Rh(NH_3)_5Cl]^{2+}$ according to the quantity employed. Sodium hexachloroiridate(III) is usually made by reduction of the iridium(IV) complex with sulphur dioxide; it may be converted into other iridium(III) complexes by very slow substitution reactions. The $[Ir(H_2O)_6]^{3+}$ ion, obtained by methods analogous to those used for the preparation of $[Rh(H_2O)_6]^{3+}$, is very easily oxidised. Substitutions in Rh(III) and Ir(III) complexes are often catalysed by reducing agents (cf. substitutions in some Co(III) complexes); intermediate formation of Rh(I) and Ir(I) compounds is believed to be involved.

The planar four-coordinated Rh(I) and Ir(I) complexes such as $RhCl(PPh_3)_3$ and *trans* $Ir(CO)Cl(PPh_3)_2$ (Vaska's compound) undergo many oxidative addition reactions. Some of these, and the carbonyls of rhodium and iridium and their derivatives, were mentioned in Sections

22.2–22.5. In addition to the carbonyls $Rh_4(CO)_{12}$ and $Rh_6(CO)_{16}$, another compound of rhodium involving metal–metal bonding is the diamagnetic dimeric acetate of rhodium(II) obtained by heating $RhCl_3.3H_2O$ with a methanolic solution of sodium acetate; this contains a Rh_2^{4+} ion and has a structure like that of chromium(II) and molybdenum(II) actetates; a Rh—Rh distance of 2.39 Å is compatible with a single bond between the metal atoms; Rh_2^{4+} has six electrons more than Mo_2^{4+}, and these occupy both π^* and the δ^* orbitals. No corresponding iridium compound is at present known.

25.10 *Palladium and platinum*

With the last pair of platinum metals the difference in chemical properties between the two elements becomes quite marked, the most stable oxidation states being II for palladium and IV for platinum. Within a given oxidation state resemblances (other than behaviour towards oxidising and reducing agents) between analogous compounds are close, however. There are considerable structural similarities between platinum(II) and palladium(II) compounds and planar low-spin nickel(II) compounds, but the extensive chemistry of octahedral and tetrahedral high-spin nickel(II) compounds has only a few parallels in palladium chemistry and none at all in platinum chemistry.

We have already mentioned palladium hydride (Section 9.6), isomerism in planar platinum(II) complexes (Section 18.6), the kinetics and mechanism of substitution in such complexes (Section 21.6), the characterisation of palladium and platinum tetracarbonyls (Section 22.2), complexes of platinum with π acceptor ligands (Section 22.6) and organometallic compounds of both elements (Chapter 23). We shall not cover much of this material again, except to add some structural information on compounds referred to in the earlier treatment of the *trans* effect; but it is desirable to complete our account of these elements by a brief systematic study of their chemistry in terms of oxidation states. Because of the much greater stability of Pd(II) than Pd(IV) we shall reverse the sequence we have followed so far in this chapter, and start with the lowest oxidation states.

Palladium(0) and platinum(0)

The simplest compounds are the tetrahedral trifluorophosphine complexes $M(PF_3)_4$ and the carbonyls $M(CO)_4$, but none of these is very stable. The triphenylphosphine complex $Pt(PPh_3)_4$ very readily dissociates in organic solvents to form $Pt(PPh_3)_3$, in which the coordination of the platinum atom is very nearly planar. This compound undergoes a wide range of oxidative addition reactions, sometimes with further elimination of triphenylphosphine, e.g.

$$Pt(PPh_3)_3 + HCl \rightarrow Pt(H)(Cl)(PPh_3)_2 + PPh_3$$

The analogous palladium compounds are similar; being rather less stable with respect to dissociation, they are particularly important in homogeneous catalysis.

Palladium(II) and platinum(II)

The only unequivocally established high-spin octahedral complex of Pd(II) is violet PdF_2, which is made by reduction of 'PdF_3' by SeF_4, and which has the rutile structure. The action of chlorine on the metal gives one of two forms of $PdCl_2$, the only binary chloride of palladium. The α-form, obtained below 550 °C, has an infinite planar chain structure

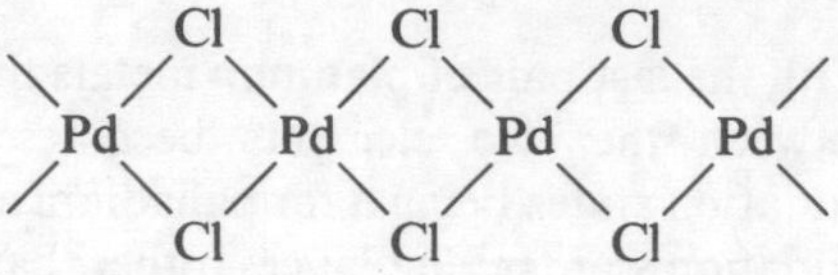

Cl Cl Cl
Pd Pd Pd Pd
Cl Cl Cl

The β-form, obtained above this temperature, is molecular Pd_6Cl_{12} and contains an octahedron of Pd atoms with a Cl atom off each edge of the Pd octahedron. So far as atomic positions are concerned this is essentially the same structure as that of the $[Nb_6Cl_{12}]^{2+}$ ion (Section 25.4), but the environment of the metal atoms in β-$PdCl_2$ is the same as that in the α-form and there is nothing to suggest Pd—Pd bonding. The red dichloride is readily soluble in water and combines with chloride ion to form the red-brown planar complex $[PdCl_4]^{2-}$, which reacts with bromide to give the more stable $[PdBr_4]^{2-}$ (typical behaviour for a class (*b*) metal ion), and with excess of ammonia to form the complex $[Pd(NH_3)_4]^{2+}$. Alkali precipitates PdO, which dissolves in perchloric acid to form $[Pd(H_2O)_4](ClO_4)_2$, the cation of which is diamagnctic and is therefore presumed to be planar. For the half-reaction

$$Pd^{2+}(aq) + 2e \rightleftharpoons Pd$$

$E^\circ = +0.98$ V (cf. the value of -0.25 V for the $Ni^{2+}(aq)/Ni$ system).

Platinum(II) compounds in general resemble those of palladium(II), but PtF_2 is not known. Platinum(II) chloride, obtained by decomposition of the tetrachloride at 350 °C, is dimorphic like palladium(II) chloride; with excess of chloride it forms $[PtCl_4]^{2-}$, which is also obtained by reduction of $[PtCl_6]^{2-}$ with hydroxylamine or oxalate. The $[Pt(H_2O)_4]^{2+}$ ion, obtained from $[PtCl_4]^{2-}$ and aqueous silver perchlorate, is kinetically much more inert than $[Pd(H_2O)_4]^{2+}$. Two very interesting compounds obtained from the tetrachloroplatinate(II) ion are Zeise's salt, $K[PtCl_3(C_2H_4)]$, which was discussed in detail in Section 23.3, and Magnus's green salt, $[Pt(NH_3)_4][PtCl_4]$. The latter, obtained by precipitation from $[Pt(NH_3)_4]Cl_2$ (which is colourless) and $[PtCl_4]^{2-}$

(which is pink), contains a chain of alternate cations and anions close enough (3.24 Å) for —Pt—Pt—Pt— interactions by d_{z^2} orbital overlap along the chain. The change from the colour of the constituent ions is believed to be associated with these interactions (cf. nickel dimethylglyoximate and the cyanoplatinates described below). In addition to $[Pt(NH_3)_4][PtCl_4]$ and *cis* and *trans* $[Pt(NH_3)_2Cl_2]$, polynuclear Pt^{II}-NH_3-Cl complexes such as

$$\left[\begin{array}{ccccc} H_3N & & Cl & & NH_3 \\ & Pt & & Pt & \\ H_3N & & Cl & & NH_3 \end{array}\right]Cl_2$$

are known. *Cis* $[Pt(NH_3)_2Cl_2]$, but not the *trans* compound, is used in chemotherapy for the treatment of cancer; it has the property of irreparably inhibiting cell division, probably by the way it binds to deoxyribonucleic acid (DNA).

Palladium and platinum both form very stable planar tetracyano complexes. Those of platinum sometimes contain only isolated $[Pt(CN)_4]^{2-}$ ions and cations (e.g. colourless $K_2[Pt(CN)_4].3H_2O$). In many, however, the $[Pt(CN)_4]^{2-}$ ions are stacked in staggered fashion (shown in Fig. 25.12) so as to give a chain of metal atoms like that in Magnus's green salt; these include yellow-green $Ba[Pt(CN)_4].4H_2O$ and violet $Sr[Pt(CN)_4].3H_2O$, in which the Pt—Pt distances are 3.32 and 3.09 Å respectively, the wave numbers of the absorptions corresponding to their colours decreasing with decrease in metal–metal distance. When potassium tetracyanoplatinate is partially oxidised with chlorine or bromine, bronze complexes of formula $K_2[Pt(CN)_4]X_{0.3}.2.5H_2O$ are obtained. These contain stacks of staggered $[Pt(CN)_4]^{2-}$ ions and isolated halide ions, the Pt—Pt distances now being only 2.88 Å (0.10 Å greater than in platinum metal); they are good one-dimensional metallic conductors (along the chains of metal atoms, of course) like polymeric sulphur nitride. The conductivity arises from the delocalisation along the chain of the extra positive charge (for 0.3 mol of X^- per mol of $K_2[Pt(CN)_4]$, 15% of the platinum must be in the tetrapositive state). Oxalate may replace cyanide as the ligand in these complexes, and other cations may replace potassium. Partial oxidation leading to cation deficiency (e.g. in $K_{1.75}[Pt(CN)_4].1.5H_2O$) also leads to metallic conductivity.

In discussing the *trans* effect in substitution in platinum(II) compounds, we pointed out that it is concerned only with rates of competing reactions, and suggested it would be better called the kinetic *trans* effect. The existence of a ground-state or thermodynamic *trans* effect is established by comparing the lengths and vibration frequencies of Pt—Cl bonds *trans* with respect to various ligands. In the series below such bond lengths are

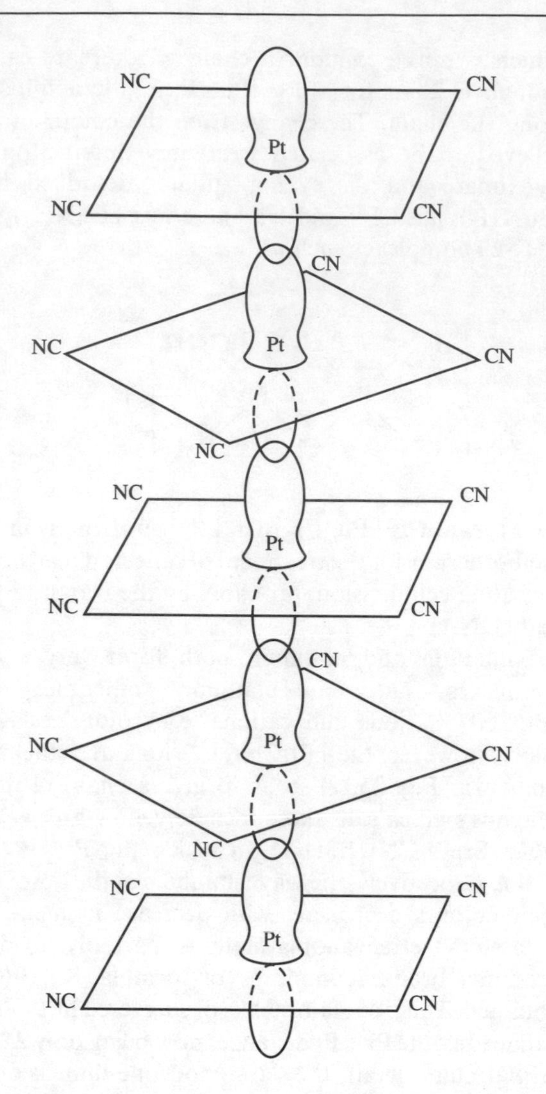

Figure 25.12
The stacking of $[Pt(CN)_4]^{2-}$ ions in coloured and conducting compounds containing the ion.

as shown:

$K_2[PtCl_4]$	$K[PtCl_3(C_2H_4)]$	*cis*$[(Me_3P)_2PtCl_2]$	Ph_2EtP, H, Pt, Cl, PPh_2Et
2.316 Å	2.327 Å	2.37 Å	2.42 Å

Further, for $K[PtCl_3(C_2H_4)]$, the Pt—Cl bond *cis* with respect to C_2H_4 has a length of 2.305 Å; the Pt—Cl stretching frequencies are 310 cm^{-1} for the 2.327 Å bond, and 331 cm^{-1} (mean of symmetric and asymmetric vibrations) for the 2.305 Å bond. Since kinetic *trans* effects of the ligands

Cl, C_2H_4, Me_3P and H are in the order $C_2H_4 > Me_3P \sim H > Cl$, it is clear that the ground-state and kinetic *trans* effects are different. This is, of course, perfectly reasonable, since substitutions in platinum(II) species are second-order reactions involving a five-coordinated metal ion in the transition state. Thus we attribute the observed *trans* effect of C_2H_4 (mainly kinetic) to withdrawal of metal $d\pi$ electron density into the π^* orbitals of the ligand, thus helping the simultaneous bonding of two negatively charged (or lone-pair containing) ligands. The *trans* effect of hydrogen, on the other hand (mainly ground-state), is the consequence of the weakening of the Pt—Cl bond as a result of electron repulsion by this atom. But clearly we have some way to go before a convincing treatment of both *trans* effects can be given.

Palladium (IV) and platinum (IV)

The only tetrahalide of palladium is PdF_4, made from the elements at 300 °C; the paramagnetic compound of empirical formula PdF_3 is actually $Pd^{II}[Pd^{IV}F_6]$, the cation having two unpaired electrons and the anion (known also in $K_2[PdF_6]$, obtained by fluorination of $K_2[PdCl_4]$) being diamagnetic. Chlorine oxidises $[PdCl_4]^{2-}$ to $[PdCl_6]^{2-}$, but the change is reversed by heating the hexachloropalladate(IV) in aqueous solution. All the platinum tetrahalides are known; $PtCl_4$ is the normal product of the chlorination of the metal, and the $[PtCl_6]^{2-}$ ion is stable in aqueous solution, from which salts of large cations $M_2[PtCl_6]$ (where $M = K^+$, NH_4^+, Rb^+ or Cs^+) may be precipitated. With alkali, $[PtCl_6]^{2-}$ yields an oxide PtO_2 which dissolves in excess to give the $[Pt(OH)_6]^{2-}$ ion. Nearly all compounds of palladium(IV) and platinum(IV) contain octahedrally coordinated metal ions. The greater kinetic inertness of platinum compounds is again noteworthy: whilst the isotructural $[PdF_6]^{2-}$ and $[PtF_6]^{2-}$ ions are both thermodynamically unstable to hydrolysis, $K_2[PdF_6]$ is decomposed by atmospheric moisture but $K_2[PtF_6]$ can be crystallised from boiling water.

Higher oxidation states

These are confined to fluorides. High-temperature fluorination of $PtCl_2$ gives tetrameric $(PtF_5)_4$, isostructural with Ru_4F_{20} etc., but by heating the metal in fluorine at 600 °C and rapidly quenching the vapour, unstable red PtF_6 is obtained. This is a very powerful oxidant, e.g.

$$O_2 + PtF_6 \rightarrow O_2^+[PtF_6]^-$$

It was the application of lattice energy and Born–Haber cycle considerations that led to the study of the interaction of xenon and PtF_6 and so opened up the chemistry of the noble gases described in Chapter 17. Palladium hexafluoride, obtained by the action of atomic fluorine on palladium powder, resembles the platinum compound.

25.11 *Silver and gold*

We saw in the last chapter that the chemistry of copper is mostly that of Cu(I) and Cu(II), with considerable emphasis on the position of the equilibrium $2Cu(I) \rightleftharpoons Cu(0)+Cu(II)$ under different conditions. Silver has only one stable oxidation state (Ag(I)), which shows no tendency to disproportionate; a few compounds of Ag(II) and a very few of Ag(III) and Ag(IV) or Ag(V) are known, but are powerful oxidising agents. Gold, on the other hand, is most stable as Au(III); Au(I) is unstable with respect to disproportionation into Au(0) and Au(III) under nearly all conditions, and Au(V) is known only in a pentafluoride and its complexes. Gold (−I) in the form of the solvated Au^- ion is produced when the metal dissolves in solutions of alkali metals in ammonia, and it is probably present in CsAu, an insulator in the solid state and a conductor when molten.

As we have emphasised earlier in this chapter, the discussion of the oxidation states of the heavy transition metals in terms of independently obtained physicochemical data is nearly always impossible owing to the absence of values for the ionisation energies and the rarity of simple ionic compounds and of simple aquo ions in solution. Data are much more plentiful for silver than for most other elements of the second transition series, and useful comparisons with copper are possible, but for gold only the first two ionisation energies have been determined (other values often quoted are estimates) and the Au^+ ion cannot be said to be present either in a solid compound or in aqueous solution. The first three ionisation energies (in kJ mol^{-1}) for copper and silver, and the first two for gold, are:

	Cu	Ag	Au
1st	745	730	890
2nd	1 960	2 070	1 970
3rd	3 580	3 360	—

Although the enthalpy of atomisation of silver (286 kJ mol^{-1}) is less than that of copper (339 kJ mol^{-1}), the greater ionic radius of the silver ion, taken in conjunction with these ionisation energies, results in silver being the more noble metal: E° is +0.80 V for Ag^+/Ag, +0.34 for Cu^{2+}/Cu and +0.52 V for Cu^+/Cu. Gold is more noble still. A comparison of factors determining E° values between silver and sodium was given in Section 7.11; another aspect of the solution chemistry of silver, the abrupt change in solubility in water between silver(I) fluoride and silver chloride, was discussed in Sections 6.8 and 7.3 and traced to a non-Coulombic contribution to the lattice energy of AgCl. Finally, the formation of ammine and halide complexes of silver was dealt with in Section 7.5, where it was shown that silver forms its most stable halide complex with iodide. To the last observation we may add that it has recently been shown that ΔH° for complexing of Ag^+ in dimethyl sulphoxide becomes more negative along the sequences Ph_3N, Ph_3P, Ph_3As; this is one of the few quantitative studies of complexing by phosphine and arsine ligands.

Silver and gold occur native and in sulphide and arsenide ores. Silver is usually worked up from the residues remaining after the isolation of copper, lead or nickel, but it may also be extracted, like gold, from all its ores by the action of sodium cyanide solution in the presence of air:

$$4M + 8CN^- + 2H_2O + O_2 \rightarrow 4[M(CN)_2]^- + 4OH^-$$

The complex cyanide thus formed is reduced to metal by zinc. Silver is not attacked by most non-oxidising acids, but it dissolves in nitric acid and is tarnished by sulphur or hydrogen sulphide. It liberates hydrogen from hot concentrated hydriodic acid owing to the formation of very stable iodo complexes. Gold is the most malleable and ductile of the metals; it is chemically unreactive, but is dissolved by hydrochloric acid in the presence of oxidising agents (owing to formation of chloro complexes) and by bromine trifluoride (with formation of $[BrF_2]^+[AuF_4]^-$). Silver and gold are both used in jewellery, and silver salts are extensively employed in photography; gold is also used in dentistry and in corrosion-resistant plating. Both metals form many useful alloys. Tertiary phosphine and thiol gold(I) complexes are used in the treatment of rheumatoid arthritis.

Although the enthalpies of atomisation of silver and gold are lower than those for earlier metals of the later transition series many metal cluster compounds of gold have been reported in recent years: most have been made by reduction of phosphine complexes of gold halides by borohydride; typical formulae are $[Au_8(PPh_3)_8](NO_3)_2$ and $[Au_{13}Cl_2(PMe_2Ph)_{10}](PF_6)_3$.

Only unstable carbonyls of silver and gold are known; they have been characterised by matrix isolation. Alkene complexes of silver(I) are like those of copper(I) and are important, but organo derivatives of silver(0) and gold(0) are unstable.

Many silver(I) salts are familiar laboratory reagents. They are nearly always anhydrous, and (except for $AgNO_3$, AgF and $AgClO_4$) are usually sparingly soluble in water. The fluoride, obtained by the action of hydrofluoric acid on the carbonate, is a valuable fluorine-exchange reagent; as we showed in Section 16.4, for a reaction such as

$$\gt C{-}Cl + MF \rightarrow \gt C{-}F + MCl$$

the effectiveness of an alkali metal fluoride as an exchange reagent increases with increase in cation size; in the case of silver(I) fluoride, the non-Coulombic contribution to the lattice energy of AgCl is an additional factor. Silver halides darken in light owing to photochemical decomposition. If the halogen produced is kept in close proximity to the finely divided metal which is also formed, the process may be reversed when the source of light is cut off, hence the use of silver chloride in photochromic spectacles. A photographic film consists of celluloid on which a layer of gelatine containing silver bromide (or iodide) in suspension has been deposited. Exposure of the film produces some submicroscopic particles of silver; on addition of an organic reducing agent, more silver bromide is reduced, the rate of reduction depending upon the intensity of illumination during the exposure period: thus parts of the film which were most strongly illuminated become darkest. It is this intensification of the latent image first formed that permits the use of short exposure times and causes silver halides to occupy their unique position in photography. Unchanged silver halide is washed out with aqueous thiosulphate, which forms the very stable complex $[Ag(S_2O_3)_2]^{3-}$, and the 'negative' is converted into a print by allowing light to pass through it on to silver bromide-containing photographic paper, which is then treated in a similar way to the negative.

Silver iodide shows a remarkably complex polymorphism. The stable form at ordinary temperature and pressure, γ-AgI, has the zinc blende structure; at high pressures this is converted into the δ-form having the sodium chloride structure, the Ag—I distance increasing from 2.81 to 3.04 Å. Between 136 and 146 °C the stable β-form has the wurtzite structure; the cooled metastable β-form is used in seeding supercooled rain-clouds (see Section 9.5 for the crystal structure of ice). Above 146 °C α-AgI becomes a fast ion electrical conductor, the conductivity at the transition temperature increasing by a factor of about 4000. In this form the iodide ions occupy positions in a caesium chloride structure, but the much smaller Ag^+ ions move freely between sites of two-, three- and four-fold coordination amongst the easily deformed iodide ions. The high temperature form of Ag_2HgI_4 shows similar behaviour.

Silver(I) oxide, precipitated by addition of alkali to silver(I) salts, is a brown solid which decomposes above 150 °C. Its aqueous suspension is alkaline, and it absorbs carbon dioxide from the atmosphere. Silver(I) oxide is, however, more soluble in aqueous alkali than in pure water, the anion $[Ag(OH)_2]^-$ being formed.

Gold(I) compounds are restricted to the chloride, bromide and iodide (all formed by controlled thermal decomposition of the gold(III) halides) and their complexes, those containing phosphorus as the donor atom (e.g. Me_3PAuCl) being the most stable. The binary halides all disproportionate on treatment with water.

Brown silver(II) fluoride is obtained by fluorinating the metal at 250 °C; it is instantly decomposed by water. It is paramagnetic, confirming the presence of a Ag^{2+} (d^9) ion, and although it is not isostructural with CuF_2 it does resemble the latter compound in having a structure based on tetragonally distorted AgF_6 octahedra. Silver(II) complexes such as $[Ag(py)_4]S_2O_8$ can be made in aqueous media by the action of powerful oxidising agents on silver(I) salts in the presence of the appropriate ligand; they are paramagnetic and usually contain square planar coordinated metal ions. The black solid of composition AgO which is precipitated when silver nitrate is warmed with persulphate solution is, however, diamagnetic, and has been shown by a neutron diffraction study to contain Ag(I) (with only two O atoms as nearest neighbours) and Ag(III) (with four O atoms as nearest neighbours). When AgO is dissolved in aqueous perchloric acid, however, the paramagnetic $[Ag(H_2O)_4]^{2+}$ ion is formed. Both AgO and Ag(II) complexes are very powerful oxidising agents; the former, for example, converts Mn(II) quantitatively to MnO_4^- in acidic solution, and $E°$ for Ag^{2+}/Ag^+ has been estimated as +2.0 V. There are a few other compounds containing silver(III), but we shall mention only the trifluoride, obtained by the action of BF_3 on $K[AgF_4]$ in HF, and the yellow complex $K[AgF_4]$ obtained by fluorination of a mixture of KCl and AgCl; this is diamagnetic and as a derivative of Ag^{3+} (d^8) presumably contains a planar $[AgF_4]^-$ anion. The orange compounds Cs_2AgF_6 and $Cs_2Ga_{0.5}Ag_{0.5}F_6$, also formed by high-temperature fluorination, are very weakly paramagnetic and diamagnetic respectively, and are therefore apparently an Ag(III) and Ag(V) mixed oxidation state complex and a derivative of Ag(V).

The action of chlorine on gold at 200 °C gives Au_2Cl_6, a diamagnetic red solid containing the planar dimeric unit

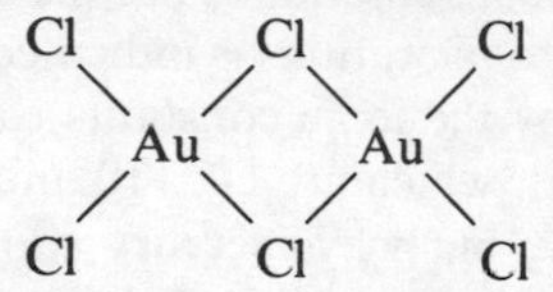

in which the Au—Cl (bridge) distances are 2.34 Å and the Au—Cl (terminal) distances 2.24 Å. In hydrochloric acid solution it forms planar $[AuCl_4]^-$, which reacts with Br^- to give $[AuBr_4]^-$ and with I^- to give AuI and I_2. A hydrated oxide $Au_2O_3.H_2O$ is precipitated from solutions of $Na[AuCl_4]$ by alkali; this reacts with excess of alkali to form the $[Au(OH)_4]^-$ ion. Partial decomposition of $CsAuCl_4$ yields diamagnetic black $CsAuCl_3$, shown by X-ray crystallography to be $Cs_2[AuCl_2][AuCl_4]$; the dark colour arises from charge-transfer between Au(I) and Au(III). Many σ bonded alkyl gold halides are obtained by interaction of gold(III) halides and lithium alkyls or Grignard reagents; gold trialkyls are unstable but form stable adducts with triphenylphosphine.

Yellow AuF_3 and red AuF_5, both of which are diamagnetic and very rapidly hydrolysed by water, may be made by the reaction sequences

$$Au \xrightarrow{BrF_3} [BrF_2][AuF_4] \xrightarrow[150\ °C]{heat} AuF_3$$

$$Au \xrightarrow[360\ °C,\ 8\ atm]{F_2,O_2} O_2[AuF_6] \xrightarrow[150\ °C]{heat} AuF_5$$

$$Au \xrightarrow[liq.\ HF]{KrF_2} KrF[AuF_6] \xrightarrow[60\ °C]{heat} AuF_5$$

Alkali metal salts of both complex fluoro anions are known. Gold(III) fluoride has a helical chain structure built up from planar AuF_4 units sharing *cis* corners; $[AuF_4]^-$ is planar; $[AuF_6]^-$ is octahedral.

25.12 Cadmium and mercury

Where definitions of transition metals are based on chemical characteristics (rather than, as in this book, on permitted combinations of quantum numbers), these metals, like zinc, are frequently considered along with the non-transition elements. Cadmium is in general very like zinc, such differences as there are being attributable to the larger sizes of the Cd atom and Cd^{2+} ion; mercury, though quite similar to cadmium, is in many respects also very like gold and thallium. In forming a stable M_2^{2+} ion, however, mercury is unique among the metals; there is evidence for both Zn_2^{2+} and Cd_2^{2+} in metal–metal halide melts, and the salt $Cd_2[AlCl_4]_2$ has been isolated recently (from molten cadmium, cadmium chloride and aluminium chloride), but neither Zn_2^{2+} nor Cd_2^{2+} can be obtained in aqueous solution. That the bond in Hg_2^{2+} is stronger than that in the other diatomic cations is indicated by the bond length (shorter than in Cd_2^{2+}) and by the force constants calculated from the Raman spectra of the M_2^{2+} ions, which are 60, 110 and 250 N m^{-1} for M = Zn, Cd and Hg respectively; but why mercury should form this ion so readily is a mystery, since it has the lowest enthalpy of atomisation of all the transition metals. It is, however, noteworthy that other polymercury cations have recently been obtained, e.g. linear Hg_3^{2+} and Hg_4^{2+} in fluoroarsenates prepared by the interaction of mercury and excess of arsenic pentafluoride in liquid sulphur dioxide.

Ionisation energies decrease regularly from zinc to cadmium, but then increase for mercury, values being given below (in kJ mol^{-1}):

	Zn	Cd	Hg
1st	906	868	1006
2nd	1733	1631	1809

There is an increase in ionic radius from Zn^{2+} (0.75 Å) to Cd^{2+} (0.95 Å); the corresponding radius for Hg^{2+} (based on the structure of HgF_2, one of the very few mercury compounds to have a typical ionic structure, in this case that of CaF_2) is 1.01 Å. The high ionisation energies of mercury may be compared with those of thallium and lead; it is sometimes said that the 6*s* inert pair effect stabilises Hg(0), Tl(I) and Pb(II), but the limitations of this kind of comment have been discussed earlier in connection with the chemistry of thallium (Section 12.1). Whatever their origin, it is clear that the abrupt increases in the first two ionisation energies far outweigh the small change in enthalpy of atomisation, and make mercury a noble metal. Values of E° for the M^{2+}/M systems are -0.76, -0.40 and $+0.85$ V respectively.

Like zinc, neither cadmium nor mercury forms a carbonyl or organo derivative of the metal in zero oxidation state; stable cyano complexes of the M^{2+} ions are obtained, however.

Since much of the chemistry of mercury is different from that of cadmium we shall now briefly consider the two elements separately; that it is profitable to do so may be taken as a sign that at the ends of the second and third transition series the consequences of the lanthanide contraction are at last becoming of minor significance.

Cadmium

Cadmium is isolated almost entirely from zinc ores; being more volatile than zinc, it concentrates in the first stage of the distillation of the metal. The distillate, a mixture of the metals and their oxides, is dissolved in acid, and cadmium is precipitated from solution by zinc owing to its less negative M^{2+}/M standard electrode potential. Its chief uses are in the manufactures of solders and low-melting alloys, but owing to its very large cross-section for the capture of thermal neutrons it is also used in control rods for nuclear reactors.

Cadmium is a reactive metal and dissolves in non-oxidising acids. Unlike zinc, however, it does not dissolve in aqueous alkali; $Cd(OH)_2$ is not amphoteric. Its only oxide, formed by heating the metal in air, is CdO which, unlike ZnO, has the sodium chloride structure; its non-stoichiometry was mentioned in Section 6.10. Cadmium fluoride, which is sparingly soluble in water, has the fluorite structure (cf. ZnF_2, rutile structure). The other halides (which are all white) are ready soluble, but their solutions contain not only Cd^{2+} and halide ions but a wide range of halo complexes: in 0.5 molar $CdBr_2$, for example, the principal species present are $CdBr^+$, $CdBr_2$ and Br^-, together with smaller quantities of Cd^{2+}, $[CdBr_3]^-$ and $[CdBr_4]^{2-}$. In contrast to zinc, the stability of halo complexes of cadmium increases from the fluoride to the iodide, i.e. Cd^{2+} is a class (*b*) cation. In the solid state $CdCl_2$, $CdBr_2$ and CdI_2 all have layer lattices whose structures and lattice energies have been discussed in Sections 6.3 and 6.7. Yellow CdS is a useful pigment and phosphor. Cadmium bromide reacts with Grignard reagents and with lithium aryls to form organo cadmium compounds which resemble the corresponding derivatives of zinc but are less reactive.

The aquo ion $[Cd(H_2O)_6]^{2+}$ is quite acidic, and dilute solutions of cadmium salts contain most of the metal in the form of solvated $CdOH^+$ or polymers thereof; in concentrated solutions Cd_2OH^{3+} is present. Complexes $[Cd(NH_3)_4]^{2+}$ and $[Cd(CN)_4]^{2-}$ (together, of course, with species containing lower ligand:metal ratios) are formed with ammonia and cyanide. Cadmium ion is capable of replacing zinc in metalloenzymes of the latter element, but since it frequently interferes with their activities it is a dangerous poison.

Mercury

The only important source of mercury is the ore *cinnabar*, HgS. When this is roasted in air at 600 °C the metal distils out:

$$HgS + O_2 \rightarrow Hg + SO_2$$

It is used in mercury cathode electrolytic cells (though to a decreasing extent), mercury vapour lamps, and a wide range of scientific apparatus; mercury compounds are employed as fungicides and in pharmacy. Soluble mercury compounds (especially methyl mercury derivatives) are, however, very poisonous. They affect both the digestive organs and the brain; the phrase 'as mad as a hatter' is believed to have originated in the use of mercury(II) nitrate in the manufacture of felt hats.

Mercury is the only metal which is a liquid at ordinary temperatures; it freezes at −39 °C and boils at 357 °C forming a monatomic vapour. The very low melting-point and low enthalpies of fusion and atomisation are only partly accounted for by the unique structure of solid mercury, a kind of distorted close packing in which each atom has six nearest neighbours at 2.99 Å and six more at 3.47 Å. Mercury has an appreciable vapour pressure at ordinary temperatures (1.3×10^{-3} mm at 20 °C), and it should always be kept in closed containers and handled in adequately ventilated places. It dissolves many metals, forming alloys known as *amalgams*; in the Na–Hg system, for example, Na_3Hg_2, NaHg and $NaHg_2$ have been characterised.

Mercury is much less reactive than cadmium, and is attacked only by oxidising acids: dilute nitric acid gives mercury(I) nitrate, $Hg_2(NO_3)_2$, whilst the concentrated acid gives mercury(II) nitrate, $Hg(NO_3)_2$. With hot concentrated sulphuric acid it forms $HgSO_4$ and SO_2. It reacts with fluorine, chlorine and bromine to give mercury(II) halides; with iodine it forms Hg_2I_2 or HgI_2 according to the relative proportions of the reactants. Mercury combines with oxygen at about 300 °C, forming the red oxide HgO, but at slightly higher temperatures the reaction is reversed – hence the formation of the metal when HgS is heated in air.

Mercury(II) compounds

Mercury(II) fluoride has the fluorite structure, but unlike most saline fluorides containing the metal in a relatively low oxidation state it is completely hydrolysed by water. The chloride and bromide are both volatile solids soluble not only in water (in which they are unionised) but also in ethanol and diethyl ether; they have molecular lattices. Mercury(II) iodide is only sparingly soluble in water; it exists as a red macromolecular form below 127 °C and a yellow molecular form above that temperature. All HgX_2 molecules are linear. Complex chlorides, bromides and iodides are formed in aqueous solution, the tetrahedral anion $[HgI_4]^{2-}$ being particularly stable. A solution of $K_2[HgI_4]$ (*Nessler's reagent*) gives a characteristic brown compound of formula $Hg_2N^+ I^-$ on treatment with ammonia. The cation in this compound is an infinite network having a structure essentially like that of cristobalite (Section 6.3); each nitrogen atom is tetrahedrally, and each mercury atom linearly, coordinated. The hydroxide $Hg_2N(OH)$ can be obtained by the action of aqueous ammonia on mercury(II) oxide:

$$2HgO + NH_3 \rightarrow Hg_2N(OH) + H_2O$$

Mercury(II) chloride combines with gaseous ammonia to form $[Hg(NH_3)_2]Cl_2$, which contains discrete linear cations; but with aqueous ammonia it forms $[Hg(NH_2)]Cl$, which contains a polymeric $—Hg—{}^+NH_2—Hg—{}^+NH_2—Hg—$ chain bent, as expected, at each nitrogen atom.

Both red and yellow forms of HgO contain infinite chain structures in which the coordination of the mercury atom is linear. Although the oxide dissolves in acids, it is only weakly basic; in aqueous solution mercury(II) salts that are ionised (e.g. the nitrate and the sulphate) are hydrolysed to a considerable extent, and many basic salts are formed. Among these is $HgO.2HgCl_2$, which is really $[O(HgCl)_3]^+Cl^-$, i.e. a substituted oxonium salt.

Mercury(II) alkyls and aryls have often been mentioned as reagents for the preparation of other organometallic compounds. They are usually made by the interaction of mercury(II) chloride and Grignard reagents, and are toxic volatile liquids or solids which, unlike the analogous zinc and cadmium compounds, do not react with air or water. This lack of reactivity is often ascribed to the weakness of the Hg—O bond that would be formed in the reaction, but it is likely that kinetic as well as thermodynamic factors are involved. Thus although the mean M—C bond dissociation energy in $M(CH_3)_2$ decreases along the series M = Zn, Cd and Hg (the values are 180, 144 and 126 kJ mol^{-1}), the first M—C bond dissociation energy is greatest for M = Hg (the values being 197, 192 and 215 kJ mol^{-1} respectively). The catalysis of the hydration of alkynes by mercury(II) salts is believed to involve ions of the type

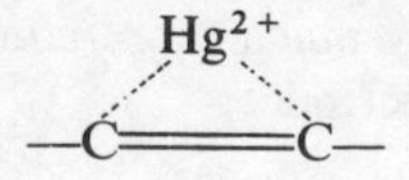

Complexes of alkenes and arenes are also known.

Mercury(I) compounds

The general method for the preparation of mercury(I) compounds is by the action of metallic mercury on mercury(II) compounds; e.g. in the preparation of Hg_2Cl_2 (calomel), $HgCl_2$ (made by sublimation from mercury(II) sulphate and sodium chloride) is heated with mercury and the sparingly soluble product is freed from $HgCl_2$ by washing with hot water.

From the measured values of E° for the half-reactions

$$Hg_2^{2+} + 2e \rightleftharpoons 2Hg \qquad E^\circ = +0.79 \text{ V}$$

$$2Hg^{2+} + 2e \rightleftharpoons Hg_2^{2+} \qquad E^\circ = +0.92 \text{ V}$$

and

$$Hg^{2+} + 2e \rightleftharpoons Hg \qquad E^\circ = +0.85 \text{ V}$$

it can be seen that Hg_2^{2+} in aqueous solution is only just stable with respect to the disproportionation

$$Hg_2^{2+} \rightleftharpoons Hg^{2+} + Hg$$

for which $K = 6 \times 10^{-3}$ at 25 °C. Reagents which form insoluble Hg^{2+} salts or stable Hg^{2+} complexes upset the equilibrium and decompose mercury(I) salts; hydroxide, sulphide and cyanide ions, for example, do so with formation of mercury and HgO, HgS and $[Hg(CN)_4]^{2-}$ respectively, and Hg_2O, Hg_2S and $Hg_2(CN)_2$ are not known. Since Hg^{2+} forms more stable complexes than the larger Hg_2^{2+}, it is not surprising that very few Hg(I) complexes are known, and that the only soluble mercury(I) salts are those containing poorly complexing anions, e.g. NO_3^- and ClO_4^-.

The binuclear nature of the mercury(I) ion in solid salts is firmly established by its diamagnetism and by X-ray diffraction: all the halides of known structure are linear molecules XHgHgX with Hg—Hg approximately 2.50 Å. In solution, the appearance of an extra line in the spectra of mercury(I) salts compared with those of the corresponding sodium salts shows that the same Hg_2^{2+} ion is present. The calomel electrode, very widely used as a standard in electrochemistry, consists of a pool of mercury (into which a platinum wire dips) in contact with a saturated solution of Hg_2Cl_2 in aqueous KCl of known concentration; the potential of the molar calomel electrode, for which the half-reaction is

$$Hg_2Cl_2 + 2e \rightleftharpoons 2Hg + 2Cl^-$$
(in M KCl)

is +0.280 V at 25 °C. Since this electrode involves no source of highly purified gas, it is very much more convenient for general use than the standard hydrogen electrode.

PROBLEMS

1 Comment on each of the following observations:
(a) The relative density of HfO_2 (9.68) is much greater than that of ZrO_2 (5.73).
(b) $K_3[Cr_2Cl_9]$ is strongly paramagnetic; $K_3[W_2Cl_9]$ is diamagnetic.
(c) The $ReCl_4$ units in the diamagnetic anion $[Re_2Cl_8]^{2-}$ are in the eclipsed configuration relative to one another but the $OsCl_4$ units in the diamagnetic anion $[Os_2Cl_8]^{2-}$ are in the staggered configuration.

(d) The Ru—N bond lengths in $[Ru(NH_3)_6]^{3+}$ and $[Ru(NH_3)_6]^{2+}$ are 2.104 and 2.144 Å respectively, and electron transfer between these ions is very fast.
(e) Unlike $[Pt(NH_3)_4][PtCl_4]$, $[Pt(C_2H_5NH_2)_4][PtCl_4]$ has an electronic absorption spectrum which is the sum of those of the constituent ions.
(f) Silver iodide is readily soluble in saturated aqueous silver nitrate solution, but silver chloride is not.
(g) When mercury(II) perchlorate is shaken with liquid mercury, the ratio [Hg(I)]/[Hg(II)] in the resulting solution is independent of the value of [Hg(II)].

2 When rhodium tribromide in the presence of diphenylmethylarsine is reduced with hypophosphorous acid, a monomeric compound **A**, containing two atoms of bromine and three molecules of the arsine per atom of rhodium, is obtained. **A** is a non-electrolyte and has a band in the infrared at 2073 cm^{-1}; the corresponding band if the complex is made using D_3PO_2 as reductant in a deuterated solvent is at 1483 cm^{-1}. Spectrophotometric tritration of **A** with bromine shows that one molecule of **A** reacts with one molecule of bromine, and by the action of excess of mineral acid on the product rhodium tribromide is regenerated.

What can you infer concerning the nature of the products formed in these changes?

3 When platinum(II) cyanide reacts with aqueous ammonia a compound of formula $Pt(CN)_2(NH_3)_2$ is obtained. Suggest three structures for this compound and indicate how you would attempt to determine which was the correct structure without recourse to X-ray diffraction methods.

When $K_2[Pt(CN)_4]$ reacts with chlorine water the product, which has the formula $K_2[Pt(CN)_4Cl_2]$, has two Raman-active CN stretching vibrations and one infrared-active CN stretching vibration which is different from the Raman-active vibrations. What does this indicate about the structure of the $[Pt(CN)_4Cl_2]^{2-}$ ion?

The reaction

$$[Pt(CN)_4Cl_2]^{2-} + Br^- \rightarrow [Pt(CN)_4ClBr]^{2-} + Cl^-$$

is catalysed by $[Pt(CN)_4]^{2-}$. Suggest a mechanism for the reaction.

4 When potassium hexachlorosmate(IV) is heated with ammonia under pressure a compound, **I**, of composition $Os_2N_9H_{24}Cl_5$ can be separated. Treatment of a solution of this compound with hydrogen iodide precipitates a compound in which three of the five chlorines of compound **I** have been replaced by iodine. One millimole of compound **I** treated with potassium hydroxide releases nine millimoles of ammonia.

The compound **I** is diamagnetic and none of the stronger absorption frequencies in the infrared spectrum appears in the Raman spectrum. Suggest a structure for **I** and account for its diamagnetism.

REFERENCES FOR FURTHER READING

Cotton, F.A. and Walton, R.A. (1982) *Multiple Bonds Between Metal Atoms*, John Wiley, New York. The definitive account of this topic.

Cotton, F.A. and Wilkinson, G. (1988) *Advanced Inorganic Chemistry*, 5th edition, Interscience Publishers, New York. Contains the definitive account in one volume of the chemistry of the metals of the second and third transition series.

Cotton, S.A. and Hart, F.A. (1975) *The Heavy Transition Elements*, Macmillan, London. Another account of the same subject with fuller coverage of individual elements than this book.

Greenwood, N.N. and Earnshaw, A. (1984) *Chemistry of the Elements*, Pergamon, Oxford. Treats metals of all three transition series together (e.g. Ti, Zr and Hf in one chapter) and is thus very useful for comparisons of the first series metals with those of later series.

Parish, R.V. (1977) *The Metallic Elements*, Longman, London. A more restricted treatment, but particularly good on oxides, halides and aqueous solution chemistry.

Porterfield, W.W. (1984) *Inorganic Chemistry*, Addison-Wesley, Reading, Massachusetts. Chapter 14 contains a very good and original account of the photochemistry of transition metal compounds, including photography.

26 *Inner transition elements: the lanthanides*

Introduction • Occurrence and separation • The metals • The tripositive oxidation state • The dipositive oxidation state • The tetrapositive oxidation state

26.1 Introduction

In this and the following chapter we complete our systematic treatment of inorganic chemistry by the discussion of the two series of inner transition elements, defined in Section 3.2 as the fourteen elements from cerium to lutetium inclusive (the lanthanides, i.e. the elements after lanthanum) and the fourteen elements from thorium to lawrencium inclusive (the actinides, i.e. the elements after actinium). Yttrium and lanthanum, which we discussed briefly in Chapter 25, closely resemble the lanthanides and occur with them, and the set of sixteen elements formed by yttrium, lanthanum and the lanthanides are often known as the *rare earth* elements. The lanthanides resemble one another much more closely than do the members of the series of ordinary transition elements discussed in the last two chapters; they have only one really stable oxidation state, and the study of their chemistry provides a unique opportunity to examine the effects of small changes in size and nuclear charge along a series of otherwise similar elements. The chemistry of the actinides is, in contrast, much more complicated, and early members of that series show a wide range of oxidation states; it is partly for this reason, and partly because their radioactivity introduces special problems into the study of their properties, that we devote a separate chapter to these elements.

Names, symbols, gaseous electronic configurations of atomic and some ionic states, and atomic and ionic radii, of lanthanum and the lanthanides (for which the general symbol Ln is often used) are assembled in Table

Table 26.1
Electronic configurations and radii of lanthanum and the lanthanides.

Atomic number	Name	Symbol	Electronic configuration*				Radii/Å	
			Ln	Ln^{2+}	Ln^{3+}	Ln^{4+}	Ln	Ln^{3+}
57	Lanthanum	La	$5d^1 6s^2$	$5d^1$	$4f^0$		1.87	1.061
58	Cerium	Ce	$4f^1 5d^1 6s^2$	$4f^2$	$4f^1$	$4f^0$	1.83	1.034
59	Praseodymium	Pr	$4f^3 6s^2$	$4f^3$	$4f^2$	$4f^1$	1.82	1.013
60	Neodymium	Nd	$4f^4 6s^2$	$4f^4$	$4f^3$	$4f^2$	1.81	0.995
61	Promethium	Pm	$4f^5 6s^2$	$4f^5$	$4f^4$		1.81	0.979
62	Samarium	Sm	$4f^6 6s^2$	$4f^6$	$4f^5$		1.80	0.964
63	Europium	Eu	$4f^7 6s^2$	$4f^7$	$4f^6$		1.99	0.950
64	Gadolinium	Gd	$4f^7 5d^1 6s^2$	$4f^7 5d^1$	$4f^7$		1.80	0.938
65	Terbium	Tb	$4f^9 6s^2$	$4f^9$	$4f^8$	$4f^7$	1.78	0.923
66	Dysprosium	Dy	$4f^{10} 6s^2$	$4f^{10}$	$4f^9$	$4f^8$	1.77	0.908
67	Holmium	Ho	$4f^{11} 6s^2$	$4f^{11}$	$4f^{10}$		1.76	0.894
68	Erbium	Er	$4f^{12} 6s^2$	$4f^{12}$	$4f^{11}$		1.75	0.881
69	Thulium	Tm	$4f^{13} 6s^2$	$4f^{13}$	$4f^{12}$		1.74	0.869
70	Ytterbium	Yb	$4f^{14} 6s^2$	$4f^{14}$	$4f^{13}$		1.94	0.858
71	Lutetium	Lu	$4f^{14} 5d^1 6s^2$	$4f^{14} 5d^1$	$4f^{14}$		1.73	0.848

*Only electrons outside the [Xe] core are indicated.

26.1. It will be noticed that, as for an ordinary series of transition elements, the electronic configurations of the ground-state atoms vary somewhat irregularly with increase in atomic number; but those of all of the tripositive ions (corresponding to what is the most stable oxidation state of all the lanthanides) are of the form $4f^n$. In other respects, however, the lanthanides differ markedly from ordinary transition elements. The splitting of the energies of the *f* orbitals in electrostatic ligand fields is very small (Δ_0 for octahedral fields, for example, is usually about 1 kJ mol^{-1}), so that crystal field stabilisation energy considerations are of little importance in lanthanide chemistry; high and variable coordination numbers are common; and only relatively unstable complexes are formed with π acceptor ligands such as carbon monoxide, cyanide ion and most organometallic groups.

The overall decrease in atomic and ionic radii from lanthanum to lutetium (the *lanthanide contraction*) has, as we have already seen, far-reaching consequences in the chemistry of the third transition series elements. This contraction is, of course, similar to that observed in an ordinary transition series, and is attributed to the same cause, the imperfect shielding of one electron by another in the same sub-shell. However, the shielding of one 4*f* electron by another is less than for one *d* electron by another, and as the nuclear charge increases along the series there is a fairly regular decrease in the size of the entire $4f^n$ sub-shell. The importance of this in methods for the separation of the lanthanides will be discussed in the following section. The ionic radii given in Table 26.1, it should be said, are for six-coordination, like the radii we have given for other ions in Section 6.4. In practice, simple compounds of the lanthanides often crystallise with quite complicated structures (e.g. in LaF_3 each La^{3+} has $7F^-$ at 2.42–2.49 Å, $2F^-$ at 2.64 Å and $2F^-$ at 3.00 Å), and observed interatomic distances increase with increase in coordination number; it is, therefore, prudent to take the values in Table 26.1 as certainly correct in a relative sense but by no means absolute ones.

The ground states of most lanthanide atoms are $[Xe]4f^n6s^2$; for a given number of electrons outside the xenon core it seems that the $4f$ orbitals increase in stability relative to the $6s$ and $5d$ orbitals as the positive charge on the species increases. In practice, of course, what ionic oxidation states are attainable depends upon the interplay of enthalpies of atomisation, ionisation energies and lattice energies in the way described in Section 16.4. In the chemistry of the lanthanides, as elsewhere, we find that higher oxidation states occur in oxides and fluorides and lower oxidation states in other halides, particularly iodides. In reviewing the oxidation states of the lanthanides we are for the moment excluding species which exhibit metallic conductivity (e.g. LaI_2, which is best formulated $La^{3+}(e^-)(I^-)_2$, and GdI, which is isostructural with ZrI) and confining attention to genuine salts which are insulators in the solid state.

When we discussed the ionisation of the transition metals (Section 20.4), it was pointed out that, so long as we are considering only d^n species, the ionisation energy increases along a series until we reach the element for which the ionisation process is $d^5 \rightarrow d^4$, then drops for $d^6 \rightarrow d^5$, and then rises again; we traced this pattern of behaviour to the fact that there is no loss of quantum mechanical exchange energy on going from d^6 to d^5, i.e. on losing an electron to form a half-filled shell. The same factor should influence the loss of an electron leading to the formation of d^0 and d^{10} ions, and it will have been noticed that Sc(III) is the only oxidation state of scandium unequivocally established as occurring under chemical conditions in mononuclear species, and that Cu(I) is the only stable unipositive cation among elements of the first transition series. Thus exchange energy considerations appear to impart a certain degree of stability (though not to the extent of overriding all other factors) to an empty, a half-filled and a full $3d$ shell. An analogous effect is found for the $4f$ shell. If we accept that the stable oxidation state of the lanthanides is the tripositive one, we can thus give an explanation for the fact that Ce^{4+} ($4f^0$) and Tb^{4+} ($4f^7$) are somewhat more stable with respect to reduction than other Ln^{4+} cations, and Eu^{2+} ($4f^7$) and Yb^{2+} ($4f^{14}$) are somewhat more stable with respect to oxidation than other Ln^{2+}-cations. However, Pr^{4+} ($4f^1$), Nd^{4+} ($4f^2$) and Dy^{4+} ($4f^8$) have also been obtained in complex fluorides, and Nd^{2+} ($4f^4$), Pm^{2+} ($4f^5$), Sm^{2+} ($4f^6$), Dy^{2+} ($4f10$) and Tm^{2+} ($4f^{13}$) in solid saline dihalides. A detailed treatment of the variation of the third ionisation energy with electronic configuration (which we shall not give here) indicates, in fact, that the discussion we have given above is an oversimplification, and that there is some evidence for extra stability being associated also with a three-quarters filled shell.

26.2 Occurrence and separation

All the lanthanides except promethium occur naturally. The most stable isotope of promethium, ^{147}Pm, is obtained as a product of the fission of heavy nuclei, and is a β^--emitter of half-life 2.6 years; most of what is known about the chemistry of the element has been worked out on milligram quantities obtained in this way and separated by ion-exchange (see below). Since yttrium (radius of Y^{3+} for six-coordination, 0.89 Å) occurs with lanthanum and the lanthanides, its separation is just like that of the lanthanides of similar ionic radii.

An important source of most of the lanthanides (which are quite widely distributed) is *monazite*, a mixed phosphate containing thorium, lanthanum, yttrium and the lanthanides. This may be decomposed by concentrated sulphuric acid; dilution followed by addition of ammonia to adjust the solution to $pH = 1$ results in the selective precipitation of thorium dioxide, which is much less soluble than the trihydroxides of the other elements. There are then two quite different types of separation possible, those based on chemical differences (a very limited group) and those based on fractionation processes (which are generally applicable).

Cerium may be separated chemically by taking advantage of the fact that it is the only lanthanide to form a tetrapositive ion stable in aqueous solution; hypochlorite oxidation converts it into Ce(IV), which is separated by carefully controlled precipitation of the dioxide, or of the tetraiodate, the other metals remaining in solution. Europium, the only one of the lanthanides to form a dipositive ion which does not decompose water rapidly, is reduced from Eu(III) to Eu(II) by zinc amalgam; the solubilities of Eu(II) salts are generally like those of salts of the alkaline earth cations (the ionic radius of Eu^{2+}, 1.10 Å, is about the same as that of Sr^{2+}, 1.16 Å), and in the dipositive oxidation state europium is easily removed by precipitation as $EuSO_4$. (Europium commonly occurs with alkaline earth metals, in fact, notably in *strontianite*, $SrCO_3$.) For the other elements, however, fractionation procedures are essential.

All these fractionation procedures depend upon the variations in two properties with change in ionic size being slightly different. The first properties utilised in this connection were solubilities in aqueous solution, which depend, as we showed in Section 7.3, on differences between lattice energies and hydration free energies. If we consider a series of isomorphous salts of general formula LnX_3, the lattice energy, which is inversely proportional to $[r(Ln^{3+})+r(X^-)]$, increases as $r(Ln^{3+})$ decreases along the series; the free energies of hydration also increase, but in a way inversely related to $r(Ln^{3+})$ and $r(X^-)$ taken separately, since in dilute solution the interactions between ions may be neglected. Thus solubilities vary regularly with decrease in cation size. The variation, however, is never very large, and many thousands of fractional crystallisations of, e.g., the bromates or double nitrates with magnesium nitrate (two of the more successful methods) are necessary to achieve a high degree of separation. Another property once utilised was fractional thermal decomposition of the nitrates, which was based on the difference in lattice energy between the nitrate and the oxide formed (obviously greatest when $r(Ln^{3+})$ is smallest). Both of these methods, however, in addition to giving only a small degree of separation, are discontinuous processes; they have now been replaced by ion-exchange and by solvent extraction, partly because these methods give better fractionation, and partly because they are suitable for continuous operation.

Typical ion-exchange resins used for this purpose are sulphonated polystyrenes such as Dowex-50, which may be denoted HR, an insoluble strong acid. When a solution containing Ln^{3+} cations is poured on to a column of the resin, the cations are absorbed, the equilibrium

$$Ln^{3+}(aq)+3HR(s) \rightleftharpoons LnR_3(s)+3H^+(aq)$$

being set up. The equilibrium distribution coefficient between the resin and the aqueous solution ($[M^{3+}(resin)]/[M^{3+}(aq)]$) is large for all the ions, but is nearly constant. The formation constants of the $EDTA^{4-}$ complexes of the Ln^{3+} ions, however, increase regularly from $10^{15.3}$ for La^{3+} to $10^{19.2}$ for Lu^{3+}, so that if a column on which all the Ln^{3+} ions have been absorbed is eluted with a dilute solution of H_4EDTA adjusted to pH 8 with ammonia, Lu^{3+} is preferentially complexed by a small margin, then Yb^{3+}, and so on. By the use of a long column, which in effect is equivalent to several thousand manual fractionations between the column and the eluting solution, 99.9 per cent pure components of the original mixture may be isolated, the progress of the separation being conveniently followed by atomic fluorescence spectroscopy. In the solvent extraction process, a dilute aqueous nitric acid solution of the lanthanides is extracted by means of a counter-flowing solution of tri-n-butyl phosphate in kerosene, and the cations of elements of higher atomic number are extracted first; a high degree of separation is again achieved.

26.3 The metals

From the solutions of salts obtained in the ion-exchange separation, the lanthanides may be precipitated as fluorides or hydroxides; the latter yield oxides on thermal decomposition, and these may be converted into chlorides by the action of ammonium chloride or carbon tetrachloride at moderate temperatures. The metals themselves may be isolated by reduction of the chlorides with sodium (used mainly for the lighter metals) or of the fluorides with magnesium or calcium, preferably in tantalum containers. The three metals that form more stable dipositive oxidation states (Sm, Eu and Yb) cannot be obtained in either of these ways, only dihalides being produced; they are prepared by electrolysis of $LnCl_3$–NaCl melts. Mixtures of all the metals for use in special steels or cigarette lighters are also made electrolytically. Neodymium is used also in magnetic alloys; individual oxides are used as phosphors, e.g. in television screens.

Lanthanum and all the lanthanides except europium crystallise in one or both of the close-packed structures; europium has the body-centred cubic structure, and the radius given for it in Table 26.1 may be adjusted to 2.04 Å for twelve-coordination. Much more important than this minor difference in structures is the fact that the atomic radii of europium and ytterbium (which have well-defined lower oxidation states) are much larger than those of the other elements, from which we may infer that in the metals they contribute fewer electrons to the metal–metal bonding than the other elements. This leads to lower enthalpies of atomisation for europium and ytterbium (177 and 152 kJ mol^{-1} respectively) than for the other lanthanides (230–430 kJ mol^{-1}). Samarium, however, which also has a low enthalpy of atomisation (206 kJ mol^{-1}) and a well-defined lower oxidation state, shows no anomaly in atomic radius. Europium and ytterbium, but not samarium, form blue solutions in liquid ammonia containing Ln^{2+} ions and solvated electrons (cf. Section 8.2).

All of the lanthanides are rather soft white metals; the heavier ones, owing to the rapid formation of an oxide coating, are kinetically less reactive than the lighter ones. Values for E° for the half-reaction

$$Ln^{3+} + 3e \rightleftharpoons Ln$$

are in the range -2.2 to -2.4 V except for Ln = Eu, for which the value is -2.0 V. This is, of course, a very small variation, and it shows that the considerable variations in enthalpy of atomisation, ionisation energies and hydration energy almost cancel. As expected, all the metals liberate hydrogen from dilute acids or steam, and burn in air, forming the oxides Ln_2O_3 except in the case of cerium, which forms CeO_2.

The metals combine with hydrogen when gently heated in the gas, forming a range of compounds between metallic (i.e. conducting) hydrides LnH_2 and essentially saline hydrides LnH_3. These compounds are potentially useful in hydrogen storage and transportation, and are now being studied intensively. Carbides Ln_3C, Ln_2C_3 and LnC_2 are formed when the metals are heated with carbon; nearly all of these exhibit metallic conductivity, and their electronic structures present interesting problems. In CeC_2, for example, the carbon–carbon bond length in the diatomic anion is 1.28 Å, compared with 1.19 Å in CaC_2; a reasonable formulation is then $Ce^{3+}C_2^{3-}$, with some overlap between the occupied antibonding orbital of the C_2^{3-} ion and the conduction band of the solid. With fluorine, cerium and terbium give tetrafluorides, the other metals trifluorides (though complex fluorides of Pr(IV), Nd(IV) and Dy(IV) can be obtained); with the other halogens, only trihalides are formed.

Metallic lanthanides do not react with carbon monoxide under the usual conditions (an observation which may be attributed to the fact that most of them in their ground states have no *d* electrons available for π bonding). However, an unstable carbonyl of neodymium, $Nd(CO)_6$, has been isolated by the matrix isolation technique, and a nitrogen complex $[(C_5Me_5)_2Sm]_2N_2$, containing a N_2 unit bonded sideways to two metal atoms, has recently been reported.

At the level of this book we should not be justified in describing large numbers of compounds of the individual lanthanide elements, and to complete our survey of these elements we shall therefore deal generally with the oxidation states of all the elements taken together, beginning in the next section with the most stable of them, the tripositive state. We may note in passing that all complexes of these metals are kinetically labile; no case of isomerism has been found.

26.4 The tripositive oxidation state

We have already discussed the ionic radii of the tripositive lanthanide cations, and in describing the separation of the elements we have indicated how the stabilities of aquo and $EDTA^{4-}$ complexes increase with increase in atomic number and concomitant decrease in (crystallographic) ionic size. The generalisation that the later lanthanides form more stable complexes may be further illustrated by reference to the hydroxides. Lanthanum hydroxide, though only sparingly soluble, is a strong base and absorbs atmospheric carbon dioxide; base strength and solubility, however, decrease as we go along the series, and $Yb(OH)_3$ and $Lu(OH)_3$ will dissolve in hot concentrated sodium hydroxide solution to form hydroxo complexes $[Yb(OH)_6]^{3-}$ and $[Lu(OH)_6]^{3-}$. Most oxides Ln_2O_3 are obtained by thermal decomposition of oxo-acid salts, but Ce, Pr and Tb give higher oxides by this method, and partial reduction of these by hydrogen is needed to obtain the sesquioxides. Mixed oxide systems involving alkaline earths and copper are now of great interest as superconductors (see Section 24.10). Fluorides, carbonates, phosphates and oxalates are sparingly soluble, but the other halides, nitrates and sulphates dissolve readily in water. Compounds such as $KCeF_4$ and $NaNdF_4$, made by fusion of alkali metal and lanthanide fluorides, are double salts and do not contain complex anions; one form of $KCeF_4$, for example, has a disordered fluorite structure.

The crystal chemistry of the lanthanides is very complicated, and we shall not attempt to give details of structures here. It is, nevertheless noteworthy that the coordination numbers of the cations decrease, as expected, with decreasing ionic radius. Thus the chlorides $CeCl_3$–$GdCl_3$ inclusive (and also $LaCl_3$) all contain nine-coordinated metal ions whose environment is like that of Re in $[ReH_9]^{2-}$ (Section 18.3), but the chlorides $TbCl_3$–$LuCl_3$ inclusive have the $AlCl_3$ layer structure in which each metal ion is octahedrally coordinated. Coordination numbers of Ln^{3+} also decrease along the series of oxides Ln_2O_3. Hydrated Dy^{3+} and Yb^{3+} in solution are eight-coordinated; coordination numbers of larger cations are probably nine.

The interpretation of the spectra of $4f^n$ ions is based on the principles outlined in Sections 3.3 and 19.6, but there are certain important differences between lanthanide and ordinary transition metal ions. For the lanthanides, spin–orbit coupling is more important than crystal field splitting, and states differing only in J values are sufficiently different in energy to be separated in the electronic spectrum. Further, since $l = 3$ for a f electron, m_l may be 3, 2, 1, 0, −1, −2 or −3, giving rise to high values of L for some f^n ions. For the configuration f^2, for example, application of Hund's rules gives the ground state (having $L = 5$, $S = 1$) as $^3H_4$. Since S, P, D, F and G states are also possible, many of them with different possible values of J, the number of theoretically possible transitions is large, even after the limitations imposed by selection rules are taken into account. Thus the spectra of lanthanide ions often contain large numbers of peaks; because the $4f$ electrons are well shielded from the surroundings of the ion, however, bands arising from f-f transitions are sharp rather than broad like d–d absorptions, and their positions are little affected by complexing. Transitions $4f \rightarrow 5d$, on the other hand, give rise to broader bands whose positions depend more on the environment.

Typical colours of lanthanide tripositive ions in aqueous solution are shown in Table 26.2; it will be seen that f^n and f^{14-n} species usually, though not invariably, have similar colours. Small quantities of various lanthanide salts are used in phosphors for television tubes and lasers because of the sharpness of their electronic spectral emissions.

The bulk magnetic moments of the lanthanide ions at the ordinary temperature are shown in Table 26.2. In general, they agree well with those calculated from formulae based on the assumption of Russell–Saunders coupling and large spin–orbit coupling constants (Section 19.6), as a consequence of which only the states of lowest J value are populated. This is not true for Eu^{3+}, however, and not quite true for Sm^{3+}. For the former, the spin–orbit coupling constant λ is about 300 cm^{-1}, only slightly greater than kT (200 cm^{-1} approximately); the ground state of the f^6 ion is $^7F_0$ (which is diamagnetic, since $J = 0$), but the states $^7F_1$ and $^7F_2$ are also populated to some extent and give rise to the observed magnetic moment. At low temperatures, as expected, the moment of the Eu^{3+} ion approaches zero. The form of the variation of magnetic moment with n in Table 26.2 arises from the operation of

Table 26.2
The colours and observed and calculated magnetic moments of La^{3} and Ln^{3+} ions (colours relate to aquo complexes).

Ion	Colour	Electronic configuration	Magnetic moment/μ_B	
			Calculated	Observed
La^{3+}	colourless	$4f^0$	0	0
Ce^{3+}	colourless	$4f^1$	2.54	2.3–2.5
Pr^{3+}	green	$4f^2$	3.58	3.4–3.6
Nd^{3+}	lilac	$4f^3$	3.62	3.5–3.6
Pm^{3+}	yellow	$4f^4$	2.68	2.7
Sm^{3+}	yellow	$4f^5$	0.84	1.5–1.6
Eu^{3+}	pale pink	$4f^6$	0	3.4–3.6
Gd^{3+}	colourless	$4f^7$	7.94	7.8–8.0
Tb^{3+}	pale pink	$4f^8$	9.72	9.4–9.6
Dy^{3+}	yellow	$4f^9$	10.63	10.4–10.5
Ho^{3+}	yellow	$4f^{10}$	10.60	10.3–10.5
Er^{3+}	rose-red	$4f^{11}$	9.57	9.4–9.6
Tm^{3+}	pale green	$4f^{12}$	7.63	7.1–7.4
Yb^{3+}	colourless	$4f^{13}$	4.50	4.4–4.9
Lu^{3+}	colourless	$4f^{14}$	0	0

Hund's third rule (Section 3.3): $J = L-S$ for a shell less than half full but $J = L+S$ for a shell more than half full; J and g_J for ground states are accordingly both larger in the second half of the series.

It is sometimes said that the lanthanide cations do not form many complexes, but since this statement refers to aqueous conditions it really only implies that water is a better ligand than other complexing agents for these cations (it is also, of course, present in much larger quantity). Thus amine complexes, which may be made in non-aqueous media, are usually decomposed by water. By far the most stable complexes of the lanthanides are those which involve polydentate chelating ligands in which oxygen is the donor atom (or, in the case of $EDTA^{4-}$, the principal donor atom). In terms of the class (*a*)–class (*b*) distinction, all the lanthanides are certainly class (*a*) species.

The molecular tris complexes of β-diketonate ions, $[Ln(RCOCHCOR)_3]$, have attracted considerable attention. Those of fluorinated diketones are often volatile, and may be used for the separation of the lanthanides by vapour-phase chromatography. Solutions of β-diketonate complexes, particularly of those of Eu^{3+} and Pr^{3+}, in organic solvents are important shift reagents in nuclear magnetic resonance spectroscopy: the effect of the magnetic field of the complex, when the latter is in contact with an organic molecule, often results in much better resolution of the components of the spectrum than is otherwise obtained.

Among the reported organic compounds of the lanthanides are the thermally stable, but air- and water-sensitive, cyclopentadienyl compounds $Ln(C_5H_5)_3$ and $(C_5H_5)Ln(CH_3)_3$. Cyano complexes are not formed in aqueous solution, but this does not exclude the possibility that they may be obtained in non-aqueous media, though none has in fact yet been characterised.

26.5 *The dipositive oxidation state*

We discussed the occurrence of the dipositive oxidation state among the lanthanides in the opening section of this chapter; we now go on to mention some aspects of the chemistry of compounds of metals in this oxidation state.

One route to compounds of dipositive lanthanides is the decomposition of the trihalides or their reduction with hydrogen or metals. Lattice energy considerations indicate that at a given temperature the reaction

$$LnX_3 \rightarrow LnX_2 + \tfrac{1}{2}X_2$$

is most likely to occur when X = I and least likely to occur when X = F. In accordance with such expectations it is observed that for Ln = Eu, thermal decompositions of the trichloride, tribromide and triiodide all give dihalides, but for Ln = Sm or Yb, only the triiodide decomposes, the trichloride and tribromide requiring reduction with hydrogen at a high temperature. Trifluorides of all three elements require the action of calcium metal at high temperatures to produce the difluorides. Low concentrations of other difluorides in solid solution in calcium fluoride may be obtained by the action of calcium vapour on solid solutions of the trifluorides in the same medium; it is by the study of such solid solutions that we obtain much of our information about Ln^{2+} species.

Another method for the preparation of lanthanide dihalides is by the solid-state reaction

$$2LnX_3 + Ln \rightarrow 3LnX_2$$

at about 500 °C – the reverse of the normal method of disproportionation of such compounds. This gives diiodides of La, Ce, Pr, Nd, Pm, Gd and Tm.

Structures are known only for a few dihalides, e.g. SmF_2 and EuF_2 are isomorphous with CaF_2, TmI_2 and YbI_2 with CdI_2. The diiodides of Sm, Eu, Tm and Yb are saline, those of La, Ce, Pr, and Gd metallic; NdI_2 exists in both forms, the insulator → metal transition being brought about by pressure.

In the series of sulphides LnS obtained by direct combination, only the Sm, Eu and Yb compounds are saline. All the monosulphides have the NaCl structure, and the Ln—S distances in the saline compounds are somewhat greater than in their neighbours in the series.

Samarium, europium and ytterbium can be reduced to the dipositive state in aqueous solution, though samarium and ytterbium require the use of alkali metal amalgams or electrolytic reduction for this purpose. The Eu(III)/Eu(II) standard potential, −0.35 V, is nearly the same as that for Cr(III)/Cr(II), and colourless Eu(II) solutions can be used for chemical studies if air is excluded; the estimated values for the Sm(III)/Sm(II) and Yb(III)/Yb(II) standard potentials, −1.5 and −1.1 V respectively, indicate that Sm(II) and Yb(II) are highly unstable with respect to oxidation even by water.

26.6 The tetrapositive oxidation state

Cerium is the only one of the lanthanides to be stable in the tetrapositive oxidation state in aqueous solution. In molar perchloric acid, in which complex formation is likely to be minimal (though not necessarily negligible), E° for the half-reaction

$$Ce^{4+} + e \rightleftharpoons Ce^{3+}$$

is +1.70 V, and yellow Ce(IV) can be obtained only by oxidation of Ce(III) with powerful oxidants such as peroxodisulphate. Cerium(IV) is thermodynamically unstable with respect to reduction by water, but the process is very slow at ordinary temperatures, and cerium(IV) sulphate (a slightly less powerful oxidant than the perchlorate owing to sulphato complex formation) is widely used as a titrimetric reagent in place of permanganate or dichromate; as a one-electron oxidant its reaction pathways are less complex than those of the two more familiar reagents. The only binary cerium(IV) compounds known are CeF_4 (obtained by heating CeF_3 in fluorine) and CeO_2 (obtained by heating $Ce(OH)_3$ in air or by oxidation of aqueous Ce(III) with hypochlorite). There are, however, several stable complexes containing Ce(IV), among them $(NH_4)_2[Ce(NO_3)_6]$, in which the nitrate ion acts as a bidentate ligand, the overall coordination of the Ce(IV) ion being icosahedral.

Praseodymium and terbium can also be oxidised in concentrated aqueous KOH—K_2CO_3, either by ozone or electrolytically, yellow Pr(IV) and red-brown Tb(IV) oxides being formed. Praseodymium forms PrO_2 when Pr_2O_3 is heated in oxygen at 500 °C and 100 atm pressure. The tetrafluoride has been obtained by the action of KrF_2 on the oxide. It decomposes at 90 °C (CeF_4 is stable up to 800 °C). Terbium forms TbF_4 and TbO_2, both obtained by oxidation of the Tb(III) compounds. All attempts to make other tetrafluorides have been unsuccessful, but the complexes $Cs_3[NdF_7]$ and $Cs_3[DyF_7]$ have been made by fluorination of a mixture of caesium chloride and the appropriate lanthanide trichloride.

The fluorides CeF_4, PrF_4 and TbF_4 (like ZrF_4) have three-dimensional structures in which the coordination of the metal atom is approximately square antiprismatic. All three compounds, together with the complex fluorides of Pr(IV), Nd(IV) and Dy(IV), are immediately hydrolysed by water. The dioxides CeO_2, PrO_2 and TbO_2 have the fluorite structure. Intermediate oxides Ln_7O_{12}, where Ln = Ce, Pr or Tb are obtained in the course of the preparations of the dioxides; these have defect lattices in which some O^{2-} vacancies in the LnO_2 fluorite lattices are compensated by the presence of Ln^{3+} ions. Many other intermediate phases are also known.

PROBLEMS

1 Use Hund's rules to derive the ground state of the Ce^{3+} ion, and calculate its magnetic moment (the spin–orbit coupling constant for Ce^{3+} is 1000 cm^{-1}, so the population of states other than the ground state may be neglected at ordinary temperature).

2 Show that the stability of a lanthanide dihalide LnX_2 with respect to disproportionation into $LnX_3 + Ln$ is greatest for X = I.

How would you attempt to show that a given compound of formula LnI_2 is a saline diiodide rather than a metallic diiodide?

3 Comment on each of the following observations:

(a) ΔH° for the formation of the $[Ln(EDTA)]^-$ complexes in aqueous solution is nearly constant and almost zero.

(b) the value of the Ce(IV)/Ce(III) standard potential, measured in molar acid, decreases along the series of acids $HClO_4$, HNO_3, H_2SO_4, HCl.

(c) $BaCeO_3$ has the perovskite structure.

REFERENCES FOR FURTHER READING

Greenwood, N.N. and Earnshaw, A. (1984) *The Chemical Elements*. Pergamon Press, Oxford. Contains a fuller account of the lanthanides than that given here.

Johnson, D.A. (1977) *Adv. Inorg. Chem. Radiochem.*, **20**, 1. The definitive review of the oxidation states of the lanthanides and their interpretation in terms of electronic configuration.

Johnson, D.A. (1982) *Some Thermodynamic Aspects of Inorganic Chemistry*, 2nd edition, Cambridge University Press. Chapter 6 contains a discussion of the thermodynamic properties of the lanthanides. Note that this author defines La—Yb inclusive as lanthanides and regards Lu as the homologue of Sc and Y; there is support for this view on chemical grounds, but it has not been generally accepted and for that reason has not been adopted in this book.

27 *Inner transition elements: the actinides*

Introduction • Isolation and general chemistry of actinium and the actinides • Actinium • Thorium • Protactinium • Uranium • Neptunium, plutonium and americium • Curium and later elements

27.1 Introduction

The existence of a second series of inner transition elements is now established beyond doubt, but in the first half of it there is a very much wider variation in chemical properties than found among the lanthanides. Thus the separation of $^{239}_{94}Pu$ from uranium, mentioned in Section 1.8, is easily achieved by conventional chemical methods, Pu(VI) being much more easily reduced to Pu(IV) than U(VI) to U(IV). In the second half of the actinide series, however, the tripositive oxidation state is the most stable one for nearly all the elements, and similarities to the lanthanides are much more apparent. The names, symbols and some properties of actinium and later elements are given in Table 27.1. Names of elements beyond lawrencium are controversial and atomic numbers (or latin equivalents) are often used instead.

All of the actinide and later elements (together with polonium, astatine, radon, francium, radium and actinium) are unstable with respect to radioactive disintegration, though the half-lives of the most abundant isotopes of thorium and uranium (^{232}Th and ^{238}U, $t_{\frac{1}{2}} = 1.4 \times 10^{10}$ and 4.5×10^{9} years respectively) are so long that for many purposes their radioactivity may be neglected. For isotopes of short half-lives, however, their radioactivity dominates the study of their chemistry. In the case of the later actinides, half-lives of only a few minutes (or even less) limit the concentrations available so that only very rapid tracer work is possible. For all the elements except thorium and uranium, radiation hazards

Table 27.1
Some properties of actinium and later elements (only those of atomic numbers 90–103 inclusive are actinides).

Atomic number	Name	Symbol	Probable electronic configuration*			Radii/Å	
			M	M^{3+}	M^{4+}	M^{3+}	M^{4+}
89	Actinium	Ac	$6d^17s^2$	$5f^0$		1.11	
90	Thorium	Th	$6d^27s^2$	$5f^1$	$5f^0$		0.99
91	Protactinium	Pa	$5f^26d^17s^2$	$5f^2$	$5f^1$		0.96
92	Uranium	U	$5f^36d^17s^2$	$5f^3$	$5f^2$	1.03	0.93
93	Neptunium	Np	$5f^46d^17s^2$	$5f^4$	$5f^3$	1.01	0.92
94	Plutonium	Pu	$5f^67s^2$	$5f^5$	$5f^4$	1.00	0.90
95	Americium	Am	$5f^77s^2$	$5f^6$	$5f^5$	0.99	0.89
96	Curium	Cm	$5f^76d^17s^2$	$5f^7$	$5f^6$	0.99	0.88
97	Berkelium	Bk	$5f^97s^2$	$5f^8$	$5f^7$	0.98	0.87
98	Californium	Cf	$5f^{10}7s^2$	$5f^9$	$5f^8$	0.98	0.86
99	Einsteinium	Es	$5f^{11}7s^2$	$5f^{10}$	$5f^9$		
100	Fermium	Fm	$5f^{12}7s^2$	$5f^{11}$	$5f^{10}$		
101	Mendelevium	Md	$5f^{13}7s^2$	$5f^{12}$	$5f^{11}$		
102	Nobelium	No	$5f^{14}7s^2$	$5f^{13}$	$5f^{12}$		
103	Lawrencium	Lr	$5f^{14}6d^17s^2$	$5f^{14}$	$5f^{13}$		
104	Rutherfordium	Rf	$5f^{14}6d^27s^2$				
105	Hahnium	Ha	$5f^{14}6d^37s^2$				

*Only electrons outside the [Rn] core are indicated.

necessitate elaborate safety precautions. Self-heating of solid compounds, even if it does not lead to thermal instability, may make structure determinations difficult, and the action of high-energy radiation on solvents (e.g. in the production of hydrogen atoms and hydroxyl radicals by their action on water) complicates the interpretation of experiments carried out in solution. At very low concentrations it is difficult to distinguish between adsorption and solid solution formation, and tracer studies involving precipitation are sometimes open to misinterpretation. All these factors make the study of the heaviest elements one of extreme technical difficulty.

The fission of heavy nuclei, and the production of nuclear energy, have been dealt with in Chapter 1, and we shall not return to these subjects here. We should, however, mention the uses of some of the actinides as light-weight power sources. Isotopes which are only α-emitters require relatively little shielding, since α-particles are easily stopped by surrounding atoms and their kinetic energy converted into heat; such heat can be used to generate electricity by means of thermopiles, and hence used in isolated situations such as satellites and heart 'pace-makers'. Among isotopes used for these purposes are ^{238}Pu ($t_{\frac{1}{2}} = 86$ years), ^{241}Am (458 years) and ^{242}Cm (162 days); the power obtained is, of course, inversely dependent upon the half-life of the isotope.

In Table 27.1 we have included elements numbers 104 and 105. Rutherfordium (number 104) has been shown to form a chloride similar in properties to hafnium tetrachloride, though less volatile, and in ion-exchange experiments to behave quite differently from the actinides. Hahnium forms a chloride and a bromide similar to niobium and tantalum pentahalides in volatility. These and later elements are therefore believed to be part of an ordinary transition series.

27.2 *Isolation and general chemistry of actinium and the actinides*

Because of the wide variation in chemical properties among members of the first half of the actinide series (about which very much more is known than about the later members), it is not satisfactory to attempt to summarise actinide chemistry in terms of oxidation states as we did for the lanthanides. In this section we shall therefore give an overall account of the properties of the actinides, but we shall then supplement this by discussing the chemistry of individual elements or of a few elements taken together.

Uranium and thorium are isolated from natural sources as described later. Actinium and protactinium are both formed, as ^{227}Ac and ^{231}Pa respectively, in the decay of ^{235}U, and may be isolated from *pitchblende* (the oxide U_3O_8); ^{227}Ac, however, is now made in milligram amounts by a (n, γ) reaction on ^{226}Ra, followed by β^--decay of the resulting ^{227}Ra. The preparation of ^{239}Np and ^{239}Pu from ^{238}U has been described in Section 1.4; lengthy irradiation of ^{239}Pu in a pile leads to the successive formation of small quantities of ^{240}Pu, ^{241}Pu, ^{242}Pu and ^{243}Pu; the last of these is a β^--emitter of half-life five hours, and decays to ^{243}Am ($t_{\frac{1}{2}}$ = 7 650 years). Further irradiation of this gives ^{244}Am, which decays to ^{244}Cm ($t_{\frac{1}{2}}$ = 18 years). Both ^{243}Am and ^{244}Cm are available on a hundred gram scale. From these, milligram amounts of ^{249}Bk, ^{252}Cf, ^{253}Es and ^{254}Es, and microgram amounts of ^{257}Fm, are also obtained by multiple neutron capture and β^--decay. Neutron capture by ^{257}Fm, however, gives an isotope (^{258}Fm) which undergoes fission within a few seconds; and for the preparation of later elements other methods must be used.

The most important of these is heavy ion bombardment in a cyclotron or similar particle-accelerating machine. The ions derived by stripping off all the electrons from ^4He, ^{11}B and ^{16}O, for example, take part in the following reactions:

$$^{253}_{99}Es + ^{4}_{2}He \rightarrow ^{256}_{101}Md + ^{1}_{0}n$$

$$^{241}_{94}Pu + ^{16}_{8}O \rightarrow ^{253}_{102}No + 4^{1}_{0}n$$

$$^{252}_{98}Cf + ^{11}_{5}B \rightarrow ^{257}_{103}Lr + 6^{1}_{0}n$$

Only minute amounts of isotopes (a few atoms in some cases) can be obtained by this method, partly because of the inaccessibility of some of the target materials and partly because of the low probability of the desired nuclear reaction. Fortunately, however, the decay properties of the products can be predicted, and by the use of ion-exchange resins, employing aqueous ammonium 2-hydroxybutyrate in particular as eluant, separations are possible, the element of highest atomic number present being eluted first; chemical studies, however, are restricted to rapid tracer experiments. Identification of the heaviest elements (now up to atomic number 109) rests entirely on decay properties.

Actinium and the actinide metals are all isolated by reduction of the anhydrous saline fluorides by lithium, magnesium or calcium at 1100–1400 °C. All are highly reactive, especially when finely divided. The action of boiling water on them, for example, gives a mixture of oxide and hydride, and combination with most non-metallic elements takes place at moderate temperatures. Hydrochloric acid attacks all the metals, but most are only slightly affected by nitric acid, probably owing to the formation of protective oxide layers; alkalis have no action. The early metals in the series crystallise in a bewildering variety of structures, many of them peculiar to one or two elements; plutonium, for instance, exists in six different forms. From americium onwards, the structures, like those of the lanthanides, are based on hexagonal close packing.

The probable ground states of the gaseous metal atoms are shown in Table 27.1. At the beginning of the series the difference in energy between the 5*f* and 6*d* orbitals is less than between the 4*f* and 5*d* orbitals for the lanthanides; later, however, the 5*f* orbitals become more stable. For tripositive and tetrapositive cations the ground-state electronic configurations are all of the general form $5f^n$, and decreases in ionic radii parallel to those observed for the lanthanide cations are found. There are many series of isomorphous compounds in actinide chemistry, e.g. dioxides, trifluorides and trichlorides.

The known oxidation states of actinium and the actinides are shown in Table 27.2. We may infer from the existence of nearly all the actinides in at least two oxidation states that successive ionisation energies probably differ by less than they do for the lanthanides. For the higher oxidation states, however, covalent bonding must certainly be involved. This may occur either because the 5*f* orbitals extend further from the nucleus than the 4*f* orbitals and are themselves involved in covalent bonding, or because the energy differences between the 5*f*, 6*d*, 7*s* and 7*p* orbitals are

Table 27.2
Oxidation states of actinium and the actinides (the most stable ones are in bold type).

Ac	Th	Pa	U	Np	Pu	Am	Cm	Bk	Cf	Es	Fm	Md	No	Lr
						2			2	2	2	2	**2**	
3			3	3	3	**3**	**3**	**3**	**3**	**3**	**3**	**3**	3	**3**
	4	4	**4**	4	**4**	4	4	4	4					
		5	5	**5**	5	5								
			6	6	6	6								
				7	7									

often within the range of chemical binding energies and hence appropriate valence states for covalently bonded species are easily attained. It is, indeed, likely that the electronic configuration for a given oxidation state sometimes depends on the environment. A simple proof that the 5*f* orbitals have greater spatial extension than the 4*f* orbitals is provided by the fine structure of the electron spin resonance spectrum of UF_3 in a CaF_2 lattice, which arises from the interaction of the electron spin of the U^{3+} ion and the F^- ions; NdF_3, the corresponding lanthanide species, shows no such effect.

The spectroscopic and magnetic properties of the actinides are very complicated, and we shall mention them only briefly in this book. The $5f \rightarrow 5f$ absorptions are weak, but they are somewhat broader and more intense, and considerably more dependent upon ligands present, than $4f \rightarrow 4f$ absorptions (i.e. there are significant crystal field stabilisation terms). The interpretation of the spectra is made difficult by the high spin–orbit coupling constants (about twice those for the lanthanides) as a result of which the Russell–Saunders coupling scheme partially breaks down. Magnetic properties show an overall similarity to those of the lanthanides in the variation of magnetic moment with the number of unpaired electrons, but values for isoelectronic species (e.g. Np(VI) and Ce(III), Np(V) and Pr(III), Np(IV) and Nd(III)) are lower for the later series, indicating partial quenching of the orbital contribution by crystal field effects.

Many stable halo, nitrato, oxalato and carbonato complexes of the actinides, in addition to complexes involving multidentate chelating ligands, may be isolated. Stable cyano and carbonyl complexes are not formed (at least under normal conditions), but many organometallic derivatives, mostly involving polyhapto cyclic ligands such as $C_5H_5^-$ and $C_8H_8^{2-}$, are known. Probably because of the availability of several kinds of metal orbital in these complexes, $(C_5Me_5)_2UCl_2$ catalyses the polymerisation of ethylene and $(C_5Me_5)_2UH_2$ activates H/D exchange in arenes. In general, therefore, we may say that the actinides show some similarities to ordinary transition elements, but their chemistry is on the whole more like that of the lanthanides, and on both physical and chemical grounds the existence of a second series of inner transition elements is firmly established.

27.3 Actinium

In all of its chemistry actinium very closely resembles lanthanum, differing from it only by being slightly more basic, as would be expected from its rather greater ionic radius (1.11 Å compared with 1.06 Å). The quantitative study of actinium compounds is complicated by radiation problems: ^{227}Ac, the most stable isotope, is an α-emitter of half-life 22 years, but its decay products, which rapidly build up, are very powerful γ-emitters, and lead to the difficulties mentioned earlier.

27.4 Thorium

The separation of thorium from monazite was mentioned in Section 26.2. The impure oxide obtained is dissolved in hydrochloric acid, and the thorium extracted by tributyl phosphate. The colourless insoluble oxide ThO_2 (CaF_2 structure), which is precipitated in neutral or even weakly acidic solution, is converted into anhydrous halides by the action of hydrogen fluoride or carbon tetrachloride at 600 °C. The fluoride is insoluble in water or alkali metal fluoride solutions, but a large number of double or complex fluorides may be made by combination of their constituents. Their structures provide an illustration of the complexity of the crystal chemistry of the actinides: $BaThF_6$ has the LaF_3 structure; $(NH_4)_3ThF_7$ and $(NH_4)_4ThF_8$ both contain infinite $[ThF_7]_n^{3n-}$ anion chains built up from ThF_9 tricapped trigonal prisms each sharing two edges, the octafluoro complex also containing an isolated F^- ion; the dodecahedral $[ThF_8]^{4-}$ anion occurs, together with extra F^- ions, in $(NH_4)_5ThF_9$. Thorium chloride, in contrast to the fluoride, is readily soluble in water and forms an octahedral complex anion $[ThCl_6]^{2-}$. Aqueous solutions of thorium salts, as expected from the high formal charge on the cation, contain hydrolysis products such as polymers of $ThOH^{3+}$, $Th(OH)_2^{2+}$, etc. Complexes such as $Th(acac)_4$ (square antiprismatic) and $[Th(NO_3)_6]^{2-}$ (shown to be icosahedral in the magnesium salt) are readily formed in aqueous media. The acetylacetonate derivative, a non-electrolyte, is volatile and soluble in organic solvents.

Thorium metal, obtained by reduction of the fluoride with calcium, or by thermal decomposition of the tetraiodide, is (like the other actinide metals) attacked by hot water, and it tarnishes in air. For the system Th^{4+}/Th, E° is -1.9 V. There is no evidence for an oxidation state other than Th(IV) in aqueous media, but by heating the metal with thorium tetraiodide ThI_2 and ThI_3 (both polymorphic) are obtained. The former is also obtained by electrochemical oxidation of thorium into an acetonitrile solution of iodine, and is a good electrical conductor presumed to be $Th^{4+}(I^-)_2(e^-)_2$; the latter is probably also a metallic halide.

27.5 Protactinium

The isotope ^{231}Pa, isolated from the decay products of ^{235}U in pitchblende, is the form of protactinium on which most work has been done. It is a sufficiently powerful α-emitter ($t_{\frac{1}{2}} = 3.2 \times 10^4$ years) to call for shielding against radiation, but the greatest difficulty associated with the study of its chemistry is the tendency of Pa(V), the stable oxidation state, to hydrolyse in solution and to be adsorbed on precipitates or even on the walls of the containing vessel. The extraction from pitchblende is a lengthy process which we shall not describe here; it ends with the precipitation of a hydrated peroxide which is decomposed to give the most important compound of protactinium, the colourless oxide Pa_2O_5, which is readily soluble in acids only in the presence of ions such as fluoride, oxalate and citrate with which it forms complexes.

The action of hydrogen fluoride on Pa_2O_5 gives a hydrated fluoride from which the water cannot be removed without decomposition, and the colourless anhydrous pentafluoride has to be made by the reaction sequence:

$$Pa_2O_5 \xrightarrow[400\,°C]{HF,H_2} PaF_4 \xrightarrow[800\,°C]{F_2} PaF_5$$

The tetrafluoride is reduced to the metal by lithium at 1300 °C. Complexes such as $KPaF_6$, Na_2PaF_7 and Na_3PaF_8 can be isolated from aqueous media; the first two of these species contain infinite chain anions consisting of PaF_8 dodecahedra each sharing two edges and PaF_9 tricapped trigonal prisms each sharing two edges, respectively; Na_3PaF_8, however, contains a cubic $[PaF_8]^{3-}$ anion. Other halides of protactinium (all coloured) include the chlorides $PaCl_5$ and $PaCl_4$ and the iodides PaI_5, PaI_4 and PaI_3; the last of these compounds is the only known Pa(III) derivative. Preparations of these substances are:

$$Pa_2O_5 \xrightarrow{SOCl_2} PaCl_5 \xrightarrow[400\,°C]{H_2} PaCl_4$$

$$Pa_2O_5 \xrightarrow[350\,°C]{AlI_3} PaI_5 \xrightarrow[400\,°C]{Al} PaI_4 \xrightarrow[350\,°C]{heat} PaI_3$$

In aqueous solution Pa(V) is reduced to Pa(IV) by zinc, $E°$ for the Pa(V)/Pa(IV) couple being about -0.1 V; solutions of Pa(IV) are rapidly oxidised by air, and have been little studied, though it has been shown that the spectrum of aqueous Pa(IV) is very like that of the isoelectronic (f^1) Ce(III) ion.

27.6 Uranium

Naturally occurring uranium consists mainly of ^{238}U (99.27 per cent) and ^{235}U (0.72 per cent), the separation of which was discussed in Section 1.7. The most important source of uranium is *pitchblende*, the oxide U_3O_8. This is dissolved by either sulphuric acid or a carbonate/bicarbonate solution above 100 °C, compressed air being blown in to act as an oxidising agent. The solution so obtained contains a sulphato or carbonato complex of the uranyl (UO_2^{2+}) cation; among several methods for the isolation of a pure uranium compound are separations involving anion-exchange resins and solvent extraction of $UO_2(NO_3)_2 . 6H_2O$ from aqueous nitric acid into diethyl ether or methyl isobutyl ketone. All separations end with the precipitation of the peroxide $UO_2(O_2) . 2H_2O$ or of the hydroxide $UO_2(OH)_2.H_2O$ and subsequent thermal decomposition to yellow UO_3 and then black U_3O_8 (a mixed U(VI) and U(IV) oxide).

The oxide UO_3 is soluble in most acids, forming solutions which contain the yellow linear $[O{=}U{=}O]^{2+}$ ion or its complexes. This ion is also found in many solid compounds, among them a series of saline oxohalides UO_2X_2, the alkaline earth 'uranates' such as $BaUO_4$ (which are therefore better described as mixed oxides), and many salts of oxo-acids, e.g. $UO_2(NO_3)_2.6H_2O$. The aqueous chemistry of the ion is discussed later.

Reduction of U_3O_8 with hydrogen at 600 °C yields another black oxide, UO_2, which has the fluorite structure. This is the key compound for the preparation of many other uranium compounds, and this aspect of the chemistry of uranium is summarised thus:

$$UO_2 \xrightarrow[500\,°C]{CCl_4} UCl_4 \xrightarrow[500\,°C]{Cl_2} U_2Cl_{10}$$

$$UO_2 \xrightarrow[550\,°C]{HF} UF_4 \xrightarrow[300\,°C]{F_2} UF_6$$

Uranium hexafluoride is a colourless volatile solid which has a vapour press of 1 atm at 56 °C and is of great importance in the separation of the uranium isotopes. It is immediately hydrolysed by water and acts as a vigorous fluorinating agent; it combines with alkali metal fluorides to give a series of ill-defined complexes. The sparingly soluble green tetrafluoride, on the other hand, is an inert solid of m.p. 1036 °C; it has a three-dimensional essentially ionic structure. Other fluorides include UF_3 (made by heating UF_4 with aluminium at 900 °C) and UF_5 (made by controlled reduction of UF_6 or from UF_6 and UF_4). An unstable hexachloride has been obtained by halogen exchange between UF_6 and BCl_3.

Uranium metal, made by reduction of the tetrafluoride with calcium or magnesium, is highly reactive. The action of hydrogen at 250 °C gives the hydride UH_3, which is also formed, together with the dioxide, by the action of boiling water on the finely divided metal.

In aqueous solution the uranyl cation is hydrolysed to some extent, forming species such as $U_2O_5^{2+}$ and $U_3O_8^{2+}$; it is a class (*a*) cation, since it is easily shown to form more stable complexes with fluoride than with other halide ions; the formation of a stable complex $[UO_2(CO_3)_3]^{4-}$ (mentioned in connection with the extraction of uranium) is also noteworthy. The reduction chemistry of the ion at $[H^+] = 1$ is shown in Fig. 27.1. From this diagram it is easily seen that UO_2^+, the first

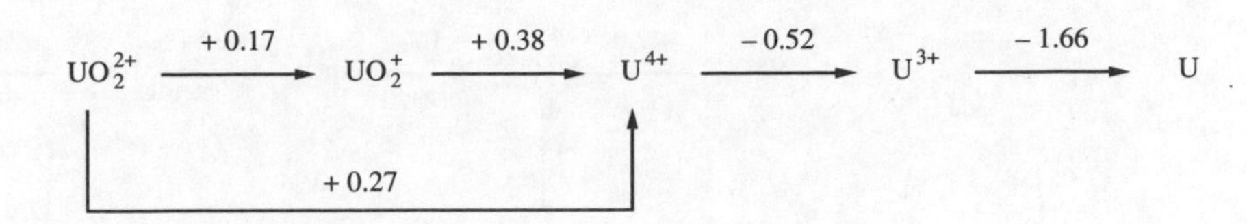

Figure 27.1
Potential diagram for uranium at $[H^+]=1$.

product of reduction of UO_2^{2+}, is somewhat unstable with respect to disproportionation into UO_2^{2+}, and U^{4+}; since protons are involved in the reaction

$$2UO_2^+ + 4H^+ \rightleftharpoons UO_2^{2+} + U^{4+} + 2H_2O$$

the position of equilibrium is pH-dependent, however; further, U(V) may be stabilised with respect to disproportionation by complexing with fluoride, the ion $[UF_6]^-$ being formed. As would be expected from Fig. 27.1, uranium metal liberates hydrogen from acids and the claret-coloured U^{3+}(aq) is a powerful reducing agent. Uranium(IV) is rapidly oxidised to UO_2^{2+} by Cr(VI), Ce(IV) or Mn(VII) but, as is often the case, oxidation by air is slow. It is interesting to note that the redox couples U^{4+}/U^{3+} and UO_2^{2+}/UO_2^+ are reversible, but UO_2^+/U^{4+} is not; the first two involve only electron-transfer, but the last one involves a structural reorganisation around the metal atom.

27.7 *Neptunium, plutonium and americium*

There are close similarities between these three elements, but a steady decrease in the stability of high oxidation states with increase in atomic number occurs, and is so important in the separation of the elements that we begin this section with the aqueous solution chemistry of the elements. Potential diagrams for $[H^+]=1$ are given in Fig. 27.2. Two unstable oxidation states have been omitted from these diagrams, Pu(VII), known only in a few salts containing the PuO_5^{3-} ion, and Am(II), known only in the solid dichloride, dibromide and diiodide; no data for them are available. The disproportionation of certain oxidation states will be discussed later; at present the main conclusions to be drawn from inspection of Fig. 27.2 are that Np(VII) will be formed only under very powerfully oxidising conditions, that all the MO_2^{2+} ions are much less stable to reduction than UO_2^{2+} (though there is only a small difference between NpO_2^{2+} and PuO_2^{2+}), that Pu^{4+} is much more easily reduced than Np^{4+}, and that all oxidation states of americium above Am(III) are powerfully oxidising. Separations of these elements from uranium (from which they are obtained in the nuclear reactor, as indicated earlier), from fission products, and from one another can thus be achieved by adjusting the solution containing them to a suitable redox potential and then using solvent extraction or precipitation to remove one element. For example the nitrates $MO_2(NO_3)_2$ may be extracted from nitric acid solution by

$$NpO_6^{5-} \xrightarrow{+2.0} NpO_2^{2+} \xrightarrow{+1.24} NpO_2^{+} \xrightarrow{+0.64} Np^{4+} \xrightarrow{+0.15} Np^{3+} \xrightarrow{-1.79} Np$$

$$NpO_2^{2+} \xrightarrow{+0.94} Np^{4+}$$

$$PuO_2^{2+} \xrightarrow{+1.02} PuO_2^{+} \xrightarrow{+1.04} Pu^{4+} \xrightarrow{+0.98} Pu^{3+} \xrightarrow{-2.00} Pu$$

$$PuO_2^{2+} \xrightarrow{+1.03} Pu^{4+} \qquad PuO_2^{2+} \xrightarrow{+1.03} Pu^{3+}$$

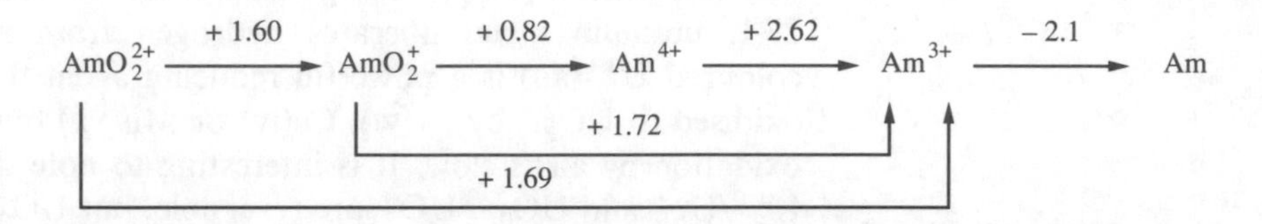

Figure 27.2
Potential diagrams for neptunium, plutonium and americium at $[H^+] = 1$.

diethyl ether, methyl isobutyl ketone or tributyl phosphate in kerosene; the nitrates $M(NO_3)_4$ may be extracted from nitric acid solution into tributyl phosphate in kerosene; or M^{4+} and M^{3+} may be precipitated as fluorides.

When ^{239}Pu ($t_{\frac{1}{2}} = 2.4 \times 10^4$ years) is prepared in the reactor by the sequence

$$^{238}U(n,\gamma) \rightarrow {}^{239}U \xrightarrow[23\ \text{min}]{-\beta^-} {}^{239}Np \xrightarrow[2.3\ \text{days}]{-\beta^-} {}^{239}Pu$$

it is almost free from ^{239}Np a few days after irradiation ceases, but ^{237}Np ($t_{\frac{1}{2}} = 2.2 \times 10^6$ years) is formed in small amount from ^{235}U by two successive (n,γ) reactions and β^--decay of the resulting ^{237}U. Small amounts of americium and heavier elements arising from multiple neutron capture by ^{239}Pu are also present, together with fission products. If the irradiated uranium is dissolved in dilute nitric acid, uranium and neptunium are present as UO_2^{2+} and NpO_2^{2+}, and plutonium as Pu^{4+}; all are extracted by tributyl phosphate in kerosene and are thus separated from other elements. Re-extraction with an aqueous solution of a mild reducing agent such as an iron(II) salt takes plutonium into the aqueous layer as Pu^{3+}, in which form it may be precipitated as the trifluoride alone or with LaF_3; separation from the latter is easily achieved after re-oxidation of the plutonium to PuO_2^{2+}, e.g. by dichromate. Separation of neptunium from uranium is effected by precipitation as NpF_4 if the uranium is present as UO_2^{2+}; selective reduction of NpO_2^{2+} can be carried out using sulphite as reductant. Many other separation schemes have been devised, but all rest on similar principles. With these elements the radiation hazards are very serious; 10^{-6} g of plutonium, for example, is a potentially lethal quantity, and all operations have to be conducted in closed systems. The effects of radiation in bringing about decompositions of solids and solvents are also important in chemical studies.

In preparation and properties the metals neptunium, plutonium and americium resemble uranium. The most stable oxides of all three elements are the dioxides obtained when the nitrates or hydroxides of the metals in any oxidation state are heated in air; however, americium dioxide is readily reduced to Am_2O_3 by hydrogen. No anhydrous trioxides MO_3 are known, but hydrated trioxides of neptunium and plutonium are obtained by the action of ozone on the tetrahydroxides. The highest oxidation states of the elements are obtained in the oxo-ions NpO_5^{3-}, PuO_5^{3-} and AmO_2^{2+}, all obtained by the use of alkaline hypochlorite, ozone or peroxodisulphate as oxidising agents; the two anions may be isolated as Ba^{2+} or $[Co(NH_3)_6]^{3+}$ salts. Neptunium(VII) and plutonium(VII) may also be obtained as the salts Li_5MO_6 by heating a mixture of Li_2O and MO_2 in oxygen at 400 °C; these M(VII) compounds are decomposed by aqueous acids with liberation of oxygen. For the Np(VII)/Np(VI) couple, E° at $[H^+]=1$ has been estimated as about +2.0 V, i.e. nearly the same as for $\frac{1}{2}S_2O_8^{2-}/SO_4^{2-}$ and Co^{3+}/Co^{2+}.

Although neptunium and plutonium form volatile hexafluorides resembling uranium hexafluoride, these compounds readily decompose to fluorine and NpF_5 and PuF_4 respectively. Americium forms only a trifluoride and a tetrafluoride. The highest chlorides are $NpCl_4$, $PuCl_3$ and $AmCl_3$; the complex $Cs_2[PuCl_6]$, however, can be obtained from CsCl, $PuCl_3$ and Cl_2 at 50 °C. Only americium forms compounds AmX_2 (X=Cl, Br, I); these are saline halides made by heating americium with the appropriate mercury(II) halide; they liberate hydrogen from water. The formation of these dihalides is of special interest since americium is the analogue of europium in the lanthanide series.

The data in Fig. 27.2 show that the ion NpO_2^+, unlike UO_2^+, is stable with respect to the disproportionation

$$2NpO_2^+ + 4H^+ \rightleftharpoons NpO_2^{2+} + Np^{4+} + 2H_2O$$

at $[H^+]=1$; PuO_2^+ is just unstable with respect to an analogous disproportionation. The closeness of the first three potentials in the reduction of PuO_2^+ is noteworthy; if PuO_2 is dissolved in excess of dilute $HClO_4$ (a non-complexing acid) at 25 °C, the solution at equilibrium contains Pu(III), Pu(IV), Pu(V) and Pu(VI). In redox systems involving plutonium, however, equilibrium is not always attained rapidly: as for uranium, couples involving only electron-transfer (e.g. PuO_2^{2+}/PuO_2^+) are rapidly reversible, but those also involving oxygen-transfer (e.g. PuO_2^+/Pu^{4+}) are not. Since hydrolysis and complex formation (the extents of which increase with increasing ionic charge, i.e.

$PuO_2^+ < PuO_2^{2+} < Pu^{3+} < Pu^{4+}$) may also complicate the situation, the study of equilibria and kinetics in solutions of plutonium compounds is a matter of great difficulty. Interesting redox reactions of neptunium and plutonium which occur in the absence of a solvent at 200–300 °C or in non-aqueous solvents include:

$$NpF_6 + 3NaF \rightarrow Na_3NpF_8 + \tfrac{1}{2}F_2$$

$$NpF_6 + CsF \rightarrow CsNpF_6 + \tfrac{1}{2}F_2$$

$$PuF_6 + 2CsF \rightarrow Cs_2PuF_6 + F_2$$

$$2CsNpF_6 \xrightarrow{\text{anhydr. HF}} Cs_2NpF_6 + NpF_6$$

In the case of americium, all oxidation states above Am(III) are powerful oxidising agents; in addition AmO_2^+ and Am^{4+} are both thermodynamically unstable with respect to disproportionation. Americium(IV) can be characterised in aqueous solution only in complexes such as $[AmF_6]^{2-}$, obtained by dissolving AmO_2 in aqueous hydrofluoric acid; but AmO_2^+, formed in the reduction of AmO_2^{2+} by chloride, disproportionates relatively slowly at low acidities.

27.8 Curium and later elements

All of these elements are of chemical interest mainly because they show how the metals in the second half of the actinide series resemble the corresponding lanthanides rather closely, and we shall deal with them only briefly. Curium is available in macroscopic quantities, but our information about the heavier elements rests on tracer studies supplemented, in the case of californium, berkelium and einsteinium, by identifications of compounds based on structure determinations by X-ray powder photography: when a few micrograms of a berkelium ion on an exchange resin is fluorinated, for example, the product has an X-ray powder diagram showing that it is isostructural with the known actinide tetrafluorides, and thus indicating that it is BkF_4. Other compounds characterised in this way are: Bk_2O_3, $BkCl_3$, BkO_2 and $Cs_2[BkCl_6]$; CfI_2, Cf_2O_3, CfO_2 and CfF_4; $EsCl_3$ and EsOCl. All are obtained by methods used for the preparations of analogous compounds of the lighter actinides. The solid-state chemistry of curium is slightly more extensive; from CmF_3 and $Cm_2(C_2O_4)_3$, which may be precipitated from aqueous solution, other compounds and the metal are obtained by the reactions shown below:

$$Cm \xleftarrow[1250\,°C]{Ba} CmF_3 \xrightarrow[400\,°C]{F_2} CmF_4$$

$$Cm_2(C_2O_4)_3 \xrightarrow[O_2/O_3]{\text{heat in}} CmO_2 \xrightarrow[600\,°C]{\text{heat}} Cm_2O_3$$

Neither curium nor berkelium shows any evidence of a dipositive oxidation state. In aqueous solution Cm^{4+} is known only as a fluoro complex; Bk^{4+} is rather more stable with respect to reduction; there is at present no evidence for a tetrapositive oxidation state in aqueous media of any of the actinides after berkelium.

The remaining actinides (fermium, mendelevium, nobelium and lawrencium) have been studied only in aqueous solution. Their behaviour on an ion-exchange column indicates that under strongly oxidising conditions all are present in the tripositive state. After reduction with magnesium, however, all except lawrencium are partially carried by insoluble difluorides, fermium least and nobelium most (californium is also carried to some extent). Tracer studies involving the action of other reducing agents indicate that the stability of the dipositive state with respect to oxidation is $Cf^{2+} < Es^{2+} < Fm^{2+} < Md^{2+} \ll No^{2+}$; No^{2+}, indeed, appears to be the more stable state of nobelium in aqueous solution. Lawrencium appears to exist only in the tripositive state. However, it may well be that later work will extend the range of known oxidation states of some of these elements.

In general, nevertheless, the similarities between the later actinides and the later lanthanides are firmly established. This is true not only of most of their observed oxidation states (as may be seen by comparing Cm and Gd, Bk and Tb, No and Yb, and Lr and Lu), but also of the spectra and magnetic properties of some compounds of isoelectronic pairs of ions (e.g. CmF_3 and GdF_3). With the demonstration mentioned earlier that rutherfordium (element number 104) is the analogue of hafnium, therefore, the delineation of the second inner transition series is complete.

PROBLEMS

1 Discuss each of the following observations:

(a) Many actinide oxides are non-stoichiometric, but few lanthanide oxides are.

(b) The ions NpO_6^{5-} and PuO_6^{5-} can be made in aqueous media only if the solution is strongly alkaline.

(c) A solution containing Pu(IV) undergoes negligible disproportionation in the presence of excess of molar H_2SO_4.

2 Suggest a method for the separation of americium from uranium, neptunium and plutonium.

What would you expect to happen when a solution of $NpO_2(ClO_4)_2$ in molar $HClO_4$ is shaken with zinc amalgam and the resulting liquid is decanted from the amalgam and aerated?

3 A solution **X** containing 21.4 g uranium(VI) dm^{-3} (25 cm^3) was reduced with zinc amalgam, decanted from the amalgam, and after being aerated for five minutes was titrated with 0.1200 molar cerium(IV) solution, 37.5 cm^3 of the latter being required for re-oxidation of the uranium to uranium(VI).

Solution **X** (100 cm^3) was then reduced and aerated as before, and treated with excess of a dilute aqueous solution of potassium fluoride. The resulting precipitate, after being dried in a platinum vessel at 300 °C, weighed 2.826 g. Dry oxygen was then passed over the precipitate at 800 °C, after which the solid reaction product weighed 1.386 g. This product was dissolved in water, and the fluoride in the solution was precipitated as lead chloride fluoride, PbClF, 2.355 g being obtained.

Deduce what you can concerning the chemical changes which took place during these experiments.

[U = 238.0; Pb = 207.2; F = 19.0; Cl = 35.5]

REFERENCES FOR FURTHER READING

Cotton, F.A. and Wilkinson, G. (1988) *Advanced Inorganic Chemistry*, 5th edition, Interscience, New York. A fuller account of the actinides, especially of their organometallic compounds.

Katz, J.J., Seaborg, G.T. and Morss, L.R. (editors) (1986) *Chemistry of the Actinide Elements*, Chapman and Hall, London. The definitive account of actinide chemistry and technology.

The Open University (1977), S304 Unit 27, *Lanthanides and Actinides*, Open University Press, Milton Keynes. Contains an interesting historical account of the recognition of the existence of the second inner transition series.

Outline answers to problems

Chapter 1

1 See calculation in Section 1.2; 2.3×10^{12} J.

2 $^{60}_{28}Ni$, $^{59}_{27}Co$; $3.1 \times 10^{-9}\ s^{-1}$.

3 Molecule must contain F in two different environments which can undergo apparent exchange readily; this suggests a T-shaped molecule and exchange by the mechanism shown for PF_5 in Fig. 1.4.

4 Assume mass of radioactive Pb can be neglected; 2.0×10^{-7} mol dm^{-3}.

5 See Section 1.2 for relationships between n, n_0, and t and between $t_{\frac{1}{2}}$ and λ; 4040 y.

6 Electrolyse a D_2O solution of any electrolyte not containing H, e.g. NaCl; pass D_2 over Li metal.
$SO_3 + D_2O \rightarrow D_2SO_4(aq)$; stir with BaO_2.
Pass CO_2 into a solution of Na_2CO_3 in D_2O.
$Mg_3N_2 + D_2O \rightarrow ND_3$;
$AlCl_3 + D_2O \rightarrow DCl$;
$ND_3 + DCl \rightarrow ND_4Cl$.
$C_6H_5MgBr + D_2O$.

7 Ratio of frequencies approx. $\sqrt{2}:1$ shows they arise from bonds containing H or D; O—H would exchange with D_2O, so anion in A must have structure

O O^-
  P
H O^-

Chapter 2

1 Although the mass of the nucleus does not appear in the simple treatment of the spectrum it does enter into the expression for the reduced mass of the atom which replaces the mass of the electron in a more accurate treatment (see Section 2.3). Ionisation energy of a one-electron species $\propto Z^2$ so I_2 for He is 4×13.6 eV and $He^{2+} + 2e \rightarrow He$ liberates 79.0 eV.

2 2.47×10^{15} Hz, 121.4 nm; 985 kJ mol^{-1}.

3 See Section 2.7.
Rotational energy levels given by

$$E = \frac{h^2}{8\pi^2 I} J(J+1)$$

so for $J = 0 \rightarrow J = 1$

$$\Delta E = h^2/4\pi^2 I$$

$$\Delta E = h\nu = \frac{h^2}{4\pi^2 I} = \frac{h^2}{4\pi^2} \cdot \frac{m_1 + m_2}{m_1 m_2 r^2}$$

Note that if h is in J s, m in kg, ν in s^{-1}, r will be in m; 1 m = 10^{10} Å. Substitution for ν, h, m_1, m_2 leads to $r = 1.13$ Å.
Vibrational energy levels are given by

$$E = (n + \tfrac{1}{2}) \frac{1}{2\pi} \sqrt{\frac{k}{\mu}}$$

so ΔE for $n=0 \rightarrow n=1$ is given by

$$\Delta E = \frac{1}{2\pi}\sqrt{\frac{k}{\mu}}$$

and $k = 4\pi^2 \nu^2 \mu$.
Conversion of $\bar{\nu}$ to ν in s^{-1} and substitution gives $k = 1.9 \times 10^3$ N m^{-1}.

4 In each case one nodal surface would have to be shown, with a change in the sign of the wave function at this surface.
Probability of finding electron at r is proportional to $\psi_r^2 r^2$, not to ψ_r^2.

Chapter 3

1 (a) 3.30 (b) 4.30. Ground state of V^+ likely to be $3d^3 4s^1$.
2 $^3P_2$, $^2P_{3/2}$, $^3F_2$, $^7S_3$, $^3F_4$.
3 (a) See Sections 2.4 and 2.10; Li^{2+} is a one-electron system so its spectrum is like that of H.
(b) Effect of increasing nuclear charge mitigated at oxygen by loss of repulsion when electron removed from doubly-occupied orbital.
(c) In each case extra electron enters valence shell; presumably less repulsion from the other electrons when shell larger; this effect apparently outweighs that of increased nuclear charge (cf. ionisation energies of alkali metals).

Chapter 4

1 Lengths should increase as electrons are added to antibonding orbitals; all paramagnetic except O_2^{2-}.
2 $^2\Sigma$, $^1\Sigma$, $^1\Sigma$, $^3\Sigma$, $^2\Pi$.
3 407 kJ mol^{-1}; should be a stronger bond in Cl_2^+ than in Cl_2.
4 (a) Some conjugation across C—C by π orbital overlap (cf. 1,3-butadiene in Section 23.4).
(b) Loss of antibonding electron when $NO \rightarrow NO^+$.
(c) See Section 4.9 for the correlation between band gap and semiconductivity.
(d) Must involve infinite three-centre two-electron bonding system analogous to that in bridge part of B_2H_6.
(e) The π overlap systems involving middle C atom must be perpendicular; so therefore must the two CXY planes, and the molecule lacks a plane or centre of symmetry.

Chapter 5

1 Tetrahedral, tetrahedral, trigonal bipyramidal, pyramidal, bent, linear, bent, regular octahedral, pentagonal bipyramidal with one position occupied by a lone pair of electrons.
2 C_{2v}, $D_{\infty h}$, C_{4v}, D_{2h}, D_{3h}, C_{2v}, D_{3h}, C_s, D_{2d}.
3 275 kJ mol^{-1}, 156 kJ mol^{-1}.
4 510 N m^{-1}, 24.9 kJ mol^{-1}. For D_2, k should be the same, but $\frac{1}{2}h\nu_0$ should be less by a factor of $\sqrt{2}$.
5 Taking the Xe—Xe bond energy as zero, 133, 133 and 128 kJ mol^{-1}; $x_{Xe} = 3.2$. Near-constancy of bond energies may imply bond order and valence state nearly constant.

Chapter 6

1 4; 5.99×10^{23} mol^{-1}.
2 622 kJ mol^{-1}.
3 (i) Born cycle based on measured enthalpy of formation and lattice energy calculated from structural data of an alkali metal hydride.
(ii) Born cycle based on estimated lattice energy (assuming CaF_2 structure with Na^{2+} somewhat smaller than Na^+).
(iii) Dissolve each in water and apply Hess's Law.
4 (a) Consider factors in Born cycle: variation in lattice energy deci-

sive for fluorides, much smaller for iodides (cf. Section 10.3).

(b) Effect of double charges on ions.

(c) $LiFeO_2$ and Li_2TiO_3 really mixed oxides $Li^+Fe^{3+}(O^{2-})_2$ and $(Li^+)_2Ti^{4+}(O^{2-})_3$ having NaCl structure like MgO; ZnO has a different structure.

(d) Write down a cycle for the reaction

$$MSO_4(s) \rightarrow MO(s) + SO_3(g)$$

MO will always have a higher lattice energy than MSO_4, but least so when M^{2+} is as large as possible, so this is the condition for ΔG° least negative.

5 (a) Successive ionisation energies less different.

(b) For small deviations from stoichiometry, $[Ni^{2+}]$ and $[O^{2-}]$ are nearly constant, so

$$K = [Ni^{3+}]^4 [\square_+]^2 / p_{O_2}$$

Since $[\square_+] = \frac{1}{2}[Ni^{3+}]$, $K \propto [Ni^{3+}]^6 / p_{O_2}$ with conductance $\propto [Ni^{3+}]$ and hence $\propto p_{O_2}^{1/6}$.

Chapter 7

1 (a) F^-, SO_4^{2-}, $[Fe(H_2O)_5OH]^{2+}$, NH_3.

(b) H_2SO_4, PH_4^+, NH_3, $VO(H_2O)^{2+}$.

2 (a) K for precipitation of CuS must be greater than K for Cu^{2+}—NH_3 complexing.

(b) A complex such as $MgOMg^{2+}$ (or a hydrate thereof) is formed.

(c) F^- must form a more stable cmplex than Cl^- or O_2^{2-} with Ti(IV).

(d) $2K^+I^- + HgI_2 \rightarrow 2K^+ + HgI_4^{2-}$; (4 anions in solution) → (3 ions in solution).

3 Evaluate ΔG° for reduction of Fe^{3+} and $[Fe(CN)_6]^{3-}$ and for complexing of Fe^{2+}; completion of cycle gives ΔG° for complexing of Fe^{3+} and hence 10^{42} for the overall formation constant of $[Fe(CN)_6]^{3-}$.

4 Precipitation of sparingly soluble CuI (solubility product 10^{-12} – see Section 7.2) drives the Cu^{2+}/I^- reaction. Copper(II) must form a very stable tartrate complex, but since tartaric acid is a weaker acid than sulphuric acid, protonation of the tartrate ion destroys this complex.

5 MO_2^+ unstable with respect to disproportionation into MO_2^{2+} and M^{4+}. M(VI) and M(V) mild oxidants, M very strong reductant, M^{3+} strong reductant. $M + H^+(aq)$ should give $M^{4+} + H_2$. O_2 should oxidise all forms of M at pH = 0 to MO_2^{2+}. $[H^+]$ should affect E° values for half-reactions involving oxygen-containing ions.

6 HCl may operate in two ways: preferential complexing of Fe^{3+} by Cl^-, and diminution of reducing power of SO_2 because of $[H^+]$ in equilibrium

$$SO_4^{2-} + 4H^+ + 2e \rightleftharpoons SO_2 + 2H_2O$$

Try effect of replacing HCl by

(a) saturated LiCl or any other very soluble chloride;

(b) $HClO_4$ or any other very strong acid that is not easily reduced.

(But see Section 15.5 for further comment.)

7 (a) (i) −1 (ii) +6 (iii) +1 (iv) $-\frac{1}{3}$ (v) 0

(b) N, 0, P, +5.

Chapter 8

1 (a) $Zn + 2NaNH_2 + 2NH_3 \rightarrow Na_2[Zn(NH_2)_4] + H_2$

$[Zn(NH_2)_4]^{2-} + 2NH_4^+ \rightarrow Zn(NH_2)_2 + 4NH_3$

$Zn(NH_2)_2 + 2NH_4I \rightarrow [Zn(NH_3)_4]^{2+} + 2I^-$

(b) $N_2O_4 \rightarrow NO^+ + NO_3^-$

$NO_3^- + H_2SO_4 \rightarrow NO_2^+ + HSO_4^- + OH^-$

$OH^- + 2H_2SO_4 \rightarrow H_3O^+ + 2HSO_4^-$

$N_2O_4 + 3H_2SO_4 \rightarrow NO^+ + NO_2^+ + H_3O^+ + 3HSO_4^-$

(c) $Ph_2C{=}CH_2 + HCl \rightarrow Ph_2C^+CH_3 + Cl^-$ to a slight extent; equilibrium then upset by

$$Cl^- + BCl_3 \rightarrow BCl_4^-$$

with increase in conductivity, but further addition of BCl_3 has no effect.

2 Prepare salts containing $AsCl_2^+$ and $AsCl_4^-$ (e.g. with $AlCl_3$ and Me_4NCl respectively), show they give conducting solutions in $AsCl_3$ and that on mixing conductivity is a minimum at 1:1 ratio.

3 H_2NNH_2, Hg_3N_2, O_2NNH_2, CH_3NH_2, $OC(NH_2)_2$, $[Cr(NH_3)_6]Cl_3$.

Chapter 9

1 DCl (by $AlCl_3/D_2O$) + $LiAlH_4$; accurate measurement of relative molecular mass or of density of water formed on combustion.

2 (a) Indicates proceeds via

$$H_2 \rightleftharpoons 2H$$
$$H + para\text{-}H_2 \rightarrow ortho\text{-}H_2 + H$$

with second step rate-determining.

(b) 3610 cm^{-1} peak due to monomer in absence of hydrogen bonding; broad peak at 3330 cm^{-1} caused by intermolecular hydrogen bonding.

(c) $MCl + HCl \rightleftharpoons M[HCl_2]$ equilibrium position governed by relative lattice energies of MCl and $M[HCl_2]$; loss of lattice energy on $M[HCl_2]$ formation least for largest cation – cf. Section 9.5.

(d) Hydrolysis of ammonium ion leads to an acidic solution of the salt.

(e) Both have wurtzite structure.

3 Measure ΔH° for reactions of hydrides and of parent metals with H_2SO_4(aq). Indicates to what extent metal–H bonding compensates H—H bonding (436 kJ mol^{-1}) if it is assumed energy needed to change metal lattice is relatively small.

Chapter 10

1 (a) Even if all the ions have primary solvation number of six, smallest hydrated ion $M(H_2O)_6^+$ with higher charge density still has greatest attraction for more water molecules, increasing effective hindrance to movement through the solution.

(b) Born cycle argument as for hydride or halide.

(c) Phase in equilibrium with the solution must change at 32 °C, e.g. hydrate → anhydrous salt. Dissolution of lower temperature phase endothermic, of higher temperature phase exothermic.

(d) Consider the stages in $M^+(aq) \rightarrow M$ (Section 7.11).

2 (a) Gives LiF + NaI (see Tables 10.2 and 10.3; ΔS° will be negligible).

(b) Consider as

$$\text{—C—Cl} + MF \rightarrow \text{—C—F} + MCl$$
$$\searrow \qquad\qquad \nearrow$$
$$\text{—C}^+ + Cl^- + M^+ + F^-$$

(all gaseous)

Only variable is $U(MCl) - U(MF)$; this will always oppose reaction, but less so when M^+ is larger.

Chapter 11

1 Born cycle: for the fluorides, substantial variation in lattice energy parallels decrease in ionisation and atomisation energies, leaving ΔH_f° fairly constant; with larger iodide ion, variation in lattice energy less, and decrease in ionisation and atomisation energies.

2 (a) See the discussion of the reaction

$$2CaF \rightarrow CaF_2 + Ca$$

in Sections 6.8 and 11.1.

(b) Dissolve each in dil. HCl, measure ΔH°, and apply Hess's Law.

Chapter 12

1 Electron counting for cluster gives $n+3$, $n+2$, $n+2$, $n+1$ and $n+3$, so an octahedron with two positions vacant, a pentagonal bipyramid with one position vacant, an octadecahedron with one position vacant, an octahedron and an icosahedron with two positions vacant. Note that for $B_{10}H_{14}$ the actual structure of the molecule could equally well have been described as derived from an icosahedron (see Fig. 12.3); by coincidence these *nido* ten-vertex and *arachno* ten-vertex fragments are the same.

2 (a) $LiAlD_4$ (from LiD, $AlCl_3$)+ BCl_3

(b) $B_2O_3 + D_2O$

(c) ND_3 (Mg_3N_2, D_2O)+B_2D_6 and heat product

(d) $B_2D_6 + CO$

(e) $B_2D_6 + C_2H_4$.

3 (a) B octet-restricted, Al not; so attack on Al by H_2O possible.

(b) $B_2H_6 \overset{\text{fast}}{\rightleftharpoons} 2BH_3$

$$BH_3 + H_2O \overset{\text{slow}}{\rightarrow} \text{products}$$

(c) Steric hindrance at B atom, not at larger Al atom.

(d) $B(OH)_3 + 2HF_2^-$

$$\rightarrow BF_4^- + 2H_2O + OH^-$$

4 (a) Steric hindrance to large group in bridges.

(b) $AlF_3 \rightarrow [AlF_6]^{3-}$; $[BF_4]^-$ more stable than $[AlF_6]^{3-}$ and AlF_3 displaced and precipitated.

(c) $GaCl_2$ contains an ion also present in $GaCl_3$/HCl and isostructural with $GeCl_4$; must be $[GaCl_4]^-$. No other Ga—Cl species there, so $GaCl_2$ must be $Ga^+[GaCl_4]^-$.

(d) Solid is $Tl^+I_3^-$. Hydrated Tl_2O_3 is very insoluble, so oxidation of Tl^+(aq) to solid Tl_2O_3 is very much easier than to Tl^{3+}(aq), and I_2 (from dissociation of I_3^-) effects this.

(e) Most likely exchange via

>Al—H—BH_3 formation.

5 **A** = B_4H_{10}, **B** = B_5H_9, **C** = RB_5H_8 with R in apical position, **D** = RB_5H_8 with R in equatorial position.

Chapter 13

1 Linear, linear, pyramidal, trigonal bipyramidal, angular, octahedral.

2 (a) KCN(aq) very alkaline owing to hydrolysis, CN^- competes unsuccessfully with OH^- for Al^{3+}.

(b) Probably contains C_3^{4-} ion or at least C_3 unit C=C=C or C—C≡C.

(c) NH_4Br an acid in NH_3 (Section 8.2).

(d) Al can replace Si in silica structure provided a cation present to maintain electrical neutrality.

(e) Si—H bond not broken in rate-determining step (presumably attack of OH^- on Si).

3 (a) $[Sn(OH)_6]^{2-} + H_2$

(b) $PbSO_4$

(c) Na_2CS_3

(d) —SiH_2O— polymers

(e) $ClCH_2SiH_3$.

4 (a) Dissolve each in concentrated HF(aq), measure ΔH°, and apply Hess's Law.

(b) Si—Si and Si—H bond energies from $\Delta H^\circ_{\text{combustion}}$ for Si_2H_6 and SiH_4 and other data; apply Pauling relationship (Section 5.7).

(c) Determine Pb(IV) by allowing it to oxidise I^- and titrating with thiosulphate (or heat with HCl, pass Cl_2 into KI(aq), and titrate).

Chapter 14

1 (a) $K^{15}NO_3 + Al$, $NaOH(aq) \rightarrow {}^{15}NH_3$; pass over Na. Oxidise the ${}^{15}NH_3$ with CuO or NaOCl. $K^{15}NO_3 + Hg$, $H_2SO_4 \rightarrow {}^{15}NO$; combine with Cl_2, $AlCl_3$.
(b) Reduce to ${}^{32}P$; treat with NaOH(aq).
${}^{32}P + I_2$; hydrolyse the product.
${}^{32}P$ + excess of S;
${}^{32}P_4S_{10} + Na_2S$.

2 (a) Tetrahedral
(b) planar (like C_2H_4 with one H missing)
(c) pyramidal at N and bent at O
(d) tetrahedral
(e) trigonal bipyramidal with apical Cl atoms.

3 (a) $P_3O_{10}^{5-}$ will give two ${}^{31}P$ signals, relative intensities 2:1; $P_4O_{13}^{6-}$ will give two signals of equal intensity.
(b) Two ${}^{19}F$ peaks of relative intensity 3:2 will coalesce at a higher temperature if rapid exchange occurs.
(c) Three possible isomers with both NMe_2 groups on same P atom, *cis* and *trans* with NMe_2 groups on different P atoms; these will give three, one and two ${}^{1}H$ peaks respectively.

4 (a) If 2Ti(III) → 2Ti(IV), N(−I) → N(−III) i.e. NH_3
(b) If 2Ag(I) → 2Ag(0), P(III) → P(V) i.e. PO_4^{3-}
(c) If 2I(0) → I(−I) twice, P(I) → P(III) → P(V) i.e. $H_3PO_2 \rightarrow H_3PO_3 \rightarrow H_3PO_4$.

5 (a) **B** = N_2O (b) **D** = N_2.

Chapter 15

1 Reverse second half-reaction, and add to first to show ΔG° negative for

$$2H_2O_2 \rightarrow O_2 + 2H_2O$$

H_2O_2 concentration 60.7 g dm^{-3}.

2 (a) Fluxional exchange via square pyramidal intermediate with lone pair: see Fig. 1.4. Exchange via a bridged dimer also possible.
(b) Reaction required is

$$SO_4^{2-} + 8H^+ + 8e \rightarrow S^{2-} + 4H_2O$$

This is assisted by very high $[H^+]$ and very low solubility of CuS.
(c) White precipitate is $Ag_2S_2O_3$, dissolves forming $[Ag(S_2O_3)_2]^{2-}$. Disproportionation of $S_2O_3^{2-}$

$$S_2O_3^{2-} + H_2O \rightarrow S^{2-} + SO_4^{2-} + 2H^+$$

brought about by removal of S^{2-} as insoluble Ag_2S.
(d) This is expected on VSEPR: Te here has 12 electrons, structure octahedral with one position occupied by lone pair.

3 (a) $S_2O_4^{2-} + 2Ag^+ + H_2O \rightarrow S_2O_5^{2-} + 2Ag + 2H^+$
(b) $S_2O_4^{2-} + 3I_2 + 4H_2O \rightarrow 2SO_4^{2-} + 6I^- + 8H^+$

4 **X** = $H_2NSO_2(OH)$.

Chapter 16

1 (a) Solution in n-hexane contains uncomplexed I_2 molecules; charge-transfer complex formed with benzene, probably O- and N-donor complexes with ethanol and pyridine; pyridine complex the most stable.
(b) Iodide complex of silver must be more stable than chloride complex.
(c) Possible products are Bu_4NCl + HI, Bu_4NH + ICl, and Bu_4NI +

HCl; for very large cation, all three salts would have about the same lattice energy; bond strength of other product the decisive factor.

(d) Hydrogen fluoride vapour is polymeric, so the hydrogen bonds are not broken on volatilisation; those in water are.

2 (a) Determine total chlorine by addition of excess of iodide and titration with thiosulphate; only HCl is a strong acid so its concentration can be determined by pH measurement – not by titration, which would upset the equilibrium. Get ΔH° from K at different temperatures.

(b) Neutralise the solution of a weighed amount of the oxide with $NaHCO_3$ and titrate the I_2; add excess of dil. HCl, and titrate again.

(c) Species with a Raman stretching frequency lower than that of Cl_2 should be formed.

No; bond order in Cl_2^- can only be 0.5, and in the decomposition

$$K^+Cl_2^- \rightarrow KCl + \tfrac{1}{2}Cl_2$$

the gain in lattice energy and the liberation of half the bond energy in Cl_2 would far outweigh the absorption of the $Cl—Cl^-$ bond energy.

3 (a) IO_4^-.

(b) $IO_4^- + 2I^- + H_2O \rightarrow IO_3^- + I_2 + 2OH^-$

$IO_3^- + 5I^- + 6H^+ \rightarrow 3I_2 + 3H_2O$.

Chapter 17

1 $[ClF_2]^-$, Br_3^-, I_3^-; $[BrF_4]^-$, I_5^-; $[SbCl_6]^{3-}$, $[IF_6]^-$; $[TeF_5]^-$, BrF_5, IF_5; $[BrF_4]^-$, I_5^-. VSEPR theory assumes bonds of two-centre two-electron type; where this assumption appears to be untrue, not a good guide (see further Section 5.2). Theory predicts a polyhedron with nine vertices, one occupied by an electron pair, for $[XeF_8]^{2-}$; actual structure is a square antiprism.

2 From ΔH° $XeF_2/I^-(aq)$ reaction and other data required. Weaker bond to Cl and higher Cl—Cl bond energy. From Born cycle assuming lattice energy of XeF about same as that of CsF.

Chapter 18

1 See Fig. 5.5.

2 (a)

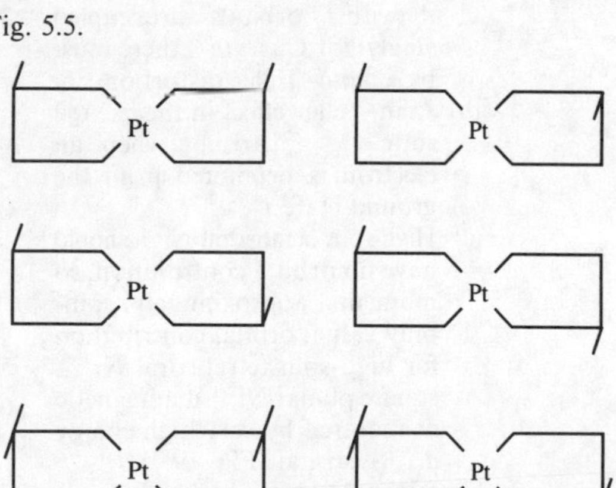

Pt Pt (Δ, Λ)

Pt Pt

(b)

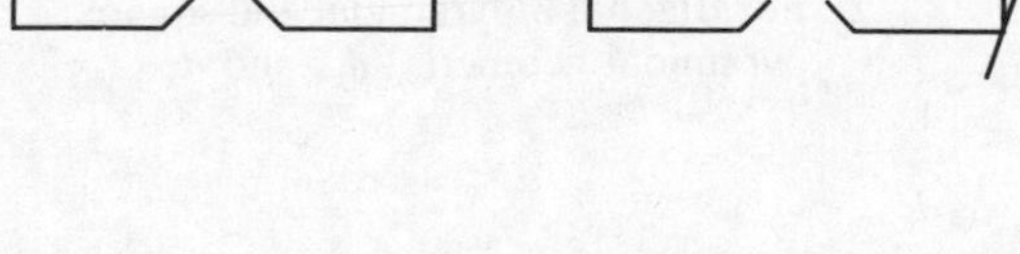

Pt Pt

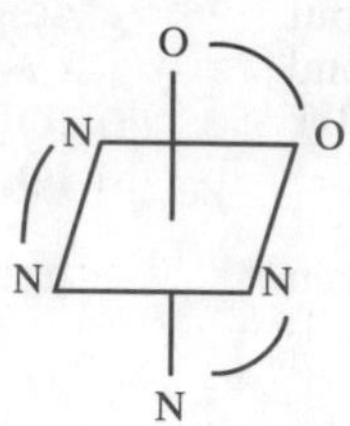

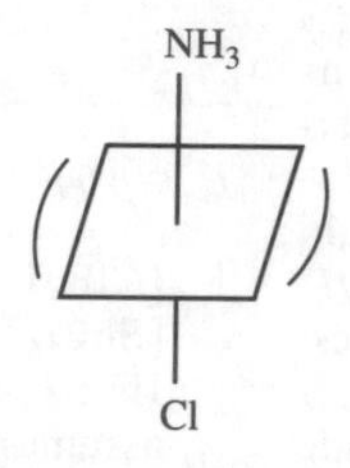

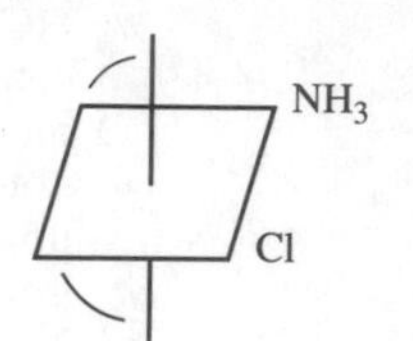

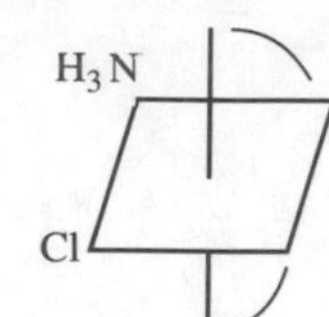

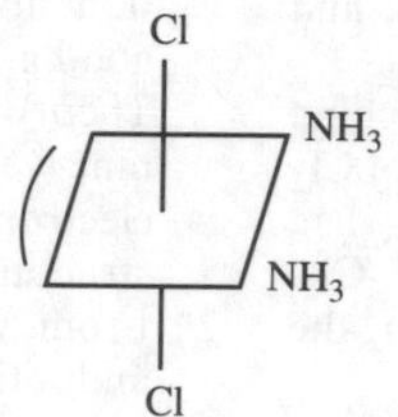

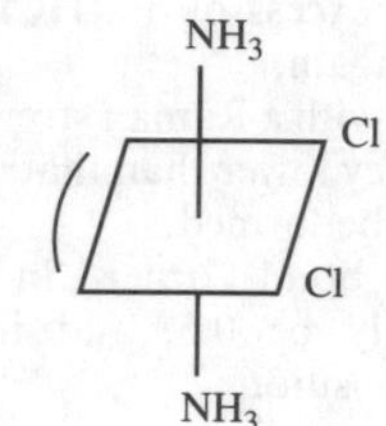

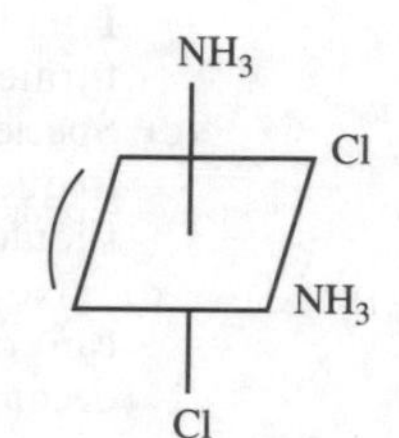

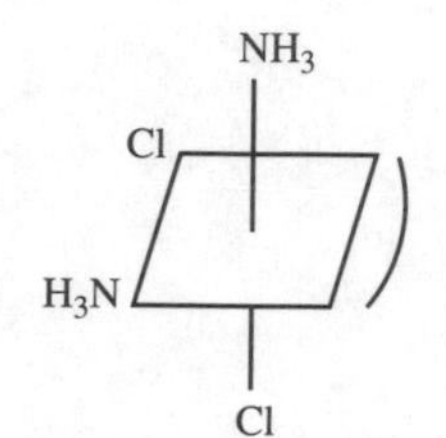

Chapter 19

1 (a) See Fig. 19.6: Co^{2+} is high-spin d^7 with t_2 orbitals all occupied singly; for Cu^{2+} (d^9) there must be a Jahn–Teller distortion.
(b) Jahn–Teller effect in the excited state $t_{2g}^3\ e_g^3$ arising when an electron is promoted from the ground state $t_{2g}^4\ e_g^2$.
(c) High-spin octahedral Ni^{2+} should have no orbital contribution, so moments approximately spin-only value; orbital contribution for high-spin tetrahedral Ni^{2+}; square planar Ni^{2+} diamagnetic as required by very high energy $d_{x^2-y^2}$ orbital (Fig. 19.7).

2 For trigonal bipyramidal and square pyramidal geometry, d_{z^2} and $d_{x^2-y^2}$ orbitals respectively are the only ones pointing directly at ligands. For Ni(II), both geometries expected to lead to diamagnetism.

3 Three (same as for a d^2 ion). *Trans* $[Co(en)F_2]^+$ has a centre of symmetry, *cis* ion has not, so Laporte rule does not apply. Charge-transfer Cl^- to Co^{3+} probably accounts for more intense colour of chloro complex; charge-transfer from F^- most unlikely.

Chapter 20

1 See Problem 7.3; 10^{35}.

2 Normal spinel ≡ tetrahedral Ni^{2+} + 2 octahedral Mn^{3+}.

Inverse spinel ≡ tetrahedral Mn^{3+} + octahedral Ni^{2+} + octahedral Mn^{3+}. One octahedral Mn^{3+} common to both, so problem reduces to tetrahedral Ni^{2+} + octahedral Mn^{3+} or octahedral Ni^{2+} + tetrahedral Mn^{3+}.

CFSE tet. Ni^{2+} + oct. Mn^{3+}
$= (0.8 \times \frac{4}{9} \times 8400) + (0.6 \times 21\,000)$
$= 15\,586\ cm^{-1}$ for normal spinel

CFSE oct. Ni^{2+} + tet. Mn^{2+}
$= (1.2 \times 8400) + (0.4 \times \frac{4}{9} \times 21\,000)$
$= 13\,813\ cm^{-1}$ for inverse spinel.

See Section 20.2 for comment: note also Jahn–Teller effect for Mn^{3+}. Although predicted normal by a small margin, $NiMn_2O_4$ is actually inverse.

3 (a) Δ(CFSE) on going from octahedral ion (aquo complex) to tetrahedral ion much less for $Co^{2+}(d^7)$ than $Ni^{2+}(d^8)$.
(b) Indicates $H_4[Fe(CN)_6]$ a weak acid so far as its fourth ionisation constant is concerned: H^+ complexing of $[Fe(CN)_6]^{4-}$ makes reduction easier.
(c) CFSE plays a minor part here: there is actually a loss of it on reduction of Mn^{3+} compared with a gain on reduction of Fe^{3+} and a loss on reduction of Cr^{3+}. Decisive factor is much larger third ionisation energy of Mn when $d^5 \rightarrow d^4$ (see Tables 20.4 and 20.5).

Chapter 21

1 (a) In $[Co(NH_3)_4CO_3]^+$, the CO_3^{2-} ion is bidentate. First step involves breaking the chelate ring, with $H_2^{18}O$ filling vacant position at Co. This is followed by protonation of the O atom of the now monodentate CO_3^{2-} as described in Section 21.2.
(b) Conjugate base dissociative (SN1) mechanism; NH_2^- in NH_3 is analogous to OH^- in H_2O (Section 8.2).
(c) Cr(II) complexes labile, Cr(III) complexes kinetically inert. Substitution catalysed by redox reaction with Cr(II)—Cl—Cr(III) intermediate; resulting Cr(II) species labile and can attack another Cr(III)Cl.

2 Two low-spin complexes of nearly equal dimensions; one low-spin complex giving a larger high-spin complex on reduction, but with help from ΔG° of reaction; one high-spin and one low-spin complex with different bond lengths.

3 (a) (i) Successive reactions with py, Br^-, NH_3.
(ii) Successive reactions with NH_3, Br^-, py.
(b) Add $[Pt(en)_2]^{2+}$ as a catalyst; see Section 21.8.

4 Common mechanism involving dissociation of one end of chelate to form a symmetrical five-coordinated intermediate. This re-forms chelate by coordination from the other carbonyl group.

Chapter 22

1 (a) Back π-bonding to CO in hexacarbonyl; where only 3CO available, more effect on these for NH_3 than for PPh_3 as other ligand, suggesting π-acceptor properties $CO > PPh_3 > NH_3$.
(b) Maximum π-bonding to relieve negative charge for Fe(−II), least for Ni(0).
(c) Substitution proceeds via a dissociative mechanism; $Ni(CO)_4$ is a derivative of $Ni(0)(d^{10})$ and 4*s*, 4*p* orbitals are filled with ligand electrons.
(d) V(0) derivative is one electron short of number for maximum π bonding. (cf. $[Fe(CN)_6]^{4-}$ and $[Fe(CN)_6]^{3-}$ Fe—C bond lengths – Section 22.9).

2 (a) Exchange with CO by dissociative mechanism preceded by faster fluxional exchange via intermediates of the form

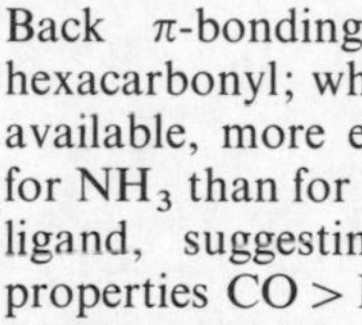

O
|
C
Co —— Co—

(b) In solvent at temperatures at which most of carbonyl is in non-bridged form, ratio of intensities of bridging CO infrared band is roughly K_{T_2}/K_{T_1}; hence $\Delta H°$.

3 Since transition metal nitrosyl derivatives usually follow the 18-electron rule, most likely structures are

$$(ON)_3Fe{-}Fe(NO)_3$$ and

$$(ON)_2Fe(\mu\text{-}NO)_2Fe(NO)_2$$

Distinguish by infrared spectroscopy (cf. bridging carbonyls).

4 For electron-counting purposes, complexes can be regarded as NO^+ complex of Co(I), with NO acting as a 3-electron donor to Co(0), and as NO^- complex of Co(III), respectively (see Section 22.8).

Chapter 23

1 For electron counting in these compounds see Section 23.1.

$(\eta^5\text{-}C_5H_5)(OC)_3Cr{-}Cr(CO)_3(\eta^5\text{-}C_5H_5)$

and

$(\eta^5\text{-}C_5H_5)(OC)_2Cr(\mu\text{-}CO)_2Cr(CO)_2(\eta^5\text{-}C_5H_5)$;

see if there are bridging carbonyl bands in the infrared spectrum.

2 A = (cycloheptatriene)Mo(CO)$_3$ B = $[(C_7H_7)Mo(CO)_3]^+$

$$Ph_3C^+BF_4^- + \mathbf{A} \rightarrow Ph_3CH + \mathbf{B}^+BF_4^-$$

Positive charge on Mo should attract electrons back from antibonding orbitals of CO and raise CO stretching frequencies.

3 (a) $(\eta^5\text{-}C_5H_5)_2$ Ni is a 20-electron system. Reaction with C_2F_4 converts it into a 18-electron system.

(b) Back donation of electrons into π^* orbital of $CH_3CH{=}CH_2$.

(c) This preserves an 18-electron configuration for Ru.

(d) $MeMn(CO)_5$ must form a 16-electron intermediate

$$MeCOMn(CO)_4$$

(presumably via a cyclic three-centre transition state) which then adds a molecule of CO to give the 18-electron system product.

Chapter 24

1 (a) Li_2TiO_3 must have the NaCl structure, i.e. it is $Li_2^+Ti^{4+}O_3^{2-}$; Li^+, Ti^{4+} and Mg^{2+} are about the same size, and $2Li^+:Ti^{4+}$ is needed for electrical neutrality.

(b) $E°$ for Ti(IV) + e $\rightleftharpoons$ Ti(III) is 0.1 V at $[H^+] = 1$, so it might be thought that in alkali no reaction with Ti^{3+} would occur. But TiO_2 is extremely insoluble and the volatility of H_2 upsets equilibrium further.

(c) The iron complexes are both low-spin, the cobalt complexes high-spin in the case of Co(II) and low-spin in the case of Co(III); see Section 21.7 for the significance of this.

(d) $[Co(NH_3)_6]^{3+}$ should, thermodynamically, certainly be decomposed by acid; sensitivity to attack by OH^- suggests a conjugate base mechanism (Section 21.3) is operative.

(e) $3KCN + CuCN \rightarrow K_3[Cu(CN)_4]$; 6 ions → 4 ions.

2 See Problem 7.6.

3 (a) Compare lattice energy from Born cycle with that interpolated from values for MnF_2 and ZnF_2 (see Section 20.2).
(b) See Problem 20.1.
(c) Born cycle. It would be reasonable to take the lattice energy as about the same as that for CoF_3, assuming both salts to be high-spin in character.

4 (a) Cr should be oxidised to Cr^{3+}, but air should have little further action.
(b) Mn should be oxidised only to Mn^{2+}, even in air; but in alkaline media, although O_2 is a much less powerful oxidant than in acid, it should oxidise $Mn(OH)_2$ to $Mn(OH)_3$. See Section 7.8 for the limitations of such predictions.

5 (a) Bond length in the O_2 ligand (see Section 10.4).
(b) Mössbauer spectrum (see Section 1.5).
(c) Show Fe^{3+}(aq) changes colour at high $[Cl^-]$ and again if $[Cl^-]$ displaced on addition of $[F^-]$.
(d) Treat it with acid, to form MnO_2 and MnO_4^-. Separate by filtration, and determine both with oxalic acid in strongly acidic (H_2SO_4) solution.

6 V(v) → V(iv); V(v) → V(ii).

Chapter 25

1 (a) Same unit cell size (effect of lanthanide contraction), but nearly twice as many nucleons in atom of Hf.
(b) W≡W bonding in $[W_2Cl_9]^{3-}$; no Cr—Cr bond in $[Cr_2Cl_9]^{3-}$.
(c) Re≣Re bond in $[Re_2Cl_8]^{2-}$; two more electrons in $[Os_2Cl_8]^{2-}$ and they pair up with the electron on each metal atom that would otherwise form the δ-bond.
(d) See Section 21.7; both are low-spin species.
(e) Bulky $C_2H_5NH_2$ ligands prevent Pt atoms of cation and anion getting close enough to interact.
(f) Complex Ag_2I^+ more stable than Ag_2Cl^+ (class (*b*) behaviour – see Section 7.7).
(g) Equilibrium involved must be

$$Hg^{2+} + Hg \rightleftharpoons Hg_2^{2+}$$

rather than

$$Hg^{2+} + Hg \rightleftharpoons 2Hg^+$$

2 Infrared shows H or D must be present, so changes are:

$$[Rh(arsine)_3Br_3] \underset{Br_2}{\overset{H_3PO_2}{\rightleftharpoons}} [Rh(arsine)_3Br_2H]$$

3

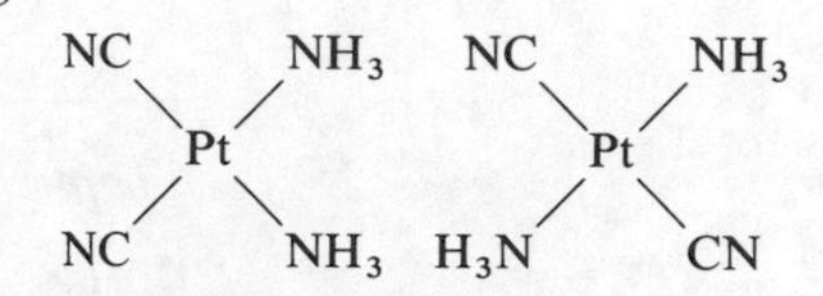

(non-electrolyte, has a dipole moment) (non-electrolyte, no dipole moment)

$[Pt(NH_3)_4][Pt(CN)_4]$ (electrolyte)

Compound $K_2[Pt(CN)_4Cl_2]$ must have *trans* structure since $Pt(CN)_4$ moiety is planar.

$$[Br—Pt—Cl—Pt—Cl]^{5-}$$

intermediate

in redox-catalysed substitution (see Section 21.8).

4 Formula of **I** indicates $7Cl^-$ from $2[OsCl_6]^{2-}$ replaced by N_9H_{24}; since all N liberated as NH_3, N_9H_{24} probably $8NH_3 + N^{3-}$. HI reaction shows $3Cl^-$ ionic, so likely formula is $[Os_2Cl_2N(NH_3)_8]Cl_3$. Vibrational spectrum shows cation has a centre of symmetry, so cation must be

$$[ClOs—N^{3-}—OsCl]^{3+}$$

$Os—N^{3-}$ bonds cannot be single, with two lone pairs of N^{3-} donated, since this would give a bent structure. Most satisfactory description is Os $d_{xy}^2 d_{xz}^1 d_{yz}^1$ with $d_{x^2-y^2}$, d_{z^2}, s, p orbitals used to bond Os to N^{3-}, Cl, $4NH_3$, sp_z linear hybrid orbitals being used by N.

Then d_{xz}, d_{yz} orbitals of each Os can overlap p_x, p_y orbitals of N atoms respectively; in each case a bonding, a non-bonding, and an antibonding orbital are formed. Two electrons from N p_x and one from each Os d_{xz} fill bonding and non-bonding orbitals, giving a π bond order of 0.5. Similar π bond from Os d_{yz} and N p_y orbitals.

Chapter 26

1 $^2F_{5/2}$; 2.54 μ_B.

2 Consider a cycle for

$$3LnX_2 \rightarrow 2LnX_3 + Ln$$

along the lines indicated for halide decomposition in Section 16.4. For a given Ln, difference in lattice energy between $3LnX_2$ and $2LnX_3$ is the governing factor: clearly least when X is largest.

Determine electrical conductivity.

3 (a) $EDTA^{4-}$ mainly an O-ligand so ΔH° for replacement of H_2O small. Large formation constants of $[Ln\ EDTA]^-$ complexes due to gain in entropy when highly charged ions combine to give singly charged species.

(b) Complexing by anion

$$Cl^- > SO_4^{2-} > NO_3^- > ClO_4^-$$

(c) A mixed oxide, not an oxo-anion salt.

Chapter 27

1 (a) Most actinides show a range of oxidation states not very different in stability; few lanthanides form more than one stable oxide.

(b) Any oxo-anion that on reduction gives a species with less oxygen bonded to the other element will be stabilised by high $[OH^-]$ because of reduction equilibria such as

$$MO_6^{5-} + 8H^+ + e \rightleftharpoons MO_2^{2+} + 4H_2O$$

(c) According to Fig. 27.2, Pu(IV)(aq) should contain substantial concentrations of disproportionation products; SO_4^{2-} must therefore stabilise Pu^{4+} much more than PuO_2^{2+}, PuO_2^+ or Pu^{3+} (consequence of higher charge).

2 Am(III) is much more difficult to oxidise than the other metals in oxidation states III and IV. So oxidise U, Np, Pu to MO_2^{2+} with e.g. Ce(IV), and precipitate AmF_3 together with CeF_3; for separation from CeF_3 oxidise Am with $S_2O_8^{2-}$ and extract (or separate Am(III) and Ce(III) by ion-exchange).

Zn amalgam should reduce Np(VI) to Np(III); O_2 at $[H^+] = 1$ should oxidise Np(III) to NpO_2^+ and some NpO_2^{2+} but oxidation might be slow.

3 U(VI) → U(IV) after aeration

$$2UF_4 + O_2 \rightarrow UF_6 + UO_2F_2$$

Index